S0-ADT-065

MATHEMATICS
with applications in management and economics

The Irwin Series in Quantitative Analysis for Business
Consulting Editor **Robert B. Fetter** Yale University

MATHEMATICS
with applications in management and economics

EARL K. BOWEN
Professor of Mathematics
Babson College

1980   Fifth edition

RICHARD D. IRWIN, INC.
Homewood, Illinois   60430

Irwin-Dorsey Limited
Georgetown, Ontario   L7G 4B3

© RICHARD D. IRWIN, INC., 1963, 1967, 1972, 1976, and 1980

All rights reserved. No part of this publication may be
reproduced, stored in a retrieval system, or transmitted,
in any form or by any means, electronic, mechanical,
photocopying, recording, or otherwise, without the prior
written permission of the publisher.

ISBN 0-256-02349-2

Library of Congress Catalog Card No. 79–88791
Printed in the United States of America

2 3 4 5 6 7 8 9 0 K 7 6 5 4 3 2 1

LEARNING SYSTEMS COMPANY—
a division of Richard D. Irwin, Inc.—has developed a
PROGRAMMED LEARNING AID
to accompany texts in this subject area.
Copies can be purchased through your bookstore
or by writing PLAIDS,
1818 Ridge Road, Homewood, Illinois   60430.

To Steven and Pamela

Preface

This text was first published in 1963 with the objectives of presenting mathematics at a level consistent with student preparation and directed specifically toward applications in management and economics. The same objectives motivate this fifth edition. The systematic presentation–explanation–example–reader exercise approach which students and teachers have approved in previous editions has been maintained. This approach provides a semiprogrammed text which is well adapted to self-study.

To limit the prerequisite for study of the text to no more than one year of secondary school algebra, I have included appendixes on the elements of mathematics for student study or reference. I should add that the first four chapters make light demands upon previous preparation, so that study of these and the appendixes can proceed simultaneously.

I will comment in some detail shortly upon the major changes in this edition. First, however, I want to refer to small changes, too numerous to list, dictated by my desire to ease student difficulties that I have observed and those which have been reported to me by others. This has been done by various means such as including a new or changed example or exercise, providing a more motivating introduction, ordering problems by degree of difficulty, and adding more steps in the algebraic development of a topic.

I have changed the format of Chapter 1 by using the slope-intercept form for finding equations of lines given the slope and the y-intercept, or the slope and a point, or two points. However, the special forms for these three cases are also shown. In Chapter 1 I have added break-even analysis from the production viewpoint to the financial-accounting viewpoint presented in earlier editions. I have also added material on row operations and reduction of a linear system to canonical form by obtaining zeros below the first longest diagonal (2.16). Chapter 3 has only minor changes, and *isolines* are the only new topic in Chapter 4.

Major rewriting occurs in Chapter 5, which now starts with an elementary example showing the how and why of the simplex procedure. This is followed by detailed step-by-step practice with simplex operations for the primal and dual problems. The chapter concludes with a new discussion of postoptimality analysis (shadow prices and right-hand-side ranges). Significant additions to Chapter 6 are the *zeros-first* method of inverting a matrix (at the end of 6.15) which eliminates the occurrence of vexing fractions during the inversion steps, and an addendum on matrix inversion using the adjoint matrix.

Chapters 7 through 11 have been rewritten almost in their entirety. Chapter 7 opens with a modern, and I think innovative, introduction to logarithms. If my classroom experience can be relied upon, this treatment makes the subject both interesting and easy to learn. The presentation includes reference to hand-held calculators (the use of which I advise strongly) but does not require calculators because an extended table of natural logarithms is included in this edition. All but a selected group of problems (marked as optional) can be solved without a calculator. The classical treatment (common logarithms, with a table) appears in the second part of the chapter, so one has the option of selecting either the modern or classical treatment.

Students have a need for simple interest calculations in their courses, so I have included this material at the beginning of Chapter 8. Also in this edition, I develop the formula for the sum of a geometric series and use this result to obtain all periodic payment formulas. Other additions include an (optional) iterative hand-calculator solution for finding the interest rate, and continuous compounding formulas for both single and periodic payments.

An objective of mine in this edition is to make the calculus material sufficiently modular to accommodate short courses and longer courses. To this end, all of the basic ideas of differential calculus are presented in Chapter 9, using only power functions. Correspondingly, the first part of Chapter 11 develops all of the basic ideas of integral calculus, using only power functions. Consequently, Chapter 1 is the only prerequisite for an introduction to differential and integral calculus. Longer courses would include Chapter 10 and the second part of Chapter 11, where exponential and logarithmic functions are encountered, and such courses would have the first part of Chapter 7 (logarithms) as a prerequisite. Additional calculus material in this edition includes sketching of polynomials, rational functions, and exponentials; detailed economic cost analysis; least-squares formula derivation and application; Lagrangian multipliers; integration by parts; and an addendum on the fundamental theorem of calculus.

The major addition to Chapter 12 (probability in the discrete case) is the use of tree diagrams and the inclusion of another table of the cumulative binomial ($n = 10$). The first part of Chapter 13 (probability in the continuous case) is new. This chapter starts by developing the idea of a density function using the uniform distribution as a simple example, shows how to convert a function to a density function and how to calculate probabilities as definite integrals, and applies this material in a discussion of the exponential density function. There is also a new section on the definition and calculation of the variance and the standard deviation by integration. I have found that the foregoing, together with numerical integration (Chapter 11), contribute to students' understanding of probability tables commonly used in statistics and the source of these tables.

The book contains approximately 1,500 numbered problems, with answers, and an additional 700 review problems that can serve for examination and lecture purposes. Answers to the review problems, worked out in detail, are available in the *Solutions Manual* available to instructors. Additionally, several hundred examples and exercises are woven into the text. To the best of my knowledge, no other text in this field provides a comparable number and variety of problems and applications.

If the entire text, including work in the appendixes, is to be completed, there is sufficient material for a three-semester sequence of courses. However, the book is structured to be adaptable to a variety of courses, from one quarter to three semesters in duration, by making parts of chapters, whole chapters and groups of chapters independent. This structure permits omissions to be made without loss of continuity or prerequisite topics. In broad terms, one quarter or one semester courses in finite mathematics can be organized from the material in Chapters 1–8 and 12; courses of corresponding duration in calculus would cover Chapters 1, 7, 9–11, and 13.

This edition has profited greatly from the careful review by Christopher J. Toy of New Hampshire College, a teacher who not only understands student difficulties in mathematics but also has devised innovative ways of lessening these difficulties. I wish to acknowledge here that I have adopted almost all of his suggestions. I wish also to acknowledge my debt to those who have taken the time to send me comments and suggestions that have contributed significantly to the improvement of this text from one edition to the next. These are: R. Andres, P. Applebaum, W. Beatty, F. Benn, T. Billesbach, G. Bloom, R. Borman, A. Brunson, R. Carlson, W. Cassidy, D. Chesnut, T. Church, D. Cleaver, C. Crell, R. Davis, W. Davis, E. Dawson, B. Dilworth, R. Dingle, D. Dixon, W. Etterbeek, J. Freigo, R. Fetter, J. Flaherty, J. Foster, R. Fox, Jr., R. Friesen, H. Frisinger, H. Fullerton, W. Furman, E. Goldstein, L. Goldstein, M. Greenberg, V. Heeren, J. Hindle, A. Ho, A. Hoffman, G. Horcutt, J. Hudson, and D. Isaacson.

Also: R. Jaffa, F. Jewett, C. I. Jones, R. J. Jones, H. King, R. Kizior, P. Latimer, R. Leezer, R. Leidig, J. Liff, S. Logan, G. Long, T. Lougheed, J. Lovell, T. Lupton, M. Malchow, E. Marrinan, Jr., P. McKeon, A. McLaury, E. Merrick, P. Merry, R. Moreland, J. Moreno, C. Murphy, D. Nichols, J. Papenfuss, R. Ralls, P. Randolph, G. Reeves, J. ReVelle, R. Salmon, F. Schwab, H. Sendek, P. Sgalla, R. Sheffield, L. Shumway, P. Siegel, B. Smith, J. Smith, W. Soule, Jr., M. Spinelli, H. Stein, D. Stoller, M. Tarrab, T. Taylor, O. Thomas, R. Tibrewalla, T. Tsukahara, E. Tyler, E. Underwood, B. Van Cor, T. Vasper, G. Waldron, B. Walker, and M. Williamson.

Additionally, I wish to acknowledge the encouragement and contributions of my colleagues at Babson: W. Carpenter, D. Kopcso, H. Kriebel, W. Montgomery, M. Riskalla, J. Saber, A. Shah, and M. Weinblatt. Finally, I thank my wife, Dorothy Holmes Bowen, for her editorial assistance and for protecting the quiet of my workplace.

January 1980   Earl K. Bowen

Contents

Chapter 9.   DIFFERENTIAL CALCULUS: POWER FUNCTIONS 402

Chapter 13. PROBABILITY IN THE CONTINUOUS CASE . 724

Appendix 1. SETS 760

Appendix 2. ELEMENTS OF ALGEBRA 773

TO THE READER

Short exercises have been woven into textual discussions to help you pick up pertinent points as they occur. For example, on page 3 you will see the following:

> **Exercise.** Taxicab fare from an airport to a nearby city B is $0.80 per mile driven, plus $2 for tolls. Let y represent fare and x the miles driven on one trip. Write the equation for y in terms of x. Answer: $y = 0.80x + 2$.

To make the best use of these exercises, you should determine the answer to the question without reference to the given answer (it would help to cover up the given answer) and then compare your answer with the given answer.

Linear equations

1

1.1 INTRODUCTION

Mathematicians, economists, statisticians, and others have applied their skills to management problems in some degree for many years, but the first concerted effort in this area occurred during World War II when these specialists were formed into Operations Analysis Groups to assist in the planning of military operations. The analysts used mathematics and statistics extensively in their studies, and the resulting recommendations were an important contribution to the war effort. Following the war, some analysts, soon joined by others, turned their attention to problems of management operations and accomplished major improvements in inventory control, quality control, warehouse location, oil industry operations, agriculture, purchasing decisions, scheduling of complex tasks such as building a shopping center, and a variety of other areas. Mathematics, old and *newly created*, coupled with innovative applications of the rapidly evolving electronic computer and directed toward management problems, resulted in a new field of study called Quantitative Methods (or

1

Management Science or Operations Research), which in time became part of the curriculum of colleges of business. The importance of quantitative approaches to management problems is now widely accepted, and a course in mathematics, with management applications, is included in the core of subjects studied by almost all management students. This text, which has been used in many hundreds of classrooms, develops mathematics in the applied context required for an understanding of the quantitative approach to management problems.

Linear relationships are the subject matter of the first five chapters of the book. From the several pleasant comments that could be made about this material, we select three. First, the mathematics has direct management applications; indeed, at this moment many electronic computers are manipulating linear relationships and printing out information to aid in making lowest-cost or highest-profit decisions. Second, Chapters 4 and 5 introduce a relatively *new* and widely used mathematical technique known as linear programming. Third, most readers will be happy to learn that the mathematics of linear relationships, as we shall develop it, is quite easy. Only a minimal background, such as that provided in the appendixes, is required to get started and to make progress.[1] Every reader should scan these appendixes. Those who feel the need should work through them systematically and refer to them as often as necessary while studying the book.

In this chapter, we first consider in some detail the algebra and geometry of linear equations in two variables, that is, equations whose graphs are straight lines. Later in the chapter, linear equations in three or more variables will be introduced to lay the foundations for analysis of systems of linear equations.

Linear equations are equations whose *terms* (the parts separated by plus, minus, and equal signs) are a constant, or a constant times *one* variable to the first power. Thus,

$$2x - 3y = 7$$

is a linear equation because it consists of the constant 7, the term $2x$, which is the constant 2 times x to the first power, and $-3y$, which is also a constant times one variable to the first power. Similarly,

$$y = \frac{1}{2}x + 3 \quad \text{and} \quad \frac{x}{3} + \frac{y}{2} = 1$$

are linear equations. However,

$$2x + 3xy = 7$$

---

[1] For a review of the minimal background, read Appendix 2. In Appendix 3 read sections A3.1–3.5 and the parts of A3.6–3.11 that deal with linear equations.

is not linear because $3xy$ is a constant times the product of two variables and

$$x^2 + y + 3x = 16$$

is not linear because of the presence of the second-power term, x^2. Linear equations arise in numerous applied situations.

Example.   Taxicab fare from an airport to nearby city A is $1.25 per mile driven, plus $0.75 for a bridge toll. Let y represent the fare and x the number of miles driven on one trip, and write the equation for y in terms of x.

Here the total fare will be

$$\text{Fare} = (1.25)(\text{number of miles driven}) + 0.75$$

or, using the specified symbols,

$$y = 1.25x + 0.75,$$

which is a linear equation.

Exercise.   Taxicab fare from an airport to a nearby city B is $0.80 per mile driven, plus $2 for tolls. Let y represent fare and x the miles driven on one trip. Write the equation for y in terms of x. Answer: $y = 0.80x + 2$.

The equations of the last example and exercise,

$$y = 1.25x + 0.75 \quad \text{and} \quad y = 0.80x + 2,$$

are linear equations in the two *variables* x and y. Such equations have graphs that are straight lines. To graph a straight line we need only to have the coordinates of two of its points. It does not matter which two, so we may arbitrarily select two values for x and obtain the corresponding values for y. For example, in

$$y = 1.25x + 0.75$$

if we choose $x = 1$, then

$$y = 1.25(1) + 0.75 = 1.25 + 0.75 = \$2$$

and we have the point $x = 1$, $y = 2$, which is designated in the conventional form (x, y) as $(1, 2)$. Again, choosing, say, $x = 5$, we find

$$y = 1.25(5) + 0.75 = 6.25 + 0.75 = \$7$$

and we have the point $x = 5$, $y = 7$, or $(5, 7)$. Plotting $(1, 2)$ and $(5, 7)$ as points, then drawing a line through them leads to Figure 1–1. The accompanying Figure 1–2 shows the graph of $y = 0.80x + 2$.

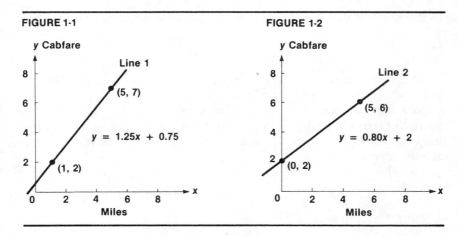

FIGURE 1-1

FIGURE 1-2

Exercise. Verify the points shown on Figure 1–2. Answer: In $y = 0.80x + 2$, if we choose $x = 0$, then $y = 2$, and if we choose $x = 5$, then $y = 6$.

Linear functions. The relationship between y and x expressed by

$$y = 1.25x + 0.75$$

is called a functional relationship because for each value of x, there is one, and only one, corresponding value for y. Notice that the expression states what y is (that is, $y =$) in terms of x and we connote this by saying y *is a function of* x and, of course, this is a linear function. If we write

$$y = \text{an expression involving } x \text{ and constants,}$$

x is called the *independent* variable and is plotted on the horizontal. The value of y depends upon what value we assign to x, and y is called the *dependent* variable, which is plotted on the vertical. Thus, when we plot points, the values of x can be chosen independently, but the corresponding values of y depend on the values chosen for x. Later in the text we will encounter situations where an expression has several letters and we will want to be able to specify which is the independent variable. This is done by writing the independent variable in parentheses next to the dependent variable. Thus, in our example, $y = 1.25x + 0.75$, we would write

$$y(x) = 1.25x + 0.75.$$

The need for identification of the independent variable can be seen if we write

$$y = kp + n$$

and state that this is a linear function because it is not clear which of the letters on the right is the independent variable. However, writing

$$y(p) = kp + n$$

specifies that p is the independent variable and, therefore, k and n are to be considered as being *constants*. In this chapter, the identity of the independent variable will be clear from the context, so we shall not use the parenthetical specification. However, *independent variable, dependent variable,* and *linear function* are words that should be a part of the reader's vocabulary.

Returning to Figures 1–1 and 1–2, observe that both lines slant upward to the right, but that line 1, on the left, rises more rapidly (is steeper) than line 2. This means that for a *given* horizontal change, the vertical change on line 1 is greater than the vertical change on line 2. We shall see that the ratio of vertical to horizontal change has numerous applied interpretations. Inasmuch as these changes are horizontal and vertical distances, we turn first to a consideration of such distances.

1.2 VERTICAL AND HORIZONTAL DISTANCES

The distance between two points is the length of the straight-line segment that joins the points. In beginning (plane) geometry and in applied mathematical work, we use a ruler or some other measuring device to determine lengths of segments. In *analytic* geometry, also called coordinate geometry, end points of line segments are specified by their x- and y-coordinates, and algebraic procedures are applied to the coordinates to find the distance between the points.

Distances on horizontal and vertical line segments play an important role in the mathematics of straight lines. Distance on a vertical segment is computed by subtracting the y-coordinates of the end points of the segment. Distance on a horizontal segment is computed by subtracting the x-coordinates of the end points of the segment. Consider, for example, a city that has avenues running east and west and streets running north and south, dividing the city into square blocks. To get from (4th St., 3d Ave.) to (10th St., 3d Ave.), we would walk on Third Avenue a distance of $10 - 4 = 6$ blocks.

Exercise. How many blocks would we walk going from (10th St., 6th Ave.) to (10th St., 13th Ave.)? Answer: Seven blocks.

We shall require that *distance* be a positive number. Hence, if the subtraction of coordinates yields a negative number, the minus sign will be discarded. Consider Figure 1–3. The segment *CD* is horizontal. The distance *CD* is found by subtraction, thus:

CD = horizontal distance = difference of x-coordinates = $9 - 3 = 6$.

If the calculation had been made as $3 - 9 = -6$, the minus sign would have been discarded.

The segment *HG* is vertical. The distance *HG* is found by subtraction:

HG = vertical distance = difference of y-coordinates = $2 - (-4) = 6$.

FIGURE 1-3

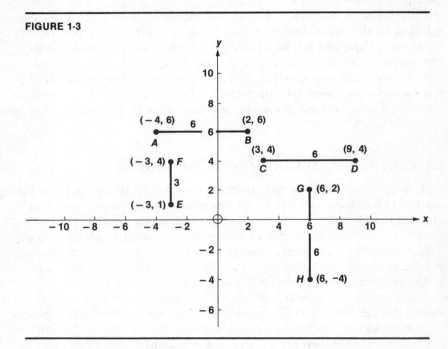

Exercise. Find the distances *AB* and *EF* on Figure 1–3. Answer: $AB = 6$ and $EF = 3$.

If the x-coordinates of the points at the end of a segment are equal, as in the case of *HG* on Figure 1–3, the segment is vertical, and if the y-coordinates of the two points are equal, as in the case for *AB*, the segment is horizontal. Thus, letting $P(a,b)$ mean the point P whose coordinates are *(a, b)*, and letting *Q(a, c)* mean the point *Q*, whose coordinates are *(a, c)*, it follows that the segment

$$P(a, b) \; Q \; (a, c)$$

is vertical. Similarly, the segment

$$R(a, b) \; S \; (c, b)$$

is horizontal.

> **Exercise.** Given $A(-2, -5)$; $B(2, 6)$; $C(3,10)$; $D(2, -5)$. If each pair of points was connected by a straight-line segment, which segment would be horizontal and which segment would be vertical? Answer: *AD* and *BD*, respectively.

Subscript notation. The symbol

$$(x_1, \; y_1)$$

is read as "x sub-one, y sub-one" or, more briefly, "x-one, y-one", and is used to designate a *particular* point. Similarly, $(x_2, \; y_2)$ denotes a different particular point. This notation is a handy way to preserve *(x, y)* to refer to the coordinates of any point, in general, with variations in the subscript serving to designate coordinates of different points. In this notation, a vertical segment would have $(x_1, \; y_1)$ and (x_1, y_2) as end points. The distance between the points would be:

$$\text{Vertical distance} = |y_2 - y_1|.$$

The vertical lines, ||, are the *absolute value* symbol which means that the difference is to be taken as positive. For example,

$$|5 - 2| = |+3| = 3$$
$$|2 - 6| = |-4| = 4$$

and, of course,

$$|-3 + 3| = |0| = 0.$$

The coordinates of the end points of a horizontal segment would be written in subscript notation as $(x_1, \; y_1)$ and $(x_2, \; y_1)$. The distance between the points would be:

$$\text{Horizontal distance} = |x_2 - x_1|.$$

1.3 PROBLEM SET 1–1

1. Given the points $A(3, 6)$, $B(-3, 6)$, $C(3, 9)$, $D(-3, 2)$, plot the points, then figure the following distances from the graph: *AB, AC, DB*.

2. Given the points $A(-5, -9)$, $B(-3, -9)$, $C(-5, 15)$, $D(12, -9)$, find the distances AB, AC, AD, BD.

3. Consider the segments AB and CD, where the coordinates are $A(p, q)$, $B(p, r)$, $C(n, q)$, and $D(m, q)$. If the two segments are extended, they will intersect in a right angle. Why?

4. Given $P_1(x_1, y_1)$, $P_2(x_2, y_2)$, $P_3(x_3, y_3)$, what relationships must exist among the coordinates if P_1P_2 is to be horizontal and P_1P_3 is to be perpendicular to P_1P_2?

5. Given P_1, P_2, P_3, P_4 with their respective coordinates in subscript form, what relationships must exist among the coordinates if P_1P_2 is to be horizontal and P_3P_4 is to be parallel to P_1P_2?

6. If cities A, B, and C are collinear (lie on the same straight line), how far would it be between B and C if:
   a) B is 60 miles east of A and C is 100 miles east of A?
   b) B is 30 miles west of A and C is 40 miles east of A?

7. If x represents sales and y represents selling expense, then $(30, 22)$ would mean $30 in sales were accompanied by $22 of selling expense. Suppose last month's figures were $(14, 10)$ and this month's are $(30, 22)$.
   a) How much did sales increase?
   b) How much did selling expense increase?
   c) Make a graph showing the points and labeling the increases.

8. What is the advantage of subscript notation for coordinates of points?

9. A section of a city is divided into square blocks by streets and avenues. How many blocks would we walk on sidewalks going from (6th St., 12th Ave.) to (1st St., 5th Ave.)?

1.4 SLANT DISTANCE FORMULA

We have been working thus far with horizontal and vertical distances and now wish to develop a formula for slant distance. Recall that the Pythagorean theorem states that the sum of the squares of the sides of a right triangle equals the square of the slant side (the hypotenuse). For example, the right triangle in Figure 1–4A has sides 3 and 4, which are a vertical and a horizontal distance, and

$$3^2 + 4^2 = 5^2.$$

That is,

$$9 + 16 = 25.$$

More generally, if the sides of the right triangle are a and b as in Figure 1–4B, we may write

$$a^2 + b^2 = c^2.$$

We want the slant distance, which is c, so we write

$$c^2 = a^2 + b^2$$

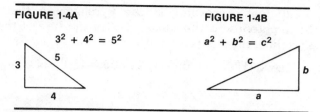

FIGURE 1-4A

$$3^2 + 4^2 = 5^2$$

FIGURE 1-4B

$$a^2 + b^2 = c^2$$

and take the square root of both sides to obtain

$$c = \sqrt{a^2 + b^2},$$

where, of course, we take the positive square root because distance is positive.

Exercise. Find the length of the hypotenuse of the right triangle whose sides are 5 and 12 units long. Answer: $c = \sqrt{169} = 13$.

Turning next to an example, Figure 1–5 shows a building whose dimensions are 80 by 90 feet. A sidewalk is to be constructed from a building exit to a point that is 200 feet on the horizontal from the lower left corner of the building. We wish to determine the length of the dotted line which represents the sidewalk. We see that the sidewalk is the hypotenuse of a right triangle whose altitude is 90 feet and whose base is a horizontal line segment of length

FIGURE 1-5

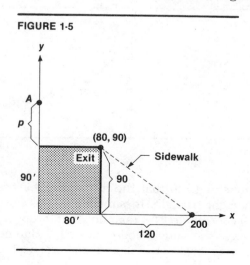

10

$(200 - 80) = 120$ feet. According to the Pythagorean theorem, the length of the hypotenuse is the square root of the sum of the squares of the sides. Hence, the desired length, L, is

$$L = \sqrt{(120)^2 + (90)^2} = \sqrt{22,500} = 150 \text{ feet.}$$

Exercise. If the sidewalk is continued to meet the vertical at point A, *a)* find the distance p from the properties of similar triangles; *b)* find the length of the extension. Answer: *a)* p is to 80 as 90 is to 120, so p is 60. *b)* $\sqrt{(60)^2 + (80)^2} = \sqrt{10,000} = 100$ feet.

Now let us develop a formula for finding the distance between two points that are on a segment not parallel to an axis by reference to the segment AB on Figure 1–6. Dashed lines have been drawn to

FIGURE 1-6

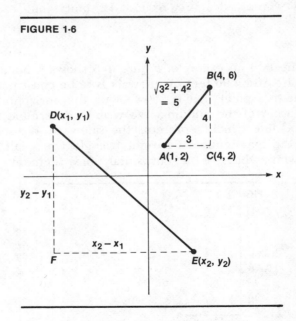

make a right angle at C, so that the distance AB is the length of the hypotenuse of a right triangle. Inasmuch as AC is horizontal, C must have the same y-coordinate as A (namely, 2). Moreover, CB is vertical, so that C must have the same x-coordinate as B (namely, 4). The horizontal distance AC is $|4-1| = 3$ and the vertical distance CB is

$|6 - 2| = 4$. Employing the Pythagorean theorem, we find that the hypotenuse, AB, is

$$\sqrt{(AC)^2 + (CB)^2} = \sqrt{3^2 + 4^2} = \sqrt{25} = 5.$$

Applying the same methodology to the segment DE on Figure 1–6, we derive the general formula:

Distance between two points $= \sqrt{(x_2 - x_1)^2 + (y_2 - y_1)^2}$.

For example, the distance between (2, 3) and (5, 2) is

$$\sqrt{(5 - 2)^2 + (2 - 3)^2} = \sqrt{9 + 1} = \sqrt{10} = 3.162.$$

1.5 PROBLEM SET 1–2

1. In each case, the given point pairs are the end points of the diagonal of a rectangle whose base is parallel to the x-axis. What are the coordinates of the other corners of the rectangle, and what is the length of the diagonal?
    a) (3, 5), (7, 8).   *b)* (−1, −2), (5, 6).
    c) (−1, −2), (−10, −14).
2. Find the distance between the points.
    a) (5, 10), (11, 18).   *b)* (0, 0), (9, 12).   *c)* (−2, −5), (3, −4).
    d) (−2, 3), (6, 9).   *e)* (3, −5), (6, −5).   *f)* (4, 7), (4, 9).
3. The coordinates of three cities are shown, in miles, on Figure A. What will be the total distance traveled if one goes from A to B, then B to C, then back to A? Note that ABC is not a right triangle.

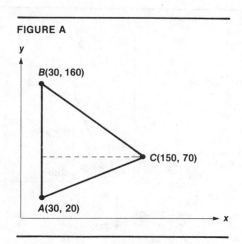

FIGURE A

4. With reference to an origin, City A is located at (3, 5), City B at (9, 13), and City C at (21, 4), all numbers being miles.
    a) Make a graph showing the positions of the cities.

  b) Compute the total distance covered traveling from *A* to *B*, then from *B* to *C*.

5. A machine shop occupies a 100-foot by 50-foot rectangular area. The electrical outlets that supply the machines are located by reference to a grid with origin at one corner of the area and the horizontal axis in the long direction of the area.
    a) What would be the coordinates of the outlet in the center of the shop?
    b) How long a line would it take to reach from the origin to the center of the shop?
    c) How long a line would it take to connect a machine at (10 feet, 5 feet) to a plug at (40 feet, 45 feet)?

1.6 SLOPE

The steepness of a ski slope, the pitch of a roof, and the steepness of the glide path of a descending airplane all are associated with the mathematical concept of the *slope* of a straight line or line segment. Numerically, the slope of a straight line is the ratio of the *rise (or fall)* to the *run* between two points on the line, where the rise or fall is the vertical separation and the run is the horizontal separation of the two points. In Figure 1–7, the slope of the segment *AB* is the ratio

$$\text{Slope} = \frac{\text{Rise}}{\text{Run}} = \frac{6}{2} = 3.$$

FIGURE 1-7

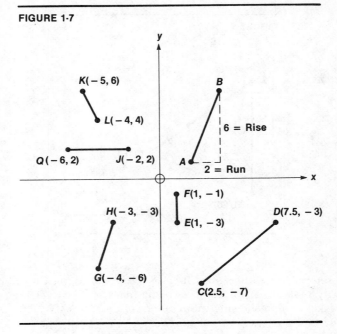

Clearly, the rise is a vertical segment and the run is a horizontal segment. Hence the slope (generally called m) of a straight line or line segment joining the points (x_1, y_1) and (x_2, y_2) is:

$$\textbf{Slope} = \textbf{m} = \frac{\text{Difference of } y\text{'s}}{\text{Difference of } x\text{'s}} = \frac{y_2 - y_1}{x_2 - x_1}.$$

Consider the segment CD in Figure 1–7:

$$m = \frac{\text{Rise}}{\text{Run}} = \frac{-3 - (-7)}{7.5 - 2.5} = \frac{4}{5} = 0.8.$$

It is important to distinguish between segments that rise to the right, such as AB in Figure 1–7, and those that fall to the right, such as KL. This is done by requiring that lines that slant upward to the right have positive slope numbers and those slanting downward to the right have negative slope numbers. This requirement will be met if we *use the same point as the starting point* when computing the rise or fall and the run. For example, the slope of $C(2.5, -7)$, $D(7.5, -3)$ computed to be 0.8 in the foregoing starting with point D for both rise and run, is also 0.8 if point C is the starting point for both rise and run. Thus,

$$m = \frac{-7 - (-3)}{2.5 - 7.5} = \frac{-4}{-5} = 0.8.$$

Exercise. See Figure 1–7. Verify that the slopes of KL and GH are, respectively, -2 and 3.

Returning to Figure 1–7, observe the horizontal segment QJ. If we substitute its coordinates into the slope formula we find:

$$m = \frac{2 - 2}{-2 - (-6)} = \frac{0}{4} = 0,$$

so the slope is zero. All horizontal segments have a slope of zero, for, as we know, the y-coordinates of two points on a horizontal segment are equal; that is, (x_1, y_1) and (x_2, y_1), with $x_1 \neq x_2$, would represent two points on a horizontal segment because both have the same y-coordinate. The slope of any such segment is

$$m = \frac{y_1 - y_1}{x_2 - x_1} = \frac{0}{x_2 - x_1} = 0.$$

The denominator of the last, $x_2 - x_1$, is not zero because $x_1 \neq x_2$. Consequently, the quotient is zero because zero divided by any nonzero number is zero.

14

Next, consider the vertical segment *EF* on Figure 1–7. From the slope formula we have:

$$m = \frac{-1 - (-3)}{1 - 1} = \frac{2}{0}.$$

The expression $2/0$ is not a number, so this vertical segment has no slope number. In general, $(x_1, y_1); (x_1, y_2)$ represent two different points on a vertical segment because they have the same x-coordinate and the slope will be

$$m = \frac{y_2 - y_1}{x_1 - x_1} = \frac{y_2 - y_1}{0}$$

so, in general, vertical segments have no slope number.[2]

Exercise. Which of the following segments is horizontal and which is vertical? *a)* The segment joining $(4, -6)$ and $(10, -6)$. *b)* The segment joining $(4, -6)$ and $(4, 10)$. Answer: *a)* is horizontal and *b)* is vertical.

Exercise. Draw three segments through the point $(4, 7)$, one with no slope number, one with $m = 0$, and one with $m = \frac{2}{3}$. (A slope of $\frac{2}{3}$ is a rise of 2 for a run of 3.)

In applications, the slope of a line segment often is interpreted as the amount of change in the vertical for a unit change (that is, a change of one) in the horizontal, and a number of such slopes have been given names.

Example. This example uses the terms *disposable income, personal consumption expenditures,* and *savings.* It will be sufficient for our purposes here to think of disposable income as the amount of income left after taxes have been paid. The part of disposable income that is placed in a bank or otherwise invested is savings, and the remainder, personal consumption expenditure, is spent on food, clothing, housing, and so on.

A line segment fitted to points whose coordinates are in the order

(disposable income, consumption expenditures)

---

[2] Saying a line has no slope number does not bring the image of a vertical line to mind. Some find it more descriptive to say a vertical line has infinite slope because a straight up-and-down line is the steepest one we can imagine. However, in using the term infinity, which is symbolized as ∞, it must be remembered that ∞ is not a number, so saying a line has infinite slope means it has no slope number.

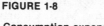

FIGURE 1-8

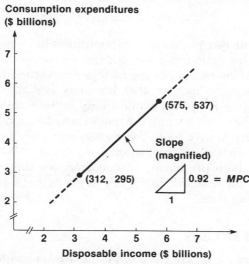

for the United States in recent years passes through

$$(312,\ 295) \quad \text{and} \quad (575,\ 537)$$

where the numbers are in billions of dollars. The slope of the segment shown on Figure 1–8 is

$$\frac{\text{Change in consumption expenditures}}{\text{Change in disposable income}} = \frac{537 - 295}{575 - 312} = \frac{242}{263} = 0.92.$$

Thus, an increase of $242 billion in consumption expenditures accompanied an increase of $263 billion in disposable income. On a proportionate basis,

$$\frac{\$242}{\$263} = \frac{\$0.92}{\$1} = 0.92,$$

so the slope of the segment, 0.92, when written with a denominator of 1, represents the change in consumption expenditures per *one* dollar of additional income. In economics, this change is called the *marginal propensity to consume*, or MPC for short, as shown on Figure 1–8. The word *marginal*, just used, means *extra*, and MPC is the extra consumption that accompanies a $1 increase in income. Income not used for consumption expenditures is *saved*. We see from the foregoing that $0.08 of each extra dollar of disposable income is saved, and 0.08 is called the *marginal propensity to save*, MPS. Clearly,

$$MPS = 1 - MPC.$$

The important bearing these two marginals have on the economic well-being of the nation can be understood by noting that savings are the source of investment in the factories and other economic activities that produce the income that becomes available to consumers. It is interesting to note that while many believe saving is a virtue, economics teaches us that when the nation has idle productive capacity, a high propensity to save (with the consequent low propensity to consume) is not a virtue for the national economy. What is needed in such times is a high propensity to consume, for this will lead to the increased demand that will bring idle capacity back into use and increase national income.

1.7 PROBLEM SET 1–3

1. If (x_1, y_1) and (x_2, y_2) are the coordinates of two points on a line,
    a) How is the rise computed?   *b)* How is the run computed?
2. *a)* What is the nature of the steepest line that can be drawn through a point?
    b) Why does such a line not have a slope number?
3. If a set of stairs rises 8 inches for every horizontal run of 12 inches, what slope number (assumed positive) describes the steepness of the stairs?
4. If x is the ground path of an airplane and y is its altitude, and at one point in time the plane is at (500, 1000) and soon after it is at (500, 0), what was the path of the airplane during the time interval?
5. A ski slope whose fall line makes a 45° angle with the horizontal is said to have a 100 percent grade. What slope number represents a 100 percent grade?
6. If the total manufacturing cost, y dollars, of producing x units of a product is $500 at 50 units output and $900 at 100 units output, and the cost-output relation is linear,
    a) What is the slope of the cost-output line?
    b) How much does the production of one unit add to total cost?
7. A line segment fitted to points whose coordinates are in the order (disposable income, personal consumption expenditures) for a nation passes through the points (32, 30), (57, 54), the numbers being in billions of dollars. What are the values, names, and interpretations of the slope and $(1 - \text{slope})$ of this line?

Compute the slope of the segment joining each point pair:

8. (0, 0), (2, 2).   9. (−3, 5), (4, −2).
10. (6, −1), (−2, 0).   11. (−3, −2), (−3, −4).

12. (2, 3), (6, 3).

13. (1/2, −1/4), (2, −1/4).

14. (−4, 7), (−4, 2).

15. (1, 1), (3, 3).

16. (2/7, −1/3), (−2/9, −2/3).

17. (0, 3), (3, 0).

18. (12, −5), (3, 6).

19. (3, −7), (3, −15).

20. (−1, −2), (−3, −4).

21. (1.6, 3.8), (−3.6, 4.2).

22. Write the formula for the slope of a line segment; then discuss the formula by means of a numerical illustration.

23. Write the expression for the slope of the segment from $P(a, -2)$ to $Q(3, -b)$ in two equivalent forms.

24. Make and label sketches showing segments having positive slope, negative slope, zero slope, and no slope number.

25. Why is division by zero excluded from arithmetic calculations? (See Appendix 2.)

Mark (T) for true or (F) for false.

26. ( ) A line that rises to the right and is almost vertical does not have a slope number.

27. ( ) The slope of the x-axis is zero.

28. ( ) The segment $P(a, b) \; Q(a, c)$ is vertical.

29. ( ) A line segment of negative slope rises to the left.

30. ( ) The slope of the y-axis is not a number.

31. ( ) A line segment that is very, very close to the vertical has a slope number that has a large absolute value.

32. ( ) No matter how large a number we may write down, there is a line segment whose slope exceeds this number.

33. ( ) A line segment contained entirely in the second quadrant necessarily has a negative slope.

34. ( ) The quadrant in which a line segment lies has no necessary relation to the sign of the slope number of the segment.

1.8 EQUATION OF A LINE: SLOPE-INTERCEPT FORM

We have been considering line segments and their slopes and now turn to the infinite extension of a segment, which is a straight line. The first linear equation we wrote in the introduction to the chapter,

$$y = 1.25x + 0.75,$$

expressed the fare paid for a taxicab ride, y, in terms of x, the number of miles driven and the constant 0.75, which was a fixed charge for a bridge toll. For example, the fares for a one-mile ride and a five-mile ride are, respectively,

$$y = 1.25(1) + 0.75 = 2$$
$$y = 1.25(5) + 0.75 = 7,$$

giving rise to the two *(x, y)* points

$$(1, 2) \quad \text{and} \quad (5, 7).$$

If we now find the slope of the line between these points as

$$\frac{y_2 - y_1}{x_2 - x_1} = \frac{7 - 2}{5 - 1} = \frac{5}{4} = 1.25,$$

we find that the slope is the coefficient of *x* in

$$y = 1.25x + 0.75.$$

> **Exercise.** If $y = 0.8x + 2$, *a)* Write the coordinates of the points where $x = 0$ and $x = 5$. *b)* Compute the slope from the points found in *(a)*. Answer: *a)* (0, 2) and (5, 6). *b)* Slope is 0.8, the coefficient of *x* in $y = 0.8x + 2$.

The last example and exercise suggest that if we write the equation of a line in the form *y* equals a constant times *x*, plus another constant, then the coefficient of *x* is the slope; that is, in

$$y = mx + b, \tag{1}$$

m is the slope. To prove this is so, suppose we take any pair of points on the line (1), *(x₁, y₁)* and *(x₂, y₂)*. Then we must prove that

$$m = \frac{y_2 - y_1}{x_2 - x_1}.$$

We start by noting that because the points are on the line (1), their coordinates must satisfy equation (1). That is, at *(x₁, y₁)*,

$$y_1 = mx_1 + b \tag{2}$$

and at *(x₂, y₂)*,

$$y_2 = mx_2 + b. \tag{3}$$

We can rewrite (2) and (3) as

$$y_1 - mx_1 = b \tag{4}$$
$$y_2 - mx_2 = b. \tag{5}$$

The left sides of (4) and (5) both equal *b*, so they are themselves equal, and we have

$$y_1 - mx_1 = y_2 - mx_2 \tag{6}$$

and, by transposing terms,

$$mx_2 - mx_1 = y_2 - y_1.$$

Factoring on the left yields

$$m(x_2 - x_1) = y_2 - y_1;$$

then, dividing both sides by $(x_2 - x_1)$, we have

$$\frac{m\cancel{(x_2 - x_1)}}{\cancel{(x_2 - x_1)}} = \frac{y_2 - y_1}{x_2 - x_1}$$

$$m = \frac{y_2 - y_1}{x_2 - x_1},$$

which is what we set out to prove. Thus, if the equation of a line is written in the form

$$y = mx + b,$$

m is the slope. Moreover, if $x = 0$, then

$$y = m(0) + b = b,$$

and $(0, b)$ is the point where the line cuts the y-axis. This point (where $x = 0$), is called the *y-intercept*. Thus, we have:

Slope-intercept form
If the equation of a line is written in the form

$$y = mx + b,$$

then m is the slope and b is the y-intercept.

Example. Write $2y + 3x = 18$ in slope-intercept form and state the value of the slope and the y-intercept.

We proceed to get y with a coefficient of 1 by itself on one side of the equation. First we transpose the term $3x$:

$$2y = -3x + 18.$$

Then, dividing both sides by 2, we have

$$\frac{2y}{2} = \frac{-3x + 18}{2}$$

$$y = -\frac{3}{2}x + 9$$

as the desired $y = mx + b$ form. The slope is $-\frac{3}{2}$ and the y-intercept is 9. The line is shown on Figure 1–9. The point $(4, 3)$ shown on the line was obtained by choosing x to be 4, arbitrarily, and computing

$$y = -\frac{3}{2}(4) + 9 = -3(2) + 9 = 3.$$

Plotting by intercepts. When a linear equation is not in slope-intercept form, the simplest points to find and use for plotting the line are its intercepts. To find the x-intercept, set $y = 0$ and solve for x;

20

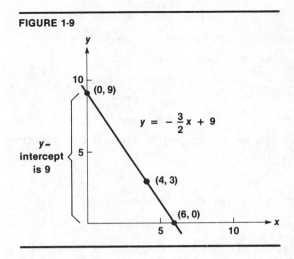

FIGURE 1-9

to find the y-intercept, set $x = 0$ and solve for y. For example, in the equation

$$2y + 3x = 18$$
x-intercept: $2(0) + 3x = 18$; $x = 6$. Point is $(6, 0)$.
y-intercept: $2y + 3(0) = 18$; $y = 9$. Point is $(0, 9)$.

This line and its intercepts are shown on Figure 1–9.

Exercise. a) Write $3y - 2x = 24$ in slope-intercept form. b) What is the slope of the line? c) What is the y-intercept? d) What is the x-intercept? e) Write the coordinates of the points found in (c) and (d). Answer: a) $y = (\frac{2}{3})x + 8$. b) $\frac{2}{3}$. c) 8. d) -12. e) $(0, 8)$ and $(-12, 0)$.

The next example illustrates one of the many interpretive applications that arise when a line is in slope-intercept form.

Example. It costs \$2500 to set up the presses and machinery needed to print and bind a paperback book. After setup, it costs \$2 per book printed and bound. Let x represent the number of books made and y the total cost of making this number of books. a) Write the equation for y in terms of x. b) State the slope of the line and interpret this number. c) State the y-intercept of the line, and interpret this number.

a) Here the total cost is y, where

$$y = 2x + 2500,$$

a linear equation in slope-intercept form.

b) The slope, 2, means that every *additional* book printed, starting with the first copy, adds $2 to total cost. Note that it would *not* be proper to say books cost $2 per copy. That is, for example, if $x = 100$ books are made, then total cost is

$$y = 2(100) + 2500 = \$2700$$

and $2700 for 100 books is

$$\frac{2700}{100} = \$27 \text{ per copy},$$

not $2 per copy. It is because of this distinction between cost per copy and the slope of the line that economists would call the slope the *marginal* cost, which is the extra cost when an *additional* copy is made.

c) The *y*-intercept, $2500, is total cost when $x = 0$ books are made. That is, at $x = 0$

$$y = 2(0) + 2500 = \$2,500.$$

This means if the machines were made ready and then it was decided not to print the book, this cost would still be incurred. It is an example of a *fixed* cost. Other examples of fixed cost are the cost of insurance on machinery whether or not it is used, and continuing rent on a building used only part time.

> *Exercise.* An agency rents cars for one day and charges $10 plus 15 cents per mile the car is driven. *a)* Write the equation for one day's rental, *y*, in terms of *x*, the number of miles driven. *b)* Interpret the slope and the *y*-intercept. *c)* What is the renter's cost per mile if a car is driven 100 miles? 200 miles? Answer: *a)* $y = 0.15x + 10$. *b)* The slope, 0.15, means that each additional mile driven adds 15 cents to total cost. The intercept, 10, is the fixed charge, which would be incurred even if the renter used the car only for a secret meeting with a friend and did not drive it out of the parking lot. *c)* $0.25 per mile and $0.20 per mile.

Linear equations in a form such as $ax + cy = d$, with *a*, *c*, and *d* constant, arise naturally in many applications. For example, suppose we have $12.60 to spend on pork and chicken. If we buy *p* pounds of pork at $0.90 per pound and *c* pounds of chicken at $0.72 per pound, our expenditures would be $0.9p + 0.72c$ dollars, and this must equal $12.60. Thus,

$$0.9p + 0.72c = 12.60.$$

In slope-intercept form, solving for p, this becomes

$$p = -\frac{0.72}{0.9}c + \frac{12.60}{0.9}, \quad \text{or} \quad p = -0.8c + 14.$$

The intercept tells us that we can buy 14 pounds of pork if we buy no chicken. The slope, -0.8, means that if we increase our purchase of chicken by one pound, we must decrease purchases of pork by 0.8 pounds. Thus, the *substitution rate* is 0.8 pounds of pork per pound of chicken.

Exercise. Solve the foregoing equation for c in terms of p in slope-intercept form, then interpret the intercept and the slope. Answer: $c = -1.25p + 17.5$. We can buy 17.5 pounds of chicken if we buy no pork. The substitution rate is 1.25 pounds of chicken per pound of pork.

1.9 STRAIGHT-LINE EQUATION GIVEN A POINT AND SLOPE

We have seen that the straight-line equation can be written directly if the slope m and a *particular* point, the y-intercept b, are given. The equation then is

$$y = mx + b.$$

If the given point is not the y-intercept, we can easily determine what this intercept is and use the last-written form.

Example. Find the equation of the line with slope 0.75 that passes through the point (8, 10).

We write first the partially complete equation

$$y = 0.75x + b, \tag{1}$$

and now we must find the value of b. Because the line passes through (8, 10), these coordinates must satisfy (1), so, substituting,

$$10 = 0.75(8) + b$$
$$10 = 6 + b$$
$$4 = b,$$

and the y-intercept is found to be $b = 4$. Hence (1) becomes the desired equation,

$$y = 0.75x + 4.$$

Exercise. Find the equation of the line that has a slope of -0.50 and passes through the point (4, 3). Answer: $y = -0.5x + 5$.

For readers who like to have separate forms for different cases, we can derive a formula easily by letting m represent the given slope and (x_1, y_1) represent the given point. The formula we want is the equation that is true not only for the point (x_1, y_1) but also for all other points (x, y) on the line. Thus, we have

$$(x_1, y_1) \text{ and } (x, y) \text{ with slope } m.$$

The slope of the line is

$$\frac{y - y_1}{x - x_1},$$

and this must equal m for all pairs of points on the line. Thus, we have:

Point-slope form: $\dfrac{y - y_1}{x - x_1} = m$, or $y - y_1 = m(x - x_1)$.

Returning to the last example, where $m = 0.75$ and the given point (x_1, y_1) is (8, 10), substitution into the slope-intercept form yields

$$\frac{y - 10}{x - 8} = 0.75$$

$$y - 10 = 0.75(x - 8)$$

$$y - 10 = 0.75x - 6$$

$$y = 0.75x + 4,$$

as before. The reader may wish to redo the last exercise using the slope-intercept form.

1.10 STRAIGHT-LINE EQUATION FROM TWO POINTS

Two points completely determine a straight line and, of course, they determine the slope of the line. Hence, we can proceed first to compute the slope, then use this value of m in

$$y = mx + b$$

together with one of the given points to find b and write the equation of the line.

Example. The total cost, y, of producing x units is a linear function. Records show that on one occasion 100 units were made at a total cost of $200, and on another occasion, 150 units were made at a total cost of \$275. Write the linear equation for total cost in terms of the number of units produced.

The information given consists of two points whose coordinates (x, y) are in the order (units made, total cost). These are:

$$(100, 200) \quad \text{and} \quad (150, 275).$$

The slope of the line is

$$m = \frac{275 - 200}{150 - 100} = \frac{75}{50} = 1.5,$$

so we have the partial answer,

$$y = 1.5x + b.$$

Picking one of the points, say (100, 200), we now determine b as follows:

$$200 = 1.5(100) + b$$
$$200 = 150 + b$$
$$50 = b.$$

Hence, the desired equation is

$$y = 1.5x + 50.$$

> **Exercise.** A publisher asks a printer for quotations on the cost of printing 1000 and 2000 copies of a book. The printer quotes \$4500 for 1000 copies and \$7500 for 2000 copies. Assume that cost, y, is linearly related to x, the number of books printed. *a)* Write the coordinates of the given points. *b)* Write the equation of the line. Answer: *a)* (1000, 4500); (2000, 7500). *b)* $y = 3x + 1500$.

To obtain a special two-point form for a straight-line equation, we designate the given points as (x_1, y_1) and (x_2, y_2). Then we let (x, y) represent any other point on the line we seek. Thus we have

$$(x_1, y_1), (x_2, y_2), (x, y).$$

The slope of the line must be the same for any pair of its points. Thus, for (x_1, y_1) and (x, y), the slope is

$$\frac{y - y_1}{x - x_1},$$

and for the given point pair, (x_1, y_1) and (x_2, y_2), the slope is

$$\frac{y_2 - y_1}{x_2 - x_1}.$$

Setting the last two expressions for the same slope equal yields:

Two-point form for a straight line:

$$\frac{y-y_1}{x-x_1}=\frac{y_2-y_1}{x_2-x_1} \text{ or } y-y_1=\left(\frac{y_2-y_1}{x_2-x_1}\right)(x-x_1).$$

To practice with this form, recall that in the last example we found the equation of the line through (100, 200) and (150, 275) was $y = 1.5x + 50$. With the two-point form we write,

$$\frac{y-200}{x-100}=\frac{275-200}{150-100}=\frac{75}{50}$$

$$\frac{y-200}{x-100}=1.5$$

$$y-200=1.5(x-100)$$

$$y-200=1.5x-150$$

$$y=1.5x+50,$$

as before. The reader may wish to redo the last exercise using the two-point form.

1.11 HORIZONTAL AND VERTICAL LINES

When the equation of a line is to be determined from two given points, it is a good idea to compare corresponding coordinates because, as we learned earlier in the chapter, if the y values are the same the line is horizontal, and if the x values are the same the line is vertical. For example, given the points

$$(3, 6) \text{ and } (8, 6)$$

we see that the line through them is horizontal because both y-coordinates are 6. This line is shown on Figure 1–10. It is clear that y is 6 at every point on this line irrespective of the value assigned to the x-coordinate, and because

$$y=6$$

describes this line, and this line only, we say $y = 6$ is the equation for the line.

If we had not noticed the equality of the y-coordinates of (3, 6) and (8, 6) and had proceeded with the slope-intercept determination of the equation by first finding

$$m=\frac{6-6}{8-3}=\frac{0}{5}=0,$$

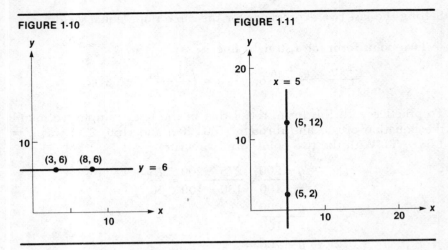

FIGURE 1-10

FIGURE 1-11

then writing the partial equation

$$y = (0)x + b,$$

substitution of one of the given points, say, (3, 6), would give

$$6 = (0)(3) + b$$
$$6 = b$$

and we would have

$$y = 6.$$

However, there is no need to proceed beyond the first step because *if the slope, m, turns out to be zero, the line is horizontal and has an equation of the form*

$$y = \text{constant},$$

where the constant is the given y-coordinate.

If the x-coordinates of the two different points are equal, as in

$$(5, 2), \quad (5, 12),$$

the line through them is vertical, as shown in Figure 1–11, and its equation is

$$x = 5.$$

If we had proceeded to apply the point-slope procedure, we would obtain

$$m = \frac{12 - 2}{5 - 5} = \frac{10}{0}.$$

We need not proceed further because *if the slope expression has a zero denominator, the line is vertical and has an equation of the form*

$$x = \text{constant},$$

where the constant is the given *x*-coordinate.

We have become accustomed to seeing both *x* and *y* in linear equations but, as we have just learned, the coefficient of *x* or the coefficient of *y* (but not both) can be zero, in which case we have a horizontal or a vertical line.

In summary:

The line through (x_1, y_1); (x_2, y_1) is $y = y_1$.

The line through (x_1, y_1); (x_1, y_2) is $x = x_1$.

Exercise. What is the equation of the line through *a)* $(-4, 7)$ and $(-4, -3)$? *b)* $(3, 6)$ and $(2, 6)$? Answer: *a)* $x = -4$. *b)* $y = 6$.

1.12 PARALLEL AND PERPENDICULAR LINES

Lines that have the same slope are parallel. Thus,

$$y = \frac{1}{2}x + 5$$

$$y = \frac{1}{2}x + 2$$

are parallel lines, as shown in Figure 1–12. We shall have occasion to be concerned about parallel lines in coming chapters where the

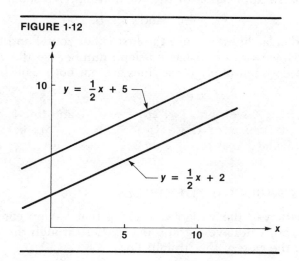

FIGURE 1-12

$y = \frac{1}{2}x + 5$

$y = \frac{1}{2}x + 2$

equation of each line expresses a condition that must be satisfied, and both conditions must be satisfied. This means that we will be seeking a point that is on *both* lines; that is, the point of intersection of the lines. Clearly, it is impossible to satisfy both conditions if the two lines are parallel because they do not intersect, and we shall say the pair of equations is inconsistent.

Exercise. Which of the following pairs of lines are parallel? *a)* $2y - 3x - 8 = 0$; $3y - 4.5x - 5 = 0$; *b)* $2x + y = 4$; $3x + 2y = 10$. Answer: The pair in *(a)*.

Perpendicular lines. If two slant lines are perpendicular, the slope of one is the negative reciprocal of the slope of the other. Thus,

$$y = \frac{2}{3}x + 5 \quad \text{and} \quad y = -\frac{3}{2}x + 10$$

are perpendicular because the negative reciprocal of the slope of the first line is

$$-\frac{1}{\frac{2}{3}} = -1(\tfrac{3}{2}) = -\tfrac{3}{2},$$

and $-\tfrac{3}{2}$ is the slope of the second line. We could as easily say that the product of the slopes of two slant lines that are perpendicular is -1. Thus, for the last pair of lines

$$m_1 m_2 = \tfrac{2}{3}(-\tfrac{3}{2}) = -1.$$

We shall have only occasional use for this relationship, so its proof is left as an exercise for the reader in the next set of problems, which also contains an application of the relationship. We should note that

$$y = 6 \quad \text{and} \quad x = 5$$

are perpendicular lines because the first is horizontal and the second is vertical. However, $x = 5$ has no slope number, so the $m_1 m_2 = -1$ relationship does not apply if the lines are not both slant lines.

Exercise. Given line 1 is $2y = 4x - 7$ and line 2 is $2y + x = 4$, to which of these is $y - 2x = 4$ *a)* parallel *b)* perpendicular? Answer: *a)* 1. *b)* 2.

1.13 LINES THROUGH THE ORIGIN

Any equation in the variables x and y that has no constant term other than zero will have a graph that passes through the origin. For example, in the case of the straight line

$$3y - 2x = 0,$$

it is obvious that $(0, 0)$, the coordinates of the origin, satisfy the equation. From an applied point of view, such lines are of interest because they are the mathematical expression of a *proportion*. That is, if we write the equation in the form

$$\frac{y}{x} = \frac{2}{3},$$

we may read the statement as "y is to x as 2 is to 3." An assumption that, say, output per man-hour is 3 units would translate to $y/x = 3$, where y is number of units of output and x is number of man-hours worked. Graphical expression of the assumption would be the straight line $y = 3x$, passing through the origin.

When considering revenue obtained from sale of a product, revenue, R, is 0 when the number of units sold, x, is 0. The revenue obtained by selling x units at a constant price of \$5 per unit would be

$$R = 5x,$$

a line through the origin.

> *Exercise.* If it costs C dollars to maintain a factory and produce x units of product, would it be reasonable to assume the cost curve goes through the origin? Answer: It would not be reasonable because at $x = 0$ (no output) costs of insurance, security, interest payments and other elements of fixed cost, which do not depend upon output, would still be incurred.

1.14 PROBLEM SET 1–4

(Note: where possible, write straight-line equations in slope-intercept form.)

Find the equation of the line passing through the given point and having the given slope:

1. $(3, 4)$, slope 3.
2. $(-5, 6)$, slope -1.
3. $(-2, 6)$, slope $-\frac{2}{3}$.
4. $(1, -4)$, slope $\frac{1}{2}$.
5. $(-3, -8)$, slope 0.
6. $(2, 7)$, slope $-\frac{1}{6}$.
7. $(-5, -8)$, slope 13.
8. $(0, 0)$, slope 0.
9. $(5, 2)$, vertical.
10. $(0, 0)$, vertical.
11. $(3, 4)$, slope 0.
12. $(4, -3)$, slope 5.

Find the equation of the line passing through each of the given pairs of points:

13. $(4, 6)$, $(-3, 7)$.
14. $(-5, 3)$, $(2, 9)$.
15. $(1, 1)$, $(3, 3)$.
16. $(-2, -4)$, $(-1, 5)$.

17. (0, 0), (2, 3).
18. (2, 4), (−3, 4).
19. (⅓, −½), (2, −3).
20. (−7, 2), (−7, −8).
21. (3, 5), (1, 4).
22. (3, 5), (−4, 5).
23. (6, 0), (10, 0).
24. (−3, −2), (4, −7).

25. Write the equation of the x-axis.
26. Write the equation of the y-axis.
27. On the line passing through (2, 3) and (−5, 6), what is the y-coordinate of the point where $x = 17$?
28. On the line of slope −2 passing through (3, 7), what is the x-coordinate of the point where $y = 17$?
29. What is the equation of the vertical line which passes through (−6, 3)?
30. a) What is the equation of the horizontal line which passes through (−6, 3)?
    b) A curve showing profit (vertical) and number of units produced and sold (horizontal) rises smoothly to a peak and then declines as we move to the right. The peak is at (100, 500). What is the equation of the tangent line at the peak? What is the significance of this equation? Hint: Make a sketch labelling the axes, showing a curve which has a mound-shaped peak.
31. a) What is the equation of the line parallel to, and 5 units above, the x-axis?
    b) What is the equation of the line parallel to, and 10 units to the left of, the y-axis?
32. As sales (x) change from $100 to $400, selling expense (y) changes from $75 to $150. Assume that the given data establish the relationship between sales and selling expense as the two change, and assume that the relationship is linear. Find the equation of the relationship.
33. As the number of units manufactured increases from 100 to 200, manufacturing cost (total) increases from $350 to $650. Assume that the given data establish the relationship between cost (y) and number of units made (x), and assume that the relationship is linear. Find the equation of the relationship.
34. If the relationship between total cost and number of units made is linear, and if cost increases by $3 for each additional unit made, and if the total cost of 10 units is $40, find the equation of the relationship between total cost (y) and number of units made (x).
35. a) If taxi fare (y) is 50 cents plus 20 cents per quarter mile, write the equation relating fare to number of miles traveled, m.
    b) The weekly earnings of a salesman are $50 plus 10% of the retail value of the goods he sells. Write the equation for earnings, E, in terms of sales volume, V. What is the slope of this line called?

Mark (T) for true or (F) for false.

36. (   ) The horizontal line through (5, 6) has the equation $x = 5$.
37. (   ) The equation of the x-axis is $y = 0$.

38. ( ) The slope of the line through (4, 6) and (5, 9) is greater than the slope of the line through (0, 0) and (1, 2).
39. ( ) The equation of the x-axis is $x = 0$.
40. ( ) The lines $x = 5$ and $x = 10$ are parallel to each other.
41. ( ) The lines $x = 5$ and $y = 10$ are perpendicular to each other.

Graph the following lines, using intercepts:

42. $x + y = 6$.
43. $3x - 2y = 12$.
44. $2x + 5y = 6$.
45. $x - y = 4$.
46. $7x + 6y - 3 = 0$.

What is the slope of each of the following lines?

47. $3x - 2y = 7$.
48. $x + y = 2$.
49. $2x - 6y = 5$.
50. $x - y = 0$.

51. *a)* A pound of food A contains 8 ounces of a nutrient, and a pound of B contains 12 ounces of the nutrient. Write the expression that must be satisfied if x pounds of A and y pounds of B are to provide 96 ounces of nutrient.
    b) What is the substitution rate of A per pound of B?
    c) What is the substitution rate of B per pound of A?

52. What is the equation of the line which has a slope of 2 and a y-intercept of −6?

53. If a straight-line equation is in the form

$$ax + by + c = 0$$

then the slope is $-a/b$; that is, the negative of the ratio of the coefficient of x to the coefficient of y. Why is this statement true?

54. If total cost is y and number of units is x, what expression represents a constant cost per unit? What equation would replace the statement that cost per unit is $3? What is the slope and what are the intercepts of the line whose equation was just written?

55. *a)* Find the equation of the line through (2, 7) which is parallel to the line $3x - 2y = 7$.
    b) Find the equation of the line through (−2, −6) which is perpendicular to the line $x = 3y - 4$.

56. A printer quotes the price of $1400 for printing 100 copies of a report and $3000 for printing 500 copies. Assuming a linear relationship, what would be the price for printing 300 copies?

57. What is the equation of the line on which the y-coordinate of any point is twice the x-coordinate?

58. Find the equation of the line through the origin which is parallel to $4x - 5y = 10$.

59. *a)* Figure A shows two perpendicular lines, l_1 and l_2. Clearly, the slope of one is positive and the other negative. Prove that angle 1 = angle

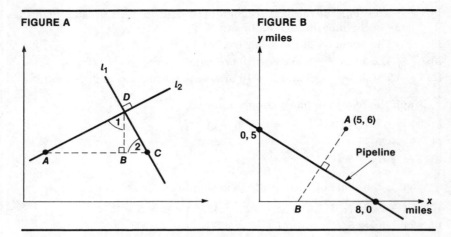

FIGURE A

FIGURE B

2. Then, using similar triangle relationships, prove that the slopes of the lines are reciprocals.

b) In what situation would slopes of perpendicular lines not be reciprocals?

60. See Figure B, which shows an existing pipeline passing through two points. Plant A is in existence at the point (5, 6) and a new plant, B, is to be located on the x-axis at a point such that the dotted pipeline which will be constructed to connect A and B to the existing line will be perpendicular to the existing line.

   a) Where should B be located?

   b) What is the advantage of having the new line meet the existing line at a right angle?

61. Prove that every line whose equation is of the form $ax + by = 0$, where a and b are any numbers (not both zero), passes through the origin.

Mark (T) for true or (F) for false.

62. ( ) The x-intercept of $3x - 2y = 12$ is -6.

63. ( ) The slope of $2x + 3y = 6$ is $-\frac{2}{3}$.

64. ( ) If the graph of the equation $ax + by + c = 0$ is to pass through the origin, c must be 0.

65. ( ) If the ratio of y to x is contant for an equation, the graph of the equation must pass through the origin.

66. ( ) $x + y = 6$ and $2x + 2y = 8$ are parallel but not identical lines.

67. ( ) $x + y = 6$ and $2x + 2y = 12$ are identical lines.

68. ( ) If the relationship between x and y can be expressed as $ax + dy = 0$, then x and y are in proportion.

69. ( ) $x = 10$ has no y-intercept.

70. ( ) Any line parallel to $x - 2y = 7$ must have a slope of $\frac{1}{2}$.

71. ( ) If y is units of output and x is man-hours worked, and if $y = 3x$, then output per man-hour is constant.

1.15 INTERPRETIVE EXERCISE: COST-OUTPUT

The purpose of this exercise is to relate the mathematical terminology of linear equations to a real-world situation. To this end, we shall assume that C is the total factory (manufacturing) cost of production of a product when Q (for quantity) units of the product are made. We assume that the relationship between C and Q is linear,[3] as shown by the line segment LM in Figure 1–13.

When making interpretations from Figure 1–13, keep in mind that a vertical distance represents a cost and a horizontal distance represents a quantity (number of units). Thus, the segment PT represents the total cost of producing OP units of product, and UM is the total cost of producing OU units of product.

FIGURE 1-13

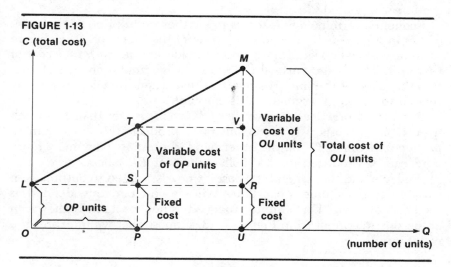

The distance OL, the vertical intercept, corresponds to the cost of operation when zero units are produced, the reasoning here being that some costs (such as insurance on the plant) exist even when no product is being made. Borrowing an accounting term, we interpret OL as *fixed cost*, that is, the component of cost that does not vary with the number of units made. In Figure 1–13, fixed cost is

$$OL = PS = UR.$$

If we make OP units of product, the total cost is PT. PT, in turn, is the sum of PS and ST. PS is fixed cost; ST we shall call the *variable*

---

[3] When the output interval is wide, the cost-output relationship may well be curved rather than straight. We shall consider the case of a curve in later chapters.

34

cost when *OP* units are made. The term variable cost refers to the component of cost that changes as the number of units produced changes. Thus, *ST* is variable cost when *OP* units are made; *RM* is the (larger) variable cost when *OU* units are made. At each level of output, total cost is the sum of fixed and variable cost.

Consider the ratio *RM/LR*. *RM* is the variable cost when *LR* (or *OU*) units are made. The ratio

Variable cost
_____
Number of units made

is the variable cost *per unit* of product made. By definition, *RM/LR* is the slope of *LM*, and the slope of a straight-line segment is constant. Consequently, variable cost per unit is constant when cost and output are linearly related.[4]

Economic terminology leads us to another description of the slope of *LM* in Figure 1–13. Economists speak of the *extra* cost (the change in total cost) when one more unit is made as the *marginal cost* of that unit. This is the vertical change for a horizontal change of one. It is the slope of the line. We may say that when total cost is linearly related to output, marginal cost is constant.

For further practice, observe that *UM* is *VM* greater than *PT*, which means that it costs *VM* more dollars to produce *OU* units than to produce *OP* units. *TV*, on the other hand, represents how many more units can be produced for *UM* dollars than for *PT* dollars.

We have said that when total cost is linearly related to output, then variable cost per unit is constant. Average cost per unit, defined as total cost over number of units produced, is not constant. Rather than argue this statement from Figure 1–13, suppose that we interpret the equation

$$y = 3x + 2,$$

letting y be total cost of producing x units. The fixed cost is $2, and the variable cost per unit (the slope) is $3. However, by substitution, we find that total cost rises from $17 to $32 if x changes from 5 to 10 units. The average cost per unit declines from 17/5 to 32/10, that is, from $3.40 to $3.20. This reduction sometimes is referred to as being a consequence of spreading fixed cost over a larger number of units.

1.16 PROBLEM SET 1–5

Mark (T) for true or (F) for false. (Assume that cost means factory cost.) Given that total cost, y, of making x units is $y = 5x + 10$:

_____

[4] If, on a cost *curve*, variable cost is divided by number of units produced, the ratio is called *average* variable cost per unit and is *not* constant.

1. (   ) The total cost of making 20 units is $110.
2. (   ) Average cost per unit is $6 if 10 units are made.
3. (   ) The marginal cost of the 11th unit is greater than $5.
4. (   ) The marginal cost of the 20th unit is $5.
5. (   ) The variable cost per unit decreases as the number of units made increases.
6. (   ) The variable cost incurred when making 10 items is $50.
7. (   ) Average cost per unit decreases as the number of units made increases.
8. (   ) Variable cost increases as the number of units made increases.
9. (   ) The marginal cost of every unit is the same.
10. (   ) The slope of the line is the variable cost per unit.
11. If the total factory cost, y, of making x units of a product is given by $y = 3x + 20$, and if 50 units are made:
    a) What is the variable cost?
    b) What is the total cost?
    c) What is the variable cost per unit?
    d) What is the average cost per unit?
    e) What is the marginal cost of the 50th unit?
12. If total factory cost, y, of making x units of a product is $y = 10x + 500$ and if 1000 units are made:
    a) What is the variable cost?
    b) What is the total cost?
    c) What is the variable cost per unit?
    d) What is the marginal cost of the last unit made?
13. A printer quotes a price of $7500 for printing 1000 copies of a book and $15,000 for printing 2500 copies. Assuming a linear relationship and that 2000 books are printed:
    a) Find the equation relating the total cost, y, to x, the number of books printed.
    b) What is the variable cost?
    c) What is the fixed cost?
    d) What is the variable cost per book?
    e) What is the average cost per book?
    f) What is the marginal cost of the last book printed?
14. If total factory cost, y, of making x units of a product is given by $y = 2x + 25$:
    a) Graph the cost-units equation.
    b) Draw a line representing fixed cost on the graph for (a).
    c) Erect a vertical line at $x = 10$, intersecting the x-axis at R, the fixed cost line at S, and the given equation at T.
    d) What are the numerical values and the interpretations of RS, ST, and RT?
15. In Figure A, the slant line represents the relation between total factory cost, C, of producing a number of units, and the number of units, Q, produced. What line segment(s), or ratios thereof, represent:

a) Fixed cost?
b) Total cost if *OA* units are made?
c) Variable cost if *OA* units are made?
d) Variable cost per unit made?
e) Average cost per unit if *OA* units are made?
f) Fixed cost per unit if *OA* units are made?
g) How many more units can be made for *AD* dollars than for *AF* dollars?
h) Marginal cost per unit?

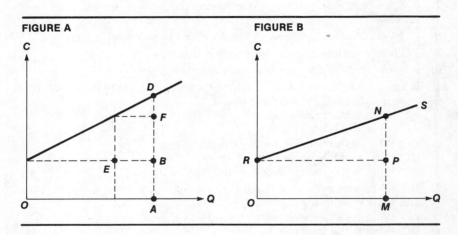

FIGURE A

FIGURE B

16. Prior to making a number of units of a certain part, a machine must be made ready, the cost incurred being called the setup cost. The total machine shop cost, *C*, of making *Q* units of the part is shown in Figure B as the segment *RS*. What interpretation would be given to:
a) *OR?*
b) *PN?*
c) *PN/OM?*
d) *MN/OM?*

1.17 COMMENT ON MODELS

We call attention here to the distinction between the mathematics of the last discussion and the real-world cost-output situation to which the mathematics was related. Nothing was said about whether costs *are* or whether costs *should be* linearly related to output. The theme of the discussion was that *if* the relationship was linear, then certain interpretations suggested themselves. In particular, mathematical analysis of the linear equation showed that the ratio of *y* to *x* decreased as *x* increased, and we interpreted this as being a consequence of spreading fixed cost over an increasing output.

The linear equation under discussion is an example of a mathematical model. In broad context, a *model* is a simplified representation of reality. Thus, for example, engineers make various models of airplane wings to test in a wind tunnel before deciding upon a final design, and architects make several small-scale models of different building designs to help a client form a mental picture of real buildings before selecting a particular design. A mathematical model is a mathematical expression, or set of expressions, that seeks to capture the essential features of a real-world situation. If this objective can be achieved, then the full force of mathematical analysis can be applied to the model, and the outcome of the analysis may have important real-world interpretations. The key idea here is that mathematical analysis is a highly developed and powerful tool, but before it can be brought to bear on a real-world problem, it is necessary to construct an adequate mathematical representation (model) of the real-world situation. We do not expect a model to be an exact replica of the real world. We hope that the model will be adequate. In any event, we must be careful not to exceed the bounds of good sense when interpreting the outcome of mathematical analysis of a model. Thus, it probably would not be realistic to assume a linear relationship between cost and output no matter how large output became, and it certainly would make no sense to extend the linear model to negative outputs.

Physical scientists have had a high degree of success in formulating mathematical models. Witness, for example, the contributions of mathematical analysis of gravitational forces to man's achievements in space. Much effort now is being directed toward model building by workers in the social sciences. The linear programming model we shall encounter a few chapters hence is a good example of a relatively recent advance that has stirred interest in many parts of the world of business and economics.

A model should *fit* actual data reasonably well. To understand what is involved here, recall an earlier example in which a straight line was fitted to points whose coordinates were in the order (disposable income, personal consumption expenditures), and the line passed through (312, 295) and (575, 537). The actual data used in determining these points are shown in Figure 1–14, which also shows the plotted points (solid) and a line drawn freehand (with the aid of a stretched thread). Clearly, no line passes through all of the points, but the freehand line comes very close to the points, and the fit is very good— much better than is the case in other situations. Essentially the same line would be obtained if we had used points for all years from 1950 to 1970 and had applied the objective techniques of fitting developed in a study of statistics.

If we let C be personal consumption expenditures and D be disposable income, then use the two-point formula with (312, 295) and (575, 537), we obtain the linear model

$$C = 0.92D + 7.96.$$

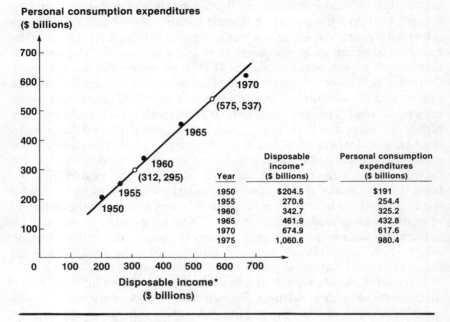

FIGURE 1-14   Linear model fitted to consumption expenditures and
disposable income in the United States. Selected years, 1950-1970.

Year	Disposable income* ($ billions)	Personal consumption expenditures ($ billions)
1950	$204.5	$191
1955	270.6	254.4
1960	342.7	325.2
1965	461.9	432.8
1970	674.9	617.6
1975	1,060.6	980.4

* Less interest paid by consumers.
Source: Economic Report of the President, 1973, 1978.

The slope, 0.92, is called the marginal propensity to consume, as we
learned earlier. What about the "intercept" or constant 7.96? Mechani-
cally, this is consumption when disposable income is zero, but we should
not infer anything about consumption when D is zero because we
have no data near zero and, in fact, D will never be zero. Moreover,
in this changing economy we should be careful about extrapolating
(going beyond the region of experience) very far. An appropriate interpretation of the constant 7.96 is that consumption expenditures are
in major part determined by income, but that at any level of income,
a part ($7.96 billion) is *independent* of income. This fixed part is called
autonomous consumption. Thus, a common model of consumption in
economics takes the form

$$C = A + MD$$

where the variables C and D are as defined earlier, the constant A is
autonomous consumption, and the constant M is the marginal propensity to consume (about 0.92 in recent years).

Having warned about extrapolating beyond the region of experience, we must now admit that a major reason for developing a model

is to make projections. Projections answer such questions as, What would consumption be if disposable income were $1061 billion, assuming the model $C = 0.92D + 7.96$ applies? Mechanical substitution leads to the projection $C = \$984$ billion. As a matter of record, disposable income did reach $1061 billion in 1975, and in that year consumption expenditures were $980 billion, so the projection was very close to the actual figure. We see that extrapolation can lead to quite accurate results, but we have no assurance of accuracy, and our confidence lessens the further we depart from the region of experience.

As a final comment, we should point out that many of the equations used in textbooks are simplistic ones, with nice whole numbers as coefficients. Students ask, quite rightly, where they come from. The answer, of course, is that they were designed by the writer of the text to facilitate learning of mathematical techniques, for such is the objective of the text. One of the purposes of the foregoing discussion of models was to indicate where real-life equations come from, and to assure the student that simplified equations used for text illustration do have real-life counterparts.

1.18 BREAK-EVEN INTERPRETATION: 1

In this section we shall consider a manufacturer who produces q units of a product and sells the product at a price of $\$p$ per unit. The symbols to be used are

$C=$ total cost of producing and selling q units.
$q=$ number of units produced and sold.
$v=$ variable cost *per unit* made, assumed to be constant.
$F=$ fixed cost, a constant.
$p=$ selling price per unit.
$R=$ total revenue received, which is the same as the dollar volume of sales.

The cost function then is given by

$$C = vq + F \qquad (1)$$

and

$$\text{Revenue} = (\text{price per unit})(\text{number of units sold})$$
$$R = pq. \qquad (2)$$

If the manufacturer is just to break even on operations, neither incurring a loss nor earning a profit, revenue (2) must equal cost (1). That is, at break-even,

$$pq = vq + F. \qquad (3)$$

40

We now may solve (3) for the production volume, q:

$$pq - vq = F$$
$$q(p - v) = F.$$

Break-even quantity: $q_e = \dfrac{F}{p - v}.$ (4)

Example. A manufacturer of Quandries has a fixed cost of $10,000 and variable cost per Quandry made is $5. Selling price per unit is $10. *a)* Write the revenue and cost equations. *b)* At what number of units will break-even occur? *c)* At what sales volume (revenue) will break-even occur?

a) The equations are:

Revenue: $R = 10q$
Cost: $C = 5q + 10,000.$

b) We can find the break-even quantity, q, by noting that

$$F = 10,000; \quad p = 10; \quad v = 5$$

and substituting these into (4) to obtain

$$q_e = \frac{10,000}{10 - 5} = \frac{10,000}{5} = 2000 \text{ units.}$$

Alternatively, we can equate the revenue and cost equations in the answer to part *(a)*, obtaining

$$10q = 5q + 10,000$$
$$5q = 10,000$$
$$q_e = \frac{10,000}{5} = 2000 \text{ units.}$$

c) At break-even, 2000 units would be produced and sold at $10 each, so the break-even sales volume would be

$$R = (10)(2000) = \$20,000.$$

> *Exercise.* A manufacturer of Queries has a fixed cost of $60,000 and variable cost is $9 per Query produced. Selling price is $12 per Query. *a)* Write the revenue and cost equations. *b)* At what number of units will break-even occur? *c)* At what sales (revenue) volume will break-even occur? Answer: *a)* $R = 12q$; $C = 9q + 60,000$. *b)* 20,000 units. *c)* $240,000.

Break-even charts. Returning to the equations of the last example,

$$R = 10q \quad \text{and} \quad C = 5q + 10,000$$

FIGURE 1-15

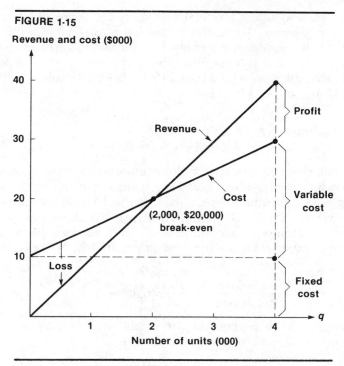

Revenue and cost ($000)

and plotting these in the usual manner, we obtain Figure 1–15. Observe that the revenue line goes through the origin and that the cost line has a *y*-intercept of $10,000. The dotted horizontal line shows that this fixed cost is constant at all levels of operation. The *total* variable cost, which is the 5*q* in

$$C = 5q + 10{,}000,$$

(not the $5 variable cost per *unit*, which is constant) is the vertical distance from the fixed cost line to the total cost line which, of course, increases as more units are produced. To the left of the break-even point (2000, $20,000), the cost line is above the revenue line and the vertical separation at any point represents loss, while to the right of break-even the vertical separation represents a profit. Break-even charts are a helpful graphic aid frequently used in planning business operations.

1.19 BREAK-EVEN INTERPRETATION: 2

Production managers and other operations executives tend to think of break-even analysis in the way it was presented in the last section.

Comptrollers and other financial managers are more likely to think in accounting terms. To illustrate this latter way of viewing break-even, we shall consider a company that purchases products and sells them at a price that is, hopefully, above the cost price. Suppose, for example, that an item which cost $130 is priced to sell at $200. The *markup* is, therefore, $70. That is,

$$\text{Cost} = \$130$$
$$\text{Retail price} = \$200$$
$$\text{Markup} = \text{Retail} - \text{Cost} = 200 - 130 = \$70.$$

From a manager's viewpoint, the dollar amounts of markup on numerous individual items, which will vary widely, are not very useful in planning and controlling operations. What is useful is the overall markup *percentage* on all items. To secure necessary comparability in financial statements, accountants use the *markup percentage on retail price* rather than on the cost price. In the example at hand, this is

$$\frac{\text{Markup}}{\text{Retail price}} = \frac{70}{200} = 0.35 \text{ or } 35\%.$$

This means that 35 percent of the retail price of $200 is markup, and the other 65 percent of $200, which is

$$0.65(200) = \$130,$$

is the cost.

We now suppose that the company in our illustration has a markup of 35 percent of retail on *all* items it purchases, so that if the firm sells x worth of merchandise, 35 percent of this amount is markup and 65 percent is cost. Thus,

$$\text{On } \$x \text{ of sales,} \quad 0.65\,x = \text{cost of goods sold.} \tag{1}$$

Next, the company incurs selling expenses, which it budgets at 10 percent of the volume of sales. Hence,

$$\text{On } \$x \text{ of sales,} \quad 0.10\,x = \text{selling expense.} \tag{2}$$

Finally, the company budgets fixed expense at $12,000, so that

$$F = \text{fixed expense} = \$12,000. \tag{3}$$

If we now let y be total cost, the sum of (1) through (3), we have

$$y = 0.65\,x + 0.10\,x + 12,000$$
$$y = 0.75\,x + 12,000. \tag{4}$$

Thus, at a sales volume of $60,000, cost will be

$$y = 0.75(60,000) + 12,000 = \$57,000$$

and profit before taxes will be

$$60,000 - 57,000 = \$3000.$$

Exercise. Compute profit if sales are $40,000. Answer: Cost would be $42,000, so a loss of $2,000 would arise.

From the foregoing example of sales levels leading to profit and loss we are led to inquire what the *break-even* level of sales would be; that is, the level of sales that will equal cost. Here, $x of sales must equal the expression for y in (4). That is,

$$x = 0.75\,x + 12,000$$
$$x - 0.75\,x = 12,000$$
$$0.25\,x = 12,000$$
$$x = \frac{12,000}{0.25}$$
$$x_e = \$48,000,$$

where x_e is the break-even sales level.

Exercise. If cost, y, is related to sales, x, by $y = 0.6\,x + 100$, show that the break-even point is $250.

If the relationship between cost and sales is written in the general slope-intercept form

$$y = mx + b$$

then the interpretations are that b represents fixed cost, mx represents variable cost, and m itself represents variable cost per dollar of sales. At break-even:

$$y = x = x_e$$

so that, by substitution:

$$x_e = mx_e + b$$
$$x_e(1 - m) = b$$
$$x_e = \frac{b}{1 - m}.$$

Hence, the break-even level occurs when sales are equal to

$$\frac{\text{Fixed cost}}{1-\text{Variable cost per dollar of sales}}.$$

Suppose that in making a budget for next year's operations, top management has set a sales goal of $200,000. Markup is to be 45 percent on retail (so cost is 55 percent of retail), and other variable cost is estimated at $0.05 per dollar of sales, so that variable cost is 0.55 + 0.05 = 0.60 per dollar of sales. Fixed cost is projected at $56,000. Then the linear cost-sales model will be

$$y = 0.6x + 56,000.$$

The break-even level of sales will be

$$x_e = \frac{56,000}{1-0.6} = 140,000.$$

The calculations show the company will make a profit if its sales exceed $140,000. For example, if sales should turn out to be $190,000, net profit would be

$$190,000 - \text{Cost} = 190,000 - [0.6(190,000) + 56,000] = 20,000.$$

It is useful to keep in mind that a budget establishes *estimates* for coming operations, and that *actual* operations are not expected to be exactly equal to the estimates. However, assuming that the linear model applies, the estimates provide the information needed to predict results (say profit) if sales vary from the estimated value.

> **Exercise.** Markup is to be 33 percent on retail and other variable cost is estimated at $0.13 per dollar of sales. Fixed cost is estimated at $4,000. *a)* What is the break-even point? *b)* Estimate profit if sales are $50,000. Answer: *a)* The equation is $y = 4,000 + 0.8x$ and the break-even point is $4000/(1 - 0.8) = \$20,000$. *b)* $6,000.

A *break-even chart* can be constructed by plotting

$$y = mx + b$$

and the line where cost equals sales, $y = x$. The point of intersection of the last two lines establishes the break-even level of sales. Figure 1–16 shows the break-even chart for the foregoing example. The break-even point is seen to be $140,000. The separation of the lines to the right of break-even indicates profit; to the left, loss.

In passing, we should note that the break-even interpretations just discussed rest upon the assumption that total cost can be separated

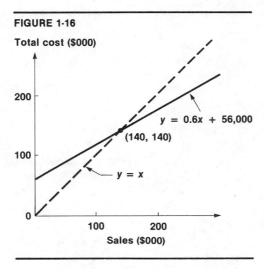

FIGURE 1-16

Total cost ($000)

$y = 0.6x + 56,000$

(140, 140)

$y = x$

Sales ($000)

into two components, one fixed and the other varying directly in proportion to sales. These assumptions often are reasonably valid for a restricted range of sales. It is not realistic, however, to assume fixed cost as constant over all ranges of sales. If sales are proving to be considerably below expected levels, management may reduce salaries or take other actions to reduce "fixed" cost. It is not our purpose here to explore managerial action, but rather again to call for exercise of judgment when interpreting mathematical models.

1.20 LINEAR DEMAND FUNCTIONS

In this section we introduce the economic concept of the demand functions for products. Usually these functions are curves rather than straight lines, but lines provide good illustrations of demand characteristics. The demand for a product is the amount, q, of the product consumers are willing and able to buy at a given price per unit, $\$p$. Price per unit and demand are related and it is conventional in economics to plot price on the vertical and demand on the horizontal. For example, suppose the demand function for a product at one point in time is

$$DD: \qquad p = -0.2q + 20 \qquad (1)$$

where q is in millions of pounds and p is in dollars per pound. The line is shown on Figure 1–17, together with the intercepts that were used to plot it and is labeled in the conventional economic manner as *DD*.

FIGURE 1-17

Price per unit

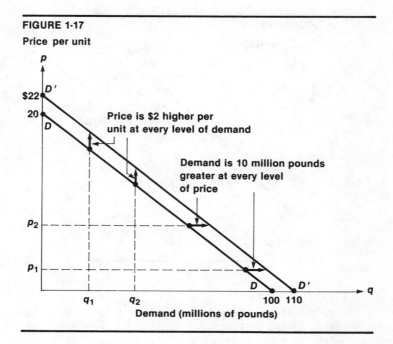

From (1), we compute that at demand $q = 25$ million pounds, $p = -0.2(25) + 20 = \$15$ per pound. Similarly, at $q = 50$, $p = \$10$ per pound. Observe that

At $p = \$15$ per pound, demand = 25 million pounds.
At $p = \$10$ per pound, demand = 50 million pounds.

Thus, higher prices are associated with lower demand, as we would expect, and this aspect of demand functions is present in (1) because the line has a negative slope and, therefore, slants downward to the right, as shown in Figure 1–17.

Our purpose in this section is to discuss the idea of a parallel *shift* in the demand function DD. The basic idea involved is that as time goes on, various factors such as introduction of competing products, changes in population or family income, or changes in individual tastes, cause the demand function to shift position. Suppose, for example, that introduction of a competing product causes the demand function to shift. And suppose the new demand function, denoted by $D'D'$, is

$$D'D': \qquad p = -0.2q + 22,$$

which is also plotted on Figure 1–17. Inasmuch as $D'D'$ is parallel to DD, the vertical separation of the lines is the same wherever it is measured. This separation is the difference in the intercepts on the vertical axis:

$$D'D' \text{ intercept} - DD\text{ intercept} = 22 - 20 = \$2.$$

We describe this *upward* shift by saying that *price per unit is $2 higher at every level of demand.* We can equally well describe the shift in terms of the constant horizontal separation of the lines, which is the difference of the intercepts on the horizontal axis:

$$D'D' \text{ intercept} - DD \text{ intercept} = 110 - 100 = 10 \text{ million pounds.}$$

This shift to the *right* is described by saying that after the shift, *demand is 10 million pounds greater at every level of price.*

> **Exercise.** Suppose that the demand functions just considered were interchanged, so that
>
> DD: $p = -0.2q + 22$, $D'D'$: $p = -0.2q + 20$,
>
> where $D'D'$ arose because of changes in incomes of families. In the manner of the foregoing italicized statements, describe: *a)* The vertical shift. *b)* The horizontal shift. Answer: *a)* Price per unit is $2 lower at every level of demand. *b)* Demand is 10 million pounds less at every price level.

The reader will notice that in the last example and exercise, the shifts were occasioned by changes in consumer demand. Mention should also be made of the fact that suppliers provide goods that consumers demand. The amount suppliers are willing to provide also is a function of price per unit, so there is a supply function that interacts with the demand function. We will return to this matter in the next chapter in our discussion of systems of linear equations.

Horizontal and vertical demand functions. Suppose the quantity of water demanded by rulers of a very wealthy desert nation remains constant when the price of water varies over a limited range. This gives rise to the notion of a vertical demand function, such as

$$DD: \qquad q = 10 \text{ billion cubic feet}$$

shown in Figure 1–18A, and we describe this by saying demand is constant at 10 billion cubic feet at different price levels. On the other hand, suppose that a coin collector will pay, say

$$p = \$5,$$

for every coin of a certain type that is offered to him. This gives rise to the horizontal demand line in Figure 1–18B. Readers will find

48

FIGURE 1-18A   FIGURE 1-18B

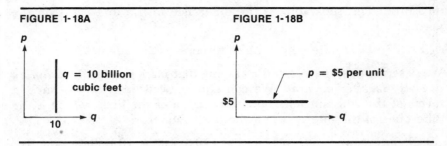

these cases discussed in economics texts under the headings *perfectly inelastic* demand, Figure 1–18A, and *perfectly elastic* demand, Figure 1–18B.

> *Exercise.* A college charges $75 per credit hour. If q is the number of credit hours supplied to students, write and interpret the equation of the supply line. Answer: $p=75$, which means that the price per credit hour is $75 no matter how many credit hours are supplied to students.

1.21 PROBLEM SET 1–6

Mark (T) for true or (F) for false.

Given that total cost incurred in producing and selling q units of a product is $C=6q+10,000$, and the q units are sold at a price of $46 per unit:

1. ( ) Total cost for 1000 units is $16,000.
2. ( ) Average cost per unit is $6 no matter how many units are made and sold.
3. ( ) Average price is $46 per unit no matter how many units are made and sold.
4. ( ) Every additional unit made and sold increases total cost by $6.
5. ( ) Variable cost is the same at different values of q.
6. ( ) When 1000 units are made and sold, variable cost will be $6,000.
7. ( ) Marginal cost is the same at all values of q.
8. ( ) Marginal cost is the same as average cost per unit.
9. ( ) A loss will be incurred if 200 units are made and sold.
10. ( ) A profit will be made if 250 units are made and sold.
11. ( ) Total cost is zero if no units are produced.
12. ( ) 250 units must be made and sold in order to break even.
13. A manufacturer has a fixed cost of $60,000 and a variable cost of $2 per unit made and sold. Selling price is $5 per unit.

    a)   Write the revenue and cost equations, using C for cost and q for number of units.

    b)   Compute profit if 25,000 units are made and sold.

    c)   Compute profit if 10,000 units are made and sold.

    d)   Find the break-even quantity.

    e)   Find the break-even dollar volume of sales (revenue).

    f)   Construct the break-even chart. Label the cost and revenue lines, the fixed cost line, and the break-even point.

14.   A manufacturer has a fixed cost of $120,000 and a variable cost of $20 per unit made and sold. Selling price is $50 per unit.

    a)   Write the revenue and cost equations, using C for cost and q for number of units.

    b)   Compute profit if 10,000 units are made and sold.

    c)   Compute profit if 1000 units are made and sold.

    d)   Find the break-even quantity.

    e)   Find the break-even dollar volume of sales (revenue).

    f)   Construct the break-even chart. Label the cost and revenue lines, the fixed cost line, and the break-even point.

15.   A company's cost function for the next 3 months is

$$C = 200,000 + 100q.$$

Find the break-even quantity if the selling price is $180 per unit.

16.   A company's cost function for the next three months is

$$C = 500,000 + 5q.$$

Find the break-even *dollar volume of sales* if the selling price is $5.50 per unit.

17.   (See Problem 15.)

    a)   What would be the company's cost if it decided to shut down operations for the next three months?

    b)   If, because of a strike, the most the company can produce is 1000 units, should it shut down? Why or why not?

18.   (See Problem 16.)

    a)   What would be the company's cost if it decided to shut down operations for the next three months?

    b)   If, because of a strike, the most the company can produce is 100,000 units, should it shut down? Why or why not?

Mark (T) for true or (F) for false.

Given that total cost, y, is related to sales volume, x, by the equation $y = 1000 + 0.2x$:

19.   ( ) Variable cost will be $200 on sales of $1000.

20.   ( ) Variable cost per dollar of sales is $0.20.

21.   ( ) Fixed cost is $1000.

22.   ( ) A loss would occur if sales were $1000.

23.   ( ) Sales of $2000 would lead to a profit of $600.

24.   ( ) The cost-sales line rises $2 for each increase of $10 in sales volume.

25. ( ) Average cost per dollar of sales is the same at various sales levels.

26. ( ) The slope of the line is interpreted as variable cost.

27. ( ) Variable cost is a constant.

28. ( ) Variable cost per dollar of sales is constant.

29. A company expects fixed cost of $22,800. Markup is to be 55 percent on retail. Variable cost in addition to costs of goods is estimated at $0.17 per dollar of sales.
    a) Find the break-even point.
    b) Write the equation relating cost and sales.
    c) What will net profit before taxes be on sales of $75,000?
    d) Make the break-even chart.

30. A company expects fixed cost of $36,000. Markup is to be 52 percent on retail, and variable cost in addition to cost of goods is estimated at $0.07 per dollar of sales.
    a) Find the break-even point.
    b) Write the equation relating sales and cost.
    c) What will net profit before taxes be on sales of $75,000?
    d) Make the break-even chart.

31. If total cost, y, is related to sales volume, x, by the equation $y = 0.47x + 29{,}786$, find:
    a) Variable cost per dollar of sales.
    b) Fixed cost.
    c) Total cost on sales of $72,000.
    d) The break-even point.
    e) Net profit before taxes on sales of $80,000.

32. If total cost, y, is related to sales volume, x, by the equation $y = 0.32x + 23{,}800$, find:
    a) Variable cost on sales of $40,000.
    b) Fixed cost.
    c) The break-even point.
    d) Net profit before taxes on sales of $30,000.

33. If variable cost per dollar of sales remains at last year's level, $0.40, but fixed cost this year is $3600 compared to $3000 last year, how much greater will this year's break-even point be than last year's?

34. Make the break-even chart for the equation in
    a) Problem 31.   b) Problem 32.

35. The demand function for a product shifts from

    $$DD: \quad p = -0.10q + 40 \quad \text{to} \quad D'D': \quad p = -0.10q + 35.$$

    Compute the horizontal and vertical shifts and write interpretive descriptions of these numbers.

36. The demand function for a product shifts from

    $$DD: \quad p = -0.05q + 40 \quad \text{to} \quad D'D': \quad p = -0.05q + 50.$$

    Compute the horizontal and vertical shifts and write interpretive descriptions of the numbers.

37. Write a descriptive interpretation of a demand function that is
    a) Vertical.   b) Horizontal.

38. The price of a particular raw material varies markedly from week to week. To be sure that it will be able to obtain this raw material, a company promises a supplier that it will buy 100 tons per week at whatever the going price is. Write the demand function implied by this promise.

39. An electric power company charges residential customers a fixed amount per kilowatt-hour of electricity used. What would be the graphical nature of the demand function in this case?

1.22 THREE-SPACE

The piece of paper or the blackboard upon which we plot points are segments of planes. On a given plane any point can be specified by an origin and two coordinates, and we refer to the points on a plane as points in space of two dimensions or, more briefly, points in *two-space*. If, now, we think of points in the interior of a room, referred to an origin, which is the point at the lower left corner of the room, three coordinates are required to locate the point. The points in the room are points in *three-space*.

Figure 1–19 illustrates coordinates for three-space. The x-, y-, and z-axes are mutually perpendicular. The x- and z-axes are in the plane of the paper, and the y-axis is to be visualized as a perpendicular coming out of the page.[5] The dotted extensions represent the negative

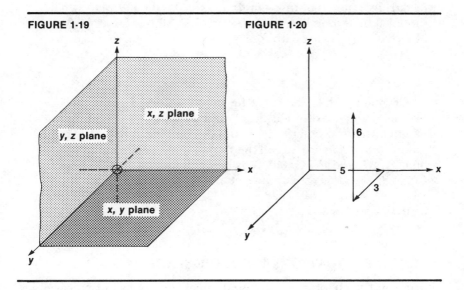

FIGURE 1-19   FIGURE 1-20

---

[5] The assignment we have chosen for the x-, y-, and z-axes is fairly standard. However, some texts have the positive direction of the y-axis into, rather than out of, the page or interchange the x- and y-axes, or both.

portions of the axes. Viewing the figure as a section of a room, the inner left corner of the room is the origin; the floor of the room is the plane of the x- and y-axes; one wall of the room is the plane of the x- and z-axes, the other is the plane of the y- and z-axes. We refer to these coordinate planes as the x, z-plane; the x, y-plane; and the y, z-plane.

The z-axis is a straight line formed by the intersection of the two walls. The x- and y-axes similarly are formed by the intersection of two coordinate planes. In passing, it is worth noting for future reference that *the intersections here discussed are particular cases of the general observation that if two planes intersect, the intersection is a complete straight line.*

Figure 1–20 illustrates the plotting of points in three-space. The point (5, 3, 6), that is:

$$x = 5, \quad y = 3, \quad z = 6,$$

is found by counting five units to the right along the x-axis, then three units out into the x, y plane in a direction parallel to the y-axis, then six units upward, parallel to the z-axis.

Exercise. Take three rectangular pieces of cardboard (3 by 5 inch cards will do) and by cutting slits and assembling, construct a model of three mutually perpendicular planes. The model should look like the interior walls, and separating floor, of a two-story house with four rooms up and four rooms down.

Exercise. The planes of the model referred to in the last exercise, when extended indefinitely, divide space into eight octants. Call the octant at the upper-front-right octant I, then going counterclockwise call the remaining octants on the upper floor II, III, and IV. Octant V is below I, and again going counterclockwise, the remaining octants on the lower side are VI, VII, and VIII. Verify that (1, −2, 3) is in octant II, (1, 2, −3) is in V, and (−1, −2, −3) is in VII.

1.23 LINEAR EQUATIONS IN THREE-SPACE

Earlier in the chapter, we saw that linear equations in two variables were represented geometrically by straight lines in two-space. The general form of such equations is $ax + by = c$. Special cases were those in which either a or b was 0, such as $y = 3$ or $x = -5$. We saw that these special cases were lines parallel to one of the coordinate axes (or perpendicular to the other). In what follows, we find that equations in three variables of the form

$$ax + by + cz = d$$

are represented geometrically in three-space as *planes,* and that special cases, where one or two of *a, b,* and *c* are zero, are represented by planes that bear parallel or perpendicular relationships to the coordinate planes and axes.

Consider the very special equation in three variables,

$$0x + 0y + z = 0,$$

which we shall refer to as the equation $z = 0$ in three-space. Graphically, any point in the *x, y*-plane has a *z*-coordinate of zero, and any point for which *z* is zero lies in the *x, y*-plane. Hence, $z = 0$ is the equation of the *x, y*-plane itself. Similarly, in three-space, $x = 0$ is the equation of the *y, z*-plane, and $y = 0$ is the equation of the *x, z*-plane. See Figure 1–21.

FIGURE 1-21

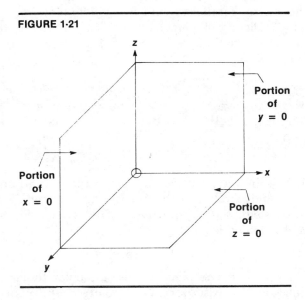

Consider next the plane that is parallel to the *y, z*-plane, and four units to the right thereof. This plane will, of course, be perpendicular to the *x, y*-plane and to the *x, z*-plane. See Figure 1–22. Clearly, *x* is 4 for any point on this plane, and any point for which *x* is 4 is on the plane. Hence, $x = 4$ is the equation of the plane. To emphasize

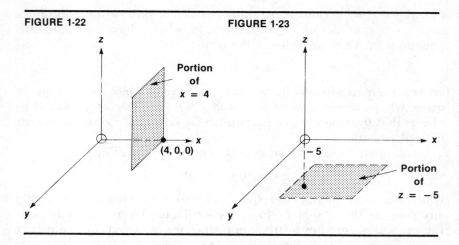

FIGURE 1-22　　　　　　　　**FIGURE 1-23**

that we are thinking in three-space, we could write the equation as $x + 0y + 0z = 4$. Similarly, in three-space, the equation $z = -5$ represents a plane that is parallel to, and five units below, the x, y-plane. See Figure 1–23. Planes, of course, extend indefinitely, so that illustrations such as those in Figures 1–22 and 1–23 represent only finite portions of planes.

> ***Exercise.*** Sketch a portion of the plane $y = 2$. Answer: This plane is parallel to, and two units "in front of," the x, z plane.

Figure 1–24 shows a segment of a plane that is perpendicular to the x, z-plane (and parallel to the y-axis) and cuts the x- and z-axes at $A(3, 0, 0)$ and $B(0, 0, 1)$, respectively. The relationship among the coordinates at A and B is

$$x + 3z = 3.$$

Moreover, inasmuch as the plane is perpendicular to the x, z-plane, the same relationship exists for every point on the plane, and the coordinates of every point on the plane will satisfy the equation. Hence:

$$x + 3z = 3$$

in three-space is the equation of a plane that is perpendicular to the x, z-plane. Similarly:

$$2z + 5y = 10$$

in three-space is a plane that is perpendicular to the y, z-plane.

FIGURE 1-24

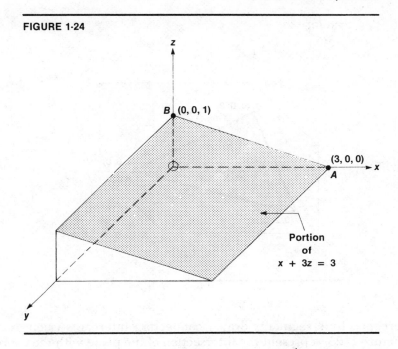

Finally, think of a plane that slices up through the floor and cuts floor and walls in the manner shown in Figure 1–25. This plane is not parallel to any of the coordinate planes. The section shown is of triangular appearance because the parts of the plane below the floor and behind the walls have not been shown. Suppose that the plane cuts the axes at the points A, B, and C, as shown. These three points completely determine a unique plane. By trial and error, we can determine that the points $(6, 0, 0)$, $(0, 0, 3)$, and $(0, 2, 0)$ satisfy the linear equation

$$x + 3y + 2z = 6.$$

We state without further demonstration that any point on the plane containing A, B, and C satisfies the last equation and that any point whose coordinates satisfy the equation will lie on the plane.

Our purpose in the last few paragraphs has been to build an intuitive basis for the statement that *every* equation of the form

$$ax + by + cz = d$$

in three-space is represented by a plane. The statement can be justified by rigorous mathematical development, but we choose here to rely upon intuition.

For sketching purposes, points such as A, B, and C on Figure 1–25 can be found by setting pairs of coordinates equal to zero to obtain

56

FIGURE 1-25

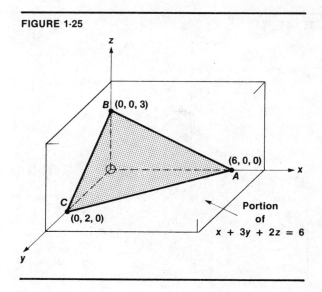

intercepts. Line segments drawn connecting intercepts (such as *AB* in Figure 1–25) represent the intersection of the plane with the coordinate plane and are called *traces;* thus, *AB* is the *x, z*-trace of the plane $x + 3y + 2z = 6$.

> ***Exercise.*** Sketch the plane $6x + y + 2z = 12$ in the manner of Figure 1–25. Answer: The intercepts are $(0, 0, 6)$, $(0, 12, 0)$, and $(2, 0, 0)$.

1.24 *n*-SPACE AND HYPERPLANES

The linear equation in four variables:

$$w + x + 2y + 3z = 12$$

is satisfied by the point $(1, 5, 0, 2)$, by $(0, 0, 0, 4)$, and by an unlimited number of other points. Each point has four coordinates (written in alphabetical order) and is a point in four-space. Inasmuch as we are limited to three-dimensional perception, the notion of four-space is purely a matter of terminology; we agree to call $(1, 5, 0, 2)$ the coordinates of a point in four-space. Again, inasmuch as the last equation is linear, like that of a plane but with more than three variables, we agree to call the expression the equation of a *hyperplane.*

Although we are limited to three-space in geometrical constructions, there is no such limit on mathematical analysis. We may speak of 6-space, 150-space, or n-space, where n is any positive integer, and we may treat equations in n variables. It should be understood that our inability to graph equations in four or more variables in no way reflects upon the realism of such equations. Consider Table 1–1. Suppose that

TABLE 1–1

Ingredient	Number of units of ingredient	Weight per unit
A	w	2
B	x	3
C	y	1
D	z	6

we wish to mix ingredients A, B, C, and D in amounts such that the total weight of the mixture will be 1000 pounds. The weight of w units of A, at two pounds per unit, is $2w$ pounds. Similar expressions hold for the weights of the other ingredients. The total weight is

$$2w + 3x + y + 6z$$

for any combination of ingredients, and the condition for the total weight is

$$2w + 3x + y + 6z = 1000.$$

If we decided to use only ingredients A and B in the mixture, the weight condition would be

$$2w + 3x = 1000.$$

Both of the last two equations describe a simple real-life situation. The fact that one is called a straight line and can be graphed, whereas the other is called a hyperplane and cannot be graphed, has no bearing on the question of realism.

> *Exercise.* If A, B, C, and D (Table 1–1) have unit costs of 2, 5, 7, and 11 cents, respectively, write the equation which specifies that the total cost of ingredients is to be \$5. Answer: $2w + 5x + 7y + 11z = 500$.

To summarize the last two sections, we state that a linear equation in two-space is a straight line. In three-space a linear equation is a plane. A linear equation in n-space, where n exceeds 3, is called a hyperplane.

1.25 SOLUTION STATEMENT WHEN VARIABLES MAY BE ARBITRARY CONSTANTS

The number of points on a straight line is without limit. For example, the equation

$$2x + y = 10$$

is satisfied by (0, 10), (1, 8), (−3, 16), and so on. For any arbitrarily stated value of x, the corresponding value of y is $10 - 2x$. We shall have occasion to state this last remark as follows: The *general* solution of the equation $2x + y = 10$ is

$$x \text{ arbitrary}$$
$$y = 10 - 2x.$$

The expression "x arbitrary," or "x is an arbitrary variable," means that arbitrary constants can be used as values of x. In the last, for example, we may arbitrarily assign x the value 1, then the value of y will be computed as 8.

Of course, we could have stated the general solution with y arbitrary; thus:

$$y \text{ arbitrary}$$
$$x = \frac{10 - y}{2}.$$

The term general solution is contrasted with a *specific* solution. Thus, (2, 6) is the specific solution when x is 2 (or when y is 6), whereas the general solution tells us how the value of one variable is computed for any (every) stated value of the other variable.

If a single linear equation has n variables, the general solution will have $n - 1$ variables (that is, all but one) arbitrary. For example:

$$2x + 3y + 4z = 12$$

has three variables. We know that if specific values are assigned to two of the variables, we can compute the proper value of the third variable so that the equation will be satisfied. If values are assigned arbitrarily to x and y, then z must be assigned the value

$$\frac{12 - 2x - 3y}{4}$$

if the equation is to be satisfied. Hence, we may write the general solution as

$$x, y \text{ arbitrary}$$
$$z = \frac{12 - 2x - 3y}{4}.$$

Obviously, the solution could be written with x and z arbitrary, or with y and z arbitrary.

From a practical point of view, the statement that a variable may be assigned values arbitrarily means that we have freedom of choice. For example, suppose that

$$x + y = 1000$$

expresses the condition that a tank which holds 1000 gallons of fluid is to be filled using x gallons of fluid A and y gallons of fluid B. The general statement of the solution as

$$x \text{ arbitrary}$$
$$y = 1000 - x$$

means that we are free to choose the value for x as long as we then take y as $1000 - x$. Inasmuch as this is an applied problem, the quantities must not be negative; hence, x is restricted to positive values from 0 to 1000, inclusive.

> *Exercise.* In the last exercise, total cost was $2w + 5x + 7y + 11z = 500$. *a)* Write the general solution with x, y, and z arbitrary. *b)* If we buy 20 pounds of B, 10 pounds of C, and 4 pounds of D, so that *(x, y, z)* are (20, 10, 4), how many pounds of A can we buy? Answer: *a)* x, y, z arbitrary, $w = 250 - 2.5x - 3.5y - 5.5z$. *b)* 143 pounds.

In later chapters we shall describe problems that involve several linear conditions on the variables and discuss methods for choosing combinations of values of the variables that will satisfy the conditions and, at the same time, be a *best* combination in the sense of maximizing a profit or minimizing a cost. In these later chapters, we shall make extensive use of arbitrary-variable terminology.

1.26 SET TERMINOLOGY

For reasons that will become clear in later applications, we have chosen to express general solutions in arbitrary-variable terminology. It may be helpful for those who are familiar with sets (see Appendix 1) to show the relationship between set and arbitrary-variable terminology.

The coordinates of a point are an example of an *ordered pair* of numbers, that is, a pair in which order has significance. In the case of coordinates, the first number of the pair is the x value and the second is the y value. An equation in two variables has ordered pairs as members of its solution set. The expression

$$\{(x, y): 2x + y = 12\}$$

is read as "the set of ordered pairs, x, y, such that $2x + y = 12$." If we interpret each ordered pair as a *point*, then the foregoing expression can be read as "the set of points such that $2x + y = 12$."

> *Exercise.* Write the symbolism that represents the solution set for the equation $y - 3x = 7$. Answer: $\{(x, y): y - 3x = 7\}$.

Clearly, the solution set for an equation such as $y - 3x = 7$ is an infinite set. Or, to put the matter in another way, the set of points on the line $y - 3x = 7$ is an infinite set of points. Similarly, the solution set of $x + 2y - 3z = 19$, which is symbolized

$$\{(x, y, z): x + 2y - 3z = 19\}$$

is an infinite set. We cannot list all the elements of the solution set. We can, however, generate as many elements of the solution set as we wish. One convenient way to do this is to solve the equation for one of the variables in terms of the other variables, in which case the equation is said to be in *explicit* form. Thus,

$$x + 2y - 3z = 19$$

is in *implicit* form because it is not solved for one of the variables, whereas

$$x = 19 - 2y + 3z$$

is in explicit form. It explains how x is computed for stated values of y and z. We may construct as many members of the solution set as we wish by giving y and z any values, selected arbitrarily, then computing x as $19 - 2y + 3z$ to complete an ordered triple that is a member of the solution set. Inasmuch as we are thinking about constructing solutions from arbitrarily selected values of y and z, we refer to these as arbitrary variables.

The terminology and symbolism in this section do not add to our ability to solve equations, but they do provide a fundamental understanding of what we mean when we refer to solutions. We do not have to come to grips with fundamentals when we deal with an equation such as $2x = 3x - 1$. We see immediately what the solution is and can prove it satisfies the equation. However, if we are asked to explain what is meant by "solution" with reference to the equation

$y = x + 6$, we must get down to fundamentals and state that the solution is the (infinite) set of ordered pairs of numbers, (x, y), such that the second number is 6 more than the first. We express the same concept symbolically by

$$\{(x, y): y = x + 6\}.$$

In this explicit form, we think of generating members of the solution set by assigning arbitrary values to x, then computing y as $x + 6$. It is clear that the statement of the general solution in terms of an arbitrary variable, shown earlier in the chapter, namely,

$$x \text{ arbitrary}$$
$$y = x + 6$$

is a satisfactory description of the solution set.

1.27 PROBLEM SET 1–7

Mark (T) for true or (F) for false:

1. ( ) Two planes may intersect at a single point.
2. ( ) $x + y = 7$ is a plane in three-space.
3. ( ) $x + y + z = 0$ is a plane in three-space.
4. ( ) In four-space, $x = 10$ would be called a hyperplane.
5. ( ) In three-space, $x = 0$ is the equation of the y-axis.
6. ( ) In three-space, $y = 0$ is the equation of the x, z plane.
7. ( ) The plane $3x + 2y = 6$ will never intersect the z-axis.
8. ( ) The plane $x - 3z = 7$ is perpendicular to the x, z plane.
9. ( ) In two-space, $y = 0$ is the equation of the y-axis.
10. ( ) In three-space, every linear equation in three variables is a plane.
11. ( ) It is not possible to graph a hyperplane.
12. ( ) Equations of hyperplanes have no applicability to real-world situations.
13. ( ) The point $(-2, -3, 5)$ is in octant III.
14. ( ) The point $(1, 3, -4)$ is in octant V.
15. ( ) The origin of coordinates lies on the plane $x + 2y + 3z = 0$.
16. ( ) The general solution for a single equation that has five variables will have four arbitrary variables.
17. ( ) There are five ways of setting up the solution for the equation described in Problem 16.
18. ( ) The plane $2x + 3y - 5z = 14$ intersects all three coordinate planes.
19. ( ) The x-intercept of the plane in Problem 18 is 7.
20. ( ) The x, y-trace of a plane is a line passing through the x- and y-intercepts of the plane.

21. Make a three-dimensional sketch showing the points:
    a) $(5, 0, 0)$.
    b) $(2, 7, 4)$.
    c) $(-3, 0, 2)$.

22. Given that $x + 3z = 6$ is the equation of a plane in three-space, rewrite the equation, including the missing variable y.

23. Make graphs showing illustrative segments of each of the following planes:
    a) $x = 10$.   b) $z = -3$.
    c) $y = 2$.   d) $x + z = 5$.
    e) $x + 2y = 4$.   f) $x + y + z = 3$.

24.

Ingredient	Number of units	Weight per unit (pounds)	Cost per unit
A	w	1	$0.50
B	x	3	0.30
C	y	2	1.20
D	z	4	0.80

   a) A mixture is to be made with total weight 500 pounds. Write the equation whose solutions are the permissible numbers of units of each ingredient in the mixture.
   b) A mixture is to be made costing $310. Write the equation whose solutions are the permissible numbers of units of each ingredient in the mixture.

25. Write in two different forms the general solution of
$$3x + 2y = 6.$$

26. Write in three different forms the general solution of
$$2x + 3y - 2z = 18.$$

27. What is meant by the statement that
$$3x - 4y = 6$$
has an unlimited number of specific solutions?

28. If the x, y-trace of the plane $3x + 4y - 6z = 12$ were considered as a straight line in two-space, what would be the equation of the line?

29. What are the coordinates of the points that establish the x, y-trace of the plane in Problem 28?

30. If $x + y = 1000$ expresses the condition that a tank that holds 1000 gallons of fluid is to be filled using x gallons of fluid A and y gallons of fluid B, show that the general solution, graphically, consists of the set of points on a line segment that is the hypotenuse of an isosceles right triangle.

31. (See Problem 30.) If A costs 20 cents a gallon and B costs 23 cents a gallon, and if the greatest amount of A that can be obtained is 600 gallons,

what combination of x and y satisfies the requirement and also leads to minimum total cost?

32. (See Problems 30 and 31.) Under what conditions would the minimum cost combination be at one of the intercepts?

33. Shingle requirements (per house) for superdeluxe, top-grade, and regular houses are, respectively, 10, 8, and 7 bundles. Write the expression for the total number of bundles of shingles needed for x superdeluxe, y top-grade, and z regular houses.

34. (See Problem 33.) A freight car can hold 560 bundles of shingles.
    a) Write the equation whose solutions contain the numbers of houses that could be shingled with one freight car load of shingles.
    b) In addition to being nonnegative, what other requirements must be satisfied by the solutions in *(a)?*

Note: Problems 35–40 relate to Section 1.26.
Translate the set expressions of Problems 35–37 into words.

35. $\{(x, y): y - 3x = 6\}$.   36. $\{(x, y, z): 2x + y - 3z = 15\}$.

37. $\{(w, x, y, z): 2w + x - y + 3z = 30\}$.

38. If the expression in Problem 35 were put into explicit form with x as the arbitrary variable, it would be $\{(x, y): y = 3x + 6\}$. Write Problem 37 with x and z arbitrary.

39. See Problem 38. Write Problem 37 with all variables but y arbitrary.

40. The general solution for Problem 36 with x and z arbitrary may be written in the form

$$x, z \text{ arbitrary}$$
$$y = 3z - 2x + 15.$$

Write the general solution for Problem 37 in this form with all variables but x arbitrary.

1.28 REVIEW PROBLEMS

1. Given the points $A(3, -4)$, $B(5, -2)$, $C(3, -2)$, find the distances
    a) AC.   b) BC.   c) AB.

2. In Problem 1, which segment is vertical and which is horizontal?

3. In each case, the given points are the end points of the diagonal of a rectangle whose sides are parallel to the axes. What are the coordinates of the other corners of the rectangle, and what is the length of the diagonal?
    a) $(0, 0)$, $(3, 4)$.   b) $(-1, 2)$, $(8, 14)$.   c) $(1, 2)$, $(3, 4)$.

4. Find the perimeter of the triangle whose vertices are $(1, 2)$, $(7, 2)$, and $(7, 10)$.

5. a) If city B is 5 miles east and 12 miles north of city A, how far is it from A to B?
    b) A section of a city has square blocks with streets perpendicular to avenues. How far would we walk (in blocks) going by sidewalk from (2nd St., 7th Ave.) to (7th St., 19th Ave.)?

c)   See b). A subway is to be built under a straight line connecting the corners. If a block is 400 feet long, how long will the subway line be in feet?

6.   If y and x represent, respectively, expenses and sales in thousands of dollars, and last week's operations are characterized as (10, 12), compared to this week's (13, 16), by how much did expenses and sales change from last week to this?

7.   If sales increased from $12 thousand to $15 thousand, the *percent* increase was (100 percent) $(15 - 12)/12 = 25$ percent.
a)   Write the expression for percent change if sales go from S_1 to S_2.
b)   What would it mean if the answer to a calculation such as that in (a) was negative?

8.   Find the slope of the segment joining the following point pairs:
a) $(-4, 7)$, $(-1, 3)$.   b) $(1, 2)$, $(5, 6)$.
c) $(0, 0)$, $(0, 5)$.   d) $(5, -1)$, $(10, -1)$.
e) $(5, 0)$, $(5, 3)$.   f) $(-2, -1)$, $(2, -4)$.
g) $(1, -3)$, $(4, -1)$.   h) $(-2, 5)$, $(3, 5)$.
i)   If personal consumption expenditures increased from $254 billion to $618 billion when disposable income (less consumer interest on loans) increased from $270 billion to $675 billion, compute the marginal propensities to consume and to save.

9.   Find the equation of the line passing through each of the point pairs in Problems 8a through 8h. State answers in slope-intercept form where possible.

10. Find the equation of the line passing through the given point and having the stated slope: State answers in slope-intercept form where possible.
a) $(1, 3)$, slope $1/5$.   b) $(0, 0)$, slope 0.
c) $(1, 1)$, slope 1.   d) $(1, 3)$, slope -2.
e) $(-2, -4)$, vertical.   f) $(-3, -4)$, slope 0.

11. What is the equation of the line parallel to and five units below the x-axis?

12. As sales, x, change from $300 to $600, selling expense changes from $250 to $400. Assume that the given data establish a linear relationship between sales and selling expense. Find the equation of the relationship in slope-intercept form.

13. If selling expense is $100 when sales are $150, and if expense increases $1 for each increase of $3 in sales, write the straight-line relationship between expense and sales in slope-intercept form.

14. If a man's weekly pay is computed at $50 (whether or not he works) plus $5 per hour worked, what equation relates weekly pay, y, to hours worked, x?

15. Explain why the lines $y = -6$ and $x = 15$ are perpendicular.

16. What is the slope and y-intercept of the line in Problem 14?

17. Graph the following using the intercepts:
a) $3x - 4y = 24$.   b) $2x + y = 10$.   c) $y = 15$.

18. Using the coefficients of x and y, determine the slope of:
a) $x - y = 4$.   b) $3x - 6y = 4$.   c) $3x + 4y = 7$.

d) An ounce of bourbon contains ½ of an ounce of alcohol and an ounce of vermouth contains ⅛ of an ounce of alcohol. Write the relation that exists if *b* ounces of bourbon and *v* ounces of vermouth are to be mixed to make a drink containing two ounces of alcohol. What is the substitution rate of bourbon per ounce of vermouth?

19. What is the equation of the line that has a slope of $-2/3$ and a *y*-intercept of -4?

20. Find the equation of the line through (3, 4) that is parallel to the line $x - 2y = 4$.

21. *a)* Find the equation of the line through $(-1, 15)$ that is perpendicular to the line $x - 2y = 4$.
    b) If an existing pipeline is described by $y = x$ and a plant at (14, 26) wishes to connect to the pipeline on a perpendicular, what is the equation of the perpendicular and how far will it be from the plant to the connecting point? Note: In the equation of the perpendicular, let $y = x$ to find the connecting point.

22. If *y* and *x* are in proportion so that *y* is to *x* as 2 is to 5, then:
    a) Write the equation relating *y* and *x*.
    b) Graph the equation in *(a)*.

23. If a company sells *x* units of a product at \$4 per unit, write the expression for *R*, the total revenue received for the *x* units. Interpret the intercepts of this line.

24. If the first of two lines has the form $ax + by + c = 0$, and the second the form $dx + ey + f = 0$, and if the first line is to pass through the origin and the second is to be parallel to the first, what relationships must exist among *a*, *b*, *c*, *d*, *e*, and *f*?

25. The *productivity* (as contrasted to *production*) of a factory is often measured by the ratio *output per labor-hour*. If a factory has a constant productivity of five units per labor-hour, what is the equation relating units of output, *y*, to number of labor-hours worked, *x*?

26. *a)* If *y* is a person's weekly pay in dollars and *x* is the number of hours the person worked, and if $y = 3.5x + 25$, interpret the numbers 3.5 and 25.
    b) A machine purchased now ($t = 0$) for \$10,000 depreciates in value by a constant amount per year for 20 years to a scrap value of \$1,000. Write the equation for *D*, the depreciated value of the machine at time *t* years. Interpret the slope and intercept of this line.

27. If total factory cost, *y*, of making *x* units of a product is given by $y = 2x + 40$, and if 100 units are made:
    a) What is the variable cost?
    b) What is total cost?
    c) What is variable cost per unit?
    d) What is average cost per unit?
    e) What is the marginal cost of the 100th unit?
    f) What is the marginal cost of the 1st unit?

28. In Figure A, the slant line represents the total factory cost, *C*, of producing a number of units, *Q*. Write an interpretation of each of the following:
    a) OT.   *b)* PL/WP.   *c)* RW/OR.   *d)* PL.   *e)* SW.

FIGURE A

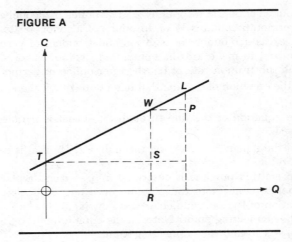

29. A manufacturer has a fixed cost of $75,000 and a variable cost of $7 per unit made and sold. Selling price is $10 per unit.
    a) Using C for cost and q for number of units made and sold, write the revenue and cost equations.
    b) Compute profit if 40,000 units are made and sold.
    c) Compute profit if 20,000 units are made and sold.
    d) Find the break-even quantity.
    e) Compute the break-even dollar volume of sales (revenue).
    f) Construct the break-even chart. Label the cost, revenue, and fixed cost lines, and the break-even point.

30. A manufacturer has a fixed cost of $40,000 and a variable cost of $1.60 per unit made and sold. Selling price is $2 per unit.
    a) Using C for cost and q for number of units made and sold, write the revenue and cost equations.
    b) Compute profit if 150,000 units are made and sold.
    c) Compute profit if 80,000 units are made and sold.
    d) Find the break-even quantity.
    e) Find the break-even dollar volume of sales (revenue).
    f) Construct the break-even chart. Label the cost, revenue, and fixed cost lines, and the break-even point.

31. A company's cost function for the next three months is

$$C = 80,000 + 12q.$$

    a) Find the break-even quantity if the company's product sells for $13.25 per unit.
    b) What will "profit" be if the company shuts down operations?
    c) If, because of a strike, the company will be able to produce only 10,000 units, should it shut down for the next three months? Why or why not?

32. A company expects fixed cost of $25,000. It plans to mark up goods 46 percent on retail, and to incur other variable costs of $0.06 per dollar of sales.
    a) Find the equation relating total cost to sales.
    b) Find the break-even point.
    c) Make the break-even chart.
    d) What will net profit before taxes be if sales are $100,000?

33. If the total cost of operations, y, is related to sales, x, by the equation $y = 10,500 + 0.58x$, find:
    a) Fixed cost.
    b) Total cost on sales of $60,000.
    c) The break-even point.
    d) Net profit before taxes on sales of $65,000.

34. If fixed cost is $10,000 and cost increases by $2 for each $3 increase in sales, find:
    a) The break-even point.
    b) The equation relating cost and sales.

35. A demand function shifts from

$$DD: \quad p = -0.2q + 50 \quad \text{to} \quad D'D': \quad p = -0.2q + 60.$$

Compute the horizontal and vertical shifts and write an interpretive description of these numbers.

36. If p, price per unit, is vertical and q, number of units demanded, is horizontal, what is the nature and interpretation of the demand functions $p = 2$ and $q = 100$?

37. In three-space, in what octant would each of the following points be?
    a) $(1, 7, 5)$.   *b)* $(-3, -5, -6)$.
    c) $(1, -3, -6)$.   *d)* $(2, 4, -3)$.

38. Make three-dimensional sketches showing the points:
    a) $(1, 3, 5)$.   *b)* $(2, 4, -6)$.   *c)* $(1, -2, 3)$.

39. Sketch the octant I portion of the plane $5x + 3y + 2z = 30$.

40. What are the coordinates of the points that establish the x, y-trace of the plane $5x + 3y + 2z = 30$?

41. If a college has w freshmen, x sophomores, y juniors, and z seniors, what is the expression for the total number of students in the college?

42. (See Problem 41). If the college wishes to enroll exactly 1500 students, write the equation whose solutions are the permissible numbers of students in each class.

43. Write the general solution of $x + 2y + 3z = 6$ in three forms.

44. Translate the following set expressions:
    a) $\{(x, y): 5x + 6y = 21\}$.   *b)* $\{(x, y, z): x + 2y + 3z = 0\}$.

45. Express the solution space for the following in set notation:
    a) $3x - 5y = 15$.   *b)* $3x - 2y + 5z - 6 = 0$.

Systems of linear equations

2

2.1 INTRODUCTION

The variables encountered in a problem may have to fulfill more than one condition. In a production problem, for example, the numbers of units of various products made will be restricted by conditions such as time available for production and money available for the purchase of raw materials. When each of the conditions can be expressed in the form of a linear equation, the mathematical description of the problem is a *system* of linear equations. The procedures for finding the values of the variables that satisfy all equations of the system *simultaneously* are the subject matter of this chapter.

Example. The break-even discussion of Chapter 1 introduced a system of two linear equations in two variables. There we saw that if sales, x, and cost, y, were related by the linear condition

$$y = 1200 + 0.4\,x$$

we could solve this equation simultaneously with

the condition $y = x$ to find where sales revenue just equals cost. If we express the 2 by 2 system as

$$y = 1200 + 0.4x$$
$$y = x,$$

y may be *eliminated* by subtracting the second equation from the first, yielding

$$0 = 1200 - 0.6x,$$

from which the break-even point is at $x = 2000$. Graphically, the simultaneous solution of the system, which is $x = y = 2000$ in this case, is the point of intersection of the graphs of the two equations.

The number of relevant variables in a system, and the number of equations representing conditions to be fulfilled, vary widely. The requirement that *all* conditions expressed by the system be satisfied sometimes cannot be met. In other cases only one set, or many sets, of values for the variables will satisfy all conditions. In this chapter we shall learn systematic procedures for solving systems with varying numbers of equations and variables, how to determine if no solution is possible, and how to express the solution if the system is satisfied by more than one set of values. We shall consider the effect of the requirement that values in the solution set not be negative, and also show how the best set can be chosen when the system has more than one solution set. Two of the five sections on applications are concerned with supply and demand analysis.

We turn first to a consideration of the relation between solutions and points of intersection.

2.2 NUMBER OF SOLUTIONS POSSIBLE

A specific solution of a system of linear equations is a set of values, one for each variable, that simultaneously satisfy all equations of the system. Geometrically, a set of values for the variables is represented by a point, and a set satisfying all the equations is represented by a point that lies on the graphs of all the equations; that is, a solution is a point of intersection of all the graphs. We shall appeal to the geometry of intersections of lines and planes to illustrate that *the number of solutions of a linear system is either zero, one, or unlimited.*

2.3 INTERSECTIONS OF STRAIGHT LINES

The graphs of linear equations in two-space are straight lines. The parts of Figure 2–1 show the intersection possibilities for such lines. If a system has two equations in two variables, either the corresponding lines intersect in a single point, or they are parallel and have no inter-

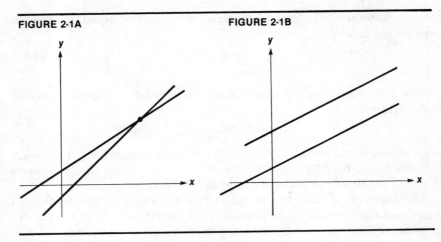

FIGURE 2-1A

FIGURE 2-1B

section point, as shown in Figures 2–1A and 2–1B. The three lines representing a system of three equations in two variables may intersect in a single point, as in Figure 2–1C, or there may be no point that lies on all three lines, as in Figures 2–1D and 2–1E. The same intersection possibilities exist if more than three lines are plotted, and we conclude that two or more different lines have one point in common or no points in common. As a special case, however, we note that if the system has two equations, and the terms of one equation are a constant multiple of the corresponding terms of the other, then both equations have the same graph. All of the (unlimited) points on one line also lie on the "other" line, so that in this special sense the system has an unlimited number of solutions.

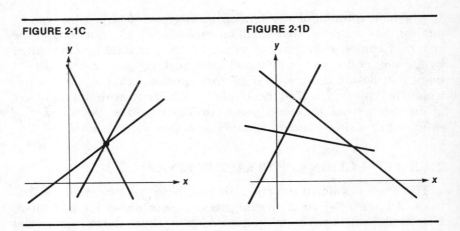

FIGURE 2-1C

FIGURE 2-1D

FIGURE 2-1E

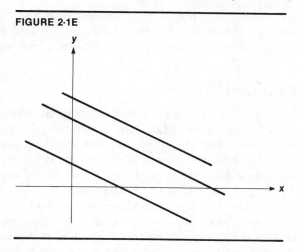

2.4 INTERSECTIONS OF PLANES

Linear equations in three-space are represented by planes. A system of two equations in three variables is represented geometrically by two planes. Any points that lie on both planes will satisfy both equations and therefore be solutions of the system. Figures 2–2A and 2–2B show that two different planes either intersect in a complete straight line *PQ*, and so have an unlimited number of points in common (Figure 2–2A), or they do not intersect at all (Figure 2–2B).

The three planes representing a system of three equations in three variables may have either zero, one, or an unlimited number of points in common. Observe Figure 2–2A, and visualize the intersection possibilities if a third plane cut into the figure. The third plane might contain the line *PQ*, so that the three planes would have an unlimited number

FIGURE 2-2A   **FIGURE 2-2B**

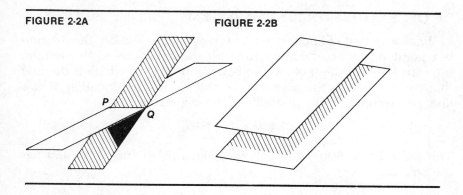

of points in common, or it might slice across PQ at an angle, hitting this line at a single point. Finally, the third plane might not cut PQ at all, so that the three planes would have no point in common.

2.5 A GENERALIZATION

Geometrical sketches cannot be made to aid us in determining the number of solutions possible when the equations of a system have more than three variables. However, intuition suggests that what has been said for systems with two variables and for systems with three variables carries over for systems with more than three variables, and intuition turns out to be correct. *The number of solutions for any linear system is either zero, one, or unlimited.* We shall see many illustrations of this generalization as we move through the chapter. When a system has zero solutions, its equations are said to be *inconsistent*. The equations of a system with one solution are said to be *consistent*, and the equations of a system with unlimited solutions are said to be *dependent*.

> *Exercise.* Imagine that two successive pages of this book are planes that meet in a straight line at the binding. Use a piece of paper as a third plane.
> *a)* Arrange the third plane so that the three planes have a single point in common.
> *b)* Arrange the third plane so that the three planes have an unlimited number of points in common.
> *c)* Find two substantially different orientations of the third plane such that the three planes have no points in common.

2.6 OPERATIONS ON LINEAR SYSTEMS

When a system of linear equations has a single solution, the solution is a point whose coordinates satisfy all the equations of the system. We start with a system of two equations, the first of which is the line that passes through the points (2, 5) and (8, 3). The equation of this line, determined by the methods of the last chapter is

$$e_1: \quad x + 3y - 17 = 0.$$

The second line passes through the points (2, 5) and (4, 2) and has the equation

$$e_2: \quad 3x + 2y - 16 = 0.$$

Notice that the point (2, 5) was used in constructing both e_1 and e_2. Hence (2, 5) satisfies both equations and is the solution of the two-equation system. We purposefully chose (2, 5) to be the solution in order to demonstrate the operations that can be performed on the equations of a system without changing the solution. We will number these operations as we proceed.

Row operations

1. *The solution is not altered if both sides of one of the equations are multiplied by a constant.*

For example, multiplying both sides of e_1 by 5 gives

$$5e_1: \ 5(x + 3y - 17) = 5(0),$$

and the solution (2, 5) satisfies this equation, as well as the unchanged e_2, because

$$5(2 + 15 - 17) = 5(0) = 0.$$

2. *The solution of a system is not changed if one of its equations is replaced by a linear combination of this and another equation in the system,* where linear combination means a constant times the given equation plus a constant times another equation.

For example, starting with the given

$$\begin{aligned} e_1: &\quad x + 3y - 17 = 0 \\ e_2: &\quad 3x + 2y - 16 = 0, \end{aligned} \tag{1}$$

let us replace e_1 by $3e_1 - 2e_2$:

$$\begin{aligned} e_3 = 3e_1 - 2e_2: &\ 3(x + 3y - 17) - 2(3x + 2y - 16) = 0 \\ &\ 3x + 9y - 51 - 6x - 4y + 32 = 0 \\ e_3: &\ -3x + 5y - 19 = 0. \end{aligned}$$

Now observe that e_3 is also satisfied by the solution (2, 5) because

$$-3(2) + 5(5) - 19 = 0.$$

Hence, the system

$$\begin{aligned} e_3: &\ -3x + 5y - 19 = 0 \\ e_2: &\quad 3x + 2y - 16 = 0 \end{aligned} \tag{2}$$

has the same solution as the original system.

3. *The solution of a system is not changed if any pair of its equations are interchanged.*

We need not illustrate the obvious fact that interchanging e_1 and e_2 in the system (1) will have no effect on the solution.

Inasmuch as the equations of a system are written horizontally, we often think of them as being *rows*, and refer to the foregoing operations as *row* operations. The manner in which row operations are used to find an unknown solution is presented next.

2.7 ELIMINATION PROCEDURE

The discussion in the last section suggests a method of attack when seeking solutions of a linear system; namely, form linear combinations of pairs of equations in a manner such that a variable is *eliminated*. For example, given the system

$$2x + 3y - 2 = 0$$
$$5x + 4y - 12 = 0$$

we see that multiplication of the first equation by 5 and the second by 2, followed by subtraction, will result in the elimination of x. Thus:

$$5(2x + 3y - 2) - 2(5x + 4y - 12) = 0$$

becomes, upon simplification:

$$7y + 14 = 0$$

from which we see that y must equal -2. Rather than forming a linear combination eliminating y in order to obtain the required value for x, we may substitute -2 for y in either of the original equations and find that x is 4. The solution of the system is $(4, -2)$.

We shall use the elimination-substitution method to solve linear systems in most of this chapter. The symbolism illustrated in the next example will be employed to keep track of the operations performed.

FIGURE 2-3

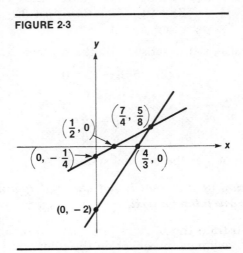

$$e_1: \ 3x - 2y = 4$$
$$e_2: \ 2x - 4y = 1.$$

Here we have given two equations, *e* sub 1 and *e* sub 2. Elimination of *y* will occur if we multiply e_1 by minus 2 and add to e_2. We obtain

$$e_3: \ -4x = -7 \qquad -2e_1 + e_2.$$

The procedure for obtaining e_3 is shown at the right of the equation. It follows from e_3 that $x = 7/4$ and substitution of this value into e_1 or e_2 yields $y = 5/8$, so the solution of the system is $(7/4, 5/8)$. The geometric nature of the system is shown in Figure 2–3, wherein the solution is seen to be the point of intersection of the two lines.

> *Exercise.* Find the solution of the following system of two equations in two unknowns, then plot the graph of the lines and label the coordinates of the point of intersection. Answer: (3, 5).
>
> $$4x - 3y = -3$$
> $$5x - \ y = 10.$$

2.8 APPLICATIONS–1

Consider a buyer who wants to combine *x* liters of Exall, which costs \$0.50 per liter, with *y* liters of Whyall, at \$0.60 per liter, to obtain 2000 liters of mixture worth \$0.53 per liter. We find two equations by noting that the total amount is $x + y$, which must equal 2000, and by noting that the value of *x* liters at \$0.50 plus *y* liters at \$0.60, which is $0.5x + 0.6y$, must equal the total value of 2000 liters at \$0.53, which is \$1,060. Hence,

$$e_1: \qquad x + \quad y = 2000$$
$$e_2: \ 0.5x + 0.6y = 1060$$

and

$$e_3: \ 0 + 0.1y = 60 \qquad -0.5e_1 + e_2$$

so that

$$y = 600 \text{ liters}$$

and from e_1

$$x = 2000 - y = 1400 \text{ liters.}$$

The buyer should mix 1400 liters of Exall with 600 liters of Whyall.

Exercise. We plan to invest x dollars in the bonds of Extron Company, which pay 7 percent interest, and y dollars in Whytron Company bonds, which pay 10 percent interest. We will invest $10,000 and require that we receive $820 interest. How much should be invested in each security? Answer: The equations are $0.07x + 0.1y = 820$ and $x + y = 10,000$. The solution is $x = \$6000$ and $y = \$4000$.

2.9 PROBLEM SET 2–1

Solve the following systems:

1. $x + y = 5$
   $2x + y = 7.$

2. $2x + 3y = 10$
   $3x + y = 1.$

3. $5x - 2y = 3$
   $2x + y = 3.$

4. $2x + 3y = 9$
   $4x - 2y = 2.$

5. $4x + 3y = 4$
   $2x + 6y = 5.$

6. $3x + 5y = 9$
   $4x + 2y = 5.$

7. If x liters of Exall, which costs $0.50 per liter, are to be mixed with y liters of Whyall, at $0.66 per liter, to obtain 1000 liters of mixture worth $0.60 per liter, how much Exall and Whyall should be used?

8. We plan to invest x dollars in Extron Company bonds, which pay 6.5 percent interest, and y dollars in Whytron Company bonds, which pay 9 percent interest. If $50,000 is to be invested and we require that $4000 interest be received, how much should be invested in each bond?

9. It takes 20 minutes and costs $2 to make one Exbee, whereas it takes 30 minutes and costs $1 to make one Whybee. If 600 minutes and $40 are available, how many Exbees and Whybees can be made?

10. It takes 10 minutes to make, and 20 minutes to paint, one Exball, whereas it takes 5 minutes to make, and 8 minutes to paint, one Whyball. If 300 minutes are available for making these products and 500 minutes are available for painting, how many Exballs and Whyballs can be made?

2.10 APPLICATIONS–2: SUPPLY AND DEMAND ANALYSIS

In the last chapter (Section 1.20), we introduced demand functions, which relate the quantity of product consumers are willing and able to buy to the unit price of the product. Linear demand functions have negative slopes because demand decreases when price increases. A supply function, on the other hand, relates the quantity producers are willing and able to supply to the unit selling price of the product, and linear supply functions have positive slopes because producers increase the supply when the selling price increases. Consider Figure 2–4, which shows supply and demand functions as the straight lines labeled SS and DD respectively. The equations are

$$e_1: \quad DD: p = -0.1q + 40 \quad \text{(negative slope, demand)}$$
$$e_2: \quad SS: p = 0.2q + 10 \quad \text{(positive slope, supply)}.$$

FIGURE 2-4

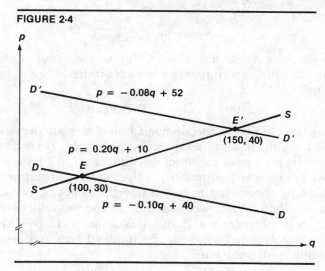

Point E, the intersection of DD and SS on Figure 2–4, is found by

$$e_3: \qquad 0 = -0.3q + 30 \qquad e_1 - e_2$$
$$0.3q = 30$$
$$q = 100 \text{ units.}$$

From e_1,

$$p = -0.1(100) + 40 = \$30 \text{ per unit}$$

so we have $E(100, 30)$, the *equilibrium* point, which shows the market price that will equate the quantity consumers are willing and able to buy with the quantity producers are willing and able to supply.

What we wish to show in this section is the effect of a change in the demand (but not the supply) function, and the effect of a change in the supply (but not the demand) function. This can be done by finding the new equilibrium point, E', after a change has occurred. For example, suppose demand changes to the function $D'D'$ shown on Figure 2–4, with supply remaining at SS. The two functions now are

$$e_1: \ D'D': p = -0.08q + 52$$
$$e_2: \ SS: \qquad p = 0.2q + 10.$$

We now find

$$0 = -0.28q + 42 \qquad e_1 - e_2$$
$$0.28q = 42$$
$$q = 150$$

and, from e_1,

$$p = -0.08(150) + 52 = 40;$$

the new equilibrium point is $E'(150, 40)$, as shown on Figure 2–4. Comparing the equilibrium points whose coordinates are in the order (quantity, price),

$$E(100, 30) \quad \text{and} \quad E'(150, 40),$$

we see that the change in the demand function is an *increase* in demand, which results in a *higher* quantity and *higher* price at the new equilibrium. In economic terminology, the demand function shifted to the right so that a larger quantity is demanded at each price level.

The foregoing shows that to determine the effect of a change in a demand or supply function, we need only to compare the two equilibrium points and interpret the changes in q and p. Thus, if there is a change in a supply function, with the demand function unchanged, and the equilibrium points are

$$E(500, 15) \quad \text{and} \quad E'(400, 20)$$

the *decrease* in supply results in a lower quantity, but a *higher* price at equilibrium. At E', Figure 2–4, note that an *increase* in demand was accompanied by an *increase* in both quantity and price, so a demand change in one direction is accompanied by quantity and price changes in the *same* direction. The effect of a supply change, however, is *inverse* to the effect just illustrated. At E' after a left shift (a decrease) in the supply function, price will be lower and quantity will be higher. At E' after a right shift in supply, price is lower and quantity is higher.

> *Exercise.* Interpret the results of the change in a supply function if the equilibrium points are $E(400, 50)$, $E'(500, 15)$. Answer: Supply has increased, and at the new equilibrium point, quantity has increased and price has decreased.

2.11 PROBLEM SET 2–2

1. *a)* Graph the supply and demand functions

$$SS: \ p = 0.1q + 8$$
$$DD: \ p = -0.5q + 50.$$

   b) Find the equilibrium point, E, for *(a)*.

   c) Plot the new demand function $D'D': \ p = -0.6q + 36$ on the graph for *(a)*.

   d) Find the new equilibrium point, E', and place its coordinates on the graph.

   e) Describe the change in equilibrium points.

2. *a)* Graph the supply and demand functions

$$SS: \ p = 0.20q + 10$$
$$DD: \ p = -0.40q + 70.$$

   b) Find the equilibrium point, E, for *(a)*.
   c) Plot the new supply function $S'S': \ p = 0.25q + 18$ on the graph for *(a)*.
   d) Find the new equilibrium point, E', and place its coordinates on the graph.
   e) Describe the change in equilibrium points.

3. (See Problem 2.) The equilibrium point for the functions in *(a)* is $E(100, 30)$. Find the new equilibrium point if the demand function changes to $D'D': \ p = -0.44q + 90$ and describe the change in the equilibrium points.

4. (See Problem 1.) The equilibrium point for the functions in *(a)* is $E(70, 15)$. Find the new equilibrium point if the supply function changes to $S'S': \ p = 0.06q + 5.2$ and describe the change in the equilibrium points.

Interpret the following supply and demand changes:

5. $E(100, 10)$ to E' $(125, 14)$ after a change in demand.
6. $E(200, 20)$ to E' $(150, 25)$ after a change in supply.
7. $E(150, 25)$ to E' $(200, 21)$ after a change in supply.
8. $E(200, 40)$ to E' $(180, 35)$ after a change in demand.
9. If either the supply or demand function (but not both) changed and the result was $E(500, 15)$ to E' $(600, 10)$, which function changed? Explain why.
10. If either the supply or demand function (but not both) changed and the result was $E(500, 15)$ to E' $(400, 10)$, which function changed? Explain why.

2.12 DEFINITION, *m* BY *n* SYSTEM

The designation *m* by *n* means that the number of equations in a system is *m* and the number of variables is *n*. The number of variables in a system is the total number appearing in the equations of the system. It is not required that all variables appear in each of the equations. As an example:

$$2x + y - \ \ z = 4$$
$$x \ \ \ \ + 2z = 7$$

has two equations and three variables. It is a 2 by 3 system.

2.13 SOLVING SYSTEMS OF LINEAR EQUATIONS

We know that linear systems will have either zero, one, or an unlimited number of solutions, and we have discussed the elimination procedure as a method of finding solutions. Attention now turns toward relating the outcome of steps in elimination to the number of solutions, and to the method of describing the general solution when the number of solutions is without limit. It will be convenient to separate the examples into the two categories, m equals n and m does not equal n.

2.14 NUMBER OF EQUATIONS EQUALS NUMBER OF VARIABLES

Systems in this category can be described as n by n systems. The solution procedure makes repeated use of linear combinations of pairs of equations.

General solution procedure, n by n

Form $(n-1)$ linear combinations of pairs of equations with each linear combination eliminating the *same* variable, making sure that every equation is used in at least one of the linear combinations. The outcome is then an $(n-1)$ by $(n-1)$ system. Repeat the procedure with the new system and continue repetition until the nature of the solution is determined.

Example.   Solve the following system of linear equations.

$$e_1: 2x + y - z = 2$$
$$e_2: x + 2y + z = 1$$
$$e_3: 3x - y + 2z = 9.$$

This is a 3 by 3 system and we proceed to reduce it to a 2 by 2 system by the following linear combinations.

$$e_4: -3y - 3z = 0 \qquad -2e_2 + e_1$$
$$e_5: -7y - z = 6 \qquad -3e_2 + e_3.$$

A linear combination of the last two equations leads to

$$e_6: 18y = -18. \qquad -3e_5 + e_4.$$

The last statement says that y must be -1. We now use backward substitution to find x and z: Setting y equal to -1 in e_5 (or e_4) yields $z = 1$. Finally, putting $y = -1$ and $z = 1$ into e_2 (or e_1 or e_3) leads to $x = 2$.

The solution of the system is the point $(2, -1, 1)$. The reader should check this solution by showing that the number triplet satisfies every one of the original three equations. The system has exactly one solution. Geometrically, we have the case of three planes that have a single point in common. The system of equations in this example does have a unique solution, and we say the equations in the system are *consistent*.

Exercise. Find the solution of the following system by use of the elimination procedure. Answer: (1, 1, 2).

$$x + y + z = 4$$
$$2x - y + z = 3$$
$$x - 2y + 3z = 5.$$

Example. Solve the following system of linear equations.

$$e_1: \quad x + y + z = 4$$
$$e_2: \quad 5x - y + 7z = 25$$
$$e_3: \quad 2x - y + 3z = 8.$$

Two linear combinations, eliminating y, lead to

$$e_4: \quad 6x + 8z = 29 \qquad e_1 + e_2$$
$$e_5: \quad 3x + 4z = 12 \qquad e_1 + e_3.$$

A linear combination of the last two equations leads to

$$e_6: \quad 0 + 0 = 5 \qquad -2e_5 + e_4.$$

The last line is a contradiction. Zero cannot equal 5. We conclude that the system has no solution. The argument here is that a solution of the system must satisfy the linear combinations. If a linear combination leads to a false statement, the system has no solution. In this case we say the equations in the system are *inconsistent*.

Exercise. Prove that the following system has no solution.

$$2x - y + 3z = 5$$
$$x + 2y - z = 6$$
$$3x + y + 2z = 8.$$

The first example of the section had a single solution, the second no solution. Let us turn next to a system that has an unlimited number of solutions and learn how to write the general solution in arbitrary-variable terminology.

Example. Solve the following system of linear equations.

$$e_1: \quad x + y + z = 4$$
$$e_2: \quad 5x - y + 7z = 20$$
$$e_3: \quad 2x - y + 3z = 8.$$

Proceeding in the usual manner, we find:

$$e_4:\ 6x + 8z = 24 \qquad e_1 + e_2$$
$$e_5:\ 3x + 4z = 12 \qquad e_1 + e_3.$$

A linear combination of the last equations yields

$$e_6:\ 0 + 0 = 0 \qquad -2e_5 + e_4.$$

The appearance of the true statement, zero equals zero, means simply that this linear combination arose from two equivalent equations. Looking back at

$$e_4:\ 6x + 8z = 24$$
$$e_5:\ 3x + 4z = 12$$

we see that e_4 could be divided by 2 and then would be identical to e_5. Any number pair satisfying one of these equations will satisfy the other. Moreover, any solution of the original system must satisfy these last equations. Solving (either) one for z yields

$$e_7:\ z = \frac{12 - 3x}{4} \quad \text{from } e_5.$$

To repeat, any solution of the original system requires that z be related to x according to the last equation.

Returning to e_1, or any of the three original equations, and replacing z by

$$\frac{12 - 3x}{4}$$

we obtain

$$e_8:\ x + y + \frac{12 - 3x}{4} = 4 \qquad e_7 \text{ into } e_1.$$

Solving the last equation for y leads to

$$e_9:\ y = \frac{4 - x}{4} \quad \text{from } e_8.$$

Both z and y have now been expressed in terms of x, and we state the *general* solution of the system as follows:

$$x \text{ arbitrary}$$
$$y = \frac{4 - x}{4}$$
$$z = \frac{12 - 3x}{4}.$$

The general solution just written shows that the number of solutions is without limits because x may be assigned any arbitrary value whatsoever. In this case, the system of equations is said to be *dependent*. The general solution provides convenient formulas for finding specific solutions. To illustrate the matter of convenience, suppose that we are asked to find the specific solutions for values of x equal to 0, 1, 2, 3, 4, 5, 6, 7, 8, and 9. We could substitute $x = 0$ in the three *original* equations, then solve to get the corresponding y, then substitute to get the corresponding z. Next, we could substitute $x = 1$ in the original equations and repeat the procedure to obtain the corresponding values for y and z. The 10 specific solutions found in this manner would require 10 separate (duplicate) manipulations of the original equations. If, on the other hand, we substitute in the formulas of the general solution, we see immediately that if $x = 0$, then $y = 1$, and $z = 3$. If $x = 1$, then $y = 3/4$, and $z = 9/4$, and so on.

The general solution also can be of use in real-world problems to determine what conditions must be satisfied if the variables in solutions are not permitted to take on negative values. In the foregoing general solution, for example, it is clear that if y and z are to be nonnegative, then x must not exceed 4; that is, x must be in the interval whose limits are 0 to 4, inclusive.

The tactics of the last example are worth reviewing. First, observation of the identity, zero equals zero, directed attention to the equations from which the statement arose. One of these equations was solved for z in terms of x. This automatically relegated x to the role of arbitrary variable, and we sought to express y also in terms of the same arbitrary variable. To do so, we returned to one of the original equations and replaced z by its x equivalent, leaving an expression which was solved for y in terms of x.

Exercise. Start by eliminating z, then make y arbitrary and prove the general solution of the system can be expressed as y arbitrary, $x = 2y + 3$, $z + y + 1$.

$$
\begin{aligned}
e_1: & \quad x - 3y + z = 4 \\
e_2: & \quad x - y - z = 2 \\
e_3: & \quad 2x - 5y + z = 7
\end{aligned}
$$

Exercise. The general solution just obtained can be checked by substitution into the original equations. For example, substitution into e_1 yields

$$(2y+3) - 3y + (y+1) = 4.$$

Removing parentheses, we find the statement is the identity

$$4 = 4.$$

Show that substitution into e_2 and e_3 also leads to identities.

Changing arbitrary variables. The arbitrary variable(s) in one general solution may be changed without solving the entire system again. For example, it may be verified that the general solution of

$$e_1:\quad x + \ y + z = 6$$
$$e_2:\ 4x - 2y + z = -9$$
$$e_3:\ 3x - \ y + z = -4$$

can be written as

$$x \text{ arbitrary}$$
$$e_4:\ y = x + 5$$
$$e_5:\ z = 1 - 2x.$$

If we wish to make y, rather than x, arbitrary, then from e_4 we write

$$e_6:\ x = y - 5$$

and then obtain z by substituting e_6 into e_5:

$$z = 1 - 2x$$
$$= 1 - 2(y - 5)$$
$$= 1 - 2y + 10$$
$$z = -2y + 11.$$

The general solution now is

$$y \text{ arbitrary}$$
$$x = y - 5$$
$$z = -2y + 11$$

Exercise. Write the general solution of the foregoing system with z arbitrary. Answer: z arbitrary; $x = (1 - z)/2$; $y = (11 - z)/2$.

2.15 MORE THAN ONE VARIABLE ARBITRARY

The following two examples provide practice with arbitrary variables and, at the same time, make clear that more than one variable may prove to be arbitrary in a general solution.

$$e_1: \quad w + x \qquad + 3z = 4$$
$$e_2: \quad w - x + 2y + z = -2$$
$$e_3: -2w + x - 3y - 3z = 1$$
$$e_4: \quad w + 2x - y + 4z = 7.$$

The attack on the 4 by 4 system starts by reducing it to a 3 by 3; thus:

$$e_5: 2w + 2y + 4z = 2 \qquad e_1 + e_2$$
$$e_6: 3w + 3y + 6z = 3 \qquad e_1 - e_3$$
$$e_7: \quad w + y + 2z = 1 \qquad 2e_1 - e_4.$$

The next step would be to eliminate another variable, reducing the 3 by 3 to a 2 by 2. However, every linear combination that eliminates a variable will lead to zero equals zero. We solve (any) one of the three equations for one variable, say w:

$$e_8: \quad w = 1 - y - 2z \quad \text{from } e_7.$$

The last statement automatically makes y and z arbitrary variables, so x must be expressed in terms of y and z. Returning to the original equations and selecting, say, the first, we have

$$e_9: (1 - y - 2z) + x + 3z = 4 \quad e_8 \text{ into } e_1.$$

Solving the last equation for x yields

$$x = y - z + 3.$$

The general solution, therefore, can be written as

$$y, z \text{ arbitrary}$$
$$x = y - z + 3$$
$$w = -y - 2z + 1.$$

> **Exercise.** Change the last-written general solution to a solution having w and z arbitrary. Answer: w, z arbitrary; $y = -w - 2z + 1$; $x = -w - 3z + 4$.

The next example shows the introduction of an arbitrary variable during the backward substitution process.

$$e_1: 2w + x + 2y + 3z = 4$$
$$e_2: \quad w + 3x + y + 5z = -2$$
$$e_3: \quad w - 2x + y - 2z = 6$$
$$e_4: 3w - x + 3y + z = 10.$$

In this case the linear combinations chosen eliminate two variables:

$$e_5: \quad -5x - 7z = 8 \qquad -2e_2 + e_1$$
$$e_6: \quad -5x - 7z = 8 \qquad -e_2 + e_3$$
$$e_7: \quad -10x - 14z = 16 \qquad -3e_2 + e_4.$$

Any linear combination of a pair of the last three equations that eliminates a variable leads to zero equals zero. Solving any of the three for x, we find

$$e_8: x = \frac{-7z - 8}{5} \quad \text{from } e_7.$$

Here, z has been delegated the role of an arbitrary variable.

$$e_9: 2w + \frac{-7z - 8}{5} + 2y + 3z = 4 \quad e_8 \text{ into } e_1$$

$$w = \frac{14 - 4z - 5y}{5}.$$

The last equation makes y arbitrary along with z. The solution of the system is

$$y, z \text{ arbitrary}$$
$$x = \frac{-7z - 8}{5}$$
$$w = \frac{14 - 4z - 5y}{5}.$$

It may be helpful to think of a solution stated in general form as a set of formulas for finding specific solutions to the system of equations. In the present case, for example, if we are asked what the solution of the system is when $y = 2$ and $z = 6$, we can compute quickly from the formulas of the general solution that x must be -10 and w must be -4. The reader may verify that $(-4, -10, 2, 6)$ satisfies every one of the original equations.

In summary, the general approach to the solution of an n by n system is to form $(n - 1)$ linear combinations of pairs of equations, each linear combination eliminating the same variable, making sure that every equation at hand is used in at least one of the linear combinations. The outcome of these eliminations is an $(n - 1)$ by $(n - 1)$ system. The reduction process is repeated until a contradiction is encountered (no solution), or until an identity such as zero equals zero is encountered (general solution required). If neither of the last two circumstances arises, the reduction will lead ultimately to a unique specific solution. The reader may wish to follow through the next examples before proceeding.

Example 1. Solve the 4 by 4 system:

e_1: $2w + 3x + y + z = 8$
e_2: $w - 2x + 3y - 2z = -7$
e_3: $3w + 7x - y + 3z = 15$
e_4: $w - 3x + 2y + z = 9.$

We proceed as follows:

e_5: $7x - 5y + 5z = 22 \quad e_1 - 2e_2$
e_6: $9x - 3y - z = -10 \quad e_1 - 2e_4$
e_7: $16x - 7y = -12 \quad e_3 - 3e_4$
e_8: $16x - 7y = -12 \quad e_7$
e_9: $52x - 20y = -28 \quad e_5 + 5e_6$

e_{10}: $\dfrac{11}{4}y = 11 \quad e_9 - \dfrac{13}{4}e_8$

$y = 4$ from e_{10}
$x = 1$ substituting $y = 4$ in e_7
$z = 7$ substituting $y = 4$, $x = 1$ in e_6
$w = -3$ substituting $y = 4$, $x = 1$, $z = 7$ in e_4.

The system is consistent and has the solution $(-3, 1, 4, 7)$ where the coordinates are in alphabetical order, (w, x, y, z).

Example 2. Solve the 2 by 2 system:

e_1: $2x + 3y = 5$
e_2: $6x + 9y = 15.$

We proceed as follows:

e_3: $0 + 0 = 0 \qquad e_2 - 3e_1$

e_4: $\qquad y = \dfrac{5 - 2x}{3} \qquad$ from e_1.

The equations in this case represent the same straight line. The system is dependent and has the general solution

$$x \text{ arbitrary}$$
$$y = \frac{5 - 2x}{3}.$$

Example 3. Solve the 3 by 3 system:

e_1: $5y - 2z = 5$
e_2: $2x + 3y + 4z = 20$
e_3: $x - y + 3z = 15.$

We proceed as follows:

e_4: $5y - 2z = -10 \quad e_2 - 2e_3$
e_5: $5y - 2z = 5 \quad e_1$
e_6: $0 - 0 = -15 \quad e_4 - e_5.$

The last statement is a contradiction, so the system is inconsistent (has no solution).

Example 4. The coefficient of one or more variables may be zero in an equation, in which case the variable is not written. Thus, in

$$e_1: \quad x + z = 2$$
$$e_2: 4x + y = 8$$
$$e_3: \ 2z + y = 3,$$

one variable is missing (has a zero coefficient) in each equation.

This simplifies the solution because we can simply rewrite e_1 (in which y is missing), then eliminate y from e_2 and e_3, as follows:

$$e_4: \quad x + \ z = 2 \qquad \text{Same as } e_1$$
$$e_5: 4x - 2z = 5 \qquad e_2 - e_3.$$

Then, eliminating z,

$$6x = 9 \qquad 2e_4 + e_5$$
$$x = \frac{9}{6} = \frac{3}{2}.$$

Hence,

$$\text{from } e_4: \ \frac{3}{2} + z = 2$$

$$z = 2 - \frac{3}{2} = \frac{4-3}{2} = \frac{1}{2}$$

and

$$\text{from } e_2: \quad 4\left(\frac{3}{2}\right) + y = 8$$
$$6 + y = 8$$
$$y = 2,$$

and the solution is $\left(\frac{3}{2}, \, 2, \, \frac{1}{2}\right)$

2.16 ZEROS BELOW THE FIRST LONGEST DIAGONAL

The elimination steps we have used in the foregoing have been the result of somewhat haphazard choices of variables to be eliminated, and while freedom of choice has the virtue of flexibility, there is also a virtue to having a standard procedure to apply in the solution of various linear systems, especially if the solution is to be programmed for execution on an electronic computer. However, we are not electronic computers, and the standard procedure described here has been designed to lessen the occurrence of fractions, which pose no difficulty for a computer but are an irritation to humans. This method will be extended in Chapter 6, and become the procedure known as Gauss-Jordan elimination.

Example 1. The section heading means that if we start with the system

$$e_1: \; 2x + y - z = 2$$
$$e_2: \; 3x - y + 2z = 9$$
$$e_3: \quad x + 2y + z = 1,$$

where the *first longest* diagonal[1] is shown by the dotted diagonal line, we will perform row operations to make the coefficients below the diagonal all zero. We shall do so in a moment, but to see the reason for doing this, we show the result first. It is

$$2x + y - z = 2$$
$$7y + z = -6$$
$$18z = 18.$$

Hence, it is clear from the last equation that $z = 1$; then from the middle equation with $z = 1$, we have $y = -1$; and then from the first equation, $x = 2$. Thus, by getting zeros below the diagonal, the solution can be obtained quickly by backward substitution.

Returning to the original system, the procedure is as follows:

1*a.* Rewrite the *first* equation.
  b. Use row operations to obtain zeros in the first *column* below the *first* equation. Thus,

$$e_4: \; 2x + y - z = 2 \qquad \text{Same as } e_1$$
$$e_5: \quad +7y + z = -6 \qquad 3e_3 - e_2$$
$$e_6: \quad +3y + 3z = 0 \qquad 2e_3 - e_1.$$

2*a.* Rewrite the first two of the last equations.
  b. Use row operations to obtain zeros in the second *column* below the *second* equation. We have

$$e_6: \; 2x + y - z = 2 \qquad \text{Same as } e_4$$
$$e_7: \qquad 7y + z = -6 \qquad \text{Same as } e_5$$
$$e_8: \qquad\qquad 18z = 18 \qquad 7e_6 - 3e_5.$$

Consequently,

from e_8: $z = 1$
from e_7: $7y + 1 = -6$, $7y = -7$, $y = -1$
from e_6: $2x - 1 - 1 = 2$, $2x = 4$, $x = 2$,

and the system is consistent, having $(2, -1, 1)$ as its solution.

---

[1] In an n by n system, this is called the *principal* or *main* diagonal.

If the system had more than three equations, we would have continued the process. Thus, steps $3a$ and $3b$ would be to rewrite the first *three* equations, then use row operations to obtain zeros in the *third* column below the *third* equation. Having done this, steps $4a$ and $4b$ would be performed in the same manner, and so on until a specific value is obtained for the last variable, or the system is discovered to be dependent or inconsistent.

Example 2.   Solve the system

$$e_1: \; 5x - y + 7z = 20$$
$$e_2: \quad x + y + \; z = 4$$
$$e_3: \; 2x - y + 3z = 8.$$

Rewriting the first equation and obtaining zeros in the first column below the first equation leads to

$$e_4: \; 5x - \; y + 7z = 20 \qquad \text{Same as } e_1$$
$$e_5: \quad\quad -6y + 2z = 0 \qquad e_1 - 5e_2$$
$$e_6: \quad\quad\; 3y - \; z = 0 \qquad 2e_2 - e_3.$$

Rewriting the first two of the last set of equations and obtaining zeros in the second column below the second equation yields

$$e_7: \; 5x - \; y + 7z = 20 \qquad \text{Same as } e_4$$
$$e_8: \quad\quad -6y + 2z = 0 \qquad \text{Same as } e_5$$
$$e_9: \quad\quad\; 0 + 0 \; = 0 \qquad e_5 + 2e_6.$$

The true statement in e_9, zero equals zero, shows the system to be dependent. When such a statement arises, we let all but one of the variables in the equation immediately above the zeros be arbitrary. Thus, with

$$y \text{ arbitrary}$$
$$\text{from } e_8: -6y + 2z = 0$$
$$2z = 6y$$
$$z = 3y$$
$$\text{from } e_7: 5x - y + 7(3y) = 20$$
$$5x = 20 + y - 21y$$
$$5x = 20 - 20y$$
$$x = 4 - 4y$$

and the solution is

$$y \text{ arbitrary}$$
$$x = 4 - 4y$$
$$z = 3y.$$

In the last example, the true statement that zero equals zero in e_9 indicated that the system was dependent and had an unlimited number of solutions. If we had encountered the contradiction that zero equals a nonzero constant, the system would be inconsistent and have no solution. The reader may wish at this time to practice the "zeros" method by applying it to problems stated in earlier exercises.

   Detaching coefficients.   The solution procedure applied to a linear system involves manipulation of the coefficients of the variables and the constants. We may, if we wish, omit repetitious writing of the letters for the variables by detaching their coefficients and writing them along with the constants. It is assumed that the constants are on the *right* side of each equation. Moreover, it is not necessary to rewrite equations that will not be used in subsequent eliminations. Thus, in Example 1 of the foregoing,

$$
\begin{aligned}
e_1&: 2x + y - z = 2 \\
e_2&: 3x - y + 2z = 9 \\
e_3&: x + 2y + z = 1,
\end{aligned}
$$

we construct the first *tableau of detached coefficients* as follows:

	x	y	z	
e_1:	2	1	-1	2
e_2:	3	-1	2	9
e_3:	1	2	1	1

where the constants (on the *right* sides of the equations) are to the right of the vertical line. We need not rewrite e_1 before proceeding to get zeros under it in the first column. The results of zeroing are

e_4:	7	1	-6		$3e_3 - e_2$
e_5:	3	3	0		$2e_3 - e_1.$

Zeroing in the second column leads to

$$18 \mid 18 \qquad 7e_5 - 3e_4$$

and the last result means

$$
\begin{aligned}
18z &= 18 \\
z &= 1.
\end{aligned}
$$

Then, the row for e_5 means

$$
\begin{aligned}
3y + 3z &= 0 \\
3y + 3(1) &= 0 \\
3y &= -3 \\
y &= -1
\end{aligned}
$$

and from any of the original rows, or the original system, we find $x = 2$. The reader may wish to practice the abbreviated detached tableau procedure by applying it to problems given in earlier exercises.

   In solving linear systems, the reader may choose to use either version of the standardized method shown in this section, or the more flexible procedure used prior to this section. The author plans to use the flexible procedure until Chapter 6, and then introduce the completed standardized procedure known as Gauss-Jordan elimination. Chapter 6 will also present the solution of linear systems by the method of determinants.

2.17 PROBLEM SET 2–3

1. Define:
   a) An m by n system of equations.
   b) An n by n system of equations.
   c) A 3 by 2 system of equations.
   d) The number of variables in a system of equations.
   e) A linear combination of two equations.

2. Write a general solution of
   a) $3x + 4y = 3$.   b) $x + 2y - 4z = 15$.

3. What developments in the solution of a system indicate
   a) No solutions?
   b) An unlimited number of solutions?

4. State the situations which may arise in the solution of a 2 by 2 system, and discuss the geometrical interpretation of each.

5. a) State the situations which may arise in the solution of a 3 by 3 system, and discuss the geometrical interpretation of each.
   b) What do the terms consistent, inconsistent, and dependent mean as applied to (a)?

6. What rules should be observed when reducing an n by n system to an $(n - 1)$ by $(n - 1)$ system?

7. Find the solution of each of the following. Plot the graph of each equation, and label the point of intersection of the lines.
   a) $x - y = 5$ b) $x - 2y = 7$
       $x + y = 1$. $2x + y = 4$.
   c) $2x + 7y - 5 = 0$
       $3x + 2y \quad = 6$.

8. Give a geometrical interpretation to the outcome of your attempt to solve
   a) $3x + 2y = 4$ b) $x - y - 4 = 0$
       $6x + 4y = 6$. $2y - 2x = -8$.

Solve the following systems:

9. $2x + y + 2z = 5$ 10. $x + z = 5$
      $x + y - z = 0$ $y + z = 3$
      $3x - 2y + z = 1$. $x - y = 2$.

11. $5x + y + z = 8$ 12. $x - 2y + z = 7$
       $x + 2y - z = 1$ $x - y + z = 4$
       $2x + y = 3$. $2x + y - 3z = -4$.

13. $\begin{aligned} x + y + z &= 10 \\ 3x - y + 2z &= 14 \\ 2x - 2y + z &= 8. \end{aligned}$

14. $\begin{aligned} x \qquad\quad &= 4 \\ x - y - z &= 7 \\ x + y + z &= 2. \end{aligned}$

15. $\begin{aligned} 2x - y + z &= 5 \\ x + 4y - 3z &= 2 \\ 3x + 3y - 2z &= 7. \end{aligned}$

16. $\begin{aligned} 2x + z &= 5 \\ x + y &= 3 \\ z - y &= 1. \end{aligned}$

17. $\begin{aligned} 2x + y - 3z &= 12 \\ x + 3y - 4z &= 6 \\ x - 2y + z &= 4. \end{aligned}$

18. $\begin{aligned} 2w + 3x + 4y + z &= 1 \\ w \qquad + 2y - z &= 3 \\ 6x + 4y + z &= 1 \\ 2w \qquad\quad - 2z &= 5. \end{aligned}$

19. $\begin{aligned} w + x - 2y + 2z &= -4 \\ w \quad - y + z &= -1 \\ 2w - 3x + y - z &= 7 \\ 3w - x - 2y + 2z &= 0. \end{aligned}$

20. $\begin{aligned} w - x - y \qquad &= -3 \\ 2x + y + z &= -1 \\ w - x \quad + z &= 0 \\ w \quad - y + z &= 3. \end{aligned}$

21. $\begin{aligned} w + z &= 0 \\ x + z &= 1 \\ y - w &= -3 \\ w + 2x + y &= 3. \end{aligned}$

22. $\begin{aligned} 2w - 5x + 3y - z &= -10 \\ w - 3x - y + z &= 4 \\ w - 5x + y + z &= -2 \\ -8x + 2y + 3z &= -3. \end{aligned}$

2.18 NUMBERS OF EQUATIONS AND VARIABLES NOT EQUAL

Cases where the *number of equations exceeds the number of variables* are not common in applications. When such systems do arise, elimination often leads to a contradiction, showing that the system has no solution. As an example, consider the system

$$\begin{aligned} e_1&: 2x + y = 7 \\ e_2&: x - 3y = 7 \\ e_3&: 2x + 4y = 10. \end{aligned}$$

We find

$$\begin{aligned} e_4&: 7y = -7 && -2e_2 + e_1 \\ e_5&: 10y = -4 && -2e_2 + e_3. \end{aligned}$$

The last two statements are contradictory because y cannot be -1 and $-4/10$ simultaneously, so the system has no solutions.

It is helpful to think of e_1, e_2, and e_3 in the foregoing as three *constraints* on the two variables, x and y. When the number of constraints exceeds the number of variables, it is not likely that the system has a solution, and the greater the number of constraints that must be satisfied, the less is the likelihood that a solution exists. Of course, we know from our discussion of geometry at the beginning of the chapter that such a system can have a single solution or an unlimited number of solutions. Thus, four planes (four constraints on three variables) can have a single point in common or a line (an unlimited number of points) in common. Examples of zero, one, and an unlimited number of solutions may be found in the next problem set.

Systems having *fewer equations than variables* arise often in practice, as we shall see later in the chapter. Such underconstrained systems have either no solutions or an unlimited number of solutions. The latter case is most common in applications and is of interest because the solution, in arbitrary-variable form, provides alternative ways of satisfying the constraints, and some alternatives may be superior to others. The geometry of an underconstrained system is illustrated by two planes (two equations in three variables), which either are parallel (no solution) or intersect in a line (unlimited number of solutions). As an example of an underconstrained system that has an unlimited number of solutions, consider the 3 by 4 system[2]

$$e_1: \quad w + x + y + 2z = 9$$
$$e_2: \quad 2w - x + y - z = 4$$
$$e_3: \quad w + 2x - y - z = 2.$$

We have:

$$e_4: \quad 3w + 2y + z = 13 \qquad e_1 + e_2$$
$$e_5: \quad 5w + y - 3z = 10 \qquad 2e_2 + e_3$$
$$e_6: \quad -7w \quad + 7z = -7 \qquad e_4 - 2e_5$$
$$w = z + 1 \qquad \text{from } e_6.$$

Resubstitution yields the general solution

$$z \text{ arbitrary}$$
$$e_7: \quad w = z + 1$$
$$e_8: \quad x = 3 - z$$
$$e_9: \quad y = 5 - 2z.$$

2.19 NONNEGATIVITY CONSTRAINTS

In applied problems, variables generally are not permitted to have negative values. If we impose this nonnegativity constraint on the foregoing example, it is clear from e_7 that w will not be negative if z is not negative. However, e_8 shows that z must be less than or equal to 3 if x is to be nonnegative, and (from e_9) less than or equal to 5/2 if y is to be nonnegative. It follows that z must be in the interval 0 to 5/2, inclusive, if all variables are to be nonnegative.

We have seen that nonnegativity has restricted the interval of permissible values for the variable z. To find corresponding restrictions for another variable, we state the solution with that variable arbitrary

---

[2] The dotted line is the appropriate diagonal for use in the zeros method of Section 2.16.

and repeat the analysis just discussed. For example, if we make y arbitrary we find from e_9 that $z = (5 - y)/2$. Substituting this z into e_7 and e_8 yields $w = (7 - y)/2$ and $x = (y + 1)/2$, respectively. Analysis of the last two sentences shows the maximum value permitted for y is 5.

> ***Exercise.***   Take w as the arbitrary variable. Using e_7, e_8, and e_9, express x, y, and z in terms of w and state the permissible range of values for w if all variables are to be nonnegative. Answer: $x = 4 - w$; $y = 7 - 2w$; $z = w - 1$. The permissible interval for w is 1 to 7/2, inclusive.

Similar analysis shows that with x arbitrary, $w = 4 - x$, $y = 2x - 1$, and $z = 3 - x$. We see that x must be at least $1/2$ if y is not to be negative and at most 3 if z is not to be negative. The permissible interval for x is $1/2$ to 3, inclusive.

When a solution has more than one variable arbitrary, the range of permissible values for one variable usually is not a fixed interval but, rather, depends upon the values assigned to the other variables. We shall consider such cases when we study linear programming in later chapters.

2.20 PROBLEM SET 2–4

Solve the following systems, which have more equations than variables:

1. $\quad x - 3y + z = 1$
    $\quad\, x + y - z = 3$
    $\quad 2x - 4y + z = 3$
    $\quad\, x - 3y + z = 10.$

2. $\quad x - 3y + z = 1$
    $\quad\, x + y - z = 3$
    $\quad 2x - 4y + z = 3$
    $\quad 4x - 6y + z = 7.$

3. $\quad 2x - y + 3z = 5$
    $\quad 3x + 2y \quad\;\; = 1$
    $\quad\, x - 3y + 6z = 8$
    $\quad 2x \qquad\; 3z = 4.$

4. $\quad 2x - y = 3$
    $\quad\, x + 2y = 4$
    $\quad 3x - 4y = 2$
    $\quad -x + 5y = 2.$

5. $\quad x + y = 2$
    $\quad x - y = 1$
    $\quad -x + 3y = 5.$

6. $\quad x + 3y + z = 6$
    $\quad -x + y + z = 2$
    $\quad 3x + y - z = 2$
    $\quad\, x + y + z = 6.$

7. $\quad x + 3y + z = 6$
    $\quad -x + y + z = 2$
    $\quad 3x + y - z = 2$
    $\quad 2x + 4y + z = 8.$

8. $\quad w + x - 3y \qquad\; = 4$
    $\quad w - 2x \qquad\; + 3z = 1$
    $\quad w - x - y + 2z = 2$
    $\quad 2w - 3x - y + 5z = 3$
    $\qquad\quad x - y - z = 5.$

Solve the following systems, which have fewer equations than variables:

9. $4x - 2y + 6z = 6$
   $-2x + y - 3z = -3.$

10. $x + 3y + z = 6$
    $-x + y + z = 2.$

11. $2x + y + z = 4$
    $6x + 3y + 3z = 10.$

12. $2x + y + z = 4$
    $6x + 3y + 4z = 15.$

13. $w + 2x - y + z = 5$
    $w - 5x + 4y - 5z = 6$
    $3w - x + 2y - 3z = 2.$

14. $w + x + z = 4$
    $w - 3x + y = 1$
    $2x + y + z = -1.$

15. $w + 2x - y + z = 4$
    $-w + 3x + z = 2$
    $2w - x - y = 2.$

16. $w - 2x + y = 1$
    $w + 3x + 2z = 7$
    $x + y + z = 3.$

17. $w + x - 2y = 0$
    $w - x + 2z = 0.$

18. See the answer for Problem 10. If nonnegativity is imposed, what are the permissible intervals for x, y, and z?

19. See the answer for Problem 14. What are the permissible intervals for the variables if nonnegativity is imposed?

20. See the answer for Problem 16. What are the permissible intervals for the variables if nonnegativity is imposed?

2.21 APPLICATIONS–3

The manner in which the topics considered in this chapter arise in applications is illustrated by examples in this section and the two sections that follow.

Example 1. A buyer wishes to combine Whyall gasoline at $0.50 per liter with Zeeall gasoline at $0.60 per liter and obtain 2000 liters of gasoline worth $0.52 per liter. How many liters of each should he buy?

Letting y and z be the respective amounts of each gasoline, we require that $y + z = 2000$ liters. The value of these amounts of gasoline is $0.50y + 0.60z$, and this must equal $1040 which is the value of 2000 liters at $0.52 per liter. Hence,

$$e_1: \quad y + z = 2000$$
$$e_2: \quad 0.5y + 0.6z = 1040.$$

Proceeding,

$$e_3: 0.1z = 40 \qquad -0.5e_1 + e_2$$

so that $z = 400$ liters of Zeeall and $y = 1600$ liters of Whyall. We note that only one set of values for x and y satisfies the constraints of this problem.

Example 2. Suppose that *in addition to the constraints of Example 1*, the buyer wants a 95 octane mixture and that Whyall and Zeeall are, respectively, 90 octane and 100 octane. This leads to

$$90y + 100z = 95(2000)$$

or, dividing by 100,

$$0.9y + z = 1900.$$

It should be clear at once that no mixture will satisfy all constraints because equal amounts of 90 and 100 octane must be mixed to obtain 95 octane, and equal numbers for y and z will not satisfy e_1 and e_2. If we proceed to write the equations

$$\begin{aligned} e_1: \quad & y + && z = 2000 \\ e_2: \quad & 0.5y + 0.6z && = 1040 \\ e_3: \quad & 0.9y + && z = 1900 \end{aligned}$$

we find

$$\begin{aligned} e_4: \quad & 0.1z = 40 && -0.5e_1 + e_2 \\ e_5: \quad & 0.1z = 100 && -0.9e_1 + e_3. \end{aligned}$$

The last two equations are a contradiction because z cannot be both 400 and 1000. We have here a case of more equations than variables and as we stated earlier in the chapter, such systems often have no solution.

Example 3.   Suppose the buyer of Example 1 also may combine Exall at \$0.40 per liter in his mixture. What choice of components does he now have for his mixture?

We now have two equations in three variables, as follows:

$$\begin{aligned} e_1: \quad & x + y + z = 2000 \\ e_2: \quad & 0.4x + 0.5y + 0.6z = 1040. \end{aligned}$$

Proceeding,

$$e_3: \quad 0.1y + 0.2z = 240 \qquad -0.4e_1 + e_2$$

so that

$$y + 2z = 2400 \quad \text{or} \quad y = 2400 - 2z.$$

Then, from e_1 we find

$$x = z - 400.$$

From the last two statements we see that if x and y are to be nonnegative, z must be at least 400 and at most 1200. The mixture choices are

$$z \text{ arbitrary in the interval 400 to 1200, inclusive.}$$
$$x = z - 400$$
$$y = 2400 - 2z.$$

In this example, the system had fewer equations than variables. As we know, if such a system has a solution, it has an unlimited number of solutions. Of course, the nonnegativity constraints can do away with some or all of the solutions.

> **Exercise.** Express the answer to Example 3 with x arbitrary, stating the permissible interval for x. Answer: $y = 1600 - 2x$, $z = x + 400$; x arbitrary in the interval 0 to 800 liters, inclusive.

In passing, we note that the highest quality gasoline is Zeeall, so if a buyer wishes to maximize quality (octane), the maximum amount of Zeeall (1200 liters) would be chosen. The other components of the mixture would then be 800 liters of Exall and 0 liters of Whyall.

2.22 APPLICATIONS–4: INTRODUCTION TO OPTIMIZATION

When a system has many solutions, some solution or solutions may be better than others because they optimize an objective such as achieving a minimum cost. Two examples follow:

Example 1. A buyer wants to mix Exall gasoline (80 octane at $0.40 per liter) with Whyall (90 octane at $0.50 per liter) and Zeeall (100 octane at $0.65 per liter) to obtain 2000 liters of 85 octane mixture at minimum cost. What amounts of each gasoline should be mixed?

Letting x, y, and z represent, respectively, the numbers of liters of Exall, Whyall, and Zeeall, the cost function, C, is

$$C = 0.4x + 0.5y + 0.65z. \tag{1}$$

The first equation to be satisfied, e_1, is that the sum x, y, and z is to be 2000. The second equation, e_2, is obtained from the condition stated for octane, which is $80x + 90y + 100z = 85(2000)$ or, dividing by 100, we have $0.8x + 0.9y + z = 1700$. Hence,

$$e_1: \quad x + \quad y + z = 2000$$
$$e_2: \ 0.8x + 0.9y + z = 1700.$$

We have

$$e_3: \ 0.2x + 0.1y = 300 \qquad e_1 - e_2$$

so that

$$y = 3000 - 2x \tag{2}$$

and from e_1

$$z = x - 1000, \tag{3}$$

so that x is arbitrary in the interval 1000 to 1500 liters. We can now express the cost function also in terms of the arbitrary variable x by substituting (2) and (3) into (1). This yields

$$
\begin{aligned}
C &= 0.40x + 0.50(3000 - 2x) + 0.65(x - 1000) \\
&= 0.40x + 1500 - x + 0.65x - 650 \\
&= 0.05x + 850.
\end{aligned}
\tag{4}
$$

From (4) it is clear that cost increases when x increases, so to minimize cost we should use the lowest possible value of x. Because x is arbitrary in the range 1000 to 1500, we choose $x = 1000$, and compute the corresponding values of y and z. We have:

$$x = 1000 \text{ liters of Exall}$$
$$y = 1000 \text{ liters of Whyall}$$
$$z = 0 \text{ liters of Zeeall.}$$

Minimum cost $= 0.05(1000) + 850 = \$900$.

> *Exercise.* *a)* Convert (2) and (3) of the foregoing example into a solution with z arbitrary. *b)* What is the permissible interval for z? *c)* Write the cost function in terms of z. *d)* What value of z will minimize cost? *e)* What is the minimum cost? Answer: *a)* z arbitrary, $x = z + 1000$, $y = 1000 - 2z$. *b)* 0 to 500. *c)* Cost $= 0.05z + 900$. *d)* $z = 0$. *e)* \$900.

In the next example, we set up the system of equations and leave optimization as an exercise for the reader.

Example 2. A mixture containing x pounds of Exboy, y pounds of Whyboy, and z pounds of Zeeboy is to be made. The mixture is to weigh five pounds and contain 1500 units of vitamin and 2500 calories. The vitamin and caloric content of the three foods is shown in Table 2–1. The problem is to determine how many pounds of each food should be in the five-pound mixture.

TABLE 2–1

Food	Number of pounds	Units of vitamin per pound	Calories per pound
Exboy	x	500	300
Whyboy	y	200	600
Zeeboy	z	100	700

The conditions of the problem require that

$$
\begin{aligned}
e_1: & \quad x + \quad y + \quad z = \quad 5 \\
e_2: & \quad 500x + 200y + 100z = 1500 \\
e_3: & \quad 300x + 600y + 700z = 2500.
\end{aligned}
$$

Proceeding, we find:

$$
\begin{aligned}
e_4: & \quad 300y + 400z = \quad 1000 & \quad 500e_1 - e_2 \\
e_5: & \quad -300y - 400z = -1000 & \quad 300e_1 - e_3 \\
e_6: & \quad 0 + 0 = \quad 0. & \quad e_4 + e_5.
\end{aligned}
$$

The system does not have a unique solution. Letting z be arbitrary:

$$
e_7: \quad y = \frac{10 - 4z}{3} \quad \text{from } e_5
$$

$$
e_8: \quad x = \frac{5 + z}{3} \quad e_7 \text{ into } e_1
$$

$$
z \text{ arbitrary.}
$$

The general solution shows that an unlimited number of different mixtures can be made that satisfy the stated conditions. For example, if we use one pound of Zeeboy ($z = 1$), then the general solution says we must use

$$
y = \frac{10 - 4}{3} = 2 \text{ pounds of Whyboy.}
$$

$$
x = \frac{5 + 1}{3} = 2 \text{ pounds of Exboy.}
$$

The reader may verify that this (2, 2, 1) mixture will weigh five pounds, and will contain 1500 units of vitamin and 2500 calories.

Applying the nonnegativity constraint to

$$
e_7: \quad y = \frac{10 - 4z}{3},
$$

we see that y will be negative if $4z$ is larger than 10, that is, if z is larger than $10/4$, which is 2.5. If z is restricted to the interval 0 to 2.5, inclusive, none of the variables will be negative. Our final solution is

$$
z \text{ arbitrary in the interval}
$$
$$
0 \text{ to } 2.5, \text{ inclusive}
$$
$$
y = \frac{10 - 4z}{3}
$$
$$
x = \frac{5 + z}{3}.
$$

Exercise. (See Example 2.) If the costs per pound of Exboy, Whyboy, and Zeeboy are, respectively, $2, $3, and $1, what is the composition and cost of the minimum-cost mixture? Answer: Mix 2.5 pounds of Exboy with 2.5 pounds of Zeeboy. Use no Whyboy. Minimum cost will be $7.50.

2.23 APPLICATIONS–5: TWO-PRODUCT SUPPLY AND DEMAND ANALYSIS

Readers planning to work through this section may wish to review the one-product discussion of Section 2.10. Here we consider the case of two products whose supply and demand expressions are interrelated. As an example consider the following expressions in which p_1 and q_1 are the price of product #1 and the quantity demanded of product #1, respectively, and similarly for the price and quantity demanded for product #2.

	Demand	*Supply*
Product #1	$p_1 = 2000 - 3q_1 - 2q_2$	$p_1 = 100 + 2q_1 + q_2$
Product #2	$p_2 = 2800 - q_1 - 4q_2$	$q_2 = 200 + 3q_1 + 2q_2.$

To achieve equilibrium, the two price expressions for each product must be equal. Hence,

$$\text{for product \#1: } 2000 - 3q_1 - 2q_2 = 100 + 2q_1 + q_2$$
$$\text{for product \#2: } 2800 - q_1 - 4q_2 = 200 + 3q_1 + 2q_2.$$

Rearranging the last two expressions, we have

$$e_1: 5q_1 + 3q_2 = 1900$$
$$e_2: 4q_1 + 6q_2 = 2600.$$

We obtain the equilibrium quantities supplied and demanded by solving the last pair of equations in the usual manner:

$$e_3: 18q_2 = 5400 \qquad 5e_2 - 4e_1.$$

Hence, $q_2 = 5400/18 = 300$. Then from e_2 we find $q_1 = 200$. Using these quantities in the original supply (or demand) expressions, we find $p_1 = \$800$ and $p_2 = \$1400$. The equilibrium prices and quantities are

$$E: \quad \begin{matrix} p_1 = \$800 & p_2 = \$1400 \\ q_1 = 200 \text{ units} & q_2 = 300 \text{ units.} \end{matrix}$$

Now suppose the supply functions remain the same but a change occurs in the demand for product #1 and, since the two products are interrelated, there is a corresponding change in the demand for #2. Then suppose the outcomes are the

New demand functions

product #1: $p_1 = 2270 - 3q_1 - 2q_2$
product #2: $p_2 = 2890 - q_1 - 4q_2$.

Equating these new demand expressions with the corresponding original supply expressions leads to

$$5q_1 + 3q_2 = 2170$$
$$4q_1 + 6q_2 = 2690.$$

Solving the last pair, we find $q_1 = 275$ and $q_2 = 265$, and using these values in the new demand or the old supply expressions yields $p_1 = \$915$, $p_2 = \$1555$. The new equilibrium is

$$E': \quad \begin{matrix} p_1 = \$915 & p_2 = \$1555 \\ q_1 = 275 \text{ units} & q_2 = 265 \text{ units}. \end{matrix}$$

Exercise. Solve the foregoing pair of equations, then carry out the substitution necessary to verify the values at E'.

Comparing E' and E, we describe what occurred in the following manner. The demand for #1 increased (from 200 to 275) and the price of this product rose (from \$800 to \$915). At the same time, there was a decrease in demand for #2 (from 300 to 265) but an increase in its price (from \$1400 to \$1555). To gain an understanding of these changes, let us think of product #1 as the standard model of an appliance and #2 as the deluxe model. Then we can see that consumer response to rising prices of appliances (both prices up) might well be to increase their demand for the standard model and decrease demand for the deluxe model, as was the outcome in this example. The outcome of supply and demand shifts depends on the nature of the products involved and relates directly to the constants in the mathematical expressions for supply and demand. Readers interested in pursuing this matter should refer to discussions of *complementary, competing* (or *substitute*), and *independent* products in economics textbooks.

2.24 PROBLEM SET 2–5

1. A mixture of pellets is to be made containing x Exbit pellets, y Whybit pellets, and z Zeebit pellets. Cost, weight and volume data for each type of pellet are shown in the table.

    Is it possible to make a mixture of these pellets at a cost of 85 cents if the mixture is to have 120 weight units and 130 volume units? If so, how many of each type of pellet should be in the mixture?

Pellet type	Number of pellets	Cost per pellet in cents	Weight units per pellet	Volume units per pellet
Exbit	x	2	1	4
Whybit	y	3	2	2
Zeebit	z	1	4	3

2. Replace the volume data in the table for Problem 1 with the numbers 1.6, 2.8, and 3.6 for Exbit, Whybit, and Zeebit, respectively. Leaving the other data and the conditions of the problem unchanged, set up and solve the resultant system of equations.

3. *a)* Foods Exboy, Whyboy, and Zeeboy are to be combined to make a seven-pound mixture that will contain 5000 calories. Exboy, Whyboy, and Zeeboy contain, respectively, 500, 1000, and 1500 calories per pound. Find the general solution, showing permissible numbers of pounds of Exboy, Whyboy, and Zeeboy that can be used in the mixture.

    b) If the costs per pound of Exboy, Whyboy, and Zeeboy are, respectively, \$1, \$3, and \$4, what is the composition and cost of the minimum-cost mixture?

4. In the table, w, x, y, and z represent tons of a certain type of steel shipped from factories to warehouses. For example, x is the number of tons shipped from factory one to warehouse two, that is, from F_1 to W_2. The numbers 40 and 60 are the total numbers of tons available at the factories, and the numbers 30 and 70 are the total tons required by the warehouses.

	W_1	W_2	
F_1	w	x	40
F_2	y	z	60
	30	70	

Note that the *marginal* total $30 + 70$ equals the *marginal* total $40 + 60$. Prove that only one of w, x, y, and z may be chosen arbitrarily by writing the general solution for w, y, and z with x arbitrary.

5. See Problem 4. Write all the equations that must be satisfied, then write the general solution of this system of equations.

	W_1	W_2	W_3	
F_1	w	x	y	90
F_2	u	v	z	60
	30	45	75	

6. Transistors are of three types, AL46TR, BM139TR, and HQ257TR. A grade A box of transistors contains 5 AL46TRs, 3 BM139TRs, and 1 HQ257TR. A grade B box contains 2 AL46TRs, 3 BM139TRs and 4 HQ257TRs. How many boxes of each grade can be filled if the total numbers of AL46TRs, BM139TRs, and HQ257TRs available are 80, 75, and 70, respectively, and all transistors are used?

7. See Problem 6. A grade C box contains 5 AL46TRs and 1 BM139TR; a grade D box contains 3 AL46TRs and 1 BM139TR; a grade E box contains 1 AL46TR and 4 BM139TRs. How many boxes of grades C, D, and E can be filled *exactly* using 44 AL46TR and 22 BM139TR transistors?

8. See Problems 6 and 7. The numbers of transistors, by type, contained in boxes of grades F, G, and H are shown in the table. Find the numbers of boxes (*x*, *y*, and *z*) that could be completely filled using all available transistors.

		Number of transistors required per box, by type		
Grade	Number of boxes	AL46TR	BM139TR	HQ257TR
F	x	1	2	2
G	y	2	7	1
H	z	1	3	1
Total available		(12)	(34)	(14)

9. Solve Problem 8 if the requirement for BM139TRs in a box of grade G is reduced from 7 to 6.

10. Solve Problem 8 if the requirement for HQ257TRs in grade F is increased from 2 to 3.

11. *a)* The table shows the numbers of hours required in each of two departments to make a unit of various products named A, B, and C. For example, product B requires 1 hour of time in Department I and 3 hours in Department II.

	Hours required per unit of product		
	A	B	C
Department I	1	1	9
Department II	1	3	7

Find the numbers of units of A, B, and C that could be made if Department I has 75 hours available and Department II has 65 hours available.

b) If profits per unit of A, B, and C are, respectively, $20, $30, and

$40, what is the maximum profit and the composition of the maximum-profit combination of outputs?

12. Solve Problem 11*a)* and *b)* if both departments have 40 hours available.

13. The demand and supply expressions for products #1 and #2 are:

Demand	*Supply*
#1: $p_1 = 1000 - 5q_1 - 4q_2$	$p_1 = 90 + 2q_1 + 3q_2$
#2: $p_2 = 900 - 2q_1 - 5q_2$	$p_2 = 120 + q_1 + 4q_2.$

a) Find the prices and quantities at equilibrium.

b) If demands change to those below, with the same supply functions, find the new equilibrium point.

Demand

#1: $p_1 = 1210 - 5q_1 - 4q_2$

#2: $p_2 = 984 - 2q_1 - 5q_2.$

14. The demand and supply expressions for products #1 and #2 are:

Demand	*Supply*
#1: $p_1 = 1500 - 4q_1 - 3q_2$	$p_1 = 400 + 3q_1 + q_2$
#2: $p_2 = 700 - q_1 - 2q_2$	$p_2 = 200 + q_1 + q_2.$

a) Find the prices and quantities at equilibrium.

b) If demands change to those below, with the same supply functions, find the new equilibrium point.

Demand

#1: $p_1 = 1400 - 3q_1 - 3q_2$

#2: $p_2 = 640 - q_1 - q_2.$

2.25 REVIEW PROBLEMS

Solve each of the following systems by the elimination method:

1. $x + 10y = 25$
    $3x - 7y = 1.$

2. $x + 2y - 3z = 11$
    $3x + 2y + z = 1$
    $2x + y - 5z = 11.$

3. $x + y - 2z = 1$
    $x - 2y + z = 7.$

4. $x + y = 6$
    $2x + y = 10$
    $x - y = 1.$

5. $3x + y + z = 4$
    $x - 2y - 3z = 0$
    $2x + 3y + 4z = 6.$

6. $2x + z = 4$
    $3y + 2z = 6$
    $4x - 3y = 2.$

7. $x + y + z = 1$
    $2x - y + 3z = 5$
    $x + 5y + 2z = 4$
    $3x - 2y + 2z = 0.$

8. $w + x - 2z = 0$
    $w - x - 2y = 0$
    $w + 2x + y - 3z = 0.$

9.
$$\begin{aligned} x - y &= 1 \\ 2x + y - 3z &= 2 \\ x + y - 2z &= 1. \end{aligned}$$

10.
$$\begin{aligned} w + x + y + z &= 0 \\ w - x + y - z &= 0 \\ 2w + x + 2y + z &= 0 \\ 3w + 2x + 3y + 2z &= 0. \end{aligned}$$

11. The table shows, for example, that one pound of food Exboy contains one ounce of nutrient P and two ounces of nutrient Q. Similar relations are shown for foods Whyboy and Zeeboy.

| | Ounces of nutrient per pound | |
Food	Nutrient P	Nutrient Q
Exboy	1	2
Whyboy	2	3
Zeeboy	3	2

If a mixture of x pounds of Exboy, y pounds of Whyboy, z pounds of Zeeboy is to contain exactly 6 ounces of P and 10 ounces of Q, find the permissible values for x, y, and z.

   a) Write the solution with x arbitrary, then with y arbitrary, and then with z arbitrary.

   b) What are the largest and the smallest permissible amounts of Exboy, Whyboy, and Zeeboy?

12. See Problem 11. If the per-pound costs of Exboy, Whyboy, and Zeeboy were 5 cents, 8 cents, and 3 cents, respectively, the cost of a mixture would be

$$M = 5x + 8y + 3z.$$

Find the composition and cost of the minimum-cost mixture.

13. See Problem 11. Suppose that the per-pound costs of Exboy, Whyboy, and Zeeboy are 10 cents, 5 cents, and 3 cents, respectively. Using the results of Problem 11, find the cost and composition of the minimum-cost mixture.

14. The table shows, for example, that a pound of food Exboy contains 2 ounces of nutrient P, 3 ounces of Q, and 1 ounce of R. Similar relations are shown for foods Whyboy and Zeeboy. If a mixture of x pounds of Exboy, y pounds of Whyboy, and z pounds of Zeeboy is to contain exactly 9 ounces of P, 13 ounces of Q, and 4 ounces of R, find the permissible values for x, y, and z.

| | Ounces per pound of: | | |
Food	Nutrient P	Nutrient Q	Nutrient R
Exboy	2	3	1
Whyboy	1	2	1
Zeeboy	1	1	0

Write the solution in arbitrary variable form, and state the permissible ranges of values for the variables.

15. See Problem 14. If the per-pound costs are 20, 10, and 5 cents for Exboy, Whyboy, and Zeeboy, respectively, find the cost and composition of the minimum-cost mixture.

16. We wish to mix x liters of Exall gasoline with y liters of Whyall and z liters of Zeeall to obtain 1000 liters of 90 octane gasoline. Costs per liter are $0.50, $0.55, and $0.65, respectively, and octane ratings are 84, 92, and 100, respectively.
    a) Write the solution in arbitrary-variable form.
    b) What is the composition and cost of the minimum-cost mixture?

17. The table shows the number of hours required to make one unit of various products (A, B, and C) in each of two departments (I and II). For example, it takes one hour of Department I time and three hours of Department II time to make one unit of product C.

| Department | Hours to make one unit of: | | |
	Product A	Product B	Product C
I.................	1	1	1
II.................	2	1	3

a) Find the numbers of units, x, y, and z of products A, B, and C that can be made if exactly 8 hours of Department I time and 14 hours of Department II time are to be utilized.
b) If the maximum possible number of units of C are made, how many hours of each department's time will be spent on making C?
c) Is it possible to utilize the hours exactly if no units of B are made? Explain.

18. The table shows, for example, that one unit of product A consumes two hours of time in Department II. Similar relations are shown for other products and departments.

| Department | Hours to make one unit of: | | | |
	Product A	Product B	Product C	Product D
I	1	2	1	2
II	2	1	0	1
III	0	1	3	2

Find the numbers of complete units, w, x, y, and z of products A, B, C, and D, respectively, which can be made if exactly 12 hours of each department's time are to be utilized. (Fractional units are not permitted in the solution).

19. The demand and supply expressions for products #1 and #2 are

	Demand	Supply

$$\#1\colon\ p_1 = 1700 - 3q_1 - q_2 \qquad p_1 = 100 + 2q_1 + q_2$$
$$\#2\colon\ p_2 = 1650 - q_1 - 2q_2 \qquad p_2 = 50 + q_1 + 2q_2.$$

a) What will be the prices and quantities at equilibrium?
b) If the demand functions change to those below and the supply functions are unchanged, find the new equilibrium point.

Demand

$$\#1\colon\ p_1 = 1720 - 3q_1 - q_2$$
$$\#2\colon\ p_2 = 1710 - q_1 - 2q_2.$$

20. The demand and supply expressions for products #1 and #2 are

	Demand	Supply

$$\#1\colon\ p_1 = 2300 - 30q_1 - 10q_2 \qquad p_1 = 180 + 5q_1 + 2q_2$$
$$\#2\colon\ p_2 = 2000 - 10q_1 - 20q_2 \qquad p_2 = 120 + q_1 + 4q_2.$$

a) What will be the prices and quantities at equilibrium?
b) If the demand functions change to those below and the supply functions remain the same, find the new equilibrium point.

Demand

$$\#1\colon\ p_1 = 4424 - 30q_1 - 10q_2$$
$$\#2\colon\ p_2 = 2708 - 10q_1 - 20q_2.$$

Systems of linear inequalities

3

3.1 INTRODUCTION

Mathematical statement of the conditions imposed by a problem often requires the use of inequalities rather than equalities. For instance, if we let x represent the number of hours a plant operates per day, x does not necessarily equal 24, but x must be less than or equal to 24. Moreover, a plant cannot operate a negative number of hours, so x must be greater than or equal to zero. Other examples of conditions leading to inequalities come easily to mind. The volume of fluid stored in a tank must be greater than or equal to zero, but less than or equal to the volume of the tank; the amount of product made is restricted by plant capacity and raw material availability. In this chapter, we shall discuss linear inequalities from an algebraic point of view and offer geometrical interpretations of systems involving one or two variables.

3.2 DEFINITIONS AND FUNDAMENTAL PROPERTIES

The symbols of inequality are $<$ and $>$. The first, $<$, means *is less than;* the second, $>$, means *is greater than*. Thus $3 > 2$ is the true statement that 3 is greater than 2, and $5 < 9$ is the true statement that 5 is less than 9. It may be of use to note that the symbols $>$ and $<$ point toward the smaller quantity. The statement $a < b$ is read "*a* is less than *b*." If we wish to state that "*a* is less than or equal to *b*," we combine the inequality sign with the equality sign and write $a \leq b$. Further practice with the symbols is afforded next:

$a \geq b$ means a is greater than or equal to b.
$a \leq 3$ means a is less than or equal to 3.
$b > 0$ means b is greater than zero; that is, b is positive.
$c < 0$ means c is negative.
$a \geq 0$ means a is not negative.

One of the fundamental properties of the real number system is the property of *order*. We say that the real numbers are ordered in the sense that either any two numbers are equal, or one is greater than the other. Thus, given any two numbers, a and b, one and only one of the following must be true:

$$a < b$$
$$a = b$$
$$a > b.$$

To say that a first number is greater than a second implies that a positive number must be added to the second number to make the sum equal the first number. For example, $5 > 2$ implies that a positive number must be added to 2 to make a sum equal to 5. In general:

$a > b$ implies that $a = b + c$, where $c > 0$.
$a < b$ implies that $a + c = b$, where $c > 0$.

The last statement that "*a* is less than *b*" implies that a positive number ($c > 0$) must be added to a to make the sum equal b.

It is a consequence of the order property of the real numbers that $-5 < -2$ and $0 > -1$. That is, inasmuch as the positive number 3 must be added to -5 to make the sum equal -2, -5 is less than -2. Again, inasmuch as the positive number 1 must be added to -1 to make the sum 0, 0 is greater than -1.

Exercise. What argument leads to the conclusion that -10 is greater than -20? (See the immediately preceding paragraph.)

The ordering of numbers can be remembered easily by reference to Figure 3–1. If the number a is to the left of the number b, then

FIGURE 3-1

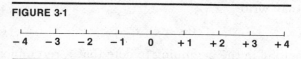

$a < b$. If the number a is to the right of the number b, then $a > b$. We see that $-1 < 0, 0 < 1, -2 > -3, 2 > -1$.

The direction in which an inequality symbol points is referred to as its *sense*, the particular use of the word being in phrases that describe inequalities as being of the *same sense* or of *opposite sense*. Thus, $x < 3$ and $x < -1$ are of the same sense, and $x \leq 5$ and $x \geq 8$ are of opposite sense.

3.3. FUNDAMENTAL OPERATIONS ON INEQUALITIES

As in the case with equalities, the same number may be added to or subtracted from both sides of an inequality, and both sides may be multiplied by, or divided by, the same *positive* (nonzero) number. However,

Multiplying or dividing both sides of an inequality by the same negative number changes the sense of the inequality.

For example:

$$i_1: \quad -3 > -6$$
$$i_2: \quad 12 < 24 \qquad i_2 \text{ is } -4\,(i_1).$$

Thus, i_1 is the true statement that -3 is greater than -6. If we multiply both sides of i_1 by -4, we change the sense to get the true statement, i_2, that 12 is less than 24. The reader will find it helpful to check the following sequence of operations.

$$
\begin{array}{lll}
i_1: & 2 > 1 & \\
i_2: & 5 > 4 & \text{Add 3 to both sides of } i_1. \\
i_3: & -30 < -24 & \text{Multiply } i_2 \text{ by } -6. \\
i_4: & 10 > 8 & \text{Divide } i_3 \text{ by } -3. \\
i_5: & -2 > -4 & \text{Subtract 12 from both sides of } i_4. \\
i_6: & -4 < -2 & \text{Read } i_5 \text{ from } \textit{right} \text{ to } \textit{left}. \\
i_7: & 2 > 1 & \text{Divide } i_6 \text{ by } -2.
\end{array}
$$

Ordinarily, we read inequalities in the usual manner, from left to right. However, as shown in i_6 of the foregoing, we may, if we wish, read from right to left. For example, just as we can write

$$6 = x \text{ or } x = 6,$$

so also we may write

$$6 < x \text{ or } x > 6.$$

It may be helpful in the beginning to note that $>$ resembles an arrowhead and we read "greater than" if we read in the direction the arrow is pointing, and "less than" if we read into the arrowpoint.

3.4 SOLVING SINGLE INEQUALITIES

Keeping the fundamental operations in mind, we solve a single inequality as shown next.

Example. Solve the following for x:

$$i_1: 3 - 2x \leq 7.$$

We proceed to get x alone on one side of the inequality.

$i_2: \quad -2x \leq \quad 4 \qquad$ Subtract 3 from both sides of i_1
$i_3: \qquad x \geq -2 \qquad$ Divide i_2 by -2.

Exercise. Solve $7 - x \geq 3$ for x. Answer: $x \leq 4$.

A single equation in two-space has a straight line as its geometric representation. A single *inequality*, on the other hand, is represented by all points on one side of a line, the line itself being included if the equality and inequality signs appear together. If we think of a straight line as dividing a plane in half, we may say that the solutions of an inequality in two-space consist of all points in a *half space*. Consider the inequality

$$2x + 3y \leq 6.$$

Our first step in representing the inequality is to draw the line specified by the *equality*

$$2x + 3y = 6.$$

The intercepts are:

$$x = 0; \ y = 2; \quad \text{point is } (0, 2)$$
$$y = 0; \ x = 3; \quad \text{point is } (3, 0).$$

The line is shown on Figure 3–2. Notice that the *origin* $(0, 0)$, satisfies the inequality statement

$$2x + 3y \leq 6$$
$$2(0) + 3(0) \leq 6$$
$$0 \leq 6.$$

FIGURE 3-2

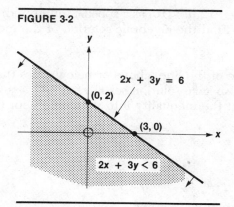

Thus the origin, which is *below* the line, is in the solution space and *if one point below a line is in the solution space, then all points below the line are in the solution space.* Hence, the solution space consists of all points on the line, or below the line, as indicated by the direction of the arrows on the line. Obviously, the solution space contains an unlimited number of points.

To solve the inequality algebraically, we change to slope-intercept form as follows:

$$2x + 3y \leq 6$$
$$3y \leq -2x + 6$$
$$y \leq -\frac{2}{3}x + 2.$$

Here, the equality part of

$$y \leq$$

means that points *on* the line are included in the solution space and the *less-than* part means points below the line are included. The algebraic solution then is

$$x \text{ arbitrary}$$
$$y \leq -\frac{2}{3}x + 2.$$

Thus, if x is assigned any arbitrary value, the last expression states the permissible values for y. For example, if $x = -12$

$$y \leq -\frac{2}{3}(-12) + 2$$
$$y \leq 8 + 2$$
$$y \leq 10$$

and we have $(x=-12, y \le 10)$ as a specific (unlimited) set of solutions.

Example. *a)* Find the algebraic solution of the inequality

$$3x - 2y < -8.$$

b) On which side of the line (above or below) does the solution space lie? *c)* Write the specific solution set if $x = 6$.

a) We first put the inequality into slope-intercept form, as follows:

$$-2y < -3x - 8$$

$$y > \frac{3}{2}x + 4$$

$$y > 1.5x + 4.$$

The general solution is, therefore,

$$x \text{ arbitrary}$$
$$y > 1.5x + 4.$$

b) The *greater-than* sign, $>$, in the last statement means the solution space is *above* the line. The line itself is *not* included because the statement specifies $y >$, not $y \ge$.

c) If $x = 6$, we have from the general solution

$$y > 1.5(6) + 4$$
$$y > 9 + 4$$
$$y > 13,$$

so the specific solution set is $(x = 6, y > 13)$. Note that $y = 13$ is not in this solution set.

> **Exercise.** Given $x - 3y \le 12$. *a)* Write the general algebraic solution. *b)* On which side of the line does the solution space lie? *c)* Write the specific solution set if $x = 9$. Answer: *a)* $y \ge (1/3)x - 4$. *b)* Above the line. *c)* $(x = 9, y \ge -1.)$

As we proceed in this chapter and the next, we shall have need to graph solution spaces. To do this, the line can be drawn from its intercepts or any two of its points and the side of the line on which the solution space lies indicated by arrows, as in Figure 3–2.

To determine the side of the line on which the solution space lies,

Either: *a)* Solve for y. Then $y >$ means above the line and $y <$ means below the line.

Or: *b)* Substitute the origin $(0, 0)$ into the inequality. If the result is a true statement, the half-space is on the same side of the plotted line as the origin. If the result is a false statement, the half-space is on the side of the plotted line which does not contain the origin. Use a point other than (0, 0) if the inequality has no constant term other than zero.

The last sentence says that if we have

$$3y + 2x \geq 0,$$

the line

$$3y + 2x = 0$$

goes through the origin, (0, 0), because

$$3(0) + 2(0) = 0,$$

and every line whose equation has zero as its only constant term passes through the origin. Both intercepts are at the origin, so to get a second point on

$$3y + 2x = 0$$

we give x an arbitrary value, say 3, and compute

$$3y + 2(3) = 0$$
$$3y = -6$$
$$y = -2.$$

The point just determined $(3, -2)$, along with $(0, 0)$, determine the line shown on Figure 3–3. Substituting (0, 0) into $3y + 2x \geq 0$ gives

FIGURE 3-3

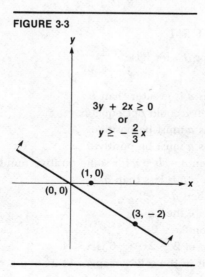

$$0 + 0 \geq 0,$$

which is true, but it does not tell us on which side of the line the solution space lies because the origin is on the line. However, if we take the point $(1, 0)$ and substitute this into the inequality we find

$$3(0) + 2(1) \geq 0$$
$$2 \geq 0,$$

which is true. From Figure 3–3, we see that $(1, 0)$ is above the line so all points in the solution space are above (or on) the line, as shown by the arrows on Figure 3–3.

We know from earlier study that

$$x = 6$$

is a *vertical* line, so the terms above and below have no meaning here. However, if we have

$$x < 6,$$

the *less than* sign, $<$, means the solution space is to the *left* of the vertical line $x = 6$. Similarly, $x > -2$ means the solution space is to the right of the vertical line $x = -2$. Of course,

$$y < 20$$

means the solution space is below the horizontal line $y = 20$, and

$$y > -12$$

means the solution space is above the horizontal line $y = -12$.

3.5 PROBLEM SET 3–1

Mark (T) for true or (F) for false.

1. ( ) $-3 > -2$.
2. ( ) $a > b$ means a is greater than b.
3. ( ) If $4 < c$, then c could not equal 5.
4. ( ) $a \leq 0$ means a must be positive.
5. ( ) $a \geq 0$ means a must be positive.
6. ( ) If $a < b$, then $a = b + c$ for some positive number c.
7. ( ) If $a > b$, then b is less than a.
8. ( ) If $-a < 2$, then $a < -2$.
9. ( ) If $a - 2 > -5$, then $a > -3$.
10. ( ) If $x \leq y$, then $2x \leq 2y$.
11. ( ) The solution of $2 - 2x > -6$ is $x < 4$.
12. ( ) The origin is in the solution space of $2x + y \leq 8$.

Solve for x:

13. $3x - 2 \leq 4$.

14. $5x + 7 \geq 22$.

15. $5 - 2x > 11$.

16. $4 - 3x < 10$.

17. $4 \leq 3 + 2x$.

18. $5 \geq 6 + 4x$.

In the following: *a)* Write the general algebraic solution by solving for y. *b)* On which side of the line does the solution space lie? *c)* Write the specific solution space for the stated value of x.

19. $3x + 2y \leq 12$; $x = 2$.

20. $2x + 5y \geq 35$; $x = 5$.

21. $7x - 5y \geq 45$; $x = 20$.

22. $3x - 2y \leq 18$; $x = 10$.

State whether or not the origin is in the solution space of each of the following:

23. $3x - 7y \leq 6$.

24. $2y - 3x \geq 4$.

25. $y - 2x \geq -1$.

26. $3x + 2y \leq -5$.

Graph the following in two-space. Indicate the side of the line on which the solution space lies by arrows.

27. $3y - 2x \leq 36$.

28. $2x - 10y \leq 20$.

29. $5x + 2y \geq 40$.

30. $2x + 3y \leq 60$.

31. $y \geq 5$.

32. $x \leq 10$.

33. $x \geq 0$.

34. $y \geq 0$.

35. $y - 3x \leq 0$.

36. $2x + 3y \leq 0$.

3.6 DOUBLE INEQUALITIES

The system of inequalities

$$x > 4$$
$$x \leq 9$$

requires that x be greater than 4 but less than or equal to 9. Hence, x is in an interval with the smaller number, 4, at the left and the larger number, 9, at the right. This may be written as

$$4 < x \leq 9.$$

Again, the statement

$$0 \leq x \leq 10$$

says that x is greater than or equal to 0 but less than or equal to 10.

Exercise. Write the statement that specifies that y is less than 0 but greater than or equal to -3. Answer: $-3 \leq y < 0$.

Two inequalities are said to be *inconsistent* if both cannot be true at the same time, and one of a pair is *redundant* if it is true automatically when the other is true. Thus, if company policy states that workers may not work more than 10 hours a day ($x \leq 10$) and the union contract specifies that workers must not work more than 8 hours a day ($x \leq 8$), company policy is redundant.

Exercise. How would the system $x < 5$ and $x > 8$ be described? Why? Answer: The inequalities are inconsistent because if x is less than 5, it cannot be greater than 8.

Double inequalities are used in describing the space between two lines. For example, the inequalities

$$i_1: \ y \geq 2x - 1$$
$$i_2: \ y \leq \ x + 2$$

are satisfied by points in the shaded space shown in Figure 3–4. The lines, which are labeled e_1 and e_2 because they are the graphs of the

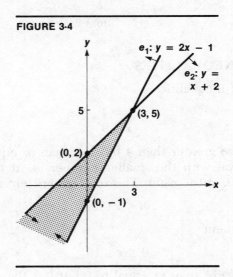

FIGURE 3-4

equality parts of the statements, are found in the usual manner to intersect at (3, 5). Hence, for any point in the solution space, the x coordinate must be less than or equal to 3.

Exercise. Suppose that we let x have the permissible value 0. What y values are permissible if $x = 0$? (See Figure 3–4). Answer: $-1 \le y \le 2$.

The exercise shows that for a given permissible value of x, the permissible values of y are those which are greater than or equal to $2x - 1$ but less than or equal to $x + 2$. That is, the solution space is completely described by

$$x \le 3; \ 2x - 1 \le y \le x + 2.$$

This last result says, for example, that if $x = -1$, then both i_1 and i_2 will be true for any y value in the closed interval

$$-3 \le y \le 1.$$

Exercise. On scratch paper, sketch the lines shown in Figure 3–4 and place arrows on them indicating the solution space for

$$y \le 2x - 1; \quad y \ge x + 2.$$

Describe the solution space algebraically. Answer: $x \ge 3$; $x + 2 \le y \le 2x - 1$.

The solution space specified by

$$i_1: \ y \le 2x - 1$$
$$i_2: \ y \le x + 2$$

is the space *below* both lines in Figure 3–4. This is the space in the lower obtuse angle formed at (3, 5). To the left of (3, 5), that is for $x < 3$, this means that points must be on or below e_1 (that is, $y \le 2x - 1$), and to the right of (3, 5) points must be on or below e_2 (that is, $y \le x + 2$). Hence, we have

$$x \le 3; \ y \le 2x - 1$$
$$x \ge 3; \ y \le \ x + 2.$$

Note in the description just given that to the right of (3, 5)

$$i_1: \ y \le 2x - 1$$

is *redundant* because it holds automatically if

120

$$i_2: \ y \leq x + 2$$

is true.

Exercise. *a*) Describe the solution space in the upper obtuse angle formed at (3, 5) in Figure 3–4. *b*) Which inequality is redundant to the right of (3, 5)? Answer: *a*) ($x \leq 3$, $y \geq x + 2$); ($x \geq 3$, $y \geq 2x - 1$). *b*) $y \geq x + 2$.

3.7 NONNEGATIVITY CONSTRAINTS

In the work to come, we shall deal often with quantities and prices of goods, and other variables that cannot take on negative values. In the case of two variables, x and y, this restriction is expressed by writing

$$x \geq 0$$
$$y \geq 0$$

along with the other inequalities at hand. Nonnegativity means that only points in the first quadrant, including the axes, are under consideration.

The system

$$x \geq 0$$
$$y \geq 0$$
$$x \leq 4$$
$$y \leq 2$$

has as its solutions the points on the boundary and inside the rectangle, Figure 3–5.

FIGURE 3-5

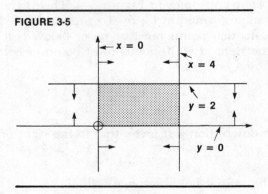

The algebraic solution of 2 by 2 systems of inequalities with nonnegativity constraints is facilitated by graphing the solution space and determining from the graph what algebraic manipulations of the inequalities will lead to the correct general solution. Consider the system

$$
\begin{aligned}
i_1: &\quad x \geq 0 \\
i_2: &\quad y \geq 0 \\
i_3: &\quad x + 2y \leq 8 \\
i_4: &\quad 7x + 4y \geq 28.
\end{aligned}
$$

The solution space is shown on Figure 3–6 and can be described algebraically by saying that if y is in the interval from 0 to 2.8, inclusive, x must be in the interval from e_4 to e_3. Thus the general solution is

$$
0 \leq y \leq 2.8
$$

$$
\frac{28 - 4y}{7} \leq x \leq 8 - 2y.
$$

FIGURE 3-6

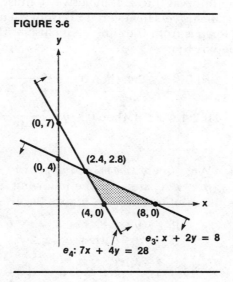

$e_3\!: x + 2y = 8$

$e_4\!: 7x + 4y = 28$

Exercise. If the solution space in Figure 3–6 is described by letting x run over the values 2.4 to 8, two statements will be required. The first is

$$
2.4 \leq x \leq 4
$$

$$
\frac{28 - 7x}{4} \leq y \leq \frac{8 - x}{2}.
$$

The second will be of the form

$$a \leq x \qquad \leq b$$
$$c \leq y \qquad \leq d.$$

Determine from Figure 3–6 the quantities that should appear in the places indicated as a, b, c, and d in the form just written.

Answer: a is 4, b is 8, c is 0, d is $\dfrac{8-x}{2}$.

Consider the next system

$$i_1: \qquad x \geq \; 0$$
$$i_2: \qquad y \geq \; 0$$
$$i_3: \quad x + 2y \leq \; 8$$
$$i_4: \; 7x + 4y \leq 28.$$

The solution space for this system is shown on Figure 3–7. We see that for x in the interval 0 to 2.4, inclusive, y must be in the interval from 0 to line e_3, inclusive. However, if x is from 2.4 up to and including 4, the interval for y is from 0 to line e_4, inclusive. Hence, the general solution may be written:

$$\text{If } 0 \leq x \leq 2.4, \text{ then } 0 \leq y \leq \frac{8-x}{2}.$$

$$\text{If } 2.4 \leq x \leq 4, \text{ then } 0 \leq y \leq \frac{28-7x}{4}.$$

Exercise. What quantities should appear in the places designated as a, b, c, d, e, and f if the following inequalities are to describe the solution space in Figure 3–7?

$$\text{If } 0 \leq y \leq 2.8, \text{ then } a \leq x \leq b.$$
$$\text{If } c \leq y \leq d, \text{ then } e \leq x \leq f.$$

Answer: a is 0, b is $\dfrac{28-4y}{7}$, c is 2.8, d is 4, e is 0, f is $8 - 2y$.

In the next chapter we will frequently encounter solution spaces similar to that in Figure 3–7. Another common type is shown in Figure 3–8, which is the solution space for

$$i_1: \; x \geq 0$$
$$i_2: \; y \geq 0$$
$$i_3: \; x + 2y \geq 8$$
$$i_4: \; 7x + 4y \geq 28.$$

FIGURE 3-7   **FIGURE 3-8**

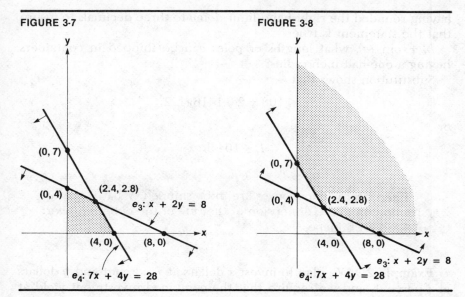

Exercise. Write the algebraic expressions for the solution space in Figure 3–8 starting with if $0 \leq x \leq 2.4$. Answer: If $0 \leq x \leq 2.4$, $y \geq (28 - 7x)/4$. If $2.4 \leq x \leq 8$, then $y \geq (8 - x)/2$. If $x > 8$, then $y > 0$.

3.8 APPLICATIONS

Example 1. The United Parcel Service in Massachusetts accepts a parcel for delivery only if its length plus girth (distance around) does not exceed 108 inches.

a) We want to ship fishing poles in cylindrical containers of radius r. What is the restriction on the length of the poles in terms of the radius of the cylindrical container, and what restrictions apply to the radius?

Here, the girth, g, is $2\pi r$, the circumference of a circle, and using L for the length, we have

$$L + 2\pi r \leq 108$$
$$L \leq 108 - 2\pi r$$

and $r > 0$. Also, $L \geq 0$, so $108 - 2\pi r \geq 0$ and $r \leq 108/2\pi$, which is 17.188 plus a bit more; that is,

$$0 < r \leq 17.188$$

having rounded the right-hand limit *down* to three decimals to insure that the statement is true.

b) From *(a),* what lengths of poles can be shipped in containers having a one-half inch radius?

Substitution shows that

$$L \le 108 - 2(3.1416)(1/2)$$

or

$$L \le 104.8.$$

Exercise. Cubical boxes are to be shipped. What is the restriction on x, the dimension of the side of the cube? Answer: $0 < x \le 21.6$ inches.

Example 2. We plan to invest x dollars at 7 percent and y dollars at i percent, and we require that the combined investment yield at least 9 percent. Express the restriction on i in terms of x and y and state the restrictions on x and y.

We have that 7 percent of x plus i percent of y must be equal to or greater than 9% of $(x + y)$. Hence, using the decimal equivalents of percents,

$$0.07x + 0.01iy \ge 0.09(x + y)$$
$$0.07x + 0.01iy \ge 0.09x + 0.09y$$
$$0.01iy \ge 0.02x + 0.09y$$
$$iy \ge \frac{0.02x + 0.09y}{0.01}$$
$$iy \ge 2x + 9y$$
$$i \ge \frac{2x + 9y}{y}.$$

Both x and y must be nonnegative. Additionally, x can be zero, but y cannot be zero. Hence,

$$x \ge 0$$
$$y > 0.$$

Exercise. If $x = \$5000$ and $y = \$4000$, what is the minimum interest rate required on the $4000? Answer: 11.5 percent.

Example 3. A mixture is to be made containing x units of food Exbit and y units of Whybit. The weight and amount of nutrient per unit of each food are shown in Table 3–1. If the mixture is to contain not less than 13 ounces of nutrient, and it is to weigh not more than 50 ounces, what combinations (numbers of units) of the foods are permissible?

TABLE 3–1

Food	Number of units in mixture	Ounces of nutrient per unit	Ounces of weight per unit
Exbit x		0.4	2
Whybit . . . . .y		0.9	3

To obtain the necessary inequalities, we find the number of ounces of nutrient in a mixture containing x units of Exbits and y units of Whybits, which is $0.4x + 0.9y$. According to the stated requirement, this sum must be 13 or more. Hence, we write

$$0.4x + 0.9y \geq 13.$$

Similarly, the weight constraint is found to be

$$2x + 3y \leq 50.$$

Taking the last two inequalities together with the nonnegativity constraints, we have the system

$$
\begin{aligned}
i_1: & \quad x \geq 0 \\
i_2: & \quad y \geq 0 \\
i_3: & \quad 0.4x + 0.9y \geq 13 \\
i_4: & \quad 2.0x + 3.0y \leq 50.
\end{aligned}
$$

The solution space for the system is shown in Figure 3–9. We see that x may take on values from 0 to 10, inclusive. For any x in this interval the value of y must be selected in the interval from line e_3 to line e_4, inclusive. Hence, the general algebraic solution is

$$0 \leq x \leq 10$$

$$\frac{13 - 0.4x}{0.9} \leq y \leq \frac{50 - 2.0x}{3.0}.$$

126

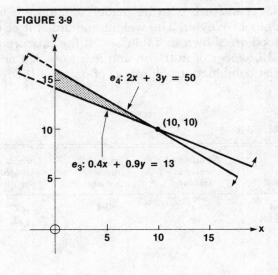

FIGURE 3-9

e_4: $2x + 3y = 50$

(10, 10)

e_3: $0.4x + 0.9y = 13$

3.9 PROBLEM SET 3–2

Make a graph showing the solution space:

1. $x \geq 0$
   $y \geq 0$
   $y \geq -1$
   $x \leq 4.$

2. $x \geq 0$
   $y \geq 0$
   $x \leq 3$
   $x \leq y.$

3. $x \geq 0$
   $y \geq 0$
   $2x + y \to 6$
   $x + 4y \leq 8.$

4. $x \leq 3$
   $y \leq 1$
   $x + 3y \geq 3.$

5. $x \geq 0$
   $y \geq 0$
   $7x + 4y \leq 28$
   $x + 2y \leq 8.$

Find solutions, if any exist:

6. $x \geq 0$
   $y \geq 0$
   $2x + 3y \leq 12$
   $x - 2y \geq 2.$

7. $x \geq 0$
   $y \geq 0$
   $4x + y \leq 12$
   $x + 5y \geq 8.$

8. $x \geq 0$
   $y \geq 0$
   $4x + y \geq 12$
   $x + 5y \leq 8.$

9. $x \geq 0$
   $y \geq 0$
   $4x + y \geq 12$
   $x + 5y \geq 8.$

10. A parcel service requires that length plus girth of parcels not exceed 108 inches. Containers with rectangular cross section (girth) of width w, height h, and length L are to be shipped.
    a) If the width is to be 16 inches, state the restriction on the height in terms of the length, and state the restrictions on the length.
    b) If the rectangular cross section is a square of side x, state the restriction on x in terms of L and state the restriction on L.

11. A mixture is to be made containing x units of food Exbit and y units of Whybit. Exbits weigh 3 ounces per unit and Whybits weigh 5 ounces per unit. Exbits contain 0.5 ounces of nutrient per unit and Whybits 1 ounce of nutrient per unit. The mixture is to have at least 8 ounces of nutrient, and the total weight is not to exceed 45 ounces. Find the general solution, showing all permissible combinations of the two foods.

12. Solve Problem 11 if the mixture is not to exceed 40 ounces in total weight.

13. Extrans and Whytrans both require the same amount of storage space per unit. Extrans cost \$3 per unit and Whytrans cost \$8 per unit. Sufficient storage space is available for 500 units (total), and \$2400 is available to spend on the items. Write the general solution, showing permissible combinations of items which may be purchased and stored without exceeding total space and money restrictions.

14. The table shows, for example, that it takes one hour in Department I and two hours in Department II to make one Exbob. Department I has up to five hours available and Department II has up to seven hours available for making Exbobs and Whybobs. Letting x be the number of Exbobs, and y the number of Whybobs made:
    a) Write the general solution showing permissible combinations of numbers of Exbobs and Whybobs which can be made in the time available.
    b) What specific combinations are possible if only complete units of product are allowed?

| | Hours required to make one | |
Department	Exbob	Whybob
I	1	1
II	2	1

15. It takes two hours to make an Exball and three hours to make a Whyball. The number of Whyballs must not be more than twice the number of Exballs. If up to 36 hours are available for making the products, find the permissible combinations of numbers of Exballs and Whyballs.

16. Exglow costs \$2 per pound and Whyglow costs \$5 per pound. If we mix x pounds of Exglow with y pounds of Whyglow, the mixture contains $x + y$ pounds. The *proportion* of Exglow in the mixture then is $x/(x + y)$. Mixtures are to be made at a total cost not exceeding \$286, and in these mixtures the proportion of Exglow must be not less than 0.2 (20 percent)

and not greater than 0.8 (80 percent). Write the general solution showing permissible combinations of Exglow and Whyglow.

17. The table shows, for example, that Department III has 190 hours available and that one unit of A requires 6 hours of Department III time. Write the general solution showing the combinations of A and B that can be made in the time available.

| Department | Available hours | Hours required to make one unit of: | |
		Product A	Product B
I	120	4	3
II	40	1	2
III	190	6	7

18. (This problem is a new interpretation of the numbers in Problem 17.) Each unit of A made contributes $6 to overhead and profit and each unit of B contributes $7. Write the general solution showing permissible combinations that can be made in the available time if total contribution to overhead and profit is to be at least $190.

| Department | Available hours | Hours required to make one unit of: | |
		Product A	Product B
I	120	4	3
II	40	1	2

3.10 REVIEW PROBLEMS

In problems 1–6: *a)* Solve for y and write the general algebraic solution. *b)* On which side of the line does the solution space lie? *c)* Write the specific solution space for the stated value of x.

1. $x - y < 5$; $x = 10$.
2. $x + y > 3$; $x = 1$.
3. $2x - 3y < -5$; $x = 2$.
4. $-2x + 5y > 2$. $x = 4$.
5. $4x - 3y > -5$; $x = 7$.
6. $x - y \leq 0$; $x = 5$.

Solve for x.

7. $2 - 3x \leq 6$.
8. $2x + 7 \geq 4$.
9. $7 - 4x \leq 0$.
10. $3 \geq 5 - 2x$.
11. $0 \leq 2x + 3$.
12. $x - 4 \geq 0$.

Write the solution, if any exists, for each of the following systems:

13. $x < 4$
$\quad x > 6.$

14. $x > -2$
$\quad x > 2.$

15. $x > 2$
$\quad x < 6.$

Make a graph showing the solution space for each of the following, assuming in each system that the nonnegativity constraints $x \geq 0$ and $y \geq 0$ prevail, and write the algebraic statement of the solution space:

16. $x - 2y \leq 0$
$\quad x + y \leq 2.$

17. $x + y \geq 2$
$\quad x + 2y \leq 4.$

18. $x - 2y \leq 0$
$\quad \cdot\, x + 2y \leq 4.$

19. $x - 2y \geq 0$
$\quad x + 2y \geq 4$
$\quad\quad\quad x \geq 3.$

20. $x + y \geq 2$
$\quad x - 2y \leq 0$
$\quad x + 2y \leq 4.$

21. $x + y \geq 2$
$\quad x - 2y \geq 0$
$\quad\quad\quad x \leq 3$
$\quad x + 2y \leq 4.$

22. *a)* We plan to invest x dollars at 6 percent and y dollars at i percent and require that the combined investment yield at least 8 percent. Express the restriction on i in terms of x and y, and state the restrictions on x and y.

   b) If $x = \$2000$ and $y = \$1000$, what is the minimum interest rate that must be obtained on the $1000?

23. A mixture of foods called Exbits and Whybits is to weigh not more than 5 pounds and contain at least 24 ounces of nutrient. Exbits contain 6 ounces of nutrient per pound and Whybits contain 4 ounces of nutrient per pound. Write the general solution showing the permissible mixtures of the two foods.

24. See Problem 23. If, in addition to the stated constraints, the mixture must contain at least 3 pounds of Exbits, write the general solution showing the permissible food mixtures.

25. See Problem 23. If, in addition to the stated constraints in Problem 23, it is required that the mixture must not cost more than 21 cents, and if Exbits cost 6 cents a pound and Whybits cost 2 cents a pound, write the general solution showing the permissible food mixtures.

26. A high-quality batch of a substance contains 3 ounces of Expop and 4 ounces of Whypop. A low-quality batch contains 1 ounce of Expop and 3 ounces of Whypop. If 70 ounces of Expop and 110 ounces of Whypop are available, write the general solution showing permissible combinations of numbers of high- and low-quality batches.

27. See Problem 26. Write the solution if a high-quality batch contributes $1 to profit and a low-quality batch contributes $2 to profit, and total profit is to be at least $65.

28. If, in addition to the conditions and facts stated in Problems 26 and 27, it is required that at least 5 percent of total output be high-quality batches, write the solution showing permissible combinations of the numbers of batches of each quality.

29. The table shows, for example, that it takes two hours in Department I, one hour in II, and one hour in III to make an Exball. If hours available

in I, II, and III are, respectively, 12, 7, and 15, write the solution showing the permissible combinations of the two products.

| Department | Hours required to make one | |
	Exball	Whyball
I	2	1
II	1	1
III	1	3

Introduction to linear programming

4

4.1 INTRODUCTION

In the real world, we are often confronted with a number of ways of accomplishing a certain objective, some ways being better in a certain sense than others. For example, there are many different combinations of foods that will provide a satisfactory diet, but some combinations are more costly than others, and we may be interested in finding the *minimum* cost of providing dietary requirements. Again, there are many combinations of products a plant can manufacture, and we may be interested in finding the combination that leads to *maximum* profit.

The variables in real-world situations are subject to restrictions that we shall call *constraints.* In the first place, it is required in most instances that the variables not take on negative values. Furthermore, certain combinations of the variables are not permissible. For example, a product mix that requires a plant to operate more than 24 hours a day obviously is not permissible, nor is it permissible to schedule output at levels exceeding capacity.

Let us suppose that the relevant variables in

131

a certain situation are x and y, and that the constraints can be expressed as a system of inequalities in these variables. Then suppose we have an expression called the *objective function*, which states how profit (or cost) is computed for specific values of x and y. The problem is to find the set of values for x and y that satisfies the constraints *and* maximizes profit (or minimizes cost). If the constraints and the objective function are linear, we have a problem in *linear programming*.

To be more specific, suppose that a textile mill buys unfinished cloth and uses 10 processes to convert it into 12 styles of finished material. The styles require varying amounts of time in each of the processes, and there is a capacity limitation on the time available for each process, making 10 capacity constraints. Further, a certain quantity (at least) of each style must be produced to satisfy customer demand, making 12 demand constraints, so we have a total of $10 + 12 = 22$ linear inequality constraints, not counting nonnegativity constraints.

By deducting raw material and processing cost from selling price, the profit contribution per yard produced for each of the 12 styles can be computed, and these profit contributions may be combined to form a linear objective function containing as its variables the unknown number of yards of each of the 12 styles to be produced. The problem is to determine the set of *positive* values for these variables that will maximize total profit, and not violate any of the 22 constraints. This set of values can then be used to establish the production schedule, which is a *program* of operations.

The situation just described involved 22 constraints on 12 variables (a 22 by 12 system), but it is by no means a large system compared to many encountered in practice. We need not be concerned about problem size, however, because large systems can be solved in a matter of seconds by entering the coefficients of the variables into one of the many computers equipped to solve such problems. In this chapter we present the fundamentals of linear programming by use of small systems, and in the next chapter we shall develop a procedure for solving larger problems.

4.2 TWO-VARIABLE CASE: AN INTRODUCTORY EXAMPLE

Table 4–1 shows that products called Extrans and Whytrans are made using the equipment of two departments, I and II. It requires one hour in each department to make an Extran, but making a Whytran takes one hour in Department I and two hours in Department II. Department I has four hours of time available, and II has six hours available. Each Extran made and sold contributes $1 to profit, and each Whytran contributes $0.50 to profit.

The problem is to determine the maximum profit that can be

TABLE 4–1

	Number of units made	Profit per unit	Hours required per unit in:	
			Department I (4 hours available)	Department II (6 hours available)
Product				
Extran x	x	\$1.00	1	1
Whytran .... y	y	0.50	1	2

achieved, keeping in mind the four- and six-hour time limitations in the departments. The profit achieved from x Extrans and y Whytrans is

(Profit per Extran)(Number of Extrans)
+ (Profit per Whytran)(Number of Whytrans)

which is

$$\$1x + \$0.5y.$$

The last is the *objective function*, which we shall name θ (theta).[1] Thus,

$$\text{profit} = \theta = x + 0.5y.$$

To make x Extrans and y Whytrans will require $x + y$ hours in I and $x + 2y$ hours in II. The time limitations therefore are

$$x + \ y \le 4$$
$$x + 2y \le 6.$$

The conditions in the foregoing, taken together with nonnegativity constraints on x and y lead to the following statement of the problem:
Find θ_{max}, subject to:

$$i_1: \qquad x \ge 0$$
$$i_2: \qquad y \ge 0$$
$$i_3: \ x + \ y \le 4$$
$$i_4: \ x + 2y \le 6$$
$$\theta = \ x + 0.5y.$$

Clearly, we can find nonnegative values of x and y (that is, values satisfying i_1 and i_2) that also satisfy i_3 and i_4. The obvious values are $x = 0$ and $y = 0$. However, if we substitute $(0, 0)$ into the objective function, we find that the profit is 0. We must make some product to achieve a profit. If we make one Extran and one Whytran ($x = 1$ and $y = 1$), we find $\theta = 1.5$ and all inequalities are satisfied.

---

[1] Z is another common name given to the objective function. We choose θ to avoid ambiguity later when z is used as a variable.

134

Exercise.   Is it possible to make two units of each product? If so, what profit will be achieved? Answer: Yes; $x = 2$, $y = 2$ satisfy all constraints in the problem. The profit will be \$3.

Exercise.   Is it possible to make three units of each product? Answer: No; $x = 3$, $y = 3$ satisfies neither i_3 nor i_4.

Our goal is not simply to list all the various profits that might be achieved but, rather, to find the *maximum* profit that can be achieved. Consider Figure 4–1.

The solution space for the inequalities i_1 through i_4 is shown on Figure 4–1 as the space bounded by lines whose intersections (which we shall call *corners*) are at O, A, B, and C. Each point in the solution space has a pair of coordinates (x, y) that, when substituted into the objective function, yield the value of θ associated with that point. For example, the point $P(1, 2)$ is in the solution space. At P, we find:

$$\theta = 1 + 0.5(2) = 2.$$

Again, point $A(0, 3)$ is in the solution space. At A:

$$\theta = 0 + 0.5(3) = 1.5.$$

FIGURE 4-1

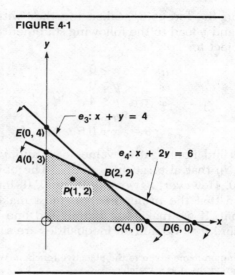

Clearly, the value of θ varies from point to point in the solution space. But observe that for a given x, we obtain the largest value of θ, for this given x, by making y as large as possible, that is, by going all the way to the boundary of the solution space. We see, therefore, that any point that will maximize θ must lie on the boundary of the solution space.

Consider the boundary segment from $A(0, 3)$ to $B(2, 2)$:

$$\text{At } A, \theta = 1.5$$
$$\text{At } B, \theta = 3.$$

This means that as we go from A toward B on the boundary AB, θ is increasing. Clearly, therefore, in seeking the maximum θ, we should go as far as possible in this direction, that is, to corner B.

Consider next the boundary segment from $B(2, 2)$ to $C(4, 0)$:

$$\text{At } B, \theta = 3$$
$$\text{At } C, \theta = 4.$$

Following the above reasoning, we move along the boundary to corner C, and then consider the segment from C to O. Obviously, θ decreases along this boundary. Whichever direction we move from C, toward O or toward B, we find θ decreases. Hence the maximum value occurs at $C(4, 0)$ and is

$$\theta_{max} = 4.$$

The solution of our problem is to make four Extrans and no Whytrans. This schedule will provide a profit of $4, the maximum possible with the stated constraints.

Exercise. Suppose that the profit function related to Figure 4–1 is

$$\theta = 2x + 3y.$$

a) Which values of x and y will now maximize profit? *b)* What is the maximum profit? Answer: *a)* $x = 2$, $y = 2$. *b)* $\theta_{max} = $10.

Example-Minimization. The function θ is the cost function,

$$\theta = 1.25x + y,$$

where x and y represent numbers of units of two products that are to be made. Find the values of x and y that minimize θ subject to the constraints

$$i_1: \ x \geq 0$$
$$i_2: \ y \geq 0$$
$$i_3: \ 2x + y \geq 10$$
$$i_4: \ x + 2y \geq 8.$$

If we write the equality parts of the last two constraints,

$$e_3: \ 2x + \ y = 10$$
$$e_4: \ \ x + 2y = 8$$

and plot these from their intercepts, the slant lines shown on Figure 4–2 are obtained. Substituting the coordinates of the origin, $(0, 0)$, into i_3 and i_4 leads to

$$2(0) + \ 0 \geq 10$$
$$0 + 2(0) \geq 8,$$

FIGURE 4-2

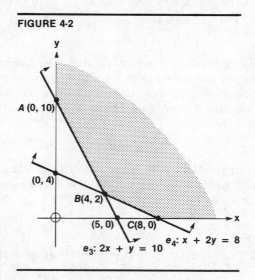

and neither of these is a true statement. The origin and other points *below* the lines are not in the solution space. Hence, the solution space lies above the slant lines as shown in Figure 4–2. Notice that this space is not enclosed but extends indefinitely above the lines in the first quadrant. This means simply that cost increases without limit as more and more units are produced. Clearly, the smaller the number of units made, the lower will be cost, and minimum cost will occur at a point along the inner boundary of the solution space. Computing total cost, θ, at the corners A, B, and C, we find

At $A(0, 10)$: $\theta = 1.25(0) + 10 = 10$
At $B(4, 2)$: $\theta = 1.25(4) + \ 2 = 7*$
At $C(8, 0)$: $\theta = 1.25(8) + \ 0 = 10.$

The asterisk on 7* shows

$$\theta_{\min} = 7 \text{ at } x = 4, y = 2.$$

Exercise. Find minimum cost for the last example if

$$\theta = 2x + 3y.$$

Answer: $\theta_{\min} = 14$ at $x = 4$, $y = 2$.

4.3 ISOLINES

The reader may be familiar with the term isobar, which is heard often in weather forecasts. An isobar is drawn on a weather map by connecting points that have the *same baro*metric pressure. We shall also use *iso* to mean *same* when we talk of *isocost* and *isoprofit* lines. Thus, in the last example, for which cost was

$$\theta = 1.25x + y \quad \text{or} \quad 1.25x + y = \theta,$$

if we wish to speak of all points *(x, y)* where cost is, say, $\theta_1 = \$5$, then

$$\theta_1: \ 1.25x + y = 5 \quad \text{or} \quad y = -1.25x + 5$$

is an isocost line. This line, plotted in the usual manner, is shown on Figure 4–3 along with the inner boundary of the solution space of Figure 4–2. At any point on $\theta_1 = 5$, cost is the same, $5. Similarly isocost lines for costs of $7, $9, and $13, which are

$$\theta_2: \ 1.25x + y = 7 \quad \text{or} \quad y = -1.25x + 7$$
$$\theta_3: \ 1.25x + y = 9 \quad \text{or} \quad y = -1.25x + 9$$
$$\theta_4: \ 1.25x + y = 13 \quad \text{or} \quad y = -1.25x + 13$$

are shown on Figure 4–3. The four isocost lines have the same slope, −1.25, so they are a set of parallel lines, and all of the isocost lines for this cost function will be parallel. More generally, the isocost for any linear cost function is a set of parallel lines. All of the first quadrant points of $\theta_4 = \$13$ are in the solution space. If we now think of laying a ruler along $\theta_4 = \$13$, then sliding toward the origin while holding it parallel to θ_4, cost is decreasing. Thus we can reduce cost to $\theta_3 = \$9$ by choosing any point on the first quadrant section of θ_3 that lies in the solution space. Sliding inward to $\theta_2 = \$7$, we see that there is just one point (4, 2) on this isocost line that is in the solution space, and sliding further inward, no points on the isocost lines will be in the solution space. Consequently, the minimum cost occurs at (4, 2) and is $\theta_2 = \$7$.

138

FIGURE 4-3

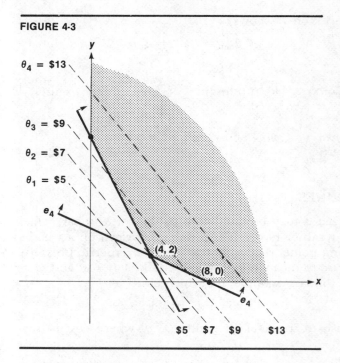

The isocost demonstration just given shows a situation where the minimum occurs at a corner of the solution space. Now suppose that we have a different cost function

$$\theta = x + 2y \quad \text{or} \quad x + 2y = \theta$$

and solve it for y to obtain

$$y = -\frac{1}{2}x + \frac{\theta}{2}.$$

We see that all isocost lines have a slope of $-(1/2)$. Observe that the constraining line

$$e_4: \ x + 2y = 8 \quad \text{or} \quad y = -\frac{1}{2}(x) + 4$$

also has a slope of $-(1/2)$, so the isocost lines are parallel to e_4 and the innermost isocost line that contains points in the solution space is the line e_4 itself. Consequently not only will the corners $(4, 2)$ and $(8, 0)$ lead to minimum cost but any point on the segment joining these two points will also lead to the same minimum cost.

If we had analyzed the maximization problem whose solution space is shown in Figure 4–1, we would have used *isoprofit* lines for the profit objective function and would be led to a result like that obtained for Figure 4–3, namely:

> *An optimum (maximum or minimum) value of the objective function in the two-variable case occurs at one corner of the solution space, or at points along a line segment joining two corner points.*

We see, therefore, that a solution of a linear programming problem does occur at a corner. However, in the case where the isolines of the objective function are parallel to a constraint line, different algebraic steps can lead to different corner points that have the same value for the objective function.

As a consequence of the foregoing, we can in principle solve a two-variable linear programming problem by changing all the constraints to equalities, finding the intersection point of each pair of inequalities to obtain all the corners, checking the corners and substituting the permissible ones into the objective function, then choosing a corner that optimizes the objective function. However, even in the simple two-variable case, this procedure involves unnecessary work because there is no reason for finding corners that are not in the solution space. For example, the axes and the constraint lines in Figure 4–2 contain the intersections (0, 0), (0, 4), and (5, 0), which are not corners of the solution space and therefore need not be calculated. An efficient method of proceeding in the two-variable case is to sketch the constraining lines, identify the solution space, and calculate only intersections that are on the boundary of the solution space.

4.4 GRAPHIC AID IN THE TWO-VARIABLE CASE

We start by considering a maximization problem in which there are four constraints in addition to the nonnegativity constraints.

Example. Given

$$
\begin{aligned}
i_1: &\quad x \geq 0 \\
i_2: &\quad y \geq 0 \\
i_3: &\quad x \leq 10 \\
i_4: &\quad y \leq 6 \\
i_5: &\quad 5x + 6y \leq 60 \\
i_6: &\quad x + 2y \leq 16.
\end{aligned}
$$

Find $\theta_{\max}$ if:

1. $\theta = x + y$
2. $\theta = 2x + 3y$
3. $\theta = x + 10y.$

We shall solve the three parts of the problem together. Consider Figure 4–4, which shows the solution space for the system of constraints. The boundary lines are plotted in the usual manner, using intercepts. Thus, changing i_5 to an equality:

$$5x + 6y = 60$$
$$\text{If } x = 0, \ y = 10$$
$$\text{If } y = 0, \ x = 12.$$

The line connecting $(0, 10)$ and $(12, 0)$ is shown as e_5 on Figure 4–4. The lines obtained by plotting equations resulting from i_1 and i_2 are, of course, the coordinate axes; those obtained from i_3 and i_4 are parallel to one of the axes.

FIGURE 4-4

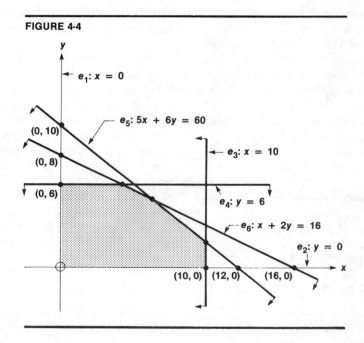

Using the symbol 1 # 4 to designate the corner where lines e_1 and e_4 intersect, and similarly for other corners, we see from Figure 4–4 that the only corners that are part of the solution space are 1 # 4, 4 # 6, 5 # 6, 3 # 5, 2 # 3, and 1 # 2. The corner 1 # 2, the origin, is of no interest. The coordinates of the remaining permissible corners are obtained in the usual manner by elimination and substitution. Thus, for 5 # 6:

$$
\begin{array}{lll}
e_5: & 5x + 6y = 60 & \\
e_6: & x + 2y = 16 & \\
e_7: & 5x + 6y = 60 & e_5 \\
e_8: & 5x + 10y = 80 & 5e_6 \\
e_9: & 4y = 20 & e_8 - e_7 \\
& y = 5 & \\
& x = 6 & \text{from } e_6.
\end{array}
$$

The corner 5 # 6 is the point (6, 5). In the case of 4 # 6:

$$e_4: \qquad y = 6$$
$$e_6: \quad x + 2y = 16.$$

Substituting $y = 6$ into e_6 yields $x = 4$, so this corner is (4, 6).

Exercise. Find the coordinates of the corner 3 # 5 and evaluate the three objective functions at this corner. Answer: See Table 4–2.

The corners, their coordinates, and values of the three objective functions are shown in Table 4–2.

TABLE 4–2

		Objective function, θ		
Corner	*Coordinates*	$x + y$	$2x + 3y$	$x + 10y$
1 # 4	(0, 6)	6	18	60
4 # 6	(4, 6)	10	26	64*
5 # 6	(6, 5)	11	27*	56
3 # 5	(10, 5/3)	35/3*	25	80/3
2 # 3	(10, 0)	10	20	10

* Indicates the maximum.

In the last example, 13 corners appear in Figure 4–4. By charting the lines, it was possible to see that only six corners were in the solution space and were permissible points at which to evaluate the objective function. Assuming that one's graph is correct, these selected corners necessarily satisfy the constraints.

Example. The minimum value of the objective function is sought in this example.

$$i_1: \qquad x \geq 0$$
$$i_2: \qquad y \geq 0$$
$$i_3: \qquad x \leq 10$$
$$i_4: \ 5x + 6y \geq 60$$
$$i_5: \quad x + 2y \geq 16.$$

Find $\theta_{\min}$ if:

$$\theta = 2x + y.$$

The solution space is shown in Figure 4–5. The minimum value of θ is 10 and occurs at the corner whose coordinates are (0, 10), as shown in Table 4–3:

142

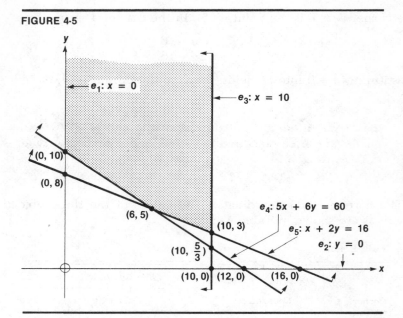

FIGURE 4-5

TABLE 4–3

Corner	Coordinates	$\theta = 2x + y$
1 # 4	$(0, 10)$	10*
4 # 5	$(6, 5)$	17
3 # 5	$(10, 3)$	23

* Indicates the minimum.

Exercise. Change the senses of i_3, i_4, and i_5 in the last example and then find θ_{min} by referring to Figure 4–5. Answer: The relevant corners are at the points $(10, 0)$, $(10, 5/3)$, and $(12, 0)$. The minimum θ, 20, occurs at the point $(10, 0)$.

4.5 PROBLEM SET 4–1

Use the graphical procedure to eliminate nonpermissible corners; then find θ maximum or minimum, as required.

1. Given:
$$x \geq 0$$
$$y \geq 0$$
$$4x + 3y \leq 24$$
$$x + 2y \leq 11.$$

2. Given:
$$x \geq 0$$
$$y \geq 0$$
$$2x + y \geq 8$$
$$6x + 10y \leq 60.$$

Find θ_{max} if:

a) $\theta = x + y$
b) $\theta = x + 3y$
c) $\theta = 3x + y$.

Find θ_{min} if:

a) $\theta = 3x + 2y$
b) $\theta = 10x + y$.

3. Given:

$$x \geq 0$$
$$y \geq 0$$
$$0.15x + 0.10y \geq 15$$
$$x + y \leq 120$$
$$0.40x \leq 32.$$

Find θ_{min} if:
$$\theta = 0.20x + 0.15y.$$

4. Given:

$$x \geq 0$$
$$y \geq 0$$
$$x \leq 9$$
$$y \leq 6$$
$$8x + 15y \leq 120$$
$$11x + 10y \leq 110.$$

Find θ_{max} if:
$$\theta = 3x + y.$$

5. Given:

$$x \geq 0$$
$$y \geq 0$$
$$x \leq 4$$
$$y \leq 4$$
$$3x + y \leq 9$$
$$x + y \leq 5.$$

Find θ_{max} if:

a) $\theta = 0.3x + 0.5y$
b) $\theta = 2x + y$
c) $\theta = 5x + y$.

6. Given:

$$x \geq 0$$
$$y \geq 0$$
$$y \leq x$$
$$x + y \leq 2$$
$$x + 3y \leq 3.$$

Find θ_{max} if:
$$\theta = x + 2y.$$

7. Given:

$$x \geq 0$$
$$y \geq 0$$
$$2x + y \geq 12$$
$$x + 2y \geq 12$$

Find θ_{min} if:
$$\theta = 2.5x + y.$$

8. Given:

$$x \geq 0$$
$$y \geq 0$$
$$4x + y \geq 16$$
$$3x + 2y \geq 24.$$

Find θ_{min} if:
$$\theta = 2x + y.$$

4.6 APPLICATION

The introductory example of Section 4.2 serves as a first application the reader may wish to review at this point. We now give another application involving two constraints in addition to nonnegativity.

Example. It takes four hours to assemble and two hours to paint an Exbox, compared to five hours to assemble and one hour to paint a Whybox. Profit is $20 per Exbox and $30 per Whybox. If available time is limited to 100 hours for assembly and 32 hours for painting, how many Exboxes and Whyboxes should be made to maximize profit? What is the maximum profit?

Letting x and y be, respectively, the number of Exboxes and Whyboxes, the information given in the example is shown in Table 4–4.

Assembly time for x Exboxes and y Whyboxes is

(4 hours per Exbox)(x Exboxes) + (5 hours per Whybox)(y Whyboxes)

TABLE 4–4

Product	Number of units made	Profit per unit	Hours required per unit Assembly (100 hours available)	Hours required per unit Painting (32 hours available)
Exbox	x	$20	4	2
Whybox	y	$30	5	1

which is

$$4x + 5y.$$

The last sum must be less than or equal to the 100 hours of total assembly time available, so we have

$$4x + 5y \leq 100.$$

Similarly, the painting constraint is $2x + y \leq 32$, and the profit function is $20x + 30y$. Hence, the problem is:

Maximize

$$\theta = 20x + 30y$$

subject to

$$
\begin{aligned}
i_1:&\quad x \geq\ \ 0 \\
i_2:&\quad\quad y \geq\ \ 0 \\
i_3:&\quad 4x + 5y \leq 100 \\
i_4:&\quad 2x +\ \ y \leq\ \ 32.
\end{aligned}
$$

The equality parts of the last two constraints are plotted from their intercepts on Figure 4–6, which shows the shaded solution space. We obtain the coordinates of the corner where the slant lines meet by writing

$$
\begin{aligned}
e_3:&\quad 4x + 5y = 100 \\
e_4:&\quad 2x +\ \ y =\ \ 32;
\end{aligned}
$$

then,

$$
\begin{aligned}
e_5:&\quad 3y = 36 \qquad e_3 - 2e_4 \\
&\quad\ \ y = 12.
\end{aligned}
$$

From e_4 we find

$$
\begin{aligned}
2x + 12 &= 32 \\
2x &= 20 \\
x &= 10,
\end{aligned}
$$

so the intersection occurs at (10, 12) as shown on Figure 4–6.

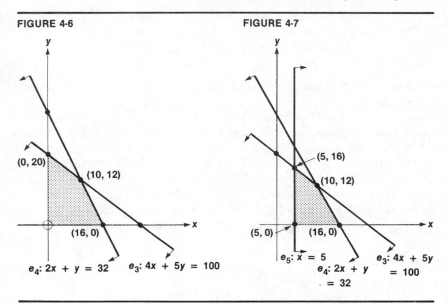

FIGURE 4-6

FIGURE 4-7

Next, evaluating $\theta = 20x + 30y$ at the corners, we find:

$$
\begin{aligned}
\text{At } (0, 0): &\quad 20(0) \ + 30(0) \ = 0 \\
\text{At } (0, 20): &\quad 20(0) \ + 30(20) = 600* \\
\text{At } (10, 12): &\ 20(10) + 30(12) = 560 \\
\text{At } (16, 0): &\ 20(16) + 30(0) \ = 320.
\end{aligned}
$$

Hence, the maximum profit is \$600 and is obtained by making no Exboxes and 20 Whyboxes.

Example. If, in the last example, there is an additional requirement that at least five Exboxes be made, what combination provides maximum profit? What is the maximum profit?

The added requirement is

$$i_5: \ x \geq 5.$$

The solution space including this requirement contains the line $x = 5$ and is shown on Figure 4–7. The corner shown as (5, 16) is the intersection of

$$
\begin{aligned}
e_3: \ 4x + 5y &= 100 \\
e_5: \ x \quad\ \ &= \ 5 \\
4(5) + 5y &= 100 \qquad e_5 \text{ into } e_3 \\
5y &= \ 80 \\
y &= \ 16.
\end{aligned}
$$

Evaluating $\theta = 20x + 30y$ at the corners, we find

$$\text{At } (5, 0): \quad 20(5) \ + 30(0) \ = 100$$
$$\text{At } (5, 16): \quad 20(5) \ + 30(16) = 580*$$
$$\text{At } (10, 12): 20(10) + 30(12) = 560$$
$$\text{At } (16, 0): \quad 20(16) + 30(0) \ = 320,$$

so the maximum profit, \$580, is obtained by making five Exboxes and 16 Whyboxes.

The next example has three constraints in addition to nonnegativity.

Example. A manufacturer makes a canned food by mixing x pounds of Exham with y pounds of Whyham. The proportion of Exham in the mixture must be at least 0.2, but not more than 0.3. Additionally, for reasons of machine-time economy, batches of at least 100 pounds of mix must be made. If costs are \$2.50 per pound of Exham and \$3 per pound of Whyham, what is the minimum-cost mixture?

The proportion of Exham in the $(x + y)$-pound mixture is

$$\frac{x}{x + y}.$$

It follows that

$$\frac{x}{x + y} \ge 0.2, \quad \frac{x}{x + y} \le 0.3, \quad \text{and} \quad x + y \ge 100.$$

Rearranging the first two of the expressions, we find

$$i_1: \ 0.8x - 0.2y \ge \ 0$$
$$i_2: \ 0.7x - 0.3y \le \ 0$$
$$i_3: \qquad x + \quad y \ge 100.$$

FIGURE 4-8

Pounds of Y

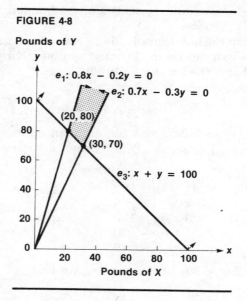

Pounds of X

Plotting the lines corresponding to the equality statements, we find the solution space as shown in the shaded area of Figure 4–8. The corners we seek are 1 # 3 and 2 # 3.

Exercise. Verify that these corners are, respectively, (20, 80) and (30, 70).

The objective function is

$$\text{Cost} = \theta = 2.5x + 3y.$$

At (20, 80), $\theta = \$290$ and at (30, 70), $\theta = \$285$. Thus, the minimum cost is $285 and is obtained by mixing 30 pounds of Exham with 70 pounds of Whyham.

4.7 PROBLEM SET 4–2

1. The table shows that it takes 5 hours in Department I and 1 hour in II to make an Extran, whereas a Whytran requires 6 hours in Department I and 2 hours in Department II. Hours of available time are 60 in I and 16 in II. Profits per Extran and per Whytran are, respectively, $1.50 and $2.00. How many units of each product should be made to maximize profit? What is the maximum profit?

Product	Number of units made	Profit per unit	Dept. I (60 hours available)	Dept. II (16 hours available)
Extrans	x	$1.50	5	1
Whytrans	y	$2.00	6	2

2. The table shows it takes 3 hours to make and 1 hour to paint an Exbob, compared to 4 hours to make and 3 hours to paint a Whybob. Profit is $1.50 per Exbob and $2.50 per Whybob. If 60 hours are available for making, and 30 hours are available for painting, how many of each should be made to maximize profit? What is the maximum profit?

Product	Units made	Profit per unit	to make (60 hours available)	to paint (30 hours available)
Exbobs	x	$1.50	3	1
Whybobs	y	$2.50	4	3

3. *a)* Exgrain and Whygrain are the two ingredients in a box of cereal which provide vitamins A and C. In terms of daily requirements for the vitamins, Exgrain has 1 vitamin A and 4 vitamin C per ounce, compared to Whygrain, which has 5 vitamin A and 2 vitamin C per ounce. Costs per ounce are 5 cents for Exgrain and 2 cents for Whygrain. A box of cereal must contain at least 9 daily requirements of vitamin A and at least 18 daily requirements of vitamin C. How many ounces of Exgrain and Whygrain should be in a box of cereal if the cost of these two ingredients is to be minimized? What is the minimum cost?

   b) If it is also required that a box contain at least 3 ounces of Exgrain, how many ounces of each grain should a box contain? What is the minimum cost?

4. *a)* To fulfill production plans for next month, it is required to schedule at least 450 hours of grinding and at least 288 hours of polishing. An Exball takes 1 hour to grind and 2 hours to polish at a cost of $1 for the operations. A Whyball takes 2 hours to grind and 1 hour to polish at a cost of $3. How many Exballs and Whyballs should be produced if the production schedule is to be fulfilled at minimum cost? What is the minimum cost?

   b) How many Exballs and Whyballs should be produced if it is also required that at least 100 Whyballs be produced? What is the minimum cost?

5. Products A and B require various times for operations on machines M1, M2, and M3. During a certain production period, M1 has 4000 minutes of time available, M2 has 6000 minutes, and M3 has 3000 minutes. Machine time per unit of A is one minute on M1, two minutes on M2, and no time on M3. Machine time per unit of B is one minute on each of the three machines. Expected profit contributions are 60 cents per unit of A made, and 75 cents per unit of B. If production during the period is limited to products A and B, how many units of each should be made to maximize profit?

6. Parts A and B are to be produced and placed in a warehouse whose capacity is 36,000 feet. Part A requires 4 feet of space per unit, B requires 3 feet. In total, 60,000 machine-hours are available for making the parts; A requires 4 machine-hours per unit, B requires 10 hours. In total, 24,000 labor-hours are available for painting the parts; A requires 2 labor-hours per unit for painting. B requires 3 labor-hours. Profit contributions are expected to be $5 per unit of A and $3 per unit of B. How many units of each should be made to maximize profit?

7. A company makes types A and B watchbands. Type A contributes 80 cents per band to profit, B contributes 60 cents. One type A band requires twice as much time to make as a type B band. If all bands were of type B, the company could make 2500 per day; however, because of material shortages, total output (both types) cannot exceed 1600 bands per day; and for the same reason the output of type A cannot exceed 800 per day, nor can the output of B exceed 1400 per day. How many of each type should be made per day to maximize profit?

8. To make one unit of product A requires three minutes in Department I and one minute in II. One unit of B requires four minutes in I and two minutes in II. Profit contributions are $5 per unit of A made, and $8 per unit of B. However, the number of units of A made must be at least as large as the number of units of B. Find the number of units of A and B that should be made to maximize profit if Department I and Department II have 150 and 60 minutes available, respectively. What is the maximum profit?

9. Suppose the profit contribution of A (see Problem 8) fell from $5 to $2 per unit. Other factors being the same, what combination would yield the maximum profit? What is this maximum profit?

10. (See Problem 8.) If, instead of requiring that the number of units of A must be at least as large as the number of units of B, it is required that at least 25 units of A be made, what then is the maximum achievable profit, and what numbers of units yield this profit?

11. (See Problem 8.) If, instead of requiring that the number of units of A must be at least as large as the number of units of B, it is required that the number of units of A must be at least three times the number of units of B and the Department II constraint is no longer binding; that is, the Department II time limitation does not affect the problem. Why is this so?

12. A quart of A costs $2 and contains 2.5 ounces of Zap. A quart of B costs $3 and contains 4 ounces of Zap. A mixture not to exceed 10 quarts, and having at least 36.4 ounces of Zap is to be made. Find the minimum-cost mixture.

13. What is the solution for Problem 12 if B is to constitute at least three-fourths of the total mixture?

4.8 THREE-VARIABLE CASE

In principle, a system in three variables can be represented graphically because linear equations in three variables are planes. However, sketching planes that are in three-dimensional space on a flat (two-dimensional) piece of paper and trying to identify the solution space is too complex to be a practical method of eliminating intersections that are not in the solution space. In the next chapter we shall present a systematic procedure that provides a permissible corner to start with and rules that show how to go to another permissible corner that *improves* the objective function. When no permissible corner improving the solution exists, the current corner provides the optimum value. Readers who plan to study Chapter 5 need only scan the remainder of this chapter to see first the need for a procedure that eliminates the extensive computation of nonpermissible corners and, second, to fix in mind the key principle that if a system has m constraints on n variables $(m < n)$ in addition to the nonnegativity constraints on the variables, then at every corner *at least* $(n - m)$ of the variables have the value zero. For example, Table 4–5 shows the corners for a system having $m = 2$ constraints (not counting nonnegativity constraints) on $n = 3$ variables, and at every corner *at least one*, that is,

$$n - m = 3 - 2 = 1$$

variable has the value zero.

Example. Solve the following three-variable linear programming problem:

Minimize

$$\theta = x + 4y + 3z$$

subject to:

$$
\begin{array}{lrcl}
i_1: & x & \geq & 0 \\
i_2: & y & \geq & 0 \\
i_3: & z & \geq & 0 \\
i_4: & 0.5x + y + 0.2z & \geq & 600 \\
i_5: & 2.0x + 3y + z & \leq & 2000.
\end{array}
$$

The problem has three variables, so each corner is the solution of a three by three system of equations. For example, writing the equality part of the first, fourth, and fifth expression, we have

Corner 1 # 4 # 5

$$
\begin{array}{lrcl}
e_1: & x & = & 0 \\
e_4: & 0.5x + y + 0.2z & = & 600 \\
e_5: & 2.0x + 3y + z & = & 2000.
\end{array}
$$

Substituting $x = 0$ into e_4 and e_5, we have

$$
\begin{array}{lrcll}
e_6: & y + 0.2z & = & 600 & \\
e_7: & 3y + z & = & 2000 & \\
e_8: & 0.4z & = & 200 & \quad e_7 - 3e_6 \\
& z & = & 500. &
\end{array}
$$

From e_6, with $z = 500$, we find $y = 500$, so the desired corner is $(0, 500, 500)$. These coordinates satisfy all inequalities, i_1 through i_5.

Corner 1 # 2 # 4

$$
\begin{array}{lrcl}
e_1: & x & = & 0 \\
e_2: & y & = & 0 \\
e_4: & 0.5x + y + 0.2z & = & 600.
\end{array}
$$

Substitution of $x = y = 0$ into e_4 yields $z = 3000$. However, $(0, 0, 3000)$ does not satisfy i_5, and so is not a permissible corner at which to evaluate the objective function.

> **Exercise.** Find and check the corner 2 # 4 # 5. Answer: $x = 2000$, $y = 0$, $z = -2000$ does not check because z is negative, contrary to constraint i_3.

In order to insure that all corners are found, we should know beforehand how many there are. This can be determined easily by the formula

$$\text{Number of corners} = \frac{c!}{n!\,(c-n)!}$$

where c is the total number of constraints (including the nonnegativity constraints) and n is the number of variables. The factorial symbol, !, means, for example,

$$5! = 5(4)(3)(2)(1)$$
$$3! = 3(2)(1).$$

For the system at hand we have

$$c = 5 \text{ constraints}$$
$$n = 3 \text{ variables}$$
$$c - n = 5 - 3 = 2,$$
$$\text{Number of corners} = \frac{5!}{3!\,2!} = \frac{5(4)(3)(2)(1)}{(3)(2)(1)(2)(1)} = 10.$$

The results of calculating and checking these ten corners are shown in Table 4–5. The desired minimum is seen to be 2,000 and occurs at $x = 400$, $y = 400$, and $z = 0$.

TABLE 4–5

Corner	Coordinates	Check	$\theta = x + 4y + 3z$
1 # 2 # 3	(0, 0, 0)	No	
1 # 2 # 4	(0, 0, 3000)	No	
1 # 2 # 5	(0, 0, 2000)	No	
1 # 3 # 4	(0, 600, 0)	Yes	2400
1 # 3 # 5	(0, 2000/3, 0)	Yes	8000/3
1 # 4 # 5	(0, 500, 500)	Yes	3500
2 # 3 # 4	(1200, 0, 0)	No	
2 # 3 # 5	(1000, 0, 0)	No	
2 # 4 # 5	(2000, 0, −2000)	No	
3 # 4 # 5	(400, 400, 0)	Yes	2000*

* Indicates the minimum.

In Table 4–5, each corner is the solution of a three by three system of equations because there are $n = 3$ variables, and each set of coordinates has *at least one zero value*. The reason for this is that, of the total number of constraints, $c = 5$, three of them are nonnegativity constraints that become

$$x = 0, \quad y = 0, \quad z = 0$$

when written as equalities. The number of remaining constraints is $n = 2$, and written as equalities these are

$$0.5x + y + 0.2z = 600$$
$$2.0x + 3y + z = 2000.$$

Clearly, any three by three system formed from the five equalities has to include at least one of the first three, $x = 0$, $y = 0$, $z = 0$, and this accounts for the presence of at least one zero in the coordinates of every corner shown in Table 4–5. In linear programming, nonnegativity constraints are always assumed to be present. Consequently, when we say that the problem has $m = 2$ constraints on $n = 3$ variables, it is to be understood that there are three nonnegativity constraints and $m = 2$ means two constraints in addition to nonnegativity. In our example,

$$n = 3, \quad m = 2, \quad n - m = 3 - 2 = 1$$

tells us that every corner will have *at least* one zero coordinate. Similarly, every corner for a problem that has $m = 4$ constraints on $n = 7$ variables will have zero values for at least three variables because

$$n - m = 7 - 4 = 3.$$

The fact just illustrated is a key principle in the procedure to be developed in the next chapter, so we highlight it here.

If a linear programming problem has m *constraints*[2] *(not counting nonnegativity constraints) on* n *variables, where* $n \geq m$, *then the coordinates of every corner of the solution space will contain at least* n − m *zeros.*

Exercise. A linear programming problem has a *total* number of constraints, $c = 16$, on $n = 10$ variables. *a)* How many nonnegativity constraints are there? *b)* What is *m*? *c)* How many zeros will every corner have? Answer: *a)* 10. *b)* $m = 6$. *c)* At least 4.

4.9 PROBLEM SET 4–3

1. Find θ_{max}, and write the coordinates of the point at which the maximum occurs if:

$\theta = 0.5x + y + 0.2z$

$x \geq 0$

$y \geq 0$

$z \geq 0$

$x + 4y + 3z \leq 1800$

$2x + 3y + z \leq 2000.$

2. Find θ_{min} if: $\theta = x + 2y + 3z$

$x \geq 0$

$y \geq 0$

$z \geq 0$

$x + 2y + 3z \geq 1200$

$2x + y + z \geq 600.$

[2] This assumes the constraints are linearly independent, that is, that no constraint is a linear combination of other constraints. If, for example, one constraint is a linear combination of two others (in the simplest case, one is a constant times another), this constraint is not counted in specifying n.

3. Find θ_{max} if: $\theta = 1.5x + 2.5y + 2.0z$

$$x \geq 0$$
$$y \geq 0$$
$$z \geq 0$$
$$x + 1.5y \qquad \leq 4.0$$
$$y + 2.5z \leq 5.0$$
$$1.5x + \quad y + 2.0z \leq 7.5.$$

4.10 APPLICATIONS: THREE VARIABLES

As we have seen, information leading to a linear programming problem may be presented in a word description of the nature of the circumstances involved and the restrictions which apply to the variables in the problem. In such a situation the first step is to extract the relevant information and set up the system of constraints.

Example. A diet (see Table 4–6) is to contain at least 10 ounces of nutrient P, 12 ounces of nutrient R, and 20 ounces of nutrient S. These nutrients are to be obtained from some combination of foods A, B, and C. Each pound of A costs 4 cents and has 4 ounces of P, 3 ounces of R, and 0 ounces of S. Each pound of B costs 7 cents and has 1 ounce of P, 2 ounces of R, and 4 ounces of S. Each pound of C costs 5 cents and has 0 ounces of P, 1 ounce of R, and 5 ounces of S. How many pounds of each food should be purchased if the stated dietary requirements are to be fulfilled at minimum cost?

The objective function is the sum of the costs of x pounds of A at 4 cents per pound, y pounds of B at 7 cents per pound, and z pounds of C at 5 cents per pound; that is:

$$\theta = 4x + 7y + 5z.$$

TABLE 4–6

Food	No. of pounds	Cost per pound	Ounces of nutrient per pound		
			P	R	S
A	x	$0.04	4	3	0
B	y	0.07	1	2	4
C	z	0.05	0	1	5

The constraints are formulated in a similar manner. Our problem becomes one of minimizing θ subject to:

$$i_1: \qquad x \geq 0$$
$$i_2: \qquad y \geq 0$$
$$i_3: \qquad z \geq 0$$
$$i_4: 4x + y \qquad \geq 10$$
$$i_5: 3x + 2y + z \geq 12$$
$$i_6: \qquad 4y + 5z \geq 20.$$

This system has 20 possible corners whose coordinates are shown in Table 4–7, together with the evaluation of θ at permissible corners. The minimum cost is 92/3 cents. The combination of foods leading to this minimum cost is 8/3 pounds of A, 0 pounds of B, and 4 pounds of C.

TABLE 4–7

Corner	Coordinates	Check	$\theta = 4x + 7y + 5z$
1 # 2 # 3	(0, 0, 0)	No	
1 # 2 # 4	(Impossible)		
1 # 2 # 5	(0, 0, 12)	No	
1 # 2 # 6	(0, 0, 4)	No	
1 # 3 # 4	(0, 10, 0)	Yes	70
1 # 3 # 5	(0, 6, 0)	No	
1 # 3 # 6	(0, 5, 0)	No	
1 # 4 # 5	(0, 10, −8)	No	
1 # 4 # 6	(0, 10, −4)	No	
1 # 5 # 6	(0, 20/3, −4/3)	No	
2 # 3 # 4	(5/2, 0, 0)	No	
2 # 3 # 5	(4, 0, 0)	No	
2 # 3 # 6	(Impossible)		
2 # 4 # 5	(5/2, 0, 9/2)	Yes	32.5
2 # 4 # 6	(5/2, 0, 4)	No	
2 # 5 # 6	(8/3, 0, 4)	Yes	92/3*
3 # 4 # 5	(8/5, 18/5, 0)	No	
3 # 4 # 6	(5/4, 5, 0)	Yes	40
3 # 5 # 6	(2/3, 5, 0)	No	
4 # 5 # 6	(20/9, 10/9, 28/9)	Yes	290/9

* Indicates the minimum.

Note that in Table 4–7 the last corner has three nonzero coordinates. This is possible because the problem has $m = 3$ constraints on $n = 3$ variables, and

$$n - m = 3 - 3 = 0.$$

The necessary presence of zero(s) in coordinates of corners arises only when n is greater than m, that is when

$$n - m > 0.$$

4.11 PROBLEM SET 4–4

1. A wholesaler has 9600 feet of space available, and $5000 he can spend to buy merchandise of types A, B, and C. Type A costs $4 per unit and requires 4 feet of storage space in the warehouse. B costs $10 per unit and requires 8 feet of space. C costs $5 per unit and requires 6 feet of space. Only 500 units of type A are available to the wholesaler. Assuming that the wholesaler expects to make a profit of $1 on each unit of A he

buys and stocks, \$3 per unit on B, and \$2 per unit on C, how many units of each should he buy and stock in order to maximize his profit?

2. A diet is to contain at least 10 ounces of nutrient P, 12 ounces of R, and 20 ounces of S. These nutrients are to be obtained from foods A, B, and C. Each pound of A costs 3 cents and contains 3 ounces of P, 2 ounces of R, and 0 ounces of S. Each pound of B costs 6 cents and contains 1 ounce of P, 3 ounces of R, and 3 ounces of S. Each pound of C costs 7 cents and contains 0 ounces of P, 1 ounce of R, and 5 ounces of S. How many pounds of each food should be purchased if the stated dietary requirements are to be met at minimum cost?

3. Products A, B, and C are sold door to door. A costs \$3 per unit, takes 10 minutes to sell (on the average), and costs \$0.50 to deliver to a customer. B costs \$5, takes 15 minutes to sell, and is left with the customer at the time of sale. C costs \$4, takes 12 minutes to sell, and costs \$1 to deliver. During any week, a salesman is allowed to draw up to \$500 worth of A, B, and C (at cost) and he is allowed delivery expenses not to exceed \$75. If a salesman's selling time is not expected to exceed 30 hours (1800 minutes) in a week, and if the salesman's profit (net after all expenses) is \$1 each on a unit of A or B and \$2 on a unit of C, what combination of sales of A, B, and C will lead to maximum profit, and what is this maximum profit?

4.12 REVIEW PROBLEMS

Solve the following two-variable linear programming problems:

1.
$$x \geq 0$$
$$y \geq 0$$
$$x + 2y \leq 40$$
$$3x + y \leq 45.$$
a) Find θ_{max} if $\theta = 2x + 3y$
b) Find θ_{max} if $\theta = 4x + y$.

2.
$$x \geq 0$$
$$y \geq 0$$
$$x + 2y \geq 40$$
$$3x + y \geq 45.$$
Find θ_{min} if $\theta = 4x + y$.

3.
$$x \geq 0$$
$$y \geq 0$$
$$x - y \geq -1$$
$$3x + 2y \leq 17$$
$$x + 4y \geq 9.$$
Find θ_{max} if $\theta = 2x + y$.

4.
$$x \geq 0$$
$$y \geq 0$$
$$x - y \geq -1$$
$$3x + 2y \geq 17$$
$$x + 4y \geq 9.$$
Find θ_{min} if $\theta = 2x + y$.

5.
$$x \geq 0$$
$$y \geq 0$$
$$x \geq 2$$
$$x + y \geq 7$$
$$3x + 4y \geq 24.$$
Find θ_{min} if $\theta = 3x + 2y$.

6. One pound of food A costs \$1 and contains 2 ounces of nutrient I and 4 ounces of nutrient II. A pound of B costs \$2 and contains 3 ounces of I and 1 ounce of II. A mixture is to contain at least 90 ounces of I and 80 ounces of II. Find the minimum-cost mixture.

7. (See Problem 6.) Suppose the mixture must weigh not more than 40 pounds. What will be the minimum-cost mixture now?

8. (See Problem 6.) What will be the minimum-cost mixture if at least 80 percent of the mixture must be food B?

9. To make one unit of product A requires three minutes each in Departments I and II. A unit of B requires two minutes in I and four minutes in II. A unit of either product contributes $1 to profit. If Departments I and II have 900 and 1200 minutes available, respectively, for making A and B, find the numbers of each that should be made to maximize profit, and find what the maximum profit is.

10. (See Problem 9.) Solve the problem if the number of units of A must be at least as great as the number of units of B.

Solve the following three-variable problems:

11.
$$x \geq 0$$
$$y \geq 0$$
$$z \geq 0$$
$$x + 3y + 4z \leq 30$$
$$x + 5y + 2z \leq 40.$$
Find θ_{max} if $\theta = x + 4y + 3z$.

12.
$$x \geq 0$$
$$y \geq 0$$
$$z \geq 0$$
$$4x + 8y + z \leq 52$$
$$8x + 28y + 3z \leq 168.$$
Find θ_{max} if $\theta = 3x + 9y + z$.

13.
$$x \geq 0$$
$$y \geq 0$$
$$z \geq 0$$
$$x + 3y + 3z \leq 50$$
$$x + 4y + 2z \leq 60$$
$$z \leq 10.$$
Find θ_{max} if $\theta = x + 4y + 2z$.

14.
$$x \geq 0$$
$$y \geq 0$$
$$z \geq 0$$
$$x + 2y + 7z \leq 21$$
$$5x + 17y + 28z \leq 140$$
$$x + 9y + 10z \leq 66.$$
Find θ_{max} if $\theta = x + 3y + 7z$.

15.
$$x \geq 0$$
$$y \geq 0$$
$$z \geq 0$$
$$x + 2y + 3z \leq 14$$
$$\frac{1}{2}x + 2y + \frac{1}{2}z \leq 10$$
$$x + 5y + 4z \leq 26.$$
Find θ_{max} if $\theta = x + 3y + 3z$.

16.
$$x \geq 0$$
$$y \geq 0$$
$$z \geq 0$$
$$x + 2y + 3z \leq 25$$
$$x + 3y + 2z \leq 30.$$
Find θ_{max} if $\theta = x + 3y + 2z$.

17. A man sells and installs products A, B, and C. The table shows, for example, that it takes three hours to sell a unit of B, four hours to install it, and net profit per unit is $40.

Product	Number of units	Selling hours per unit	Installation hours per unit	Profit per unit
A	x	1	1	$10
B	y	3	4	40
C	z	2	1	10

During a 38-hour week, the man allots no more than 18 hours to selling and no more than 20 hours to installation. Find the combination of numbers of units of A, B, and C that would yield maximum profit.

18. The table shows, for example, that two labor-hours are needed to sell a unit of B, three labor-hours to deliver it, and three labor-hours to install it. Profit per unit of B is $60.

Product	Number of units	Selling hours per unit	Delivery hours per unit	Installation hours per unit	Profit per unit
A	x	1	1	2	$20
B	y	2	3	3	60
C	z	3	2	5	40

If 220 labor-hours are available, of which not more than 50 are to be used for selling, not more than 60 for delivery, and not more than 110 for installation, find the combination of A, B, and C that yields maximum profit and state what this maximum is.

19. The table shows, for example, that in making a unit of B, three minutes are required for stamping, 13 minutes for forming, and five minutes for painting, and each unit of B contributes $4 to profit.

Product	Number of units	Stamping	Forming	Painting	Profit per unit
A	x	1	3	1	$1
B	y	3	13	5	4
C	z	2	2	5	2

(*Minutes per unit for* spans the Stamping, Forming, and Painting columns.)

Minutes available for stamping, forming, and painting are 40, 144, and 70, respectively. Find the combination of numbers of units of A, B, and C that leads to maximum profit. (Note: Assume that fractional units are permissible and that a fractional unit contributes its fraction to profit.)

Linear programming:
The simplex method

5

5.1 INTRODUCTION

The graphic solution of two-variable linear programming problems in Chapter 4 serves well to introduce the fundamental concept of optimizing a linear objective function subject to linear constraints. However, applied problems contain numerous constraints involving many variables and cannot be solved by graphic procedures. Solution by determination of all corners as carried out for the simple three-variable example at the end of Chapter 4, while conceptually possible, would require a tremendous amount of computation, most of which would serve only to eliminate corners that are not in the solution space. The efficient way to solve an applied problem is to start with a corner that is in the solution space, then move to another corner in the solution space that *improves* the objective function, continuing the process of moving to another corner until no corner can be found that will further improve the objective function. The process thus works only with corners of the solution space and stops at the optimum corner, where no further improvement can be made in the objective function. The process to which we have just referred is called the *simplex method* of solving linear programming problems.

This chapter starts with an example showing the general idea of the simplex procedure, then provides practice in carrying out each of the simplex steps. Following this practice, the full simplex procedure is again demonstrated as it applies to maximization problems. Next we show how to change a minimization problem into a maximization problem by applying the dual theorem, so that at this point the simplex method can be applied to solve both maximization and minimization problems. The last part of the chapter presents analyses that can be made after the optimal simplex solution has been obtained. These analyses provide information (shadow prices, for example) that can be more important than the optimal solution itself.

5.2 FUNDAMENTAL PROCEDURES AND TERMINOLOGY

The introductory illustration of Chapter 4 will serve as our starting point, and we shall repeat it here.

Example.   Table 5–1 shows Extrans and Whytrans are made using the equipment of Departments I and II. It requires 1 hour in each department to make an Extran, but making a Whytran takes 1 hour in Department I and 2 hours in Department II. Department I has 4 hours of available time, and II has 6 hours of available time. Each Extran made contributes $1 to profit and each Whytran contributes $0.50 to profit. Determine how many Extrans and Whytrans should be made to maximize profit, and compute the maximum profit.

TABLE 5–1

Product	Number of units made	Profit per unit	Hours required in: Department I (4 hours available)	Department II (6 hours available)
Extrans x		$1.00	1	1
Whytrans .... y		$0.50	1	2

We form the linear programming problem as follows:
Maximize

$$\theta = x + 0.5y$$

subject to:

$$x \geq 0;\ y \geq 0$$
$$i_2:\ x + y \leq 4$$
$$i_3:\ x + 2y \leq 6. \tag{1}$$

Slack variables. The first step in our procedure changes the inequalities in the problem statement into equalities. Thus,

$$x + y \le 4$$

requires that a *positive* quantity may have to be added to the left side to increase its value to 4. Let us call this quantity r and write

$$x + y + r = 4.$$

For example, if we do not make any Extrans or Whytrans,

$$x = y = 0$$

and

$$r = 4.$$

Remembering that the inequality at hand is the Department I constraint, the last says that if we do not make any Extrans or Whytrans, the 4 hours of available time in I are not used. This unused time is called *slack* time, and the nonnegative variable r is called a *slack* variable.

In a similar manner, we convert

$$x + 2y \le 6$$

to

$$x + 2y + s = 0$$

by adding the nonnegative slack variable, s.

Inasmuch as slack time does not produce any product, the slack variables contribute nothing to profit, so

$$\theta = x + 0.5y + (0)r + (0)s$$

and we shall not bother to write the slack variables in the objective function, θ. We now formulate the problem as follows:

$$
\begin{aligned}
&x \ge 0; \quad y \ge 0; \quad r \ge 0; \quad s \ge 0 \\
&e_1: \ \theta = x + 0.5y \\
&e_2: \ x + \ y + r = 4 \\
&e_3: \ x + 2y + s = 6.
\end{aligned}
\tag{2}
$$

The original problem, (1), had *two* constraints (i_2 and i_3), not counting nonnegativity constraints, which have been changed to the equalities e_2 and e_3. However, in the change we introduced two more variables, r and s, so that we now have $m = 2$ constraints on $n = 4$ variables (x, y, r, and s). According to the principle illustrated and stated in Chapter 4, Section 4.8, in any solution of the system at least $n - m$ of the variables must have the value zero. Here,

$$n - m = 4 - 2 = 2,$$

so at least two of the variables must be zero in a solution of the last-written system, (2), and the question is which two are zero in the optimal solution.

We start our search by making sure that we have a permissible solution, which we now refer to as a *feasible* solution. The obvious starting point when all constraints are $\leq$ is the origin, where $x = y = 0$. To be clear on this point, refer to Figure 5-1. Any point on the boundary of the shaded space, or within it, is a *feasible* solution, and the corner $(0, 0)$, which we shall now call a *vertex* of the solution space, is feasible.

Our next step is to write (2) with the zero-valued variables (x and y) as arbitrary variables. Thus,

$$x + y + r = 4$$

becomes

$$r = 4 - x - y.$$

Similarly,

$$x + 2y + s = 6$$

becomes

$$s = 6 - x - 2y$$

FIGURE 5-1

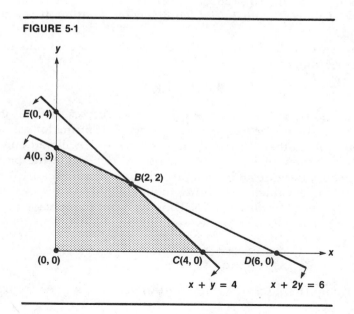

and θ is already expressed in terms of x and y. We have

$$
\begin{aligned}
e_4:\ \theta &= \quad\ \ x + 0.5y \\
e_5:\ r &= 4 - x - \quad y \\
e_6:\ s &= 6 - x - \quad 2y.
\end{aligned}
\tag{3}
$$

Now, remembering that the variables on the *right* side are to take the value zero, we have the feasible solution

$$
x = y = 0; \quad r = 4; \quad s = 6; \quad \theta = 0.
\tag{4}
$$

Now for more terminology. The variables that are *in* the solution, the ones on the *left* side, r and s, are called *basic* variables. The basic variables, all of them, form the *basis*. Thus, currently, r is in the basis and s is in the basis. Together, r and s are the basis. The variables that are *out of* the solution, the ones on the right that have the value zero, are called nonbasic variables, that is, x and y are nonbasic variables at the present time. The solution, (4), which is *on the boundary* of the feasible space in Figure 5–1 (it is the vertex at the origin) is called a *basic feasible* solution to distinguish it from feasible solutions that are not on the boundary. Our interest centers on the vertices, so we confine attention to the basic feasible solutions at the vertices.

The basic feasible solution (4), in which $x = y = 0$, means, of course, that no Extrans or Whytrans are produced, so Department I has $r = 4$ hours of unused time (slack time) and II has $s = 6$ hours of slack. No product is made, so profit $\theta = 0$. *This vertex serves the important purpose of providing an initial basic feasible solution.* It is a starting point.

What we plan to do next is decide which vertex we should move to next. Examining

$$
\theta = (1)x + 0.5y,
$$

which currently is zero, we note that if we *increase* x from its zero value, then every increase of 1 in x (every Extran made) will add \$1 to profit, but another Whytran (increasing y by 1) will add only \$0.50 to profit. We now plan to bring *one* new variable into the basis (and take one out). Because an Extran provides more profit than a Whytran, it is reasonable to choose to bring in all the Extrans we can, leaving the number of Whytrans at zero. Thus, we want to increase x *as much as possible*. If we look at (3), we see

$$
\begin{aligned}
e_5:\ r &= 4 - x - \ y \\
e_6:\ s &= 6 - x - 2y.
\end{aligned}
$$

Covering up the y terms (which remain at zero), we see from e_5 that

$$
r = 4 - x,
$$

and x must not be increased to more than 4, because if it is r will become negative. Similarly, from e_6 with $y = 0$, we see that

$$s = 6 - x,$$

and x cannot be increased to more than 6. The controlling condition therefore is that x not be increased by more than 4 (which guarantees it will not be increased by more than 6). Thus, we found from θ that x should be brought *into* the basis and from the last analysis that it should be brought in at e_5 so that r goes out of the basis. Consequently, we solve e_5 for x to obtain

$$e_7\colon\ x = 4 - y - r,$$

so x now is a basic variable and r is nonbasic along with y. We now must change e_4 and e_6 in (3) so that r and y become the nonbasic variables. We start with

$$e_6\colon\ s = 6 - x - 2y.$$

We have, by substituting e_7 into the last,

$$s = 6 - (4 - y - r) - 2y$$
$$= 6 - 4 + y + r - 2y$$
$$e_8\colon\ s = 2 - y + r.$$

Then, converting θ, we start with

$$e_4\colon\ \theta = x + 0.5y$$

and substitute from e_7 to obtain

$$\theta = (4 - y - r) + 0.5y$$
$$e_9\colon\ \theta = 4 - 0.5y - r.$$

Our new system is

$$
\begin{aligned}
e_7\colon&\ x = 4 - y - r \\
e_8\colon&\ s = 2 - y + r \\
e_9\colon&\ \theta = 4 - 0.5y - r,
\end{aligned}
\tag{5}
$$

in which x and s are the basic variables. The variables y and r are nonbasic and have the value zero. The solution at this juncture (setting the nonbasic variables y and r equal to 0) is

$$x = 4 \quad s = 2; \quad y = r = 0; \quad \theta = 4. \tag{6}$$

This is a basic feasible solution and appears as the vertex $C(4, 0)$ on Figure 5–1. Thus, our procedure started with the obvious basic feasible solution, the origin, and moved to another basic feasible solution that increased the objective function.

We again look at the objective function, which is now

$$e_9\colon\ \theta = 4 - 0.5y - r,$$

and note that both nonbasic variables (y and r on the right) have *negative* coefficients. We should not increase them from their current values of zero because to do so will *decrease* the value of the objective function. There is no way to increase θ so the current solution is optimal and the maximum value of θ is

$$\theta_{max} = \$4 \text{ at } x = 4 \text{ and } y = 0.$$

Hence, 4 Extrans and no Whytrans should be made. From Table 5–1 we see that if 4 Extrans are made (and zero Whytrans) the Extrans will use up all 4 available hours in Department I, so there is no slack in I, and $r = 0$, as the solution (6) states. However, only 4 of the 6 hours in II will be used, so the unused time is 2 hours, which is the value $s = 2$ in the solution.

The procedure illustrated in the foregoing is summarized in what follows, using the word constraint to refer to constraints other than nonnegativity constraints.

1. Introduce a positive slack variable to change each $\leq$ constraint to an equality.
2. Write the first system with the slacks as the basic variables (on the left).
3. *a)* If θ can be increased by bringing in a new basic variable and taking one out, determine which variable is to be brought in. Go to Step 4.
   b) If at any juncture θ cannot be increased (all coefficients negative), the present solution is optimal. Stop.
4. Determine which variable is to come out.
5. Manipulate the equations to bring in the new variable, taking one out (see Step 4). Write the new system. Return to Step 3.

Step 5 turns out to be the most laborious when the number of variables and constraints increases, so we shall first develop a mechanical procedure for manipulating solutions to bring one variable in and take one out. Then we shall proceed step by step to develop and practice the rules used to perform Steps 1 through 5 in the foregoing. We present no problem set here because the purpose of this section has been only to introduce fundamental procedures and definitions.

5.3 CHANGING BASIC VARIABLES: ONE VARIABLE NONBASIC

Prior to this chapter, we have expressed unlimited solution sets in terms of arbitrary variables. As we have already said, in linear programming arbitrary variables are assigned the value zero in solutions and called nonbasic, so we shall use this term for our present work. If we have the two by three system of equalities,

$$e_1: \; 2x + y + \; z = 12$$
$$e_2: \; \quad x + y + 2z = 18, \tag{1}$$

we may eliminate y by subtraction. Thus,

$$e_3: \; x - z = -6 \qquad e_1 - e_2.$$

If we now choose x to be basic and z nonbasic,

$$e_4: x = -6 + z.$$

Substituting e_4 into e_1 yields

$$2(-6 + z) + y + z = 12$$
$$-12 + 2z + y + z = 12$$
$$e_5: \qquad\qquad\qquad y = 24 - 3z.$$

Thus, we have the general solution

$$z \text{ nonbasic}$$
$$e_4: \; x = -6 + z \tag{2}$$
$$e_5: \; y = 24 - 3z,$$

where x and y are the basic variables and form the *basis*. We now want to develop a set of rules for taking one variable out of the basis and bringing another in; that is, for making a variable other than z nonbasic. Suppose we want to make y the nonbasic variable. Then from

$$e_5: \quad y = 24 - 3z$$
$$3z = 24 - \; y$$
$$e_6: \quad z = \; 8 - \frac{1}{3}y.$$

Putting e_6 into e_4 we have

$$x = -6 + \left(8 - \frac{1}{3}y\right)$$
$$e_7: \; x = 2 - \frac{1}{3}y$$

and our result is

$$y \text{ nonbasic}$$
$$e_6: \; z = 8 - \frac{1}{3}y \tag{3}$$
$$e_7: \; x = 2 - \frac{1}{3}y.$$

The constants that appear on the right side of (3) obviously are related to the constants in (2). We want to find in a general way what this relationship is. To this end, we designate the constants by a, b, c, and d, carry out the procedures to change variables, then observe the forms in which the constants appear in the outcome. Thus, if a solution is

$$z \text{ nonbasic}$$
$$e_8\colon\ y = a + bz \tag{4}$$
$$e_9\colon\ x = c + dz$$

and if we wish to make y nonbasic and z basic, we find from e_8

$$y - bz = a$$
$$-bz = a - y$$
$$e_{10}\colon \qquad z = -\frac{a}{b} + \frac{1}{b}y.$$

Then, substituting e_{10} into e_9, we have

$$x = c + d\left(-\frac{a}{b} + \frac{1}{b}y\right)$$
$$x = \left(c - \frac{ad}{b}\right) + \frac{d}{b}y.$$

In the last, we put the terms in parentheses over the common denominator, b, to obtain

$$e_{11}\colon\ x = \frac{cb - ad}{b} + \frac{d}{b}y,$$

so the outcome is

$$y \text{ nonbasic}$$
$$e_{10}\colon\ z = -\frac{a}{b} + \frac{1}{b}y \tag{5}$$
$$e_{11}\colon\ x = \frac{cb - ad}{b} + \frac{d}{b}y.$$

To show the relationships between the constants in (5) and those in (4) so that we can formulate rules for going from (4) to (5) and omit the detail just used, we adopt the following *tableau* form for writing the starting solution, (4).

$$\begin{array}{ccc} & & z \\ e_7\colon\ y & \quad a \quad & b* \\ e_8\colon\ x & \quad c \quad & d. \end{array}$$

The tableau form omits the equality symbols and places the variable z at the head of the column so that attention may be directed toward the constants a, b, c, and d. The appearance of z as a *column* head means that the solution now has z as nonbasic, and the *row* variables y and x are basic. The asterisk (*) on b^* means we propose to rewrite the solution by interchanging y and z which are the row and column intersecting at b^*. Thus, y is to become nonbasic, and z is to become basic. The element with the * is called the *pivot* element.

The tableau forms of (4) and the solution (5) are shown next side by side in (6) for comparison so that we can see how (5) is obtained from (4).

$$
\begin{array}{cc}
 & z \\
y & a \quad b^* \\
x & c \quad d.
\end{array}
\qquad
\begin{array}{ccc}
 & z & y \\
z & -\dfrac{a}{b} & \dfrac{1}{b} \\
x & \dfrac{cb-ad}{b} & \dfrac{d}{b}.
\end{array}
\tag{6}
$$

The relationship between the tableaus is stated in the form of four rules which tell how to obtain the numbers in the right tableau from those of the left tableau.

1. Replace the pivot element by its reciprocal. Thus, b^* in the left tableau goes to $1/b$ in the right tableau.
2. Replace the elements in the pivot column by the quotient

$$\frac{\text{Element}}{\text{Pivot}}.$$

Thus, d in the left tableau becomes d/b on the right.

3. Replace the elements of the pivot row by

$$-\frac{\text{Element}}{\text{Pivot}}.$$

Thus, a in the left tableau goes to $-a/b$ on the right.

4. Replace elements not in pivot row or pivot column by their pivot cross product. The pivot cross product for an element is

$$\frac{(\text{Element})(\text{pivot}) - \text{product of opposite diagonal elements}}{\text{Pivot}}.$$

To clarify the meaning of "opposite diagonal," note that four elements in

$$
\begin{array}{cc}
a & b^* \\
c & d
\end{array}
$$

form a rectangle (in this case a square). One diagonal goes from c to the pivot b, and the "opposite" diagonal goes from a to d.

In the present case, c is in neither pivot row nor column. Its pivot cross product is

$$\frac{cb - ad}{b},$$

which is shown in the appropriate location in the right tableau of (6).

Comparison of the last two tableaus in (6) shows that these four rules carry the one on the left to the one on the right. We shall now illustrate by applying these rules to the system (1) at the beginning of the section

$$e_1:\ 2x + y + \ z = 12$$
$$e_2:\ \ x + y + 2z = 18.$$

The first solution, shown in (2), was

$$z \text{ nonbasic}$$
$$x = -6 + \ z$$
$$y = \ 24 - 3z$$

which, in tableau form, is

$$
\begin{array}{ccc}
 & & z \\
x & -6 & 1 \\
y & 24 & -3*.
\end{array}
$$

The asterisk shows that we plan to make y nonbasic instead of z. Interchanging y and z, the form of the next tableau will be

$$
\begin{array}{ccc}
 & & y \\
x & ? & ? \\
z & ? & ?.
\end{array}
$$

Following the rules, we have:

1. Replace pivot element, -3, by $1/-3$.
2. Replace pivot column element, 1, by $1/-3$.
3. Replace pivot row element, 24, by $-(24/-3) = 8$.
4. Replace -6 by its pivot cross product
$$\frac{(-6)(-3) - (24)(1)}{-3} = 2.$$

The second tableau is

$$
\begin{array}{ccc}
 & & y \\
x & 2 & -\dfrac{1}{3} \\
 & & \\
z & 8 & -\dfrac{1}{3}.
\end{array}
$$

The equation reproduction of the last tableau is

$$y \text{ nonbasic}$$

$$x = 2 - \frac{1}{3}y$$

$$z = 8 - \frac{1}{3}y.$$

The result is the system (3) obtained earlier by manipulating equations. As another example, we shall show that

		z					v
u	5	−2			u	−3	2
v	4	−1*	goes to		z	4	−1.

That is, the solution

z nonbasic		v nonbasic
$u = 5 - 2z$		$u = -3 + 2v$
$v = 4 - \ \ z$	goes to	$z = \ \ 4 - \ \ v.$

The details of the change in tableaus are:

1. Pivot, −1, goes to reciprocal, $1/-1 = -1$.
2. Pivot column element −2 goes to

$$\frac{\text{Element}}{\text{Pivot}} = \frac{-2}{-1} = 2.$$

3. Pivot row element 4 goes to

$$-\frac{\text{Element}}{\text{Pivot}} = -\frac{4}{-1} = 4.$$

4. Element not in pivot row or column, 5, goes to

$$\text{Pivot cross product} = \frac{(5)(-1) - 4(-2)}{-1} = -3.$$

As a final example before the problem set, the reader may verify that

		y					w
w	3	7*			y	$-\dfrac{3}{7}$	$\dfrac{1}{7}$
			goes to				
x	5	6			x	$\dfrac{17}{7}$	$\dfrac{6}{7}.$

That is, in equation form:

y nonbasic	w nonbasic
$w = 3 + 7y$	$y = -\dfrac{3}{7} + \dfrac{1}{7}w$

$$\text{goes to}$$

$x = 5 + 6y$	$x = \dfrac{17}{7} + \dfrac{6}{7}w$

5.4 PROBLEM SET 5–1

Mark (T) for true or (F) for false:

1. (   ) The column in which the pivot element occurs must have a variable at its head.

2. (   ) Elements in the pivot row of the first tableau are replaced by element over pivot.

3. (   ) Elements in the pivot column of the first tableau are replaced by the negative of element over pivot.

4. (   ) The pivot cross product is computed only for elements not in the pivot row or the pivot column.

5. (   ) The replacement for p in $\begin{matrix} p & 2^* \\ 3 & 6 \end{matrix}$ would be $-\dfrac{p}{2}$.

6. (   ) The replacement for p in $\begin{matrix} 3 & 6^* \\ 2 & p \end{matrix}$ would be $\dfrac{p}{6}$.

7. (   ) The replacement for p in $\begin{matrix} 3 & 6 \\ 2 & p^* \end{matrix}$ would be $-\dfrac{1}{p}$.

8. (   ) The replacement for p in $\begin{matrix} p & 6^* \\ 3 & 2 \end{matrix}$ would be $\dfrac{2p-18}{6}$.

9. (   ) The replacement for 4 in $\begin{matrix} 4 & 2^* \\ -1 & 1 \end{matrix}$ would be 3.

10. (   ) The replacement for 4 in $\begin{matrix} 3 & -2^* \\ 4 & 1 \end{matrix}$ would be $\dfrac{11}{2}$.

Write the numbers that would appear in the second tableau:

11. $\begin{matrix} 2 & 1 \\ 6 & 4^* \end{matrix}$

12. $\begin{matrix} 3 & 4^* \\ 1 & 2 \end{matrix}$

13. $\begin{matrix} 5 & -1^* \\ -3 & 2 \end{matrix}$

14. $\begin{matrix} -\dfrac{1}{2} & 2 \\ -5 & 3^* \end{matrix}$

15. $\begin{matrix} \dfrac{1}{4} & -\dfrac{1^*}{3} \\ -2 & 1 \end{matrix}$

Convert the equations into first tableau form; then find the second tableau. Finally, write the equations specified by the second tableau.

16. $x = 3 - 2y$
    $w = 5 - 3y^*$.

17. $v = 10 + 4z^*$
    $y = 5 - z$.

18. $y = 8 - 2x$
    $w = 2 - x^*$.

19. $u = 5 + v^*$
    $w = 1 - v$.

20. $x = 3 + 2z$
    $y = 4 - 3z^*$.

5.5 MORE THAN ONE NONBASIC VARIABLE

The pivot rules developed in the last section apply without alteration when the number of variables is increased. Consider the system

$$w, z \text{ nonbasic}$$
$$x = 3 + 2w + 4z$$
$$y = 5 + w - 3z.$$

To make x nonbasic instead of z, we start with the tableau

		w	z
x	3	2	4*
y	5	1	-3

making the intersection of the x row and z column the pivot. To obtain the second tableau:

1. Pivot element 4 is replaced by reciprocal, $1/4$.
2. Pivot column element -3 is replaced by

$$\frac{\text{Element}}{\text{Pivot}} = \frac{-3}{4}.$$

3. Pivot row elements 3 and 2 are replaced by the negative of

$$\frac{\text{Element}}{\text{Pivot}}, -\frac{3}{4} \text{ and } -\frac{1}{2}, \text{ respectively.}$$

4. Elements 5 and 1 are replaced by their pivot cross products.

$$5 \text{ is replaced by } \frac{(5)(4) - (3)(-3)}{4} = \frac{29}{4}$$

$$1 \text{ is replaced by } \frac{(1)(4) - (2)(-3)}{4} = \frac{5}{2}.$$

The second tableau is

$$\begin{array}{ccc} & w & x \\ z & -\dfrac{3}{4} & -\dfrac{1}{2} & \dfrac{1}{4} \\ y & \dfrac{29}{4} & \dfrac{5}{2} & -\dfrac{3}{4} \end{array}$$

which represents the equation system

$$w,\ x \text{ nonbasic}$$

$$z = -\frac{3}{4} - \frac{1}{2}w + \frac{1}{4}x$$

$$y = \frac{29}{4} + \frac{5}{2}w - \frac{3}{4}x.$$

As a more extensive example, let us start with the following tableau representation of a system with five nonbasic variables:

$$\begin{array}{ccccccc} & & v & w & x & y & z \\ r & 2 & 4 & 1 & -3 & 1 & 3 \\ s & 1 & 3 & -3 & 5^* & -2 & -3 \\ t & 3 & 1 & 2 & 6 & 4 & 2. \end{array}$$

The asterisk shows we intend to make s nonbasic in place of x. The form of the second tableau, with pivot row and column entries, is:

$$\begin{array}{ccccccc} & & v & w & s & y & z \\ r & ? & ? & ? & -\dfrac{3}{5} & ? & ? \\ x & -\dfrac{1}{5} & -\dfrac{3}{5} & \dfrac{3}{5} & \dfrac{1}{5} & \dfrac{2}{5} & \dfrac{3}{5} \\ t & ? & ? & ? & \dfrac{6}{5} & ? & ?. \end{array}$$

All the elements indicated by a question mark are obtained as pivot cross products. Thus the first three elements in the first row of the initial tableau, 2, 4, and 1, are replaced as follows:

$$2 \text{ goes to} \qquad \frac{(2)(5) - (1)(-3)}{5} = \frac{13}{5}$$

$$4 \text{ goes to} \qquad \frac{(4)(5) - (3)(-3)}{5} = \frac{29}{5}$$

$$1 \text{ goes to} \qquad \frac{(1)(5) - (-3)(-3)}{5} = -\frac{4}{5}.$$

Similar calculations lead to the following completed tableau:

	v	w	s	y	z	
r	$\dfrac{13}{5}$	$\dfrac{29}{5}$	$-\dfrac{4}{5}$	$\dfrac{3}{5}$	$-\dfrac{1}{5}$	$\dfrac{6}{5}$
x	$-\dfrac{1}{5}$	$-\dfrac{3}{5}$	$\dfrac{3}{5}$	$\dfrac{1}{5}$	$\dfrac{2}{5}$	$\dfrac{3}{5}$
t	$\dfrac{9}{5}$	$-\dfrac{13}{5}$	$\dfrac{28}{5}$	$\dfrac{6}{5}$	$\dfrac{32}{5}$	$\dfrac{28}{5}$.

5.6 PROBLEM SET 5–2

1. Given:

	y	z	
w	1	2	1
x	5	3	-2.

a) Make w nonbasic in place of y.
b) Make x nonbasic in place of y.
c) Make w nonbasic in place of z.
d) Make x nonbasic in place of z.

2. Given:

	w	x	y	
u	3	2	1	2
v	5	-1	2	-3.

a) Make u nonbasic in place of x.
b) Make v nonbasic in place of w.
c) Make u nonbasic in place of w.
d) Make u nonbasic in place of y.
e) Make v nonbasic in place of x.
f) Make v nonbasic in place of y.

3. Given

	v	w	x	y	
r	4	-2	4	0	6
s	6	2	$-2*$	4	0
t	8	0	-6	2	-4.

a) Write the second tableau.
b) Write the equation systems corresponding to the given tableau and the second tableau.

4. Given:

$$w, x, \text{ and } y \text{ nonbasic}$$
$$u = 4 + 3w + 2x + \ y$$
$$v = 2 - \ w + 3x - 2y.$$

a) Write the tableau for the given system of equations.
b) Construct a second tableau which makes v nonbasic in place of x, and write the equation system corresponding to the second tableau.

5.7 INTRODUCING THE OBJECTIVE FUNCTION

If we have a system of equations and an objective function, θ, and the system has a general solution in terms of nonbasic variables, θ

can also be expressed in terms of the nonbasic variables, as shown in the next example:

$$x = 24 - 2z$$
$$y = 12 + z$$
$$\theta = x + 3y + z.$$

By substitution for x and y into θ, we obtain θ in terms of the nonbasic variable z, as follows:

$$\theta = (24 - 2z) + 3(12 + z) + z$$
$$= 24 - 2z + 36 + 3z + z$$
$$\theta = 60 + 2z.$$

If we enter θ into the tableau, we may manipulate it in the usual manner as basic variables are changed. The current tableau is

		z
θ	60	2
x	24	$-2*$
y	12	1.

Pivoting on $-2*$, we obtain the new tableau

		x
θ	84	-1
z	12	$-\dfrac{1}{2}$
y	24	$-\dfrac{1}{2}$

which corresponds to the equations

$$\theta = 84 - x.$$
$$z = 12 - \frac{1}{2}x.$$
$$y = 24 - \frac{1}{2}x.$$

5.8 SLACK VARIABLES

The programming problems we wish to discuss involve systems of inequalities. In this section, we shall see how to convert a system of inequalities into a system of equalities by means of *slack* variables.

Consider the inequality

$$x + 3y \leq 10$$

where we require that both x and y be nonnegative; that is, the variables must be equal to or greater than zero. We define the slack variable, r, by the statement

$$x + 3y + r = 10$$

r also being nonnegative; that is, r takes up the slack between $(x + 3y)$ and 10. The original inequality is satisfied by various number pairs, one of which is

$$x = 1$$
$$y = 2.$$

For this pair the inequality is the true statement $7 \leq 10$. The slack in this instance is 3; that is, if the number pair is placed into the equality $x + 3y + r = 10$ the value of r will be 3.

Again, the inequality is satisfied by

$$x = 1$$
$$y = 3$$

because substitution leads to the true statement $10 \leq 10$. In this instance, however, there is no slack, and r would take the value zero.

We shall work with slack variables in programming problems. The manner in which the problems will be set up is illustrated next. Suppose that we have to maximize

$$\theta = 2x + 3y$$

subject to the inequality constraints

$$x + 3y \leq 10$$
$$2x + y \leq 6.$$

The inequalities will be converted to equalities by means of slack variables. The appearance of the problem will then be

$$\theta = 0 + 2x + 3y$$
$$x + 3y + r = 10$$
$$2x + y + s = 6.$$

All variables (including slack variables) are to be nonnegative. We shall solve the equations for the slack variables, then make a tableau. Thus:

$$\theta = 0 + 2x + 3y$$
$$r = 10 - x - 3y$$
$$s = 6 - 2x - y$$

leads to the tableau

		x	y
θ	0	2	3
r	10	−1	−3
s	6	−2	−1.

The solution of the programming problem will be found by manipulating tableaus in the usual manner, following a set of rules for determining which element is to be the pivot.

5.9 PROBLEM SET 5–3

Construct the tableau corresponding to each of the following programming situations:

1. $\theta = 2x + y$, subject to:
$$3x + 2y \leq 20$$
$$x + y \leq 12.$$

2. $\theta = x + 2y + z$, subject to:
$$x + y + z \leq 8$$
$$2x + 3y + z \leq 12$$
$$x + 2y + 3z \leq 10.$$

3. $\theta = 3w + 2.8x + y + 2z$, subject to:
$$2w + x + 2y + 3z \leq 150$$
$$2x + y + 2z \leq 100$$
$$3w + x + 4y \leq 200.$$

5.10 THE SIMPLEX PROCEDURE

Linear programming problems can be solved by a procedure called the *simplex* method. The word simplex is a mathematical term used in n-space geometry and is not a synonym for *simple*. The important attribute of the simplex method is that it provides a step-by-step solution of linear programming problems in which only basic feasible solutions (vertices) are investigated and each step brings us closer to the optimal value of the objective function. The steps in the simplex procedure have already been broadly outlined in Section 5.2, and in this section we shall describe them in detail. At any step, we can read the current solution (not necessarily optimal) from the tableau. We can determine also whether the solution is optimal; if it is not, we select the proper pivot and construct the next tableau, continuing in this manner until the optimum is reached.

5.11 READING THE SOLUTION; IDENTIFYING THE OPTIMUM

The solution at any step is read from the tableau by letting all nonbasic variables be zero, and assigning the numbers in the first column (the constants) as values of the corresponding basic variables. For example, suppose that we have the tableau

		r	z	y
θ	10	-1	2	1
x	8	-2	2	2
s	12	2	-6^*	1
t	6	4	-1	3.

Disregard the pivot element for the moment. In the tableau the solution is

$$\theta = 10$$
$$x = 8$$
$$s = 12$$
$$t = 6$$
$$r = z = y = 0.$$

A solution is optimal if the numbers in the θ row *under the nonbasic variables* are all zero or negative. The specified numbers in the last tableau are -1, 2, and 1, so the solution is not optimal. A pivot must be selected and the next tableau formed.

5.12 SELECTION OF THE PIVOT

The pivot column is the nonbasic variable column that has the largest positive number in the θ row. In the last tableau the pivot will be in the z column.

Now, divide each *negative* element of the pivot column into the corresponding constant in the first column. Select as the pivot row that row for which the quotient is smallest in *absolute value* (that is, smallest disregarding sign). In the present case the quotients are

$$s \text{ row: } \frac{12}{-6} = -2$$

$$t \text{ row: } \frac{6}{-1} = -6.$$

The former of these two numbers, -2, is the smaller in absolute value, so we select the s row as the pivot row. As the foregoing tableau shows, the pivot element is -6^* at the intersection of the s row and the z column. The next step would be to form a new tableau in the usual manner, making s nonbasic in place of z. Rather than proceed further with this example, we shall start a new one at the beginning and carry it all the way through. The example is arranged as a set of exercises so that the reader can obtain practice in the methodology. When the reader comes to an exercise, he or she should cover the following material (which contains the answer to the exercise) and do the exercise on a scratch pad before proceeding. Consider the following linear programming problem:

Maximize

$$\theta = x + y$$

subject to:

$$3x + 2y \le 34$$
$$x + 2y \le 18$$
$$x + 6y \le 48$$
$$x \le 10.$$

Exercise. Introduce the slack variables *r*, *s*, *t*, and *u* and convert the given system into a system of equalities.

The system of equalities is found to be

$$3x + 2y + r = 34$$
$$x + 2y + s = 18$$
$$x + 6y + t = 48$$
$$x \quad\quad + u = 10.$$

Exercise. Solve each of the last equations for the slack variable and write the resultant equations and the objective function in terms of *x* and *y*.

The set of equations is found to be

$$\theta = 0 + x + y$$
$$r = 34 - 3x - 2y$$
$$s = 18 - x - 2y$$
$$t = 48 - x - 6y$$
$$u = 10 - x + 0y.$$

Exercise. Construct the first tableau.

The first tableau is found to be

		x	y
θ	0	1	1
r	34	-3	-2
s	18	-1	-2
t	48	-1	-6
u	10	-1	0.

The solution here is not optimal because the θ row has a positive number, one, under both x and y. We may choose either the x or y column as the pivot column. (The rule is to choose the column having the largest positive number in the θ row.)

Exercise. Take x as the pivot column, then find the pivot row.

To find the pivot row, we form the following quotients for the *negative* numbers in the x column:

$$r \text{ row: } \frac{34}{-3}$$

$$s \text{ row: } \frac{18}{-1}$$

$$t \text{ row: } \frac{48}{-1}$$

$$u \text{ row: } \frac{10}{-1}.$$

The smallest quotient, disregarding sign, is $10/-1$, so we take u as the pivot row. The pivot element is the -1 at the intersection of the u row and x column.

> **Exercise.**   Place an asterisk at the intersection of the u row and x column of the tableau. Manipulate the tableau in the usual manner to make u nonbasic in place of x and write the second tableau.

The second tableau is found to be

		u	y
θ	10	-1	1
r	4	3	-2
s	8	1	-2
t	38	1	-6
x	10	-1	0.

> **Exercise.**   Write the solution exhibited by the second tableau.

The solution at this point is

$$\theta = 10$$
$$r = 4$$
$$s = 8$$
$$t = 38$$
$$x = 10$$
$$u = y = 0.$$

Exercise. Is the present solution optimal?

The solution is not optimal because the θ row has a positive element under a nonbasic variable, namely, the one under y. The y column becomes the new pivot column.

Exercise. Take y as the pivot column and find the pivot row.

To find the pivot row we form the quotients

$$r \text{ row:} \frac{4}{-2}$$

$$s \text{ row:} \frac{8}{-2}$$

$$t \text{ row:} \frac{38}{-6}$$

x row: None (why?).

In absolute value the r row has the smallest quotient, and so becomes the pivot row. The pivot element is the -2 at the intersection of the r row and y column.

Exercise. Place an asterisk next to the pivot element, -2, of the last tableau, then calculate and write the third tableau.

The third tableau is found to be

		u	r
θ	12	$\frac{1}{2}$	$-\frac{1}{2}$
y	2	$\frac{3}{2}$	$-\frac{1}{2}$
s	4	-2	1
t	26	-8	3
x	10	-1	0.

Exercise. Write the solution exhibited by the third tableau. Is this solution optimal?

The current solution from the last tableau is

$$\theta = 12$$
$$y = 2$$
$$s = 4$$
$$t = 26$$
$$x = 10$$
$$u = r = 0.$$

The solution is not optimal because the nonbasic variable column u has a positive number, $1/2$, in the θ row.

Exercise. Find the pivot element of the third tableau, then carry out the manipulations and write the fourth tableau.

The pivot is found to be the -2 at the intersection of the s row and u column. The fourth tableau is

		s	r
θ	13	$-\dfrac{1}{4}$	$-\dfrac{1}{4}$
y	5	$-\dfrac{3}{4}$	$\dfrac{1}{4}$
u	2	$-\dfrac{1}{2}$	$\dfrac{1}{2}$
t	10	4	-1
x	8	$\dfrac{1}{2}$	$-\dfrac{1}{2}$.

In the last tableau, all elements in the θ row under the variables are negative. The present solution is optimal. It is

$$\theta_{max} = 13$$
$$x = 8$$
$$y = 5$$
$$u = 2$$
$$t = 10$$
$$s = r = 0.$$

The answer to the original problem is that the maximum value of θ is 13, and it occurs when x is 8 and y is 5. The number pair (8, 5) satisfies all the original inequalities and maximizes θ. The significance of the complete solution can be seen by referring to the original inequality system: at (8, 5),

$$3x + 2y \leq 34 \text{ becomes } 24 + 10 \leq 34.$$

There is no slack in the last statement, so the slack variable r is zero, as shown in the complete solution. Similar demonstration shows why slack variable s is also zero.

On the other hand, at (8, 5),

$$x + 6y \leq 48 \text{ becomes } 8 + 30 \leq 48$$

showing a slack of 10 in this inequality. This is the meaning of $t = 10$ in the complete solution. Similar demonstration shows why $u = 2$ in the complete solution.

Observe the increase in θ from tableau to tableau in the foregoing example. At the start, θ was zero. It increased successively to 10, to 12, and finally to the maximum, 13. The progress of the solution can be seen graphically from Figure 5–2, which shows the solution space for

FIGURE 5-2

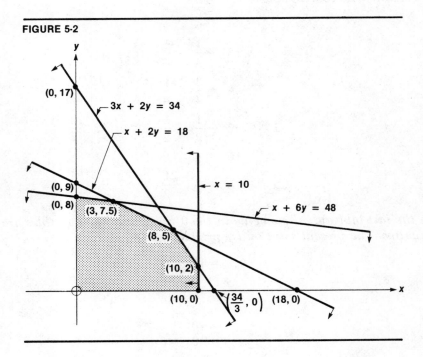

$$3x + 2y \le 34$$
$$x + 2y \le 18$$
$$x + 6y \le 48$$
$$x \le 10$$
$$x \ge 0$$
$$y \ge 0.$$

Observing the corners of the solution space, we see that the simplex method started with the corner $(0, 0)$. The next tableau moved us to the corner $(10, 0)$, the next to $(10, 2)$, and the final tableau to $(8, 5)$, where θ reaches its maximum. Figure 5–2 shows that the simplex progression goes from one basic feasible solution (a vertex) to another that increases θ, and avoids infeasible vertices such as $(0, 9)$ that are not on the boundary of the shaded solution space.

Before turning to the problem set, we review the simplex method with an example that yields the optimum after one step:

Maximize

$$\theta = 2w + x + z$$

subject to:

$$w + 2x + 2z \le 20$$
$$2w + 3x + z \le 18.$$

The usual nonnegativity restrictions apply to all variables. Introducing and solving for slack variables, we have

$$\theta = \qquad 2w + x + z$$
$$r = 20 - w - 2x - 2z$$
$$s = 18 - 2w - 3x - z.$$

The first tableau is

		w	x	z
θ	0	2	1	1
r	20	-1	-2	-2
s	18	-2^*	-3	-1

The pivot column is w because it has the largest positive number in the θ row. The absolute value of $18/-2$ is smaller than that of $20/-1$, so s is the pivot row. The next tableau is

		s	x	z
θ	18	-1	-2	0
r	11	$\dfrac{1}{2}$	$-\dfrac{1}{2}$	$-\dfrac{3}{2}$
w	9	$-\dfrac{1}{2}$	$-\dfrac{3}{2}$	$-\dfrac{1}{2}$.

No element in the θ row under a variable is positive, so the solution is optimal. It is $\theta_{max} = 18$.

5.13 PROBLEM SET 5–4

In each of the following, find the maximum value of θ by the simplex method, and state the values of the variates which make θ maximum. It is to be assumed that all variables must be zero or positive.

1. $\theta = x + 3y + 2z$
$x + 2y + 3z \leq 50$
$x + 3y + 5z \leq 60$.

2. $\theta = x + 4y + 3z$
$x + 3y + 4z \leq 30$
$x + 5y + 2z \leq 40$.

3. $\theta = x + 4y + 6z$
$x + 3y + 6z \leq 48$
$x + 6y + 3z \leq 90$
$x + 9y + 10z \leq 137$.

4. $\theta = x + 6y + z$
$x + 5y + 2z \leq 30$
$x + 7y \leq 40$
$2x + y + 3z \leq 70$.

5. $\theta = 3x + 7y + 6z$
$x + y + z \leq 4$
$x + y \leq 3$.

6. $\theta = 1.5x + 2.5y + 2z$
$1.5x + y + z \leq 7.5$
$x + y + z \leq 4.0$
$y + 2.5z \leq 5.0$.

7. $\theta = 3x + 5y + 3z$
$2x + 3y + 6z \leq 50$
$3x + 4y + z \leq 40$
$3x + 5y + 2z \leq 20$.

8. $\theta = 2x + 3y + z$
$2x + y + 3z \leq 10$
$x + 3y + 2z \leq 20$.

9. $\theta = x + 2y + z$
$2x + y + 3z \leq 12$
$x + 2y \leq 6$
$2x + z \leq 4$.

10. $\theta = x + y + z$
$x + 2y + z \leq 12$
$2x + y + z \leq 20$
$x + y + 3z \leq 15$.

11. A wholesaler has 9600 feet of space available, and $5000 he can spend to buy merchandise of types A, B, and C. Type A costs $4 per unit and requires 4 feet of storage space in the warehouse. B costs $10 per unit and requires 8 feet of space. C costs $5 per unit and requires 6 feet of space. Only 500 units of type A are available to the wholesaler. Assuming that the wholesaler expects to make a profit of $1 on each unit of A he buys and stocks, $3 per unit on B, and $2 per unit on C, how many units of each should he buy and stock in order to maximize his profit?

12. Products A, B, and C are sold door to door. A costs $3 per unit, takes 10 minutes to sell (on the average), and costs $0.50 to deliver to a customer. B costs $5, takes 15 minutes to sell, and is left with the customer at the time of sale. C costs $4, takes 12 minutes to sell, and costs $1 to deliver. During any week, a salesman is allowed to draw up to $500 worth of A, B, and C (at cost) and he is allowed delivery expenses not to exceed $75. If a salesman's selling time is not expected to exceed 30 hours (1800 minutes) in a week, and if the salesman's profit (net after all expenses) is $1 each on a unit of A or B and $2 on a unit of C, what combination of sales of A, B, and C will lead to maximum profit, and what is this maximum profit?

5.14 MINIMIZATION BY MAXIMIZING THE DUAL

Our discussion to this point has involved only maximization problems with $\leq$ constraints. In this section we shall learn how to change a minimization problem with $\geq$ constraints into a maximization problem with $\leq$ constraints so that the solution method already developed can be applied to either type of problem. Thus, we shall:

Solve a "less than or equal max" by the methods already shown but change a "greater than or equal min" to a "less than or equal to max" and then solve by the method already shown.

The original "greater than or equal to min" will be called the *primal* problem. The "less than or equal to max" to which we change will be called the *dual* problem. We first illustrate how the primal is converted to the dual, then give the necessary rules for reading the solution of the primal from the solution of the dual.

Example. Write the dual of the following problem.
Minimize

$$\theta = 2x + 3y$$

subject to:

$$x, y \geq 0$$
$$i_1: \; 3x + 2y \geq 12$$
$$i_2: \; 7x + 2y \geq 20.$$

First, observe that in i_1, if $3x + 2y$ is, say, 15, the inequality is satisfied because 15 is 3 *more* than 12; that is, there is a *surplus* of 3. Hence, to convert i_1 and i_2 to equalities, *positive* variables would be *subtracted*, and these are called *surplus* variables rather than slack variables. We denote the surplus variables here by r and s, and indicate them next to their respective constraints, as follows:

Primal
$$x, y \geq \; 0$$
$$r: \; 3x + 2y \geq 12$$
$$s: \; 7x + 2y \geq 20$$
Minimize: $2x + 3y = \theta$.

To form the *dual* problem, which is a maximization problem with $\leq$ constraints, we look at the first column (x's) in the primal, make 3 the coefficient of r and 7 the coefficient of s to obtain

$$3r + 7s.$$

Next, change to $\leq$, use the coefficient of x in θ as the right-hand constant, and write

$$3r + 7s \leq 2.$$

Now repeat the process with the second *(y)* column to obtain

$$2r + 2s \leq 3.$$

Finally, the new objective function is formed in a similar manner from the right-hand column of constants. It is

$$12r + 20s = \theta.$$

The dual problem is:

Dual

Maximize

$$\theta = 12r + 20s$$

subject to:

$$r, s \geq 0$$
$$x: \ 3r + 7s \leq 2$$
$$y: \ 2r + 2s \leq 3.$$

Notice that the dual is a "less than or equal max" problem and has x and y as *slack* variables, as indicated at the left of the last two inequalities. We offer another illustration for practice.

Example. Write the dual of the following problem.

Primal

Minimize:

$$\theta = 5x + 8y + 6z$$

subject to:

$$x, y, z \geq 0$$
$$2x + \ y + 4z \geq 4$$
$$2x + \ 3y + \ z \geq 2$$
$$4x + 10y + 3z \geq 6.$$

We rewrite the primal for clarity, as follows:

Primal

$$x, y, z \geq 0$$
$$r: \ 2x + \ y + 4z \geq 4$$
$$s: \ 2x + \ 3y + \ z \geq 2$$
$$t: \ 4x + 10y + 3z \geq 6$$
$$\overline{ 5x + \ 8y + 6z = \theta,}$$

where r, s, and t are surplus variables. Attaching the column of x coefficients to r, s, and t, and using the θ coefficient in the x column as the constant, leads to the dual $\leq$ constraint

$$2r + 2s + 4t \leq 5.$$

In a similar manner, the second and third columns of the primal lead to the corresponding constraints in the dual. The new objective function is formed from the column of constants on the right. We have:

Dual

$$r, s, t \geq 0$$

$$
\begin{aligned}
x: \ & 2r + 2s + \ 4t \leq 5 \\
y: \ & \ r + 3s + 10t \leq 8 \\
z: \ & 4r + \ s + \ 3t \leq 6 \\
\hline
\text{Maximize: } & 4r + 2s + \ 6t = \theta.
\end{aligned}
$$

In the dual, x, y, and z become the slack variables.

> **Exercise.** *a)* Construct the dual of the following using r and s as surplus variables: Minimize $\theta = 4x + 3y$ subject to $x, y \geq 0$; $2x + y \geq 10$; $x + 5y \geq 20$. *b)* What are the slack variables in the dual? Answer: *a)* Maximize $\theta = 10r + 20s$ subject to $r, s \geq 0$; $2r + s \leq 4$; $r + 5s \leq 3$. *b)* x and y.

We have formed the dual of a primal "greater than or equal min" problem because the dual is a "less than or equal to max" problem that can be solved by the simplex procedure developed earlier in the chapter. We should mention, however, that a "less than or equal to max" can be considered to be the primal. Then the dual is a "greater than or equal min," and it is clear that whichever problem is the primal, the dual of its dual is the original primal problem. In any event, the dual theorem proves that all of the information contained in the solution of the primal (dual) is present in the solution of the dual (primal). The development of the dual theorem is outside the scope of this text, so we shall state without proof that part of the theorem which we want to use.

If a minimization problem with $\geq$ constraints has an optimum (minimum) value, this value equals the maximum of the dual problem.

Thus, to solve a primal "greater than or equal to min" we shall set up the dual and find its maximum in the usual manner. The dual theorem then assures us that this maximum of the dual is the minimum of the primal.

Example. Solve the following:
Minimize

$$\theta = 3x + 5y$$

subject to:

$$x, y \geq 0$$
$$i_1: \ 2x + 5y \geq 50$$
$$i_2: \ \ x + 2y \geq 24.$$

Here we have a two-variable problem that can be solved quickly by the graphical procedure of Chapter 4. The solution is

$$\theta_{min} = 60 \quad \text{at} \quad x = 0, \ y = 12.$$

To solve by maximizing the dual, we first set up the dual as follows:

Dual

Maximize

$$\theta = 50r + 24s$$

subject to:

$$r, s \geq 0$$
$$2r + \ s \leq 3$$
$$5r + 2s \leq 5.$$

As stated earlier, x and y now become the slacks in the last two constraints, so we have

$$2r + \ s + x = 3$$
$$5r + 2s + y = 5,$$

which we solve for the slacks, and list with θ, as follows:

$$\theta = \ \ \ \ \ 50r + 24s$$
$$x = 3 - \ \ 2r - \ \ s$$
$$y = 5 - \ \ 5r - \ 2s.$$

We now proceed to maximize θ. The starting tableau is

		r	s
θ	0	50	24
x	3	-2	-1
y	5	-5^*	-2.

Pivoting on -5^* leads to the second tableau:

		y	s
θ	50	-10	4
x	1	$\dfrac{2}{5}$	$-\dfrac{1}{5}$
r	1	$-\dfrac{1}{5}$	$-\dfrac{2^*}{5}$.

Pivoting again on the last starred element leads to:

Maximum of Dual

	y	r	
θ	60	-12	-10
x	$\dfrac{1}{2}$	$\dfrac{1}{2}$	$\dfrac{1}{2}$
s	$\dfrac{5}{2}$	$-\dfrac{1}{2}$	$-\dfrac{5}{2}$

In the last we see that θ cannot be increased further, so we have:

For the *dual:* $\theta_{max} = 60$.

The dual theorem states that the minimum of the primal must equal the maximum of the dual. Hence,

For the *primal:* $\theta_{min} = 60$.

At the start of the example we found by graphical means that

$$\theta_{min} = 60 \quad \text{at} \quad x = 0, \, y = 12.$$

The values for x and y at the minimum can also be obtained from the solution of the dual by the following rules.

To obtain the values of the variables in the solution of the primal, set the variables in the first column of the optimal dual tableau equal to zero and set the values of the variables in the column headings equal to the negative of the number under them in the θ row. The nonbasic variables in the dual are the basic variables in the primal.

Applying the rules, we have $\theta_{min} = 60$ and

$$\begin{aligned} x &= 0 \\ s &= 0 \\ y &= 12 \\ r &= 10. \end{aligned} \tag{1}$$

To see the full significance of the complete solution (1), recall the original primal problem, which contained

$$\begin{array}{lll} i_1\text{:} & 2x + 5y \geq 50 & \text{(Surplus variable is } r) \\ i_2\text{:} & x + 2y \geq 24 & \text{(Surplus variable is } s). \end{array}$$

The solution (1) has $x = 0$, $y = 12$. Placing these in i_1 yields

$$\begin{aligned} 2(0) + 5(12) &\geq 50 \\ 60 &\geq 50, \end{aligned}$$

so i_1 is satisfied and has a *surplus* of 10; that is, 60 is 10 more than 50. This surplus is the value $r = 10$ in the solution (1). Similarly, placing $x = 0$, $y = 12$ into i_2 yields

$$0 + 2(12) \geq 24$$
$$24 \geq 24,$$

and there is no surplus, which is the meaning of $s = 0$ in the solution (1).

In Section 4.10, we stated a minimization problem in three variables and proceeded to solve it by the laborious method of finding all 20 intersections, eliminating 15 vertices as infeasible, then determining the optimum from the remaining five basic feasible solutions. We shall now rewrite the problem and solve it by maximizing the dual.

Example. Three variables. A diet is to contain at least 10 ounces of nutrient R, 12 ounces of nutrient S, and 20 ounces of nutrient T. These nutrients are to be obtained by some combination of foods A, B, and C. Each pound of A costs 4 cents and has 4 ounces of R, 3 of S, and 0 of T. Each pound of B costs 7 cents and has 1 ounce of R, 2 of S, and 4 of T. Each pound of C costs 5 cents and has 0 ounces of R, 1 of S, and 5 of T. How many pounds of A, B, and C should be combined to provide the required amount of nutrient at minimum cost?

Letting x, y, and z represent, respectively, the number of pounds of A, B, and C, the data of the problem lead to the following:

Minimize

$$\theta = 4x + 7y + 5z$$

subject to:

$$
\begin{array}{lll}
& x, y, z \geq 0 & \\
r\colon\ 4x + y\ \ \ \ \ \ \geq 10 & \text{(Nutrient R constraint)} & \\
s\colon\ 3x + 2y + z \geq 12 & \text{(Nutrient S constraint)} & \text{(2)} \\
t\colon\ \ \ \ \ \ \ 4y + 5z \geq 20 & \text{(Nutrient T constraint),} &
\end{array}
$$

where r, s, and t are surplus variables in the primary, and x, y, and z will become slack variables in the dual.

We first write the dual, as follows:

Dual

Maximize

$$\theta = 10r + 12s + 20t$$

subject to:

$$
\begin{array}{ll}
r, s, t \geq 0 & \\
4r + 3s + (0)t \leq 4 & \text{(Slack is } x) \\
r + 2s + 4t \leq 7 & \text{(Slack is } y) \\
0(r) + s + 5t \leq 5. & \text{(Slack is } z)
\end{array}
$$

Introducing x, y, and z as slacks, we change the constraints to

$$
\begin{array}{l}
4r + 3s + (0)t + x = 4 \\
r + 2s + 4t + y = 7 \\
0(r) + s + 5t + z = 5.
\end{array}
$$

Next, solving for the slacks, we find

$$x = 4 - 4r - 3s - (0)t$$
$$y = 7 - r - 2s - 4t$$
$$z = 5 - 0(r) - s - 5t$$

and set up the initial tableau

		r	s	t
θ	0	10	12	20
x	4	-4	-3	0
y	7	-1	-2	-4
z	5	0	-1	-5^*.

Pivoting on -5^* produces

		r	s	z
θ	20	10	8	-4
x	4	-4^*	-3	0
y	3	-1	$-\dfrac{6}{5}$	$\dfrac{4}{5}$
t	1	0	$-\dfrac{1}{5}$	$-\dfrac{1}{5}$.

Pivoting next on -4^* leads to

		x	s	z
θ	30	$-\dfrac{5}{2}$	$\dfrac{1}{2}$	-4
r	1	$-\dfrac{1}{4}$	$-\dfrac{3^*}{4}$	0
y	2	$\dfrac{1}{4}$	$-\dfrac{9}{20}$	$\dfrac{4}{5}$
t	1	0	$-\dfrac{1}{5}$	$-\dfrac{1}{5}$.

The last pivot on $(-3/4)^*$ gives the optimal tableau

		x	r	z
θ	$\dfrac{92}{3}$	$-\dfrac{8}{3}$	$-\dfrac{2}{3}$	-4
s	$\dfrac{4}{3}$	$-\dfrac{1}{3}$	$-\dfrac{4}{3}$	0
y	$\dfrac{7}{5}$	$\dfrac{2}{5}$	$\dfrac{3}{5}$	$\dfrac{4}{5}$
t	$\dfrac{11}{15}$	$\dfrac{1}{15}$	$\dfrac{4}{15}$	$-\dfrac{1}{5}$.

(3)

Following the rules for writing the minimum of the primal from the last tableau (which provides the maximum of the dual), we have

$$\text{Minimum cost of mixture: } \theta_{min} = \frac{92}{3} \text{ cents;}$$

Contents of minimum-cost mixture

$$x = \frac{8}{3} \text{ pounds of food A}$$

$$y = 0 \text{ pounds of food B}$$

$$z = 4 \text{ pounds of food C.}$$

This solution is the same as that obtained in Section 4.10. The important point to note, however, is that the method of Section 4.10 is an unthinkable procedure to apply to problems with numerous variables and constraints. The simplex method is a general procedure that can solve any linear programming problem efficiently. It is true, of course, that an extensive amount of arithmetic has to be done in applying the simplex method to large problems, but the fact that the method applies the *same* steps to go from one tableau to another means that these steps can be carried out by an electronic computer that performs thousands of arithmetic steps in a fraction of a second. Consequently, large applied problems are solved by the simplex method on high-speed computers.

5.15 PROBLEM SET 5–5

Convert each of the primal problems into its dual and write the *first* tableau for the dual. It is not necessary to carry out iterations.

1. Given $\theta = 5x + 3y$, find θ_{min} subject to

$$4x + 2y \geq 20$$
$$6x + y \geq 10.$$

2. Given $\theta = 4x + 2y$, find θ_{min} subject to

$$3x + 2y \geq 10$$
$$x + 2y \geq 20.$$

3. Given $\theta = x + y + z$, find θ_{min} subject to

$$3x + 3y + z \geq 2$$
$$2x + y \geq 4$$
$$4x + 2y + 2z \geq 6.$$

4. Given $\theta = 2x + 3y + z$, find θ_{min} subject to

$$x + y + z \geq 10$$
$$2x + 3y + z \geq 15$$
$$3x + y + 2z \geq 20.$$

If the maximum of the dual has the following (partial) tableau, what is the minimum of the primal, and at what vertex does the minimum occur?

5.

		x	y
θ	10	-3	-2
r	4		
s	1		

6.

		x	t	z
θ	50	-5	-1	-7
r	4			
y	1			
s	3			

Find the minimum value of θ by maximizing the dual.

7.
$$x \geq 0$$
$$y \geq 0$$
$$4x + 3y \geq 24$$
$$x + 2y \geq 11$$
$$3x + 2y = \theta.$$

8.
$$x \geq 0$$
$$y \geq 0$$
$$3x + 2y \geq 24$$
$$x + y \geq 9$$
$$5x + 4y = \theta.$$

9.
$$x \geq 0$$
$$y \geq 0$$
$$z \geq 0$$
$$x + 2y + 3z \geq 1200$$
$$2x + y + z \geq 600$$
$$x + y + 3z = \theta.$$

10.
$$x \geq 0$$
$$y \geq 0$$
$$z \geq 0$$
$$x + 2y - z \geq 2$$
$$2x + 3y + 8z \geq 6$$
$$18x + 30y + 4z = \theta.$$

11. A quart of Expop cost $2 and contains 2 ounces of Zip. A quart of Whypop costs $3 and contains 6 ounces of Zip. A mixture of the two is to contain at least 4 quarts and at least 12 ounces of Zip. How much Expop and Whypop should be used to obtain the minimum-cost mixture? What is the minimum cost?

12. A diet is to contain at least 18 ounces of nutrient R, 24 ounces of nutrient S, and 36 ounces of T. The nutrients are to be obtained from some combination of foods Excal, Whycal, and Zeecal. Each pound of Excal costs 18 cents and contains, respectively, 1, 2, and 3 ounces of R, S, and T. Each pound of Whycal costs 12 cents and contains, respectively, 1, 3, and 2 ounces of R, S, and T. Each pound of Zeecal costs 6 cents and contains, respectively, 1, 1, and 2 ounces of R, S, and T. How many pounds of each of the foods should be combined to obtain the minimum cost diet? What is the minimum cost?

5.16 SHADOW PRICES: RIGHT-HAND SIDE RANGES

The following illustration will provide the background for introducing the concept of shadow prices.

Example. The Fancy Paint Company specializes in applying new finishes to automobiles. It does two types of finish, one called Extone and the other called Whytone. Each job requires painting and drying. Painting time is restricted to 10 hours each day and drying time is restricted to 20 hours. It takes 2 hours to paint and 1 hour to dry an Extone finish, compared to 1 hour to paint and 3 hours to dry a Whytone finish. Profit is $100 per Extone and $60 per Whytone. How many

TABLE 5–2

Finish type	Number of jobs	Profit per job	Hours per job to	
			Paint (10 hours available)	Dry (20 hours available)
Extone x		$100	2	1
Whytone .... y		$ 60	1	3

of each type of finish job should be scheduled each day if profit is to be maximized?

The information in the problem is summarized in Table 5–2.

We have the following problem:

Maximize

$$\theta = 100x + 60y$$

subject to:

$$x, y \geq 0$$
$$i_1: \ 2x + \ y \leq 10 \qquad \text{(Painting time).} \tag{1}$$
$$i_2: \ \ \ x + 3y \leq 20 \qquad \text{(Drying time).}$$

Inserting slack variables, we have

$$\theta = 100x + 60y$$
$$e_1: 2x + \ y + r = 10 \qquad \text{(Painting time).} \tag{2}$$
$$e_2: \ \ \ x + 3y + s = 20 \qquad \text{(Drying time).}$$

The initial tableau is

$$
\begin{array}{c c c c}
 & & x & y \\
\theta & 0 & 100 & 60 \\
r & 10 & -2^* & -1 \\
s & 20 & -1 & -3.
\end{array}
\tag{3}
$$

Pivoting on the -2^*, we obtain the second tableau

$$
\begin{array}{c c c c}
 & & r & y \\
\theta & 500 & -50 & 10 \\
x & 5 & -\dfrac{1}{2} & -\dfrac{1}{2} \\
s & 15 & \dfrac{1}{2} & -\dfrac{5^*}{2}.
\end{array}
\tag{4}
$$

The next pivoting operation provides the optimal tableau

$$
\begin{array}{c c c c}
 & & r & s \\
\theta & 560 & -48 & -4 \\
x & 2 & -\dfrac{3}{5} & \dfrac{1}{5} \\
y & 6 & \dfrac{1}{5} & -\dfrac{2}{5}.
\end{array}
\tag{5}
$$

The maximum profit is $560, and can be attained by doing 2 Extone and 6 Whytone finishes per day. The complete solution

$$\theta_{max} = \$560; \ x = 2; \ y = 6; \ r = 0; \ s = 0$$

shows that both slack variables, r and s, are zero. If we recall the initial inequality constraints

$$i_1: 2x + y \leq 10 \qquad \text{(Painting time; slack is } r)$$
$$i_2: \quad x + 3y \leq 20 \qquad \text{(Drying time; slack is } s)$$

we see that

$$x = 2 \text{ Extone finishes}$$
$$y = 6 \text{ Whytone finishes}$$

satisfy i_1 and i_2 as equalities. That is, at $x = 2$, $y = 6$

$$i_1: 2(2) + 6 \leq 10$$
$$i_2: \quad 2 + 3(6) \leq 20,$$

and we have no slack painting or drying time. We shall say that when a slack variable is zero in an optimal solution, the associated constraint is *binding.* The binding constraints in the present case mean simply that no more Extone or Whytone finishes can be done in a day because all of the available resources (painting and drying time) have been used. Consequently, the only way to increase output is to increase the amount of the painting and/or drying time resources. By way of contrast, a slack that turns out to be greater than zero is associated with a nonbinding constraint because the corresponding available resource is not fully utilized.

Effect of changing a right-hand side constant in a binding constraint. The Fancy Paint Company of our example, noting that the binding constraints on resources limit the number of jobs that can be done, and consequently limit profit per day, wishes to know the effect of enlarging its painting facility. To determine this effect, suppose that the amount of painting time is increased by an amount c hours. That is, the amount of painting time available is to be changed from 10 hours to $10 + c$ hours. The setup (2) now becomes

$$\theta = 100x + 60y$$
$$e_1: 2x + y + r = 10 + c \qquad \text{(Painting time, slack is } r) \qquad (6)$$
$$e_2: \quad x + 3y + s = 20 \qquad \text{(Drying time, slack is } s)$$

and the initial tableau is

		x	y
θ	0	100	60
r	$10 + c$	-2^*	-1
s	20	-1	$-3.$

(7)

Note that -2^* is taken as the pivot in (7). This implies that

$$\left| \frac{10 + c}{-2} \right| \leq \left| \frac{20}{-1} \right|,$$

because if $(10 + c)/(-2)$ were larger in absolute value than $20/(-1)$, s rather than r would go out of the basis. We shall see on page 198 that c is indeed less than or equal to 30 and why this must be so.

The next tableau is

		r	y
θ	$500 + 50c$	-50	10
x	$5 + \dfrac{c}{2}$	$-\dfrac{1}{2}$	$-\dfrac{1}{2}$
s	$15 - \dfrac{c}{2}$	$\dfrac{1}{2}$	$-\dfrac{5^*}{2}$.

$$(8)$$

The next pivot leads to the optimal tableau (9).

		r	s
θ	$560 + 48c$	-48	-4
x	$2 + \dfrac{3}{5}c$	$-\dfrac{3}{5}$	$\dfrac{1}{5}$
y	$6 - \dfrac{c}{5}$	$\dfrac{1}{5}$	$-\dfrac{2}{5}$.

$$(9)$$

In (9) we see that if the painting resource is changed by c hours, then the optimal solution is

$$\theta = 560 + 48c;$$

that is, θ is changed by $48c$. Similarly,

$$x = 2 + \frac{3}{5}c, \quad y = 6 - \frac{c}{5},$$

that is, x (the number of Extones) changes by $(3/5)c$ and y (the number of Whytones) changes by $-c/5$.

Shadow price. The optimal value in (9) is

$$\theta = 560 + 48c,$$

where c is the change in the number of painting time hours available. Now suppose painting time is increased from the original 10 hours to 11 hours, so that $c = 1$. Profit would then increase by \$48 to

$$\theta = 560 + 48(1) = \$608.$$

Thus, an hour of painting time has a value of \$48, and this value is called the *shadow price* of an hour of painting time. It is also called the *marginal value* of an hour of painting time because it is the additional profit when the available amount of the resource is increased by *one* unit. Finally, it may be referred to as the *imputed price* of an hour of painting time because the original problem statement did not provide a dollar value for painting time, and the $48 per hour was imputed from the analysis.

In the foregoing, we obtained the consequences of a change of c hours in painting time by solving the entire problem again. We now wish to show that the optimal tableau with the change (9) can be obtained easily from the optimal tableau (5) in the original solution without repeating the solution. To this end, (5) and (9) are shown next.

		r	s			r	s	
θ	560	-48	-4	θ	$560 + 48c$	-48	-4	
(5) x	2	$-\dfrac{3}{5}$	$\dfrac{1}{5}$	x	$2 + \dfrac{3}{5}c$	$-\dfrac{3}{5}$	$\dfrac{1}{5}$	(9)
y	6	$\dfrac{1}{5}$	$-\dfrac{2}{5}$	y	$6 - \dfrac{c}{5}$	$\dfrac{1}{5}$	$-\dfrac{2}{5}$	

Remember that we are considering a change in painting time hours, where the slack variable is r. In (5), refer to the r column. Now *subtract* c times the entry in the r column from the first column entry. In the first row this gives:

$$560 - (-48)c = 560 + 48c.$$

In the second row we have,

$$2 - \left(-\frac{3}{5}\right)c = 2 + \frac{3}{5}c$$

and in the third row

$$6 - \left(\frac{1}{5}\right)c = 6 - \frac{c}{5}.$$

The results just obtained are the new first column entries in (9). Other entries in (9) are the same as those in (5). We have the following rule and definition:

Less than or equal max problem

To determine the effect of a change in the right-hand side constant of a binding constraint (a constraint whose slack is non-basic) by c, subtract c times the slack column entries in the optimal tableau from the first column entries. The shadow price of the resource that has this slack variable is the negative of the slack column entry in the θ row. The shadow price of a slack that is in the basis is zero.

Note the last sentence. The reason for this statement is that a slack variable that is not zero identifies a resource that is not fully utilized, so adding to the available amount of that resource would not increase profit. We repeat that a shadow price exists only for a resource for which the constraint is binding; that is, a resource whose slack variable is nonbasic in the optimal tableau.

Right-hand-side ranges. The rule just stated applies *if the change, c, does not change the basis.* Clearly, if a change of c caused x or y to be negative in (9), the simplex procedure would not have produced (9), because adherence to the simplex rules insures that basic variables will not be negative. Hence, it follows that in (9), the values of x and y must be greater than or equal to zero. That is,

$$x = 2 + \frac{3}{5}c \geq 0 \quad \text{and} \quad y = 6 - \frac{c}{5} \geq 0.$$

Working with x we find

$$2 + \frac{3}{5}c \geq 0$$

$$\frac{3}{5}c \geq -2$$

$$c \geq -2\left(\frac{5}{3}\right)$$

$$c \geq -\frac{10}{3}. \tag{10}$$

Working with y we have

$$6 - \frac{c}{5} \geq 0$$

$$-\frac{c}{5} \geq -6$$

$$\frac{c}{5} \leq 6$$

$$c \leq 30. \tag{11}$$

Taking the (10) and (11) together gives

$$-\frac{10}{3} \leq c \leq 30.[1] \tag{12}$$

Recalling again that we are working with the slack variable r, which is the slack in the painting time constraint

---

[1] Recall that the selection of the pivot in (7) implied that $c \leq 30$. We see here that this must be so if the basis is not to change.

$$2x + y + r \leq 10,$$

the result (12) means that we can decrease the right-hand side by as much as 10/3 to

$$10 - \frac{10}{3} = \frac{20}{3}$$

and increase it by as much as 30 to

$$10 + 30 = 40$$

without changing the variables that are in the basis. Thus the range of right-hand values for painting time is from 20/3 to 40. In summary, then, the shadow price of an hour of painting time is $48 for available painting times in the range 20/3 to 40 hours.

The price we have found in the shadows, $48 per hour of painting time, is an important value because it shows that profit per day can be increased by $48 for each hour of increased painting time made available. It follows that it would be profitable to increase painting facilities if the cost per additional hour of daily painting time is less than $48.

> **Exercise.** Refer to the optimal tableau (5) of the foregoing example. *a)* What is the shadow price of an hour of drying time? *b)* What are the limits of the change, c, in the range of right-hand side values for the drying time constraint? *c)* What is the range of right-hand side values for the drying time constraint? Answer: *a)* $4 per hour of drying time. *b)* $-15 \leq c \leq 10$. *c)* From 5 to 30 hours.

We have determined shadow prices and right-hand value ranges for a "less than or equal max" problem. The rules that apply in the case of a "greater than or equal to min" are different because we solve such problems by changing to the dual and maximizing the dual. The differences to be noted are, first, instead of slack for resources we now have surplus for requirements. Second, the change to the dual results in a transposition of rows and columns. Third, as we learned earlier in reading the primal solution from the dual, sign changes occur. To show how the analysis of the optimum dual tableau is carried out, recall the nutrient example near the end of Section 5.14, which was
Minimize

$$\theta = 4x + 7y + 5z$$

subject to:

$$x, y, z \geq 0$$

$$
\begin{array}{ll}
4x + y \geq 10 & \text{(Nutrient R requirement, surplus is } r) \\
3x + 2y + z \geq 12 & \text{(Nutrient S requirement, surplus is } s) \\
4y + 5z \geq 20 & \text{(Nutrient T requirement, surplus is } t).
\end{array}
\tag{13}
$$

In this example, a diet has minimal requirements for quantities of nutrients R, S, and T, which are to be obtained from x pounds of food A, y pounds of B, and z pounds of C. The objective function represents the total cost of these foods. After changing to the dual and proceeding with pivoting operations, we obtained the following in (3) of Section 5.14.

Optimal dual tableau

	x	r	z	
θ	$\dfrac{92}{3}$	$-\dfrac{8}{3}$	$-\dfrac{2}{3}$	-4
s	$\dfrac{4}{3}$	$-\dfrac{1}{3}$	$-\dfrac{4}{3}$	0
y	$\dfrac{7}{5}$	$\dfrac{2}{5}$	$\dfrac{3}{5}$	$\dfrac{4}{5}$
t	$\dfrac{11}{15}$	$\dfrac{1}{15}$	$\dfrac{4}{15}$	$-\dfrac{1}{5}$

(14)

From (14), we obtain the solution to the primal as

$$\theta_{\min} = \frac{92}{3} \quad \text{at} \quad x = \frac{8}{3}, \; r = \frac{2}{3}, \; z = 4, \; s = y = t = 0.$$

Thus, there is a *surplus* of $r = 2/3$ ounces in the nutrient R requirement, but the nutrient S and nutrient T constraints are *binding*. Their shadow prices are the entries in the first *column*. Thus,

$$\text{Nutrient S:} \quad \text{Shadow price} = \frac{4}{3} \text{ cents per ounce}$$

$$\text{Nutrient T:} \quad \text{Shadow price} = \frac{11}{15} \text{ cents per ounce.}$$

The shadow price of an ounce of nutrient R is zero because there is a surplus of R; that is, the constraint is not binding.

Next we state the rule for determining the right-hand-side ranges of binding constraints, and the definition of shadow prices for minimization problems.

Greater than or equal min problem
To determine the effect of changing the right-hand side of a binding constraint by c in the optimal dual tableau, add c times

the surplus variable row entries to the θ row. The shadow price of a binding requirement is the value in the first column of the row for the associated surplus variable. The shadow price of a requirement whose surplus variable is in the basis is zero.

Applying the rule just stated to (14), we obtain the following for a change of c in the right-hand side of the nutrient S requirement (surplus is s).

	x	r	z
θ	$\dfrac{92}{3}+\dfrac{4}{3}c$ $\quad -\dfrac{8}{3}-\dfrac{1}{3}c$	$-\dfrac{2}{3}-\dfrac{4}{3}c$	$-4+0c$
s	$\dfrac{4}{3}$ $\quad -\dfrac{1}{3}$	$-\dfrac{4}{3}$	0
y	$\dfrac{7}{5}$ $\quad \dfrac{2}{5}$	$\dfrac{3}{5}$	$\dfrac{4}{5}$
t	$\dfrac{11}{15}$ $\quad \dfrac{1}{15}$	$\dfrac{4}{15}$	$-\dfrac{1}{5}$

$$(15)$$

Right-hand-side ranges from the dual. As (15) shows, the change in the right-hand side of a binding requirement constraint affects the θ row entries in the optimal dual tableaus. For this tableau to remain optimal, the entries under x, r, and z *must not be positive.* That is,

$$i_1: \quad -\frac{8}{3}-\frac{1}{3}c \le 0$$

$$i_2: \quad -\frac{2}{3}-\frac{4}{3}c \le 0$$

$$i_3: \quad -4+0c \le 0.$$

The last, i_3, always is true. However, we have

$$i_1: \quad -\frac{8}{3}-\frac{1}{3}c \le 0$$

$$8+\ \ c \ge 0$$

$$c \ge -8.$$

Similarly,

$$i_2: \quad -\frac{2}{3}-\frac{4}{3}c \le 0$$

$$2+4c \ge 0$$

$$4c \ge -2$$

$$c \ge -\frac{1}{2}.$$

Examining the last two consequences

$$c \geq -\frac{1}{2}; \quad c \geq -8$$

we see that $c \geq -8$ is redundant because if c is greater than minus one-half, it is automatically greater than minus eight. Hence, the nutrient S constraint remains binding if the requirement for S is *decreased* by any amount not exceeding $1/2$ ounce, or is *increased* by any amount whatsoever.

Recalling that the requirement for nutrient S in (13) is 12 ounces, the range for this requirement is

$$\left(12 - \frac{1}{2}\right) \text{ to any value,}$$

which we may write as

11.5 to ∞ ounces.

In summary, the shadow price of an ounce of S is $4/3$ cents and the right-hand side range for the nutrient S requirements is 11.5 to ∞ ounces, where ∞ means any positive amount whatever.

> **Exercise.** From (14): *a)* What is the shadow price of an ounce of nutrient *T*, which has a requirement of 20 ounces? *b)* What is the range of change, *c*, in the right-hand side of the nutrient T requirement? *c)* What is the range of right-hand-side values of the nutrient T requirement? Answer: *a)* $11/15$ cents per ounce. *b)* $-20 \leq c \leq 5/2$. *c)* 0 to 22.5 ounces.

The determinations of shadow prices and right-hand-side ranges *after* the optimal tableau has been obtained are examples of *post-optimality* analysis. This analysis can be more important than the optimal solution itself because it may indicate a way to improvement as, for example, when the shadow price of a constraining resource is high enough to make it profitable to incur the expense of increasing the available amount of that resource. In closing this section, we should mention that there are other post-optimality analyses that can be made to determine, for example, the effect of changes in the coefficients of the objective function. We must leave these and numerous other topics in linear programming for texts devoted entirely to the subject of mathematical programming.

5.17 CONCLUDING COMMENTS

In this chapter we have introduced three of the central concepts of linear programming—the simplex procedure, the dual theorem, and post-optimality analysis. Before moving on to other topics, it is worthwhile to take a broad look at linear programming problems as they occur and are solved in practice. The first characteristic of applied problems is that they are large; that is, they have numerous constraints involving many variables. For example, a problem with 20 constraints (not counting negativity) on 10 variables would be a relatively small applied problem. Now, if we append a slack (or surplus) variable to each constraint the number of variables is 10 plus 20, and the resultant system has 20 equations in 30 variables. Obviously, the sensible way to solve even this small problem is to construct a computer program that carries out the number crunching required in manipulation of the coefficients and constants to go from one tableau to the next. Such computer programs have been written, and many readers will find that their university computer center has one available on the computer all the time. The user needs only to supply the computer with the coefficients and constants of the constraints, the type of constraint ($\leq$, $=$, or $\geq$), and the coefficients of the objective function. The computer then will perform the calculations and start printing out the results very quickly—say in less than a minute for 20 constraints on 10 variables.

A second characteristic of applied problems is the possible occurrence of a *mix* of $\leq$, $\geq$, and $=$ constraints. We have applied the simplex method in this chapter to problems having all $\geq$, or all $\leq$ constraints. It is not difficult to change a problem with a mix into one with all $\leq$ or all $\geq$ constraints and adapt it to solution by the simplex method, although it would carry us too far into this specialized area to show here how this is done. The important fact is that a mix of constraints can be changed into a "less than or equal max" problem or a "greater than or equal min" problem and solved by the simplex procedure.

Next we note that in contrast to problems in this book, all of which have optimum solutions, an applied problem may be *infeasible;* that is, the problem does not have a solution. On the other hand a problem may turn out to have an *unbounded* solution as would be the case, for example, if the constraints did not limit the production of a product that contributes to profit. Finally, it may have occurred to the reader that in carrying out the pivoting procedure to get, say, the sixth tableau from the fifth, the sixth might turn out to be the same as the fifth, so continuation would bounce us back and forth endlessly between the fifth and sixth tableaus. This situation as well as infeasibility and unboundedness, can occur in applied problems. However, it turns out that the simplex procedure can be adapted to handle them and that the following statement can be shown to be true.

The simplex procedure can be adapted to solve any linear programming problem using a finite number of pivoting operations. It determines the optimum if one exists and identifies infeasibility or unboundedness.

It might be inferred from the last statement that any linear programming problem can be solved on a computer. However, we should point out that a given problem can be too large to solve on a particular computer, and that approximations arising when the results of computer calculations are rounded can have a significant effect; for example, a computer may indicate incorrectly that a problem is infeasible. Thus, while the computer is a powerful and indispensable machine for solving linear programming problems, a given computer may fail to solve a given problem correctly. The best safeguard against such a failure is to obtain the advice of a member of the computer staff who knows the characteristics of both the computer system and the linear programming "package" the system uses.

5.18 PROBLEM SET 5–6

1. A glass manufacturer makes x Exglasses, y Whyglasses, and z Zeeglasses, each glass made contributing to profit. Resources available are 80 manhours of moulding time (slack is r), 40 manhours of grinding time (slack is s), and 24 manhours of polishing time (slack is t). Suppose the optimal profit-maximizing tableau is that given below, where θ is dollars of profit.

		s	y	t
θ	1500	−30	−3	−50
x	100	−20	−2	−25
r	5	1	3	−5
z	80	−40	−5	20

a) How many of each type of glass should be made to maximize profit? What is the maximum profit?

b) Find the shadow price of any resource for which the constraint is binding and state the meaning of this price.

c) Find the right-hand-side ranges for the constraints in (b).

2. (All dollar values are in thousands of dollars.) A contractor plans to build x Superdeluxe, y Deluxe, and z Standard houses, each of which contributes to θ, the profit function. Resources available are the amounts of money a bank will loan for land purchase, building, and landscaping. Maximum amounts that will be loaned are $100 for land (slack is r), $500 for building (slack is s), and $40 for landscaping (slack is t). Suppose the optimal tableau is as follows:

		r	z	t
θ	125	$-\dfrac{1}{5}$	-2	$-\dfrac{1}{20}$
x	4	1	$-\dfrac{2}{3}$	-2
y	9	3	$\dfrac{1}{4}$	$-\dfrac{1}{3}$
s	5	-1	0	$\dfrac{1}{2}$

a)   How many of each type of house should be built to maximize profit? What is the maximum profit?

b)   Find the shadow price of any resource for which the constraint is binding and state the meaning of this price.

c)   Find the right-hand-side ranges for the constraints in (b).

3.   A producer has to make at least 20 A-Trans (surplus is r) and at least 18 B-Trans (surplus is s). Machines known as Exmills and Whymills make known numbers of A-Trans and B-Trans per hour. The Exmill is to be run for x hours and the Whymill for y hours at known costs in hundreds of dollars per hour. The objective is to minimize the cost of making at least the required number of A-Trans and B-Trans. The optimal dual tableau was found to be the following:

		x	y
θ	7	-3	-4
s	$\dfrac{1}{4}$	$\dfrac{1}{4}$	$-\dfrac{1}{2}$
r	$\dfrac{1}{8}$	$-\dfrac{3}{8}$	$\dfrac{1}{4}$

a)   How many hours should be scheduled for the Exmill and the Whymill? What is the minimum cost?

b)   Find the shadow price for any requirement for which the constraint is binding and state the meaning of this price.

c)   Find the right-hand-side ranges for the constraints in (b).

4.   A contractor rents X-Trucks and Y-Trucks to pick up and deliver gravel and crushed rock to a road construction site. Requirements are at least 18,000 tons of gravel (surplus is r) and at least 10,000 tons of crushed rock (surplus is s). The trucks can carry known amounts of gravel and crushed rock at a known cost, in dollars per day, for renting each type of truck. X-Trucks are to be rented for x days and Y-Trucks for y days. The goal is to mimimize pick up and delivery cost for at least the total requirements. The optimal dual tableau was found to be the following:

		x	y
θ	9200	-60	-20
s	$\dfrac{2}{5}$	$-\dfrac{1}{50}$	$\dfrac{1}{100}$
r	$\dfrac{1}{5}$	$\dfrac{3}{100}$	$-\dfrac{1}{50}$.

a)   For how many days should each truck be rented to minimize cost? What is the minimum cost?

b)   Find the shadow price for any requirement for which the constraint is binding and state the meaning of this price.

c)   Find the right-hand-side ranges for the constraints in (b).

5.   The table shows, for example, that making one Extran requires 1 minute for stamping, 3 minutes for forming and 1 minute for painting, and contributes $1 to profit.

		Minutes per unit for			
Product	Number of units	Stamping	Forming	Painting	Profit per unit
Extran x		1	3	1	$1
Whytran ....y		3	10	5	$4
Zeetran ....z		2	5	5	$2

Minutes available for stamping, forming, and painting are, respectively, 48, 150, and 70.

   a) How many units of each product should be made to maximize profit? What is the maximum profit?
   b) Find the shadow price of any resource for which the constraint is binding and state the meaning of this price.
   c) Find the right-hand-side range for constraints in (b).

6. A chemical firm has three processes, Exmix, Whymix, and Zeemix, for producing A-oil, B-oil, and C-oil. It has an order to supply 10 barrels of A-oil, 40 barrels of B-oil, and 20 barrels of C-oil. In one hour, Exmix can make 1 barrel of A-oil, 3 of B-oil, and 2 of C-oil. In one hour Whymix can make 3 barrels of A-oil, 10 of B-oil, 5 of C-oil. In one hour Zeemix can make 1 barrel of A-oil, 5 of B-oil, and 5 of C-oil. Cost, per hour to run the processes are: $96 for Exmix; $300 for Whymix; $140 for Zeemix.

   a) How long should each process be run to produce at least the amounts required by the order at minimum cost? What is the minimum cost?
   b) Find the shadow price of any requirement for which the constraint is binding and state the meaning of this price.
   c) Find the right-hand-side ranges for the constraints in (b).

5.19 REVIEW PROBLEMS

Solve by the simplex procedure.

1. Given $\theta = x + 6y + 10z$, find θ_{max} subject to $x, y, z \geq 0$ and

$$x + 5y + 10z \leq 150$$
$$3x + 20y + 25z \leq 500$$
$$x + 15y + 15z \leq 300.$$

2. Given $\theta = x + 6y + 3z$, find θ_{max} subject to $x, y, z \geq 0$ and

$$x + 5y + 4z \leq 39$$
$$2x + 13y + 5z \leq 90$$
$$x + 2y + 3z \leq 40.$$

3. Given $\theta = 2x + 3y + z$, find θ_{max} subject to $x, y, z \geq 0$ and

$$3x + 2y + z \leq 8$$
$$2x + 3y + z \leq 10$$
$$5x + 3y + 2z \leq 17.$$

4. Given $\theta = x + 4y + 7z$, find θ_{max} subject to $x, y, z \geq 0$ and

$$x + 3y + 7z \le 70$$
$$x + 5y + 5z \le 100$$
$$x + 10y + 8z \le 190.$$

5. Given $\theta = 3x + 2y + z$, find θ_{max} subject to x, y, $z \ge 0$ and

$$3x + y + z \le 35$$
$$2x + 10y + 3z \le 140$$
$$4x + 4y + z \le 50.$$

6. Given $\theta = x + 10y + 3z$, find θ_{max} subject to x, y, $z \ge 0$ and

$$x + 9y + 3z \le 240$$
$$3x + 35y + z \le 800$$
$$x + 12y + 5z \le 300.$$

7. Given $\theta = 2x + y + 2z$, find θ_{max} subject to x, y, $z \ge 0$ and

$$x + y + 2z \le 8$$
$$2x + y + z \le 10$$
$$3x + y + 3z \le 15.$$

8. Given $\theta = 3x + y + 7z$, find θ_{max} subject to x, y, $z \ge 0$ and

$$2x + y + 7z \le 21$$
$$17x + 5y + 28z \le 140$$
$$9x + y + 10z \le 66.$$

Solve Problems 9–14 by the dual procedure.

9. Given $\theta = 6x + 24y + 12z$, find θ_{min} subject to x, y, z, ≥ 0 and

$$x + 2y + z \ge 1$$
$$x + y + 3z \ge 2$$
$$x + 3y + z \ge 3.$$

10. Given $\theta = 2x + y + z$, find θ_{min} subject to x, y, $z \ge 0$ and

$$2x + 3y + 2z \ge 10$$
$$x + 4y + 3z \ge 20.$$

11. Given $\theta = 12x + 6y + 4z$, find θ_{min} subject to x, y, $z \ge 0$ and

$$2x + y + 2z \ge 1$$
$$x + 2y \ge 2$$
$$3x + z \ge 1.$$

12. Given $\theta = 8x + 10y + 15z$, find θ_{min} subject to x, y, $z \ge 0$ and

$$x + 2y + 3z \ge 2$$
$$x + y + z \ge 1$$
$$2x + y + 3z \ge 2.$$

13. Given $\theta = 2x + 3y + z$, find the *minimum* value of θ subject to x, y, $z \ge 0$ and

$$x + y + z \ge 10$$
$$2x + 3y + z \ge 15$$
$$3x + y + 2z \ge 20.$$

14. Given $\theta = 5x + 7y + 6z$, find θ_{min} subject to $x, y, z \geq 0$ and

$$\begin{aligned} x \quad &+ z \geq 3 \\ x + y \quad &\geq 2 \\ y + z &\geq 1. \end{aligned}$$

15. Quality X gasoline cost $0.40 per liter and has an octane rating of 70. The per-liter costs and octane ratings are $0.50 and 90 for Y, $0.60 and 100 for Z. We wish to buy at least 1000 liters and mix them, requiring that the mixture have an octane rating of at least 85. Find the minimum cost and the numbers of liters to be mixed to obtain this minimum.

16. A toy manufacturer makes x Extoys, y Whytoys, and z Zeetoys a day, each of which contributes to the profit function θ. Resources available are 100 hours of cutting time (slack is r), 40 hours of assembly time (slack is s), and 80 hours of painting time (slack is t). The table below gives the tableau for the solution of the dollar-profit maximum problem.

		r	s	z
θ	800	-60	-30	-5
x	200	-50	40	-2
t	10	1	1	1
y	300	-20	-50	3

   a) How many of each type of toy should be made to maximize profit? What is the maximum profit?
   b) Find the shadow price of any resource for which the constraint is binding and state the meaning of this price.
   c) Find the right-hand-side ranges for the constraints in (b).

17. A papermaking company has two processes, Express and Whypress, each of which can make bond paper and standard paper. The hourly cost, in dollars, of operating each process is known, and the hourly output of each type of paper for each process is known. The company has an order to supply 500 cases of bond paper (surplus is r) and 1000 cases of standard (surplus is s). The cost of meeting these requirements is to be minimized. Suppose the optimal dual tableau is as follows, where x and y represent hours of operation for the Express and the Whypress:

		x	y
θ	900	-3	-5
r	20	$\dfrac{1}{6}$	$-\dfrac{1}{2}$
s	12	$-\dfrac{1}{5}$	1

   a) For how many hours should each process be operated? What is the minimum cost?
   b) Find the shadow price for any requirement for which the constraint is binding and state the meaning of this price.
   c) Find the right-hand-side ranges for each of the constraints in (b).

Compact notation: Vectors, matrices, and summation

6

6.1 INTRODUCTION

Brevity in mathematical statements is achieved through the use of symbols. Thus the brief expression $247 \div 793$ takes the place of what would be a very lengthy statement if one chose to write a complete description of the steps involved in the long division. The price paid for brevity, of course, is the effort spent in learning the meaning of the symbol.

In this chapter, we shall learn the symbols for matrices and vectors and apply them in the statement and solution of input-output problems and other problems involving linear systems. Then we shall introduce the summation symbol and show its application in linear systems and in statistics.

6.2 MATRICES AND VECTORS

Numerical data arranged in a form that we shall come to call a matrix are very common in everyday life. For example, suppose that a company has six gasoline stations, three in region # 1 and three in region # 2. January sales volume, in thousands of gallons, is shown for each station in Table 6–1.

TABLE 6–1
Tiger Oil Company: Sales in thousands of gallons, January

Station	Region #1	Region #2
#1	10	15
#2	12	18
#3	8	12

We note that sales of Station #1 in Region #2 were 15 thousand gallons. The position of the *entry* (also called *element*) 15 is at the intersection of the first row and second column, and we may symbolize the entry as a_{12} (read "a sub one two")[1] where the first subscript refers to the row and the second to the column. To avoid having to specify which of a pair of subscripts specifies the row and which the column, we shall *always* use the row, column order convention.

Exercise. In Table 6–1, *a)* what are a_{11} and a_{32}? *b)* Write the symbol for the entry 8. Answer: *a)* 10 and 12. *b)* a_{31}.

If we keep in mind that rows are stations and columns are regions, we can omit the stub and caption of the table and write the *matrix:*

Sales matrix by station by region

$$\begin{pmatrix} 10 & 15 \\ 12 & 18 \\ 8 & 12 \end{pmatrix}.$$ [1]

Definition: A matrix is a rectangular array of numbers. Matrices are enclosed in grouping symbols such as parentheses or brackets.

If we use m by n to mean a matrix of m rows by n columns, m by n is called the *order* of the matrix or, more descriptively, the *shape* of the matrix. Thus, a 5 by 4 matrix has the shape of a rectangle 5 (rows) down by 4 (columns) across.

Definition: A 1 by n matrix, such as the 1 by 3 matrix (4 7 6), is called a *row vector,* and an n by 1 matrix, such as the 2 by 1 matrix

---

[1] A comma is required to avoid ambiguity if the number of rows or columns exceeds nine. For example, an entry in the 12th row 3rd column would be $a_{12,3}$.

$$\binom{5}{7},$$

is called a *column vector*. Entries in row and column vectors often are referred to as the *components* of the vectors. The matrix [1] may be thought of as consisting of two column vectors or three row vectors.

> *Exercise.* Write the first row vector in the matrix, [1]. Answer: (10   15).

Next, suppose Tiger Oil Company sells one grade of gasoline but, due to transportation costs, prices vary in the two regions as shown in Table 6–2.

TABLE 6–2
Tiger Oil Company: Regional price variation

Region	Price per gallon
#1	$1.10
#2	$1.20

We can represent the price matrix as a column or a row vector. Thus,

Price vector by region

$$\binom{1.10}{1.20} \qquad \text{or} \qquad (1.10 \quad 1.20). \qquad\qquad [2]$$

6.3 PRODUCT OF A NUMBER AND A MATRIX

In matrix algebra, an ordinary number is called a *scalar*. To multiply a matrix by a scalar, we multiply each entry in the matrix by the scalar. For example

$$12\begin{pmatrix} 1 & 3 \\ 4 & 2 \end{pmatrix} = \begin{pmatrix} 12 & 36 \\ 48 & 24 \end{pmatrix}.$$

To see the origin of the word scalar, think of the entries 1, 3, 4, 2 in the left matrix as being in feet. Scaling these to inches requires that each entry be multiplied by 12. Similarly, Tiger Oil Company may wish to set February sales quotas that are 10% higher than actual January sales. The quotas will then be 1.1 times the January sales matrix, [1]. Thus,

212

$$\begin{array}{cc}
[1] & [3] \\
\text{January Sales} & \text{February Quota}
\end{array}$$

$$1.1 \begin{pmatrix} 10 & 15 \\ 12 & 18 \\ 8 & 12 \end{pmatrix} = \begin{pmatrix} 11 & 16.5 \\ 13.2 & 19.8 \\ 8.8 & 13.2 \end{pmatrix}.$$

Exercise. Tiger Oil Company's price vector is (1.10 1.20) in dollars per gallon. What would be the scaling factor and the resultant price vector if prices are to be in dollars per 1000 gallons? Answer: The scaling factor is 1000. Hence, 1000 (1.10 1.20) = (1100 1200).

6.4 ADDITION AND SUBTRACTION OF MATRICES

Matrices are added or subtracted by adding or subtracting *corresponding* entries. Since entries must correspond, the matrices involved must have the same shape. We cannot compute

$$\begin{pmatrix} 3 & 1 & 5 \\ 2 & 4 & 7 \end{pmatrix} + \begin{pmatrix} 5 & 2 \\ -3 & 6 \end{pmatrix}$$

because the first is 2 by 3 and the second is 2 by 2. However,

$$\begin{pmatrix} 3 & 1 \\ 2 & 4 \end{pmatrix} + \begin{pmatrix} 5 & -6 \\ 2 & 0 \end{pmatrix} = \begin{pmatrix} 8 & -5 \\ 4 & 4 \end{pmatrix}$$

and

$$\begin{pmatrix} 5 & 8 \\ -1 & 3 \end{pmatrix} - 2\begin{pmatrix} 1 & 0 \\ 3 & -2 \end{pmatrix} = \begin{pmatrix} 3 & 8 \\ -7 & 7 \end{pmatrix}.$$

Exercise. Compute

$$3\begin{pmatrix} 4 & -1 \\ 2 & 5 \end{pmatrix} - 2\begin{pmatrix} 1 & 2 \\ -3 & 4 \end{pmatrix}. \quad \text{Answer:} \begin{pmatrix} 10 & -7 \\ 12 & 7 \end{pmatrix}.$$

Returning to the affairs of Tiger Oil Company, recall its February sales quota matrix, [3]. This is the middle matrix in the following. The left is the new matrix, [4], of *actual* February sales, and the difference is matrix [5] which shows the deviation of actual February sales from quota.

[4]   [3]   [5]

February Sales   February Quota   Deviation from Quota

$$\begin{pmatrix} 12 & 15.5 \\ 13 & 20 \\ 8.5 & 12.9 \end{pmatrix} - \begin{pmatrix} 11 & 16.5 \\ 13.2 & 19.8 \\ 8.8 & 13.2 \end{pmatrix} = \begin{pmatrix} 1 & -1 \\ -0.2 & 0.2 \\ -0.3 & -0.3 \end{pmatrix}.$$

From [5] we see, for example, that Station #1, Region #2 sales were 1 thousand gallons below quota.

> *Exercise.* In [5], interpret the entry a_{21}. Answer: Station #2, Region #1 sales were 0.2 thousand gallons below quota.

6.5 MULTIPLICATION OF MATRICES

Matrix multiplication is a specialized form of multiplication devised for very practical reasons. Consider the following:

$$(3 \quad 2)\begin{pmatrix} 5 \\ 4 \end{pmatrix} = 23.$$

We obtain the product, 23, as the sum of 3(5) + 2(4). To fix this in mind, we may place our left index finger on the 3 and our right index finger on the 5. Multiply to obtain 15. Then move the left finger *across* to the 2 and the right finger *down* to the 4, multiply to obtain 8, then add this to 15 to get 23. A product obtained in this manner is called the *inner product*. Observe that the inner product of a row vector by a column vector is an ordinary number, a scalar. As another example,

$$(2 \quad -3 \quad 1 \quad 0)\begin{pmatrix} 4 \\ -2 \\ 3 \\ -1 \end{pmatrix} = 8 + 6 + 3 + 0 = 17.$$

> *Exercise.* Evaluate
> $$(3 \quad -1)\begin{pmatrix} 4 \\ 2 \end{pmatrix}.$$ Answer: 10.

It is clear that if the "left finger across, right finger down" technique is to be defined there must be as many numbers across as there are down. That is, the first matrix must have as many columns as the second has rows. In our last example we had a 1 by 4 multiplied by a 4 by 1. If we indicate this as (1 by 4) (4 by 1), observe that the two middle numbers are the same, 4. When the shapes of two matrices are indicated in this manner and the middle numbers are the same, the two are said to be *conformable for multiplication*. Alternatively, we may say that two matrices are conformable for multiplication if the number of columns in the first is the same as the number of rows in the second. In the next example, we multiply a 2 by 2 and a 2 by 3.

$$\begin{pmatrix} 1 & 2 \\ 3 & 4 \end{pmatrix} \begin{pmatrix} 5 & 6 & 7 \\ 8 & 9 & 10 \end{pmatrix} = \begin{pmatrix} 21 & 24 & 27 \\ 47 & 54 & 61 \end{pmatrix}.$$

The calculations leading to the rightmost matrix are:

First row: $1(5) + 2(8) = 21$
$1(6) + 2(9) = 24$
$1(7) + 2(10) = 27.$

Second row: $3(5) + 4(8) = 47$
$3(6) + 4(9) = 54$
$3(7) + 4(10) = 61.$

Exercise. Compute the following product:

$$\begin{pmatrix} 1 & 2 & 3 \\ 4 & 5 & 6 \end{pmatrix} \begin{pmatrix} 0 & 1 \\ 2 & 3 \\ 4 & 5 \end{pmatrix}. \qquad \text{Answer: } \begin{pmatrix} 16 & 22 \\ 34 & 49 \end{pmatrix}.$$

As an example of the applied meaning of matrix multiplication, let us multiply the gallons sold matrix (entries in thousands of gallons) by the price per thousand gallons vector for the Tiger Oil Company.

Station	Region #1	#2		Region	Dollars per thousand gallons
#1	10	15		#1	1100
#2	12	18		#2	1200
#3	8	12			

The product is

Station	Total dollar volume of sales, both regions
#1	$10(1100) + 15(1200) = \$29,000$
#2	$12(1100) + 18(1200) = \$34,800$
#3	$8(1100) + 12(1200) = \$23,200.$

Thus, if Mr. Jones owns both #1 stations, his dollar volume of sales is 10 thousand gallons at $1100 per thousand in Region #1, plus 15 thousand gallons at $1200 per thousand in Region #2 for a total of $29,000 in both regions. Similarly, the owner of the #2 stations grosses $34,800 and the owner of the #3 stations grosses $23,200.

Matrix addition (subtraction) has the associative and commutative properties of ordinary algebra. Matrix multiplication has the associative property, but *the commutative property does not hold in general for matrix multiplication.* For example,

$$\begin{pmatrix} 1 & 2 \\ 3 & 4 \end{pmatrix} \begin{pmatrix} 5 & 6 \\ 7 & 8 \end{pmatrix} = \begin{pmatrix} 19 & 22 \\ 43 & 50 \end{pmatrix},$$

but if we interchange the factor matrices we find

$$\begin{pmatrix} 5 & 6 \\ 7 & 8 \end{pmatrix} \begin{pmatrix} 1 & 2 \\ 3 & 4 \end{pmatrix} = \begin{pmatrix} 23 & 34 \\ 31 & 46 \end{pmatrix}.$$

Thus, if we *premultiply* by

$$\begin{pmatrix} 1 & 2 \\ 3 & 4 \end{pmatrix}$$

we get one result, but if we *postmultiply* we get a different result.

Exercise. What do we obtain if we

a) premultiply $\begin{pmatrix} 1 \\ 2 \end{pmatrix}$ by $(3 \quad 4)$,

b) if we postmultiply $\begin{pmatrix} 1 \\ 2 \end{pmatrix}$ by $(3 \quad 4)$?

Answer: a) 11.   b) $\begin{pmatrix} 3 & 4 \\ 6 & 8 \end{pmatrix}$.

We *can* design matrices that commute with each other in multiplication. For example,

$$\begin{pmatrix} 1 & 0 \\ 0 & 1 \end{pmatrix} \begin{pmatrix} 2 & 3 \\ 4 & 5 \end{pmatrix} = \begin{pmatrix} 2 & 3 \\ 4 & 5 \end{pmatrix} \begin{pmatrix} 1 & 0 \\ 0 & 1 \end{pmatrix} = \begin{pmatrix} 2 & 3 \\ 4 & 5 \end{pmatrix}.$$

However, as we have seen, the commutative property does not hold in general.

We repeat the suggestion that the reader keep in mind that an m by n matrix can be postmultiplied by an n by p matrix (they are conformable), and the result will be an *m* by *p* matrix. Thus, we can multiply a 5 by 8 and an 8 by 11, and the product matrix will be 5 by 11.

Exercise. *a)* What will be the shape of the product matrix if a 5 by 7 is postmultiplied by a 7 by 6? *b)* What happens if we premultiply the 5 by 7 by the 7 by 6? Answer: *a)* 5 by 6. *b)* The matrices are not conformable for multiplication.

6.6 UNIT (IDENTITY) MATRIX

A *unit* or *identity* matrix is one that has the same number of rows and columns, with 1 as the element in each position on the main diagonal (upper left to lower right) and 0 as the element in all other positions. Thus,

$$I = \begin{pmatrix} 1 & 0 & 0 \\ 0 & 1 & 0 \\ 0 & 0 & 1 \end{pmatrix}$$

is a 3 by 3 unit matrix. Inasmuch as all unit matrices have the same number of rows and columns, we may describe the last also by calling it the unit matrix of order 3.

The unit matrix of order 2 is

$$I = \begin{pmatrix} 1 & 0 \\ 0 & 1 \end{pmatrix}.$$

The most important property of the unit matrix is illustrated by the statements

$$AI = A \quad \text{and} \quad IA = A.$$

That is, the product of any given matrix and the unit matrix is the given matrix itself. In other words, the unit matrix behaves like the number 1 in ordinary arithmetic where we say

$$(a)(1) = (1)(a) = a.$$

The reader may verify the unity property of *I* by carrying out the following multiplication in which the first written matrix, *I*, when multiplied by the second matrix, yields the second matrix as the product.

$$\begin{pmatrix} 1 & 0 & 0 \\ 0 & 1 & 0 \\ 0 & 0 & 1 \end{pmatrix} \begin{pmatrix} 2 & 3 & 4 \\ -1 & -2 & 0 \\ 5 & 2 & -3 \end{pmatrix} = \begin{pmatrix} 2 & 3 & 4 \\ -1 & -2 & 0 \\ 5 & 2 & -3 \end{pmatrix}.$$

Exercise. Given

$$A = \begin{pmatrix} a & b \\ c & d \end{pmatrix}$$

verify that $AI = A$ and that $IA = A$.

6.7 PROBLEM SET 6–1

Perform the following operations:

1. $(2\ 3\ 4) + (1\ -2\ 3)$.

2. $\begin{pmatrix} 1 \\ -3 \end{pmatrix} + \begin{pmatrix} -3 \\ 5 \end{pmatrix}$.

3. $\begin{pmatrix} 2 \\ 4 \end{pmatrix} + \begin{pmatrix} 5 \\ 7 \end{pmatrix} + \begin{pmatrix} 3 \\ 2 \end{pmatrix}$.

4. $(6\ -1\ 2) - (3\ -2\ 4) + (5\ -1\ 6)$.

5. $3\begin{pmatrix} 7 \\ 2 \end{pmatrix}$.

6. $-2(5\ -9\ 3)$.

7. $(2\ 7)\begin{pmatrix} 3 \\ 5 \end{pmatrix}$.

8. $(1\ -3\ 6\ 2)\begin{pmatrix} 2 \\ 1 \\ 2 \\ -3 \end{pmatrix}$.

9. $\begin{pmatrix} 3 & 1 & 2 \\ 1 & 4 & 1 \end{pmatrix} + \begin{pmatrix} 1 & -5 & -2 \\ -3 & 2 & 4 \end{pmatrix}$.

10. $\begin{pmatrix} 4 & 7 \\ 1 & 3 \end{pmatrix} + \begin{pmatrix} 1 & -2 \\ 6 & -3 \end{pmatrix} - \begin{pmatrix} 2 & -4 \\ 5 & 7 \end{pmatrix}$.

11. $4\begin{pmatrix} 1 & -3 & 2 \\ 5 & 1 & -3 \end{pmatrix} - 3\begin{pmatrix} 2 & 5 & -3 \\ 1 & 2 & -1 \end{pmatrix}$.

12. $\begin{pmatrix} 1 & 2 \\ 3 & 5 \end{pmatrix}\begin{pmatrix} 3 & 0 & 5 \\ 1 & -2 & 0 \end{pmatrix}$.

13. $(5\ 6\ 7)\begin{pmatrix} 1 & 2 \\ 0 & 3 \\ 3 & 1 \end{pmatrix}$.

14. $\begin{pmatrix} 3 & 0 \\ -5 & 4 \end{pmatrix}\begin{pmatrix} 1 & 6 \\ 2 & 0 \end{pmatrix}$.

15. $\begin{pmatrix} 2 & 1 & 1 & 0 \\ 1 & 3 & 0 & 2 \\ -1 & -2 & 1 & 4 \end{pmatrix}\begin{pmatrix} 5 & 6 \\ 1 & 1 \\ 2 & 3 \\ 0 & -1 \end{pmatrix}$.

16. $\begin{pmatrix} 1 & -1 & 2 \\ 2 & 0 & 3 \end{pmatrix}\begin{pmatrix} 4 & 0 & 1 & 2 \\ -1 & 3 & 0 & 2 \\ 1 & 1 & 2 & 3 \end{pmatrix}$.

17. $\begin{pmatrix} 1 & 0 & 0 \\ 0 & 1 & 0 \\ 0 & 0 & 1 \end{pmatrix}\begin{pmatrix} 1 & 4 \\ 2 & 5 \\ 3 & 6 \end{pmatrix}$.

18. $\begin{pmatrix} 1 & 2 & 3 \\ 3 & 2 & 1 \end{pmatrix}\begin{pmatrix} 1 & 0 & 0 \\ 0 & 1 & 0 \\ 0 & 0 & 1 \end{pmatrix}$.

19. Interest at the rates 0.06, 0.07, and 0.08 is earned on respective investments of $3000, $2000, and $4000. *a)* Express the total amount of interest earned as the product of a row vector by a column vector. *b)* Compute the total interest by matrix multiplication.

20. Two canned meat spreads, Regular and Superior, are made by grinding beef, pork, and lamb together. The numbers of pounds of each meat in a 15 pound batch of each brand are as follows:

	Pounds of		
Brand	*Beef*	*Pork*	*Lamb*
Superior	8	2	5
Regular	4	8	3

a) Suppose we wish to make 10 batches of Superior and 20 of Regular. Multiply the meat matrix in the table and the batch vector (10 20) and interpret the result.

b) Suppose that the per pound prices of beef, pork, and lamb are $2.50, $2.00, and $3.00, respectively. Multiply the price vector and the meat matrix and interpret the results.

6.8 MATRIX SYMBOLS

We have used a bold face capital letter for a matrix. Thus, A is a matrix. Readers may use a wavy underscore, $\underset{\sim}{A}$, to designate a matrix. Similarly, bold face small letters, such as b, will be used to designate a vector. Subscripted small letters will be used to symbolize the entries in a matrix. Thus, a_{23} is the entry at the intersection of the 2nd row and 3rd column. The 3 by 4 matrix A will then mean

$$A = \begin{pmatrix} a_{11} & a_{12} & a_{13} & a_{14} \\ a_{21} & a_{22} & a_{23} & a_{24} \\ a_{31} & a_{32} & a_{33} & a_{34} \end{pmatrix}.$$

We shall refer to A as the *compact* form and to the righthand expression as the *expanded matrix form.*

Exercise. Write the expanded matrix form of the 2 by 3 matrix B. Answer:

$$B = \begin{pmatrix} b_{11} & b_{12} & b_{13} \\ b_{21} & b_{22} & b_{23} \end{pmatrix}.$$

The expanded form of the 1 by 4 vector c is

$$c = (c_1 \quad c_2 \quad c_3 \quad c_4)$$

and similarly for a column vector.

6.9 LINEAR EQUATIONS IN MATRIX FORM

To secure the advantage of matrix representation of equations, we shall use the same letter with varying subscripts to designate different variables. That is, instead of using x, y, and z as three variables, we shall use x_1, x_2, and x_3. Consider the linear equation

$$3x_1 + 2x_2 = 7$$

and observe that the technique of matrix multiplication makes it possible to express the equation as

$$(3 \quad 2)\begin{pmatrix} x_1 \\ x_2 \end{pmatrix} = 7.$$

Exercise. Write the linear equation specified by

$$(5 \quad 1 \quad 4)\begin{pmatrix} x_1 \\ x_2 \\ x_3 \end{pmatrix} = 10.$$

Answer: $5x_1 + x_2 + 4x_3 = 10$.

Similarly, the equation

$$a_1 x_1 + a_2 x_2 + a_3 x_3 = b$$

is

$$(a_1 \quad a_2 \quad a_3)\begin{pmatrix} x_1 \\ x_2 \\ x_3 \end{pmatrix} = b$$

and the general linear equation in n variables,

$$a_1 x_1 + a_2 x_2 + a_3 x_3 + \cdots + a_n x_n = b$$

becomes

$$(a_1 \quad a_2 \quad a_3 \cdots a_n)\begin{pmatrix} x_1 \\ x_2 \\ x_3 \\ \vdots \\ x_n \end{pmatrix} = b.$$

We can now see the advantage of compact form. Namely, all the information in the expanded vector expression just written can be summarized by stating simply that the equation is

$$ax = b$$

and that a is 1 by n. *We do not have to state the shape of x because conformability requires that this vector be n by 1.*

Exercise. If p is 1 by 2 and we have $py = q$, *a)* what is the shape of y? *b)* Write the expanded vector form of the equation. *c)* Write the usual algebraic form. Answer: *a)* y must be 2 by 1.

b) $(p_1 \ p_2) \begin{pmatrix} y_1 \\ y_2 \end{pmatrix} = q.$ *c)* $p_1 y_1 + p_2 y_2 = q.$

Exercise. If a customer buys u_1 units of product #1, u_2 units of #2, and so on to u_g units of product #g at unit prices of, respectively, p_1, p_2, . . . , p_g, write the expression for the total cost, t, in usual algebraic form and in compact form. What are the shapes of p and u? Answer: $p_1 u_1 + p_2 u_2 + \cdots + p_g u_g = t$; $pu = t$; p is 1 by g, and u is g by 1.

It is now easy to express any system of linear equalities or inequalities in matrix form. For example,

$$3x_1 + 2x_2 + x_3 = 5$$
$$2x_1 + \ x_2 - x_3 = 4$$

becomes

$$\begin{pmatrix} 3 & 2 & 1 \\ 2 & 1 & -1 \end{pmatrix} \begin{pmatrix} x_1 \\ x_2 \\ x_3 \end{pmatrix} = \begin{pmatrix} 5 \\ 4 \end{pmatrix}.$$

The general 2 by 3 system (2 equations, 3 variables) is

$$a_{11} x_1 + a_{12} x_2 + a_{13} x_3 = b_1$$
$$a_{21} x_1 + a_{22} x_2 + a_{23} x_3 = b_2,$$

where now we need double subscripts to specify the row and column position of a coefficient. The foregoing system can be described in compact form as

$$Ax = b, \text{ where } A \text{ is 2 by 3.}$$

Note again that if A is 2 by 3, the vector x must be 3 by 1 and the product vector, b, is 2 by 1.

> **Exercise.** If $cy \leq d$ represents a 3 by 2 system of $\leq$ inequalities, write *a)* the expanded matrix form and *b)* the usual algebraic form of the system.
>
> Answer: *a)* $\begin{pmatrix} c_{11} & c_{12} \\ c_{21} & c_{22} \\ c_{31} & c_{32} \end{pmatrix} \begin{pmatrix} y_1 \\ y_2 \end{pmatrix} \leq \begin{pmatrix} d_1 \\ d_2 \\ d_3 \end{pmatrix}.$ *b)* $\begin{aligned} c_{11}\, y_1 + c_{12}\, y_2 &\leq d_1 \\ c_{21}\, y_1 + c_{22}\, y_2 &\leq d_2 \\ c_{31}\, y_1 + c_{32}\, y_2 &\leq d_3. \end{aligned}$

The general m by n system,

$$
\begin{aligned}
a_{11} x_1 + a_{12} x_2 + a_{13} x_3 + \cdots + a_{1n} x_n &= b_1 \\
a_{21} x_1 + a_{22} x_2 + a_{23} x_3 + \cdots + a_{2n} x_n &= b_2 \\
&\ \vdots \\
a_{m1} x_1 + a_{m2} x_2 + a_{m3} x_3 + \cdots + a_{mn} x_n &= b_m
\end{aligned}
$$

can be specified simply as

$$Ax = b \qquad \text{where } A \text{ is } m \text{ by } n.$$

Checking the shapes again, we see that if A is m by n, the vector x is n by 1, and the product vector b is m by 1.

> **Exercise.** In the p by q system $Cy = d$, what are the shapes of C, y, and d? Answer: C is p by q, y is q by 1, and d is p by 1.

Finally, let us state the linear programming problem of the last chapter in compact form. We have m constraints on n variables, an m by n system. Further, all n variables must be greater than or equal to zero. Using 0 to mean a vector or matrix whose entries are all zeros, the statement $x \geq 0$ gives the nonnegativity requirement. The objective function, $\theta = c_1 x_1 + c_2 x_2 + \cdots + c_n x_n$ can be written compactly as cx. The objective is to maximize cx subject to the m by n system of constraints $Ax \leq b$. The problem may be stated as
Maximize

$$cx$$

subject to:

$$Ax \leq b \quad \text{(where } A \text{ is } m \text{ by } n\text{)}$$

$$x \geq 0.$$

Analyzing the shapes of the matrices, we reason that A is m by n, so x is n by 1, and b is m by 1. Since in cx the vector c premultiplies x, c is 1 by n.

Exercise. Given that the components of c are 1, 3, and 2; the components of b are 10, 20, and 30; x is an appropriate vector, and

$$A = \begin{pmatrix} 5 & 3 & 4 \\ 2 & 4 & 2 \\ 1 & 1 & 5 \end{pmatrix}.$$

Write the linear programming problem
Maximize

$$cx$$

subject to:

$$Ax \leq b$$
$$x \geq 0$$

first in expanded vector form, then in ordinary algebraic form.

Answer: The expanded vector form is
Maximize

$$(1 \quad 3 \quad 2) \begin{pmatrix} x_1 \\ x_2 \\ x_3 \end{pmatrix}$$

subject to:

$$\begin{pmatrix} 5 & 3 & 4 \\ 2 & 4 & 2 \\ 1 & 1 & 5 \end{pmatrix} \begin{pmatrix} x_1 \\ x_2 \\ x_3 \end{pmatrix} \leq \begin{pmatrix} 10 \\ 20 \\ 30 \end{pmatrix}$$

$$\begin{pmatrix} x_1 \\ x_2 \\ x_3 \end{pmatrix} \geq \begin{pmatrix} 0 \\ 0 \\ 0 \end{pmatrix}.$$

The ordinary algebraic form is
Maximize

$$x_1 + 3x_2 + 2x_3$$

subject to:

$$5x_1 + 3x_2 + 4x_3 \leq 10$$
$$2x_1 + 4x_2 + 2x_3 \leq 20$$
$$x_1 + x_2 + 5x_3 \leq 30$$
$$x_1 \geq 0$$
$$x_2 \geq 0$$
$$x_3 \geq 0.$$

It is often useful in the solution of linear programming problems to convert an inequality constraint into an equality by adding a new nonnegative variable (called a *slack* variable and designated by s) to the constraint. For example, $2x_1 + 3x_2 \leq 15$ might be converted to $2x_1 + 3x_2 + s_1 = 15$ by addition of the slack variable s_1, where, of course, $s_1 \geq 0$. Consider the following linear programming problem, which contains two constraints (other than nonnegativity) and, therefore, employs two slack variables to convert the constraints into equalities.

Maximize

$$3x_1 + 2x_2$$

subject to:

$$3x_1 + 4x_2 + s_1 = 20$$
$$x_1 + 2x_2 + s_2 = 12$$
$$x_1 \geq 0$$
$$x_2 \geq 0$$
$$s_1 \geq 0$$
$$s_2 \geq 0.$$

The reader may verify that the problem at hand can be written in expanded matrix form as:

Maximize

$$(3 \quad 2)\begin{pmatrix} x_1 \\ x_2 \end{pmatrix}$$

subject to:

$$\begin{pmatrix} 3 & 4 \\ 1 & 2 \end{pmatrix}\begin{pmatrix} x_1 \\ x_2 \end{pmatrix} + \begin{pmatrix} s_1 \\ s_2 \end{pmatrix} = \begin{pmatrix} 20 \\ 12 \end{pmatrix}$$

$$\begin{pmatrix} x_1 \\ x_2 \end{pmatrix} \geq \begin{pmatrix} 0 \\ 0 \end{pmatrix}$$

$$\begin{pmatrix} s_1 \\ s_2 \end{pmatrix} \geq \begin{pmatrix} 0 \\ 0 \end{pmatrix}.$$

In compact matrix notation the last is of the form:
Maximize

$$cx$$

subject to:

$$Ax + s = b$$
$$x \geq 0$$
$$s \geq 0.$$

The transpose of a matrix. The transpose of a matrix is obtained by writing its rows as columns. For example, if we take

$$A = \begin{pmatrix} 1 & 2 \\ 3 & 4 \end{pmatrix}$$

and write its rows as columns, we have

$$A^t = \begin{pmatrix} 1 & 3 \\ 2 & 4 \end{pmatrix},$$

where A^t is read "A transpose". Similarly, the transpose of a column (row) vector is a row (column) vector. Thus,

$$\begin{pmatrix} 1 \\ 2 \\ 3 \end{pmatrix}^t = (1 \quad 2 \quad 3).$$

The transpose is sometimes used in writing linear systems or linear programming problems *to make all vectors in a statement be column vectors.* We did not do this in the foregoing. For example, the statement

$$\text{maximize } 3x_1 + 2x_2,$$

when expressed in vector form is

$$\text{maximize } (3 \quad 2) \begin{pmatrix} x_1 \\ x_2 \end{pmatrix},$$

where $(3, 2)$ is a row vector. If we want all vectors to be column vectors, the statement would be

$$\text{maximize } \begin{pmatrix} 3 \\ 2 \end{pmatrix}^t \begin{pmatrix} x_1 \\ x_2 \end{pmatrix}.$$

This point need not concern us here but is mentioned because readers may encounter it in other writings. We shall, however, have a more important use of the transpose in a later section of this chapter.

6.10 PROBLEM SET 6–2

Write the following in expanded matrix form:

1. $2x_1 + 3x_2 = 5$
   $x_1 + 2x_2 = 3.$

2. $x_1 + 2x_2 + y_1 = 10$
   $2x_1 + 3x_2 + y_2 = 12.$

3. $3x_1 + x_3 - x_4 \leq 5$
   $2x_1 + x_2 - 5x_4 \leq 10$
   $x_2 + 3x_3 + x_4 \leq 8.$

Write the following in usual algebraic form:

4. $\begin{pmatrix} 3 & 1 & 2 \\ 1 & 4 & 1 \end{pmatrix} \begin{pmatrix} x_1 \\ x_2 \\ x_3 \end{pmatrix} = \begin{pmatrix} 5 \\ 4 \end{pmatrix}.$

5. $\begin{pmatrix} 2 & 3 \\ 4 & 6 \\ 1 & 7 \end{pmatrix} \begin{pmatrix} x_1 \\ x_2 \end{pmatrix} = \begin{pmatrix} 5 \\ 10 \\ 6 \end{pmatrix}.$

6. $(x_1 \quad x_2) \begin{pmatrix} 3 & 5 \\ 1 & 4 \end{pmatrix} = \begin{pmatrix} 2 \\ 6 \end{pmatrix}.$

7. $\begin{pmatrix} p_{11} & p_{12} & p_{13} & p_{14} \\ p_{21} & p_{22} & p_{23} & p_{24} \\ p_{31} & p_{32} & p_{33} & p_{34} \end{pmatrix} x = q$, if x and q are appropriate vectors.

8. $\begin{pmatrix} 2 & 1 & 5 \\ 4 & 6 & 2 \end{pmatrix} \begin{pmatrix} x_1 \\ x_2 \\ x_3 \end{pmatrix} + \begin{pmatrix} y_1 \\ y_2 \end{pmatrix} = \begin{pmatrix} 10 \\ 5 \end{pmatrix}.$

9. Maximize $(3 \quad 2 \quad 4) \begin{pmatrix} x_1 \\ x_2 \\ x_3 \end{pmatrix}$ subject to $\begin{pmatrix} 2 & 1 & 4 \\ 3 & 5 & 2 \end{pmatrix} \begin{pmatrix} x_1 \\ x_2 \\ x_3 \end{pmatrix} \leq \begin{pmatrix} 10 \\ 15 \end{pmatrix}$

   and $x \geq 0.$

10. Maximize cx
    subject to: $Ax + s = b$
    $$x \geq 0$$
    $$s \geq 0,$$

    where

    $$b = \begin{pmatrix} 12 \\ 15 \\ 10 \\ 20 \end{pmatrix}, c = (2 \ 5 \ 4 \ 3), A = \begin{pmatrix} 1 & 3 & 2 & 0 \\ 0 & 1 & 3 & 2 \\ 5 & 3 & 0 & 2 \\ 1 & 4 & 6 & 0 \end{pmatrix},$$

    and s and x are appropriate vectors.

11. The coefficient matrix of a system of equalities, A, is p by q. The variables,

y, and the constants, g, are vectors with the proper number of components.

a) What are the proper shapes for y and for g?

b) Write the system in compact matrix notation.

6.11 APPLICATIONS–1: MARKOV CHAINS

To set the stage for this section, suppose that a restaurant chain notes that 30 percent of the dinners it sells each week are beef dinners, and 70 percent are other dinners. The chain manager has a special arrangement for volume buying of beef at relatively low prices, and would like to raise the proportion of beef dinners sold. He carries out a promotional campaign to increase beef sales and collects the information shown in Table 6–3.

TABLE 6–3
Transition Proportions,
one week to the next week

One week	Next week	
	Beef	Other
Beef	0.8	0.2
Other	0.6	0.4

The matrix of proportions is called the *transition matrix*. The number 0.8 means that 80 percent of those buying beef dinners one week buy beef dinners again the next week. Similarly, 20 percent of those buying beef one week buy other dinners the next week.

> **Exercise.** Interpret the second row of Table 6–3. Answer: 60 percent of those buying other dinners one week change to beef dinners the next week, and the remaining 40 percent buy other dinners again the next week.

The state of affairs at the beginning of the section was 30 percent beef and 70 percent other. We shall call the vector (0.3 0.7) the *state* vector. Writing the state vector to the left of the transition matrix, we have

$$
(0.3 \text{ beef} \quad 0.7 \text{ other}) \quad \begin{array}{c} \text{Beef} \\ \text{Other} \end{array} \begin{pmatrix} \overset{\text{Beef}}{0.8} & \overset{\text{Other}}{0.2} \\ 0.6 & 0.4 \end{pmatrix}.
$$

If we wish to find what proportion will buy beef after a one-week transition, we note that 80 percent of the 30 percent who bought beef one week will buy it the next and an additional 60 percent of the 70 percent who bought other dinners one week will buy beef the next week. The sum is $0.3(0.8) + 0.7(0.6) = 0.66$, and we found this by the usual inner product matrix multiplication procedure. In the same manner $0.3(0.2) + 0.7(0.4) = 0.34$ is the proportion buying other dinners next week, and the new state vector is (0.66 beef 0.34 other). In brief,

$$(0.3 \quad 0.7) \begin{pmatrix} 0.8 & 0.2 \\ 0.6 & 0.4 \end{pmatrix} = (0.66 \quad 0.34).$$

Now suppose the promotion activity is maintained and the transition matrix remains constant from week to week. Then the state vector for week #1, (0.66 0.34) can be used as a premultiplier of the transition matrix to obtain the state vector for week #2.

> *Exercise.* What proportions will buy beef and other in week #2?
>
> Answer: $(0.66 \quad 0.34) \begin{pmatrix} 0.8 & 0.2 \\ 0.6 & 0.4 \end{pmatrix} = (0.732 \quad 0.268),$
>
> which is 73.2 percent beef and 26.8 percent other dinners.

If we continue the chain, we may examine the successive state vectors as shown in Table 6–4.

TABLE 6–4
Successive state vectors

Week	Beef	Other
Beginning	(0.30	0.70)
#1	(0.66	0.34)
#2	(0.732	0.268)
#3	(0.7464	0.2536)
#4	(0.74928	0.25072)
#5	(0.749856	0.250144)

Observe that the components of each state vector sum to 1, as must be the case because they are proportions of a whole. Also, for the same reason, the rows of any transition matrix must each sum to 1. Note also that the state vector appears to be approaching (0.75 0.25).

What would happen if we used this state vector as a multiplier of the transition matrix? We would obtain

$$(0.75 \ 0.25) \begin{pmatrix} 0.8 & 0.2 \\ 0.6 & 0.4 \end{pmatrix} = (0.75 \ 0.25).$$

We see that the transition of (0.75 0.25) leads to the same state, (0.75 0.25), and we call this vector the *steady state*. The actual calculation of successive state vectors will never yield exactly (0.75 0.25), so it is a matter of importance to learn how to find the steady state by a method other than tabulating successive state vectors and guessing the steady state from the sequence of results. In our problem, let us call the steady state $(v_1 \ v_2)$. Then it must be true that

$$(v_1 \ v_2) \begin{pmatrix} 0.8 & 0.2 \\ 0.6 & 0.4 \end{pmatrix} = (v_1 \ v_2),$$

where, of course, $v_1 + v_2 = 1$. The matrix multiplication yields

$$
\begin{array}{lll}
e_1: 0.8v_1 + 0.6v_2 = v_1 & \quad & -0.2v_1 + 0.6v_2 = 0 \\
e_2: 0.2v_1 + 0.4v_2 = v_2 & \text{or} & \ \ 0.2v_1 - 0.6v_2 = 0.
\end{array}
$$

Clearly, e_1 and e_2 are the same equation, so we need only one of them. Taking e_1 with $v_1 + v_2 = 1$ we have

$$
\begin{array}{lll}
e_1: & -0.2v_1 + 0.6v_2 = 0 & \\
e_3: & v_1 + \quad v_2 = 1 & \\
e_4: & \quad 0.8v_2 = 0.2 & 0.2e_3 + e_1.
\end{array}
$$

From the last statement, $v_2 = 0.25$ and from e_3, v_1 is 0.75, so we have the steady state vector (0.75 0.25) that we obtained by guessing from the sequence in Table 6–4.

Exercise: Find the steady state for the transition matrix

$$\begin{pmatrix} 0.5 & 0.5 \\ 1 & 0 \end{pmatrix}.$$

Answer: (2/3 1/3).

Next, let us suppose that once a customer buys beef he is so satisfied that he will buy beef the next time. This results in a 1 in the transition matrix. Thus,

$$
\begin{array}{cc}
& \text{Beef} \quad \text{Other} \\
(0.3 \text{ beef} \quad 0.7 \text{ other}) \quad \begin{array}{c} \text{Beef} \\ \text{Other} \end{array} & \begin{pmatrix} 1.0 & 0.0 \\ 0.6 & 0.4 \end{pmatrix}.
\end{array}
$$

Inasmuch as 60 percent of those buying other meals change to beef, then continue to buy beef, it is reasonable to expect that ultimately all customers will buy beef and the steady state will be (1 0). In this situation, beef has absorbed all the business, and the chain leading to this steady state is called an *absorbing Markov chain.*

Exercise. Verify that the steady state for the last written beef-other matrix is (1 0).

We have used 2 by 2 transition matrices to illustrate this section. Clearly, the methodology can be applied to n by n matrices. All that would change is the complexity of the calculations, so we shall not pursue this topic further. We should mention that Markov was a mathematician whose name has been given to processes of the type discussed in this section; namely, processes in which the future state is completely determined by the present state and not at all by the way in which the present state arose. Finally, we should call attention to the fact that the transition matrices in our discussion have been constant as we changed from state to state, and this fact can be made explicit by referring to the processes as *stationary* Markov processes.

6.12 PROBLEM SET 6–3

1. Eager and Beaver Companies each have 50 percent of the market for a product. Because of a promotion campaign, buyers are switching between Eager and Beaver according to the following transition matrix.

$$\begin{array}{cc} & \begin{array}{cc} \text{Eager} & \text{Beaver} \end{array} \\ \begin{array}{c} \text{Eager} \\ \text{Beaver} \end{array} & \begin{pmatrix} 0.6 & 0.4 \\ 0.5 & 0.5 \end{pmatrix}. \end{array}$$

   a) What do the numbers 0.6 and 0.4 mean?
   b) What will be the market shares after the first and second transitions?
   c) What are the steady state market shares?

2. At a point in time 95 percent of the population were spenders of copper pennies and 5 percent were savers. Because of the increasing value of pennies, only 30 percent of the spenders remain spenders, and 10 percent of the savers become spenders.
   a) What will be the (spender saver) state vector after one transition?
   b) What is the steady state vector?

3. At a point in time, 1 percent of the population use a drug and 99 percent do not. In a year, $\frac{1}{10}$ of one percent of nonusers become users, but all users remain users.
   a) What will be the percent of users and nonusers after one transition?
   b) What is the steady state?

4. Carry out the multiplication and interpret the result.

$$
\begin{array}{c}
\text{State} \\
\begin{array}{cc} \#1 & \#2 \\ (a & b) \end{array}
\end{array}
\begin{array}{c}
\begin{array}{cc} \#1 & \#2 \end{array} \\
\begin{array}{c} \#1 \\ \#2 \end{array}
\begin{pmatrix} 1 & 0 \\ 0 & 1 \end{pmatrix}.
\end{array}
$$

5. The following chain is cyclical, meaning that it returns periodically to the same state. Write the successive state vectors until the initial one, (0.6 0.3 0.1) reappears.

$$
(0.6\ \ 0.3\ \ 0.1)
\begin{pmatrix}
0 & 0 & 1 \\
1 & 0 & 0 \\
0 & 1 & 0
\end{pmatrix}.
$$

6.13 SQUARE MATRICES

An n by n matrix has the same number of rows and columns and is called a square matrix. We may specify a square matrix simply by stating n. Thus, a 3 by 3 matrix is a square matrix of order 3. In the remainder of this chapter, all matrices except vectors will be square matrices.

6.14 ROW OPERATIONS

Row operations on a matrix consist either of multiplying (dividing) a row by a nonzero constant, adding a multiple of one row to another row, or interchanging two rows.[2] For example, given the matrix

$$
A = \begin{pmatrix} 1 & 2 \\ 3 & 9 \end{pmatrix},
$$

we may multiply the first row by -2 to give another matrix

$$
\begin{pmatrix} -2 & -4 \\ 3 & 9 \end{pmatrix}.
$$

Or, dividing the second row of A by 3 leads to the matrix

$$
\begin{pmatrix} 1 & 2 \\ 1 & 3 \end{pmatrix}.
$$

If we multiply the first row of A by -3 and add the result to the second row (leaving the first row unchanged), we have

$$
\begin{pmatrix} 1 & 2 \\ 0 & 3 \end{pmatrix}.
$$

---

[2] See also Chapter 2, Section 6.

Exercise. What row operation performed on

$$\begin{pmatrix} 2 & 5 \\ 6 & 13 \end{pmatrix}$$

will lead to another matrix which will have a 0 in place of the 6, and what will the new matrix be? Answer: Multiply the first row of the given matrix by −3 and add to the second row, obtaining

$$\begin{pmatrix} 2 & 5 \\ 0 & -2 \end{pmatrix}.$$

If we have a statement of matrix equality, such as

$$AB = C$$

and we perform the same row operations on A and C (not on B and C) we are led to new statements of equality. For example,

$$\begin{pmatrix} 1 & 2 \\ 0 & 3 \end{pmatrix}\begin{pmatrix} 2 & -4 \\ -1 & 5 \end{pmatrix} = \begin{pmatrix} 0 & 6 \\ -3 & 15 \end{pmatrix}$$

is a true statement. If we multiply the first row of the leftmost matrix by 3 and the first row of the rightmost matrix by 3 we obtain

$$\begin{pmatrix} 3 & 6 \\ 0 & 3 \end{pmatrix}\begin{pmatrix} 2 & -4 \\ -1 & 5 \end{pmatrix} = \begin{pmatrix} 0 & 18 \\ -3 & 15 \end{pmatrix},$$

which the reader may verify is also a true statement. Or, starting over, if we take the leftmost and rightmost matrices and add twice the first row to the second row, we obtain

$$\begin{pmatrix} 1 & 2 \\ 2 & 7 \end{pmatrix}\begin{pmatrix} 2 & -4 \\ -1 & 5 \end{pmatrix} = \begin{pmatrix} 0 & 6 \\ -3 & 27 \end{pmatrix},$$

which computation will prove to be a true statement.

Exercise. Starting with the original matrices in the foregoing, select the leftmost and rightmost matrices and multiply the second row by −2 and add to the first. Write the new statement and verify that it is an equality.

In summary, we have

1. Row operations on a matrix are:
   a) Multiplying a row by a constant or dividing a row by a nonzero constant.
   b) Adding a multiple of one row to a multiple of another row.
   c) Interchanging two rows.
2. The matrix relationship

$$AB = C$$

remains true if the same row operations are performed on A and C, leaving B unchanged.

6.15 THE INVERSE OF A MATRIX

In ordinary algebra, we write

$$a(a^{-1}) = (a^1)(a^{-1}) = a^{1-1} = a^0 = 1.$$

Thus,

$$aa^{-1} = 1,$$

and a^{-1} is called the multiplicative inverse of a because the product just written is 1. For example, 3^{-1} is the inverse of 3 because

$$3(3^{-1}) = 3 \cdot \frac{1}{3} = 1.$$

In matrix algebra, we have an analogous operation illustrated by

$$\begin{pmatrix} 1 & 2 \\ 2 & 3 \end{pmatrix}\begin{pmatrix} -3 & 2 \\ 2 & -1 \end{pmatrix} = \begin{pmatrix} 1 & 0 \\ 0 & 1 \end{pmatrix} = I.$$

That is, the product of the matrices at the left is the unit (identity) matrix I, of order 2. If we call the leftmost matrix A, that is,

$$A = \begin{pmatrix} 1 & 2 \\ 2 & 3 \end{pmatrix},$$

we define the second matrix as the *multiplicative inverse*, A^{-1}, and write

$$A^{-1} = \begin{pmatrix} -3 & 2 \\ 2 & -1 \end{pmatrix}.$$

Thus,

$$\begin{pmatrix} 1 & 2 \\ 2 & 3 \end{pmatrix}\begin{pmatrix} -3 & 2 \\ 2 & -1 \end{pmatrix} = I$$
$$\quad A \qquad\qquad A^{-1} \quad = I$$

Definition. Two matrices are inverses of each other if their product is the identity matrix, *I*.

Observe that the definition says *inverses of each other*. This means that either matrix can be called *A* and the other is A^{-1} or, more to the point, we may write the product as AA^{-1} or $A^{-1}A$ and still obtain

$$AA^{-1} = A^{-1}A = I.$$

For example, in the foregoing, we have

$$A^{-1}A = \begin{pmatrix} -3 & 2 \\ 2 & -1 \end{pmatrix} \begin{pmatrix} 1 & 2 \\ 2 & 3 \end{pmatrix} = \begin{pmatrix} 1 & 0 \\ 0 & 1 \end{pmatrix} = I.$$

Exercise. Verify that the following matrices are inverses of each other.

$$\begin{pmatrix} 13 & 3 \\ 4 & 1 \end{pmatrix} \begin{pmatrix} 1 & -3 \\ -4 & 13 \end{pmatrix}.$$

Answer: Multiplication of the two matrices leads to the unit matrix

$$\begin{pmatrix} 1 & 0 \\ 0 & 1 \end{pmatrix}.$$

We shall see later that not every matrix has an inverse. However, if the inverse of *A* does exist, it is unique: that is, a matrix that has an inverse has exactly one inverse. The uniqueness of the inverse may not seem obvious, so let us see what would happen if in addition to A^{-1}, *A* had another inverse, *B*. Then, according to the definition of the inverse, the product *AB* would have to be *I*; that is,

$$AB = I.$$

Now if we multiply both sides of this expression by A^{-1} we have

$$A^{-1}AB = A^{-1}I.$$

Noting that on the left $A^{-1}A$ can be replaced by *I*, then *IB* can be replaced by *B*, and on the right $A^{-1}I$ can be replaced by A^{-1}, we see that

$$B = A^{-1}.$$

So the supposedly different inverse, *B*, turns out to be the original inverse, A^{-1}. We shall make use of the uniqueness of the inverse a little later when we will ask what the expression (?) must be to make the following a true statement:

234

$$I = (?)A.$$

The answer will be that (?) must be A^{-1}.

Gauss-Jordan inversion. We are now ready to attack the main objective of this section, which is to compute the inverse of a given matrix, if one exists. After we have learned computational procedures, we shall discover that they are directly applicable to the solution of n by n systems of linear equations. We present first the method attributed to Gauss and Jordan, which is the method we shall use as we continue our work. We shall also present two other procedures for carrying out matrix inversion—the central procedure involved in the solution of linear systems of equations.

Briefly, the Gauss-Jordan method starts by writing the given matrix at the left, and the corresponding unit matrix next to it, at the right. Then select and carry out row operations which will convert the given matrix into the unit matrix, and *apply the same operations to the matrix at the right*. When the left (given) matrix becomes the unit matrix, the matrix on the right will be the desired inverse. To illustrate, let us find the inverse of the matrix

$$\begin{pmatrix} 3 & 2 \\ 1 & 1 \end{pmatrix}.$$

We start by writing the given matrix to the left and the unit matrix to the right; thus,

$$\left(\begin{array}{cc|cc} 3 & 2 & 1 & 0 \\ 1 & 1 & 0 & 1 \end{array} \right).$$

To change the left matrix into the unit matrix will require several steps (which we perform on both matrices). We start by getting a 1 in the upper left corner. Divide the first row above by 3 to obtain

$$\left(\begin{array}{cc|cc} 1 & \dfrac{2}{3} & \dfrac{1}{3} & 0 \\ 1 & 1 & 0 & 1 \end{array} \right).$$

Next, get a 0 in the second row, first column, of the left matrix by multiplying the first row above by -1 and adding to the second row. We have

$$\left(\begin{array}{cc|cc} 1 & \dfrac{2}{3} & \dfrac{1}{3} & 0 \\ 0 & \dfrac{1}{3} & -\dfrac{1}{3} & 1 \end{array} \right).$$

Multiply the second row, just above, by 3 to obtain a 1 in the 2nd row, 2nd column of the left matrix. Thus,

$$\left(\begin{array}{cc|cc} 1 & \frac{2}{3} & \frac{1}{3} & 0 \\ 0 & 1 & -1 & 3 \end{array}\right).$$

Finally, multiply the second row, just above, by $-\frac{2}{3}$ and add to the first to obtain a zero in the 1st row, second column of the left matrix. This gives

$$\left(\begin{array}{cc|cc} 1 & 0 & 1 & -2 \\ 0 & 1 & -1 & 3 \end{array}\right).$$

We now have the identity matrix at the left, so the inverse of

$$\left(\begin{array}{cc} 3 & 2 \\ 1 & 1 \end{array}\right)$$

is the matrix at the right; namely,

$$\left(\begin{array}{cc} 1 & -2 \\ -1 & 3 \end{array}\right).$$

Exercise. Apply the procedure of the last example and find the inverse of the matrix

$$A = \left(\begin{array}{cc} 7 & 3 \\ 2 & 1 \end{array}\right).$$

Answer: The inverse is the matrix

$$A^{-1} = \left(\begin{array}{cc} 1 & -3 \\ -2 & 7 \end{array}\right).$$

We next illustrate the computation of the inverse of a matrix of order 3, the left matrix in the following:

$$\left(\begin{array}{ccc|ccc} 2 & 3 & 1 & 1 & 0 & 0 \\ 1 & 4 & 2 & 0 & 1 & 0 \\ 5 & 6 & 4 & 0 & 0 & 1 \end{array}\right).$$

Divide the first row above by 2 to obtain

$$\left(\begin{array}{ccc|ccc} 1 & \frac{3}{2} & \frac{1}{2} & \frac{1}{2} & 0 & 0 \\ 1 & 4 & 2 & 0 & 1 & 0 \\ 5 & 6 & 4 & 0 & 0 & 1 \end{array}\right).$$

Multiply the first row above by -1 and add to the second row to obtain

$$\begin{pmatrix} 1 & \dfrac{3}{2} & \dfrac{1}{2} & \bigg| & \dfrac{1}{2} & 0 & 0 \\ 0 & \dfrac{5}{2} & \dfrac{3}{2} & \bigg| & -\dfrac{1}{2} & 1 & 0 \\ 5 & 6 & 4 & \bigg| & 0 & 0 & 1 \end{pmatrix}.$$

Multiply the first row above by -5 and add to the third row to obtain

$$\begin{pmatrix} 1 & \dfrac{3}{2} & \dfrac{1}{2} & \bigg| & \dfrac{1}{2} & 0 & 0 \\ 0 & \dfrac{5}{2} & \dfrac{3}{2} & \bigg| & -\dfrac{1}{2} & 1 & 0 \\ 0 & -\dfrac{3}{2} & \dfrac{3}{2} & \bigg| & -\dfrac{5}{2} & 0 & 1 \end{pmatrix}.$$

The tactics we have followed are these: first, get a 1 in the first column, first row, then use combinations of this row with each of the other rows to get zeros in the first columns of these rows. We now repeat these tactics, starting by getting a 1 in the second column of the second row, then using this row to get zeros in the second columns of the other rows.

Divide the second row by 5/2 to obtain the new second row

$$0 \quad 1 \quad \dfrac{3}{5} \quad \bigg| \quad -\dfrac{1}{5} \quad \dfrac{2}{5} \quad 0.$$

If we multiply this new second row by $-3/2$ and add to the first row, then multiply the new second row by $3/2$ and add to the third row, we obtain the new matrices

$$\begin{pmatrix} 1 & 0 & -\dfrac{2}{5} & \bigg| & \dfrac{4}{5} & -\dfrac{3}{5} & 0 \\ 0 & 1 & \dfrac{3}{5} & \bigg| & -\dfrac{1}{5} & \dfrac{2}{5} & 0 \\ 0 & 0 & \dfrac{12}{5} & \bigg| & -\dfrac{14}{5} & \dfrac{3}{5} & 1 \end{pmatrix}.$$

Finally, we repeat the tactics by getting a 1 in the third row of the third column and then using this row to get zeros in the third column of the other rows. The new third row, obtained by dividing the previous third row by 12/5, is

$$0 \quad 0 \quad 1 \quad \left| \quad -\frac{7}{6} \quad \frac{1}{4} \quad \frac{5}{12} \right. .$$

If we multiply this new third row by 2/5 and add to the first row, then multiply the new third row by −3/5 and add to the second row we find

$$\begin{pmatrix} 1 & 0 & 0 & \bigg| & \dfrac{1}{3} & -\dfrac{1}{2} & \dfrac{1}{6} \\[2mm] 0 & 1 & 0 & \bigg| & \dfrac{1}{2} & \dfrac{1}{4} & -\dfrac{1}{4} \\[2mm] 0 & 0 & 1 & \bigg| & -\dfrac{7}{6} & \dfrac{1}{4} & \dfrac{5}{12} \end{pmatrix},$$

which has the matrix I to the left and the desired inverse to the right.

> **Exercise.** Multiply the original matrix of the last illustration by the inverse and verify that the product is I.

Zeros first.[3]   The Gauss-Jordan method proceeds to obtain ones on the main diagonal and zeros for the off-diagonal elements. A variation on this procedure is to obtain the off-diagonal zeros first and, following this, obtain ones on the diagonal. This *zeros-first* variation can simplify the work involved in hand calculation by avoiding fractions until the last step.

Example.   By the zeros-first method, find the inverse of

$$A = \begin{pmatrix} 2 & 3 \\ 4 & 7 \end{pmatrix}.$$

We start in the usual manner by writing

$$\begin{pmatrix} 2 & 3 & \bigg| & 1 & 0 \\ 4 & 7 & \bigg| & 0 & 1 \end{pmatrix}.$$

Multiplying the first row by −2 and adding to the second row, we have

$$\begin{pmatrix} 2 & 3 & \bigg| & 1 & 0 \\ 0 & 1 & \bigg| & -2 & 1 \end{pmatrix}.$$

We now have a zero as the lower off-diagonal element and proceed to obtain a zero as the upper off-diagonal element. Multiplying the second row by −3 and adding to the first yields

---

[3] Suggested by Professor Van Cor of New Hampshire College.

$$\begin{pmatrix} 2 & 0 & | & 7 & -3 \\ 0 & 1 & | & -2 & 1 \end{pmatrix}.$$

The off-diagonal elements now are zero, and ones are obtained on the main diagonal by dividing the first row by 2 to give

$$\begin{pmatrix} 1 & 0 & | & \dfrac{7}{2} & -\dfrac{3}{2} \\ 0 & 1 & | & -2 & 1 \end{pmatrix}.$$

The desired inverse is

$$A^- = \begin{pmatrix} \dfrac{7}{2} & -\dfrac{3}{2} \\ -2 & 1 \end{pmatrix}.$$

As another example, we find the inverse of

$$A = \begin{pmatrix} 3 & 2 & -1 \\ 1 & -1 & 0 \\ 2 & 0 & -1 \end{pmatrix}$$

by first writing

$$\begin{pmatrix} 3 & 2 & -1 & | & 1 & 0 & 0 \\ 1 & -1 & 0 & | & 0 & 1 & 0 \\ 2 & 0 & -1 & | & 0 & 0 & 1 \end{pmatrix}.$$

To get zeros in the first column, under the 3, we perform the following row operations:

Replace row two by row one minus 3 times row two.
Replace row three by 2 times row one minus 3 times row three.
We have

$$\begin{pmatrix} 3 & 2 & -1 & | & 1 & 0 & 0 \\ 0 & 5 & -1 & | & 1 & -3 & 0 \\ 0 & 4 & 1 & | & 2 & 0 & -3 \end{pmatrix}.$$

Next:

Replace row three by 4 times row two minus 5 times row three.

$$\begin{pmatrix} 3 & 2 & -1 & | & 1 & 0 & 0 \\ 0 & 5 & -1 & | & 1 & -3 & 0 \\ 0 & 0 & -9 & | & -6 & -12 & 15 \end{pmatrix}.$$

Then:

Replace row two by row three minus 9 times row two.
Replace row one by row three minus 9 times row one.

$$\begin{pmatrix} -27 & -18 & 0 & \bigm| & -15 & -12 & 15 \\ 0 & -45 & 0 & \bigm| & -15 & 15 & 15 \\ 0 & 0 & -9 & \bigm| & -6 & -12 & 15 \end{pmatrix}.$$

Next:

Replace row one by 2 times row two minus 5 times row 1.

$$\begin{pmatrix} 135 & 0 & 0 & \bigm| & 45 & 90 & -45 \\ 0 & -45 & 0 & \bigm| & -15 & 15 & 15 \\ 0 & 0 & -9 & \bigm| & -6 & -12 & 15 \end{pmatrix}.$$

Finally,

Divide row one by 135; divide row two by -45; divide row three by -9. We obtain

$$\begin{pmatrix} 1 & 0 & 0 & \bigm| & \dfrac{1}{3} & \dfrac{2}{3} & -\dfrac{1}{3} \\[2ex] 0 & 1 & 0 & \bigm| & \dfrac{1}{3} & -\dfrac{1}{3} & -\dfrac{1}{3} \\[2ex] 0 & 0 & 1 & \bigm| & \dfrac{2}{3} & \dfrac{4}{3} & -\dfrac{5}{3} \end{pmatrix}$$

and A^{-1} is the right matrix in the foregoing product.

The foregoing illustrates how the zeros-first method postpones the introduction of fractions until the final step. Most readers will find this an advantage for hand calculation, and we recommend its use. Matrix inversion on high-speed computers, on the other hand, is carried out by programs following the Gauss-Jordan method. In an addendum, Section 6.33, we shall show how to invert a matrix by the use of determinants, cofactors, and the adjoint matrix.

Inversion of large matrices is a task best left for electronic computers. However, we should practice the procedure long enough to understand that even though the arithmetic is tedious, the basic methodology is not complicated. To see why the method works, consider the true statement

$$A = IA.$$

Suppose that we carry out row operations on the left of the equal sign to change the left matrix to I, and maintain the equality by applying the same operations to the first matrix on the right of the equal sign (which starts out as I). The end result will be I on the left and something times A on the right; thus,

$$I = (?)A.$$

We showed earlier that the inverse of a matrix is unique. Hence, there is only one appropriate entry for (?) in the foregoing. It is A^{-1}. It follows that if we write a given matrix A with I to its right, then change the left to I by row operations which are applied also to the right, the end result will be A^{-1} on the right.

Not every matrix has an inverse. For example, consider the matrix

$$\begin{pmatrix} 1 & 1 \\ 2 & 2 \end{pmatrix}.$$

If we set this matrix up with the unit matrix to the right, we have

$$\left(\begin{array}{cc|cc} 1 & 1 & 1 & 0 \\ 2 & 2 & 0 & 1 \end{array}\right).$$

The indicated row operation is to multiply the first row by -2 and add to the second row. We find

$$\left(\begin{array}{cc|cc} 1 & 1 & 1 & 0 \\ 0 & 0 & -2 & 1 \end{array}\right).$$

No row operations for the left matrix can be found that will provide a 1 in the lower left corner and a 0 in the upper right. The given matrix has no inverse and is said to be *singular*. In the matrix context, the word singular does not mean one or single but, rather, connotes that the matrix is peculiar in that it has no inverse.

Interchanging rows: Removing a constant factor. Interchange of rows is a permissible matrix operation that may be used for convenience, or that may be necessary to maintain a matrix in the form required for steps in the inversion process. As for convenience, note that

$$\left(\begin{array}{cc|cc} 3 & 5 & 1 & 0 \\ 1 & 2 & 0 & 1 \end{array}\right)$$

is a bit simpler to work with if the rows are interchanged to give:

$$\left(\begin{array}{cc|cc} 1 & 2 & 0 & 1 \\ 3 & 5 & 1 & 0 \end{array}\right)$$

because now we have a 1 as the upper left diagonal element. On the other hand, in

$$\left(\begin{array}{cc|cc} 0 & 1 & 1 & 0 \\ 2 & 3 & 0 & 1 \end{array}\right)$$

we have an unwanted zero as the upper left diagonal element. Interchanging rows, we have

$$\left(\begin{array}{cc|cc} 2 & 3 & 0 & 1 \\ 0 & 1 & 1 & 0 \end{array}\right)$$

and can proceed to the inverse in the usual manner. Thus, add minus 3 times row two to row 1:

$$\left(\begin{array}{cc|cc} 2 & 0 & -3 & 1 \\ 0 & 1 & 1 & 0 \end{array}\right).$$

Then, divide row one by 2:

$$\left(\begin{array}{cc|cc} 1 & 0 & -\dfrac{3}{2} & \dfrac{1}{2} \\ 0 & 1 & 1 & 0 \end{array}\right)$$

and we have that

$$\text{if } \mathbf{A} = \begin{pmatrix} 0 & 1 \\ 2 & 3 \end{pmatrix}, \quad \text{then} \quad \mathbf{A}^{-1} = \begin{pmatrix} -\dfrac{3}{2} & \dfrac{1}{2} \\ 1 & 0 \end{pmatrix}.$$

Finally, a common factor may be removed in order to simplify the numbers occurring in the inversion. For example, the expression

$$\left(\begin{array}{ccc|ccc} -27 & -18 & 0 & -15 & -12 & 15 \\ 0 & -45 & 0 & -15 & 15 & 15 \\ 0 & 0 & -9 & -6 & -12 & 15 \end{array}\right),$$

which we encountered a page or two earlier can be simplified if rows one, two, and three are divided, respectively, by −3, −15, and −3 to give

$$\left(\begin{array}{ccc|ccc} 9 & 6 & 0 & 5 & 4 & -5 \\ 0 & 3 & 0 & 1 & -1 & -1 \\ 0 & 0 & 3 & 2 & 4 & -5 \end{array}\right).$$

We leave it to the reader to decide whether or not it is worth the time to rewrite the factored matrix or work with the former one.

6.16 PROBLEM SET 6–4

Find the inverse of each of the following, if one exists.

1. $\begin{pmatrix} 7 & 3 \\ 2 & 1 \end{pmatrix}.$ 2. $\begin{pmatrix} 9 & 4 \\ 2 & 1 \end{pmatrix}.$

3. $\begin{pmatrix} 2 & 2 \\ 3 & 5 \end{pmatrix}.$ 4. $\begin{pmatrix} 1 & -1 \\ -1 & 2 \end{pmatrix}.$

5. $\begin{pmatrix} 2 & 2 \\ 6 & 6 \end{pmatrix}.$ 6. $\begin{pmatrix} 3 & 3 \\ 2 & 2 \end{pmatrix}.$

7. $\begin{pmatrix} 1 & 3 \\ 2 & 0 \end{pmatrix}.$ 8. $\begin{pmatrix} 2 & 5 \\ 3 & 4 \end{pmatrix}.$

9. $\begin{pmatrix} 0 & 1 \\ 2 & 3 \end{pmatrix}.$

10. $\begin{pmatrix} 0 & 3 \\ 2 & 5 \end{pmatrix}.$

11. $\begin{pmatrix} 2 & 8 & -11 \\ -1 & -5 & 7 \\ 1 & 2 & -3 \end{pmatrix}.$

12. $\begin{pmatrix} 2 & 2 & 3 \\ 0 & 1 & 1 \\ 4 & 0 & 3 \end{pmatrix}.$

13. $\begin{pmatrix} 1 & 1 & 1 \\ 1 & 1 & 1 \\ 2 & 2 & 2 \end{pmatrix}.$

14. $\begin{pmatrix} 0 & -1 & 1 \\ -1 & 1 & 2 \\ 1 & 0 & -2 \end{pmatrix}.$

15. $\begin{pmatrix} 2 & 1 & 4 \\ 3 & 0 & 2 \\ 1 & 2 & 3 \end{pmatrix}.$

16. $\begin{pmatrix} 1 & -1 & 0 \\ 2 & 1 & 3 \\ 3 & 0 & 3 \end{pmatrix}.$

6.17 APPLICATIONS–2: MATRIX SOLUTION OF LINEAR SYSTEMS

In this section we shall show how the solution of an n by n system of linear equations is accomplished by matrix inversion. To set the stage, consider first the manner in which a simple linear equation such as

$$2x = 3$$

is solved for x. To bring out what we have in mind, two procedures using slightly different symbols are presented side by side.

Regular solution symbols	*Inverse solution symbols*
$2x = 3$	$2x = 3$
$\dfrac{1}{2}(2x) = \left(\dfrac{1}{2}\right)(3)$	$(2^{-1})(2x) = (2^{-1})(3)$
$(1)x = \dfrac{1}{2}(3)$	$(1)x = (2)^{-1}(3)$
$x = \dfrac{1}{2}(3).$	$x = (2)^{-1}(3).$

Notice on the right that the solution was obtained by multiplying both sides of the original equation by the multiplicative inverse of 2, which is 2^{-1}. More generally, if we start with

$$ax = b$$

then multiply both sides of the equation by the inverse, a^{-1}, to obtain

$$(a^{-1})(a^1 x) = (a^{-1})b$$

we have

$$(a^{-1+1})x = (a^{-1})b$$
$$a^0 x = a^{-1}b$$
$$(1)x = a^{-1}b$$
$$x = a^{-1}b.$$

The solution of a linear system proceeds in a manner analogous to that just illustrated for a single equation, except that now we multiply both sides of the matrix form of the system by the inverse of the coefficient matrix. To see how this works out, consider the 2 by 2 system

$$2x_1 + 3x_2 = 17$$
$$x_1 + 2x_2 = 10.$$

The system can be written in expanded matrix form as

$$\begin{pmatrix} 2 & 3 \\ 1 & 2 \end{pmatrix} \begin{pmatrix} x_1 \\ x_2 \end{pmatrix} = \begin{pmatrix} 17 \\ 10 \end{pmatrix}.$$

If we multiply both sides of the last equation by the inverse of the coefficient matrix, which is

$$\begin{pmatrix} 2 & 3 \\ 1 & 2 \end{pmatrix}^{-1},$$

we have

$$\begin{pmatrix} 2 & 3 \\ 1 & 2 \end{pmatrix}^{-1} \begin{pmatrix} 2 & 3 \\ 1 & 2 \end{pmatrix} \begin{pmatrix} x_1 \\ x_2 \end{pmatrix} = \begin{pmatrix} 2 & 3 \\ 1 & 2 \end{pmatrix}^{-1} \begin{pmatrix} 17 \\ 10 \end{pmatrix}.$$

Inasmuch as the product of a matrix and its inverse is the unit matrix, we may rewrite the last equation as

$$\begin{pmatrix} 1 & 0 \\ 0 & 1 \end{pmatrix} \begin{pmatrix} x_1 \\ x_2 \end{pmatrix} = \begin{pmatrix} 2 & 3 \\ 1 & 2 \end{pmatrix}^{-1} \begin{pmatrix} 17 \\ 10 \end{pmatrix}.$$

The left member now is the matrix

$$\begin{pmatrix} x_1 \\ x_2 \end{pmatrix}$$

so we may write

$$\begin{pmatrix} x_1 \\ x_2 \end{pmatrix} = \begin{pmatrix} 2 & 3 \\ 1 & 2 \end{pmatrix}^{-1} \begin{pmatrix} 17 \\ 10 \end{pmatrix}.$$

The last equation shows that x_1 and x_2 (the elements of the solution vector) can be computed as soon as the inverse of the coefficient matrix is at hand. The reader may verify that the desired inverse is

$$\begin{pmatrix} 2 & 3 \\ 1 & 2 \end{pmatrix}^{-1} = \begin{pmatrix} 2 & -3 \\ -1 & 2 \end{pmatrix}.$$

Hence the solution of the original system is given by

$$\begin{pmatrix} x_1 \\ x_2 \end{pmatrix} = \begin{pmatrix} 2 & -3 \\ -1 & 2 \end{pmatrix} \begin{pmatrix} 17 \\ 10 \end{pmatrix},$$

that is, after computing the right member,

$$\begin{pmatrix} x_1 \\ x_2 \end{pmatrix} = \begin{pmatrix} 4 \\ 3 \end{pmatrix},$$

so that $x_1 = 4$ and $x_2 = 3$.

In general, if A is the matrix of coefficients and b is the column vector of constants, the matrix expression for an n by n system of linear equations is

$$Ax = b.$$

If both sides of the last equation are multiplied by the inverse of the coefficient matrix, A^{-1}, we have

$$A^{-1}Ax = A^{-1}b,$$

from which it follows that

$$x = A^{-1}b.$$

The key to the solution of the system is A^{-1}, the inverse of the coefficient matrix. Once A^{-1} has been found, the solution vector, x, can be computed by premultiplying b, the vector of constants, by A^{-1}. Consider the system

$$\begin{aligned} 2x_1 + 2x_2 + 3x_3 &= 3 \\ x_2 + x_3 &= 2 \\ x_1 + x_2 + x_3 &= 4. \end{aligned}$$

The coefficient matrix is

$$A = \begin{pmatrix} 2 & 2 & 3 \\ 0 & 1 & 1 \\ 1 & 1 & 1 \end{pmatrix}$$

and the reader may verify that the inverse of this matrix is

$$A^{-1} = \begin{pmatrix} 0 & -1 & 1 \\ -1 & 1 & 2 \\ 1 & 0 & -2 \end{pmatrix}.$$

The vector of constants is

$$b = \begin{pmatrix} 3 \\ 2 \\ 4 \end{pmatrix}.$$

The solution vector, x, is found as the product $A^{-1}b$, which is

$$\begin{pmatrix} x_1 \\ x_2 \\ x_3 \end{pmatrix} = A^{-1}b = \begin{pmatrix} 0 & -1 & 1 \\ -1 & 1 & 2 \\ 1 & 0 & -2 \end{pmatrix} \begin{pmatrix} 3 \\ 2 \\ 4 \end{pmatrix} = \begin{pmatrix} 2 \\ 7 \\ -5 \end{pmatrix}.$$

We find that $x_1 = 2$, $x_2 = 7$, and $x_3 = -5$. The reader may verify that these values satisfy all three original equations. It is worth pointing out that once A^{-1} has been computed, we have in effect solved *all* linear systems that have A as the coefficient matrix. In the case of the present system, for example, if we were to change the constants on the right from 3, 2, and 4 to, say, 0, 1, and 2, respectively, the solution of the system would be the foregoing A^{-1} multiplied by the new vector of constants; that is,

$$\begin{pmatrix} x_1 \\ x_2 \\ x_3 \end{pmatrix} = \begin{pmatrix} 0 & -1 & 1 \\ -1 & 1 & 2 \\ 1 & 0 & -2 \end{pmatrix} \begin{pmatrix} 0 \\ 1 \\ 2 \end{pmatrix} = \begin{pmatrix} 1 \\ 5 \\ -4 \end{pmatrix}$$

so that $x_1 = 1$, $x_2 = 5$, $x_3 = -4$ would be the solution of the system

$$\begin{aligned} 2x_1 + 2x_2 + 3x_3 &= 0 \\ x_2 + x_3 &= 1 \\ x_1 + x_2 + x_3 &= 2. \end{aligned}$$

Exercise. If the system

$$\begin{aligned} 7x_1 + 3x_2 &= 5 \\ 2x_1 + x_2 &= 7 \end{aligned}$$

is expressed in matrix form as $Ax = b$, what are A, x, and b; what is A^{-1}? Answer: A is the coefficient matrix

$$\begin{pmatrix} 7 & 3 \\ 2 & 1 \end{pmatrix},$$

x is the solution vector

$$\begin{pmatrix} x_1 \\ x_2 \end{pmatrix},$$

b is the vector of constants

$$\begin{pmatrix} 5 \\ 7 \end{pmatrix},$$

and the inverse of the coefficient matrix is found by computation to be

$$A^{-1} = \begin{pmatrix} 1 & -3 \\ -2 & 7 \end{pmatrix}.$$

> **Exercise.** The solution of the foregoing exercise is $x = A^{-1}b$. Substitute the proper numbers into $A^{-1}b$ and compute the solution vector, x. Answer:
>
> $$x = \begin{pmatrix} x_1 \\ x_2 \end{pmatrix} = \begin{pmatrix} 1 & -3 \\ -2 & 7 \end{pmatrix}\begin{pmatrix} 5 \\ 7 \end{pmatrix} = \begin{pmatrix} -16 \\ 39 \end{pmatrix}.$$

As an application of the material we have just studied, suppose a company makes liquid products Primeoil, Midoil, and Lastoil, which contain different amounts of additives A_1, A_2, and A_3 per gallon as shown in Table 6–5.

TABLE 6–5

Liquid	Gallons made	Pounds of additive per gallon		
		A_1	A_2	A_3
Primeoil	x_1	1	1	1
Midoil	x_2	1	2	1
Lastoil	x_3	2	0	1
Available additive, end of week		a_1	a_2	a_3

The additives deteriorate if not used within a week, so each Saturday the company schedules production of $(x_1 \; x_2 \; x_3)$ gallons of Primeoil, Midoil, and Lastoil to use up the additives on hand. These amounts vary from week to week and are shown as $(a_1 \; a_2 \; a_3)$ in Table 6–5. If we schedule $(x_1 \; x_2 \; x_3)$ gallons of the liquids, the pounds of additive A_1 used will be $x_1(1) + x_2(1) + x_3(2)$, and this should equal a_1, the amount available. Hence, $x_1 + x_2 + 2x_3 = a_1$. Combining this condition with the conditions on additives A_2 and A_3 leads us to the system

$$\begin{aligned} x_1 + x_2 + 2x_3 &= a_1 \\ x_1 + 2x_2 \quad\;\; &= a_2 \\ x_1 + x_2 + x_3 &= a_3. \end{aligned}$$

The coefficient matrix is

$$A = \begin{pmatrix} 1 & 1 & 2 \\ 1 & 2 & 0 \\ 1 & 1 & 1 \end{pmatrix}$$

and it may be verified that the inverse is

$$A^{-1} = \begin{pmatrix} -2 & -1 & 4 \\ 1 & 1 & -2 \\ 1 & 0 & -1 \end{pmatrix}.$$

Hence, the solution vector is

$$\begin{pmatrix} x_1 \\ x_2 \\ x_3 \end{pmatrix} = \begin{pmatrix} -2 & -1 & 4 \\ 1 & 1 & -2 \\ 1 & 0 & -1 \end{pmatrix} \begin{pmatrix} a_1 \\ a_2 \\ a_3 \end{pmatrix}.$$

Suppose that on a given Saturday the amounts of additives available are $a_1 = 20$ pounds of A_1, $a_2 = 30$ pounds of A_2, and $a_3 = 20$ pounds of A_3. To use up these additives, the production schedule should be

$$\begin{pmatrix} x_1 \\ x_2 \\ x_3 \end{pmatrix} = \begin{pmatrix} -2 & -1 & 4 \\ 1 & 1 & -2 \\ 1 & 0 & -1 \end{pmatrix} \begin{pmatrix} 20 \\ 30 \\ 20 \end{pmatrix} = \begin{pmatrix} 10 \\ 10 \\ 0 \end{pmatrix},$$

which is 10 gallons of Primeoil, 10 gallons of Midoil, and no Lastoil.

Exercise. If the additives available at a week's end are $(a_1\ a_2\ a_3) = (80\ 100\ 70)$, what should the production schedule be? Answer: $(x_1\ x_2\ x_3) = (20\ 40\ 10)$.

Of course, it may not be possible to schedule production to use all the additives available on a given Saturday. For example, if $(a_1\ a_2\ a_3) = (40\ 60\ 50)$, the solution vector is $(x_1\ x_2\ x_3) = (60\ 0\ -10)$, and the value $x_3 = -10$ is not possible. In such a case we might choose to maximize either the total amount of liquid made or the total amount of additive used, depending upon cost considerations, and we would apply the linear programming techniques of Chapters 4 and 5 to determine the optimum production schedule. In this context, it should be noted that the manipulations carried out in the solution of linear systems in earlier chapters are operations on matrices. In this book, however, we have chosen to make our major study of such problems (Chapters 1–5) not depend upon an ability to use matrix algebra, and we shall not recast these problems in matrix form.

Finally, it should be repeated that matrix inversion is defined only for square matrices. If the reader wishes to work with m by n systems where $m \neq n$, or if he encounters a singular square matrix (one with no inverse), he should refer to Chapter 2 to find the solution for the system, if one exists.

6.18 PROBLEM SET 6–5

1. Consider the system

$$8x_1 + 5x_2 = 2$$
$$3x_1 + 2x_2 = 1.$$

   a) Relating the system to $Ax = b$, what is A?, x?, b?
   b) Compute A^{-1}.
   c) Write $x = A^{-1}b$ in expanded matrix form.
   d) Compute the solution from part (c).
   e) What would be the solution vector if the elements in the vector of constants, 2 and 1, were changed to each of the following:
      (1) 1, 0. (2) 0, 1. (3) 1, 1. (4) 3, 4. (5) −3, 1.

2. Answer parts (a) through (d) of Problem 1 for the system

$$4x_1 + 3x_2 = 2$$
$$9x_1 + 7x_2 = 3.$$

   e) What would be the solution vector if the elements in the vector of constants, 2 and 3, were changed to each of the following:
      (1) 1, 0. (2) 0, 1. (3) 1, 1. (4) 2, 1. (5) −1, 2.

3. Answer parts (a) through (d) of Problem 1 for the system

$$6x_1 + 8x_2 = 3$$
$$2x_1 + 3x_2 = 1.$$

   e) What would be the solution vector if the elements in the vector of constants, 3 and 1, were changed to each of the following:
      (1) 1, 1. (2) 0, 1. (3) 1, 0. (4) 2, 3. (5) −1, 1.

4. Given the system

$$8x_1 - 7x_2 = b_1$$
$$-5x_1 + 5x_2 = b_2$$

   find the missing elements to complete the following equation;

$$\begin{pmatrix} x_1 \\ x_2 \end{pmatrix} = \begin{pmatrix} & \\ & \end{pmatrix} \begin{pmatrix} b_1 \\ b_2 \end{pmatrix}.$$

5. Answer parts (a) through (d) of Problem 1 for the system

$$3x_1 \qquad\quad + 5x_3 = 3$$
$$2x_1 + 2x_2 + 5x_3 = 7$$
$$x_2 + \ x_3 = 2.$$

   e) What would be the solution vector if the elements in the vector of constants, 3, 7, and 2, were changed to each of the following:
      (1) 1, 2, 1. (2) 2, 3, 4. (3) 1, 0, −1. (4) 1, 6, 2.
      (5) 2, 10, 1.

6. Answer parts (a) through (d) of Problem 1 for the system

$$7x_1 + 3x_2 \qquad\quad = 1$$
$$3x_2 + 5x_3 = 2$$
$$x_1 + \ x_2 + \ x_3 = 3.$$

e) What would be the solution vector if the elements in the vector of constants were changed from 1, 2, and 3 to each of the following:
  (1)   1, −1, 1.   (2)   2, 3, 0.   (3)   −1, 2, −2.   (4)   10, −1, 1.
  (5)   0, −1, 0.

7. Given the system

$$2x_1 + 2x_2 + 3x_3 = b_1$$
$$x_2 + x_3 = b_2$$
$$4x_1 + 3x_3 = b_3$$

find the missing elements to complete the following equation:

$$\begin{pmatrix} x_1 \\ x_2 \\ x_3 \end{pmatrix} = \begin{pmatrix} & & \\ & & \\ & & \end{pmatrix} \begin{pmatrix} b_1 \\ b_2 \\ b_3 \end{pmatrix}.$$

8. Given the system

$$x_1 + 2x_2 + x_3 = b_1$$
$$2x_1 + x_2 + x_3 = b_2$$
$$3x_1 + 2x_3 = b_3$$

a) Verify by multiplication that the inverse of the coefficient matrix is

$$A^{-1} = \begin{pmatrix} -\dfrac{2}{3} & \dfrac{4}{3} & -\dfrac{1}{3} \\ \dfrac{1}{3} & \dfrac{1}{3} & -\dfrac{1}{3} \\ 1 & -2 & 1 \end{pmatrix}.$$

b) Compute the solution vector if the b's are, respectively,
  (1)   3, 0, 3.   (2)   6, 3, 0.

9. Given the system

$$x_1 + x_2 + x_3 + x_4 = b_1$$
$$x_1 + 2x_2 + 2x_3 + 2x_4 = b_2$$
$$x_1 + 2x_2 + 3x_3 + 3x_4 = b_3$$
$$x_1 + 2x_2 + 3x_3 + 4x_4 = b_4$$

a) Verify by multiplication that the inverse of the coefficient matrix is

$$\begin{pmatrix} 2 & -1 & 0 & 0 \\ -1 & 2 & -1 & 0 \\ 0 & -1 & 2 & -1 \\ 0 & 0 & -1 & 1 \end{pmatrix}.$$

b) What is the solution vector if the b's are, respectively,
  (1)   1, 1, 1, 1.   (2)   1, 0, 1, 0.

10. The table shows that we plan to make x_1 Primers. The machine hours used in making a Primer are 1 hour on machine M_1 and 1 hour on M_3. H_1 is the total of hours available on M_1. Other entries in the table have corresponding interpretations.

  a) Set up the system of equations which must be solved if all available machine hours are to be used in making $(x_1\ x_2\ x_3)$ units of the products.

Product	Units made	Machine hours per unit on machine		
		M_1	M_2	M_3
Primer	x_1	1	0	1
Middler	x_2	0	1	2
Laster	x_3	2	3	0
Total hours available		H_1	H_2	H_3

b) Set up the coefficient matrix and find its inverse.
What numbers of units can be made if the numbers of available hours, $(H_1 \ H_2 \ H_3)$, are:
c) (160 80 200)?   *d)* (400 400 400)?   *e)* (320 192 256)?

6.19 APPLICATIONS–3: INPUT-OUTPUT ANALYSIS

Before presenting the formal terminology of this section, it will be helpful to introduce the basic ideas by an example.

Example. Fran's business is producing electric power. Jon owns oil wells and his business is producing oil. Jon uses some of his oil to supply heat for processing his crude oil. He also sells some to Fran to use to run her electric power generators, and he supplies oil to the general public. Similarly, Fran uses some of her power output in her plant operations, sells some to Jon to run his machinery, and supplies some to the general public. Jon's output is barrels of oil and Fran's is kilowatts of electricity, so in order to have a common measure of outputs we shall use their dollar values. Jon uses 20 percent of the value of his own production, supplies 60 percent of the value of Fran's production, and also supplies $140 worth of oil to the public. Fran, on the other hand, uses 10 percent of the value of her own output, supplies 30 percent of the value of Jon's output, and satisfies public demand for $150 worth of electricity. The problem is to find the output (measured in dollar value) that Jon and Fran should schedule to just equal their own and each other's needs and public demand.

We shall let $\$x_1$ be (the value of) Jon's output and $\$x_2$ be (the value of) Fran's output. The data given for the problem are organized in Figure 6–1. To be in balance, Jon's output of $\$x_1$ must contribute the 20% of the output he uses himself, as shown by the arrow marked with the amount $0.2x_1$. Additionally, Jon's output makes up 60% of the value of Fran's power, as shown by the $0.6x_2$ arrow going to Fran. Finally, Jon has to supply $140 to the public. In summary,

Distribution of (the value of) Jon's output

$0.2\,x_1$ to Jon
$0.6\,x_2$ to Fran
$\$140$ to the public
Total: $0.2\,x_1 + 0.6\,x_2 + 140.$

FIGURE 6-1

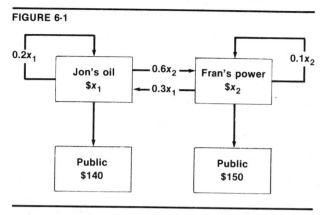

To be in balance, (the value of) Jon's output of x_1 must just equal the total distributed. Hence,

$$x_1 = 0.2\,x_1 + 0.6\,x_2 + 140$$

or

$$e_1: \quad 0.8x_1 - 0.6x_2 = 140.$$

> **Exercise.** Write the equation e_2 (corresponding to e_1) for balancing Fran's output with its distribution. Answer: e_2: $-0.3x_1 + 0.9x_2 = 150$.

Writing e_1 from the foregoing and e_2 from the exercise, we have

$$e_1: \quad 0.8x_1 - 0.6x_2 = 140$$
$$e_2: \quad -0.3x_1 + 0.9x_2 = 150.$$

Then,

$$e_3: \qquad\qquad 1.8x_1 = 720 \qquad\qquad 3e_1 + 2e_2$$
$$x_1 = \$400.$$

Substituting $x_1 = \$400$ into e_2 gives

$$-0.3(400) + 0.9x_2 = 150$$
$$0.9x_2 = 150 + 0.3(400)$$
$$0.9x_2 = 270$$
$$x_2 = \$300.$$

Our answer is that

$$\text{Jon's output} = x_1 = \$400$$
$$\text{Fran's output} = x_2 = \$300$$

will just meet their own needs and public demand. The distributions are shown next.

	$x_1 = Jon's\ \$400\ output$	$x_2 = Fran's\ \$300\ output$
To Jon:	$0.2x_1 = \$\ 80$	$0.3x_1 = \$120$
To Fran:	$0.6x_2 = \ \ 180$	$0.1x_2 = \ \ 30$
To public:	140	150
Total:	$400	$300

We now turn to a formal development of the input-output model. Such a model is used to determine how much each industry has to produce to just satisfy its own needs, interindustry (Jon and Fran) requirements, and final demand (public, nonindustry demand). To develop the model in a simple fashion, we assume that an economy has only two industries, No. 1 and No. 2, whose outputs are measured in dollars and are designated as x_1 and x_2, respectively. Final demands for the outputs of industries No. 1 and No. 2 will be symbolized by d_1 and d_2 dollars, respectively. In addition to supplying output to satisfy final demand, each industry must satisfy interindustry requirements for its output. That is, No. 1 must provide enough output to satisfy its requirement for its own output and to satisfy the industry No. 2 requirement for the output of No. 1. Similarly, No. 2 must satisfy its requirement for its own product and the industry No. 1 requirement for the output of industry No. 2.

We start with interindustry requirements. Suppose that each dollar of industry No. 1 output requires $0.40 of its own output and $0.10 of industry No. 2 output. Similarly, each dollar of industry No. 2 output requires $0.50 of industry No. 1 output and $0.20 of industry No. 2 output. These interindustry requirements are summarized in the *technological matrix* in the following manner:

Technological Matrix

	User No. 1	User No. 2
Producer No. 1	0.4	0.5
Producer No. 2	0.1	0.2

The first column under *user* states that each dollar of industry No. 1 output requires $0.4 of industry No. 1 output and $0.1 of industry No. 2 output.

Exercise. What is the meaning of the second *user* column of the foregoing technological matrix? Answer: Each dollar of industry No. 2 output requires $0.5 of industry No. 1 output and $0.2 of industry No. 2 output.

Recalling that total outputs of No. 1 and No. 2 are x_1 and x_2, respectively, it follows that industry No. 1 must produce $0.4x_1$ to satisfy its own requirement and $0.5x_2$ to satisfy the industry No. 2's requirement for the output of industry No. 1. Thus, for interindustry requirements, No. 1 must produce $0.4x_1 + 0.5x_2$.

Exercise. How much must industry No. 2 produce to satisfy interindustry requirements? Answer: $0.1x_1 + 0.2x_2$.

In addition to satisfying interindustry requirements, output must also be sufficient to fill final demands, d_1 and d_2. Suppose that final demands for outputs of No. 1 and No. 2 are $90 and $200, respectively. If equilibrium is to be achieved, industry No. 1 must produce $0.4x_1 + 0.5x_2$ to meet interindustry demand, plus 90 to satisfy final demand. That is, its output, x_1, must be such that

$$x_1 = 0.4x_1 + 0.5x_2 + 90.$$

Exercise. What must x_2 be to meet interindustry demand and final demand? Answer: $x_2 = 0.1x_1 + 0.2x_2 + 200$.

We now have the linear system

$$e_1: \quad x_1 = 0.4x_1 + 0.5x_2 + 90$$
$$e_2: \quad x_2 = 0.1x_1 + 0.2x_2 + 200$$

or

$$e_1: \quad 0.6x_1 - 0.5x_2 = 90$$
$$e_2: \quad -0.1x_1 + 0.8x_2 = 200.$$

Exercise. Find the values of x_1 and x_2 which satisfy the foregoing system by the elimination procedure. Answer: $x_1 = 400$, $x_2 = 300$.

To interpret the answer in the exercise, let us substitute numbers into the original equation, e_1. We find

$$400 = 160 + 150 + 90.$$

This means that industry No. 1 produces \$400 of output, uses \$160 of this itself, supplies industry No. 2 with \$150, and satisfies final demand of \$90.

> **Exercise.** What is the disposition of the output of industry No. 2? Answer: Industry No. 2 output is \$300. It uses \$60 of this itself, supplies \$40 to industry No. 1, and satisfies final demand of \$200.

In the circumstances we have described, all requirements are met, and all output is used. We have solved the input-output problem for the given levels of final demand. But suppose that the levels of final demand change, what then will be the required outputs, given that the technological matrix (that is, the interindustry demand relationships) is constant? We could, of course, repeat the solution for each set of demands, but it is more efficient to employ the matrix solution discussed in the previous section of this chapter. First, let us display the relevant information in the following manner:

| | User | | Final | Total |
Producer	No. 1	No. 2	demand	output
No. 1	a_{11}	a_{12}	d_1	x_1
No. 2	a_{21}	a_{22}	d_2	x_2

The *user* columns constitute the technological matrix, A. Thus,

$$A = \begin{pmatrix} a_{11} & a_{12} \\ a_{21} & a_{22} \end{pmatrix}.$$

The next column is the (final) demand vector, D. Thus,

$$D = \begin{pmatrix} d_1 \\ d_2 \end{pmatrix}$$

and the last column is the total output vector, X. Thus,

$$X = \begin{pmatrix} x_1 \\ x_2 \end{pmatrix}.$$

In the technological matrix, a_{ij} represents the amount of industry i's output required in the output of \$1 by industry j. Thus, for example, if we had a 9 by 9 matrix, a_{37} would represent the amount of industry No. 3 output required in the output of \$1 by industry No. 7.

> ***Exercise.*** Interpret a_{21}. Answer: This is the amount of industry No. 2 product required in the output of \$1 by industry No. 1.

Following our introductory example, we may write

$$x_1 = a_{11} x_1 + a_{12} x_2 + d_1$$
$$x_2 = a_{21} x_1 + a_{22} x_2 + d_2.$$

The last may be rearranged to give

$$(1 - a_{11}) x_1 - a_{12} x_2 = d_1$$
$$-a_{21} x_1 + (1 - a_{22}) x_2 = d_2.$$

Changing now to matrix-vector form, we have

$$\begin{pmatrix} 1 - a_{11} & - a_{12} \\ - a_{21} & 1 - a_{22} \end{pmatrix} \begin{pmatrix} x_1 \\ x_2 \end{pmatrix} = \begin{pmatrix} d_1 \\ d_2 \end{pmatrix}.$$

In the last equation, the leftmost matrix is the same as $I - A$, where I is the identity matrix. That is,

$$\begin{pmatrix} 1 & 0 \\ 0 & 1 \end{pmatrix} - \begin{pmatrix} a_{11} & a_{12} \\ a_{21} & a_{22} \end{pmatrix} = \begin{pmatrix} 1 - a_{11} & - a_{12} \\ - a_{21} & 1 - a_{22} \end{pmatrix}.$$

We may therefore write the matrix-vector form compactly as

$$(I - A)X = D.$$

We can now solve for the solution (output) vector, X, by pre-multiplying both sides by the inverse, $(I - A)^{-1}$ to obtain

$$X = (I - A)^{-1}D.$$

The last statement shows that the core mathematical problem in input-output computations is the calculation of the inverse, $(I - A)^{-1}$, given the technological matrix, A. Let us demonstrate by repeating the example with which this section started. The relevant data are:

Producer	User No. 1	No. 2	Final demand	Total output
No. 1	0.4	0.5	90	x_1
No. 2	0.1	0.2	200	x_2

The technological matrix is

$$A = \begin{pmatrix} 0.4 & 0.5 \\ 0.1 & 0.2 \end{pmatrix}$$

or, in fractional form,

$$A = \begin{pmatrix} \dfrac{2}{5} & \dfrac{1}{2} \\ \dfrac{1}{10} & \dfrac{1}{5} \end{pmatrix}.$$

Subtracting A from the identity matrix yields

$$\begin{pmatrix} 1 & 0 \\ 0 & 1 \end{pmatrix} - \begin{pmatrix} \dfrac{2}{5} & \dfrac{1}{2} \\ \dfrac{1}{10} & \dfrac{1}{5} \end{pmatrix} = \begin{pmatrix} \dfrac{3}{5} & -\dfrac{1}{2} \\ -\dfrac{1}{10} & \dfrac{4}{5} \end{pmatrix}.$$

We proceed in the manner of the last section to find $(I - A)^{-1}$.

$$\left(\begin{array}{cc|cc} \dfrac{3}{5} & -\dfrac{1}{2} & 1 & 0 \\ -\dfrac{1}{10} & \dfrac{4}{5} & 0 & 1 \end{array} \right).$$

Divide row 1 by 3/5:

$$\left(\begin{array}{cc|cc} 1 & -\dfrac{5}{6} & \dfrac{5}{3} & 0 \\ -\dfrac{1}{10} & \dfrac{4}{5} & 0 & 1 \end{array} \right).$$

Multiply row 1 by 1/10, add to row 2

$$\left(\begin{array}{cc|cc} 1 & -\dfrac{5}{6} & \dfrac{5}{3} & 0 \\ 0 & \dfrac{43}{60} & \dfrac{1}{6} & 1 \end{array} \right).$$

Divide row 2 by 43/60:

$$\left(\begin{array}{cc|cc} 1 & -\dfrac{5}{6} & \dfrac{5}{3} & 0 \\ 0 & 1 & \dfrac{10}{43} & \dfrac{60}{43} \end{array} \right).$$

Multiply row 2 by 5/6, add to row 1:

$$\begin{pmatrix} 1 & 0 & \Big| & \dfrac{80}{43} & \dfrac{50}{43} \\[2ex] 0 & 1 & \Big| & \dfrac{10}{43} & \dfrac{60}{43} \end{pmatrix}.$$

Hence we have

$$(I - A)^{-1} = \begin{pmatrix} \dfrac{80}{43} & \dfrac{50}{43} \\[2ex] \dfrac{10}{43} & \dfrac{60}{43} \end{pmatrix}$$

or, after factoring,

$$(I - A)^{-1} = \frac{10}{43}\begin{pmatrix} 8 & 5 \\ 1 & 6 \end{pmatrix}.$$

The solution vector, $X = (I - A)^{-1}D$ then is

$$X = \frac{10}{43}\begin{pmatrix} 8 & 5 \\ 1 & 6 \end{pmatrix}\begin{pmatrix} 90 \\ 200 \end{pmatrix}$$

so that

$$x_1 = \frac{10}{43}(720 + 1000) = 400$$

$$x_2 = \frac{10}{43}(90 + 1200) = 300$$

and we have the same solution found earlier in the chapter. The matrix solution is more efficient, however, because with it we may now find quickly the solution vector for varying final demand levels. For example, if final demand changes to 215 for industry No. 1 and 430 for No. 2, we have

$$X = \frac{10}{43}\begin{pmatrix} 8 & 5 \\ 1 & 6 \end{pmatrix}\begin{pmatrix} 215 \\ 430 \end{pmatrix}.$$

Exercise. Compute the elements of the output vector for the demand levels just stated. Answer: $x_1 = 900$, $x_2 = 650$.

Matrix inversion with a 3 by 3 matrix is a tiresome and error-prone calculation when carried out by hand, and we shall not add arithmetical complications by enlarging our two-industry economy. The methodology is perfectly general, however, and applies to economies with hundreds or even thousands of industries. Input-output analysis is a thriving offshoot of mathematical economics, which has been nurtured by the computational power of modern large-scale computers.

Before turning to the Problem Set, let us review our two-industry input-output analysis methodology and present a final example. We have as given the values for the technological matrix, A, and the values for final demand vector, D. We proceed as follows:

1. Compute $I - A$; that is, subtract the technological matrix from the identity matrix.
2. Compute the inverse, $(I - A)^{-1}$.
3. Write the output (solution) vector as $X = (I - A)^{-1}D$.
4. Compute the components, x_1 and x_2, of the output vector X by matrix multiplication.

Suppose that final demands for the outputs of industries No. 1 and No. 2 are 330 and 550, respectively. One dollar of industry No. 1 output requires $\frac{1}{6}$ dollars of its own output and $\frac{1}{3}$ dollars of industry No. 2 output. Finally, \$1 of industry No. 2 output requires $\frac{1}{4}$ dollars of its own output and $\frac{1}{2}$ dollars of industry No. 1 output. the given information is tabulated next.

Producer	User No. 1	User No. 2	Final demand	Total output
No. 1	$\frac{1}{6}$	$\frac{1}{2}$	330	x_1
No. 2	$\frac{1}{3}$	$\frac{1}{4}$	550	x_2

The technological matrix is

$$A = \begin{pmatrix} \dfrac{1}{6} & \dfrac{1}{2} \\ \dfrac{1}{3} & \dfrac{1}{4} \end{pmatrix}.$$

Subtracting A from the identity matrix yields

$$(I - A) = \begin{pmatrix} 1 - \dfrac{1}{6} & 0 - \dfrac{1}{2} \\ 0 - \dfrac{1}{3} & 1 - \dfrac{1}{4} \end{pmatrix} = \begin{pmatrix} \dfrac{5}{6} & -\dfrac{1}{2} \\ -\dfrac{1}{3} & \dfrac{3}{4} \end{pmatrix}.$$

We omit the steps in the calculation of the inverse of $(I - A)$. The exercise following verifies that this is

$$(I - A)^{-1} = \frac{1}{11}\begin{pmatrix} 18 & 12 \\ 8 & 20 \end{pmatrix}.$$

Exercise. Verify that the foregoing is the correct inverse by showing that $(I - A)(I - A)^{-1} = I$. (We omit showing the steps but pause to note that it is wise to check an inverse by this procedure).

Finally, the output vector is

$$X = \frac{1}{11}\begin{pmatrix} 18 & 12 \\ 8 & 20 \end{pmatrix}\begin{pmatrix} 330 \\ 550 \end{pmatrix} = \frac{110}{11}\begin{pmatrix} 18 & 12 \\ 8 & 20 \end{pmatrix}\begin{pmatrix} 3 \\ 5 \end{pmatrix}.$$

Hence,

$$x_1 = 10(54 + 60) = 1140, \quad \text{and} \quad x_2 = 10(24 + 100) = 1240.$$

6.20 PROBLEM SET 6–6

1. Given the following:

Producer	User No. 1	No. 2	Final demand	Total output
No. 1	$\frac{1}{5}$	$\frac{3}{10}$	d_1	x_1
No. 2	$\frac{3}{5}$	$\frac{1}{10}$	d_2	x_2

a) What is the technological matrix, A?
b) What does the first column of A mean?
c) What is $(I - A)$?
d) What is $(I - A)^{-1}$?
e) Write the solution equation in matrix-vector form, using the elements found for $(I - A)^{-1}$.
f) Compute the elements x_1 and x_2 of the output vector if final demand is 270 for industry No. 1 and 405 for No. 2.
g) Compute the elements of the output vector if final demand is 540 for industry No. 1 and 810 for No. 2.

2. Each dollar of industry No. 1 output requires $\frac{1}{10}$ dollars of its own output and $\frac{3}{10}$ dollars of industry No. 2 output. Each dollar of industry No. 2

output requires ⅗ dollars of industry No. 1 output and ⅕ dollars of its own output. Let d_1, d_2 be final demands and x_1, x_2 total outputs.

a) Compute the matrix which completes the equation

$$\begin{pmatrix} x_1 \\ x_2 \end{pmatrix} = \begin{pmatrix} \quad \end{pmatrix} \begin{pmatrix} d_1 \\ d_2 \end{pmatrix}.$$

b) Find the elements of the output vector if $d_1 = 81$ and $d_2 = 135$.

c) Find the elements of the output vector if the demands in (b) double.

d) Find the elements of the output vector if $d_1 = 216$ and $d_2 = 171$.

3. Each dollar of industry No. 1 output requires ⅙ dollars of its own output and ⅗ dollars of industry No. 2 output. Each dollar of industry No. 2 output requires ¼ dollars of its own output and ¾ dollars of industry No. 1 output. Find the outputs of industry No. 1 and industry No. 2 if demands for final products are as follows: (Be careful in setting up the technological matrix).

a) $d_1 = 210$ and $d_2 = 420$.

b) $d_1 = 420$ and $d_2 = 630$.

4. Suppose that we have three industries whose total outputs are to be x_1, x_2, and x_3, respectively. Final demands are d_1, d_2, and d_3.

a) What is the meaning of a_{ij} in the general case?

b) What is the meaning of a_{13} in the three-industry case?

c) Write the elements of the technological matrix for the three industries.

d) Write the expanded matrix-vector expression for the output vector. That is, complete the equation

$$\begin{pmatrix} x_1 \\ x_2 \\ x_3 \end{pmatrix} = \begin{pmatrix} \quad \end{pmatrix} \begin{pmatrix} \quad \end{pmatrix}.$$

6.21 SUMMATION SYMBOL

The letter Σ (sigma) is the mathematical symbol for summation. The expression

$$\sum_{j=1}^{3} j$$

is read as "sigma j, j going from 1 to 3" and means to insert 1 for j, then 2 for j, then 3 for j, and *sum* the results. Thus:

$$\sum_{j=1}^{3} j = 1 + 2 + 3 = 6.$$

Similarly:

$$\sum_{j=2}^{5} j = 2 + 3 + 4 + 5 = 14$$

$$\sum_{j=1}^{4} 2j = 2(1) + 2(2) + 2(3) + 2(4) = 20$$

$$\sum_{j=1}^{3} (j-1) = (1-1) + (2-1) + (3-1) = 3$$

$$\sum_{j=2}^{3} j^3 = 2^3 + 3^3 = 8 + 27 = 35.$$

We shall have frequent occasion for indicating the sum of n terms, without specifying a particular value for n. Consider, for example, the sum of the first n integers. In expanded form, we would indicate this sum by

$$1 + 2 + 3 + \cdots + n$$

where the three dots are read "and so on" and mean that the unwritten terms of the series are formed according to the same rule that applies to the first written terms; that is, in the present case, each number is formed by adding one to the preceding number. In compact summation notation the sum of the first n integers is

$$\sum_{j=1}^{n} j$$

because, by definition, this symbol expands to

$$\sum_{j=1}^{n} j = 1 + 2 + 3 + \cdots + n.$$

6.22 PROBLEM SET 6–7

Find the numerical values of

1. $\sum_{p=4}^{7} p.$

2. $\sum_{q=1}^{n} q$, if n is 5.

3. $\sum_{u=6}^{9} u^2.$

4. $\sum_{j=1}^{5} 3j.$

5. $\sum_{p=1}^{3} p^3.$

Express in summation notation:

6. The sum of the first q integers.
7. The sum of the squares of the first n integers.
8. $1 + 8 + 27 + 64.$
9. $(1 + 2) + (2 + 2) + (3 + 2) + (4 + 2).$
10. $3(1) + 3(2) + 3(3) + 3(4) + 3(5).$

6.23 SUMMATION ON INDICES: LINEAR EQUATIONS

As a first step toward expressing equations in compact summation form, we choose one letter and tag it with different subscripts to indicate different variables, or choose one letter and tag it with subscripts to indicate different constants. For example, x_1, x_2, x_3 can be used to represent three different variables, and c_1, c_2, c_3 to represent three different constants.

Now, consider the symbol

$$\sum_{j=1}^{3} x_j.$$

As before, the statement means to replace j first by 1, then by 2, then by 3, and sum the results. We see that

$$\sum_{j=1}^{3} x_j = x_1 + x_2 + x_3.$$

Similarly:

$$\sum_{j=1}^{5} a_j = a_1 + a_2 + a_3 + a_4 + a_5$$

$$\sum_{j=1}^{n} x_j = x_1 + x_2 + x_3 + \cdots + x_n.$$

It is clear that

$$\sum_{j=1}^{3} a_j x_j = a_1 x_1 + a_2 x_2 + a_3 x_3$$

$$\sum_{j=1}^{n} a_j x_j = a_1 x_1 + a_2 x_2 + \cdots + a_n x_n.$$

Next, consider the expression

$$\sum_{j=1}^{2} a_j x_j = b$$

where the a_j and b are constants. In expanded form, the expression becomes

$$a_1 x_1 + a_2 x_2 = b$$

which is a linear equation in the two variables, x_1 and x_2. It is clear that any linear equation can be written in compact summation notation as

$$\sum_{j=1}^{n} a_j x_j = b$$

because any linear equation is of the form to which the last expression expands; namely:

$$a_1 x_1 + a_2 x_2 + \cdots + a_n x_n = b.$$

For example, the linear equation

$$5x_1 + 2x_2 + 4x_3 = 6$$

is of the stated form with

$$n = 3$$
$$a_1 = 5$$
$$a_2 = 2$$
$$a_3 = a_n = 4$$
$$b = 6.$$

6.24 PROBLEM SET 6–8

Write the expanded form of:

1. $\displaystyle\sum_{j=1}^{5} y_j.$ 2. $\displaystyle\sum_{j=1}^{3} c_j x_j.$

3. $\displaystyle\sum_{j=1}^{n} b_j y_j.$ 4. $\displaystyle\sum_{j=1}^{5} p_j x_j = 10.$

5. $\displaystyle\sum_{j=1}^{n} a_j x_j = c.$

Express in compact summation form:

6. $x_1 + x_2 + x_3 + x_4.$
7. $a_1 x_1 + a_2 x_2 + a_3 x_3 + a_4 x_4 + a_5 x_5 + a_6 x_6 = b.$
8. $c_1 x_1 + c_2 x_2 + \cdots + c_9 x_9.$
9. $a_1 x_1 + a_2 x_2 + \cdots + a_q x_q.$
10. Identify n, b, and each of the a's in

$$\sum_{j=1}^{n} a_j x_j = b$$

with a number in the equation

$$x_1 + 2x_3 + 5x_4 = 7.$$

6.25 SUMMATION FORM FOR SYSTEMS

We have seen how to express a single linear equation in terms of the summation symbol by tagging variables with a subscript. To extend the notation to systems of equations, another subscript is required to identify the different equations of the system. Consider the following expression:

$$\sum_{j=1}^{3} a_{ij} x_j = b_i \qquad i = 1, 2.$$

The symbolism means to insert $i = 1$ first, obtaining

$$\sum_{j=1}^{3} a_{1j} x_j = b_1.$$

Then insert $i = 2$ in the original summation, obtaining

$$\sum_{j=1}^{3} a_{2j} x_j = b_2.$$

The expression with $i = 1$ expands to

$$\sum_{j=1}^{3} a_{1j} x_j = a_{11} x_1 + a_{12} x_2 + a_{13} x_3 = b_1.$$

The expression with $i = 2$ expands in the same manner. Thus the original expression

$$\sum_{j=1}^{3} a_{ij} x_j = b_i \qquad i = 1, 2$$

means the following system of two linear equations in three variables:

$$a_{11} x_1 + a_{12} x_2 + a_{13} x_3 = b_1$$
$$a_{21} x_1 + a_{22} x_2 + a_{23} x_3 = b_2.$$

The first number in each double subscript names the equation; the second names the variable. Thus, a_{21} is in the second equation, and it is the coefficient of variable number one, that is, x_1. Similarly, a_{49} would be in the fourth equation, and would be the coefficient of x_9 in that equation.

Exercise. Write the system of equations specified by

$$\sum_{j=1}^{2} c_{ij} x_j = b_i \qquad i = 1, 2, 3.$$

Answer: The system has three equations, each of the form $c_{i1} x_1 + c_{i2} x_2 = b_i$. The three equations can be generated by letting i equal 1, then 2, then 3.

Consider the linear system

$$a_{11} x_1 + a_{12} x_2 + a_{13} x_3 + a_{14} x_4 = b_1$$
$$a_{21} x_1 + a_{22} x_2 + a_{23} x_3 + a_{24} x_4 = b_2$$
$$a_{31} x_1 + a_{32} x_2 + a_{33} x_3 + a_{34} x_4 = b_3.$$

Note that the beginning subscript on all a's in an equation and the subscript on b are the same as the equation number. For example, the first equation has ones in the positions just described. If we let i represent the equation number, it follows in the present case that i must take on the values 1, 2, and 3 to generate the three equations. Next, letting j be the number indicating the variable, it follows that each equation is a summation with j going from 1 to 4, inclusive. The foregoing system of equations, therefore, is represented by

$$\sum_{j=1}^{4} a_{ij} x_j = b_i \qquad i = 1, 2, 3.$$

The first equation of the system is generated from the last expression by letting i be 1, and then constructing the sum for j going from 1 through 4. The other two equations are generated in the same manner after letting i be 2, then 3.

Exercise. How would a system of three equations in five variables be symbolized in sigma notation? Answer: The symbols would be the same as those written in the immediately foregoing expression, except the upper limit on j would be 5 rather than 4.

We are now ready to symbolize any system of linear equations. Suppose that we have m linear equations in n variables, an m by n system. To indicate this system in expanded form, we would write

$$a_{11} x_1 + a_{12} x_2 + \cdots + a_{1n} x_n = b_1$$
$$a_{21} x_1 + a_{22} x_2 + \cdots + a_{2n} x_n = b_2$$
$$\vdots \qquad \vdots \qquad \qquad \vdots \qquad \vdots$$
$$a_{m1} x_1 + a_{m2} x_2 + \cdots + a_{mn} x_n = b_m.$$

In compact summation form the system is simply

$$\sum_{j=1}^{n} a_{ij} x_j = b_i \qquad i = 1, 2, \ldots, m.$$

6.26 LINEAR PROGRAMMING PROBLEMS IN SUMMATION NOTATION

The linear programming problems of the last chapter consisted of a system of linear inequalities with nonnegativity constraints, and an objective function which was to be maximized. The nonnegativity constraints can be expressed as

$$x_j \geq 0 \text{ for all } j.$$

The objective function is of the form

$$\theta = c_1 x_1 + c_2 x_2 + \cdots + c_n x_n$$

where the letter c is adopted to represent the constants in this function. In symbols:

$$\theta = \sum_{j=1}^{n} c_j x_j.$$

The system of linear inequalities can be expressed in the same manner as a system of equalities. Hence the compact summation form of the linear programming problems of the last chapter is

Maximize

$$\sum_{j=1}^{n} c_j x_j$$

subject to:

$$\sum_{j=1}^{n} a_{ij} x_j \le b_i \qquad i = 1, 2, \ldots, m$$

$$x_j \ge 0 \qquad \text{for all } j.$$

6.27 PROBLEM SET 6–9

1. Write in expanded form:

$$\sum_{j=1}^{4} a_{ij} x_j = b_i \qquad i = 1, 2.$$

2. Write in expanded form:
   Maximize

$$\sum_{j=1}^{3} c_j x_j$$

subject to:

$$\sum_{j=1}^{3} a_{ij} x_j \le b_i \qquad i = 1, 2, \ldots, 4$$

$$x_j \ge 0 \qquad \text{for all } j.$$

3. A system has p linear equations in q variables.
   a) Write the system in expanded form, using . . . where necessary.
   b) Write the system in compact summation form.
4. A linear programming maximization problem has p "less than or equal" constraints on q variables.
   a) Write the statement of the problem in expanded form, using . . . where necessary.
   b) Write the problem in compact summation form.
5. Express the following in compact summation notation:

$$a_{11} x_1 + a_{12} x_2 + \cdots + a_{18} x_8 = b_1$$
$$a_{21} x_1 + a_{22} x_2 + \cdots + a_{28} x_8 = b_2$$
$$\vdots \qquad \vdots \qquad \qquad \vdots \qquad \vdots$$
$$a_{61} x_1 + a_{62} x_2 + \cdots + a_{68} x_8 = b_6.$$

6. In the system

$$2x_1 + 3x_2 + 4x_4 = 20$$
$$x_1 + 9x_3 + 8x_4 = 50$$
$$5x_1 + 7x_3 + 6x_4 = 100,$$

   a) What is m? What is n?
   b) What symbol would represent the following constants which appear in the system: 2, 3, 8, 9, 7, 6?
   c) What constants in the system correspond to a_{13}, b_1, a_{31}, a_{22}, a_{32}?

7. We buy x_1 units of item number one at d_1 dollars per unit, x_2 units of item two at d_2 dollars per unit, and so on. Write the expression for the total cost of 10 different items:
   a) In expanded form.
   b) In compact summation notation.

8. Repeat parts (a) and (b) of Problem 7 if we buy p different items.

6.28 APPLICATIONS OF Σ IN STATISTICS

Statistical formulas which describe the computations to be performed on data make extensive use of the summation operation and its symbol. Suppose that we have tabulated sales data for eight weeks, as follows:

Week i	Sales in $ thousands S_i
1	10
2	12
3	8
4	9
5	14
6	11
7	12
8	12

Total sales for the eight weeks would be

$$\sum_{i=1}^{8} S_i = \$88 \text{ thousand.}$$

Often, the index of summation is not specified and in such cases we simply assume that the summation runs over all the data. Thus, $\Sigma\ S$

means the sum of weekly sales for as many weeks *(n)* as are entered in the table. Viewed in this tabular light, $\Sigma \, S$ is an instruction to sum the data numbers whose name is S. If the values of S are in a column, the instruction $\Sigma \, S$ means to sum the column headed S.

The arithmetic (pronounced arith-met'ic) average is the ordinary average found by dividing the sum of the data items by the number of items. We find average weekly sales to be

$$\frac{\text{Sum of weekly sales}}{\text{Number of weeks}} = \frac{88}{8} = \$11 \text{ thousand.}$$

In general, if we have n items of sales data, S, we may denote the arithmetic average as $\overline{S}$ (read S *bar*) and compute it as

$$\overline{S} = \frac{\sum_{i=1}^{n} S_i}{n} \text{ or, more briefly, } \overline{S} = \frac{\Sigma \, S}{n}.$$

Exercise. If your grades on four examinations are 70, 85, 95, 80, and g represents grades, what formula denotes your average grade? What is n? Compute this average. Answer: $\overline{g} = (\Sigma \, g)/n; \; n = 4; \; \overline{g} = 330/4 = 82.5$.

Squares and products of variables play an important role in statistics. Common forms are:[4]

$$\Sigma \, x^2 = \text{The sum of the squares of the values of } x.$$
$$(\Sigma \, x)^2 = \text{The square of the sum of the } x\text{'s.}$$
$$\Sigma \, xy = \text{The sum of the } xy \text{ products.}$$
$$(\Sigma \, x)(\Sigma \, y) = \Sigma \, x \, \Sigma \, y = \text{The product of the sum of the } x\text{'s times the sum of the } y\text{'s.}$$

To illustrate, suppose that x represents sales of a company during a given month and y represents expenses during the same month. Table 6–6 shows sales-expense data for four months to the left of the double divider. To the right of the double divider we have computed extensions headed x^2, y^2, and xy. The column headed x^2 contains the squares of the respective values of x, similarly for the column headed y^2. The xy column contains the xy products for each data pair. The number of data pairs (which is the number of rows in the table) is the number n and here $n = 4$.

---

[4] Sometimes large X is used for the data numbers and small x is defined to be the difference between a data number and the average of all the data numbers; that is, $x = X - \overline{X}$. Here we use small x to represent a data number, not the difference $X - \overline{X}$.

TABLE 6–6

x	y	x^2	y^2	xy
5	3	25	9	15
10	5	100	25	50
6	4	36	16	24
8	4	64	16	32
29	16	225	66	121

We see that $\Sigma\ x^2 = 225$, $\Sigma\ y^2 = 66$, and $\Sigma\ xy = 121$. Note carefully, however, that

$$(\Sigma\ x)^2 = (29)^2 = 841.$$

Exercise. Find $(\Sigma\ y)^2$ and $\Sigma\ y\ \Sigma\ x$ from Table 6–6.
Answer: $(\Sigma\ y)^2 = (16)^2 = 256$, $\Sigma\ x\ \Sigma\ y = (29)(16) = 464$.

Fitting a straight line to observational data. The sales *(x)* and expense *(y)* data of Table 6–6 are plotted on Figure 6–2. We call these *observational* data because they represent actual numbers recorded from sales records, so the points of Figure 6–2 are not obtained from an equation and obviously do not all lie on a straight line. However, the points indicate a linear *tendency,* which we wish to describe *approximately* by a straight-line equation. Our desire is to run a straight line through the data, and have this line come as close as possible (in a sense to be defined shortly) to the points taken as a group. The criterion commonly applied to specify the *best-fitting* line is called

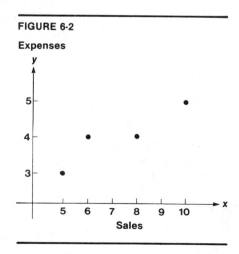

FIGURE 6-2

Expenses

the *least-squares* criterion and is developed in a later chapter (see Index reference to *least squares*). In brief, the criterion states that the best-fitting line is the one for which the sum of the squared errors has the smallest possible value, where each error is the vertical distance between an observed point and the line. We need not concern ourselves here with the details of the criterion because all we wish to do is show that an understanding of summation notation makes it possible to carry out the statistical procedure for determining the best-fitting line even though we may not have statistical training.

According to the least-squares criterion, the line that best fits the points has an equation of the form

$$y = mx + b$$

where the slope, m, and the intercept, b, are computed from the data by means of the formulas

$$m = \frac{n \Sigma\, xy - \Sigma\, x\, \Sigma\, y}{n \Sigma\, x^2 - (\Sigma\, x)^2},$$

$$b = \frac{\Sigma\, y - m \Sigma\, x}{n}.$$

From Table 6–6, we have that

$$
\begin{aligned}
n &= 4 & \Sigma\, xy &= 121 \\
\Sigma\, x &= 29 & \Sigma\, x\, \Sigma\, y &= 464 \\
\Sigma\, y &= 16 & \Sigma\, x^2 &= 225 \\
& & (\Sigma\, x)^2 &= 841.
\end{aligned}
$$

Substituting, we find the slope of the best fitting line to be

$$\frac{4(121) - (29)(16)}{4(225) - (29)^2} = \frac{484 - 464}{900 - 841} = \frac{20}{59} = 0.34.$$

With this value for m, we find the y-intercept to be

$$b = \frac{16 - (0.34)(29)}{4} = 1.54.$$

The desired equation then is

$$y = 0.34x + 1.54.$$

The last example illustrates the fact that an understanding of the meaning of the summation symbol makes the techniques of statistical analysis available to the non-statistician.

Exercise. If we let $x = 0$ and $x = 10$ in the last equation, the corresponding values of y are 1.54 and 4.94, respectively. On Figure 6–1, plot lightly, with pencil, the points (0, 1.54) and (10, 4.94) and verify that the line joining them does indeed come close to the points.

6.29 PROPERTIES OF THE Σ OPERATION

An elementary, but important, property of summation is illustrated by

$$\sum_{i=1}^{3} 5 = 5 + 5 + 5 = 15.$$

We see here that the quantity being summed is the *constant*, 5, when $i = 1$, $i = 2$, and $i = 3$. We could write more directly

$$\sum_{i=1}^{3} 5 = 3(5) = 15.$$

Exercise. Compute the value of $\sum_{i=1}^{8} 2$. Answer: $8(2) = 16$.

In general, if c is any *constant*, then

$$\sum_{i=1}^{n} c = nc.$$

The word *constant* here means that the expression is independent of the index of summation.

Exercise. What is $\sum_{j=1}^{m} p$? Answer: mp.

The point of the last exercise is that the expression p is constant with respect to the index of summation.

A second property of the summation operation is that it may be applied term by term to an expression.

To illustrate,

$$\sum_{i=1}^{3} (x_i + y_i) = x_1 + y_1 + x_2 + y_2 + x_3 + y_3$$

$$= x_1 + x_2 + x_3 + y_1 + y_2 + y_3$$

$$\sum_{i=1}^{3} (x_i + y_i) = \sum_{i=1}^{3} x_i + \sum_{i=1}^{3} y_i.$$

The property obviously applies whatever the value of n; that is,

$$\sum_{i=1}^{n} (x_i + y_i - z_i) = \sum_{i=1}^{n} x_i + \sum_{i=1}^{n} y_i - \sum_{i=1}^{n} z_i.$$

Finally, a constant factor of the quantity being summed can be taken outside the summation symbol. Thus,

$$\Sigma\, 3x_i = 3\, \Sigma\, x_i \text{ and, in general, } \Sigma\, cx_i = c\, \Sigma\, x_i$$

where c is a constant with respect to the index of summation.

Summarizing the properties just discussed, we have

1. $\displaystyle\sum_{i=1}^{n} cx_i = c \sum_{i=1}^{n} x_i;$ or $\Sigma\, cx_i = c\, \Sigma\, x_i.$

2. $\displaystyle\sum_{i=1}^{n} c = nc;$ or $\Sigma\, c = nc,$ assuming c is constant with respect

    to the index, which goes from 1 to n.

3. $\displaystyle\sum_{i=1}^{n} (x_i + y_i - z_i) = \sum_{i=1}^{n} x_i + \sum_{i=1}^{n} y_i - \sum_{i=1}^{n} z_i,$

    or

    $\Sigma\, (x_i + y_i - z_i) = \Sigma\, x_i + \Sigma\, y_i - \Sigma\, z_i.$

For practice with these properties, let us compute

$$\Sigma\, (4x + 2)$$

for the following data: $x = 1, 2,$ and 3. We may write

$$\Sigma\, (4x + 2) = \Sigma\, 4x + \Sigma\, 2 = 4\, \Sigma\, x + 3(2)$$

where, in the last term, the factor 3 is the value of n for the three data items. Now, $\Sigma\, x$ is $1 + 2 + 3 = 6$, so

$$4\, \Sigma\, x + 3(2) = 4(6) + 6 = 30.$$

Alternatively, we could have performed the calculation as shown in Table 6–7,

TABLE 6–7

x	$4x + 2$
1	6
2	10
3	14
	30

which shows that $\Sigma\, (4x + 2) = 30$.

Exercise. Given the data y: 5, 7, 2, 9, compute $\Sigma\ (2y - 1)$ by the two procedures just illustrated. Answer: 42.

Continuing, let us find an alternate expression for

$$\Sigma\,(x - \bar{x})^2.$$

The expression implies that we have a set of n values for the variable x, and we have computed the average of these n values, $\bar{x}$. Hence, the values of x vary, but $\bar{x}$ is a constant. Squaring the expression leads to

$$\Sigma\,(x - \bar{x})^2 = \Sigma(x^2 - 2x\bar{x} + \bar{x}^2) = \Sigma\,x^2 - \Sigma\,2x\bar{x} + \Sigma\,\bar{x}^2.$$

Inasmuch as $\bar{x}$ is a constant, we may write the last as

$$\Sigma\,x^2 - 2\bar{x}\Sigma\,x + n\bar{x}^2.$$

Substituting from the definition of $\bar{x}$,

$$\bar{x} = \frac{\Sigma\,x}{n},$$

we find

$$\Sigma\,x^2 - 2\,\frac{\Sigma\,x}{n}\left(\Sigma\,x\right) + n\frac{(\Sigma\,x)^2}{n^2} = \Sigma\,x^2 - \frac{(\Sigma\,x)^2}{n}.$$

To illustrate what has just been proved, consider the data $x = 4$, 6, 11. Here, $\bar{x} = 21/3 = 7$, and $n = 3$. Direct calculation of $\Sigma\,(x - \bar{x})^2$ is shown in Table 6–8.

TABLE 6–8

x	$(x - \bar{x})$	$(x - \bar{x})^2$
4	−3	9
6	−1	1
11		16
$\bar{x} = 7$	4	26

We find that by direct calculation, $\Sigma\,(x - \bar{x})^2 = 26$. The alternate computation requires only $\Sigma\,x^2$ and $(\Sigma\,x)^2$, as shown next.

x	x^2
4	16
6	36
11	121
21	173

$$\Sigma\,x^2 - \frac{(\Sigma\,x)^2}{n} = 173 - \frac{(21)^2}{3} = 173 - 147 = 26.$$

Both the direct and the alternate method of computation lead to the same result, 26. In this example, as in others which arise in statistics, the advantage of the alternate method of computation is that it is simpler than the direct calculation.

As an alternative way of providing practice in the use of Σ properties, let us specify that

$$\sum_{i=1}^{n} (2x_i - 3)$$

is to be expressed in terms of n and $\bar{x}$. Expanding, we have $\Sigma(2x_i - 3) = \Sigma(2x_i) - \Sigma(3) = 2\Sigma x_i - 3n = 2n\bar{x} - 3n$, as specified. Observe the substitution of $n\bar{x}$ for Σx_i to obtain the result in the specified form. As another example, let us express

$$\sum_{i=1}^{n} x_i (y_i - \bar{y})$$

in terms of $\Sigma x_i y_i$, n, Σx_i, and Σy_i. Expanding, we find

$$\Sigma x_i y_i - \Sigma y_i \bar{y} = \Sigma x_i y_i - \bar{y}\Sigma x_i = \Sigma x_i y_i - \frac{(\Sigma y_i)(\Sigma x_i)}{n}$$

as specified.

> **Exercise.** The average of the x_i's is $\bar{x}$. If a is added to each x_i, the new average will be $\Sigma(x_i + a)/n$. Express the new average in terms of the old average and a. Interpret the outcome. Answer: The new average is $\bar{x} + a$, which means that if the constant a is added to each x_i, the average will be increased by a.

6.30 THE VARIANCE AND STANDARD DEVIATION

Conventionally in statistics, $\bar{x}$ represents the average of a *sample* of items from an overall group called the *population*. The symbol μ (mu) is used to represent the average of the population. Thus, if 10 students in a class of 30 averaged 75 percent on a test and the class average was 80 percent, we would use $\bar{x} = 75$ for the sample of 10 students, but $\mu = 80$ percent for the population of 30 students. Similarly, the measures we shall discuss in this section are represented by s^2 and s in samples and σ and σ^2 for populations, where σ is the small Greek letter sigma. Our reason for using σ and μ in this section is that we plan to develop in the next section an important result concerning the variation of the numbers in a *population* about the population average.

The *variance* of a population of n data items, symbolized by σ^2, is defined as

$$\sigma^2 = \frac{\sum\limits_{i=1}^{n} (x_i - \mu)^2}{n}.$$

The standard deviation of the population, σ, is the square root of the variance. To use this formula for a given set of x_i's, we must first find the average, $\mu = \Sigma x_i / n$, in the usual manner as shown at the foot of the first column of the next table.

x_i	$x_i - \mu$	$(x_i - \mu)^2$
4	−5	25
13	4	16
10	1	1
$\Sigma x_i = 27$		$42 = \Sigma(x_i - \mu)^2$
$\mu = 27/3 = 9.$		

Then, in the second column, we compute the deviation of each x_i from the average, $\mu = 9$. Squaring the second column entries leads to the third column entries. The variance is the average of the third column,

$$\sigma^2 = \frac{42}{3} = 14.$$

The standard deviation is $\sigma = \sqrt{14} = 3.74$.

> *Exercise.* Compute the variance and standard deviation of the two-item population 4, 10. Answer: Variance, σ^2, is 9; standard deviation, σ, is 3.

The standard deviation is a measure of variability. Admittedly, this measure has little intuitive appeal. However, it has important practical applications, which we shall see in the next section and in a later chapter on probability.

6.31 TCHEBYSHEFF'S THEOREM

To set the stage for this section, suppose that a college student learns that the 30 students who graduated last year with a major in a certain subject area are receiving annual salaries averaging $25 thousand, with a standard deviation of $3 thousand. By itself, the average salary seems quite impressive, but the student realizes that very different groups of numbers could lead to this average. For example, the majority of salaries could be considerably below $25 thousand, with a relatively small group of very high salaries pulling the average up; or the salaries may spread more or less evenly over a wide interval; or salaries may cluster very closely around $25 thousand. Clearly, these circumstances have different implications to the student considering the subject area as a major, and it would be helpful to know whether or not the individual salaries cluster closely about the average. The standard deviation is a measure of variability around the average, and our objective is to show how to use the standard deviation to make specific statements, such as, "at least 75 percent of all the salaries are in the interval $19 to $31 thousand," based upon the given average and standard deviation. The method is:

a) Arbitrarily pick any number greater than or equal to 1 and call this number k. For example, suppose $k = 2$.

b) Form the interval $\mu \pm k$(standard deviation). In our example, this is $25 \pm 2(\$3)$ thousand, or $19 to $31 thousand.

c) Compute the fraction $1 - 1/k^2$. For our example, $1 - 1/2^2 = 3/4$.

d) Make the statement, "At least $1 - 1/k^2$ of the numbers in the population are in the interval $\mu \pm k$(standard deviation)." Our result is "At least $3/4$ of the salaries are in the interval $19 to $31 thousand."

If we had started with the multiple $k = 3$ in step a), the interval would be b) $25 \pm 3(\$3)$ or $16 to $34 thousand. In c) we would find $1 - 1/k^2 = 1 - 1/3^2 = 1 - 1/9$ or $8/9$. The statement d) would be "At least $8/9$ of the numbers in the population are in the interval $16 to $34 thousand."

Exercise. If we start with the multiplier $k = 2.5$ in a), what would be the resulting statement in d)? Answer: "At least 0.84 or 84 percent of the salaries are in the interval $17,500 to $32,500."

The average and the standard deviation play important roles in a world where summarization is needed to aid in comprehension of large masses of data. The procedure we have just illustrated is important because it applies to any population of numerical data. The procedure is justified by the theorem we shall now prove.

Tchebysheff's Theorem: At least $1 - 1/k^2$ of the data items in a

population are in the inclusive interval that extends k standard devia-
tions on either side of the population mean, μ.

We start the proof by writing the definition of the variance,

$$\sigma^2 = \frac{\sum\limits_{i=1}^{n}(x_i - \mu)^2}{n}. \tag{1}$$

Now if we pick an interval extending $k\sigma$ on either side of the average, μ, a data item either is in this interval or it is outside the interval. Note in particular that for an item outside the interval, the deviation $(x_i - \mu)$ is larger absolutely than $k\sigma$, so for such an item, $(x_i - \mu)^2 > (k\sigma)^2$. Now suppose the number of items inside the interval is m, so the remaining $n - m$ are outside. Let us separate the two groups by writing (1) in two parts. Thus,

$$\sigma^2 = \frac{\sum\limits_{i=1}^{m}(x_i - \mu)^2 + \sum\limits_{i=m+1}^{n}(x_i - \mu)^2}{n}. \tag{2}$$

Observe that all the individual terms in the sums in the numerator are squares, so none can be negative. Consequently, if we discard the first sum of squares in the numerator, we will have a smaller value on the right (or the same value if the discarded amount is zero). It follows that

$$\sigma^2 \geq \frac{\sum\limits_{i=m+1}^{n}(x_i - \mu)^2}{n}. \tag{3}$$

Now remember that there are $n - m$ terms in the numerator sum in (3), and for each of the x_i,

$$(x_i - \mu)^2 > (k\sigma)^2$$

because we are using only items that are more than $k\sigma$ from μ. Consequently, if we replace each $(x_i - \mu)^2$ in (3) by $(k\sigma)^2$ we are using smaller values at the right, and it remains true that

$$\sigma^2 \geq \frac{\sum\limits_{i=m+1}^{n}(k\sigma)^2}{n}.$$

Inasmuch as $(k\sigma)^2$ is a constant, and we are summing $n - m$ of these constants, we have

$$\sigma^2 \geq \frac{(n - m)(k\sigma)^2}{n}.$$

Cancelling the σ^2, and writing $(n - m)/n$ as $1 - m/n$, we have

$$1 \geq \left(1 - \frac{m}{n}\right)k^2 \quad \text{or} \quad \frac{1}{k^2} \geq 1 - \frac{m}{n},$$

from which

$$\frac{m}{n} \geq 1 - \frac{1}{k^2}.$$

Now, since m is the number of items in the interval and n is the total number of items, m/n is the proportion of items in the interval. Hence, the last statement says that the proportion in the inclusive interval $\mu \pm k\sigma$ is at least (greater than or equal to) $1 - 1/k^2$, which proves the theorem.

It is often stated that an average is misleading because it is not *typical* or *representative* as illustrated by the comment, Sure, the average salary is $25,000, but that does not mean much to me because it is not representative. What is implied here, at least in part, is that salaries vary so greatly that the average is not a reliable indicator for an individual. The moral of this section is that the standard deviation provides a means of assessing representativeness, and the theorem just proved shows a way of using the standard deviation to this end that applies to any population of data.

To close this section let us practice the use of the theorem by stating an example in a form slightly different from that used in the introductory examples.

Example. If the average and standard deviation of monthly salaries paid in a certain job category are $1200 and $100, respectively, what can we say about the proportion of workers earning between $800 and $1600, inclusive? Here we note that the interval $\mu \pm k\sigma$ is $1200 \pm 4(100)$, so $k = 4$. Hence, $1 - 1/k^2 = 1 - 1/16 = 15/16$ or 0.9375, so that at least 93.75 percent of all workers earn between $800 and $1600, inclusive.

> *Exercise.* From the data of the foregoing example, what can we say about the proportion of workers earning between $1000 and $1400, inclusive? Answer: At least 75 percent have salaries in the interval.

6.32 PROBLEM SET 6–10

1. Given the values $x = 2, 3,$ and 7, compute the following:
   a) $\Sigma\, x$. b) $\Sigma\, x^2$. c) $\bar{x}$. d) $\Sigma\, (x - \bar{x})^2$.

   e) The square of the average of the x's.

   f) The average of the squares of the x's.

2. Given the values $y = 5, 7, 12$, compute the following:

   a) $\Sigma y.$ *b)* $\Sigma y^2.$ *c)* $\bar{y}.$ *d)* $\Sigma (y - \bar{y})^2.$

   e) The square of the average of the y's.

   f) The average of the squares of the y's.

3. Compute the following from the tabulated values of x and y:

   a) $\Sigma xy.$ *b)* $\Sigma x \Sigma y.$ *c)* $\Sigma x^2.$

   d) $(\Sigma x)^2.$ *e)* $\Sigma (x - \bar{x})(y - \bar{y}).$

x	y
4	7
5	8
9	10
2	3

4. Compute the following from the tabulated values of x and y.

   a) $\Sigma xy.$ *b)* $\Sigma x \Sigma y.$ *c)* $\Sigma x^2.$

   d) $(\Sigma x)^2.$ *e)* $\Sigma (x - \bar{x})(y - \bar{y}).$

x	y
1	2
3	4
5	6

5. *a)* Plot the following points on a graph.

x	y
3	1
4	3
5	6
6	6
7	9

   b) Find the equation of the best fitting line, $y = mx + b$, where

$$m = \frac{n \Sigma xy - \Sigma x \Sigma y}{n \Sigma x^2 - (\Sigma x)^2}, \quad \text{and} \quad b = \frac{\Sigma y - m \Sigma x}{n}.$$

   c) Sketch the line found in *(b)* on the graph in *(a)*.

6. Carry out the instructions in 5*(a)*, 5*(b)*, and 5*(c)* for the data

x	y
3	4
12	16
6	8
9	12

.

7. The *coefficient of linear correlation* is a measure of the degree of relationship between the two variables. It is designated as r, and its value is always in the interval -1 to $+1$. Values near -1 or $+1$ indicate high correlation, and values near zero indicate low correlation. The coefficient is computed from the formula

$$r = \frac{n \Sigma xy - \Sigma x \Sigma y}{\sqrt{[n \Sigma x^2 - (\Sigma x)^2][n \Sigma y^2 - (\Sigma y)^2]}}.$$

The sign of the square root is $+$. The slope, m, of the best fitting line is positive when r is positive, and negative when r is negative. That is, r is positive when y tends to increase as x increases and negative when y tends to decrease as x increases.

If the observed points, (x, y), all lie on a straight line rising to the right (m positive), r will be 1. Verify this statement by computing r for the following points, all of which lie on the line $y = 2x + 5$.

x	y
0	5
2	9
3	11

8. (See Problem 7). Compute the correlation coefficient from the following observed values of x and y.

x	y
-2	0
-1	0
0	1
1	1
2	3

Compute the average, variance and standard deviation for 9–11.

9. x_i: 19, 23, 17, 21, 20.

10. x_i: 10, 15, 6, 12, 7, 10.

11. x_i: 8, 9, 11, 1, 2, 3, 4, 8, 8.

12.   Given the values $x = 3, 4, 8,$
  a)   Show that $\Sigma\,(x - \bar{x}) = 0.$
  b)   Prove that for *any* group of data, the sum of the deviations of the data items from their average is 0. That is, prove

$$\sum_{i=1}^{n} (x_i - \bar{x}) = 0.$$

13.   *a)*   Prove that

$$\sum_{i=1}^{n} (x_i - 5)^2 = \sum_{i=1}^{n} x_i^2 - 10n\bar{x} + 25n.$$

  b)   Illustrate *(a)* with the data $x = 5, 4, 9.$

14.   *a)*   If $\bar{x}$ is the average of the x_i, express the following in terms of $\bar{x}$, n, and b, where b is a constant.

$$\frac{\sum_{i=1}^{n} (x_i - b)}{n}.$$

  b)   If we have 1000 numbers whose average is 60 and we subtract 7 from each of these numbers, what will be the average of the resultant group of 1000 numbers?

15.   Express in terms of Σx_i^2, $\bar{x}$, and n:

$$\sum_{i=1}^{n} (x_i + 5)(x_i - 3).$$

16.   A professor reports that the average grade on a test was 72 percent, with a standard deviation of 3 percentage points. Using the multiple $k = 5$ in the Tchebysheff procedure, what statement can be made about the class grades?

17.   The average of the straight-time hourly wages of a group of workers is \$5 with a standard deviation of \$0.40. What statement can be made about the proportion of workers who earn between \$4 and \$6 per hour, inclusive?

6.33 ADDENDUM: MATRIX INVERSION BY ADJOINT MATRIX (OPTIONAL)

We have learned how to invert matrices by the Gauss-Jordan method and the similar zeros-first procedure. In this optional section we show another method for the important process of matrix inversion. This method requires the use of *determinants, cofactors,* and the *adjoint* of a matrix.

The determinant of a matrix. The determinant of a *square* matrix is a *number* (a scalar) calculated from the elements of the matrix by operations that are developed next.

Definition (2 by 2): If $A = \begin{pmatrix} a_{11} & a_{12} \\ a_{21} & a_{22} \end{pmatrix}$, then

$$\text{Determinant of } A = \det A = |A| = \begin{vmatrix} a_{11} & a_{12} \\ a_{21} & a_{22} \end{vmatrix} = a_{11}a_{22} - a_{12}a_{21}.$$

Thus, the determinant of a 2 by 2 matrix is the product of the main (top left to lower right) diagonal elements minus the product of the opposite diagonal elements. For example, if

$$A = \begin{pmatrix} 2 & 3 \\ 1 & 4 \end{pmatrix},$$

then

$$|A| = \begin{vmatrix} 2 & 3 \\ 1 & 4 \end{vmatrix} = 2(4) - 3(1) = 8 - 3 = 5.$$

Exercise. Find $|X|$ if $X = \begin{pmatrix} 5 & 1 \\ -2 & 3 \end{pmatrix}$. Answer: 17.

Minors of a matrix. Consider the 3 by 3 matrix

$$A = \begin{pmatrix} 2 & 1 & 3 \\ 3 & 2 & 1 \\ 2 & 3 & 1 \end{pmatrix}$$

and observe that the first row and first column have been lined out leaving

$$\begin{pmatrix} 2 & 1 \\ 3 & 1 \end{pmatrix}.$$

The determinant of this reduced matrix is called a *minor* and it is

$$\text{Minor of row 1, column 1} = |M_{11}| = \begin{vmatrix} 2 & 1 \\ 3 & 1 \end{vmatrix} = 2(1) - 1(3) = -1.$$

Specifically, the subscripts on

$$|M_{11}|$$

mean this is the minor obtained by deleting row 1 and column 1. In a similar manner, the minor obtained by deleting row 1 and column 2 is

$$| \boldsymbol{M}_{12} | = \begin{vmatrix} 3 & 1 \\ 2 & 1 \end{vmatrix} = 3(1) - 1(2) = 1.$$

Exercise. For **A** in the foregoing, find: *a)* $| \boldsymbol{M}_{13} |$. *b)* $| \boldsymbol{M}_{23} |$. Answer: *a)* 5. *b)* 4.

Cofactors (signed minors). A cofactor is a minor preceded by a *plus (minus)* sign if the sum of the subscripts is *even (odd)*. For example, the cofactor, c_{12}, of the minor $| \boldsymbol{M}_{12} |$ is

$$c_{12} = - | \boldsymbol{M}_{12} |$$

because the sum of the subscripts is odd. Similarly,

$$c_{13} = + | \boldsymbol{M}_{13} |$$

because the sum of the subscripts is even (the plus sign need not be written, of course). For the matrix **A** of the foregoing, we have

$$c_{12} = - | \boldsymbol{M}_{12} | = - \begin{vmatrix} 3 & 1 \\ 2 & 1 \end{vmatrix} = -[3(1) - 1(2)] = -1.$$

Note that a cofactor is a *scalar*.
Similarly,

$$c_{33} = | \boldsymbol{M}_{33} | = + \begin{vmatrix} 2 & 1 \\ 3 & 2 \end{vmatrix} = 2(2) - 1(3) = 1.$$

Exercise. For the matrix **A** of the foregoing, find: *a)* c_{23}. *b)* c_{22}. Answer: *a)* −4. *b)* −4.

Computing a determinant: Expansion by cofactors. The determinant of 3 by 3 or larger square matrices can be computed by cofactor expansion. This is done by *expansion on a row* or *expansion on a column*. Either procedure may be chosen. To expand on a row, each element in the row is multiplied by its cofactor and the results are summed. For example, if

$$A = \begin{pmatrix} 2 & 1 & 3 \\ 3 & 2 & 1 \\ 2 & 3 & 1 \end{pmatrix}$$

we may expand by, say, the first row to obtain

$$2c_{11} + 1c_{12} + 3c_{13},$$

which, with appropriate signs, is

$$2 \mid M_{11} \mid - 1 \mid M_{12} \mid + 3 \mid M_{13} \mid$$

or

$$2 \begin{vmatrix} 2 & 1 \\ 3 & 1 \end{vmatrix} -1 \begin{vmatrix} 3 & 1 \\ 2 & 1 \end{vmatrix} +3 \begin{vmatrix} 3 & 2 \\ 2 & 3 \end{vmatrix}$$
$$= 2 \ (-1) \ -1 \ (1) \ \ \ +3 \ (5)$$
$$= -2 - 1 + 15$$
$$= 12.$$

Hence,

$$\mid A \mid = 12.$$

As we have said, the row or column upon which the expansion is based is a matter of choice. For example, expanding A on the second *column* leads to

$$1c_{12} \quad\quad + 2c_{22} \quad\quad +3c_{32} \tag{1}$$
$$= -1 \mid M_{12} \mid +2 \mid M_{22} \mid -3 \mid M_{32} \mid \tag{2}$$

$$= -1 \begin{vmatrix} 3 & 1 \\ 2 & 1 \end{vmatrix} +2 \begin{vmatrix} 2 & 3 \\ 2 & 1 \end{vmatrix} -3 \begin{vmatrix} 2 & 3 \\ 3 & 1 \end{vmatrix}$$
$$= -1 \ (1) \ \ \ +2 \ (-4) \ -3 \ (-7)$$
$$= -1 \quad\quad\quad -8 \quad\quad +21$$
$$= 12. \tag{3}$$

as before.

> ***Exercise.*** Compute $\mid A \mid$ of the foregoing by expanding on the third row. *a)* Show Step (1) of the immediate foregoing. *b)* Show Step (2) of the foregoing. *c)* What is the value of $\mid A \mid$? Answer: *a)* $2c_{31} + 3c_{32} + 1c_{33}$. *b)* $2 \mid M_{31} \mid -3 \mid M_{32} \mid +1 \mid M_{33} \mid$. *c)* $2(1-6) - 3(2-9) + 1(4-3) = 12$.

In passing, we note that inasmuch as any row or column may be chosen for expansion, it will simplify calculations to choose the row or column that has the greatest number of zeros, if the matrix has any zero elements.

The cofactor matrix. We express the cofactor matrix as C. It is the matrix whose elements are the cofactors of a given matrix. For example, if we have

$$A = \begin{pmatrix} 2 & 1 & 3 \\ 3 & 2 & 1 \\ 2 & 3 & 1 \end{pmatrix}$$

and replace each of its elements by the cofactor of that element we obtain the cofactor matrix. Thus, the element in the first row and first column of C is c_{11}, where

$$c_{11} = |M_{11}| = \begin{vmatrix} 2 & 1 \\ 3 & 1 \end{vmatrix} = -1.$$

Proceeding across the first row we find

$$c_{12} = -|M_{12}| = - \begin{vmatrix} 3 & 1 \\ 2 & 1 \end{vmatrix} = -(3-2) = -1.$$

$$c_{13} = |M_{13}| = \begin{vmatrix} 3 & 2 \\ 2 & 3 \end{vmatrix} = (9-4) = 5.$$

Turning to the second row of A we compute

$$c_{21} = -|M_{21}| = - \begin{vmatrix} 1 & 3 \\ 3 & 1 \end{vmatrix} = -(1-9) = 8.$$

$$c_{22} = |M_{22}| = \begin{vmatrix} 2 & 3 \\ 2 & 1 \end{vmatrix} = (2-6) = -4.$$

$$c_{23} = -|M_{23}| = - \begin{vmatrix} 2 & 1 \\ 2 & 3 \end{vmatrix} = -(6-2) = -4.$$

Exercise. Find the third row elements c_{31}, c_{32}, c_{33} of the cofactor matrix. Answer: $c_{31} = (1-6) = -5$; $c_{32} = -(2-9) = 7$; $c_{33} = (4-3) = 1$.

Combining the foregoing calculations, we have the cofactor matrix

$$C = \begin{pmatrix} -1 & -1 & 5 \\ 8 & -4 & -4 \\ -5 & 7 & 1 \end{pmatrix}.$$

The adjoint matrix. If we now write the rows of C as columns, giving the *transpose*, C^t, of C, we have

$$C^t = \begin{pmatrix} -1 & 8 & -5 \\ -1 & -4 & 7 \\ 5 & -4 & 1 \end{pmatrix}.$$

This matrix is called the *adjoint* of the original matrix A from which it was formed. Thus,

Adjoint of $A = $ adj $A = C^t$

where C is the matrix whose elements are the cofactors of the elements of A.

Matrix inversion by adjoint matrix. We now have all the details needed in computing the inverse of a matrix. In brief symbolism

$$A^{-1} = \frac{1}{|A|}(\text{adj } A).$$

Reviewing the past calculations, for

$$A = \begin{pmatrix} 2 & 1 & 3 \\ 3 & 2 & 1 \\ 2 & 3 & 1 \end{pmatrix}$$

we first computed the value of the determinant and found

$$|A| = 12.$$

Next we constructed the cofactor matrix and transposed it to obtain the adjoint matrix

$$\text{adj } A = \begin{pmatrix} -1 & 8 & -5 \\ -1 & -4 & 7 \\ 5 & -4 & 1 \end{pmatrix}.$$

Consequently, the inverse of A is

$$A^{-1} = \frac{1}{|A|}(\text{adj } A) = \frac{1}{12}\begin{pmatrix} -1 & 8 & -5 \\ -1 & -4 & 7 \\ 5 & -4 & 1 \end{pmatrix}.$$

To demonstrate that the foregoing is the correct inverse, we compute

$$A^{-1}A = \frac{1}{12}\begin{pmatrix} -1 & 8 & -5 \\ -1 & -4 & 7 \\ 5 & -4 & 1 \end{pmatrix}\begin{pmatrix} 2 & 1 & 3 \\ 3 & 2 & 1 \\ 2 & 3 & 1 \end{pmatrix}$$

$$= \frac{1}{12}\begin{pmatrix} 12 & 0 & 0 \\ 0 & 12 & 0 \\ 0 & 0 & 12 \end{pmatrix} = \begin{pmatrix} 1 & 0 & 0 \\ 0 & 1 & 0 \\ 0 & 0 & 1 \end{pmatrix} = I.$$

The fact that

$$A^{-1}A = I$$

proves the matrices are inverses. In summary:

To compute the inverse of a square matrix, A,

1. Compute the value of the determinant of $|A|$ by expanding on a row or column.

2. Compute the cofactor matrix of **A** by forming a matrix whose elements are the cofactors of the elements of **A**.
3. Write the transpose of the cofactor matrix by writing its rows as columns. This is the adjoint matrix, adj **A**.
4. The inverse of **A** is

$$A^{-1} = \frac{1}{|A|}(\text{adj } A).$$

Example. Find A^{-1} if

$$A = \begin{pmatrix} 3 & 4 \\ 2 & 5 \end{pmatrix}.$$

We first compute the determinant of **A**,

$$|A| = 3(5) - 4(2) = 15 - 8 = 7.$$

The cofactors are the single elements

$$c_{11} = +5, \quad c_{12} = -2, \quad c_{21} = -4, \quad c_{22} = 3$$

so the cofactor matrix is

$$C = \begin{pmatrix} 5 & -2 \\ -4 & 3 \end{pmatrix}.$$

Writing the rows of **C** as columns gives the transpose of **C** which is the adjoint matrix

$$\text{adj } A = C^{t} = \begin{pmatrix} 5 & -4 \\ -2 & 3 \end{pmatrix}$$

and

$$A^{-1} = \frac{1}{|A|}(\text{adj } A) = \frac{1}{7}\begin{pmatrix} 5 & -4 \\ -2 & 3 \end{pmatrix}.$$

To verify the result, we compute

$$A^{-1}A = \frac{1}{7}\begin{pmatrix} 5 & -4 \\ -2 & 3 \end{pmatrix}\begin{pmatrix} 3 & 4 \\ 2 & 5 \end{pmatrix} = \frac{1}{7}\begin{pmatrix} 7 & 0 \\ 0 & 7 \end{pmatrix} = \begin{pmatrix} 1 & 0 \\ 0 & 1 \end{pmatrix} = I.$$

Exercise. Find A^{-1} by the adjoint method if $A = \begin{pmatrix} 2 & 2 \\ 3 & 4 \end{pmatrix}$.

Answer: $A^{-1} = \frac{1}{2}\begin{pmatrix} 4 & -2 \\ -3 & 2 \end{pmatrix}.$

Example. Compute A^{-1} if

$$A = \begin{pmatrix} 2 & 3 & 1 \\ 0 & 8 & 5 \\ 1 & 2 & 1 \end{pmatrix}.$$

We first compute the determinant of A by expanding on the first column, which has a zero element.

$$|A| = 2 \begin{vmatrix} 8 & 5 \\ 2 & 1 \end{vmatrix} - 0 + 1 \begin{vmatrix} 3 & 1 \\ 8 & 5 \end{vmatrix}$$
$$= 2(8 - 10) + (15 - 8) = -4 + 7 = 3.$$

Next we compute the elements of the cofactor matrix

$$c_{11} = + \begin{vmatrix} 8 & 5 \\ 2 & 1 \end{vmatrix} = (8 - 10) = -2$$

$$c_{12} = - \begin{vmatrix} 0 & 5 \\ 1 & 1 \end{vmatrix} = -(0 - 5) = 5$$

$$c_{13} = + \begin{vmatrix} 0 & 8 \\ 1 & 2 \end{vmatrix} = (0 - 8) = -8$$

$$c_{21} = - \begin{vmatrix} 3 & 1 \\ 2 & 1 \end{vmatrix} = -(3 - 2) = -1$$

$$c_{22} = + \begin{vmatrix} 2 & 1 \\ 1 & 1 \end{vmatrix} = (2 - 1) = 1$$

$$c_{23} = - \begin{vmatrix} 2 & 3 \\ 1 & 2 \end{vmatrix} = -(4 - 3) = -1$$

$$c_{31} = + \begin{vmatrix} 3 & 1 \\ 8 & 5 \end{vmatrix} = (15 - 8) = 7$$

$$c_{32} = - \begin{vmatrix} 2 & 1 \\ 0 & 5 \end{vmatrix} = -(10 - 0) = -10$$

$$c_{33} = + \begin{vmatrix} 2 & 3 \\ 0 & 8 \end{vmatrix} = (16 - 0) = 16.$$

The cofactor matrix is

$$C = \begin{pmatrix} -2 & 5 & -8 \\ -1 & 1 & -1 \\ 7 & -10 & 16 \end{pmatrix}.$$

Writing the rows of C as columns, we obtain the transpose

$$\text{adj } A = C^t = \begin{pmatrix} -2 & -1 & 7 \\ 5 & 1 & -10 \\ -8 & -1 & 16 \end{pmatrix}.$$

The inverse of A is

$$A^{-1} = \frac{1}{|A|} (\text{adj } A) = \frac{1}{3} \begin{pmatrix} -2 & -1 & 7 \\ 5 & 1 & -10 \\ -8 & -1 & 16 \end{pmatrix}.$$

Finally to prove the result, we compute

$$A^{-1} A = \frac{1}{3} \begin{pmatrix} -2 & -1 & 7 \\ 5 & 1 & -10 \\ -8 & -1 & 16 \end{pmatrix} \begin{pmatrix} 2 & 3 & 1 \\ 0 & 8 & 5 \\ 1 & 2 & 1 \end{pmatrix}$$

$$= \frac{1}{3} \begin{pmatrix} 3 & 0 & 0 \\ 0 & 3 & 0 \\ 0 & 0 & 3 \end{pmatrix} = \begin{pmatrix} 1 & 0 & 0 \\ 0 & 1 & 0 \\ 0 & 0 & 1 \end{pmatrix} = I.$$

Inversion by computation of the adjoint matrix can be extended to larger matrices by the procedure shown in the foregoing. Thus, if A is a 4 by 4 matrix, the cofactor matrix will have 16 elements found by evaluating the determinants of sixteen 3 by 3 matrices. Clearly, the arithmetic becomes extensive for inversion of large matrices and because of this, and because no new principles are involved, we leave this task for the computer.

Singular matrices. The statement

$$A^{-1} = \frac{1}{|A|} (\text{adj } A)$$

shows that A^{-1} does not exist if

$$\det A = |A| = 0.$$

Thus, if $|A|$ equals zero, the matrix is singular. This result is useful because if we compute $|A|$ at the beginning and find it to be zero, we know there is no point in carrying out calculations to invert A because it has no inverse.

Example. Find A^{-1} if

$$A = \begin{pmatrix} 1 & 1 & 1 \\ 2 & -1 & -1 \\ 4 & 1 & 1 \end{pmatrix}.$$

First we compute $|A|$ by expanding on the first row.

$$|A| = +1 \begin{vmatrix} -1 & -1 \\ 1 & 1 \end{vmatrix} - 1 \begin{vmatrix} 2 & -1 \\ 4 & 1 \end{vmatrix} + 1 \begin{vmatrix} 2 & -1 \\ 4 & 1 \end{vmatrix}$$

$$= \quad (-1+1) - 1 (2+4) + 1 (2+4)$$

$$= \quad 0 \quad - \quad 6 \quad + \quad 6$$

$$= 0.$$

Hence, A is singular, so A^{-1} does not exist.

6.34 PROBLEM SET 6–11

Find the inverse of each of the following, if one exists.

1. $\begin{pmatrix} 8 & 4 \\ 3 & 2 \end{pmatrix}$.

2. $\begin{pmatrix} 6 & 1 \\ -1 & -2 \end{pmatrix}$.

3. $\begin{pmatrix} 1 & 1 \\ 2 & 4 \end{pmatrix}$.

4. $\begin{pmatrix} 3 & 5 \\ 2 & 4 \end{pmatrix}$.

5. $\begin{pmatrix} 2 & 3 \\ 4 & 6 \end{pmatrix}$.

6. $\begin{pmatrix} -1 & -3 \\ 4 & 12 \end{pmatrix}$.

7. $\begin{pmatrix} 1 & 2 & 3 \\ 3 & -2 & 1 \\ -2 & 1 & -3 \end{pmatrix}$.

8. $\begin{pmatrix} 5 & 4 & 1 \\ 2 & -3 & 0 \\ -1 & 6 & -2 \end{pmatrix}$.

9. $\begin{pmatrix} 1 & 2 & 1 \\ 4 & 5 & -3 \\ 3 & 4 & -2 \end{pmatrix}$.

10. $\begin{pmatrix} 2 & 2 & 3 \\ 0 & 1 & 1 \\ 1 & 1 & 1 \end{pmatrix}$.

11. $\begin{pmatrix} 1 & 2 & 3 \\ -1 & 5 & 2 \\ 2 & 4 & 6 \end{pmatrix}$.

12. $\begin{pmatrix} 3 & -6 & -1 \\ 4 & -6 & -2 \\ -4 & 8 & 2 \end{pmatrix}$.

13. $\begin{pmatrix} 0 & 1 & 2 \\ 3 & 4 & 0 \\ 5 & 0 & 6 \end{pmatrix}$.

14. $\begin{pmatrix} 1 & -1 & 0 \\ 2 & 1 & 3 \\ 1 & 0 & 1 \end{pmatrix}$.

6.35 REVIEW PROBLEMS

Perform the following operations:

1. $(1\ 4\ 6) + (3\ -1\ 0) - 5(3\ 5\ -4)$.

2. $\begin{pmatrix} 8 \\ 2 \end{pmatrix} - 3\begin{pmatrix} 4 \\ 6 \end{pmatrix} + 2\begin{pmatrix} 1 \\ 5 \end{pmatrix}$.

3. If p and q are two-component column vectors, express the compact vector statement $px + qy = 0$
   a) In expanded vector form.
   b) In usual algebraic form.

4. If x, y, and r are three-component row vectors, express the compact vector statement $x + y = r$
   a) In expanded vector form.
   b) In usual algebraic form.

5. If x and 0 are four-component column vectors, express the compact vector statement $x \geq 0$
   a) In expanded vector form.
   b) In usual algebraic form.

6. $\begin{pmatrix} 3 & 2 & 0 \\ 1 & 5 & 4 \end{pmatrix} + \begin{pmatrix} 1 & -3 & 6 \\ 2 & 4 & 5 \end{pmatrix} - 2\begin{pmatrix} 1 & 1 & 3 \\ 4 & 0 & -1 \end{pmatrix}$.

7. $\begin{pmatrix} 3 & -2 \\ 1 & 7 \end{pmatrix} - \begin{pmatrix} -3 & 4 \\ 6 & 0 \end{pmatrix} - 3\begin{pmatrix} 1 & 2 \\ 1 & 3 \end{pmatrix} + 5\begin{pmatrix} 1 & -1 \\ -2 & 2 \end{pmatrix}.$

8. $(1 \ 0 \ 2)\begin{pmatrix} 2 & -1 & 3 & 5 \\ 0 & 2 & -2 & 1 \\ 1 & 4 & 2 & 3 \end{pmatrix}.$

9. $\begin{pmatrix} 1 & 3 & 2 & 2 \\ 2 & 0 & 1 & -1 \\ 0 & 1 & 2 & 3 \end{pmatrix}\begin{pmatrix} 3 & 4 \\ 2 & 0 \\ -1 & 5 \end{pmatrix}.$

10. $\begin{pmatrix} 2 & 3 \\ 1 & -1 \end{pmatrix}\begin{pmatrix} 1 & 3 & 5 \\ 2 & 4 & 6 \end{pmatrix}.$

11. $\begin{pmatrix} 4 & 1 & -1 \\ 2 & 0 & 3 \\ 0 & -2 & 4 \end{pmatrix}\begin{pmatrix} 0 & 1 & -1 \\ 1 & -1 & 0 \\ -1 & 0 & 1 \end{pmatrix}.$

12. The numbers of trees, bushes, and shrubs used in landscaping small, medium, and large lots are shown as the requirements matrix in the table. Costs per unit are shown at the left.

Unit cost	Item	Requirements matrix Number needed to landscape a		
		Small lot	Medium lot	Large lot
$30	Red maple.....................	0	1	2
20	Hard maple...................	1	1	1
20	Yew..........................	2	4	5
40	Arbovitae.....................	0	2	2
50	Spruce	0	2	3
20	Rhododendron	2	3	6
10	Laurel........................	2	2	4
20	Azalea........................	2	5	8

a) A contractor orders plantings for six small, 10 medium, and five large lots. Multiply the planting vector and the requirements matrix and interpret the results.

b) Multiply the cost vector and the requirements matrix and interpret the results.

Write the following in expanded matrix form:

13.
$$\begin{aligned} x_1 + 2x_2 \quad\;\; + x_4 &= 12 \\ 3x_2 - x_3 + 2x_4 &= 6 \\ 5x_1 - x_2 + x_3 - x_4 &= 7. \end{aligned}$$

14.
$$\begin{aligned} x_1 + 2x_2 &\leq 18 \\ 2x_1 + 3x_2 &\leq 24 \\ -x_1 \quad\;\; &\leq 0 \\ -x_2 &\leq 0. \end{aligned}$$

15.
$$\begin{aligned} x_1 + 2x_2 + s_1 &= 10 \\ x_1 + 3x_2 + s_2 &= 15. \end{aligned}$$

Write the following in usual algebraic form:

16. $\begin{pmatrix} 1 & 2 \\ 3 & 4 \\ 5 & 5 \end{pmatrix} \begin{pmatrix} x_1 \\ x_2 \end{pmatrix} = \begin{pmatrix} 7 \\ 8 \\ 9 \end{pmatrix}$.

17. $\begin{pmatrix} 3 & 0 & 1 \\ 0 & 2 & 4 \\ 2 & 3 & 0 \end{pmatrix} \begin{pmatrix} x_1 \\ x_2 \\ x_3 \end{pmatrix} \leq \begin{pmatrix} 14 \\ 10 \\ 12 \end{pmatrix}$.

18. $\begin{pmatrix} 3 & 1 \\ 2 & 2 \\ 1 & 4 \end{pmatrix} \begin{pmatrix} x_1 \\ x_2 \end{pmatrix} + \begin{pmatrix} s_1 \\ s_2 \\ s_3 \end{pmatrix} = \begin{pmatrix} 20 \\ 30 \\ 40 \end{pmatrix}$.

19. Maximize $(2\ 3) \begin{pmatrix} x_1 \\ x_2 \end{pmatrix}$ subject to $\begin{pmatrix} 1 & 1 \\ 2 & 1 \\ 1 & 2 \end{pmatrix} \begin{pmatrix} x_1 \\ x_2 \end{pmatrix} \leq \begin{pmatrix} 10 \\ 15 \\ 20 \end{pmatrix}$ and $x \geq 0$.

20. Minimize $(1\ 2\ 3\ 4) \begin{pmatrix} x_1 \\ x_2 \\ x_3 \\ x_4 \end{pmatrix}$ subject to $\begin{pmatrix} 1 & 0 & 2 & 3 \\ 2 & 1 & 0 & 4 \end{pmatrix} \begin{pmatrix} x_1 \\ x_2 \\ x_3 \\ x_4 \end{pmatrix} \geq \begin{pmatrix} 25 \\ 30 \end{pmatrix}$ and $x \geq 0$.

21. In a section of the country, 35 percent live in urban areas and 65 percent in rural areas at a point in time. Urban-rural movement each year is described by the transition matrix

	Urban	Rural
Urban	0.8	0.2
Rural	0.4	0.6

What will be the proportions in urban and rural areas after
a) One year? b) Two years? c) Three years?
d) What is the steady state?

22. Brand X has 25 percent of the market and other brands share the rest of the market. Because of a promotional effort, 50 percent of those buying other brands shift to Brand X each month, while 70 percent of those buying Brand X continue to buy this brand.
a) What percent of the market will Brand X have after one month?
b) What is the steady state?

Find the inverse of each of the following matrices (if an inverse exists):

23. $\begin{pmatrix} 0 & 1 \\ 1 & 0 \end{pmatrix}$.

24. $\begin{pmatrix} 3 & 1 \\ 0 & 2 \end{pmatrix}$.

25. $\begin{pmatrix} 1 & 2 \\ 3 & 3 \end{pmatrix}$.

26. $\begin{pmatrix} 1 & 1 \\ 2 & 1 \end{pmatrix}$.

27. $\begin{pmatrix} 5 & 3 \\ 2 & 2 \end{pmatrix}$.

28. $\begin{pmatrix} 1 & 2 \\ 3 & 4 \end{pmatrix}$.

29. $\begin{pmatrix} 1 & 0 & 1 \\ 0 & 1 & 2 \\ 2 & 3 & 0 \end{pmatrix}$.

30. $\begin{pmatrix} 1 & 2 & -1 \\ 1 & 0 & 1 \\ 0 & 5 & -5 \end{pmatrix}$.

31. $\begin{pmatrix} 1 & 0 & 2 \\ 0 & 1 & 3 \\ 1 & 2 & 0 \end{pmatrix}$.

32. $\begin{pmatrix} 0 & 0 & 1 \\ 0 & 1 & 0 \\ 1 & 0 & 0 \end{pmatrix}.$ 33. $\begin{pmatrix} 60 & 30 & 20 \\ 30 & 20 & 15 \\ 20 & 15 & 12 \end{pmatrix}.$ 34. $\begin{pmatrix} 1 & 1 & 1 \\ 2 & 1 & 1 \\ 1 & 2 & 2 \end{pmatrix}.$

35. $\begin{pmatrix} 1 & 2 & 1 \\ 2 & 1 & 1 \\ 1 & 2 & 2 \end{pmatrix}.$ 36. $\begin{pmatrix} 2 & 4 & 1 \\ 3 & 1 & 2 \\ 0 & 5 & 6 \end{pmatrix}.$ 37. $\begin{pmatrix} 2 & 0 & 1 \\ 3 & 1 & 3 \\ 0 & 1 & 4 \end{pmatrix}.$

38. Consider the system

$$7x_1 + 11x_2 = 3$$
$$5x_1 + 8x_2 = 5.$$

   a) Relating the system to $Ax = b$, what is A?, x?, b?
   b) Compute A^{-1}.
   c) Write $x = A^{-1}b$ in expanded matrix form.
   d) Compute the solution from part (c).
   e) What would be the solution vector if the elements in the vector of constants, 3 and 5, were changed to
      (1) 1, 1.   (2) 0, 1.   (3) 1, 0.   (4) 2, −3.

39. Answer parts (a) through (d) of Problem 38 for the system

$$2x_1 + 4x_2 = 4$$
$$5x_1 + 6x_2 = 8.$$

   e) What would be the solution vector if the elements in the vector of constants, 4 and 8, were changed to
      (1) 4, 12.   (2) 12, 16.   (3) 20, 40.   (4) 0, 4.

40. Answer parts (a) through (d) of Problem 38 for the system

$$3x_1 \qquad + 2x_3 = 1$$
$$5x_1 + 2x_2 + 5x_3 = 2$$
$$x_2 + x_3 = 3.$$

   e) What would be the solution vector if the elements in the vector of constants, 1, 2, and 3 were changed to
      (1) 1, 1, 1.   (2) 1, 0, 1.   (3) 1, 1, 0.

41. Answer parts (a) through (d) of Problem 38 for the system

$$2x_1 + 3x_2 - 15x_3 = 3$$
$$-5x_1 - 7x_2 + 35x_3 = 2$$
$$3x_1 + 4x_2 - 21x_3 = 1.$$

   e) What would be the solution vector if the elements in the vector of constants, 3, 2, and 1, were changed to
      (1) 1, 1, 1.   (2) 0, 1, 1.   (3) 1, 1, 0.

42. The table shows that x_1 units of product X_1 are to be made and that a unit of X_1 requires 1 hour of time on machine M_1, 3 hours on M_2, and 2 hours on M_3. Machine M_1 has H_1 hours of available time. Similar interpretations apply to the other entries.

Product	Units to be made	Hours of machine time per unit of product on		
		M_1	M_2	M_3
X_1 .... x_1		1	3	2
X_2 .... x_2		2	1	3
X_3 .... x_3		3	2	1
Total machine hours available		H_1	H_2	H_3

Find the numbers of units that can be made if the machine time availability vector $(H_1\ H_2\ H_3)$ is

a) (180 540 360).   b) (180 180 180).   c) (270 360 360).

43. Given the system

$$3x_1 + 2x_2 = b_1$$
$$4x_1 + 3x_2 = b_2$$

find the missing elements to complete the following equation:

$$\begin{pmatrix} x_1 \\ x_2 \end{pmatrix} = \begin{pmatrix} \quad \end{pmatrix}\begin{pmatrix} b_1 \\ b_2 \end{pmatrix}.$$

44. Given the system

$$2x_1 + x_2 + x_3 = b_1$$
$$x_1 + x_2 + x_3 = b_2$$
$$x_1 + 2x_2 + x_3 = b_3$$

find the missing elements to complete the following equation:

$$\begin{pmatrix} x_1 \\ x_2 \\ x_3 \end{pmatrix} = \begin{pmatrix} \quad \end{pmatrix}\begin{pmatrix} b_1 \\ b_2 \\ b_3 \end{pmatrix}.$$

45. Consider the system

$$x_1 + x_2 + x_3 = 3$$
$$2x_1 + 3x_2 + 2x_3 = 5$$
$$x_1 + 2x_2 + x_3 = 2.$$

a) Show that the matrix of coefficients does not have an inverse.
b) Find the solution of the system by the methods of Chapter 2.
c) Change the constant in the first equation from 3 to 4.
   (1) Does the change in the constant have an effect upon the matrix of coefficients?
   (2) Does the change have an effect upon the solution in part (b)?
   (3) Our example illustrates what it means if the matrix of coefficients in an n by n system has no inverse. What does it mean if the coefficient matrix has no inverse?

46. a) Write the technological matrix from the following input-output data.

Producer	User		Final demand	Total output
	No. 1	No. 2		
No. 1...........	$\frac{1}{5}$	$\frac{3}{10}$	d_1	x_1
No. 2...........	$\frac{1}{5}$	$\frac{1}{10}$	d_2	x_2

b) What does the second column of the technological matrix mean?

c) What is $(I - A)$?   *d)* What is $(I - A)^{-1}$?

e) Write the equation for the output vector X in matrix-vector form using the computed elements of $(I - A)^{-1}$ and the elements x_1, x_2, d_1, d_2.

f) Compute the elements x_1 and x_2 of the output vector if final demand for industry No. 1 is 180 and for industry No. 2 is 216.

g) Compute the elements x_1 and x_2 of the output vector if final demand is 36 for industry No. 1 and 108 for industry No. 2.

47. Each dollar of industy No. 1 output requires $\frac{1}{10}$ dollars of its own output and $\frac{1}{2}$ dollars of industry No. 2 output. Each dollar of industry No. 2 output requires $\frac{3}{5}$ dollars of industry No. 1 output and $\frac{1}{5}$ dollars of its own output. Let d_1, d_2 be final demand and x_1, x_2 be total output, respectively, for industries No. 1 and No. 2.

a) Compute the matrix that completes the following equation:

$$\begin{pmatrix} x_1 \\ x_2 \end{pmatrix} = \begin{pmatrix} & \\ & \end{pmatrix} \begin{pmatrix} d_1 \\ d_2 \end{pmatrix}.$$

b) Find the elements x_1 and x_2 of the output vector if final demand is 315 for industry No. 1 and 504 for industry No. 2.

c) Find the elements x_1 and x_2 of the output vector if final demand is 63 for industry No. 1 and 189 for No. 2.

Find the numerical values of:

48. $\sum_{n=3}^{6} (n^2 - n)$.   49. $\sum_{i=1}^{5} 2i^2$.   50. $\sum_{p=3}^{7} (2p + 1)$.

Express in summation notation:

51. The sum of the first 100 positive integers.

52. The sum of the 5th power of the first 10 positive integers.

53. The sum of the first 50 positive even integers.

54. The sum of the first 50 odd integers.

Write in expanded form:

55. $\sum_{j=10}^{15} c_j y_j$.   56. $\sum_{j=10}^{10+n} c_j x_j$.   57. $\sum_{i=m}^{m+n} a_i x_i$.

58. $\displaystyle\sum_{i=1}^{4} x_i = 25.$ 59. $\displaystyle\sum_{j=1}^{3} 2x_j = 7.$ 60. $\displaystyle\sum_{j=1}^{6} c_j x_j = 15.$

Express in compact summation form:

61. $a_1 x_1 + a_2 x_2 + a_3 x_3.$
62. $a_1 x_1 + a_2 x_2 + \cdots + a_9 x_9.$
63. $a_1 x_1 + a_2 x_2 + \cdots + a_n x_n.$
64. $a_1 x_1 + a_2 x_2 + \cdots + a_n x_n + a_{n+1} x_{n+1} + a_{n+2} x_{n+2} + \cdots + a_{n+m} x_{n+m}.$
65. Identify m, c, and each of the b's in

$$\sum_{j=1}^{m} b_i y_i = c$$

with a number in the equation

$$5y_1 + 3y_2 - y_3 + 4y_6 = 12.$$

66. Write the following system in expanded form

$$\sum_{j=1}^{3} a_{ij} x_j = b_i \qquad i = 1, 2, 3, 4.$$

67. Write in expanded form

Maximize

$$\sum_{j=1}^{2} c_j x_j$$

subject to:

$$\sum_{j=1}^{2} a_{ij} x_j \leq b_i \qquad i = 1, 2$$

$$x_j \geq 0 \text{ for all } j.$$

68. A linear system has p equations in p variables.
   a) Write the system in expanded form, using $\cdots$ where necessary.
   b) Write the system in compact summation form.
69. A linear programming minimization problem has 15 "greater than or equal" constraints on 25 variables.
   a) Write the problem in expanded form using $\cdots$ to shorten the expressions.
   b) Write the problem in compact summation form.
70. Consider the following system and its compact representation

$$\begin{aligned}
3x_1 + 2x_2 + 5x_3 &= 6 \\
x_1 + 7x_2 - x_3 &= 10 \\
4x_2 - 8x_3 &= 2 \\
6x_1 + 9x_2 &= 11.
\end{aligned}$$

$$\sum_{j=1}^{n} a_{ij} x_j = b_i \quad i = 1, 2, \ldots, m.$$

   a) What is m? What is n?

b) What symbol would represent the following constants which appear in the system: 3, 7, 10, 9, 11?

c) What constants in the system correspond to a_{23}, a_{31}, b_4, a_{33}, a_{12}?

71. To find the average of a group of n numbers, we add up the numbers and divide by n; thus, the average of the three numbers 2, 7, 6 is

$$\frac{2+7+6}{3} = 5.$$

a) Write the expression for the average of any three numbers, x_1, x_2, and x_3 in expanded form and in compact summation form.

b) Write the expression for the average of any n numbers in expanded form (using $\cdots$) and in compact summation form.

72. To compute the weighted average of a group of numbers, each number, x_i, is multiplied by its weight, w_i; the products so obtained are summed and the sum is divided by the sum of the weights to obtain the average.

a) Given the numbers 3 and 5 with weights 2 and 8, respectively, compute the weighted average.

b) Write the compact formula for the weighted average of n numbers, where each number, x_i, is weighted by w_i.

c) Write the formula in (b) in expanded form.

73. Compute the following for the values $x = 2, 3, 7$.

a) $\Sigma\ x$.   b) $\Sigma\ x^2$.   c) $\bar{x}$.   d) $\Sigma\ (x - \bar{x})^2$.

e) Compute the square of the average of the x's.

f) Compute the average of the squares of the x's.

74. a) Plot the following points on a graph.

x	y
4	2
0	1
6	4
2	3
6	2

b) Find the equation of the best fitting line, $y = mx + b$, where

$$m = \frac{n\ \Sigma\ xy - \Sigma\ x\ \Sigma\ y}{n\ \Sigma\ x^2 - (\Sigma\ x)^2} \quad \text{and} \quad b = \frac{\Sigma\ y - m\ \Sigma\ x}{n}.$$

c) Sketch the line found in (b) on the graph in (a).

d) Find the coefficient of linear correlation for the data in (a) by evaluating the formula

$$\frac{n\ \Sigma\ xy - \Sigma\ x\ \Sigma\ y}{\sqrt{[n\ \Sigma\ x^2 - (\Sigma\ x)^2][n\ \Sigma\ y^2 - (\Sigma\ y)^2]}}.$$

75. The *harmonic* average (or harmonic mean) of a set of numbers is the reciprocal of the average (arithmetic) of the reciprocals of the numbers.

It is used on occasion to average ratios. The formula for the harmonic mean, h, is

$$\frac{1}{\dfrac{\sum\limits_{i=1}^{n}(1/x_i)}{n}} = \frac{n}{\sum\limits_{i=1}^{n}(1/x_i)}.$$

Compute the harmonic mean of $x = 2, 3, 4$.

Find the average, variance, and standard deviation:

76. x_i: 1, 2, 7, 5, 10, 5.   77. x_i: 57, 69, 51, 63, 60.

78. If μ and σ^2 are the average and variance of the x_i and we multiply each x_i by the constant a and compute a new mean and variance,
   a) Express the new mean in terms of μ and a.
   b) Express the new variance and standard deviation in terms of μ and a.
   c) The numbers x_i: 2, 6, 0, 4, 3 have $\mu_1 = 3$ and $\sigma_1^2 = 4$. If we triple these numbers what will be the new average, variance, and standard deviation?

79. Express

$$\sum_{i=1}^{n} x_i(x_i - \mu)$$

   a) in terms of Σx_i^2, n, and μ;
   b) in terms of Σx_i^2, Σx_i, and n.

Natural and common logarithms

7

7.1 INTRODUCTION

This chapter contains both a modern and a standard treatment of logarithms. The modern treatment emphasizes, but does not require, the solution of problems involving logarithms by means of a hand-held calculator. The standard treatment applies a table of logarithms to the base 10 to solve problems and provides notes on how these problems are solved with a hand-held calculator. To avoid undue repetitions, the definitions and rules of logarithms are included in the modern treatment only. This treatment appears first, in Sections 7.2 through 7.9. All readers should study these sections because the definitions and rules of logarithms are applied in the standard treatment in the latter part of the chapter, and in later chapters of the book. The modern treatment has been designed so that it can be studied even if a calculator is not available and, of course, a calculator is not required for the standard treatment. Consequently, *a hand-held calculator is not required* for study of this chapter, but it is strongly recommended that one be used because at the touch of a key it will provide, almost instantly, a logarithm that is much more accurate than logarithms available in most tables.

The modern treatment provides all we need on logarithms in later work. The standard treatment, therefore, can be omitted. We have included it because of the high probability that readers will encounter topics using this treatment in other courses. To be suitable for work in this chapter, a calculator must

1. Provide the logarithm of a number.
2. Provide the inverse of the logarithm of a number.
3. Raise a number to a power.
4. Have floating point display of digits.[1]

At the time of this writing, a calculator with the foregoing capabilities can be purchased for about $20. By way of contrast, 15 or 20 years ago students purchased a slide rule to do these operations, and paid about $40 for a device that is much harder to use than a hand-held calculator, and much less accurate.

7.2 THE NEED FOR LOGARITHMS

In the early part of this century, and prior thereto, logarithms were extensively used to ease calculational difficulties arising in a variety of fields including finance, navigation, and engineering. At the present time such calculations are made accurately and rapidly by large computers and small special-purpose computers. However, computers have not outmoded logarithms. For one thing, computers routinely find logarithms and use them in calculations. For example, calculations like

$$1000(1.005)^{60}$$

are performed millions of times a day by banks' data-processing computers, and the computers routinely apply logarithms in carrying out such calculations.

A second reason that logarithms have not been outmoded is that they are *necessary* in the solution of certain types of equations. Third, as we shall see, the *rules* of logarithms continue to play an important role in mathematical analysis.

Logarithms are necessary in the solution of the equation

---

[1] If 0.012345 is multiplied by 0.054321 on a calculator having a ten-digit display with floating point, the displayed product will be 6.70592745 $-$ 04. The $-$04 means the decimal point should be moved four places to the left, so the number is 0.000670592745. Similarly 123456 times 654321 will be displayed as 8.077985338   10, and the 10 at the right, which is +10, means the decimal point should be moved ten places to the right to give 80,779,853,380. Calculators automatically convert to floating point when very large numbers or small decimals are displayed.

$$10^x = 5,$$

which requires that we find the power of 10 that equals 5. We know that

$$10^0 = 1$$
$$10^1 = 10,$$

so x is between 0 and 1, but this is not a sufficiently accurate answer. The answer is found easily by referring to a table of logarithms that have the 10 in 10^x as their "base" number. Such a table, computed by mathematicians, lists numbers and the power of 10 that equals each of the numbers. This power of 10 for a number, N, is called the *logarithm of N to the base 10*. Base 10 logarithms are also called *common* logarithms. In symbols, the common logarithm of a number N is written

$$\log_{10} N.$$

Table 7–1 provides a few logarithms, correct to six decimal places, for our use here.

TABLE 7–1

Number N	Logarithm to the base 10 $\log_{10} N$ (common logs)
1	0.000 000
1.02	0.008 600
1.04	0.017 033
1.06	0.025 306
1.08	0.033 424
2	0.301 030
3	0.477 121
4	0.602 060
5	0.698 970
6	0.778 151
7	0.845 098
8	0.903 090
9	0.954 243
10	1.000 000
16	1.204 120

Returning to

$$10^x = 5$$

we see from Table 7–1 that the power of 10 that equals 5, that is, $\log_{10} 5$, is

$$x = \log_{10} 5 = 0.698\ 970.$$

Exercise. Solve for x: $10^x = 8$. Answer: 0.903 090.

Next, in Table 7–1, observe that

$$\log_{10} 4 \ = 0.602 \ 060$$
$$\log_{10} 16 = 1.204 \ 120,$$

and we see that the logarithm of 16 is two times the logarithm of 4; that is,

$$\log_{10} 16 = 2(0.602060) = 2 \log_{10} 4 = 1.204120.$$

This result occurs because 16 is 4 to the second power, and a *rule of logarithms,* which we will prove later, states that *the logarithm of a number to a power equals the power times the logarithm of the number.* That is,

$$\log_{10} 16 = \log_{10} 4^2 = 2 \log_{10} 4.$$

Similarly,

$$\log_{10} 5^4 = 4 \log_{10} 5$$
$$\log_{10} 7^3 = 3 \log_{10} 7,$$

and so on.

Exercise. From Table 7–1: *a)* Compute 3 times $\log_{10} 2$. *b)* The result in *a)* is the logarithm of what number? Answer: *a)* 0.903090. *b)* 8, which is 2^3.

We introduced the foregoing rule to illustrate the point that logarithms are needed in the solution of practical problems. To pursue this point, suppose that Jan has $1000 to invest and finds that interest can be earned at the rate of 6 percent, compounded annually. Jan wants to know how many years, n, it will take at this interest rate for $1000 to grow to $2000. A friend who knows some formulas from finance (which we study in the next chapter) says the answer is the solution of the equation

$$1000(1.06)^t = 2000.$$

Here the unknown is an exponent, and logarithms are needed to determine the value of the exponent. First, dividing both sides by 1000, we have

$$(1.06)^t = 2.$$

Next, take the logarithms of both sides of the equation. Thus:

$$\log_{10} (1.06)^t = \log_{10} 2.$$

Now apply the rule of logarithms stated a little earlier to obtain

$$t \log_{10} (1.06) = \log_{10} 2.$$

$$t = \frac{\log_{10} 2}{\log_{10} (1.06)}.$$

We find the required logarithms in Table 7–1 and have

$$t = \frac{0.301\ 030}{0.025\ 306}$$

$$= 11.9,$$

so it will take 11.9 years, or about 12 years, for Jan's investment of $1000 to grow to $2000.

Note: From this point onward, we shall simplify writing $\log_{10} N$ by agreeing that log shall always mean $\log_{10}$.

Definition: log N means $\log_{10} N$.

Example: Solve for x: $3^x = 8$.
As before, we take the logarithm of both sides to obtain

$$\log 3^x = \log 8.$$

Apply the rule for the power giving

$$x \log 3 = \log 8$$

$$x = \frac{\log 8}{\log 3}.$$

Inserting logarithms from Table 7–1, we find

$$x = \frac{0.903090}{0.477121}$$

$$= 1.893.$$

Exercise.   Solve for x: $9^x = 2$. Answer: 0.31546.

Note: We shall return to base ten (common) logarithms later in the chapter. Until then we shall use the natural logarithm system introduced in the next section.

7.3 NATURAL (BASE *e*) LOGARITHMS

One of the most important constants in mathematics is the number π, which is

$$\pi = 3.14159$$

correct to five decimal places. Inasmuch as π has infinitely many decimal places, it cannot be written exactly using a finite number of digits. This poses no problem, however, because its value is known, or can be computed, to as many places as required for the accuracy needed in applications. What was just stated about the characteristics of π also applies to a second very important mathematical constant, *e*. Correct to six decimals, this constant is

$$e = 2.718\,282.$$

Some may recall that 22/7 is a rough approximation of π. A somewhat rougher approximation of *e* is 19/7. The reason the unlikely number *e* was selected as the base of a system of logarithms and the logarithms are called *natural* logarithms, will be clear when we move into study of the calculus. For present purposes, natural logarithms are simply numbers calculated for us that can be found in a table or taken from a hand calculator. Table X at the end of the book provides natural logarithms of some numbers for our use. We shall use the symbol *ln* (pronounced as is the name Lynn) to designate logarithms to the base *e*.

Definition: $\ln N = \log_e N.$

From Table X we find that

$$\ln 6.7 \ = 1.90211$$
$$\ln 0.1 \ = -2.30259$$
$$\ln 6.14 = 1.81482.$$

In the last, note that the 1.8 at the beginning of the row in Table X applies to all numbers in that row, so space is saved by writing 1.8 only once at the beginning. Then note that the row containing

$$\ln 6.05$$

starts with .1.7 and *0006 is entered in the .05 column. The * means to increase 1.7 to 1.8. Thus,

$$\ln 6.05 = 1.80006.$$

Similarly,

$$\ln 54 = 3.98898$$
$$\ln 55 = 4.00733.$$

Exercise.   Find the following: *a)* ln 4.5. *b)* ln 0.2. *c)* ln 9.26. *d)* ln 0.06. *e)* ln 9.8. *f)* ln 27. *g)* ln 197. *h)* ln 603. Answer: *a)* 1.50408. *b)* −1.60944. *c)* 2.22570. *d)* −2.81341. *e)* 2.28238. *f)* 3.29584. *g)* 5.28320. *h)* 6.40192.

The problems we solved earlier with common (base 10) logarithms can be solved in the same manner with natural (base *e*) logarithms.
Example.   Solve for *n* by natural logarithms:

$$(1.06)^n = 2.$$

We write

$$\ln(1.06)^n = \ln 2$$
$$n \ln(1.06) = \ln 2$$
$$n = \frac{\ln 2}{\ln 1.06}$$
$$n = \frac{0.69315}{0.05827}$$
$$n = 11.9.$$

Exercise.   Solve $9^x = 2$ for *x* by natural logarithms. Answer: 0.31547.

7.4 LOGARITHMS FROM HAND-HELD CALCULATORS

Two decades ago, a "fully automatic" desk calculator sold for about $1500. Fully automatic meant that the calculator could perform not only addition and subtraction, but also multiplication and division. The calculator itself was about the size and weight of a large typewriter. Its mechanism was an ingenious series of gears, springs, and levers which clattered away during operation—especially the time-consuming operation of division. Today we can purchase for about $20 a shirt-pocket sized calculator weighing a few ounces that does all that the earlier $1500 machine did and much more, and does it silently. In this text we shall refer to these compact devices as hand calculators to distinguish them from large-scale systems operated by universities, business firms, and others for data processing, which includes processing of alphabetic and other characters as well as numbers. These large-scale systems we shall refer to as computers.

Hand calculators are available with various options. In this book the important options are the operations of taking a logarithm and raising to a power. Calculators that have a logarithm key can be expected to have a power key, but this should be checked before purchase. Currently, some calculators have both a natural logarithm and a common (base 10) logarithm key. Others have only one such key, which probably is the natural logarithm but, again, this should be checked. Readers should refer to their calculator instruction booklet to learn how to obtain a logarithm and raise a number to a power, or simply ask someone who knows. We will not attempt here to give instructions for operating specific calculators because the instructions would be outdated before this book is in print.

In work with logarithms, it makes no difference (until we get into calculus) whether we work with the common or natural system. If, for some reason, we have a logarithm to one base and want to find the corresponding logarithm to the other base, the conversion is simple. It is:

Ln to Log Conversion

$$\ln N = (\ln 10) \log N = 2.30259 \log N.$$
$$\log N = (\log e) \ln N = 0.434294 \ln N.$$

> *Exercise.* Table 7–1 shows log 5 = 0.698970 and Table X shows ln 5 = 1.60944. *a)* How can ln 5 be computed from log 5? *b)* How can log 5 be computed from ln 5? Answer: *a)* ln 5 = (2.30259) log 5 = 2.30259(0.698970) = 1.60944. *b)* log 5 = (0.434294) ln 5 = (0.434294)(1.60944) = 0.698970.

7.5 PROBLEM SET 7–1

Solve the following for x by reference to Table 7–1 or by use of a hand calculator.

1. $2^x = 3.$
2. $3^x = 2.$
3. $10^x = 5.$
4. $10^x = 6.$
5. $2^x = 1.08.$
6. $5^x = 1.02.$
7. $(1.06)^x = 3.$
8. $(1.08)^x = 2.$
9. $10^x = 7.$
10. $10^x = 3.$

Apply the power rule and Table 7–1 to find the following:

11. log 125.
12. log 27.
13. log 81.
14. log 625.

Solve the following for x by reference to Table X or by use of a hand calculator.

15. $(20)^x = 100$.   16. $(4)^x = 500$.

17. $(0.5)^x = 8$.   18. $(0.25)^x = 10$.

19. $(1.08)^x = 3$.   20. $(1.03)^x = 5$.

7.6 DEFINITION OF LOGARITHMS

The need for logarithms in certain calculations was demonstrated in the previous sections of this chapter. Logarithms also play an important role, independent of their calculational use, in applied mathematical analysis. To understand this role, it is necessary to be thoroughly familiar with the definition of a logarithm and the rules that apply to operations with logarithms. We start by recalling that the logarithm of a given number is the exponent of the power to which another number, the base, must be raised to equal the given number. If we write

$$2^3 = 8,$$

then 3 is the power to which 2, the base, must be raised to equal 8 and we say the logarithm of 8 to the base 2 equals 3. Thus,

$$2^3 = 8 \quad \text{means} \quad \log_2 8 = 3,$$

and conversely. The two statements are called *inverse* forms. Similarly

$$\log_5 125 = 3 \quad \text{means} \quad 5^3 = 125$$

$$\log_2 \left(\frac{1}{2}\right) = -1 \quad \text{means} \quad 2^{-1} = \frac{1}{2}$$

$$\log_b M = x \quad \text{means} \quad b^x = M.$$

Example. Write $2^5 = 32$ and $\log_{10} N = x$ in inverse form.
For the first part,

$$2^5 = 32,$$

we identify the base as 2. The exponent, 5, equals the logarithm, which is the logarithm of 32. Hence,

$$\log_2 32 = 5.$$

The arrows below show a system for remembering how to do the second part, starting with the base.

$$\log_{10} N = x \quad \text{means} \quad 10^x = N.$$

308

Exercise. Write the following in inverse form. *a)* $\log_{10} 100 = 2$. *b)* $\log_a N = x$. *c)* $(0.5)^2 = 0.25$. *d)* $x^2 = 10$. *e)* $a^z = M$. Answer: *a)* $10^2 = 100$. *b)* $a^x = N$. *c)* $\log_{0.5}(0.25) = 2$. *d)* $\log_x 10 = 2$. *e)* $\log_a M = z$.

7.7 RULES OF LOGARITHMS

Logarithms, as we stated, are exponents and therefore must obey rules corresponding to exponent rules. Examples of exponent rules are

$$x^2 (x^3) = x^{2+3} = x^5$$

$$\frac{x^4}{x^2} = x^{4-2} = x^2$$

$$(x^3)^2 = x^{3(2)} = x^6.$$

In brief statement, the foregoing say that when the bases are the same, exponents of the powers are added in multiplication, subtracted in division, and multiplied when a power is raised to a power. The corresponding rules for logarithms are:

Logarithm rules

Product: $\log ab = \log a + \log b$.

Quotient: $\log\left(\dfrac{a}{b}\right) = \log a - \log b$.

Power: $\log a^b = b(\log a)$.

Warning. Logarithms of sums and differences *cannot* be taken term by term. That is, log *(a + b) cannot* be written as log *a* + log *b*, and similarly for log *(a − b)*.

Thus, the logarithm of a product is the sum of the logarithms of the factors, the logarithm of a quotient is the logarithm of the numerator minus the logarithm of the denominator, and the logarithm of a number to a power is the power times the logarithm of the number. These rules apply whatever the base of the logarithms is. To prove the product rule, assume the base is some number *c,* and let

$$\log_c a = x \quad \text{so that} \quad c^x = a$$
$$\log_c b = y \quad \text{so that} \quad c^y = b.$$

If we multiply *a* by *b,* at the right, the result is the same as multiplying the equivalent values, c^x and c^y. Hence,

$$(c^x)(c^y) = ab.$$
$$c^{x+y} = ab.$$

Writing the inverse of the last statement gives

$$\log_c ab = x + y.$$

Now, remembering that $x = \log_c a$ and $y = \log_c b$, we have

$$\log_c ab = \log_c a + \log_c b$$

which proves the product rule. Similar proofs can be written for the quotient and power rules, but we shall leave these for exercises in the next problem set.

Example. Assume that for some base, log $a = 0.3000$, log $b = 2.8642$, and log $c = 1.7642$. Find

$$\log \left(\frac{a^3 \, b^{1/2}}{c} \right).$$

The purpose of this example is to fix the rules of logarithms in mind. If we apply the product and quotient rules, then the power rule, we have

$$\log \left(\frac{a^3 \, b^{1/2}}{c} \right) = \log a^3 + \log b^{1/2} - \log c$$

$$= 3 \log a + \frac{1}{2} \log b - \log c$$

$$= 3(0.3000) + \frac{1}{2}(2.8642) - 1.7642$$

$$= 0.9000 + 1.4321 - 1.7642$$

$$= 0.5679.$$

Exercise. Assume that, for some base, log $x = 0.5$, log $y = 1.5$, and log $z = 3$. Compute the value of

$$\log \frac{xy}{z^{1/3}}.$$

Answer: 1.

Example. Rewrite the following using the log symbol only once:

$$\log a - 2 \log b + \frac{1}{2} \log c.$$

We proceed to use the power rule first, then apply the product and quotient rules.

$$\log a - 2 \log b + \frac{1}{2} \log c = \log a - \log b^2 + \log c^{1/2}$$

$$= \log \left(\frac{ac^{1/2}}{b^2}\right).$$

Exercise. Rewrite $2 \log x - \log y - 0.5 \log z$ using the log symbol only once. Answer: $\log \left(\dfrac{x^2}{yz^{0.5}}\right)$.

Points to remember. The base, b, of a system of logarithms can be any positive number except 1. For such a number

$$\log_b b = 1; \quad \text{that is,} \quad b^1 = b$$
$$\log_b 1 = 0; \quad \text{that is,} \quad b^0 = 1.$$

Thus, for example, remembering that $\log N$ means $\log_{10} N$ and $\ln N = \log_e N$,

$$\log 10 = 1; \quad \ln e = 1; \quad \log 1 = 0; \quad \ln 1 = 0.$$

These points are worthy of special note because they arise often in applications.

Example. If $y = x(x - \ln x)$, compute y when $x = 1$.

Substituting $x = 1$, we find

$$y = 1(1 - \ln 1)$$
$$= 1(1 - 0)$$
$$= 1.$$

Example. Solve for x:

$$e^{-0.1x} = 0.2.$$

Inasmuch as the base of this exponential is e, taking the natural logarithm of both sides of the equation is the simplest way to proceed. Thus,

$$\ln(e^{-0.1x}) = \ln 0.2$$
$$-0.1x \ln e = \ln 0.2,$$

and $\ln e = 1$, so

$$-0.1x = -1.60944$$
$$x = \frac{-1.60944}{-0.1}$$
$$= 16.0944.$$

Exercise. *a)* Evaluate ln $(x^2 - 3)$ when $x = 2$. *b)* Solve $0.5x$ log $10 = 2$ for x. Answer: *a)* 0. *b)* $x = 4$.

Log$_b$ N requires that N be positive. A glance at Table X shows no logarithms are given for 0 or for negative numbers, and an attempt to obtain logarithms for such numbers on a hand calculator will lead to an error signal. This is because the base of a logarithm system, which must be positive, when raised to any power, is always positive. For example, for base 10,

$$\log_{10} N = x \quad \text{means} \quad 10^x = N.$$

If we examine various values of 10^x, we can see that these values are always positive. Thus,

$$10^2 = 100$$
$$10^0 = 1$$
$$10^{-1} = 0.1$$
$$10^{-5} = 0.00001,$$

and, in general,

$$10^x > 0$$

$$10^{-x} = \frac{1}{10^x} > 0.$$

Hence, for every value of x in

$$\log_{10} N = x,$$

N must be strictly positive; that is, N cannot be zero or negative.

7.8 GRAPH OF $y = \ln x$

To sketch the graph of

$$y = \ln x,$$

we start at $x = 1$, where $y = \ln 1 = 0$, and move to the right (x increasing) to

$$x = 2 \quad \text{and} \quad y = \ln 2 = 0.69,$$

then to

$$x = 3, \qquad y = \ln 3 = 1.10,$$

and so on. We see that as we move to the right, ln x increases, and the curve of $y = \ln x$ rises. If we move to the left of $x = 1$ to

$$x = 0.5, \quad y = \ln 0.5 = -0.69$$

then to

$$x = 0.1, \quad y = \ln 0.1 = -2.30,$$

and so on, the curve of $y = \ln x$ falls ever further below the x-axis. Figure 7–1 illustrates the behavior of $y = \ln x$. The curve rises slowly, but always rises, to the right of $x = 1$. Recalling that x cannot be zero

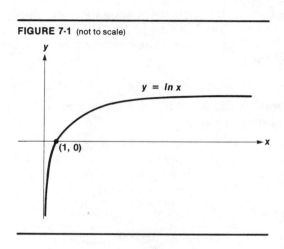

FIGURE 7-1 (not to scale)

$y = \ln x$

(1, 0)

or negative, we conclude that to the left of $x = 1$ the curve falls indefinitely as x gets closer to zero, but never touches the y-axis (where $x = 0$). It is not important to graph $y = \ln x$ carefully, but it is useful to know its characteristics, including the intercept point, (1, 0).

7.9 APPLICATION OF INVERSE NATURAL LOGARITHMS

By definition of a natural logarithm,

$$\log_e N = \ln N = x \quad \text{means} \quad N = e^x.$$

That is, for example, if

$$\ln N = 2$$

then

$$N = e^2 = (2.718282)^2$$
$$= 7.38906$$

to 5 places of decimals. Again, if

$$\ln N = -2$$

then

$$N = e^{-2} = \frac{1}{e^2}$$

$$= \frac{1}{(2.718282)^2}$$

$$= 0.13534.$$

Positive and negative powers of e can be obtained from a hand calculator having an e^x key by simply entering the value of x and depressing the key. Keys on some calculators perform two operations, and a special switch key of some sort is provided to select the operation desired. Thus, entering 2 on a calculator with a key named both LN and INV (or equivalent) will provide ln 2 with the switch in one position and e^2 with the switch in the other position. Values of e^x and e^{-x} can also be found in tables such as Table IX at the back of the book. More extensive tables are available but, for speed and accuracy, a hand calculator is superior to a table.

Exercise. From Table IX or by hand calculator, find: *a)* e^3. *b)* e^{-3}. *c)* $e^{0.05}$. *d)* $e^{-0.1}$. Answer: (Accurate to 5 decimal digits.) *a)* 20.08554. *b)* 0.04979. *c)* 1.05127. *d)* 0.90484.

Occasionally it is helpful to use the symbol antiln (x) instead of e^x.

Definition: antiln $(x) = e^x$.

Ln and antiln are inverse operations in the sense that one annuls the effect of, or undoes, the other. For example, if we find

$$\ln 2 = 0.69315$$

and take the antiln, thus,

$$\text{antiln } (\ln 2) = \text{antiln } (0.69315)$$
$$= e^{0.69315}$$
$$\text{antiln } (\ln 2) = 2.$$

Rule: antiln $(\ln x) = x$.

Example. One dollar has been deposited in a bank account. At time t years later, the expression relating S, the amount in the account, to the time t is

$$\ln S = 0.08t.$$

Find the amount in the account at time $t = 20$ years.

Substituting $t = 20$, we have

$$\ln S = 0.08(20) = 1.6.$$

Hence,

$$\text{antiln} (\ln S) = \text{antiln} (1.6)$$
$$S = e^{1.6} = \$4.95.$$

Example. (Optional—Hand calculator.) A $1000 deposit in a bank account grows to $2000 in 10 years at an interest rate of i, where i must be computed from

$$1000(1 + i)^{10} = 2000.$$

Find i.

We first divide by 1000, giving

$$(1 + i)^{10} = 2.$$

An easy way to proceed is to get $(1 + i)$ by raising both sides of the equation to the 0.1 power; that is, by taking the tenth root. Thus,

$$[(1 + i)^{10}]^{0.1} = 2^{0.1}$$
$$(1 + i)^1 = 2^{0.1}$$

because, on the left, we multiply exponents when a power is raised to a power. Consequently,

$$1 + i = 2^{0.1}$$
$$i = 2^{0.1} - 1.$$

With a hand calculator we find 2 to the 0.1 power is 1.07177. Hence,

$$i = 1.07177 - 1 = 0.07177.$$

An alternate way of solving

$$(1 + i)^{10} = 2$$

is to take the logarithms of both sides of the equation. Thus,

$$\ln (1 + i)^{10} = \ln 2$$
$$10 \ln (1 + i) = 0.69315$$
$$\ln (1 + i) = 0.069315.$$

The last, in inverse form, says

$$1 + i = e^{0.069315}$$

and by hand calculator e to the 0.069315 power is 1.07177, so

$$1 + i = 1.07177$$
$$i = 1.07177 - 1 = 0.07177$$

as before.

The last example illustrates a commonly occurring problem in finance, that of determining an unknown interest rate. The answer 0.07177, or 7.177 percent, means that $1000 at this interest rate, applied in the compounding procedure described in the next chapter, will grow to $2000 in 10 years.

Exercise. (Optional—Hand calculator.) *a)* If $\ln S = 3 + 0.1t$, find S when $t = 5$. *b)* Find i if $10,000(1 + i)^{10} = 30,000$. Answer: *a)* 33.11545. *b)* 0.11612.

Looking ahead. Any positive number except 1 can be used as the base for a system of logarithms. The base 10 (common logarithms) has been used widely in the past because it fits the decimal system and, as we explain later in the chapter, this makes possible a logarithm table that is more concise than tables of logarithms with any other base. However, tables of logarithms generally available give less accurate results than those which can be obtained more quickly by the touch of a key on a hand calculator. Consequently logarithm tables are referred to less and less frequently as time goes on, and it is quite possible that a decade hence they will not appear at the end of mathematics texts. Until that time, however, we shall include tables because readers will encounter them in other courses and one objective of this text is to prepare students for other courses. Our recommendation is that readers become familiar with the tables, *but use hand calculators to obtain logarithms.* If this is done, logarithms are available for more numbers than are included in the tables, they will be easier to obtain and work with, and, most important, they will be more accurate. The writer plans to present natural logarithms of numbers, obtained from a hand calculator, accurate to five places of decimals except where the needs of the problem at hand would be served better by something other than five places of decimals. In a later section of the chapter treating the use of a table of common (base 10) logarithms, accuracy of calculations will be restricted by limitations of the table.

The next section shows how the value of e and the values of natural logarithms are computed. The section is optional because it is not a prerequisite for study of other parts of the book.

7.10 COMPUTING e AND NATURAL LOGARITHMS (OPTIONAL)

A hand calculator is a black box that determines logarithms by some process known to only a relatively small number of people, and logarithms found in a table are computed by methods known to only a few. This is as it should be because one person cannot know all fields.

However, it is not difficult to learn one process for computing logarithms and a reader who learns the process can properly claim to be "one in a million" because, in fact, very few of the millions of people who use logarithms have any idea about where they come from. We are going to illustrate computation of natural logarithms, so let us first learn how their base, e, can be calculated to any desired degree of accuracy.

The mathematical definition of e, developed in a later chapter, says that e can be approximated to any degree of accuracy from the expression

$$\left(\frac{n+1}{n}\right)^n$$

by using a sufficiently large value of n. Unfortunately, sufficiently large is very large indeed if high accuracy is required. For example, the writer's calculator shows that for $n = 10,000$

$$\left(\frac{10000 + 1}{10000}\right)^{10000} = 2.7181$$

to four decimal places, whereas e is 2.7183 correct to four decimal places. To obtain an alternate method of approximating e,

$$\left(\frac{n+1}{n}\right)^n$$

can be expanded by the binomial theorem, and mathematicians have proved from this expansion that

$$e = 1 + \frac{1}{1!} + \frac{1}{2!} + \frac{1}{3!} + \frac{1}{4!} + \frac{1}{5!} + \frac{1}{6!} + \frac{1}{7!} + \frac{1}{8!} + \cdots$$

where the three dots $\cdots$ mean to continue indefinitely. That is, the more terms included in the sum, the closer the sum is to e. The meaning of the factorial symbol, !, is illustrated by

$$4! = 4 \text{ factorial} = 4 \cdot 3 \cdot 2 \cdot 1 = 24$$
$$3! = 3 \text{ factorial} = 3 \cdot 2 \cdot 1 \quad = 6.$$

The endless sum just presented is the infinite-series definition of e. Computing the factorials, we have

$$e = 1 + \frac{1}{1} + \frac{1}{2} + \frac{1}{6} + \frac{1}{24} + \frac{1}{120} + \frac{1}{720} + \frac{1}{5040} + \frac{1}{40320} + \cdots$$
$$= 2.7182788 + \cdots.$$

Rounded to five places of decimals, the last is 2.71828 and is the value of e correct to five places of decimals. Including three more terms yields 2.718281828, which is *e* correct to nine places of decimals. Lest the reader be led to think the last set of digits is a repeating decimal (note the 1828, 1828 sequence) we point out that this is not so, and correct to 12 places of decimals, *e* is 2.7182 8182 8459. In any event, the series approximation leads quickly to a value of *e* as accurate as we need in any application.

Mathematicians have also derived a set of series approximation formulas for computing natural logarithms of numbers. One of them is

$$\ln (1 + x) = x - \frac{x^2}{2} + \frac{x^3}{3} - \frac{x^4}{4} + \frac{x^5}{5} - \frac{x^6}{6} \cdots; \ 0 < x \le 1,$$

where the signs of the terms alternate. This series will settle in (converge) on a value for a logarithm only if *x* is positive and not greater than 1, but other formulas and the rules of logarithms can be applied to obtain logarithms for all numbers. To illustrate a logarithm calculation, we compute

$$\ln (1.2) = \ln (1 + 0.2).$$

This is

$$\ln (1 + x)$$

with *x* = 0.2. Hence,

$$\ln 1.2 = 0.2 - \frac{(0.2)^2}{2} + \frac{(0.2)^3}{3} - \frac{(0.2)^4}{4} + \frac{(0.2)^5}{5} - \frac{(0.2)^6}{6} + \cdots$$
$$\doteq 0.18232,$$

where $\doteq$ is read "equals approximately". The reader may verify by hand calculator or from Table X that the value calculated is ln 1.2 correct to five places of decimals.

> ***Exercise.*** *a)* Approximate ln (1.01) using only the first two terms of the foregoing series. *b)* By how much does the result in *a)* differ from the value 0.00995 for ln (1.01) in Table X? Answer: *a)* 0.00995. *b)* No difference.

After the logarithm of one number, *N*, has been computed by the series approximation, logarithms of all powers of *N* can be computed by simple multiplication because

$$\ln N^x = x \ln N.$$

Similarly, after the logarithms of two numbers have been computed, logarithms of their product and quotient can be found by simple addition and subtraction. That is,

$$\ln MN = \ln M + \ln N$$

$$\ln \frac{M}{N} = \ln M - \ln N.$$

Example. Two logarithms, correct to eight decimal places, have been computed. They are

$$\ln 4 = 1.3862\ 9436$$
$$\ln 10 = 2.3025\ 8509.$$

Compute the following, correct to five decimals, from the values of $\ln 4$ and $\ln 10$: *a)* $\ln 64$. *b)* $\ln 40$. *c)* $\ln 0.04$. We have

a) $64 = 4^3$; $\ln 64 = 3 \ln 4$
$$= 3(1.3862\ 9436)$$
$$= 4.1588\ 8308$$
$$= 4.15888.$$

b) $40 = 4(10)$; $\ln 40 = \ln 4 + \ln 10$
$$= 1.3862\ 9436 + 2.3025\ 8509$$
$$= 3.6888\ 7945$$
$$= 3.68888.$$

c) $0.04 = \dfrac{4}{100} = \dfrac{4}{10^2}$;

$$\ln(0.04) = \ln\left(\frac{4}{10^2}\right) = \ln 4 - 2 \ln 10$$
$$= 1.3862\ 9436 - 2(2.3025\ 8509)$$
$$= -3.2188\ 7582$$
$$= -3.21888.$$

Exercise. Given $\ln 4 = 1.3862\ 9436$ and $\ln 10 = 2.3025\ 8509$, show how the following can be computed, and carry out the computation correct to five decimal digits. *a)* $\ln 1000$. *b)* $\ln 0.01$. *c)* $\ln 25$. Answer: *a)* $\ln 1000 = 3 \ln 10 = 6.90776$. *b)* $\ln (0.01) = \ln 1 - 2 \ln 10 = -4.60517$. *c)* $\ln 25 = 2 \ln 10 - \ln 4 = 3.21888$.

It is clear from the foregoing that only a selected group of logarithms has to be calculated by the series approximation. With these and the rules of logarithms, a table can be constructed.

7.11   PROBLEM SET 7–2

Note: Remember that $\log x = \log_{10} x$ and $\ln x = \log_e x$.
Write the following in inverse form.

1. $3^2 = 9$.

2. $2^5 = 32$.

3. $\log_3 N = x$.

4. $\log_2 N = y$.

5. $\left(\dfrac{1}{2}\right)^3 = 0.125$.

6. $(0.2)^2 = 0.04$.

7. $\log_{16} 4 = \dfrac{1}{2}$.

8. $\log_{27} 3 = \dfrac{1}{3}$.

9. $2^{-3} = 0.125$.

10. $3^{-2} = \dfrac{1}{9}$.

11. $\log_5 0.04 = -2$.

12. $\log_2 0.0625 = -4$.

13. $\log 100 = 2$.

14. $\ln 20 = 2.9957$.

15. $\ln 10 = 2.3026$.

16. $\log 10 = 1$.

17. $\log 0.5 = -0.3010$.

18. $\ln 0.5 = -0.6931$.

What is the value of x in each of the following?

19. $\ln e = x$.

20. $\log 10 = x$.

21. $\log_2 4 = x$.

22. $\log_3 27 = x$.

23. $\log_7 7 = x$.

24. $\log_{0.5} (0.5) = x$.

25. $\log_4 2 = x$.

26. $\log_{64} 4 = x$.

27. By taking natural logarithms and antilogarithms of both sides, prove that $x = 2$ if

$$e^{\ln x} = 2.$$

28. By taking common logarithms and antilogarithms of both sides, prove that $x = 3$ if

$$10^{\log x} = 3.$$

29. What value must x have if

$$10^{\log 5} = x?$$

30. What value must x have if

$$e^{\ln 4} = x?$$

31. Find the value of x:

$$7^{\log_7 6} = x.$$

32. Find the value of x:

$$4^{\log_4 5} = x.$$

33. Rewrite the following using the ln symbol only once.

$$\frac{\ln x}{0.5} - 2 \ln y + 3 \ln z.$$

34. Rewrite the following using the log symbol only once.

$$\frac{\log x}{0.2} + \frac{1}{2}\log y - 2 \log z.$$

Given that $\log_a x = 4.2$, $\log_a y = 1.4$, $\log_a z = -1.2$, compute the value of each of the following:

35. $\log_a (xyz)$.

36. $\log_a \sqrt{z}$.

37. $\log_a \left(\frac{x}{z}\right)$.

38. $\log_a \left(\frac{xy}{z}\right)$.

39. $\log_a (z^{3/2})$.

40. $\log_a z^2$.

41. $\log_a \sqrt{xy}$.

42. $\log_a (xz)^{1/3}$.

43. $\frac{\log_a x}{\log_a z}$.

44. $\frac{\log_a (xy)}{\log_a z}$.

Using the rules of logarithms, write the following in a different, but equivalent form.

45. $2 \ln x + 3 \ln y$.

46. $3 \ln x - 2 \ln y$.

47. $0.2 \ln x^5$.

48. $\frac{1}{2} \ln x^4$.

49. $\frac{\ln y}{x}$.

50. $\left(\frac{1}{y}\right) \ln x$.

Solve for x.

51. $\ln x = 0.42$.

52. $\ln x = 1.2$.

53. $\ln x = -1.1$.

54. $\ln x = -0.01$.

55. $\ln(1 + x) = 0.3$.

56. $\ln(2 + x) = 1$.

57. $\ln(0.5x - 2) = -0.5$.

58. $\ln(1 - 0.1x) = -0.65$.

(Optional—Hand calculator) Use the compound interest formula below to solve the following:

$$S = P(1 + i)^t.$$

59. Compute t if $S = 5000$, $P = 2000$, and $i = 0.085$.
60. Compute t if $S = 10,000$, $P = 5000$, and $i = 0.12$.
61. Compute i if $S = 4000$, $P = 1000$, and $t = 10$.
62. Compute i if $S = 2500$, $P = 1000$, and $t = 5$.

(Optional) Compute the following by summing five terms of the series approximation,

$$\ln(1 + x) = x - \frac{x^2}{2} + \frac{x^3}{3} - \frac{x^4}{4} + \frac{x^5}{5} \cdots 0 < x \leq 1.$$

63. $\ln 1.1$.

64. $\ln(1.3)$.

7.12 COMMON (BASE 10) LOGARITHMS

Note: In this and following sections on base 10 logarithms our purpose is to illustrate the use of the base 10 table of logarithms. Consequently, we shall not take logarithms from a hand calculator.

Selection of 10 as the base leads to the most efficient method of tabulating logarithms. Actually only part of the logarithm of a number, known as the *mantissa*, is tabulated and the other part, called the *characteristic*, is determined by the user by a procedure derived from a rule of logarithms. Mantissas are provided in Table I at the back of the book. Mantissas are *positive decimals*, and a decimal point should be inserted when a mantissa is written. In Table I, the first two digits in a number are written in the left *margin* and the third digit is in the top *margin*. The *field*, or body, of the table contains the mantissas. Table 7–2 shows a part of Table I.

TABLE 7–2

N	0	1	2	3	4	5	6	.
10	0000	0043	0086	0128	0170	0212	0253	.
.	.	.	.	.	.	.	.	.
36	5563	5575	5587	5599	5611	5623	5635	.
37	5682	5694	5705	5717	5729	5740	5752	.
38	5798	5809	5821	5832	5843	5855	5866	.
.	.	.	.	.	.	.	.	.
99	9956	9961	9965	9969	9974	9978	9983	.

In Table 7–2, the digit sequence 372 has .5705 as its mantissa. Moreover any number that differs from 372 only in the position of its decimal point, such as 0.372, 37.2, 3720, 0.00372, and so on, has the same mantissa.

> *Exercise.* Find the mantissas of the logarithms of *a)* 105. *b)* 1.05. *c)* 0.0105. *d)* 10. *e)* 1. *f)* 38. Answer: *a)*, *b)*, and *c)*: .0212. *d)* and *e)*: .0000. *f)* .5798.

Standard position of the decimal point. Reading a number from left to right, the standard position of the decimal point, indicated by ‸, is after the first nonzero digit. For example, the numbers

$$48.7; \quad 0.0526; \quad 9;$$

marked with standard position are

$$4{‸}8.7; \quad 0.05{‸}26; \quad 9{‸}.$$

Rule for characteristic. The characteristic of the logarithm of a number is the *count* of the number of places *from* standard position *to* the actual position of the decimal point. The characteristic is positive (negative) if the actual point is to the right (left) of standard position.

Thus in

$$4\underset{\smile}{.}8.7$$

$$+1$$

the actual position of the point is 1 place to the *right* of standard position and the characteristic is +1. In

$$0.\underset{\smile}{0}5.26,$$

$$-2$$

the actual point is 2 places to the *left* of standard position and the characteristic is −2. For the number 9, or

$$9_{\wedge},$$

the actual point is in standard position and the characteristic is 0.

> *Exercise.* Write the characteristics of the logarithms of *a)* 374. *b)* 0.374. *c)* 3.8. *d)* 10. *e)* 1. *f)* 0.000590. Answer: *a)* +2. *b)* −1. *c)* 0. *d)* 1. *e)* 0. *f)* −4.

Standard form of a logarithm. The standard form of a base 10 logarithm is the combination of the characteristic and the positive decimal mantissa with positive characteristics written at the beginning and negative characteristics at the end.

For example, to write

$$\log 38.2,$$

we find the mantissa is 0.5821. The characteristic is +1. Hence,

$$\log 38.2 = 1.5821.$$

However, in writing

$$\log 0.000382,$$

which also has 0.5821 as its mantissa, the characteristic is −4, so it is written at the end. Thus,

$$\log 0.000382 = 0.5821 - 4.$$

It would be incorrect to write the last as −4.5821 because the mantissa part, 0.5821, is positive. This is why negative characteristics are written at the end. Some readers may be familiar with the practice of using +6 − 10, which is −4, and writing log 0.000382 = 6.5821 − 10. It does not matter which of the two methods is followed, but if a base 10 logarithm table is used, the decimal part of a logarithm must be positive.

Note: On a hand calculator with a base 10 key, we find

$$\log 0.000382 = -3.4179.$$

This number is

$$\log 0.000382 = .5821 - 4 = -3.4179.$$

The calculator does not work with a table of logarithms, so there is no need for separating characteristic and mantissa. On calculators having only a natural logarithm key, a base 10 logarithm can be found from

$$\log N = \frac{\ln N}{\ln 10} = 0.43429 \ln N.$$

Table 7–3 shows some examples of determining characteristics and writing the logarithms of numbers.

TABLE 7–3

Number N	Standard position	Characteristic	Mantissa from Table I	Log N
200	2⌄00.	+2	.3010	2.3010
20	2⌄0.	+1	.3010	1.3010
2	2⌄	0	.3010	0.3010
0.2	0.2⌄	−1	.3010	.3010 − 1
55.6	5⌄5.6	1	.7451	1.7451
0.00258	0.002⌄58	−3	.4116	.4116 − 3
9240	9⌄240.	+3	.9657	3.9657
0.0003780	0.0003⌄780	−4	.5775	.5775 − 4
10	1⌄0.	+1	.0000	1.0000
1	1⌄	0	.0000	0.0000

Exercise. Write the common logarithms of the following numbers. *a)* 25. *b)* 0.25. *c)* 2.5. *d)* 7250. *e)* 0.0123. Answer: *a)* 1.3979. *b)* .3979 − 1. *c)* 0.3979. *d)* 3.8603. *e)* .0899 − 2.

The rule of characteristics is a consequence of rules of logarithms. The mantissas tabulated in Table I are the actual logarithms of numbers 1 through 9.99. Thus, the entry for 20, which is 0.3010, is actually

$$\log 2 = 0.3010.$$

Now, remembering that

$$\log 10 = \log_{10} 10 = 1,$$

we can find log 20 by a rule of logarithms, as follows:

$$
\begin{aligned}
\log(20) = \log(10)(2) &= \log 10 + \log 2 \\
&= 1 + \log 2 \\
&= 1 + .3010 \\
&= 1.3010.
\end{aligned}
$$

That is, log 20 is log 2 plus a characteristic of 1.
   Again,

$$
\begin{aligned}
\log 0.2 = \log \frac{2}{10} &\\
&= \log 2 - \log 10 \\
&= .3010 - 1,
\end{aligned}
$$

and

$$
\begin{aligned}
\log 0.02 = \log \frac{2}{100} = \log \frac{2}{10^2} &\\
&= \log 2 - \log 10^2 \\
&= \log 2 - 2 \log 10 \\
&= 0.3010 - 2,
\end{aligned}
$$

and so on for any sequence of digits containing 2, and zeros which serve to position the decimal point. A table of natural logarithms, or logarithms to any base other than 10, requires separate entries for the logarithms of 200, 20, 2, 0.2, 0.002, and so on. Consequently, base 10 leads to a more compact table than is possible with any other base. However, to apply a base 10 table, the user must supply part of the logarithm.

7.13 PROBLEM SET 7–3

Find the logarithms of the following numbers:

1. 34.6.	2. 3.46.	3. 0.346.
4. 0.0346.	5. 34,600.	6. 0.903.
7. 1.01.	8. 258.	9. 0.01.
10. 0.0125.	11. 10^3.	12. 10^{-1}.

13.	$10^{0.786}$.	14.	10^{132}.	15.	$10^{2.3456}$.
16.	2340.	17.	0.00478.	18.	1.
19.	9.87.	20.	604.	21.	. 0.032.
22.	0.000007.	23.	24,500.	24.	6.
25.	0.1.	26.	10.	27.	2.03.
28.	568.	29.	0.000101.	30.	3.28.
31.	0.0001.	32.	57.	33.	5.
34.	16.	35.	1.6.	36.	1240.
37.	111.	38.	15.6.	39.	12.
40.	0.04.	41.	0.139.	42.	0.17.
43.	1.75.	44.	829.0.	45.	0.0566.
46.	155,000.	47.	0.002.	48.	0.137.
49.	5.27.	50.	0.00000000478.		

7.14 ANTILOGARITHMS

In the procedure described in the last section, we were given a number N and instructed to find the logarithm of the number. In the inverse process, we are given the logarithm of a number and instructed to find the number. For example, given

$$\log N = 1.5821,$$

we have the mantissa, 0.5821, and the characteristic, 1, and we want to know what number, N, has this mantissa and characteristic. To write this number, we need first to find its sequence of digits by working backwards in Table I. Having this sequence of digits, we mark standard position, then use the characteristic to find the actual position of the decimal point. To obtain the digit sequence in N if

$$\log N = 1.5821$$

we search the *field* (as contrasted to the left and top margins) for the mantissa. The reader should now refer to Table 7–2 in the foregoing (or Table I at the back of the book) and note that the mantissa 5821 appears in the row with the marginal number 38 and the column with the heading 2. We write the digit sequence 382 and mark standard position

$$3{\scriptstyle\wedge}82$$

and in

$$\log N = 1.5821$$

the characteristic $+1$ tells us the actual point is one place to the *right* of standard position. Thus, the number sought is

$$3{\scriptsize\frown}8{,}2 = 38.2.$$

$$+1$$

Antilog and log are inverse operations and one annuls the effect of the other. Thus, if

$$\log N = 1.5821,$$

then

$$\text{antilog}(\log N) = \text{antilog}(1.5821)$$
$$N = \text{antilog}(1.5821)$$
$$= 38.2$$

as demonstrated earlier.

Example. Find antilog$(0.8476 - 2)$.

In Table I, the mantissa, 8476 is found in the field. It is in the 70 row and the 4 column, so the desired digit sequence is 704. Writing this with standard position marked, we have

$$7{\scriptsize\frown}04$$

and the characteristic, -2, tells us the actual point is two places to the left of standard position. Hence,

$$\text{antilog}(0.8476 - 2) = 0.0704.$$

Exercise. *a)* Find antilog(2.7042). *b)* Find antilog$(0.8274 - 1)$. *c)* Find antilog(0.9542). Answer: *a)* 506. *b)* 0.672. *c)* 9.

Antilogs on a hand calculator. If we have

$$\log N = 0.3010,$$

then

$$N = \text{antilog}(0.3010).$$

Moreover, the inverse form of

$$\log_{10} N = 0.3010$$

is

$$N = 10^{0.3010}.$$

Hence,

$$N = \text{antilog}(0.3010) = 10^{0.3010}.$$

This states that

$$\text{antilog}(0.3010)$$

can be found by raising 10 to the 0.3010 power. In general:

Rule: Antilog $x = 10^x$.

Example. (Optional—Hand calculator). Find *a)* Antilog 1.3078. *b)* Antilog(0.1274 − 1).

For *a)* we have

$$\text{antilog}(1.3078) = 10^{1.3078} = 20.3142.$$

Turning to *b)*, we recall that calculators do not use a logarithm table and 0.1274 − 1 must be changed to

$$0.1274 - 1 = -0.8726.$$

Then

$$\begin{aligned}
\text{antilog}(0.1274 - 1) &= \text{antilog}(-0.8726) \\
&= 10^{-0.8726} \\
&= 0.13409.
\end{aligned}$$

Exercise. (Optional—Hand calculator). Find *a)* antilog(0.2345). *b)* antilog(3.0572). *c)* antilog(0.2896 − 3). Answer: *a)* 1.71593. *b)* 1,140.78. *c)* 0.00194805.

7.15 PROBLEM SET 7–4

Find the following.

1. antilog(1.4472).
2. antilog(2.4472).
3. antilog(.9031 − 2).
4. antilog(.9031 − 1).
5. antilog(.8451).
6. antilog(.4624).
7. antilog(4.4116).
8. antilog(3.7388).
9. antilog(.6096 − 3).
10. antilog(.4843 − 4).
11. antilog(8.2405 − 10).
12. antilog(9.2405 − 10).
13. antilog(3.0253).
14. antilog(5.0128).
15. antilog(.2742).
16. antilog(.8500).
17. antilog(0).
18. antilog(0.0000).
19. antilog(.0043).
20. antilog(.0086).
21. antilog(.0086 − 3).
22. antilog(.0043 − 2).
23. antilog(1.7042 − 2).
24. antilog(2.4997 − 3).
25. antilog(4.0334).
26. antilog(3.0170).

7.16 SUMMARY

Both logarithms and antilogarithms appear in the work to come and, at the beginning, the reader will have to be sure in each case that the proper procedure is followed. In summary, the procedures are:

1. To find log N (example: Find log 0.573):
   a) Locate the three-digit sequence of N, which is 573 in the example, in the *margins* of Table I and write down the corresponding mantissa from the *field* of the table, inserting a decimal point at the front. For the example, this is .7582.
   b) Find the characteristic of N. For 0.573, or 0.5ᴧ73, this is −1. Now write log N. For the example, this is

$$\log 0.573 = 0.7582 - 1.$$

2. To find antilog x (example: Find antilog[2.7427]):
   a) Locate the decimal part of x in the *field* of Table I and copy down the corresponding digit sequence from the margins. For the example, this sequence is 553.
   b) Starting from the standard position in the sequence, locate the actual decimal point by moving the number of places specified by the whole number part of x. For the example, this is +2, so the desired number is 5ᴧ53 with the decimal two places to the right of ᴧ. Hence,

$$\text{antilog}(2.7427) = 553.$$

Example. Find log 0.8000 and antilog(0.8).

We choose this example to emphasize the point that there is no way of telling from the number at hand (0.8 in both of the foregoing) which procedure is to be applied. It is the instruction *log* or *antilog* that specifies the procedure. Thus,

$$\log(0.8000) = \log 0.8 = .9031 - 1,$$

but

$$\text{antilog}(0.8) = \text{antilog}(0.8000) = 6.31.$$

The two procedures are mixed up in the next problem set. By working these problems, the reader will become confident that the proper one is carried out.

7.17 PROBLEM SET 7–5

Find the following.

1. log(3860).
2. antilog(1.8370).
3. antilog(3.9609).
4. log(0.0057).

5. log(.9400).   6. antilog(.73).
7. antilog(.94).   8. log(.73).
9. log(2).   10. antilog(3).
11. antilog(2).   12. log(3).
13. log(1.000).   14. antilog(1).
15. antilog(.29).   16. log(0.2900).

7.18 INTERPOLATION

We can find the mantissas of logarithms of any 3-digit number such as 52.6 in Table I, and the mantissas of such numbers are accurate to four places of decimals. Mantissas for 4-digit numbers can be approximated by *interpolation*. For example,

$$\log 52.64$$

is between log 52.60 and 52.70, so we may write

Number   Logarithm

$$0.10 \begin{bmatrix} 0.04 \begin{bmatrix} 52.60 & 1.7210 \\ 52.64 & 1. \end{bmatrix} x \\ 52.70 & 1.7218 \end{bmatrix} 0.0008$$

The numbers on the arrows are found by subtracting the number at the tail of the arrow from the number at the arrowhead. We find the value of x by setting up the proportion

$$\frac{x}{0.0008} = \frac{0.04}{0.10}.$$

Hence,

$$x = \frac{0.04}{0.10}(0.0008)$$
$$= 0.4(0.0008)$$
$$= 0.00032$$
$$= 0.0003,$$

where, in the last step, we have rounded to the fourth place of decimals because no mantissa can be expected to be accurate to more places than the tabulated mantissas, which are given to only four decimal places.

Note the arrowhead associated with x in our setup of the interpolation. In the direction of this arrowhead, logarithms are *increasing*, so x must be *added* to the number at the *tail* of the arrow. Thus,

$$\log 52.64 = 1.7210 + 0.0003 = 1.7213.$$

Exercise. Find log 297.6 by interpolating in Table I. Answer: 2.4728 + 0.0008 = 2.4736.

Abbreviated interpolation setup. Here, we write only the digit sequence for the number and the digits of the mantissa, as shown next in finding log 52.64.

Number   *Mantissa*

$$10 \begin{bmatrix} 4 & \begin{bmatrix} 5260 \\ 5264 \\ 5270 \end{bmatrix} \end{bmatrix} \quad \begin{bmatrix} 7210 \\ \\ 7218 \end{bmatrix} x \end{bmatrix} 8$$

$$\frac{x}{8} = \frac{4}{10}$$

$$x = \frac{4}{10}(8)$$

$$= 3.2$$

$$= 3,$$

rounding the last so that no digits appear to the right of the decimal point. We now have, for the mantissa digits,

$$7210 + 3 = 7213.$$

But, of course, the decimal point must be inserted, so we have

$$\text{mantissa} = 0.7213.$$

The characteristic of log 52.64 is 1. Hence,

$$\log 52.64 = 1.7213.$$

Mental interpolation. In the last (and any) setup the ratio of the numbers for the left brackets is always a certain number of tenths. In the case of 52.64, which is between 52.60 and 52.70, this is 4 tenths (the ending digit in 56.24) or 0.4. Therefore x is 0.4 times the difference in the mantissas, which appear in Table I as

$$7210 \qquad 7218.$$

This difference is $18 - 10 = 8$, so

$$x = 0.4(8) = 3.2 \text{ or } 3$$

and

$$\log 52.64 = 1.7213.$$

With practice, the foregoing can be done mentally or by a small amount of pencil work. The tabular setup is a safer way to proceed, but it is unduly time consuming if many interpolations are made. As another example of mental interpolation, we find

$$\log 0.001868.$$

We need 8 tenths (.8) of the tabular difference. Locating the mantissas for 186 and 187 (which are 2695 and 2718), we place a finger on them and subtract mentally to obtain 23. Then (mentally, or by jotting it down) eight-tenths of 23 is 18.4 or 18. Adding this to the smaller mantissa gives 2713. Then, inserting the decimal point and appending the characteristic, we have

$$\log 0.00186 = .2713 - 3.$$

Exercise. By mental interpolation in Table I, find: *a)* log 4.137. *b)* log 0.1084. Answer: *a)* 0.6167. *b)* 0.0350 − 1.

Rounding in interpolation. We shall follow the practice of rounding 5 or more up and less than 5 down. For example:

Number	*Rounded to 4 decimal places*
0.86123	0.8612
0.86124	0.8612
0.86125	0.8613
0.86126	0.8613
0.861249	0.8612

Note that for the number in the last line, 0.861249, we look at the 4th digit, then at the next digit, which is *less than 5,* so the rounded number is 0.8612.

Interpolation in antilogarithms. If the 4-digit mantissa wanted in finding an antilogarithm is not given exactly in the field of Table I, we copy the two mantissas that straddle the given mantissa, together with the corresponding marginal numbers. For example, in determining

$$\text{antilog } 1.4050,$$

we seek the mantissa 4050 in the field of Table I. It is between 4048 and 4065, which have corresponding marginal digits of 254 and 255, respectively. That is, antilog 1.4050 is between

$$25.4 \quad \text{and} \quad 25.5.$$

Writing the foregoing in a tabular setup, we have

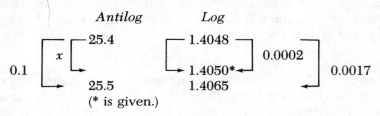

$$\begin{array}{ccc} & Antilog & Log \\ & 25.4 & 1.4048 \\ 0.1 & x & 1.4050* \quad 0.0002 \\ & 25.5 & 1.4065 \quad 0.0017 \\ & (* \text{ is given.}) & \end{array}$$

Hence,

$$\frac{x}{0.1} = \frac{0.0002}{0.0017}$$

$$x = (0.1)\left(\frac{0.0002}{0.0017}\right)$$

$$x = 0.012$$
$$= 0.01$$
$$\text{antilog}(1.4050 = 25.4 + 0.01 = 25.41.$$

In abbreviated setup, the last would be

$$\begin{array}{cc} Antilog & Mantissa \\ (Marginal\ digits) & (Field) \\ \\ 1 \quad x \quad \begin{matrix} 254 \\ ? \\ 255 \end{matrix} & \begin{matrix} 4048 \\ 4050 \quad 2 \\ 4065 \end{matrix} \quad 17 \end{array}$$

$$\frac{x}{1} = \frac{2}{17}$$

$$x = \frac{2}{17}$$

$$= 0.12$$

$$= 0.1,$$

which says the digit sequence we want, ? in the setup, is one-tenth of the way between 254 and 255, which is 2541. Hence, applying the characteristic 1, we have

$$\text{antilog}(1.4050) = 25.41.$$

Writing can be reduced to a minimum by simply writing down the given mantissa and the two that straddle it, in order. These are

$$4048 \quad 4050 \quad 4065.$$

Subtract the first from the second, then subtract the first from the third and divide the smaller by the larger to yield

$$\frac{2}{17} = 0.12$$

$$= 0.1.$$

This means that the marginal number corresponding to the first mantissa, which is 254, has a 1 in the next place. As before, we write 2541, apply the characteristic, and have

$$\text{antilog}(1.4050) = 25.41.$$

All we need in this interpolation is the single next digit, which is the ratio of the differences rounded to the first decimal place.

Exercise. Find *a)* Antilog(0.9368). *b)* Antilog(.8422 − 1). Answer: *a)* 8.646. *b)* 0.6953.

We have gone into interpolation in some detail because it can be applied in tables other than tables of logarithms. In passing, we should point out that our procedure is called *linear* interpolation because it approximates by assuming the graph of the logarithm curve between two points is a segment of a straight line. Figure 7–1 earlier in the chapter shows that $y = \log x$ is not a straight line, so interpolated results are necessarily approximations. However, for two points close together, the curved section is close to a straight line, especially at the right (large values of *x*). If *x* is near 1, where the curve departs most from linearity, linear approximation will have the most error. The reader can see this matter clearly by looking at the first and last rows of Table I. In the first row, entries change by larger amounts from column to column than is the case with entries in the last row. Consequently, interpolation in the first row is subject to the larger errors.

7.19 PROBLEM SET 7–6

Evaluate:

1. log 24.65.
2. log 89.85.
3. log 0.03175.
4. log 3272.
5. log 0.004932.
6. antilog 0.3762.
7. antilog 0.5950.
8. antilog 0.4860.
9. antilog (0.9070 − 2).
10. antilog (8.7438 − 10).
11. log 0.6080.
12. antilog 0.6080.
13. log 1.3670.
14. antilog 1.3670.
15. antilog (0.3670 − 3).
16. log 1.0350.

17.	antilog 1.0350.	18.	log 347.8.
19.	antilog 0.6120.	20.	log 0.4777.
21.	antilog 0.4777.	22.	log 4.8620.
23.	antilog 4.8620.	24.	log 0.003727.
25.	antilog $(0.9232 - 3)$.	26.	log 5.623.
27.	antilog $(2.8493 - 3)$.	28.	log 10.05.
29.	antilog 3.7417.	30.	antilog $(0.3742 - 3)$.

7.20 COMPUTATIONS: BASE 10

The product rule for logarithms states

$$\log ab = \log a + \log b.$$

If we take the antilogarithms of both sides of the last we have

$$\text{antilog}(\log ab) = \text{antilog}(\log a + \log b).$$

Now antilog and log are inverses, so that one annuls the other. For example, antilog(log 2) = 2, and antilog(log x) = x. Applying this to the left of the last expression, and remembering that logs and antilogs of a sum or difference (on the right) *cannot* be taken term by term, we have the computational rule

$$ab = \text{antilog}(\log a + \log b).$$

This says the product ab can be obtained by finding log a and log b, adding them, and then finding the antilog of the sum. Similarly, the quotient rule for logarithms says

$$\log \frac{a}{b} = \log a - \log b.$$

Hence,

$$\text{antilog}\left(\log \frac{a}{b}\right) = \text{antilog}(\log a - \log b)$$

and

$$\frac{a}{b} = \text{antilog}(\log a - \log b).$$

Therefore, a quotient can be computed by finding the logarithm of the numerator, subtracting from this the logarithm of the denominator, and then finding the antilogarithm of the difference. Finally, from the power rule,

$$\log a^b = b \log a$$
$$\text{antilog}[\log (a^b)] = \text{antilog}(b \log a)$$
$$a^b = \text{antilog}(b \log a),$$

which says a base to a power can be evaluated by multiplying the power by the logarithm of the base, then finding the antilogarithm of the product. In summary, we have:

Rules for calculation

$$ab = \text{antilog}(\log a + \log b)$$

$$\frac{a}{b} = \text{antilog } (\log a - \log b)$$

$$a^b = \text{antilog}(b \log a).$$

Example.   Compute x by logarithms, if

$$x = \frac{(22.4)(0.137)}{(6.08)(0.214)}.$$

We write

$x = \text{antilog}[\log 22.4 + \log 0.137 - \log 6.08 - \log 0.214]$
$ = \text{antilog}[1.3502 + (0.1367 - 1) - (0.7839) - (0.3304 - 1)]$
$ = \text{antilog}[1.3502 + 0.1367 - 1 - 0.7839 - 0.3304 + 1]$
$ = \text{antilog } (0.3726)$
$ = 2.358.$

Exercise.   Compute $x = 857(0.0236)/(0.458)$ by logarithms. Show the numbers used. Answer: $x = \text{antilog}[2.9330 + (0.3729 - 2) - (0.6609 - 1)] = \text{antilog}(1.6450) = 44.16.$

Converting to proper form.   If we compute

$$\frac{0.416}{27.3} = \text{antilog}[\log 0.416 - \log 27.3]$$

$$\phantom{\frac{0.416}{27.3}} = \text{antilog}[(0.6191 - 1) - 1.4362]$$

$$\phantom{\frac{0.416}{27.3}} = \text{antilog}(-1.8171)$$

we cannot find the antilog using Table I because in -1.8171 *all* of this number is negative and mantissas *must* be positive. To change to the proper form, we add to -1.8171 the smallest whole number that will make the result positive, which is 2, then correct for this change by subtracting the same number. That is,

$$-1.8171 = (-1.8171 + 2.0000) - 2$$
$$ = 0.1829 - 2.$$

The last has a positive decimal part so it is a proper form. Returning to the problem,

$$\text{antilog}(-1.8171)$$
$$= \text{antilog}(0.1829 - 2)$$
$$= 0.01524.$$

Exercise. Compute $(0.0257)/(9.7)$ showing the logarithms used. Answer: $\text{antilog}[(0.4099 - 2) - (0.9868)] = \text{anti-}$
$\log(-2.5769) = \text{antilog}(0.4231 - 3) = 0.002649.$

Example. Compute $(13.8)^{-2/3}$ by logarithms.
By the power rule for logarithms, we have

$$(13.8)^{-2/3} = \text{antilog}[-\tfrac{2}{3} \log(13.8)]$$
$$= \text{antilog}[-\tfrac{2}{3}(1.1399)]$$
$$= \text{antilog}\left(-\frac{2.2798}{3}\right)$$
$$= \text{antilog}(-0.75993)$$
$$= \text{antilog}(-0.7599) \qquad \text{(Improper form)}$$
$$= \text{antilog}[(-0.7599) + 1 - 1] \qquad \text{(Change form)}$$
$$= \text{antilog}(0.2401 - 1) \qquad \text{(Proper form)}$$
$$= 0.1738.$$

Example. Compute $(0.0417)^{3/4}$ by logarithms.
By the power rule,

$$(0.0417)^{3/4} = \text{antilog}[\tfrac{3}{4} \log(0.0417)]$$
$$= \text{antilog}[\tfrac{3}{4} (0.6201 - 2)]$$
$$= \text{antilog}[\tfrac{3}{4} (-1.3799)]$$
$$= \text{antilog}\left(\frac{-4.1397}{4}\right)$$
$$= \text{antilog}(-1.034925)$$
$$= \text{antilog}(-1.0349) \qquad \text{(Improper form)}$$
$$= \text{antilog}[(-1.0349 + 2) - 2] \qquad \text{(Change form)}$$
$$= \text{antilog}(0.9651 - 2) \qquad \text{(Proper form)}$$
$$= 0.09228.$$

Exercise. Compute by logarithms, showing the logarithms.
a) $(9.5)^{-0.2}$. b) The cube root of $225 = (225)^{1/3}$. Answer: a)
$\text{antilog}[-0.2(0.9777)] = \text{antilog}(0.8045 - 1) = 0.6376.$ b)
$\text{antilog}[(1/3)(2.3522)] = \text{antilog}(0.7841) = 6.083.$

Example.   Solve for x: $2^x = 8$.

The solution of this *exponential* equation is $x = 3$, because 2 to the third power equals 8. To illustrate how such equations are solved by logarithms, we first take the logarithm of both sides:

$$\log(2^x) = \log 8.$$

Applying the power rule for logarithms gives

$$x \log 2 = \log 8$$

$$x = \frac{\log 8}{\log 2}$$

$$= \frac{0.9031}{0.3010}$$

$$= 3.0003$$

$$= 3.000.$$

The result has been rounded as shown because the numbers 0.9031 and 0.3010 are accurate only to four digits. We call attention to the fact that the solution was obtained by *dividing* 0.9031 by 0.3010. Students sometimes confuse

$$x = \frac{\log 8}{\log 2},$$

which says to divide one logarithm by another, with the *different* expression,

$$\log\left(\frac{8}{2}\right)$$

which is found by subtraction as $\log 8 - \log 2$.

Example.   Solve for x: $200(1.25)^x = 500$.

We first divide by 200, giving

$$(1.25)^x = \frac{500}{200} = 2.5.$$

Then, taking logarithms,

$$x \log(1.25) = \log 2.5$$

$$x = \frac{\log 2.5}{\log 1.25}$$

$$= \frac{0.3979}{0.0969}$$

$$= 4.106.$$

Example. Solve for x: $50(1 + x)^{10} = 20$.
Dividing by 50 gives

$$(1 + x)^{10} = \frac{20}{50} = 0.4.$$

Next, taking logarithms,

$$\log(1 + x)^{10} = \log (0.4)$$
$$10 \log(1 + x) = 0.6021 - 1$$
$$\log(1 + x) = \frac{0.6021 - 1}{10}$$
$$= \frac{-0.3979}{10}$$
$$= -0.03979$$
$$\log(1 + x) = -0.0398 \qquad \text{(Improper form)}$$
$$= (-0.0398 + 1) - 1$$
$$\log(1 + x) = 0.9602 - 1. \qquad \text{(Proper form)}$$

Now, taking the antilog of both sides,

$$\text{antilog}[\log(1 + x)] = \text{antilog}(0.9602 - 1)$$
$$1 + x = \text{antilog}(0.9602 - 1)$$
$$1 + x = 0.9124$$
$$x = 0.9124 - 1$$
$$x = -0.0876.$$

Exercise. Solve for x by logarithms. *a)* What must x be if $8^x = 64$? *b)* Solve $8^x = 64$ by logarithms. *c)* Solve for x: $40(1 + x)^5 = 24$. Answer: *a)* 2. *b)* 2. *c)* −0.0972.

7.21 PROBLEM SET 7–7

Carry out the following computations by logarithms.

1. $(362)(2.58)$.
2. $(0.00716)(329)$.
3. $(3.654)(12.96)$.
4. $(28.4)(0.0512)$.
5. $(0.018)(0.00389)$.
6. $(53.45)(2.719)$.
7. $\dfrac{527}{315}$.
8. $\dfrac{2.43}{816}$.
9. $\dfrac{5.247}{0.0821}$.
10. $\dfrac{0.3472}{0.0516}$.

11. $\dfrac{217.3}{4.296}$.

12. $(2.31)^5$.

13. $(0.627)^{1/4}$.

14. $(0.0345)^3$.

15. $(10)^{1/3}$.

16. $(2446)^{1/2}$.

17. $(0.03)^{-1/2}$.

18. $(0.426)^{-0.7}$.

19. $\dfrac{(27.5)(3.64)}{19.1}$.

20. $\dfrac{(0.317)(316.2)}{107.6}$.

21. $(0.0745)^{-2/3}$.

22. $(425)(0.1567)$.

23. $\dfrac{(2.3)^{1/2}}{(5.6)^{2/3}}$.

24. $\dfrac{3.67}{814.3}$.

25. $(432)^{-4/5}$.

26. $(0.045)^{-2/3}$.

27. $(0.562)^{3/4}$.

28. $(85)^{0.2}$.

29. $(259)^{1/3}$.

30. $(0.0856)^{1/4}$.

Solve the following for x by logarithms.

31. $2^x = 25$.

32. $5^x = 12$.

33. $4^{-x} = 3$.

34. $3^{-x} = 2$.

35. $(0.5)^x = 5$.

36. $(0.2)^x = 8$.

37. $20^x = 5$.

38. $50^x = 10$.

39. $(1 + x)^5 = 2$.

40. $(1 + x)^{10} = 4$.

41. $(1 + x)^5 = 0.2$.

42. $(1 + x)^{10} = 0.4$.

43. $2(5^x) = 10$.

44. $4(3^x) = 8$.

45. $10(1 + x)^5 = 30$.

46. $20(1 + x)^{10} = 40$.

47. $20(1 + x)^5 = 10$.

48. $60(1 + x)^8 = 18$.

49. $1 + x = \text{antilog}(0.1220)$.

50. $\left(\dfrac{2}{x}\right)^{-5} = 10$.

7.22 REVIEW PROBLEMS

1. What is the value of e accurate to six decimal places?
2. State the definition of the natural logarithm of a number N.

Find the following:

3. $\ln 8.46$.

4. $\ln 33$.

5. $\ln 0.08$.

6. $\ln 0.1$.

7. $\ln 2.27$.

8. $\ln 3.36$.

9. $\ln 58$.

10. $\ln 8.18$.

11. $\ln 425$.

12. $\ln 786$.

13. $\ln e$.

14. $\ln 1$.

15. In $\log_b N$, why must N be positive?
16. Sketch a graph showing the characteristics of $y = \ln x$.

Find the following.

17. $e^{0.25}$.

18. $e^{0.01}$.

19. $e^{-0.25}$.

20. $e^{-0.01}$.

21. $e^{1.6}$. · 22. $e^{2.3}$.

23. e^{-2}. · 24. e^{-3}.

Find the following.

25. antiln(0.17). · 26. antiln(0.08).

27. antiln(−1.3). · 28. antiln(−2.1).

29. antiln(−0.01). · 30. antiln(1).

Write the following in inverse form.

31. $5^2 = 25$. · 32. $2^3 = 8$.

33. $\log_7 49 = 2$. · 34. $\log_5 125 = 3$.

35. $27^{1/3} = 3$. · 36. $64^{1/2} = 8$.

37. $\log_8 2 = \tfrac{1}{3}$. · 38. $\log_{81} 3 = \tfrac{1}{4}$.

39. $\ln e = 1$. · 40. $\log 1 = 0$.

41. $7^0 = 1$. · 42. $5^1 = 5$.

What is the value of x in each of the following?

43. $\log_2 32 = x$. · 44. $\log_3 81 = x$.

45. $\log_{0.2}(0.04) = x$. · 46. $\log_{0.2} 5 = x$.

47. $\log_3 x = 3$. · 48. $\log_2 x = 4$.

49. $\ln x = 1$. · 50. $\log x = 0$.

51. $e^{\ln e} = x$. · 52. $10^{\log 10} = x$.

53. $3^{\log_3 5} = x$. · 54. $5^{\log_5 4} = x$.

55. If, for base a, $\log_a x = 0.6$, $\log_a y = 1.8$, $\log_a z = 1.2$, compute the value of

$$\log_a \left[\frac{x^2}{(yz)^{1/3}} \right].$$

56. Using the logarithms of Problem 55, compute the value of

$$\log_a \left[\frac{z^{1/4}}{x^{1/3} y^{1/6}} \right].$$

57. Rewrite the following using the ln symbol only once.

$$2 \ln x - \frac{1}{2} \ln(y^2) + \ln z.$$

58. Rewrite the following using the ln symbol only once.

$$3 \ln(x^{1/3}) - \frac{1}{2} \ln y - 2 \ln z.$$

Solve the following for x.

59. $\ln x = 0.84$. · 60. $\ln x = 2.4$.

61. $\ln x = -2.3$. · 62. $\ln x = -0.08$.

63. $\ln(1 + x) = 1.2$. · 64. $\ln(1 + x) = -0.5$.

Solve the following for x.

65. $(6.75)^x = 4$.
66. $(25)^x = 127$.
67. $(0.23)^x = 0.15$.
68. $(0.55)^x = 1.8$.
69. $3^{-x} = 2$.
70. $4^{-x} = 8$.
71. $(0.2)^{-x} = 0.7$.
72. $(0.8)^{-x} = 0.5$.
73. $\left(\dfrac{4}{x}\right)^{-5} = 0.26$.
74. $\left(\dfrac{x}{5}\right)^{2.1} = 8$.
75. $e^x = 25$.
76. $10^x = 80$.
77. $x^{10} - 1 = 0$.
78. $x^{10} - 2 = 0$.
79. $(1 + x)^{15} = 2.5$.
80. $(1 + x)^{-10} = 0.5$.
81. $(1 + x) = $ antiln (0.14)
82. $(1 + x) = $ antiln (0.1239).
83. $x[1 + (1.01)^{50}] = 5$.
84. $2500[1 + (1.02)^{-x}] = 3000$.
85. State the definition of the common logarithm of a number, N.
86. Why is $\log(0.001)$ equal to -3?
87. Why cannot N be zero in $\log N$?
88. Why cannot N be negative in $\log N$?

Carry out the following computation by means of common (base 10) logarithms.

89. $(451)(3.62)$.
90. $(0.00815)(527)$.
91. $\dfrac{648}{267}$.
92. $\dfrac{5.26}{927}$.
93. $\dfrac{6.452}{0.0765}$.
94. $\dfrac{0.5681}{0.0437}$.
95. $\dfrac{(32.6)(2.59)}{18.7}$.
96. $\dfrac{(0.415)(327.4)}{128.7}$.
97. $(20)^{1/3}$.
98. $(4262)^{1/2}$.
99. $(0.025)^{-1/2}$.
100. $(0.507)^{-0.6}$.

Mathematics of finance

8

8.1 INTRODUCTION

Just about everyone becomes involved in transactions where interest rates affect the amount of money to be paid or received. For many, the largest such transaction is the purchase of a home. As we shall learn, a person who borrows $50,000 for such a purchase and cancels the debt by making monthly payments over a period of 30 years, will pay back about $145,000 if the interest rate is 9 percent per year, compounded monthly. Not very many years ago, when the interest rate was about 7 percent, the corresponding payback would have been about $120,000. The size of these paybacks serves to show both the effect of compound interest and the effect of changes in interest rates. In this chapter, we first look at simple interest calculation, then turn to the major objective of the chapter, which is to develop and apply formulas for financial transactions that involve compound interest calculations. In so doing, we shall ask and answer questions such as the following, for various interest rates and frequencies of compounding:

a) If $5000 is deposited in a bank account now, what will be the amount in the account ten years from now?

b) How much must be deposited in a bank account now if the amount in the account five years from now is to be $10,000?

c) If $1000 is added to an account every year, to what amount will the account grow in fifteen years?

d) How much must be deposited in an account each year if the amount in the account at the end of ten years is to be $15,000?

e) What sum deposited now will provide an income of $10,000 per year each year for the next 20 years?

f) If $30,000 is borrowed now to pay for a home, how much must be paid back each month if the debt is to be cancelled in 20 years?

In the last part of the chapter we shall answer questions like the foregoing if interest is compounded instantaneously (continuously) rather than at discrete points in time such as the end of each day, month, four months, six months, or one year.

Extensive tabulations are available to aid in carrying out the calculations involved in financial tabulations. Tables II through VII at the end of the book will suffice for our purposes. Additionally, we shall provide optional examples and exercises in which the calculations are carried out on a hand calculator. We recommend that the reader do this optional work (or do all problems and exercises) with a hand calculator because even the most extensive tabulations provide a limited cross-section of interest rates and time periods. However, it is important to understand that the chapter has been designed so that it is possible to learn the subject matter and do practice calculations using only the tables provided in this book. Also, we should point out that compound interest calculations involve exponents and, as a consequence, logarithms play a role in some of these calculations.

8.2 SIMPLE INTEREST AND THE AMOUNT

Interest rates are generally quoted in percentage form and, for use in calculations, must be converted to the equivalent decimal value by dividing the percentage by 100; that is, by moving the decimal point in the percentage two places to the left. For example,

$$r = 8\tfrac{1}{4}\% = 8.25\% = 0.0825.$$

Unless otherwise stated, a *quoted rate is a rate per year*. Thus, $1 at 8% means that interest of $0.08 will be earned in a year, and $100 at this rate provides

$$100(0.08) = \$8$$

of interest in one year. Interest on $100 at 8% for 9 months is interest for 9/12 year; that is,

$$\text{Interest} = \underset{\uparrow}{100} \quad \underset{\uparrow}{(0.08)} \quad \underset{\uparrow}{\left(\frac{9}{12}\right)} = \$6.00.$$

$$\text{Interest} = (\text{Principal}) \ (\text{rate}) \ (\text{time in years}).$$

The last line introduces the following definitions, which apply in simple interest calculations:

Definitions:

$I =$ Interest, in dollars.
$P =$ Principal, the sum of money on which interest is being earned.
$r =$ Rate of interest per year.
$t =$ Time, number of years.

Simple interest formula: $I = Prt$.

Thus, interest on $600 at 7½ percent for 10 months is computed using $P = 600$, $r = 0.075$, and $t = 10/12$ years. This is

$$I = 600(0.075)\left(\frac{10}{12}\right) = \$37.50.$$

Exercise. Compute the interest on $480 at 6¼ percent for 9 months. Answer: $22.50.

The simple interest formula can be solved easily for any of its variables.

Example. Find the interest rate if $1000 earns $45 interest in 6 months.

Here, $I = 45$, $P = 1000$, $t = 6/12 = 0.5$. Hence,

$$I = Prt$$
$$45 = 1000\,(r)(0.5)$$
$$45 = 500\,r$$
$$\frac{45}{500} = r$$
$$0.09 = r.$$

To obtain the percent rate, the decimal rate, $r = 0.09$, is multiplied by 100. Thus,

$$r = 0.09 = 100(0.09)\% = 9\%.$$

The *yield* on the common stock of a company is a percent obtained by dividing the amount (called the *dividend*) that a shareholder receives per share of stock held by the price of a share of the stock. Thus, yield is like an interest rate with the dividend analogous to the interest for $t = 1$ year and the price per share analogous to the principal. A stock market report showing

Gn Food 1.64   32¼

means that at the time of the quotation, a share of General Foods stock sold for $32.25 and the annual dividend was estimated to be $1.64 per share.

> **Exercise.** Compute the yield for General Foods from the market report in the foregoing. Answer: 5.09%.

The amount: If interest on $1000 at 9 percent for 8 months is computed as

$$I = Prt = 1000(0.09)\left(\frac{8}{12}\right) = \$60$$

and the interest is added to the principal, the sum is called the *amount*, S. Thus,

$$S = 1000 + 1000(0.09)\left(\frac{8}{12}\right)$$
$$= 1000 + 60$$
$$= \$1060.$$

Definition: **Amount:** $S = P + Prt = P(1 + rt)$.

Example. Find the amount if $20,000 is borrowed at 6 percent for 3 months.

Here, 3 months is $3/12 = 1/4$ of a year, so $t = 1/4$. Hence,

$$S = 20,000\left[1 + 0.06\left(\frac{1}{4}\right)\right]$$
$$= 20,000(1 + 0.015)$$
$$= \$20,300.00.$$

> **Exercise.** Find the amount of $5,000 at 10 percent for 9 months. Answer: $5,375.00.

The amount formula can be solved easily for any one of its variables.

Example. Jan received $50 for a diamond at a pawn shop and a month later paid $53.50 to get the diamond back. Find the percent interest rate.

Here,

$$P = \$50, \ S = \$53.50, \ n = \frac{1}{12} \text{ year.}$$

Therefore,

$$53.50 = 50 \left[1 + r\left(\frac{1}{12}\right) \right].$$

Dividing both sides by 50 yields

$$\frac{53.50}{50} = 1 + \frac{r}{12}.$$

Multiplying both sides by 12, we find

$$12\left(\frac{53.50}{50}\right) = 12 + r$$
$$12.84 = 12 + r$$
$$0.84 = r.$$

The last is the *decimal* rate. The percent rate is 100 times the decimal rate

$$0.84 = 100(0.84)\% = 84\%.$$

> **Exercise.** Fran has placed $500 in an employees' savings account that pays eight percent simple interest. How long will it be, in months, until the investment amounts to $530? Answer: 3/4 of a year = 9 months.

8.3 SIMPLE DISCOUNT: PRESENT VALUE

In the last exercise, $500 now amounts to $530 nine months from now if the interest rate is 8 percent. In reverse, we say that the *present value* of $530 receivable in 9 months is $500 now if the interest rate is 8 percent. This present value is analogous to a principal, so we shall denote it by P. Inasmuch as

$$S = P(1 + rt)$$

we obtain the simple present value formula by dividing both sides by $(1 + rt)$. Thus,

$$\frac{S}{1+rt} = P.$$

Definition: **Present value:** $P = \dfrac{S}{1+rt}$.

Thus, the present value of $530 receivable 9 months from now if interest is at 8 percent is

$$P = \frac{530}{1 + 0.08\left(\dfrac{9}{12}\right)}$$

$$= \frac{530}{1.06}$$

$$= \$500.$$

> **Exercise.** How much will Fran have to invest now in the employees' 8 percent savings account in order to have $600 a year from now? Answer: $600/1.08 = \$555.56$.

It is important in this chapter to keep in mind that a future amount of money, S, is worth less than S now. The sense of this is that certainly, in a business transaction, a person who promises to pay back $1000 to a lender at some time in the future cannot expect to receive as much as $1000 now.

> **Exercise.** Find the present value of $1000 at 9 percent due 8 months from now. Answer: $943.40.

8.4 BANK DISCOUNT

In many loans, the interest charge is computed not on the amount the borrower receives, but on the amount that is repaid later. A charge for a loan computed in this manner is called the *bank discount*, and the amount the borrower receives is called the *proceeds* of a loan. Proceeds begins with P and it is an amount received now. The future amount to be paid back is S. If $1000 is borrowed at 12 percent for 6 months, the borrower receives the proceeds, P, and pays back $S =$ 1000. The proceeds will be $1000 minus the interest on $1000. This will be

$$P = 1000 - 1000(0.12)\left(\frac{6}{12}\right)$$

$$
\begin{array}{cccc}
\uparrow & \uparrow & \uparrow & \uparrow \\
P = S & - S & (d) & (t)
\end{array}
$$

$$= 1000 - 60$$

$$= \$940.$$

In the foregoing, the interest rate was designated as d and is called the *bank discount rate*.

Definition: Proceeds: $P = S(1 - dt)$.

The example just completed shows that the borrower receives $940 but pays interest on the future sum $1000. If the borrower wants to receive proceeds of $P = \$1000$, then

$$1000 = S\left[1 - 0.12\left(\frac{6}{12}\right)\right]$$

$$1000 = S(1 - 0.06)$$

$$1000 = 0.94\,S$$

$$\frac{1000}{0.94} = S$$

$$\$1,063.83 = S.$$

Thus, the borrower who wants $1000 now will pay back $1,063.83 six months from now.

> **Exercise.** *a)* A borrower signs a note promising to pay a bank $5000 ten months from now. How much will the borrower receive if the discount rate is 8.4 percent? *b)* How much would the borrower have to repay in order to receive $5000 now? Answer: *a)* $4,650. *b)* $5,376.34.

Proceeds are an amount received now with repayment in the future, so they are analogous to present value, which was discussed in the last section. However, proceeds are not equal to present value because the proceeds from a future obligation to pay are always *less than* the present value of that obligation if, of course, the same rate of interest is used in both calculations. In the next set of problems, proceeds should be computed when the interest rate is stated as a *discount rate* or a *bank discount rate* and present value should be computed for similar problems where the interest rate is given without the qualifier, discount.

8.5 PROBLEM SET 8–1

Find *a)* the interest and *b)* the amount for each of the following principals for the stated simple interest rate and time period.

1. $500; 7%; 1 year.
2. $1,000; 8%; 1 year.
3. $1,000; 9%; 6 months.
4. $2,000; 6%; 6 months.
5. $100; 36%; 4 months.
6. $500; 24%; 3 months.
7. $200; 12%; 18 months.
8. $500; 18%; 16 months.
9. $5,000; 24%; 3 years.
10. $4,000; 30%; 2 years.
11. How many months will it take until the interest on $900 at 12 percent will be $135?
12. A credit card holder has owed the credit card company $200 for a month and receives a bill containing an interest charge of $3. Find the interest rate.
13. Compute the yield of New England Electric Company stock from the following stock market report.

NEngEl   1.94   22

14. How much must be deposited in an account paying 7 percent if interest of $100 is to be earned in 24 months?
15. How many months will it take at 8 percent interest for $2000 to grow to an amount of $2400?
16. Fran deposits $1000 in an employees' savings account at 6 percent. How many months will it be until the amount in the account is $1100?
17. Dan buys a TV set priced at $500 and is to pay this amount, plus interest, 3 months later. The total bill was $520. Compute the interest rate.
18. At what rate of interest will an investment of $1000 for 2 years grow to the amount of $1100?
19. Find the present value of $460 receivable 18 months from now if the interest rate is 10 percent.
20. Find the present value of $1000 receivable 2 years from now if the interest rate is 8.5 percent.
21. How much will Sam have to invest now in an employees' savings account at 7 percent in order to have $1000 in the account 18 months from now?
22. (See Problem 21.) How much would Sam have to invest if the interest rate was 10 percent?
23. Find the proceeds of a $2000, 18 month loan from a bank if the discount rate is 12 percent.
24. Find the proceeds of a $500, 9 month loan from a bank if the discount rate is 9 percent.
25. Dan wants $2000 now from a bank, to be repaid 18 months from now. How much will the repayment be if the discount rate is 15 percent?
26. Fran signs a note promising to pay a bank $1000 ten months from now and receives $900. Find the discount rate.

8.6 COMPOUND INTEREST AND AMOUNT

To see how compound interest works and develop a formula for computing the compound amount, suppose \$5000 is invested at 10 percent interest compounded each year. The amount at the end of the first year would be

$$S_1 = 5000 + 5000(0.10)(1)$$
$$= 5000 + 500$$
$$= \$5500.$$

This \$5500 becomes the principal at the beginning of the second year, and the amount at the end of the second year is

$$S_2 = 5500 + 5500(0.10)(1)$$
$$= 5500 + 550$$
$$= \$6050.$$

Thus, in the second year, interest is earned on not only the \$5000 invested, but also on the \$500 of interest earned in the first year. This common practice of computing interest on interest is called *compounding* interest.

To obtain a formula for computing the compound amount, we will use *i as the interest rate per period*. It will suffice for the moment to think of the period as being a year, and we will adjust our formula accordingly later when the period is something other than a year.

Definition: $i =$ interest rate per period.

Assuming, then, that the period is 1 year, a principal of \$P will amount to

$$S_1 = P(1 + i)^1$$

at the end of the first year. At the beginning of the second year, $P(1 + i)$ becomes the new beginning principal, which is multiplied by $(1 + i)$ to find the compound amount at the end of the second year. Thus,

$$S_2 = P(1 + i)(1 + i) = P(1 + i)^2.$$

Hence, at the beginning of the third year, the new principal is $P(1 + i)^2$, and to obtain the compound amount at the end of the third year, this must be multiplied by $(1 + i)$. Thus,

$$S_3 = P(1 + i)^2(1 + i) = P(1 + i)^3.$$

Similarly, the compound amount at the end of 10 years would be

$$S_{10} = P(1 + i)^{10}$$

and, in general, at the end of n years, the compound amount will be

$$S_n = P(1 + i)^n.$$

Conventionally, the subscript n on S_n is not written.

Definition: **Compound amount:** $S = P(1 + i)^n$.

The last is the *compound amount of $P for n periods at an interest rate of i per period.*

Example.　Find the compound amount of $1000 at 7 percent for 10 years.

We have

$$S = 1000(1 + 0.07)^{10}$$
$$= 1000(1.07)^{10}$$
$$= 1000(1.96715)$$
$$= \$1967.15.$$

The calculation of $(1.07)^{10}$ can be done quickly on a calculator or the desired result can be obtained from Table II at the back of the book. The title of this table is the *Compound Amount of $1.* The reader should verify that the entry for 7 percent, 10 periods, is 1.96715. This is the compound amount of $1, so the compound amount of $1000 is 1000 times 1.96715.

Exercise.　If $500 is invested at 6 percent compounded annually, what will be the amount 30 years later? Answer: $2871.75.

8.7 THE CONVERSION PERIOD

Quoted interest rates are rates *per year* if not accompanied by a qualifying statement such as 1½ percent per month. In the absence of a qualifier, the quoted annual rate is called the *nominal* rate and it is symbolized by j.

Definition: **Nominal rate** = rate per year = j.

Although the quoted nominal rate is per year, it is common practice to compound interest more frequently than once a year. Many banks compound interest on savings accounts on a daily basis, or 365 times a year. In other transactions, interest is compounded monthly, quarterly, or semiannually.

Conversion	Number of conversions per year, m
Daily	365
Monthly	12
Quarterly	4
Semiannually	2

Definition: **Number of conversions per year** $= m$.

Example. Find the compound amount of $500 at 8 percent compounded quarterly for 10 years.

There are four quarters in one year and

$$4(10) = 40$$

quarters in 10 years. Thus we multiply the number of years by the number of conversions per year. We shall now use this number of periods (40 quarters) as n in the compound amount formula.

Next, inasmuch as the quoted 8 percent is a nominal or per-year rate, we divide 8 percent by the number of conversions per year to obtain

$$i = \frac{8\%}{4} = 2\% \text{ per period.}$$

From Table II, at $i = 2\%$ and $n = 40$, we find

Compound amount of $1 = 2.20804.$

Hence, for $500, we have

$$S = 500(2.20804)$$
$$= \$1104.02.$$

On a hand calculator, the corresponding computation would be

$$S = 500\left(1 + \frac{0.08}{4}\right)^{10(4)}$$
$$= 500(1.02)^{40}$$
$$= 500(2.208040)$$
$$= \$1104.02.$$

In summary, *when interest is compounded more often than once a year:*

$$n = (\text{number of years})(\text{number of conversions per year})$$

$$i = \frac{\text{nominal rate per year}}{\text{number of conversions per year}} = \frac{j}{m}.$$

Exercise. If $800 is invested at 6 percent compounded semiannually, what will be the amount in 5 years? Answer: $1075.13.

Example. (Optional—Hand calculator). Compute the compound amount of $5000 at 7 percent compounded monthly for 10 years.
Here we have

$$-n = (10 \text{ years})(12 \text{ months per year}) = 120.$$
$$r = \frac{0.07}{12}.$$

Hence,

$$S = 5000 \left(1 + \frac{0.07}{12}\right)^{120}$$

$$= 5000(2.009 \quad 661 \quad 299). \text{ (See } Note \ 1 \text{ following.)}$$
$$= \$10,048.31. \text{ (See } Note \ 2 \text{ following.)}$$

Note 1: Hand calculator accuracy. Like any computing device, hand calculators have limited accuracy. In a *single* elementary operation $(+, -, \times, \div)$ the maximum error is ± 1 in the rightmost digit that can be displayed on the calculator; that is, for example, the eighth digit on a calculator that displays 8 digits, not including the exponent on floating point (see Index for reference) calculators. If a sequence of elementary operations is performed, the error can increase. The logarithm of a number may have an error as much as ± 3 in the rightmost digit. Raising to a power can lead to errors that affect more digits than the rightmost one, and this is an important consideration in financial computations where large powers arise. In the last example, for instance, the 120th power obtained on the writer's 10 digit display calculator is correct only to seven digits. Below it is compared with the number correct to ten digits obtained from a computer:

Hand calculator result: 2.009   661   299
Correct result: 2.009   661   377.

Note 2: Accuracy in multiplication and division. Suppose we compute

$$(25.4)(36.532) = 927.9128.$$

Now, if 25.4 is a number accurate to only three digits, the product cannot be accurate to more than three digits and should be rounded to 928. In a sequence, or chain, of multiplications and divisions, *the chain is only as strong as its weakest link.* Consequently, the result of such a computation should be rounded to the number of accurate digits in the least accurate figure used in the calculation. But note that *beginning zeros are not included in the count of the number of accurate digits.* For example, if the 342 in 0.00342 are accurate, the number has three accurate digits, and

$$(0.00342)(284.56) = 0.9731952 = 0.973$$

where 0.973 is the properly rounded product.

> *Exercise.* (Optional—Hand calculator). A bank pays 5.75 percent compounded daily on 90-day notice accounts. If $500 is deposited in such an account, what will be the amount in 90 days? (Use 365 days per year.) Answer: $507.14.

8.8 FINDING THE TIME, *n*

In the formula

$$S = P(1 + i)^n,$$

we can determine *n* if *S*, *P*, and *i* are given; that is, we can find how long it will take for *P* dollars deposited now at *i* percent to grow to an amount of *S* dollars.

Example. At 8 percent compounded annually, how many years will it take for $2000 to grow to $3000?

We have

$$3000 = 2000(1 + 0.08)^n$$

$$\frac{3000}{2000} = (1.08)^n$$

$$1.5 = (1.08)^n.$$

Taking the logarithm of both sides, using natural logarithms because our table of natural logarithms is more accurate than the common logarithm table, we write

$$\ln 1.5 = \ln(1.08)^n = n \ln(1.08).$$

From Table X,

$$0.40547 = n(0.07696)$$

$$\frac{0.40547}{0.07696} = n$$

$$5.27 = n,$$

so *n* is a bit more than five and a quarter years.

Exercise. Using Table X or a hand calculator, find how many years it will take at 6 percent compounded annually for $1000 to grow to $2000. Answer: About 11.9 years.

8.9 FINDING THE INTEREST RATE

Example. At what interest rate compounded annually will a sum of money double in 10 years?

Here, double means $100 grows to $200, $25 grows to $50, $1 grows to $2, and so on, and the time required is the same in any doubling. We shall use

$$P = \$1, \quad S = \$2.$$

Hence, in

$$S = P(1 + i)^n$$

or

$$P(1 + i)^n = S$$
$$(1)(1 + i)^{10} = 2$$
$$(1 + i)^{10} = 2.$$

Then

$$\ln (1 + i)^{10} = \ln 2$$
$$10 \ln(1 + i) = \ln 2$$
$$10 \ln(1 + i) = 0.69315$$
$$\ln(1 + i) = \frac{0.69315}{10}$$
$$\ln(1 + i) = 0.069315.$$

To obtain $(1 + i)$, we must take the antilogarithm of both sides. Thus,

$$\text{antiln } [\ln(1 + i)] = \text{antiln}(0.069315)$$
$$(1 + i) = \text{antiln}(0.069315).$$

On a hand calculator, antiln (0.069315) is found simply by raising e to the 0.069315 power to obtain 1.07177. This is $(1 + i)$, so $i = 0.07177$ or 7.177 percent. If Table X is used, it will be necessary to interpolate, as follows:

1. In the *field* of Table X, find the entries that straddle 0.069315. These are 0.06766 and 0.07696. Write down 0.069315, these field entries and their corresponding *marginal* numbers, which are, respectively, 1.07 and 1.08. We have

356

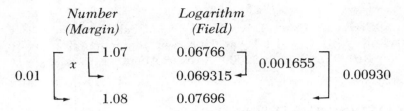

	Number (Margin)		Logarithm (Field)		

$$
0.01
\begin{bmatrix}
x
\begin{bmatrix}
1.07 \\
\\
1.08
\end{bmatrix}
\end{bmatrix}
\begin{matrix}
0.06766 \\
0.069315 \\
0.07696
\end{matrix}
\begin{bmatrix}
0.001655
\end{bmatrix}
\quad 0.00930
$$

2. Compute x by the proportion

$$\frac{x}{0.01} = \frac{0.001655}{0.00930}$$

$$x = 0.01\left(\frac{0.001655}{0.00930}\right)$$

$$= 0.00178$$

$$= 0.0018.$$

The last result was rounded because linear interpolation is approximate and we cannot in general expect interpolation results from Table X to be accurate beyond the fourth decimal place.

3. Note that the numbers (margin) *increase* in the direction of the x arrowhead, so *add* x to the number at the tail of the arrow to obtain

$$x = 1.07 + 0.0018$$

$$= 1.0718.$$

Now, returning to our problem,

$$(1 + i) = \text{antiln}\ (0.069315)$$

$$1 + i = 1.0718$$

$$i = 1.0718 - 1$$

$$= 0.0718$$

$$i = 7.18\%.$$

Taking logarithms on hand calculators gives more accurate results than tabular interpolation, and calculator operation is much easier than interpolation.

Exercise. By interpolation or hand calculator, find the rate of interest that, compounded annually, will result in tripling a sum of money in 10 years. Answer: 11.61 percent by interpolation and 11.612 percent by hand calculator.

8.10 PROBLEM SET 8–2

Find the compound amounts at the stated nominal interest rate compounded annually (once a year). (Compute by Table II or hand calculator.)

1. $200; 20 years; 5%.
2. $300; 10 years; 6%.
3. $400; 40 years; 8%.
4. $500; 15 years; 7%.

Find the compound amount using the appropriate interest rate and number of periods.

5. $150; 8 years; 8% compounded quarterly.
6. $250; 3 years; 12% compounded monthly.
7. $600; 20 years; 8% compounded semiannually.
8. $1000; 10 years; 16% compounded quarterly.
9. How many years will it take at 7 percent compounded annually for $5000 to amount to $20,000?
10. How many years will it take for a sum of money to double at 10 percent compounded annually?
11. By interpolation or hand calculator, find the rate of interest compounded annually at which a sum of money will double in 20 years.
12. By interpolation or hand calculator, find the rate of interest compounded semiannually at which $5000 will grow to $12,000 in eight years.

The following are optional. Use a hand calculator.

13. A bank pays 5.25 percent compounded daily on certificate accounts running for 6 years. Using 365 days per year, compute the amount of a deposit of $5000 for 6 years.
14. A bank pays 5.25 percent compounded daily on certain accounts. Find the amount of a deposit of $2000 for 45 days.
15. How many years will it take at 9 percent compounded annually for $5000 to grow to $10,000?
16. At what rate of interest compounded annually will $1000 grow to $5000 in 10 years?

8.11 COMPOUND DISCOUNT: PRESENT VALUE

Dividing both sides of the compound amount formula

$$S = P(1 + i)^n$$

by $(1 + i)^n$ leads to

$$\frac{S}{(1 + i)^n} = P \quad \text{or} \quad P = \frac{S}{(1 + i)^n}.$$

By the definition of a negative exponent,

$$P = \frac{S}{(1+i)^n} = S(1+i)^{-n}.$$

Definition: **Present value:** $P = S(1+i)^{-n}$
Compound discount factor $= (1+i)^{-n}$.

Values of the compound discount factor are available in Table III, which has the title *Present Value of $1*. In this, as in all the financial tables, the tabular entry is multiplied by the number of dollars specified in the problem at hand. On a hand calculator, of course, it is as easy to divide by $(1+i)^n$ as it is to multiply by $(1+i)^{-n}$, but multiplication is easier for handwork without a calculator.

Example. What is the present value of $2500 payable 4 years from now at 8 percent compounded quarterly?

Here, the amount 4 years hence is $S = 2,500$. With quarterly compounding,

$$n = (4 \text{ periods per year})(4 \text{ years}) = 16 \text{ periods}.$$

$$i = \frac{0.08}{4} = 0.02.$$

Hence,

$$P = 2500(1 + 0.02)^{-16}$$
$$= 2500(0.728446),$$

where 0.728446 is the compound discount factor taken from Table III. Multiplication gives

$$P = \$1821.115.$$

There is no generally applied rule for rounding a monetary number to the last cent, and we shall follow the practice of rounding a digit up when the next digit is 5 or more. Thus,

$$P = \$1821.12.$$

A hand calculator shows the last result is, more accurately, $1821.1145, which would round to $1821.11.

Exercise. What is the present value of $4000 payable in 20 years at 8 percent compounded semiannually? Answer: $833.16.

Example. (Optional—Hand calculator). How much must be deposited now in an account paying 7.3 percent compounded daily in order to have just enough in the account 3 years from now to make $10,000 available for investment in a business enterprise?

With one day as a period,

$$n = (365 \text{ days per year})(3 \text{ years}) = 1095 \text{ periods}.$$

$$i = \frac{0.073 \text{ per year}}{365 \text{ conversions per year}}.$$

Hence,

$$P = 10000 \left(1 + \frac{0.073}{365}\right)^{-1095}$$

$$= \frac{10000}{\left(1 + \dfrac{0.073}{365}\right)^{1095}}$$

$$= \frac{10000}{1.2448040}$$

$$= \$8033.39.$$

Exercise. (Optional—Hand calculator). How much must be deposited now in an account paying 8 percent compounded monthly in order to have just enough in the account 5 years from now to make a $10,000 down payment on a home? Answer: $6712.10.

8.12 PROBLEM SET 8–3

Compute the present value for the following using Table III or a hand calculator.

1. $1000 at 8% compounded annually, due in 20 years.
2. $2000 at 7% compounded annually, due in 10 years.
3. $5000 at 10% compounded semiannually, due in 5 years.
4. $4000 at 12% compounded monthly, due in 3 years.
5. What sum of money deposited now at 8% compounded quarterly will provide just enough money to pay a $1000 debt due seven years from now?
6. What sum of money invested now at 12% compounded monthly will provide just enough to pay a debt of $2500 due in 3 years?
7. If output per manhour increases by 5% compounded annually and is currently 100 units per manhour, what was output per manhour 5 years ago?
8. An account bearing interest at 6 percent compounded semiannually was established 10 years ago. The account balance now is $9030.55. What was the initial amount when the account was established?

Optional: Compute the following with a hand calculator.

9. Find the present value of $1000 due in 2 years at 8 percent compounded daily. (365 days in a year.)
10. Find the present value of $2000 due in 10 years at 9 percent compounded monthly.

8.13 GEOMETRIC SERIES

In the calculations we have made thus far, one sum of money was involved, and we found the future amount or present value of this single sum. In many transactions, a series of equal sums is deposited or paid periodically over time. Examples of this are saving $100 per month to accumulate a desired amount of money, or paying $350 per month to pay off a mortgage on a home. In compound interest computations, the formulas for transactions involving periodic payments or receipts are derived from the formula for the sum of a geometric series. An example of such a series is

$$1 + 2 + 4 + 8 + 16.$$

This is a *unit* geometric series because the first term is one. The second term is 2 times the first, the third is 2 times the second, the fourth is two times the third, and so on.

Definition: The unit geometric series is the sum of a sequence of terms in which the first term is 1 and each successive term is r times the previous term.

We next state and illustrate the formula for the sum of such a series, and then will show how it arises.

Formula: Sum of n terms of a unit geometric series:

$$S_n = \frac{r^n - 1}{r - 1}.$$

The series written above has five terms, and its sum, computed longhand, is

$$S_5 = 1 + 2 + 4 + 8 + 16 = 31.$$

The number r is the multiplier, which gives a new term from the previous term; that is, r is the constant ratio of two successive terms. In the example, $r = 2$. Hence, by formula,

$$S_5 = \frac{r^5 - 1}{r - 1} = \frac{2^5 - 1}{2 - 1} = \frac{32 - 1}{1} = 31,$$

as before.

Example. Compute the sum of the first 10 terms of

$$1 + 1.06 + (1.06)^2 + (1.06)^3 + \cdots .$$

Here, the ratio r is 1.06. Hence,

$$S_{10} = \frac{(1.06)^{10} - 1}{1.06 - 1} = \frac{1.79085 - 1}{1.06 - 1} = \frac{0.79085}{0.06} = 13.181$$

to three decimal places. Observe that the power

$$(1.06)^{10} = 1.79085$$

can be found in Table II with $i = 6$ percent, $n = 10$, or it can be obtained from a hand calculator.

Exercise. *a)* By formula, compute the sum of $1 + 3 + 3^2 + 3^3 + 3^4$. *b)* By Table II or hand calculator, compute the sum of the first 20 terms of $1 + (1.08) + (1.08)^2 + (1.08)^3 + \cdots$. Answer: *a)* 121. *b)* 45.762.

Source of the geometric formula. If we write

$$S_5 = 1 + 2 + 2^2 + 2^3 + 2^4,$$

observe that there are $n = 5$ terms, but the exponent of the last term is $n - 1 = 4$. Hence,

$$S_n = 1 + r + r^2 + r^3 \ldots + r^{n-2} + r^{n-1}$$

is the correct general expression for the sum of n terms of the unit geometric series. Next, we multiply S_n by r to obtain

$$rS_n = r(1 + r + r^2 + r^3 \ldots + r^{n-2} + r^{n-1})$$
$$= r + r^2 + r^3 + \ldots + r^{n-2} + r^{n-1} + r^n.$$

Now subtract S_n from rS_n, thus,

$$rS_n - S_n = r + r^2 + r^3 \ldots + r^{n-2} + r^{n-1} + r^n$$
$$- (1 + r + r^2 + r^3 \ldots + r^{n-2} + r^{n-1})$$
$$= r^n - 1.$$

Hence,

$$rS_n - S_n = r^n - 1.$$

Factoring out S_n on the left gives

$$S_n (r - 1) = r^n - 1,$$

so

$$S_n = \frac{r^n - 1}{r - 1},$$

which is the formula stated at the beginning of the section.

8.14 ORDINARY ANNUITIES: AMOUNT

An *ordinary* annuity is a series of equal periodic payments in which each payment is made at the *end* of the period. In our work we shall use the following symbols:

n = Number of periods
i = Interest rate per period
R = Payment per period
S_n = Amount of the annuity after n payments.
s_n = Amount of an annuity of $1 per period for n periods.

Let us start with a payment of $1 at the end of each period for five periods, and an interest rate of $i = 0.02$ per period. Figure 8–1

FIGURE 8-1

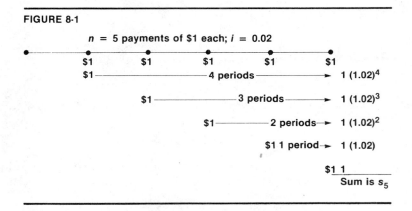

shows the payments for this annuity. Note that the payments are shown at the *end* of each period because this is an *ordinary* annuity. Therefore, the first payment draws interest for four periods and, by the compound amount formula, contributes

$$1(1.02)^4$$

to the sum s_5. The second payment draws interest for three periods so its contribution to s_5 is

$$1(1.02)^3,$$

and similarly for the other payments. The last payment occurs *at* the point in time where s_5 is evaluated and so contributes $1, but no interest, to s_5. Adding the terms of s_5 from the bottom up in Figure 8–1, we have

$$s_5 = 1 + (1.02)^1 + (1.02)^2 + (1.02)^3 + (1.02)^4.$$

The last is a geometric progression with $n = 5$ terms, $r = 1.02$, so

$$s_5 = \frac{(1.02)^5 - 1}{1.02 - 1} = \$5.20.$$

In general, if the interest rate is i per period and n periodic payments are made, then the amount at the end is

$$s_n = \frac{(1 + i)^n - 1}{(1 + i) - 1} = \frac{(1 + i)^n - 1}{i}$$

where, we recall, s_n means the payment is \$1 per period. For a payment of \$$R$ per period we simply multiply s_n by R to obtain

Amount of ordinary annuity: $S_n = R \left[\dfrac{(1 + i)^n - 1}{i} \right].$

Table IV at the back of the book entitled *Amount of \$1 per Period* provides values of s_n. To apply the table when the payment is R per period, we multiply the appropriate entry by R.

Example. If \$100 is deposited in an account at the end of every quarter for the next five years, how much will be in the account at the time of the final deposit if interest is 8 percent compounded quarterly?

We have

$$n = (5 \text{ years})(4 \text{ quarters per year}) = 20 \text{ periods,}$$
which means 20 quarterly payments.

$$i = \frac{8\%}{4} = 2\%$$

$$R = \$100.$$

From Table IV, with $i = 2\%$, $n = 20$, we find that \$1 per period would amount to \$24.29737. Hence, with $R = \$100$ per period

$$S_{20} = 100(24.29737) = \$2429.74.$$

Observe that 20 payments of \$100 each amount to \$2000. The \$2429.74 amount is this \$2000 plus interest for varying lengths of time on all payments except the last one.

> **Exercise.** Five hundred dollars is deposited in an account at the end of each six-month period for eight years. Find the amount in the account after the last deposit has been made if interest is earned at the rate of 10 percent compounded semiannually. Answer: \$11,828.75.

Example. (Optional—Hand calculator). If \$100 is deposited in an account each month for 10 years and the account earns 7 percent compounded monthly, how much will be in the account after the last deposit is made?

We have

$$n = 10 \text{ years (12 months per year)} = 120 \text{ periods}$$

$$r = \frac{0.07}{12}$$

$$R = \$100.$$

Hence,

$$S_{120} = 100 \left[\frac{\left(1 + \dfrac{0.07}{12}\right)^{120} - 1}{\dfrac{0.07}{12}} \right]$$

$$= 100 \left[\frac{1.009661}{0.005833333} \right]$$

$$= 100(173.0848)$$

$$= \$17,308.48.$$

Note: In the numerator of the last fraction, only the seven digits the writer knew to be accurate were included, and the denominator was written with seven accurate digits (not counting beginning zeros).

> **Exercise.** (Optional—Hand calculator). Five hundred dollars is to be deposited in an account at the end of each six-month period for 25 years. Find the amount in the account after the last deposit is made if interest is computed at 6 percent compounded semiannually. Answer: \$56,398.43.

8.15 ORDINARY ANNUITY: PRESENT VALUE

We have learned earlier that the present value of any amount due n periods from now is found by multiplying the amount by the compound discount factor,

$$(1 + i)^{-n}.$$

If we multiply the future amount of an annuity of \$1 per period, s_n, by the compound discount factor we have the present value of an annuity of \$1 per period, which is symbolized by a_n. Hence,

$$a_n = s_n(1 + i)^{-n}$$

$$= \left[\frac{(1 + i)^n - 1}{i}\right](1 + i)^{-n}$$

$$= \left[\frac{(1 + i)^n(1 + i)^{-n} - 1(1 + i)^{-n}}{i}\right]$$

$$= \left[\frac{(1 + i)^0 - (1 + i)^{-n}}{i}\right]$$

$$a_n = \left[\frac{1 - (1 + i)^{-n}}{i}\right].$$

Values for a_n are given in Table V, entitled *Present Value of $1 per Period*, at the back of the book. The present value of an ordinary annuity of $R per period is R times the tabular entry.

Present value of ordinary annuity: $A_n = R\left[\dfrac{1 - (1 + i)^{-n}}{i}\right].$

Present value annuity calculations arise when we wish to determine what lump sum must be deposited in an account now if this sum and the interest it earns are to provide equal payments for a stated number of periods, with the last payment making the account balance zero.

Example. What sum deposited now in an account earning 4 percent interest compounded quarterly will provide quarterly payments of $1000 for 10 years, the first payment to be made three months from now?

Here we have

$$n = (10 \text{ years})(4 \text{ quarters per year}) = 40$$

$$i = \frac{4\%}{4} = 1\%$$

$$a_{40} = 32.83469 \text{ (From Table V)}.$$

Hence,

$$A_n = 1000(32.83469) = \$32,834.69.$$

Exercise. *a)* A sum of money invested now at 6 percent compounded semiannually is to provide payments of $1500 every six months for eight years, the first payment due six months from now. How much should be invested? *b)* How much interest will the investment earn? Answer: *a)* $18,841.65. *b)* $5158.35.

Example. (Optional—Hand calculator). *a)* The directors of a company have voted to establish a fund that will pay a retiring accountant, or his estate, $1000 per month for the next ten years, the first payment to be made a month from now. How much should be placed in the fund if it earns interest at 7 percent compounded monthly? *b)* How much interest will the fund earn during its existence?

Here, for part *(a)*

$$n = (10 \text{ years})(12 \text{ months per year}) = 120$$

$$i = \frac{0.07}{12}$$

$$R = 1000.$$

Hence,

$$A_{120} = 1000 \left[\frac{1 - \left(1 + \dfrac{0.07}{12}\right)^{-120}}{\dfrac{0.07}{12}} \right]$$

$$= 1000 \left[\frac{0.5024037}{0.005833333} \right]$$

$$= 1000 \, [86.12635]$$

$$= \$86,126.35.$$

For part *(b)* we have

$$(120 \text{ payments})(\$1000 \text{ per payment}) = \$120,000.$$
$$\text{Interest earned} = \$120,000 - \$86,126.35 = \$33,873.65.$$

Exercise. (Optional—Hand calculator). Answer *(a)* and *(b)* of the last example if the interest rate is 8 percent compounded monthly. Answer: *a)* $82,421.48. *b)* $37,578.52.

8.16 PROBLEM SET 8–4

Find the amount of the following ordinary annuities.

1. $500 per month for 3 years at 12% compounded monthly.
2. $1000 every 3 months for 10 years at 8% compounded quarterly.
3. $2000 per year for 20 years at 7% compounded annually.
4. $2500 a year for 34 years at 8% compounded annually.

Find the present value of the following ordinary annuities.

5. $400 every three months for 5 years at 8% compounded quarterly.

6. $1000 per month for 3 years at 12% compounded monthly.

7. $2000 every six months for 18 years at 8% compounded semiannually.

8. $1000 every year for 40 years at 6% compounded annually.

9. When Kathy was born, her parents decided to deposit $500 every six months thereafter for 15 years in an account earning 6 percent compounded semiannually. How much will be in the account after the last deposit is made?

10. Greg has $100 deducted from his salary at the end of each month and invested in an employees' fund that, because of company contributions, pays 12 percent interest compounded monthly. How much will Greg's account amount to when he retires three years from now after receiving his last salary check?

11. A company offers its salesmen a bonus of $500 per quarter for 3 years. To win a bonus, a salesman must have sold at least $1 million worth of the company's products in the period January 1 through December 31, and the first bonus payment is made at the end of the first quarter following. The company funds each bonus on December 31 by a lump sum deposit in a bank account that pays 8 percent compounded quarterly, and the bank sends out the bonus checks. *a)* What total sum is received by each bonus winner? *b)* How much does it cost the company to fund each bonus?

12. A college alumni club has decided to establish a scholarship fund that will provide grants of $5000 a year for 25 years, the first grants to be made a year from now. *a)* What should be the sum placed in the fund if interest on it is earned at the rate of 8 percent compounded annually? *b)* What is the total amount of scholarship aid the fund will provide over its life?

(Optional) Do the following on a hand calculator.

13. Find the amount of an ordinary annuity of $1000 per quarter for 25 years at 8 percent compounded quarterly.

14. Find the present value of an ordinary annuity of $1000 per month for 10 years at 7 percent compounded monthly.

8.17 AMORTIZATION

In many financial transactions, a current obligation is discharged by making a series of payments in the future. After the last payment, the obligation ceases to exist—it is dead—and it is said to have been *amortized* by the payments. Prominent examples of amortization are loans taken to buy a home or a car and amortized over a period of 20 to 30 years in the case of a home mortgage and over 2 or 3 years in the case of a car purchase loan. Given the amount of the loan (the current principal, P), the number of periods (n), and the interest rate (i), the quantity to be calculated is R, the amount of the periodic payment. Figure 8–2 illustrates the nature of the problem. The n payments of R dollars each constitute an ordinary annuity whose pres-

FIGURE 8-2

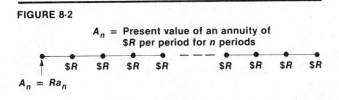

A_n = Present value of an annuity of $R per period for n periods

$A_n = Ra_n$

ent value is A_n, and we have learned that A_n is R times the present value of an annuity of \$1, which is $R(a_n)$. This present value must equal P, the current principal amount borrowed. Thus,

$$A_n = Ra_n = P$$

$$R = \frac{P}{a_n} = P\left(\frac{1}{a_n}\right).$$

Therefore, to find the periodic payment needed to amortize a debt of P, we need only to divide P by a_n, or multiply by $1/a_n$, the reciprocal of a_n. The complete formula for the amortization payment is

$$R = \frac{P}{a_n} = \frac{P}{\left[\dfrac{1 - (1 + i)^{-n}}{i}\right]}.$$

The denominator on the right is a simple fraction, so we may invert it and multiply by the numerator to obtain

Amortization payment: $R = P\left[\dfrac{i}{1 - (1 + i)^{-n}}\right] = P\left(\dfrac{1}{a_n}\right).$

Values of $1/a_n$ are provided in Table VI, entitled *Per-Period Equivalent of \$1 Present Value*, at the back of the book.

Example. Sam borrowed \$5000 to buy a car. He will amortize the loan by monthly payments of R each over a period of 3 years. *a)* Find the monthly payment if interest is 12 percent compounded monthly. *b)* Find the total amount Sam will pay.

For part *(a)* we have,

$$P = 5000$$

$$n = (3 \text{ years})(12 \text{ months per year}) = 36$$

$$i = \frac{12\%}{12} = 1\% \text{ per month.}$$

From Table VI with $i = 1$ percent, $n = 36$, we find $1/a_n$ is 0.033214. Hence,

$$R = 5000(0.033214) = \$166.07.$$

b) Sam pays $166.07 a month for 36 months. The total paid will be

$$36(166.07) = \$5978.52,$$

of which $978.52 is interest.

> ***Exercise.*** (See the foregoing example.) Sam paid for his car, but more than once was late in making payments because of financial reverses, one being an accident that badly damaged the car. He now wishes to buy another car and receives a 3-year loan for $5000, but the interest charge is 24 percent compounded monthly. *a)* What will Sam's monthly payment be for the new car? *b)* How much interest will he pay on this loan? Answer: *a)* $196.17 from Table VI; more accurately, by calculator, $196.16. *b)* $2062.12; more accurately, $2061.91.

Example. (Optional—Hand calculator.) A company has borrowed $50,000 at 10 percent compounded quarterly. The debt is to be amortized by equal payments each quarter over 15 years. *a)* Find the quarterly payment. *b)* How much interest will be paid?

a) We have $P = 50{,}000$, $i = 0.10/4$, and $n = 15(4) = 60$. Hence,

$$R = 5000 \left[\frac{\dfrac{0.10}{4}}{1 - \left(1 + \dfrac{0.10}{4}\right)^{-60}} \right]$$

$$= 50000(0.032353396)$$

$$= 1617.6698$$

$$= \$1617.67 \text{ per quarter.}$$

b) Payments for 60 quarters will be

$$60(1617.6698) = \$97{,}060.19.$$

Interest paid will be

$$97{,}060.19 - 50{,}000 = \$47{,}060.19.$$

> ***Exercise.*** (Optional—Hand calculator). A real estate developer borrows $100,000 at 8 percent compounded monthly. The debt is to be discharged by monthly payments for the next 6 years. *a)* Find the monthly payment. *b)* How much interest will be paid? Answer: *a)* $1753.32. *b)* $26,239.04; more accurately, $26,239.33.

Mortgage payments. In a typical home purchase transaction, the home-buyer pays part of the cost in cash and borrows the remainder needed, usually from a bank. The bank is said to have a *mortgage* on the home. The buyer amortizes the mortgage by periodic payments over a period of time. Typically, payments are monthly and the time period is long—30 years is not unusual. A mortgage at 9.25 percent monthly for 30 years means

$$n = (12)(30) = 360 \text{ periods},$$

$$i = \frac{9.25\%}{12} = \frac{0.0925}{12} = 0.007\ 708\ 333\ 333.$$

Because large sums of money are involved, mortgage calculations require carrying 10 or more accurate digits. Consequently, tables for $1/a_n$ for varying interest rates and mortgage durations are too lengthy to include in this book. We have, however, calculated entries for selected rates and durations on a computer, and these are provided in Table XI at the back of the book.

Example. A $70,000 home is to be purchased by paying $10,000 in cash and a $60,000 mortgage for 30 years at 9.75 percent compounded monthly. *a)* Find the monthly payment on the mortgage. *b)* What will be the total amount of interest paid?

In Table XI, all entries have been computed for monthly compounding, so it is not necessary to calculate the rate per period. We simply enter the table at the nominal rate $j = 9.75$ percent,

$$n = 30 \text{ years}(12 \text{ months per year}) = 360 \text{ months}$$

and find

$$0.008\ 591\ 5441.$$

a) For a principal amount of $60,000, the monthly payment is

$$R = 60000(0.008\ 591\ 5441) = \$515.49265,$$

or, rounded, $515.49 per month.

b) The total amount paid in 360 months will be

$$360(515.49265) = \$185,577.35.$$

Interest paid will be

$$185,577.55 - 60,000 = \$125,577.35.$$

The amount of interest just calculated is very large, but it must be remembered that the buyer has a home to live in for 30 years and, moreover, real estate values have risen, and are predicted to continue to rise as time goes on. If the experience of the past 30 years is repeated, the value of the home when the mortgage has been paid off could easily be three times its purchase cost.

> **Exercise.** A manufacturer has a 15-year, 8.5 percent, mortgage for $100,000 on a building. Payments are made monthly. *a)* Find the monthly payment. *b)* How much interest will be paid? Answer: *a)* $984.74. *b)* $77,253.12.

Amortization schedules. When a debt is amortized, part of each payment is interest on the balance outstanding, and the remainder is used to reduce the balance outstanding. The largest interest charge occurs at the first payment because then interest is due on the entire principal and, of course, the smallest reduction of principal occurs at the time of the first payment. To see how this works, recall the $60,000, 9.75 percent, 30-year mortgage of the last example. The monthly payment is $515.49. At the time the first payment is due, one month has passed and the interest on the $60,000 is

$$I = 60{,}000(0.0975)\left(\frac{1}{12}\right) = \$487.50.$$

The amount applied to reduce the balance outstanding is

$$515.49 - 487.50 = \$27.99,$$

so the new balance is

$$60{,}000 - 27.99 = \$59{,}972.01.$$

It is clear that the first payment is almost entirely absorbed by the interest charge, with little left to reduce the balance still owed. Each month, of course, the interest charge decreases and the reduction of the amount owed increases. However, in financial jargon, early payments do not increase the homeowner's equity very much.

Continuing to the second payment, the beginning balance is now $59,972.01, and we proceed as follows:

Beginning balance owed		$59,972.01
Payment	$515.49	
Interest charge $(59{,}972.01)(.0975)\left(\frac{1}{12}\right)$:	487.27	
Reduction of balance owed	28.22	28.22
Ending balance owed		$59,943.79

When the process just described is repeated for the entire period of the loan (360 months for the example) and the results tabulated, the table is called an *amortization schedule*. The process is simple, but it must be repeated many times, and the chore is best left to a computer that can be programmed to do the process once and then instructed to repeat it over and over again, printing out a line of entries in an amortization schedule at the end of each repetition.

> **Exercise.** For the $100,000, 15-year, 8.5 percent mortgage of the last exercise, the monthly payment is $984.74. For the first and second payments, find the *a)* Interest charge. *b)* Reduction in balance owed. *c)* Ending balance owed. Answer: First month: *a)* $708.33. *b)* $276.41. *c)* $99,729.59. Second month: *a)* $706.42. *b)* $278.32. *c)* $99,451.27.

8.18 SINKING FUNDS

A sinking fund is a fund into which periodic payments are made in order to accumulate a specified sum at some point in the future. For example, a corporation that obtains money needed to expand by selling $1 million worth of bonds payable in 10 years must pay interest to bond holders (usually semiannually) while the bonds mature and, at maturity 10 years later, pay $1 million to redeem the bonds. To be sure that the $1 million is available 10 years hence, the corporation may set up a sinking fund to accumulate this amount. The problem is to determine R, the required periodic payment into the sinking

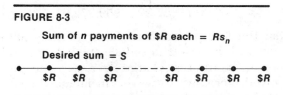

FIGURE 8-3

Sum of *n* payments of $R each = Rs_n

Desired sum = S

R R R ----- R R R R

fund. As shown on Figure 8–3, the *n* payments constitute an ordinary annuity of R per period, and the amount of this annuity is

$$S_n = Rs_n.$$

It is required that S_n be the desired sum, S. That is,

$$Rs_n = S.$$

Hence, the periodic payment is

$$R = \frac{S}{s_n} = S\left(\frac{1}{s_n}\right)$$

$$= S\left[\frac{1}{\dfrac{(1+i)^n - 1}{i}}\right]$$

Inverting the simple fraction in the denominator, and multiplying this by the numerator, 1, we have:

$$\textbf{Sinking fund payment: } R = S\left(\frac{1}{s_n}\right) = S\left[\frac{i}{(1+i)^n - 1}\right].$$

Table VII at the back of the book, entitled *Per-Period Equivalent of $1 Future Value*, provides values for $1/s_n$.

Example. How much should be deposited in a sinking fund at the end of each quarter for 5 years to accumulate $10,000 if the fund earns 8 percent compounded quarterly?

Table VII, for

$$i = \frac{8\%}{4} = 2\%$$

$$n = (5 \text{ years})(4 \text{ quarters per year}) = 20 \text{ periods}$$

shows

$$\frac{1}{s_n} = 0.0411567.$$

Hence,

$$R = 10000(0.0411567) = \$411.567.$$

Thus, the quarterly payment is, rounded, $411.57. Over the life of the sinking fund, the sum of the deposits will be

$$20(411.567) = \$8,231.34.$$

This sum, plus interest earned, will provide the desired $10,000.

> *Exercise.* Instead of creating a sinking fund at a bank, a company has its controller create a reserve fund in the company's accounts to which a contribution is made each quarter. By so doing, the company can realize a return of 12 percent compounded quarterly. What should be the amount transferred quarterly to the reserve fund to accumulate $10,000 in 5 years? Answer: $372.16.

Example. (Optional—Hand calculator.) A company wants to accumulate $100,000 to purchase replacement machinery 8 years from now. To accomplish this, equal semiannual payments are made to a fund that earns 7 percent compounded semiannually. Find the amount of each payment.

We have

$$S = 100,000$$

$$i = \frac{0.07}{2} = 0.035$$

$$n = 8(2) = 16.$$

From

$$R = S\left[\frac{i}{(1+i)^n - 1}\right]$$

we have

$$R = 100,000\left[\frac{0.035}{(1.035)^{16} - 1}\right]$$

$$= 100,000(0.04768483)$$

$$= \$4768.48.$$

Exercise. (Optional—Hand calculator.) A company issues $1 million of bonds and sets up a sinking fund at 8 percent compounded quarterly to accumulate $1 million 15 years hence to redeem the bonds. Find the quarterly payment to the sinking fund. Answer: $8,767.97.

8.19 PROBLEM SET 8–5

1. What payment at the end of each month for 2 years will discharge a current debt of $1000 if the interest charge on the debt balance at any time is 12 percent compounded monthly?

2. What payment at the end of each 6-month period for 10 years will discharge a current debt of $2500 if the interest charge on the debt balance is 10 percent compounded semiannually.

3. A company borrows $100,000 at 12 percent compounded semiannually. The debt is amortized by making equal payments at the end of each six months for 7 years.
   a) Find the amount of each payment.
   b) How much of the first payment is for interest, and by how much does it reduce the balance owed?
   c) How much of the second payment is for interest, and by how much does it reduce the balance owed?

4. Fran borrowed $6000 at 24 percent compounded monthly to buy a car. The debt is to be discharged by equal payments at the end of each month for 3 years.

a) Find the amount of each payment.

b) How much of the first payment is for interest, and by how much does it reduce the balance owed?

c) How much of the second payment is for interest, and by how much does it reduce the balance owed?

5. Exall Corporation has taken out a $1,500,000, 25-year mortgage on its new office building, with interest at 9 percent compounded monthly.

   a) Find the monthly payment.

   b) How much of the first payment is for interest, and by how much does it reduce the balance owed?

6. The Smiths have taken out a $35,000, 30-year mortgage on their home, with interest at 8.75 percent compounded monthly.

   a) Find the monthly payment.

   b) How much of the first payment is for interest, and by how much does it reduce the balance owed?

7. What amount should be deposited at the end of each quarter in a sinking fund earning 8 percent compounded quarterly if the amount in the fund after 4 years is to be $90,000?

8. What amount should be deposited at the end of each 6-month period in a sinking fund earning 6 percent compounded semiannually if the amount in the fund after 15 years is to be $75,000?

9. Whyall Corporation has decided to transfer a sum of money to a reserve account at the end of each year to accumulate $100,000 to be used to replace machinery 10 years from now. How much should be transferred each year if interest at 8 percent compounded annually is credited to the reserve?

10. The Joneses are going to deposit a sum of money at the end of each six-month period in an account earning 8 percent compounded semiannually in order to accumulate $15,000 for a downpayment on a home eight years from now. What should be the amount of each deposit?

(Optional—Hand calculator.) Compute the following:

11. Exall Corporation has borrowed $5,000,000 at 7 percent compounded semiannually. The debt is discharged by equal payments at the end of each six-month period for 30 years. Find the amount of each payment.

12. Whyall Corporation has issued $10 million worth of bonds to obtain money now for expanding its corporate activities. To redeem the bonds, which fall due in 30 years, Whyall will transfer an amount to a reserve fund at the end of each six-month period. How much should be transferred if the account earns 7 percent compounded semiannually?

8.20 SUMMARY OF FINANCIAL TABLES

The purpose of this section and the next problem set is to provide practice in identifying the type of problem at hand so that the proper table or tables will be selected for the solution. We will use time diagrams to analyze problems, and the reader is encouraged to do the same.

Example. How much should be deposited now in an account earning 8 percent compounded quarterly if the amount in the account 10 years from now is to be $10,000?

The time diagram, Figure 8–4, demonstrates that the $10,000 is a

FIGURE 8-4

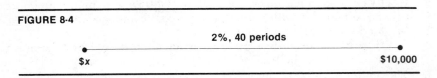

future amount, and the unknown amount, x, is the present value of the future amount. Table III should be selected, yielding

$$x = 10000(0.452890) = \$4528.90.$$

Example. How much will be accumulated at the end of 10 years by depositing $1000 at the end of each six-month period in an account paying 6 percent compounded semiannually?

Figure 8–5 shows the given $1000 deposits at the end of each six-

FIGURE 8-5

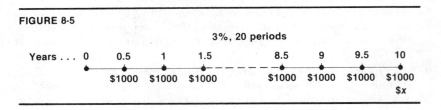

month period, with the unknown amount, x, at the end of 10 years. Thus, x is the (future) amount of $1,000 per period, and Table IV applies. We have

$$x = 1000(26.87037) = \$26,870.37.$$

Example. Smith borrows $5000 from a bank at 12 percent compounded monthly and promises to discharge the debt by equal payments at the end of each month for 3 years. Find the amount of each payment.

FIGURE 8-6

1%, 36 periods

Months ...	0	1	2	3		33	34	35	36
	$5000	$x	$x	$x		$x	$x	$x	$x

Figure 8–6 shows the present debt is $5000, and the x payments are the per-period equivalent of $5000 present value, so Table VI applies and we find

$$x = 5000(0.033214) = \$166.07.$$

Example. Sue will start college six months from now and her parents have decided to establish a bank account now to provide $2500 every six months for tuition payments. If eight tuition payments are to be made and the account earns 6 percent compounded semiannually, how much should be deposited?

Figure 8–7 shows the unknown deposit, x, to be made now, is

FIGURE 8-7

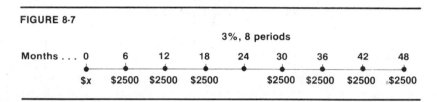

3%, 8 periods

the present value of $2500 per period, so Table V applies and we have

$$x = 2500(7.01969) = \$17,549.23.$$

Example. If the Gross National Product now is $1,683.5 billion and is growing at the compound annual rate of 6 percent, what will the GNP be 20 years from now?

Figure 8–8 shows that we want the compound amount of the single value, $1,683.5 billion. We have, from Table II,

$$x = 1683.5(3.20714) = \$5,399.2 \text{ billion.}$$

FIGURE 8-8

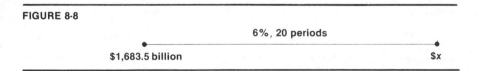

6%, 20 periods

Example. Sam plans to buy a car two years from now and decides to accumulate $3000 to help pay for it by having a deduction made from his monthly salary at the end of each month and deposited in an employees' savings account that, because of employer contributions, earns 12 percent interest compounded monthly. How much will be deducted each month?

FIGURE 8-9

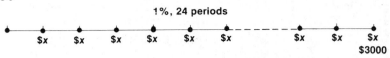

Figure 8–9 shows the amount Sam desires as a future amount. The periodic payment, x, needed to accumulate this amount is the per period equivalent of $3000 future value. Table VII applies, and we find

$$x = 3000(0.0370735) = \$111.22.$$

8.21 MULTISTEP PROBLEMS

The solution of a problem may require more than one step and involve more than one interest rate, as illustrated by the following examples.

Example. Sam wants to determine how much he should deposit in an account now at 8 percent compounded quarterly so that the amount in the account 10 years from now will provide an income of $5000 every six months for 12 years, with the first $5000 to be received in 10½ years. Sam estimates that 10 years from now he should be able to earn 6 percent compounded semiannually on the account when it is used to provide his semiannual income of $5000. How much should Sam deposit now?

Figure 8–10 shows the structure of the problem. Sam's goal is to

FIGURE 8-10

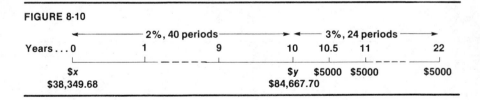

have y in his account at the point shown, where y must be the then present value of $5000 per period. Applying Table V, we find

$$y = 5000(16.93554) = \$84,677.70,$$

as shown in Figure 8–10. The current deposit now, at time 0, is the present value of y. Hence, from Table III, at 2 percent, 40 periods, we have

$$x = 84{,}677.70(0.452890) = \$38{,}349.68.$$

More accurately, by hand calculator, the current deposit should be $38,349.72.

The last example is an illustration of a *deferred annuity;* that is, an annuity purchased now with the annuity payments being deferred so that they start at a time later than would be the case for an ordinary annuity.

Example. Fran borrows $2000 from Silverbank and signs a note promising to discharge the debt with interest at 12 percent compounded monthly at a maturity date two years from now. Six months later, Silverbank needed more cash and sold Fran's note to Goldbank. Goldbank computed the maturity amount of Fran's note, and gave Silverbank the present value of this amount, computed at 8 percent compounded quarterly. How much did Silverbank receive?

Figure 8–11 shows the structure of the problem. The maturity value

FIGURE 8-11

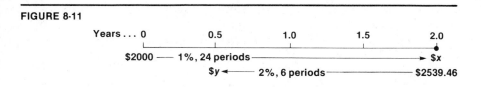

of the note, $x, is the compound amount of $2000, which, using Table II, is

$$x = 2000(1.26973) = \$2539.46.$$

When Goldbank buys the note, it still has 1½ years or 6 quarters until maturity, so the amount Silverbank receives, $x, is the present value of $y = \$2539.46$, which, using Table III, is

$$y = 2539.46(0.887971) = \$2254.97.$$

More accurately, by hand calculator, $y = \$2254.98$. This example illustrates the *compound discounting of a note.* Simple interest calculations are frequently used to discount notes (see Index reference to *bank discount*), but we shall not use simple interest in this section.

Example. Sam wishes to provide himself, or his estate, with an income of $5,000 every six months, starting 15½ years from now and continuing for 20 years. He deposits $25,000 in the account now, and he has a guaranteed inheritance of $10,000, which he will receive 10 years from now and add to the account. He knows these sums will not provide the income he wants, so he plans to make periodic deposits to the account at the end of every six months for 15 years

FIGURE 8-12

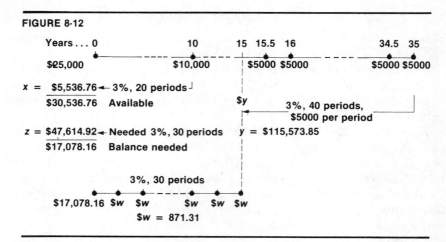

to make up the difference. How much should the periodic deposits be if all interest is computed at 6 percent compounded semiannually?

Figure 8–12 brings all amounts back to time zero. Here Sam has available $25,000 plus x, the present value of the $10,000 inheritance, which is

$$x = 10,000(0.553676) = \$5536.76,$$

so the total amount available now is $30,536.76. The deferred annuity Sam wants to establish has a principal of $\$y$, and it is an annuity of $5000 for 40 periods (20 years) at 3 percent per period. Hence, using Table V,

$$y = 5000(23.11477) = \$115,573.85.$$

The present value of $\$y$ is the amount Sam needs. This is, using Table III with $n = 30$ and $i = 0.03$,

$$z = (115,573.85)(0.411987) = \$47,614.92.$$

The annuity Sam must provide by his additional payments to the account must make up the difference between the amount needed and the amount available, which is

$$47614.92 - 30536.76 = \$17,078.16.$$

Consequently, Sam needs to know the per-period equivalent of $17,078.16 present value which, using Table VI with $n = 30$ and $i = 0.03$ is

$$w = 17078.16(0.051019) = \$871.31.$$

Hence, Sam should deposit $871.31 to the account every six months for the next 15 years.

8.22 PROBLEM SET 8–6

Solve the following by use of the appropriate tables at the back of the book. All periodic payments are in the form of ordinary annuities.

1. If $100 is deposited at the end of every six months for five years at 6 percent compounded semiannually, what will be the amount in the account after the last deposit?

2. Sue borrowed $7000 at 12 percent compounded monthly for three years to buy a car. How much will she have to pay at the end of each month to discharge the debt?

3. At 8 percent compounded quarterly, what will be the amount of a current deposit of $5,000 in 10 years?

4. How much should be deposited at the end of each year to an account earning 8 percent compounded annually in order to accumulate $10,000 at the time of the last deposit 9 years from now?

5. What sum of money deposited now at 8 percent compounded annually will grow to $10,000 in 20 years?

6. How much should be deposited now at 7 percent compounded annually to provide an income of $20,000 at the end of each year for the next 22 years?

7. The population of a town now is 52,000. Population five years ago is unknown, but it is estimated that population has increased at the rate of 6 percent compounded annually. Estimate population five years ago.

8. Sam borrowed $4000 at 24 percent compounded monthly to pay for construction of a garage. The debt is to be discharged by payments at the end of each month for 30 months. Find the amount of the monthly payment.

9. Jill has $250 taken from her salary at the end of each quarter and deposited in an employees' fund that earns 8 percent compounded quarterly. What will be the amount in the account after the last deposit is made 5 years from now?

10. The board of directors of a company has voted to establish a fund that will provide a retiring executive with an income of $5000 at the end of each quarter for 10 years. The fund will be invested in the company, which earns 12 percent compounded quarterly. Find the amount that should be invested.

11. The real estate tax on a piece of property now is $2000 per year. If taxes increase at the rate of 5 percent compounded annually, what will the tax on this property be 10 years from now?

12. Sam wants to accumulate $10,000 for a downpayment on a home 8 years from now. He will do this by making a deposit at the end of each quarter in an account earning 8 percent compounded quarterly. How much should he deposit each quarter?

13. A note for $3000 with interest at 12 percent compounded monthly is payable 40 months from now. Find the (then present) value of the note 19 months from now if this value is computed at 8 percent compounded quarterly.

14. Sue has purchased $20,000 worth of securities earning 10 percent compounded semiannually. Ten years from now, she plans to use the securities and interest to establish an account earning 7 percent compounded annually, and to exhaust this account by equal withdrawals at the end of each year for 5 years. How much will each withdrawal be?

15. How much should be deposited now at 8 percent compounded semiannually to make possible equal withdrawals of $5000 at the end of each year for 5 years, the first withdrawal to be made 10 years from now. Interest during the withdrawal period is to be 7 percent compounded annually.

16. Nine years from now, Sam wants to have an amount available to deposit to an account that earns 6 percent compounded annually. This account is to provide Sam with an income of $10,000 at the end of each year for ten years. To accomplish this, Sam invests in an eight-year bank certificate that pays eight percent compounded semiannually, and he will use this certificate, plus its interest, to establish his income account. What should the principal value of the certificate be?

17. Fran will make 20 equal semiannual deposits to an account earning 8 percent compounded semiannually then, after the last deposit, she will use the amount in the account to establish an ordinary annuity earning 6 percent compounded annually which will provide her with $10,000 at the end of each year for 5 years. How much should Fran's semiannual deposit be?

18. During a three-year period when his business was prospering, Jack was able to deposit $1000 at the end of each month in an account earning 12 percent compounded monthly. The business slackened, and Jack could not continue the deposits. Moreover, the interest rate on his accumulated deposits fell to 8 percent compounded quarterly and remained at this level for 10 years, at which time Jack decided to exhaust the account by withdrawing equal amounts at the end of every six months for five years. The interest rate remained at 8 percent compounded semiannually over the time of the withdrawals. How much did Jack withdraw every six months?

19. (All interest rates are 7 percent compounded annually.) Jill wishes to provide herself, or her estate, an income of $10,000 at the end of each year for 10 years. She will make a lump sum deposit when the account is established and add $3000 at the end of each year for 12 years. The income is to start at the end of the year following the year in which the last deposit was made. Compute the required lump sum deposit.

8.23 FINDING THE INTEREST RATE: PERIODIC PAYMENTS

An unknown interest rate can be approximated by interpolation in the periodic payment tables, or by hand calculator. The hand calculator provides more accurate results but, as we shall see, the periodic payment formulas cannot be solved for i, and a series of successive approximations is required.

Example.   Sam purchased a piece of property by making a down-payment and signing an agreement to pay the remaining amount, $5000, by making equal year-end payments of $700 for ten years. By interpolation, find the rate of annually compounded interest that Sam is paying.

In this problem, we know that $700 is the per-period equivalent of $5000 present value. To use the Table VI, we need the corresponding tabular entry for the periodic equivalent of $1 present value. To convert, note that

Entry is the periodic equivalent of $1 present value.
$700 is the periodic equivalent of $5000 present value.

Hence,

$$\frac{\text{Entry}}{700} = \frac{1}{5000}.$$

$$\text{Entry} = 700\left(\frac{1}{5000}\right) = 0.14.$$

We now enter Table VI at the row $n = 10$ and look across this row to find the two entries that straddle 0.14. These are 0.135868 (at 6 percent) and 0.142378 (at 7 percent). The interpolation is set up as follows:

Interest rate	Entry

$$0.01 \left[\begin{array}{c} x \left[\begin{array}{c} 0.06 \\ \\ 0.07 \end{array} \right. \end{array} \right. \quad \begin{array}{c} 0.135868 \\ 0.14 \\ 0.142378 \end{array} \left. \begin{array}{c} \\ \end{array} \right] 0.004132 \quad \left. \right] 0.00651$$

$$\frac{x}{0.01} = \frac{0.004132}{0.00651}$$

$$x = 0.01\left(\frac{0.004132}{0.00651}\right)$$

$$= 0.006347$$

$$= 0.006.$$

We shall round interpolated interest rates to the third decimal place, as shown. The desired approximate interest rate is

$$0.06 + 0.006 = 0.066 = 6.6\%.$$

Shortly, we shall find that the interest rate, accurate to four digits, is 6.637 percent.

We note in passing that Table V also could be used in the foregoing example. That is, the given information states that $5000 is the present value of $700 per period. Hence,

Entry is the present value of $1 per period.
$5000 is the present value of $700 per period.

$$\frac{\text{Entry}}{5000} = \frac{1}{700}.$$

$$\text{Entry} = 5000\left(\frac{1}{700}\right).$$

$$= 7.14286.$$

The desired interest rate now is determined by interpolating in Table V to find the interest rate for $n = 10$ that corresponds to the entry 7.14286. The result, 6.6 percent, is the same as that obtained from Table VI.

Exercise. At what interest rate, compounded annually, will an ordinary annuity of $100 per year amount to $4,000 in 22 years? Answer: 5.3%.

Hand calculator approximation. (Optional). Returning to the last example, where $700 a year for ten years is the periodic equivalent of $5000 present value, we may substitute $P = 5000$, $R = 700$, and $n = 10$ into the formula for the per-period equivalent of $5000 present value

$$P = R\left[\frac{i}{1 - (1 + i)^{-n}}\right]$$

to obtain

$$700 = 5000\left[\frac{i}{1 - (1 + i)^{-10}}\right].$$

This formula cannot be solved explicitly for i. What we shall do is rearrange the terms to make an equation that has zero on one side. First, dividing by 5000, we have

$$\frac{700}{5000} = \frac{i}{1 - (1 + i)^{-10}}$$

$$0.14 = \frac{i}{1 - (1 + i)^{-10}}.$$

Multiplying both sides by the denominator on the right gives

$$0.14[1 - (1 + i)^{-10}] = i$$
$$0.14 - 0.14(1 + i)^{-10} = i$$
$$0.14 - 0.14(1 + i)^{-10} - i = 0.$$

Multiplying both sides by minus one, we have

$$-0.14 + 0.14(1 + i)^{-10} + i = 0,$$

or, rearranging the last,

$$0.14(1 + i)^{-10} - 0.14 + i = 0. \tag{1}$$

How the terms are arranged does not matter, but we want zero on one side. The writer's preference is to have the power term first with a positive coefficient.

What we plan to do now is choose an arbitrary value for i, substitute it into the equation and see how the left side compares to zero. We know interest rates are in the neighborhood of 5 percent, so we shall start with $i = 0.05$ and compute

$$0.14(1 + 0.05)^{-10} - 0.14 + 0.05 = -0.00405.$$

We see that $i = 0.05$ does not satisfy the equation because it leads to a remainder of -0.00405 rather than zero on the right, so we must choose $i = 0.04$ or $i = 0.06$ and try again. We could analyze the equation to determine whether the minus sign means that $i = 0.05$ is too small (in which case the next trial would be $i = 0.06$). However, it is simpler here to pick either one and compare the result with -0.00405 to learn what $+$ and $-$ remainders mean. We choose $i = 0.04$. Substituting into (1) and carrying out the calculation, we find

$$0.14(1.04)^{-10} - 0.14 + 0.04 = -0.00542.$$

This is further from zero than -0.00405, so we moved in the wrong direction. Now we know that a minus remainder means too small and a plus remainder means too large.

Next we show the earlier result for $i = 0.05$, and the result for $i = 0.06$. We shall not write the equation (1) any longer, but the reader can check what follows by substituting the interest rates into (1).

Rate, i	Remainder	i is
0.05	−0.00405	too small
0.06	−0.00182	"
0.07	+0.00117	too large

We want the remainder to be zero, or close to zero. The foregoing calculations show the remainder is minus at $i = 0.06$ and plus at 0.07, so the value of i where the remainder is zero is between 0.06 and 0.07. We proceed with the calculation using the midpoint, $i = 0.065$.

Rate, i	Remainder	i is
0.065	−0.000418	too small
0.066	−0.000115	"
0.067	+0.000195	too large

The system should be clear by now. The sign change in the remainders, just calculated, means i is between 0.066 and 0.067, so we try the midpoint, 0.0665, and continue in this manner, as shown next.

Rate, i	Remainder	i is
0.0665	+0.0000393	too large
0.0664	+0.00000828	"
0.0663	−0.0000227	too small
0.06635	−0.00000720	"
0.06636	−0.00000410	"
0.06637	−0.00000101	"
0.06638	+0.00000209	too large
0.066375	+0.000000539	"

Inasmuch as the foregoing table shows that 0.06637 is too small but 0.066375 is too large, the value of i is

$$i = 0.06637 = 6.637 \text{ percent},$$

accurate to four digits. The foregoing procedure can be used to solve equations other than the one at hand. The idea is simply to use trial values of the unknown in steps or increments, locate two values between which the remainder changes sign, then try the midpoint of these two values and proceed. More sophisticated approximation procedures, available in books on *numerical analysis,* can be used to obtain a solution more rapidly, but the "brute force" method we used will satisfy our needs.

> ***Exercise.*** (Optional—Hand calculator.) *a)* Write the formula for the amount of an ordinary annuity of $100 per year having an amount of $4000 in 22 years and rewrite this formula as an equation with zero on one side. *b)* By hand calculator, compute the interest rate accurate to three significant digits. Answer: *a)* $(1 + i)^{22} - 40i - 1 = 0$. *b)* 0.0532 or 5.32 percent. More accurately, $i = 5.3208$ percent.

8.24 THE EFFECTIVE RATE OF INTEREST

Because of lack of comparability, it is hard to judge whether interest quoted at 8 percent compounded semiannually results in more or less interest than would be the case if the rate was 7.9 percent compounded monthly. To make the comparison possible, we change both to their equivalent annual rates, and these equivalents are called *effective* rates. For example, $1 at 8 percent compounded quarterly for *one* year, would amount to

$$S = 1\left(1 + \frac{0.08}{4}\right)^4$$

$$= (1.02)^4$$

$$= 1.08243,$$

which is the same as the amount of $1 at a rate of 0.08243, or 8.243 percent for one year. Similarly, by calculator, $1 at 7.9 percent compounded monthly for *one* year would amount to

$$S = 1\left(1 + \frac{0.079}{12}\right)^{12}$$

$$= 1.08192,$$

which is equivalent to the amount of $1 at a rate of 8.192 percent for one year.

In general, at nominal (annual) rate j compounded m times a year, $1 grows to

$$S = (1)(1+i)^m, \quad i = \frac{j}{m},$$

in one year. At the *effective* rate, r, $1 grows to

$$S = 1 + r$$

in a year. Hence,

$$1 + r = (1+i)^m$$
$$r = (1+i)^m - 1.$$

Effective Rate of i Compounded m Times a Year

$$r = (1+i)^m - 1; \quad i = \frac{j}{m}.$$

Example. Find the effective rate of 24 percent compounded monthly. Here, as usual,

$$i = \frac{24\%}{12} = 2\%$$

$$m = 12 \text{ months in a year.}$$

Consequently,

$$r = (1 + 0.02)^{12} - 1$$
$$= 1.26824 - 1$$
$$= 0.26824$$
$$= 26.824\%.$$

Table II provides

$$(1 + i)^m$$

for some values of i and, of course, this power can be found easily on a hand calculator for any i and m in common use.

8.25 PROBLEM SET 8–7

Approximate the following by interpolation accurate to two significant digits. (Beginning zeros are not significant digits.)

1. At what rate of interest compounded annually will payments of $2000 at the end of each year for eight years amount to $20,000?
2. At what rate of interest compounded annually will $50,000 be accumulated by payments of $2000 at the end of each year for 15 years?
3. At what rate of interest will a principal of $20,000 provide payments of $2000 at the end of each year for 20 years?
4. At what rate of interest compounded annually will $5000 payments at the end of each year for 11 years be provided by a principal investment now of $40,000?

(Optional—Hand calculator.) By successive approximation, find the interest rate accurate to three significant digits for:

5. Problem 1.
6. Problem 2.
7. Problem 3.
8. Problem 4.

Find the effective interest rate for each of the following by table or hand calculator.

9. 8%, compounded quarterly.
10. 10%, compounded semiannually.
11. 12%, compounded monthly.
12. 16%, compounded quarterly.

(Optional—Hand calculator.) Find the effective interest rate for each of the following:

13. 10%, compounded monthly.
14. 7%, compounded quarterly.
15. 18%, compounded monthly.
16. 9%, compounded quarterly.
17. 9%, compounded monthly.
18. 9%, compounded daily (365 days per year).

8.26 CONTINUOUS (INSTANTANEOUS) COMPOUNDING

Before the appearance of modern high-speed data processing computers, calculation and recording of interest for even a few thousand bank accounts was too costly in time to be done at frequent intervals. Consequently, the common practice was to compound quarterly—once every three months—and depositors who withdrew money from their accounts between interest dates did not receive interest for the time between the last interest calculation and the date of withdrawal. Now, with high-speed computers, it is possible to calculate and add interest whenever any account transaction takes place. A common practice currently is to compute interest from the day of deposit to the day of withdrawal, with interest compounded daily. The number of compoundings in a year would then be 365, or 366 for a leap year. However, banks may count a year as twelve 30-day months, 360 days for a year.

We start on our way toward continuous, or instantaneous, compounding by recalling the number e, which is the base of the system of natural logarithms (Chapter 7). This constant, like the constant π, has an unending sequence of digits and therefore cannot be expressed exactly by any finite sequence of digits. However, e can be expressed as accurately as needed for any applied problem. To twelve places of decimals,

$$e = 2.7182\ 8182\ 8459+.$$

The value of e can be computed to any desired degree of accuracy by calculating

$$\left(1 + \frac{1}{m}\right)^m$$

using a *sufficiently large* value of m. For example, with $m = 200{,}000$

$$\left(1 + \frac{1}{200{,}000}\right)^{200{,}000} = 2.7182\ 8182\ 8,$$

which is e correct to 9 decimal places. The important point to note is that the expression

$$\left(1 + \frac{1}{m}\right)^m$$

can be interpreted as the compound amount, S, of \$1 at 100 percent interest ($i = 1.00$) for one year, compounded m times a year. Thus, for example, \$1 at 100% compounded monthly for a year would yield

$$S = 1\left(1 + \frac{1}{12}\right)^{12} = \$2.61.$$

FIGURE 8-13

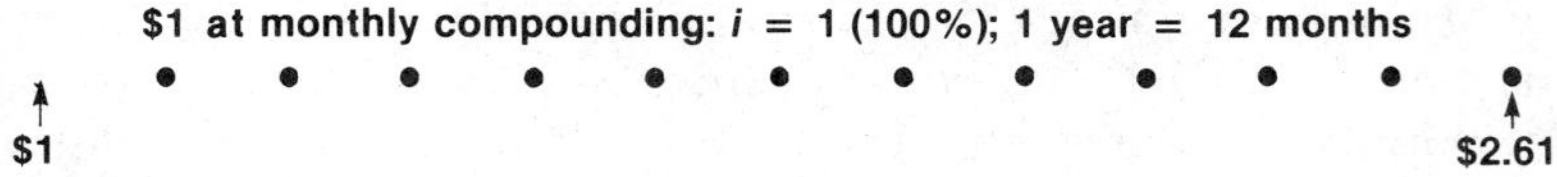

$1 at monthly compounding: i = 1 (100%); 1 year = 12 months

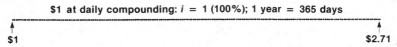

$1 $2.61

FIGURE 8-14

$1 at daily compounding: i = 1 (100%); 1 year = 365 days

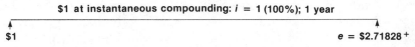

$1 $2.71

FIGURE 8-15

$1 at instantaneous compounding: i = 1 (100%); 1 year

$1   e = $2.71828 +

The last is illustrated on Figure 8–13, which shows 12 discrete points (that is, separated points) representing the 12 compoundings that make $1 at 100% grow to $2.61 in a year.

If we compound $1 at 100% for a year with daily compoundings, $m = 365$ and

$$S = 1\left(1 + \frac{1}{m}\right)^m = 1\left(1 + \frac{1}{365}\right)^{365} = \$2.71$$

as shown on Figure 8–14, which has a discrete set of 365 points representing the daily compoundings. Note that in the result, $2.71, we have the first two digits of e. Note also that the discrete set of 365 points is beginning to look like a *continuous* line; that is a line drawn without lifting the pencil from the paper. Now, going one step further, suppose we compound every hour for a year so that

$$m = (365 \text{ days}) \, (24 \text{ hours per day}) = 8760.$$

Then

$$S = \left(1 + \frac{1}{m}\right)^m = \left(1 + \frac{1}{8760}\right)^{8760} = 2.7181,$$

a result which has the first four digits of e. It might be possible with the aid of a magnifying glass and a very sharp point to make a figure like Figure 8–14 showing these 8,760 compoundings, but if the magnifying glass were removed, the result would appear to be a continuous line without gaps, but the set of points is discrete, and the gaps are there.

To approach the core concept of instantaneous compounding, we now contemplate compounding every minute ($m = $ 525,600 compoundings per year), every second ($m = $ 31,536,000 compoundings a year), and so on, with the duration of the compounding period getting smaller and the number of compounding points increasing. There is no limit to the number of compounding periods but, as it grows, the duration of a period approaches *zero,* and this leads us to the concept of a *point* in time of zero duration, which is connoted by the word *instant.* The nature of time in the real world forces us to attach meaning to a point in time—an instant—because, for example, the time from 9 A.M. to 11 A.M. goes through all times in the interval, and at some point the time must be precisely 10 o'clock. We cannot, of course, mark a clock dial in instants. However, real time does not progress in discrete jumps—it is *continuous.* A real time line, then, is not a discrete set of points, but a continuous line with no gaps. It is a line we represent on paper by drawing a pencil along a straight edge without removing the pencil from the paper, as shown on Figure 8–15, which represents instantaneous compounding. As this figure shows, instantaneous compounding of $1 at 100% for one year yields a compound amount of $e. For present purposes, we state that

e is the sequence of digits generated by calculating

$$\left(1 + \frac{1}{m}\right)^m$$

using larger and larger values of m.

Of course, the letter *m* is arbitrary, and we can say equally well that *e* is generated by

$$\left(1 + \frac{1}{x}\right)^x \quad \text{or} \quad \left(1 + \frac{1}{p}\right)^p$$

as *x,* on the left, or *p,* on the right, becomes larger and larger.

We have demonstrated that $1 at 100% for 1 year, compounded continuously (instantaneously), yields $e. If we have $1 at nominal rate *j* per year for *t* years, the amount is

$$S = 1e^{jt}.$$

as shown in the footnote.[1] If $P rather than $1 is the principal, we have:

---

[1] The compound amount of $1 at rate *j* compounded *m* times a year for *t* years is

$$\left(1 + \frac{j}{m}\right)^{mt}. \tag{1}$$

Now let $p = m/j$ so that $j/m = 1/p$ and $m = pj,$ and substitute these into (1) giving

$$\left(1 + \frac{1}{p}\right)^{pjt} = \left[\left(1 + \frac{1}{p}\right)^p\right]^{jt}. \tag{2}$$

Now as *m* becomes larger and larger, *p,* which is *m/j,* becomes larger and larger, the bracketed expression in (2) generates *e,* and we obtain $e^{jt}.$

Compound amount with continuous compounding:

$$S = Pe^{jt}$$

Note: We shall use the symbol t to be a number of years. This number can be a fraction or a decimal. We do this to distinguish t from n because n has been defined to be a *whole* number of periods.

Example. Find the compound amount of $500 at 8 percent compounded continuously for 9 years and 3 months.

Remembering that t is the number of years, we find

$$t = 9 + \frac{3}{12} = 9.25 \text{ years.}$$

Hence,

$$\begin{aligned}
S &= 500e^{0.08(9.25)} \\
&= 500e^{0.74} \\
&= 500(2.0959) \\
&= \$1047.95.
\end{aligned}$$

The power of e just used, 2.0959, was taken from Table IX at the back of the book, and is accurate to only five digits. Consequently, the answer should be rounded to five digits, giving $1048.0. Using a more accurate value of $e^{0.74}$ taken from a hand calculator, the answer is $1047.97, to the nearest cent.

> *Exercise.* Find the amount of $200 at 6 percent compounded continuously for 30 months. Answer: $232.36 using Table IX; $232.37, accurate to the nearest cent.

If both sides of the formula

$$S = Pe^{jt}$$

are divided by the exponential term, we have

$$\frac{S}{e^{jt}} = P,$$

or, using the negative exponent

$$Se^{-jt} = P.$$

This is the present value, P, of a future amount, S.

Present value with continuous compounding: $P = Se^{-jt}$

> *Exercise.* How much must be deposited now in an account earning 7.5 percent compounded continuously if the amount in the account eight years from now is to be $10,000? Answer: $5488 by Table IX. $5488.12 accurate to the nearest cent.

8.27 EFFECTIVE RATE: CONTINUOUS COMPOUNDING

At a nominal rate $j = 0.08$, \$1 compounded continuously for one year amounts to

$$S = e^{0.08} = 1.0833,$$

so the interest earned is

$$1.0833 - 1 = 0.0833 \text{ or } 8.33\%.$$

In general

Effective rate of nominal rate j: $r = e^j - 1$.

> *Exercise.* Find the effective rate of 10 percent compounded continuously. Answer: 10.52 percent.

Current banking practice. (Optional—Hand calculator.) There are legal limits on the interest rates banks can offer for various types of accounts. In the last decade, these limits have not compared favorably with interest rates that can be obtained from non-bank investments and, in an effort to attract more deposits, some banks have adopted not only the ultimate in compounding, continuous compounding, but also the *modified* year,

$$\frac{365}{360}$$

which is greater than one. The writer has before him a newspaper advertisement that states that the effective rate on 8 percent is 8.45 percent. To obtain this result, we replace the exponent in

$$e^{0.08} - 1$$

by

$$\frac{365}{360}(0.08), \text{ or } 0.08111,$$

to obtain

$$\begin{aligned} r &= e^{0.08111} - 1 \\ &= 1.08449 - 1 \\ &= 0.08449. \end{aligned}$$

The last number, rounded, is 0.0845 or 8.45 percent.

> *Exercise.* (Optional—Hand calculator.) Using the modified year, what is the effective rate of 6 percent compounded continuously? Answer: 6.27 percent.

In passing, we note that going from daily to continuous compounding contributes very little to interest earned, but the modified year makes a significant contribution when large sums of money are involved. *We shall use one year as* t = 1 *unless specifically instructed to use the modified year.*

We return now to the nominal rate j compounded continuously, for which the effective rate is

$$r = e^j - 1.$$

If we solve this for j we have the nominal rate, which, compounded continuously, yields a *given* effective rate, r. Rearranging the terms of the last equation, we have

$$e^j = 1 + r.$$

Taking the natural logarithm of both sides,

$$\ln e^j = \ln(1 + r).$$

By the power rule for logarithms,

$$j \ln e = \ln(1 + i),$$

where $\ln e = 1$. Hence,

Continuous j equivalent of r effective: $j = \ln(1 + r).$

Example. A bank states that the effective interest on savings accounts that earn continuous interest is 7 percent. Find the nominal rate.

Here,

$$r = 0.07,$$

so

$$
\begin{aligned}
j &= \ln(1 + r) \\
&= \ln(1.07) \\
&= 0.06766 \text{ or } 6.766\%,
\end{aligned}
$$

where $\ln(1.07)$ was taken from Table X at the back of the book.

> *Exercise.* What nominal rate compounded continuously gives an effective rate of 8 percent? Answer: 7.696 percent.

8.28 FINANCIAL FORMULAS: CONTINUOUS COMPOUNDING

In this section we shall present without development the continuous compounding counterparts of the formulas developed earlier in the chapter for periodic compounding. For periodic payments of $R each, payment is made at the end of the period, and between payments the amount in the account is compounded continuously.

TABLE 8–1
Continuous compounding formulas
(t = number of years; j = nominal rate; m = number of periods per year)

Quantity	Formula
1. Amount	$S = Pe^{jt}$
2. Present value	$P = Se^{-jt}$
3. Amount of $R per period	$S_n = R\left[\dfrac{e^{jt}-1}{e^{j/m}-1}\right]$
4. Present value of $R per period	$A_n = R\left[\dfrac{1-e^{-jt}}{e^{j/m}-1}\right]$
5. Per-period equivalent of $P present value	$R = P\left[\dfrac{e^{j/m}-1}{1-e^{-jt}}\right]$
6. Per-period equivalent of $S future value	$R = S\left[\dfrac{e^{j/m}-1}{e^{jt}-1}\right]$

Example. How much must be deposited at the end of each quarter for 5 years at 8 percent compounded continuously to accumulate $10,000 at the time of the last deposit?

Here we have to find the per-period equivalent of $10,000 future value, so Formula 6 of Table 8–1 applies with $t = 5$, $j = 0.08$, $m = 4$, and $S = 10,000$.

$$\frac{j}{m} = \frac{0.08}{4} = 0.02; \quad jt = 0.08(5) = 0.4.$$

Substituting into Formula 6 yields

$$R = 10000\left[\frac{e^{0.02}-1}{e^{0.4}-1}\right]$$

$$= 10000\left[\frac{1.0202-1}{1.4918-1}\right]$$

$$= 10000\left[\frac{0.0202}{0.4918}\right]$$

$$= \$410.74$$

$$= \$411.$$

Because of the limited accuracy of the table used (Table IX), the last answer was rounded to three significant digits. Table IX will suffice to provide practice with the formulas but, for accuracy, a hand calcula-

396

tor should be used. In this case, the answer by hand calculator, accurate to the nearest cent, happens to be $410.74.

Exercise. How much should be deposited now at 6 percent compounded continuously to provide payments of $2000 at the end of each six months for 8 years? Answer: (Formula 4 of Table 8–1.) Using Table IX and rounding to the third digit gives $25,000. By hand calculator, the answer is $25,035.13.

8.29 PROBLEM SET 8–8

Note: Problems 1–28 can be solved by table or hand calculator.

Find the amount of the following deposits if interest is compounded continuously.

1. $1000; 6%; 5 years.   2. $500; 8%; 10 years.
3. $5000; 8%; 4 years and 6 months.
4. $4000; 6%; 5 years and 8 months.

Find the present value of the following if interest is compounded continuously.

5. $800; 8.5%; due in 10 years.
6. $2500; 9.5%; due in 12 years.
7. $1000; 12%; due in 9 months.
8. $3000; 10%; due in 18 months.

The following are nominal rates. Find the effective rate if interest is compounded continuously.

9. $j = 0.05$.   10. $j = 0.06$.
11. $j = 0.07$.   12. $j = 0.08$.

What nominal rate compounded continuously will yield the following effective rates?

13. $r = 0.12$.   14. $r = 0.09$.
15. $r = 0.05$.   16. $r = 0.10$.

17. Sam deposits $500 at the end of each six months for 10 years in an account earning 6 percent. How much will be in the account after the last payment? (Use continuous compounding.)
18. Fran has an opportunity to lend a growing company a sum of money and earn 10 percent interest. The company will discharge its debt to Fran by sending her equal amounts every six months for five years. How much should Fran lend if she wants to receive $2000 every six months? (Use continuous compounding.)
19. Bill wants to accumulate $5000 to buy a boat five years from now by making deposits at the end of each quarter to an account earning 8 per-

cent. How much should Bill deposit each quarter? (Use continuous compounding.)

20. How much will a deposit of $5000 grow to in 20 years at 6.8 percent interest compounded continuously?

21. Jan has lent a new company $10,000 at 12 percent. The company will discharge its debt by sending Jan checks of equal amounts, one each month, for 6 years. How much will Jan receive each month? (Use continuous compounding).

22. How much should be deposited now at 8.4 percent compounded continuously if the amount in the account 10 years from now is to be $10,000?

23. George has $100 deducted from his salary at the end of each month and invested in an employees' fund that earns 12 percent interest. How much will he have after the last payment 10 years from now? (Use continuous compounding.)

24. How much invested now at 8 percent will provide an income of $1,000 per quarter for the next seven years? (Use continuous compounding.)

25. Sam invests $10,000 in a bank account paying 7.6 percent compounded continuously for 15 years. How much will the account amount to at the end of this time?

26. A company wants to accumulate $15,000 to replace a machine five years from now. To do this, equal payments are to be made at the end of each six-month period to an account earning 10 percent interest. Find the periodic payment. (Use continuous compounding.)

27. How much must be deposited now in an account paying 7.5 percent interest compounded continuously if the amount in the account six years from now is to be $7500?

28. Bill has borrowed $5000 at 24 percent to purchase a car. He will discharge the debt by equal end-of-the-month payments for three years. Find the payment. (Use continuous compounding.)

The following are optional problems for solution on a hand calculator.

29. History tells us that Peter Minuit purchased Manhattan Island in New York from the Indians for $24 about 360 years ago. If the $24 had been invested at 5 percent compounded continuously, what would be its amount after 360 years?

30. How much should be deposited now in an account paying 7.6 percent compounded continuously if the account is to grow to $10,000 in eight years?

31. What is the effective rate of 7.9 percent compounded continuously?

32. What nominal rate compounded continuously will yield an effective rate of 8.22 percent?

33. A bank offers 7.6 percent compounded continuously and uses the modified year, 365/360. Find the effective rate.

34. What sum of money deposited at 7.5 percent compounded continuously will yield an income of $1500 at the end of each month for six years?

35. How much should be deposited at the end of each month for 10 years to accumulate $20,000 if the account earns 8 percent compounded continuously?

36. Bill borrowed $6000 at 18 percent compounded continuously to buy a car. He will discharge the debt by equal end-of-the-month payments for three years. Find the amount of the payment.

37. Fran deposits $150 at the end of each month for 6 years in an account earning 7.2 percent compounded continuously. What will be the amount in the account after the last deposit?

8.30 REVIEW PROBLEMS

1. Compute the simple interest at 9 percent on $500 for 15 months.
2. Compute the amount of $1500 at 8 percent simple interest for 10 months.
3. How many months will it take for the simple interest on $2000 at 7 percent to be $175?
4. A credit card holder has owed the card company $400 for one month and receives a bill for $405. Find the simple rate of interest.
5. Compute the yield of the stock of Detroit Edison Company from the following stock market report.

DetEdis 1.52 15½

6. How much must be deposited in an account paying 8.5 percent simple interest if $210 interest is to be earned in 30 months?
7. At what rate of simple interest will $500 grow to $560 in nine months?
8. How many months will it take at 8.4 percent simple interest for $2000 to grow to $2280?
9. Find the present value of $1500 due two years from now with simple interest at 10 percent.
10. How much will Fran have to deposit now in an employee's savings account earning 10 percent simple interest in order to have $2500 in the account two and one-half years from now?
11. Find the proceeds of a $1000, 2-year loan from a bank if the simple bank discount rate is 18 percent.
12. Bill receives $2000 from a bank now, to be repaid in 30 months. If the bank discount rate is 12 percent, how much will Bill have to pay back?
13. Sam signs a note promising to pay a bank $3000 in two years. If Sam receives $2400 when he signs the note, what is the bank's discount rate?
14. Find the amount of $10,000 at 12 percent compounded monthly for 3 years and 4 months.
15. Find the amount of $4000 at 8 percent compounded quarterly for 7 years.
16. Find the amount of $5000 at 7 percent compounded annually for 20 years.
17. An employees' Credit Union Fund pays interest at 12 percent compounded monthly. How much will an employee who invests $2500 now have three years from now?
18. How many years will it take at 8 percent compounded annually for $2,000 to grow to $10,000?

19. How many years will it take for a sum of money to double at 9 percent compounded annually?

20. Find the rate of interest compounded annually at which $1000 will grow to $2500 in 10 years.

21. A company sold 125,000 machines five years ago and this year sold 350,000 machines. Find the annually compounded percentage rate of increase in sales over the five-year period.

22. Find the present value of $5000 payable 10 years from now with interest at 10 percent compounded semiannually.

23. The population of a town now is about 500,000. If population has increased at the rate of 7 percent compounded annually, estimate the population 20 years ago.

24. What sum invested now at 8 percent compounded quarterly will grow to $10,000 in 9 years?

25. Find the amount of an ordinary annuity of $500 per quarter for 10 years at 8 percent compounded quarterly.

26. How much should be deposited now in an account earning 6 percent compounded semiannually to provide an income of $5000 at the end of each six-month period for 17½ years?

27. Sam has borrowed $12,000 at 12 percent, compounded monthly, from a bank. He will discharge the debt by end-of-the-month payments for the next 30 months. Find the monthly payment.

28. *a)* Whyall Corporation has taken out a $5 million, 30-year, 10 percent mortgage on its new manufacturing facility. How much will Whyall pay each month to discharge this mortgage?

    b) How much of the first payment is for interest, and by how much does it reduce the balance owed?

    c) How much of the second payment is for interest, and by how much does it reduce the balance owed?

29. *a)* The Browns have taken out a $50,000, 20-year, 8 percent mortgage on their home. How much will they pay each month to discharge this mortgage?

    b) How much of the first payment is for interest, and by how much does it reduce the balance owed?

    c) How much of the second payment is for interest, and by how much does it reduce the balance owed?

30. Sue plans to accumulate $8000 for a trip around the world five years from now. Find how much she should deposit to the trip fund at the end of each quarter if the fund earns 8 percent compounded quarterly.

31. Sam deposits $250 at the end of every 3 months for 10 years to an account earning 8 percent compounded quarterly. How much will he have in the account after the last deposit?

32. A note for $3000 with interest at 12 percent compounded monthly is payable 30 months from now. Find the (then present) value of the note 12 months from now if this value is computed at 8 percent compounded quarterly.

33. Bill promises to pay off a current debt of $10,000 at 7 percent compounded

annually by making equal end-of-the-year payments for the next ten years. Compute the payment.

34. How much should be deposited now at 7 percent compounded annually to make possible withdrawals of $7500 at the end of each year for 10 years, the first withdrawal to be made 15 years from now? Interest during the withdrawal period is 6 percent compounded annually.

35. If fuel is now consumed by a plant at the rate of 25,000 barrels a year, and the rate increases by 7 percent compounded annually, what will yearly fuel consumption be 15 years from now?

36. Whyall Corporation is setting up a reserve account earning 10 percent compounded semiannually. At the end of each six-month period, equal sums are transferred to this account. The purpose of the account is to provide $20,000 after the last transfer eight years from now to be used to replace equipment. Compute the semiannual transfer to the account.

37. (All interest at 7 percent compounded annually.) A business man has an obligation to pay $8000 four years from now, and another obligation to pay $12,000 nine years from now. The money needed is to be provided by setting up an account into which equal deposits are made at the end of each year for nine years. Find the annual deposit.

38. Sam wants to buy a boat for $10,000 nine years from now. How much should he deposit now at 8 percent compounded quarterly in order to have the desired amount available?

39. A company will need 1500 barrels of fuel at the end of this month, and its needs will increase at the rate of one percent per month (12 percent compounded monthly). The company plans to take advantage of a currently favorable price situation and purchase a three-year supply of fuel. How many barrels should be purchased?

40. How much should be deposited now in an account at 7 percent compounded annually to provide an income of $10,000 at the end of each year for 22 years?

41. (All interest compounded annually.) Sue has an investment that will pay her $20,000 plus interest at 5 percent in ten years. She can obtain the present value of this investment, computed at 6 percent, and deposit this amount in an account paying 8 percent. If she does so, what will be the amount of the account in 10 years?

42. (Approximate by interpolation in the tables or by hand calculator. State answer accurate to two significant digits.) At what rate of interest compounded annually will payments of $1000 per year for 15 years discharge a current debt of $10,000?

43. (See instructions, Problem 42.) At what rate of interest compounded annually will a series of end-of-the-year deposits of $2500 amount to $41,000 in 12 years?

44. (Optional—Hand calculator.) Compute the percent in Problem 42 accurate to three significant digits.

45. (Optional—Hand calculator.) At what rate of interest compounded annually will a series of end-of-the-year deposits of $1000 amount to $25,000 in 15 years? State answer accurate to three significant digits.

46. By table or hand calculator, find the effective interest rate for 12 percent compounded quarterly.

47. By table or hand calculator, find the effective interest rate of 24 percent compounded monthly.

48. (Optional—Hand calculator.) Find the effective interest rate for 6 percent compounded daily using 365 days per year.

49. Find the amount of $5000 for 9 years at 7 percent compounded continuously.

50. How much should be deposited now at 8 percent compounded continuously to have $10,000 seven years from now?

51. What is the effective rate of 12 percent compounded continuously?

52. What nominal rate of interest compounded continuously will yield an effective rate of 12 percent?

53. Fran wants to have $4000 available three years from now. She makes equal deposits at the end of each quarter in an employees' fund earning 12 percent compounded continuously. Compute the amount of the deposit.

54. Bill has lent a company $20,000 at 14 percent compounded continuously. The company will discharge the debt by sending Bill a check at the end of each six-month period for ten years. Compute the amount of the check.

55. How much should be deposited now in an account earning 6.6 percent compounded continuously in order to have $12,000 fifteen years from now?

56. Sam has an opportunity to invest in an income account at 12 percent compounded continuously. The amount invested, plus interest, comes back to Sam in the form of a check at the end of each month for eight years. How much should Sam invest if he wants the monthly income check to be $1000?

57. How much will a deposit of $2500 amount to in 10 years at 8.3 percent compounded continuously?

58. Sue deposits $400 at the end of each six-month period for 12 years in an account earning 8 percent compounded continuously. What will the amount in the account be after the last payment?

Differential calculus:
Power functions

9

9.1 INTRODUCTION

In this chapter we present differential calculus mainly as a tool that can be applied to solve optimization problems; that is, for example, to determine how to maximize profit or minimize cost. We shall also present one interpretation of the derivative as a rate of change, but the majority of rate interpretations will be found in Chapter 10 along with further topics in differential calculus and optimization. The objective of this chapter is to provide all the basic concepts of differential calculus that we need and do so with a minimum of prerequisite study. Consequently, only algebra and a study of straight lines and slopes (Chapter 1) are a prerequisite for this chapter. To make this minimum of prerequisite study possible, we shall work with a set of expressions known as power functions. In the next chapter, we introduce and apply exponential and logarithmic functions that have the material in the first part of Chapter 7 (the modern treatment of logarithms) as an additional prerequisite.

9.2 WHY STUDY CALCULUS?

The answer to the question posed has several parts, from which we select the part most readers will find significant; namely, that the tools of calculus can be used to solve applied problems. The purpose of this section is to document the last statement by first posing a simple applied problem and showing how it can be solved approximately by 'brute force' without calculus. Then we shall explain what tools are needed to obtain an *exact* solution in a *simple* manner. The tools, of course, are the procedures of calculus.

Example.   The United Parcel Service operates a fleet of trucks that pick up and deliver packages. UPS will not handle packages whose length plus girth[1] exceeds 108 inches.

Exall Company ships a fresh product that requires as much ventilation as possible, and its boxes are made of a perforated material. To obtain maximum ventilation, Exall wants the boxes to have as large a surface area as possible. The problem then is to find what box dimensions will utilize the entire 108 inches allowed *and* provide maximum surface area. The box is to be rectangular, with square bases, as shown in Figure 9–1. The length of the box is L, and its girth is $4x$. Thus,

$$\text{Length plus girth} = L + 4x.$$

FIGURE 9-1

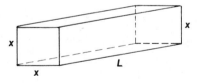

When all 108 inches are utilized,

$$L + 4x = 108$$

so that

$$L = 108 - 4x. \tag{1}$$

---

[1] For example, the length of this book is the vertical (longer) dimension of the cover, and the girth is the distance around the book in the horizontal direction.

To find the expression for the surface area, which is to be maximized, we note first that the area of the square base is $(x)(x)$, and there are two bases. Hence,

$$\text{Area of bases} = 2x^2.$$

The area of the rectangle forming one side is $(x)(L)$, and there are four such sides, so that

$$\text{Side area} = 4xL.$$

Hence,

$$\text{Total area} = A = 2x^2 + 4xL.$$

If (1) is substituted into the last expression, we have

$$A = 2x^2 + 4x(108 - 4x)$$
$$= 2x^2 + 432x - 16x^2$$
$$A = 432x - 14x^2. \tag{2}$$

We want to find the value of x that makes A as large as possible. This value can then be substituted into (1) to get the value of L. In the 'brute force' approach to the solution, we let $x = 1$, then $x = 2$, 3, 4, and so on. For each value of x, the area is computed by (2) and tabulated. For example,

$$x = 1: A = 432(1) - 14(1)^2 = 418 \text{ square inches.}$$
$$x = 2: A = 432(2) - 14(2)^2 = 808 \text{ square inches.}$$

We see that $x = 2$ provides a larger area than $x = 1$, so we try $x = 3$.

$$x = 3: A = 432(3) - 14(3)^2 = 1{,}170.$$

Continuing in this manner, we obtain the results shown in Table 9–1.

TABLE 9–1

x	Area	x	Area	x	Area
1	418	9	2754	15.1	3331.06
2	808	10	2920	15.2	3331.84
3	1170	11	3058	15.3	3332.34
4	1504	12	3168	15.4	3332.56
5	1810	13	3250	15.5	3332.5*
6	2088	14	3304	15.41	3332.5666
7	2338	15	3330	15.42	3332.5704
8	2560	16	3328*	15.43	3332.5714
				15.44	3332.5696*

* Marks a point in a sequence where area decreases.

Observe that as x increases area increases until the point marked with * at the bottom of the fourth column. When x went from 15 to 16, area decreased from 3330 to 3328. We conclude that the maximum area occurs near $x = 15$. In the right section of Table 9–1, we find that $x = 15.1$ gives a larger area than $x = 15$, so we continue by steps of 0.1 as shown, then change to steps of 0.01 after the *. Clearly, we could proceed to steps of 0.001 and so on to get a more exact value for x but we have demonstrated that this time-consuming procedure is a 'brute force' attack on the problem. We leave the foregoing work with the conclusion that the area is maximized when x is approximately 15.4 inches.

The calculus method of solving this problem can be understood by reference to the graph of

$$A = 432x - 14x^2$$

shown in Figure 9–2. Observe that the graph is mound-shaped and the largest area possible is the verticle height of the peak of the mound.

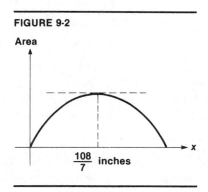

FIGURE 9-2

Area

$\dfrac{108}{7}$ inches

Note also that the line tangent to the curve at the peak of the mound is horizontal, so this tangent has a slope of zero. In this chapter, we shall learn that the tangent to

$$A = 432x - 14x^2$$

is horizontal when

$$28x = 432$$

$$x = \frac{432}{28}$$

$$= \frac{108}{7},$$

exactly, or 15.4286 inches, correct to four places of decimals. The corresponding value for *L*, computed from $L = 108 - 4x$ is 324/7 inches exactly, or 46.2857 inches, correct to four decimals. The question, of course, is how we obtained $28x = 432$, and that is the question we shall answer in this chapter. In the course of developing the tools needed to answer the question, and in the application of these tools, functional notation will be used extensively, so we shall review functions before proceeding.

9.3 FUNCTIONAL NOTATION

In Chapter 1, the expression

$$y = mx + b$$

was called the slope-intercept form of the equation of a straight line. The quantities *m* (the slope) and *b* (the *y*-intercept) for any *specific* line are constants, as in

$$y = 2x + 5,$$

whereas *x* and *y* are variables. Thus, even though

$$y = mx + b$$

contains four letters *(y, m, x, b)*, we know from the context that *x* and *y* are to be considered as the variables. The first advantage of functional notation is to ensure that the variable in any expression is stated specifically. To do this for $y = mx + b$, we write

$$y(x) = mx + b,$$

and read $y(x)$ as "the *y* function of *x*" or, more briefly, "*y* of *x*." In the symbol $y(x)$, *x* is called the *independent* variable. Similarly, in

$$C(L) = \frac{NF}{L} + \frac{iL}{2},$$

L is the independent variable and, this being the case, *N*, *F*, and *i* are constants as are *m* and *b* in

$$y(x) = mx + b.$$

If we write

$$g(x) = 3x^3 - 2x + 10,$$

g is the name of the function, *x* is the independent variable, and $g(x)$ represents values of the function. To find the value of the function when $x = 2$, written as $g(2)$, we evaluate

$$g(x) = 3x^3 - 2x + 10$$

at $x = 2$ to find

$$g(2) = 3(2^3) - 2(2) + 10$$
$$= 30.$$

Exercise. Given $h(q) = 2q - 10/q$, find: *a)* $h(1)$. *b)* $h(5)$. *c)* $h(100)$. Answer: *a)* -8. *b)* 8. *c)* 199.9.

The foregoing exercise shows a second advantage of functional notation; namely, brevity of statement. That is, the instruction "Find h(1)" replaces the longer instruction "Find the value of the expression when the variable equals 1." For emphasis, we note that if

$$f(k) = k^2 - 2kg$$

then

$$f(5) = 25 - 10g,$$

but if, for the same expression, g is the independent variable as in

$$h(g) = k^2 - 2kg,$$

then

$$h(5) = k^2 - 10k.$$

Exercise. If $p(q) = q^2 - r^2 + 5$ and $h(r) = q^2 - r^2 + 5$, what is: *a)* $p(2)$? *b)* $h(2)$? Answer: *a)* $9 - r^2$. *b)* $q^2 + 1$.

It is helpful to think of a function as a rule that specifies how to find the value of the function for a stated value of the independent variable. For example, given

$$f(x) = x^2 - 6,$$

the rule, $x^2 - 6$, says to square the value of the independent variable and then subtract 6. Thus, $f(5)$ is obtained by squaring 5, then subtracting 6, and the resulting function value is 19. Similarly,

$$f(x + a)$$

will be obtained by squaring $(x + a)$ and subtracting 6, so

$$f(x + a) = (x + a)^2 - 6$$
$$= x^2 + 2ax + a^2 - 6.$$

Example. Find $g(a) - g(x - a)$ if $g(x) = x^2 + 10$.

We find

$$g(a) = a^2 + 10$$
$$g(x - a) = (x - a)^2 + 10.$$

Hence,

$$g(a) - g(x - a) = (a^2 + 10) - [(x - a)^2 + 10]$$
$$= a^2 + 10 - (x - a)^2 - 10$$
$$= a^2 - (x^2 - 2ax + a^2)$$
$$= a^2 - x^2 + 2ax - a^2$$
$$= 2ax - x^2.$$

Exercise. Find $f(x + a) - f(x)$ if $f(x) = x^2 - 3$. Answer: $2ax + a^2$.

A function must be single-valued. To avoid ambiguity in the meaning of *the* value of a function, we require that a function shall have one, and only one, value for each permissible value of the independent variable. In particular, in

$$f(x) = x^{1/2} = \sqrt{x}$$

we define $f(x)$ to be the *positive* square root of x. That is,

$$f(4) = 4^{1/2} = 2.$$

This does not mean that

$$y^2 = 4$$

has only $y = 2$ as a solution, because $y^2 = 4$ is a *conditional equality*, not a function, and both $y = 2$ and $y = -2$ satisfy the conditional equality.

What has been said for

$$f(x) = x^{1/2}$$

applies also to any other fractional power that has an even denominator, such as

$$g(x) = x^{3/4}, \ h(x) = x^{-5/6}.$$

That is, the function value for such even roots is the positive value. Note also that for even roots, x must not be negative, so only positive numbers (or zero if the exponent is positive) are permissible values for x. An odd root, such as

$$p(x) = x^{2/3}$$

raises no question of ambiguity because such roots are single valued. Thus,

$$p(8) = 8^{2/3}$$
$$= (8^{1/3})^2$$
$$= (2)^2$$
$$= 4$$

and

$$p(-8) = (-8)^{2/3}$$
$$= (-8^{1/3})^2$$
$$= (-2)^2$$
$$= 4.$$

We note in passing that while odd roots of negative numbers are permissible, we shall not use them in our work.

Exercise. For

$$g(x) = \frac{x^{3/2}}{32} - 16x^{-1/2} + 2x^{1/3},$$

find: *a)* $g(64)$. *b)* $g(-1)$. Answer: *a)* 22. *b)* $x = -1$ is not permissible.

9.4 DELTA NOTATION

Calculus has been described as the mathematics of change because it was invented to solve problems involving rates of change. In this section we show how to find the general expression for the change in the value of a function when its independent variable changes value. Conventionally, the symbol Δ, delta, is taken to mean *the change in.* Thus, Δx, read as "delta x," is the change in x; that is, the change in the value of the independent variable. Similarly, $\Delta f(x)$, read as "delta f of x," means the change in the value of the function, $f(x)$, when x changes by Δx. In summary:

Δ **means the change in**
Δx **means the change in x**
$\Delta f(x)$ **means the change in $f(x)$ when x changes by Δx.**

Example. Find the expression for $\Delta g(z)$ if

$$g(z) = z^2 + 3z.$$

Here, z is the independent variable, so $\Delta g(z)$ is the change in $g(z)$ when z changes by Δz. We can picture the change in z as

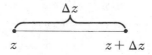

so that z changes to $z + \Delta z$. The change in $g(z)$ is therefore

$$\Delta g(z) = g(z + \Delta z) - g(z)$$
$$= [(z + \Delta z)^2 + 3(z + \Delta z)] - (z^2 + 3z)$$
$$= [z^2 + 2z(\Delta z) + (\Delta z)^2 + 3z + 3(\Delta z)] - z^2 - 3z$$
$$= \quad\quad 2z(\Delta z) + (\Delta z)^2 + 3(\Delta z)$$
$$= \Delta z[2z + (\Delta z) + 3].$$

The last expression says, for example, that if z changes from, say, 5 to 5.5, so that

$$\Delta z = 5.5 - 5 = 0.5,$$

then $g(z)$ will change by

$$\Delta g(z) = 0.5[2(5) + 0.5 + 3]$$
$$= 0.5(13.5)$$
$$= 6.75.$$

This may be verified as follows:

$$g(z) = z^2 + 3z$$
$$g(5.5) = (5.5)^2 + 3(5.5) = 46.75$$
$$g(5) = 5^2 + 3(5) \quad\quad = 40$$
$$\Delta g(z) = g(5.5) - g(5) = 46.75 - 40 \quad = 6.75.$$

Attention should be called to the fact that in the little figure at the beginning of the example, $z + \Delta z$ was shown at the right of z. This will be the case if Δz is positive, but $z + \Delta z$ would be to the left of z for negative values of Δz. However, the expression we derived for $\Delta g(z)$ is correct for both positive and negative values of Δz. We may now state:

Definition. $\Delta f(x) = f(x + \Delta x) - f(x).$

Exercise. *a)* Find the algebraic expression for $\Delta g(y)$ if $g(y) = 2y^2 - 5$. *b)* Find $\Delta g(y)$ if y goes from 2 to 3 by substitution into the answer for *a)*. *c)* Compute $\Delta g(y)$ as $g(3) - g(2)$. Answer: *a)* $2(\Delta y)[2y + \Delta y]$. *b)* 10. *c)* 10.

As an application to help fix the foregoing in mind, recall that in Chapter 1 the marginal cost of a unit of production was defined as the change in total cost when that unit is produced. Thus, if the total cost of making g gallons of Exall is $C(g)$, then the marginal cost of the 10th gallon is Δg when g changes from 9 to 10. This is

$$C(10) - C(9),$$

which is the total cost of 10 gallons minus total cost of 9 gallons. In general, the marginal cost of the gth gallon is

$$\Delta C(g) = C(g) - C(g - 1).$$

Now suppose we have the cost function

$$C(g) = 1000 + 5g + 0.01g^2$$

and we want the expression for the marginal cost of the gth gallon;

$$
\begin{aligned}
C(g) - C(g-1) \\
&= [1000 + 5g + 0.01g^2] - [1000 + 5(g-1) + 0.01(g-1)^2] \\
&= 1000 + 5g + 0.01g^2 - [1000 + 5g - 5 + 0.01(g^2 - 2g + 1)] \\
&= 1000 + 5g + 0.01g^2 - 1000 - 5g + 5 - 0.01g^2 + 0.02g - 0.01 \\
&= 5 + 0.02g - 0.01 \\
&= 4.99 + 0.02g.
\end{aligned}
$$

Thus,

$$\text{marginal cost of } g\text{th gallon} = 4.99 + 0.02g.$$

Exercise. For the cost function of the last example, find the marginal cost of the *a)* Tenth gallon. *b)* Fiftieth gallon. Answer: *a)* \$5.19. *b)* \$5.99.

In the answer to the last exercise, note that the marginal cost of the fiftieth gallon is greater than that of the tenth gallon. In Chapter 1, where *linear* cost functions were considered, marginal cost was the *constant* slope of a straight line. The cost function of the exercise has a second power term $(0.01g^2)$ and is a curve. After we have discussed slope as it applies to a curve, we will see the reason for the increase in marginal cost observed in the answer to the last exercise.

9.5 PROBLEM SET 9–1

1. If $f(x) = 3x - 2$, find the value of, or the algebraic expression for:
   a) $f(3)$.
   b) $f(-2)$.
   c) $f(a)$.
   d) $[f(a)]^2$.
   e) $f(ab)$.
   f) $f(3y + 4)$.
   g) $f(x + 1)$.
   h) $f(x + 1) - f(x)$.

2. If $h(x) = x^2 + 3x$, find the value of, or the algebraic expression for:
   a) $h(2)$.
   b) $h(-3)$.
   c) $h\left(\dfrac{1}{2}\right)$.
   d) $h\left(\dfrac{2}{a}\right)$.
   e) $h(x + 0.5)$.
   f) $h(a + 1)$.
   g) $h(a - 1)$.
   h) $h(x) - h(x - 1)$.

3. *a)* If $g(x) = x^2y - y^2$, write the expression for $g(a)$.
   b) If $f(y) = x^2 - y^2$, write the expression for $f(a)$.

4. *a)* If $h(y) = 2x + 5y$, write the expression for $h(3)$.
   b) If $p(x) = 2x + 5y$, write the expression for $p(3)$.

5. If $p(x) = 2x^{-1} - 3x^{-2}$, write the value of, or the algebraic expression for:
   a) $p(2)$.
   b) $p(3)$.
   c) $p(a)$.
   d) $p(x + 1)$.

6. If $f(x) = 2x^{1/3} + 3x^{-2/3}$, find the value of:
   a) $f(1)$.
   b) $f(64)$.
   c) $f\left(\dfrac{1}{8}\right)$.

7. What is the meaning of:
   a) Δx?   b) $\Delta f(x)$?

8. Given $f(x) = x^2 - 3x + 5$, find $\Delta f(x)$ if x changes from 2 by the amount $\Delta x = 0.5$.

9. Given $f(x) = 2x^2 - 10x + 8$, find $\Delta f(x)$ if x changes from 0 by the amount $\Delta x = 0.1$.

10. Find the expression for $\Delta f(x)$ if $f(x) = 10 - 3x$.

11. Find the expression for $\Delta g(x)$ if $g(x) = mx + b$.

12. Find the expression for:
    a) $\Delta f(x)$ if $f(x) = x^2$.
    b) $\Delta g(x)$ if $g(x) = 2x^2 - 3x + 5$.

13. Find the expression for:
    a) $\Delta f(x)$ if $f(x) = 2x^2$.
    b) $\Delta g(x)$ if $g(x) = x^2 + 2x - 10$.

14. If the total cost of making g gallons of Exall is $C(g)$ dollars and

$$C(g) = 100 + 3g + 0.01g^2,$$

   a) Find the expression for the marginal cost of the gth gallon, which is $C(g) - C(g-1)$.
   b) Find the marginal cost of the tenth gallon.
   c) Find the marginal cost of the fiftieth gallon.

15. If the total cost of making g gallons of Whyall is $C(g)$ dollars and

$$C(g) = 50 + g + 0.1g^2$$

   a) Find the expression for the marginal cost of the gth gallon, which is $C(g) - C(g-1)$.
   b) Find the marginal cost of the tenth gallon.
   c) Find the marginal cost of the fiftieth gallon.

9.6 LIMITS

Limits are the core concept in the development of calculus. We shall not pursue the theory of limits in this text, but in this section the reader will become familiar enough with the limit concept to use it when needed as we proceed. We start by observing that the function

$$g(x) = \frac{x^2}{x}$$

does not have a value at $x = 0$, because at $x = 0$ the ratio is the meaningless expression $0/0$, and we shall say $g(x)$ is *not defined* at $x = 0$. Readers who think this ratio has the value 1 should look up the Index reference to discussion of *zero*. There it is shown why we cannot divide any number, including zero, by zero. To emphasize this exception, we can write

$$g(x) = \frac{x^2}{x}; \quad x \neq 0.$$

It is true, though, that

$$g(x) = \frac{x^2}{x} = x; \quad x \neq 0.$$

That is, $g(x) = x$ for any value of x except zero. Because this is true, it is correct to write

$$g(x) \text{ is close to zero if } x \text{ is close to zero}$$

or

$$g(x) \text{ approaches zero if } x \text{ approaches zero.}$$

We describe this behavior of $g(x)$ by saying $g(x)$ approaches zero as a limiting value as x approaches zero and write

limit of $g(x)$ as x approaches zero is zero.

The last is expressed symbolically as

$$\lim_{x \to 0} g(x) = 0.$$

Thus, in the case of $g(x)$ we are able to state that a limit exists. By way of contrast, consider

$$f(x) = \frac{x}{x^2}; \quad x \neq 0.$$

Like the earlier $g(x)$, $f(x)$ at $x = 0$ becomes the meaningless expression $0/0$ and $f(x)$ is not defined at $x = 0$. It is true, of course, that

$$f(x) = \frac{x}{x^2} = \frac{1}{x}; \quad x \neq 0,$$

that is, $f(x) = 1/x$ for any value except $x = 0$. But $f(x)$, unlike $g(x)$, does not have a limit as x approaches zero. For example, if x takes on the sequence of values

$$1, \ 0.1, \ 0.01, \ 0.001,$$

and so on, approaching zero, then

$$f(1) \qquad = \frac{1}{1} \qquad = 1$$

$$f(0.1) \qquad = \frac{1}{0.1} \qquad = 10$$

$$f(0.01) \qquad = \frac{1}{0.01} = 100$$

$$f(0.001) = \frac{1}{0.001} = 1000,$$

and so on, and this last sequence of values does not approach a limiting value but, instead, becomes larger and larger as x gets closer and closer to zero. Similarly, if we took $x = -1/2, -1/3, -1/4, -1/5$, and so on, as a sequence approaching zero, then

$$f\left(-\frac{1}{2}\right) = \frac{1}{-\frac{1}{2}} = -2$$

$$f\left(-\frac{1}{3}\right) = \frac{1}{-\frac{1}{3}} = -3$$

$$f\left(-\frac{1}{4}\right) = \frac{1}{-\frac{1}{4}} = -4$$

$$f\left(-\frac{1}{5}\right) = \frac{1}{-\frac{1}{5}} = -5$$

and so on. Again, the sequence $-2, -3, -4, -5$, continues indefinitely and does not approach a limit. Hence,

$$\lim_{x \to 0} \frac{x}{x^2} \quad \text{does not exist.}$$

Comparing

$$g(x) = \frac{x^2}{x}; \quad x \neq 0$$

$$f(x) = \frac{x}{x^2}; \quad x \neq 0,$$

we see that they are alike at $x = 0$ in that both become the meaningless ratio, $0/0$. However, they have the important difference that $g(x)$ has a limit as x approaches zero and $f(x)$ does not. The distinction is important because, as we shall see, the central definition of the differential calculus leads to $0/0$, but we can attach a meaningful interpretation to the expression that leads to $0/0$ if this expression has a limit.

The limits we shall use in the major part of the chapter are easily found. We start by stressing the real meaning of the limit concept, but the reader will quickly learn how to find the limits without formally applying the concept. The real meaning of the concept is embodied in the proper interpretation of $x \to a$. This is that x approaches, *but never equals*, a.

Example. Find $\lim_{x \to 2}(x + 3)$.

Keeping in mind that $x \to 2$ means x must not equal 2, we set up any sequence of x values approaching 2, and compute and write the corresponding values of the function $x + 3$. Thus, for example,

$$x: \quad 1.9 \quad 1.99 \quad 1.999 \quad 1.9999 \to 2$$
$$x + 3: \quad 4.9 \quad 4.99 \quad 4.999 \quad 4.9999 \to ?$$

Hopefully, it is obvious that the ? at the right of the sequence for $(x + 3)$ approaches 5, so that

$$\lim_{x \to 2} (x + 3) = 5.$$

We said hopefully in the last sentence because we do not plan to go into the detail needed to prove this formally and rigorously. Again, hopefully, it is obvious that the limit of $(x + 3)$ is 5 for all sequences of values of x approaching 2.

The reader will have observed that the limit just written, 5, is the value of the function $(x + 3)$ when $x = 2$ and wonder why the sequence approach was taken. The answer, of course, was stated at the beginning. It is that the real meaning of the limit concept is preserved only if $x \to a$ is taken to mean x is approaching, but does not equal, a. While this meaning is important to preserve, the writer admits that, as a practical matter, many limits encountered in this book can be obtained by replacing $x \to a$ by $x = a$.

Exercise. Find $\lim_{x \to 3} (2x - 5)$. Answer: 1.

Example. Find the limit if it exists:

$$\lim_{x \to 2} \frac{x}{x - 2}.$$

First note that when $x = 2$, the fraction becomes 2/0, which is meaningless, so the expression is not defined at $x = 2$. We now write a sequence of values of x approaching 2, find the corresponding sequence of values for the fraction, examine the latter sequence, and try to determine whether or not it has a limit. A convenient sequence is

$$x: \quad 2.1, \quad 2.01, \quad 2.001, \quad 2.0001, \quad \cdots \to 2$$
$$\frac{x}{x - 2}: \quad \left(\frac{2.1}{0.1}\right), \quad \left(\frac{2.01}{0.01}\right), \quad \left(\frac{2.001}{0.001}\right), \quad \left(\frac{2.0001}{0.0001}\right), \ldots$$
$$= \quad 21, \quad 201, \quad 2001, \quad 20001, \quad \cdots \to ?$$

This sequence of values for the fraction grows without limit, and the same would be true for any sequence of fractions whose numerator is nearly 2, but whose denominators get ever closer to zero. Hence,

$$\lim_{x \to 2} \frac{x}{x-2} \text{ does not exist.}$$

We would encourage the reader to do the last example mentally. To do this, look at the numerator of

$$\frac{x}{x-2}$$

and observe that if x is very close to 2, the numerator is close to 2. The denominator, however, gets closer and closer to zero as x approaches 2. The key point to keep in mind now is that a fraction whose numerator is a number that is not zero and whose denominator is close to zero has a *large* (absolute) value, and this value becomes ever larger the closer the denominator is to zero. By thinking in this manner, the reader can quickly see, without writing sequences, that as $x \to 2$, the fraction at hand gets ever larger and a limit does not exist.

Example. Find the limit if it exists:

$$\lim_{x \to 2} \frac{x^2 - 4}{x - 2}.$$

Observe that if $x = 2$, the fraction is $0/0$, which is meaningless, and the expression is not defined at $x = 2$. However, we can factor the numerator to obtain

$$\lim_{x \to 2} \frac{x^2 - 4}{x - 2} = \lim_{x \to 2} \frac{(x+2)(x-2)}{x-2}.$$

If x is not 2, and $x \to 2$ says it is not, we may cancel and obtain

$$\lim_{x \to 2} \frac{(x+2)(x-2)}{x-2} = \lim_{x \to 2} (x+2) = 4.$$

Exercise. Find the limit if it exists.

a) $\lim_{x \to 5} \dfrac{x^2 - 25}{x - 5}.$ b) $\lim_{x \to 1} \dfrac{x+1}{x^2-1}.$ Answer: a) 10.

b) Limit does not exist.

Example. Find the limit if it exists:

$$\lim_{a \to b} (3a + 2b).$$

We note that for any sequence of values of a, approaching b, $3a$ will approach $3b$ and $3a + 2b$ will approach $3b + 2b$. Hence,

$$\lim_{a \to b} (3a + 2b) = 3b + 2b = 5b.$$

Example. Find the limit if it exists:

$$\lim_{b \to a} \frac{a^2 - b^2}{a - b}.$$

Factoring leads to

$$\lim_{b \to a} \frac{a^2 - b^2}{a - b} = \lim_{b \to a} \frac{(a + b)(a - b)}{a - b}$$

$$= \lim_{b \to a} (a + b)$$

$$= 2a.$$

Note that in this example b approaches a, whereas in the previous example a approaches b.

Exercise. Find the limit if it exists. *a)* $\lim_{y \to x} (2x + y)$.

b) $\lim_{x \to y} \frac{x^2 - y^2}{x - y}$. Answer: *a)* $3x$. *b)* $2y$.

Limit theorems. We now state and illustrate limit theorems we shall refer to from time to time as our work progresses. These theorems are intuitively reasonable, but their proof requires attention to details that the writer does not consider to be appropriate for this text. The following are true if all of the indicated limits exist.

1. If k is any constant, $\lim_{x \to a} k = k$.

   Here note that $x \to a$ specifies that x is changing and approaching a. The constant k does not change (it does not involve x). Thus, the limit of a constant is the constant.

   Examples. $\lim_{x \to 5} 10 = 10$. $\lim_{a \to b} c = c$.

2. $\lim_{x \to a} k f(x) = k \lim_{x \to a} f(x)$.

   That is, a constant factor, here k, may be placed inside or outside the limit symbol.

   Example. $\lim_{x \to a} 3x^2 = 3(\lim_{x \to a} x^2) = 3(a^2) = 3a^2$.

3. $\lim_{x \to a} [f(x) \pm g(x)] = \lim_{x \to a} f(x) \pm \lim_{x \to a} g(x).$

That is, the limit of a sum or difference is the sum or difference of the limits. Or, we can say that the limit of an expression can be taken term by term.

Example. $\lim_{x \to a} (x^2 - 2x + 3) = \lim_{x \to a} x^2 - 2 \lim_{x \to a} x + \lim_{x \to a} 3$

$$= a^2 - 2a + 3.$$

4. $\lim_{x \to a} [f(x)g(x)] = [\lim_{x \to a} f(x)] [\lim_{x \to a} g(x)].$

That is, the limit of a product is the product of the limits.

Example. $\lim_{x \to 2} (x + 3)(x - 2) = [\lim_{x \to 2} (x + 3)] [\lim_{x \to 2} (x - 2)]$

$$= [5] [0]$$

$$= 0.$$

5. $\lim_{x \to a} \left[\dfrac{f(x)}{g(x)} \right] = \dfrac{\lim\limits_{x \to a} f(x)}{\lim\limits_{x \to a} g(x)},$ if $\lim_{x \to a} g(x) \neq 0.$

That is, the limit of a quotient is the quotient of the limits if the denominator limit is not zero.

Example. $\lim_{x \to 3} \dfrac{(x + 5)}{(x - 2)} = \dfrac{\lim\limits_{x \to 3} (x + 5)}{\lim\limits_{x \to 3} (x - 2)} = \dfrac{8}{1} = 8.$

6. $\lim_{x \to a} [f(x)]^n = [\lim_{x \to a} f(x)]^n.$

That is, the limit of a power of $f(x)$ is the power of the limit of $f(x)$.

Example. $\lim_{x \to 3} (x - 1)^5 = [\lim_{x \to 3} (x - 1)]^5 = 2^5 = 32.$

9.7 PROBLEM SET 9–2

Find the limit for each of the following if it exists.

1. $\lim_{x \to 1} (x^2 + 2x - 2).$

2. $\lim_{x \to 2} (x^3 - 5x^2 - 1).$

3. $\lim_{x \to b} ax^2.$

4. $\lim_{b \to a} a^2b^2.$

5. $\lim_{x \to 1} \dfrac{x^2 - 1}{x + 1}.$

6. $\lim_{x \to 1} \dfrac{x^2 + 1}{x + 1}.$

7. $\lim_{x \to 2} \dfrac{x + 5}{x - 2}.$

8. $\lim_{x \to 1} \dfrac{x^2 + 1}{x - 1}.$

9. $\lim_{x \to a} (x - 1)^{1/3}.$

10. $\lim_{x \to 24} (x + 1)^{1/2}.$

11. $\lim\limits_{b \to a} (3a + 5b)$.

12. $\lim\limits_{a \to b} (5a - b)$.

13. $\lim\limits_{x \to 3/2} \dfrac{4x^2 - 9}{2x - 3}$.

14. $\lim\limits_{x \to 5/3} \dfrac{9x^2 - 25}{3x - 5}$.

15. $\lim\limits_{x \to 0} \dfrac{x^4}{x^3}$.

16. $\lim\limits_{x \to 0} \dfrac{x^3}{x^4}$.

17. $\lim\limits_{x \to 1/2} \dfrac{2x + 1}{2x - 1}$.

18. $\lim\limits_{x \to 1/3} \dfrac{3x - 1}{3x + 1}$.

19. $\lim\limits_{\Delta x \to 0} \dfrac{(\Delta x)(2x + 3\Delta x)}{\Delta x}$.

20. $\lim\limits_{\Delta x \to 0} \dfrac{(\Delta x)[3x^2 + 2x(\Delta x)]}{\Delta x}$.

21. $\lim\limits_{a \to 0} \dfrac{\dfrac{1}{2 + a} - \dfrac{1}{2}}{a}$.

22. $\lim\limits_{a \to 0} \dfrac{\dfrac{1}{5 + a} - \dfrac{1}{5}}{a}$.

9.8 THE DIFFERENCE QUOTIENT

The function

$$g(x) = mx + b$$

is a *linear* function represented graphically by a straight line. Thus,

$$f(x) = 0.8x + 3$$

is a linear function. Conventionally when graphing functions, the function values, $f(x)$, are on the vertical, and the independent variable values are on the horizontal. The graph of $f(x)$, which has a slope of 0.8 and a vertical intercept of 3, is shown in Figure 9–3. If the indepen-

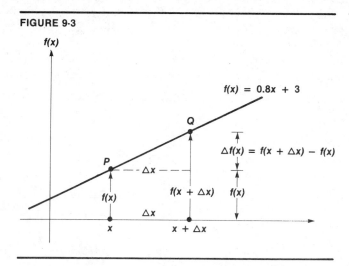

FIGURE 9-3

$f(x)$

$f(x) = 0.8x + 3$

$\Delta f(x) = f(x + \Delta x) - f(x)$

dent variable changes from x to $x + \Delta x$, the value of the function changes from $f(x)$ to $f(x + \Delta x)$, so the change in the function is

$$\Delta f(x) = f(x + \Delta x) - f(x),$$

as shown on Figure 9–3. The slope of the line can be determined from any two of its points, such as P and Q. It is the vertical change divided by the horizontal change; that is, the change in the function divided by the change in x, which is

$$\frac{\Delta f(x)}{\Delta x}.$$

This expression is called the *difference quotient*. Thus, the difference quotient is simply the slope of a straight line. For our function,

$$f(x) = 0.8x + 3$$

we find

$$\begin{aligned}\Delta f(x) = f(x + \Delta x) - f(x) &= [0.8(x + \Delta x) + 3] - [0.8x + 3] \\ &= 0.8x + 0.8\Delta x + 3 - 0.8x - 3 \\ &= 0.8(\Delta x).\end{aligned}$$

Hence, the difference quotient is

$$\frac{\Delta f(x)}{\Delta x} = \frac{0.8(\Delta x)}{\Delta x} = 0.8.$$

The result, a constant, is not surprising, of course, because $f(x)$ is a straight line and we know that a given line has the same slope number for any pair of its points.

Exercise. If $g(x) = mx + b$, what will the difference quotient be? Answer: m, the slope of the line specified by $g(x)$.

Consider next the function x^3, a section of which is shown on Figure 9–4. A line, called a *secant* line, has been drawn through two points,

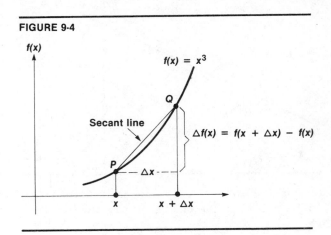

FIGURE 9-4

f(x)

f(x) = x³

Q

Secant line

$\Delta f(x) = f(x + \Delta x) - f(x)$

P

Δx

x   x + Δx

P and Q, on the curve. The difference quotient for this pair of points we shall call m_s, where m stands for slope and s for secant. We have

$$m_s = \frac{\Delta f(x)}{\Delta x} = \frac{f(x + \Delta x) - f(x)}{\Delta x} = \frac{(x + \Delta x)^3 - x^3}{\Delta x}. \tag{1}$$

First we find $(x + \Delta x)^3$:

$$
\begin{aligned}
(x + \Delta x)^3 &= (x + \Delta x)(x + \Delta x)^2 \\
&= (x + \Delta x)[x^2 + 2x(\Delta x) + (\Delta x)^2] \\
&= x^3 + 2x^2(\Delta x) + x(\Delta x)^2 + x^2(\Delta x) + 2x(\Delta x)^2 + (\Delta x)^3 \\
&= x^3 + 3x^2(\Delta x) + 3x(\Delta x)^2 + (\Delta x)^3.
\end{aligned}
$$

Hence,

$$
\begin{aligned}
m_s &= \frac{(x + \Delta x)^3 - x^3}{\Delta x} \\
&= \frac{x^3 + 3x^2(\Delta x) + 3x(\Delta x)^2 + (\Delta x)^3 - x^3}{\Delta x} \\
&= \frac{3x^2 \Delta x + 3x(\Delta x)^2 + (\Delta x)^3}{\Delta x} \\
&= \frac{\Delta x[3x^2 + 3x(\Delta x) + (\Delta x)^2]}{\Delta x} \\
&= 3x^2 + (\Delta x)(3x + \Delta x).
\end{aligned}
$$

Here the difference quotient, m_s, is not constant as it is in the case of a straight line but depends, instead, upon the size of Δx and the particular value of x at hand. In other words, the slope of a secant line for two points on the *curve* in Figure 9–4 depends upon the points chosen. For a given Δx, m_s will be larger as we move to the right on the curve, which means, using a suggestive term, that the curve gets steeper as we move to the right.

> ***Exercise.*** Find the expression for the difference quotient for $f(x) = x^2$. Answer: $2x + \Delta x$.

9.9 DEFINITION OF THE DERIVATIVE

We now have the tools needed to obtain the rules of differential calculus. The problem at hand can be thought of as that of determining the slope of a line tangent to a curve at *a* point on the curve. This is a new problem because until now it has been necessary to have *two* points to compute a slope and here we have only one point. But we also have the function that determines the shape of the graph, and this shape governs the slopes of the tangents.

422

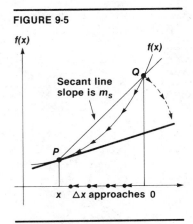

FIGURE 9-5

We start, as shown in Figure 9–5, by taking the stated point on a curve, P, and a nearby point, Q. The secant line through these points has the slope m_s. We will take the slope of the secant line,

$$m_s = \frac{\Delta f(x)}{\Delta x},$$

as an approximation of the slope of the tangent line. Clearly, the closer Q is to P, the more closely will the secant slope approximate the tangent slope. Thus, if we move Q down the curve toward P, as indicated by the arrowheads on the curve, so that Q *approaches* P, secant slopes will change and *approach* the slope of the tangent line at P.[2] Thus, as Q approaches P,

$$\text{secant line} \to \text{tangent line at } P$$

$$\text{secant slope} \to \text{tangent slope at } P$$

$$m_s = \frac{\Delta f(x)}{\Delta x} \to \text{tangent slope at } P.$$

The next step is to find a mathematical procedure that can be used to make Q approach P. Returning to Figure 9–5, note the points at the arrowheads on the x-axis. The distance from x to one of these points is a value of Δx. We generate these arrowhead points by letting Δx approach zero. The corresponding points vertically above on the curve then move from Q toward P as Δx approaches zero. Hence, as

---

[2] The reader will find it helpful to construct a curve on a full sheet of paper, and use a ruler to sketch a tangent line at a point, P. Then take a piece of thread and stretch it across P and another point Q to represent the secant line on Figure 9–5. Now, holding the thread fixed at point P, rotate the stretched thread so that it cuts the curve at points successively closer to P. Observe that as the thread, which represents the secant line, moves downward, it approaches the tangent line.

$\Delta x \rightarrow$ zero

$Q \rightarrow P$

$\dfrac{\Delta f(x)}{\Delta x} \rightarrow$ tangent slope at P.

The smaller Δx is, the closer the secant slope is to the tangent slope at P. As a consequence, it is intuitively reasonable to define the tangent slope at P to be the limit of the secant slope as Δx approaches zero.

$$\text{Tangent slope} = \lim_{\Delta x \to 0} \frac{\Delta f(x)}{\Delta x}.$$

The function derived when this limit is obtained is called the *derivative* of $f(x)$ and is symbolized as $f'(x)$.

Definition: $f'(x) =$ the derivative of $f(x)$

$$f'(x) = \lim_{\Delta x \to 0} \frac{\Delta f(x)}{\Delta x} = \lim_{\Delta x \to 0} \left[\frac{f(x + \Delta x) - f(x)}{\Delta x} \right].$$

The definition just stated is the fundamental definition of differential calculus, and all rules for finding derivatives are developed by starting from this definition. Its statement requires the use of functions, delta procedures, and limits, which we have studied earlier in the chapter. It is worth noting in particular that the need for the limit concept arises because the secant slope in Figure 9–5 is

$$m_s = \frac{\Delta f(x)}{\Delta x} = \frac{f(x + \Delta x) - f(x)}{\Delta x}.$$

If Δx is set equal to zero, m_s becomes

$$m_s = \frac{f(x + 0) - f(x)}{0} = \frac{f(x) - f(x)}{0} = \frac{0}{0},$$

so m_s is not defined at $\Delta x = 0$. What has happened here, of course, is that $\Delta x = 0$ means that on Figure 9–5, points P and Q are the same point and the slope of a line cannot be determined from one of its points. However, as we mentioned in the discussion of limits, if we let $\Delta x \rightarrow 0$, rather than set $\Delta x = 0$, and a limit exists, then this limit has an important interpretation; namely, the limit is the slope of a tangent line at a (one) point on the curve. Thus, the problem that is solved by finding $f'(x)$ is that of finding the expression that gives the slope of tangent lines when only one point on such a line is specified.

Before going on to an example, we state the following definition in order to shorten statements.

Definition: *The slope of a curve at a point means the slope of a line tangent to a curve at a point.*

Example. Find the slope of $f(x) = x^3$ at the point where $x = 0.5$.

First we must find the expression for the derivative, $f'(x)$. This will be done in steps that we shall number so that the reader can refer to them.

1. Find $f(x + \Delta x)$. For $f(x) = x^3$

$$\begin{aligned}
f(x + \Delta x) &= (x + \Delta x)^3 \\
&= (x + \Delta x)(x + \Delta x)^2 \\
&= (x + \Delta x)[x^2 + 2x(\Delta x) + (\Delta x)^2] \\
&= x^3 + 2x^2(\Delta x) + x(\Delta x)^2 + x^2(\Delta x) + 2x(\Delta x)^2 + (\Delta x)^3 \\
&= x^3 + 3x^2(\Delta x) + 3x(\Delta x)^2 + (\Delta x)^3.
\end{aligned}$$

2. Set up the difference quotient. This is

$$\begin{aligned}
\frac{\Delta f(x)}{\Delta x} &= \frac{f(x + \Delta x) - f(x)}{\Delta x} \\
&= \frac{x^3 + 3x^2(\Delta x) + 3x(\Delta x)^2 + (\Delta x)^3 - x^3}{\Delta x} \\
&= \frac{3x^2(\Delta x) + 3x(\Delta x)^2 + (\Delta x)^3}{\Delta x} \\
&= \frac{\Delta x[3x^2 + 3x(\Delta x) + (\Delta x)^2]}{\Delta x} \\
&= 3x^2 + 3x(\Delta x) + (\Delta x)^2.
\end{aligned}$$

3. Find $f'(x)$, which is the limit of the difference quotient as $\Delta x \to 0$, if a limit exists. Here we have

$$f'(x) = \lim_{\Delta x \to 0} [3x^2 + 3x(\Delta x) + (\Delta x)^2] = 3x^2.$$

FIGURE 9-6

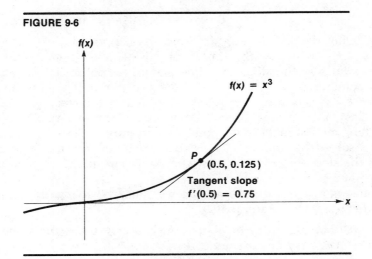

$f(x)$

$f(x) = x^3$

P
(0.5, 0.125)
Tangent slope
$f'(0.5) = 0.75$

x

Observe in the last line that Δx is the changing quantity and the terms with Δx as a factor vanish as $\Delta x \to 0$. However, the first term does not involve Δx, so it will not change, and

$$f'(x) = 3x^2.$$

The last line says the slope of

$$f(x) = x^3$$

at any point is computed as three times the square of the x-coordinate of the point. Thus, for the point where $x = 0.5$,

$$f'(0.5) = 3(0.5)^2 = 3(0.25) = 0.75.$$

Figure 9–6 shows $f(x)$ and the tangent line at the point P where

$$x = 0.5; \; f(x) = f(0.5) = (0.5)^3 = 0.125.$$

> **Exercise.** Given $f(x) = x^2$, follow the foregoing steps and write the results of *a)* Step 1. *b)* Step 2. *c)* Step 3. *d)* Compute the slope of $f(x)$ at the point where $x = 1$. Answer: *a)* $x^2 + 2x(\Delta x) + (\Delta x)^2$. *b)* $2x + \Delta x$. *c)* $2x$. *d)* 2.

We now return to the introductory example in Section 9.2. There the problem was that of finding box dimensions that provided a box of maximum surface area. The problem, illustrated in Figure 9–2, was to find x so that the slope of the tangent to

$$A = A(x) = 432x - 14x^2$$

was zero. This horizontal tangent marked the peak (highest value) of the area function in Figure 9–2. Applying the three-step procedure to find $A'(x)$, we have

1. $A(x + \Delta x) = 432(x + \Delta x) - 14(x + \Delta x)^2$
$$= 432 + 432(\Delta x) - 14[x^2 + 2x(\Delta x) + (\Delta x)^2]$$
$$= 432 + 432(\Delta x) - 14x^2 - 28x(\Delta x) - 14(\Delta x)^2.$$

2. $$\frac{\Delta A(x)}{\Delta x} = \frac{A(x + \Delta x) - A(x)}{\Delta x}$$

$$= \frac{[432x + 432(\Delta x) - 14x^2 - 28x(\Delta x) - 14(\Delta x)^2] - [432x - 14x^2]}{\Delta x}$$

$$= \frac{432(\Delta x) - 28x(\Delta x) - 14(\Delta x)^2}{\Delta x}$$

$$= \frac{\Delta x[432 - 28x - 14(\Delta x)]}{\Delta x}$$

$$= 432 - 28x - 14(\Delta x).$$

3. $A'(x) = \lim_{\Delta x \to 0} [432 - 28x - 14(\Delta x)]$

$= 432 - 28x,$

where, as before, terms with Δx as a factor vanish when $\Delta x \to 0$. Now we wish to know where the slope of the curve is zero. This is where $A'(x) = 0$.

$$A'(x) = 0 \quad \text{where} \quad 432 - 28x = 0$$
$$432 = 28x$$
$$\frac{432}{28} = x$$
$$\frac{108}{7} = x,$$

which is the result stated in Section 9.2.

We shall pause now for a brief problem set that will serve to fix the fundamental definition of the derivative in mind and then return to develop a simple rule that can be applied to take derivatives quickly.

9.10 PROBLEM SET 9–3

Apply the three steps of the delta method to determine $f'(x)$ for each of the following, and find the slope of the curve of $f(x)$ for the stated value of x.

1. $f(x) = 3x + 2$, at $x = 1$.
2. $f(x) = 2 - 0.5x$, at $x = 1$.
3. $f(x) = x^2 - 2x - 1$, at $x = 1$.
4. $f(x) = 3x^2 - 12x + 2$, at $x = 3$.
5. $f(x) = \dfrac{1}{x}$, at $x = 2$.
6. $f(x) = \dfrac{1}{x^2}$, at $x = -1$.

The definition of the derivative of $f(x) = x^2$ is

$$f'(x) = \lim_{\Delta x \to 0} \frac{(x + \Delta x)^2 - x^2}{\Delta x}.$$

Write the corresponding definition of the derivative of the following. Do not continue beyond writing the definition.

7. $f(x) = x^5 - 2x^4$.
8. $f(x) = 3x^6 + 2x^3$.
9. $f(x) = x^{1/3}$.
10. $f(x) = x^{1/2}$.

9.11 THE SIMPLE POWER RULE

There is a simple rule that can be applied to find the derivative of any power of x. To see where the rule comes from, we state some results obtained earlier. Namely,

$$\text{for } f(x) = x^2, \quad \frac{f(x + \Delta x) - f(x)}{\Delta x} = 2x + \Delta x;$$

for $f(x) = x^3$, $\dfrac{f(x + \Delta x) - f(x)}{\Delta x} = 3x^2 + \Delta x(3x + \Delta x)$.

Omitting the algebra, we state the further results that:

$$\text{for } f(x) = x^4, \frac{f(x + \Delta x) - f(x)}{\Delta x} = 4x^3 + \Delta x(\quad)$$

$$\text{for } f(x) = x^5, \frac{f(x + \Delta x) - f(x)}{\Delta x} = 5x^4 + \Delta x(\quad),$$

where, in the last two lines, $\Delta x(\quad)$ means that all remaining terms have Δx as a factor and hence will vanish in the limit,

$$f'(x) = \lim_{\Delta x \to 0} \frac{f(x + \Delta x) - f(x)}{\Delta x}.$$

Thus, taking the limit of each of the foregoing, we find that:

$$\text{if } f(x) = x^2, \ f'(x) = 2x;$$
$$\text{if } f(x) = x^3, \ f'(x) = 3x^2;$$
$$\text{if } f(x) = x^4, \ f'(x) = 4x^3.$$

We see that in each case,

$$f'(x) = (\text{power})(x \text{ to the power minus one}),$$

and this can be proved to be true for x to any power that is a constant, for all values of x where the derivative is defined. To see the need for the last clause, which implies a restriction on values of x, observe that

$$\text{if } f(x) = x^{1/2},$$

then

$$f'(x) = \frac{1}{2} x^{1/2 - 1}$$

$$= \frac{1}{2} x^{-1/2}$$

$$= \frac{1}{2x^{1/2}}$$

and note that $f'(x)$ is not defined at $x = 0$ because

$$f'(0) = \frac{1}{2(0)^{1/2}} = \frac{1}{0}$$

is not defined. *To avoid endless repetition, we shall follow convention and omit the necessary qualification that a derivative rule holds only for values of the independent variable where the derivative is defined.*

Simple power rule: If $f(x) = x^n$, $f'(x) = nx^{n-1}$.

Example. Find $f'(x)$ if: *a)* $f(x) = x^{10}$. *b)* $f(x) = x^{2/3}$. *c)* $f(x) = \dfrac{1}{x^3}$.

We have the following:

a) $n = 10$, so $f'(x) = 10x^{10-1} = 10x^9$.

b) $n = \dfrac{2}{3}$, so $f'(x) = \dfrac{2}{3}x^{2/3-1} = \dfrac{2}{3}x^{-1/3} = \dfrac{2}{3x^{1/3}}$.

c) $f(x) = \dfrac{1}{x^3} = x^{-3}$, so $n = -3$. Hence, $f'(x) = -3x^{-3-1} = -3x^{-4} = -\dfrac{3}{x^4}$.

> ***Exercise.*** Find $g'(x)$ for each of the following. *a)* $g(x) = x^7$.
> *b)* $g(x) = x^{6.5}$. *c)* $g(x) = x^{4/3}$. *d)* $g(x) = x^{1/2}$. *e)* $g(x) = x$. *f)*
> $g(x) = \dfrac{1}{x}$. Answer: *a)* $7x^6$. *b)* $6.5x^{5.5}$. *c)* $\dfrac{4}{3}x^{1/3}$. *d)* $\dfrac{1}{2x^{1/2}}$. *e)* 1. *f)* $-\dfrac{1}{x^2}$.

We have called the rule at hand the *simple* power rule to distinguish it from the *power function rule* we shall develop later and apply to cases where the base of the power is a function of x other than simply x itself. Thus, the derivative of

$$f(x) = x^9$$

is found by the simple power rule, but the derivative of

$$g(x) = (3x^2 + 2x - 7)^9$$

will be found later by the power function rule.

9.12 $\dfrac{d}{dx}$ NOTATION AND RULES OF OPERATIONS

The two-part statement,

$$\text{if } f(x) = x^3, \text{ then } f'(x) = 3x^2,$$

is awkward, and it will be convenient to use a symbol that means to take the derivative with respect to an independent variable.

Definition: $\dfrac{d}{dx}$ *means take the derivative with respect to* x.

$\dfrac{df(x)}{dx}$ *means* $f'(x)$.

Thus, the symbol d/dx is an instruction and should *never* be read as d over d times x. Instead, say "the derivative with respect to x." The symbol may be written apart from the function, as in

$$\frac{d}{dx} f(x),$$

or it may appear as

$$\frac{df(x)}{dx}.$$

We now can make simple one-part statements such as

$$\frac{d}{dx} x^3 = 3x^2; \quad \frac{d(x^4)}{dx} = 4x^3.$$

Next we state the rules that govern operations with derivatives, and give an example of each. We shall prove the first two rules and leave the third one for the reader to prove in the coming problem set.

1. If k is any constant, $\dfrac{d}{dx}(k) = 0$.

    That is, the derivative of any constant is zero. Thus, if we have

    $$f(x) = 10, \frac{df(x)}{dx} = \frac{d}{dx}(10) = 0.$$

    Graphically, this means simply that $f(x) = 10$, or $f(x) = k$, is a *horizontal* straight line, and the slope of a horizontal line is zero.

    Proof: *a)*   If $f(x) = k$, meaning $f(x)$ is the same no matter what the value of x is, then $f(x + \Delta x) = k$.

    b) $\dfrac{f(x + \Delta x) - f(x)}{\Delta x} = \dfrac{k - k}{\Delta x} = 0.$

    c) $f'(x) = \lim\limits_{\Delta x \to 0} (0) = 0$, because the limit of a constant is the constant.

2. $\dfrac{d}{dx}[kf(x)] = k\dfrac{df(x)}{dx} = kf'(x).$

    That is, a constant factor (multiplier) remains in the derivative or, we may say, a constant factor may be placed inside or outside of the derivative symbol.

    Examples: $\dfrac{d}{dx}(2x) = 2\dfrac{d}{dx}(x) = 2(1) = 2.$

    $$\frac{d}{dx}(3x) = 3\frac{d}{dx}(x) = 3.$$

    $$\frac{d}{dx}(3x^2) = 3\frac{d}{dx}(x^2) = 3(2x) = 6x.$$

Proof: Let $g(x) = kf(x)$ and apply steps a, b, and c of the foregoing to $g(x)$.

a) $g(x + \Delta x) = kf(x + \Delta x)$.

b) $\dfrac{g(x + \Delta x) - g(x)}{\Delta x} = \dfrac{kf(x + \Delta x) - kf(x)}{\Delta x}$

$$= \dfrac{k[f(x + \Delta x) - f(x)]}{\Delta x}.$$

c) $\displaystyle \lim_{\Delta x \to 0} \dfrac{g(x + \Delta x) - g(x)}{\Delta x} = \lim_{\Delta x \to 0} \dfrac{k[f(x + \Delta x) - f(x)]}{\Delta x}$

$$= k \lim_{\Delta x \to 0} \dfrac{f(x + \Delta x) - f(x)}{\Delta x}$$

$$= kf'(x),$$

which is rule number two. Note in the next to the last step that we applied the limit theorem which states that a constant factor may be placed inside or outside of the limit symbol.

3. $\dfrac{d}{dx}[f(x) \pm g(x)] = \dfrac{df(x)}{dx} \pm \dfrac{dg(x)}{dx}$.

That is, the derivative of an expression may be taken term by term. This follows from the limit theorem which says the limit may be taken term by term. (See the last problem in Problem Set 9.4.)

Example. $\dfrac{d}{dx}(5x^4 + 3x^2 + 2x + 7)$

$$= \dfrac{d}{dx}(5x^4) + \dfrac{d}{dx}(3x^2) + \dfrac{d}{dx}(2x) + \dfrac{d}{dx}(7)$$

$$= 5(4x^3) + 3(2x) + 2(1) + 0$$

$$= 20x^3 + 6x + 2.$$

Summary of rules

1.   The derivative of a constant is zero.
2.   The derivative of a constant times a function equals the constant times the derivative of the function.
3.   The derivative of an expression may be taken term by term.

Because the point seems unclear to some students, we call special attention to the fact that

$$f(x) = kx \text{ is } f(x) = kx^1.$$

By the power rule,

$$\dfrac{df(x)}{dx} = k(1)x^{1-1} = kx^0 = k.$$

That is, for example,

$$\frac{d}{dx}\left(\frac{x}{3}\right) = \frac{1}{3}, \frac{d}{dx}(5x) = 5, \frac{d}{dx}(0.2x) = 0.2, \text{ and } \frac{d}{dx}(-2x) = -2.$$

Note also that in

$$\frac{d}{dx}(3x^2 + ax - b),$$

the derivative is *with respect to x*. This means that other letters are treated as constants because x is the independent variable. Thus,

$$\frac{d}{dx}(3x^2 + ax - b) = 6x + a.$$

Again,

$$\frac{d}{dq}(3q^2 - pq + 4r) = 6q - p.$$

Exercise. Find: *a)* $\dfrac{d}{dx}(2x^{3/2} - 4x^3 + 5x - 10)$. *b)* $\dfrac{d}{dp}(2q^3 - 4p^2q^2 + 3p)$. Answer: *a)* $3x^{1/2} - 12x^2 + 5$. *b)* $-8pq^2 + 3$.

9.13 MAXIMA AND MINIMA: QUADRATIC FUNCTIONS

Functions of the general form

$$f(x) = ax^2 + bx + c, \ a \neq 0$$

are called *quadratic* functions. Examples are

$$f(x) = 3x^2 - 4x + 10$$
$$f(x) = x^2 - 16$$
$$f(x) = 4x - 12x^2$$
$$f(x) = x^2.$$

FIGURE 9-7

Vertical parabolas: $f(x) = ax^2 + bx + c$

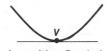

a is positive: Parabola opens upward; vertex is lowest point

a is negative: Parabola opens downward; vertex is highest point

---

Graphs of quadratics are curves called *vertical parabolas,* which may open either upward or downward as shown in Figure 9–7. As stated on the graphs, the parabola opens upward (downward) if the x^2 term is positive (negative). Thus,

$$f(x) = 3x^2 \qquad \text{opens upward;}$$
$$g(x) = 5 - 2x^2 \text{ opens downward.}$$

The highest, or lowest, point on a vertical parabola is called the *vertex.* In Figure 9–7, note that the line tangent to a parabola at its vertex is horizontal, so the slope of the parabola at the vertex is zero. The vertex is of significance because it is the maximum or minimum value the quadratic function can attain. To find the vertex, we first find the slope, $f'(x)$, then set $f'(x) = 0$ and solve for x to determine where the slope is zero.

   Example.   For $f(x) = 3x^2 - 24x + 60$: *a)* Find the coordinates of the vertex. *b)* Is this a maximum or minimum point? Why? *c)* Sketch the curve.

   a) We start with

$$f'(x) = 6x - 24.$$

Then

$$f'(x) = 0 \quad \text{where} \quad \begin{aligned} 6x - 24 &= 0 \\ 6x &= 24 \\ x &= 4. \end{aligned}$$

The value of the function when $x = 4$ is

$$\begin{aligned} f(4) &= 3(4)^2 - 24(4) + 60 \\ &= 48 - 96 + 60 \\ &= 12, \end{aligned}$$

so the vertex is the point (4, 12).

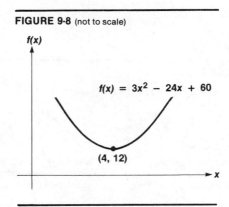

FIGURE 9-8 (not to scale)

f(x)

$f(x) = 3x^2 - 24x + 60$

(4, 12)

x

b) The parabola opens upward because the term $3x^2$ is positive. Hence, the vertex is a minimum, and the minimum value of $f(x)$ is 12.

c) All that is needed to *sketch* the function is to mark the vertex, (4, 12), and show a parabola opening upward from this point. No other points are required and the sketch need not be to scale. See Figure 9–8.

Exercise.   For $f(x) = 20x - x^2 - 80$: *a)* Find the coordinates of the vertex. *b)* Is this a maximum or minimum? *c)* Sketch the graph of $f(x)$. Answer: *a)* (10, 20). *b)* Maximum. *c)* The graph is a parabola opening downward from the vertex point (10, 20).

Parabolic cost functions: Marginal cost.   Suppose the total cost, in dollars, of producing g gallons of Galloil can be represented on the interval $0 \le g \le 150$ by the cost function

$$C(g) = 0.1g^2 + 2g + 30.$$

This function is sketched as the vertical parabola shown on Figure 9–9. In Chapter 1, and earlier in this chapter, we determined the marginal cost of, say, the 51st gallon as

$$C(51) - C(50);$$

that is, the additional cost incurred when the 51st gallon is produced. In Figure 9–9, this will be the slope of the secant line joining the points where $g = 50$ and $g = 51$. However, we should realize that a gallon is an arbitrary unit and we could as easily deal with quarts, in which case the interval from 50 to 51 gallons becomes 200 to 204

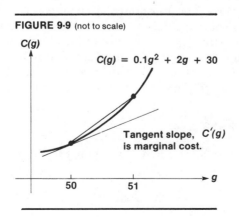

FIGURE 9-9 (not to scale)

$C(g)$

$C(g) = 0.1g^2 + 2g + 30$

Tangent slope, $C'(g)$ is marginal cost.

50   51 g

quarts. Now we could compute the marginal cost of the 201st, 202d, 203d, and 204th quarts. These again would be secant slopes, all of which will be different because our cost function curves, and all of which will be different from the marginal cost of the 51st gallon, $C(51) - C(50)$. Again, quarts are an arbitrary unit, and we could think in terms of pints, in which case each of the 8 pints in the 51st gallon would have a different marginal cost. The way out of the ambiguity caused by the fact that measuring units are arbitrary is to define marginal cost *at a point;* that is, to use the slope of the tangent line, $C'(g)$, as marginal cost.

Definition: *If $C(x)$ is the total cost of producing x units of a product, $C'(x)$ is the point marginal cost.*

Hereafter in this text, marginal cost will mean point marginal cost.
Example. If $C(g) = 0.1g^2 + 2g + 30$, find the marginal cost at 50 and 51 units of output.
Here,

$$\text{Marginal cost} = C'(g) = 0.2g + 2.$$
$$C'(50) = 0.2(50) + 2 = \$12 \text{ per gallon.}$$
$$C'(51) = 0.2(51) + 2 = \$12.20 \text{ per gallon.}$$

Exercise. If the total cost of producing p pounds of Poundum is $C(p) = 0.0015p^3 - 0.9p^2 + 200p + 60,000$, compute marginal cost at outputs of *a)* 100 pounds. *b)* 200 pounds. *c)* 300 pounds. Answer: *a)* \$65 per pound. *b)* \$20 per pound. *c)* \$65 per pound.

The major purpose of this chapter is to solve optimization problems by finding a point on a curve where the derivative (the slope of the tangent line) is zero. The foregoing discussion not only provided the calculus definition of marginal cost, but also showed that nonzero values of derivatives have important interpretations. Marginal cost is one such interpretation, and we shall present a detailed analysis of cost by the use of derivatives near the end of this chapter. Additional interpretations of the derivative appear in the next chapter.

9.14 PROBLEM SET 9–4

In Problems 1 through 18, find $f'(x)$. Do not leave negative exponents in answers.

1. $f(x) = 2 + k$.   2. $f(x) = x$.   3. $f(x) = \dfrac{x}{2}$.

4. $f(x) = 2x + 3.$

5. $f(x) = \dfrac{x}{3} + 4.$

6. $f(x) = \dfrac{2}{3} - \dfrac{3x}{2}.$

7. $f(x) = 3x^2 + 2x - 5.$

8. $f(x) = \dfrac{x^3}{3} - \dfrac{x^2}{2} + x + 12.$

9. $f(x) = 0.01x^2 + 2x + 100.$

10. $f(x) = 0.5x^3 - \dfrac{x^2}{2} + 7.$

11. $f(x) = mx + b.$

12. $f(x) = ax^2 + bx + c.$

13. $f(x) = \dfrac{2}{x} - \dfrac{1}{x^2}.$

14. $f(x) = \dfrac{1}{2x} + \dfrac{1}{x^3}.$

15. $f(x) = 2x^{3/2} - 4x^{1/2} + 3x^{2/3} + 2x - 7.$

16. $f(x) = 2x^{1/3} + x^{4/3} + 5x^{1.2} + x - 1.$

17. $f(x) = \dfrac{1}{3x} + \dfrac{2}{x^{1/2}}.$

18. $f(x) = \dfrac{3}{x^{1/3}} - \dfrac{12}{x^{1/4}}.$

Find each of the following:

19. $\dfrac{d}{dx}(3x^2 - 2x + 5).$

20. $\dfrac{d}{dy}(10y^2 - 4x + 7).$

21. $\dfrac{d}{dz}(az + b).$

22. $\dfrac{d}{dx}(3pw^2 - 2p^3).$

23. $\dfrac{d}{dh}(kh^2 - ah^{1/3} + 5ak).$

24. $\dfrac{d}{dm}\left(\dfrac{a}{m} - 3m^2 + 5a^3\right).$

Find the slope of the tangent to each of the following curves at the indicated value of x.

25. $f(x) = 3;\ x = 1.$

26. $f(x) = -5;\ x = 4.$

27. $f(x) = 2x + 6;\ x = -0.5.$

28. $f(x) = 3 - 4x;\ x = -3.$

29. $f(x) = 3x^2 - 2x + 5;\ x = 0.5.$

30. $f(x) = 10x - 2x^2 + 3;\ x = 2.5.$

31. $f(x) = 3x + \dfrac{12}{x};\ x = 2.$

32. $f(x) = 12x^{1/2} - 3x;\ x = 9.$

33. $f(x) = 8x^{1/3} + x;\ x = 8.$

34. $f(x) = \dfrac{1}{x^2} - \dfrac{1}{x} + x;\ x = 2.$

Find the value of x for which the slope is zero.

35. $f(x) = 10 - 3x^2 + 3x.$

36. $f(x) = 0.2x^2 - 40x + 50.$

37. $f(x) = 3x^{1/3} - 4x;\ x > 0.$

38. $f(x) = 3x^{1/2} - 2x.$

39. $f(x) = x + \dfrac{9}{x};\ x > 0.$

40. $f(x) = 3x + \dfrac{48}{x};\ x > 0.$

For Problems 41 through 44, find the maximum or minimum value of $f(x)$ and describe the graph of $f(x)$.

41. $f(x) = 3x^2 + 5.$

42. $f(x) = -2x^2 + 10.$

43. $f(x) = 20x - 2x^2 - 40.$

44. $f(x) = 3x^2 - 18x + 31.$

45. The profit made by a company that produces and sells x barrels of Exall is $P(x)$ dollars, where

$$P(x) = 100x - 0.01x^2 - 120,000.$$

a) What should x be if profit is to be maximized?
b) What is the amount of the maximum profit?
c) How do we know b) is the maximum profit?

46. The average cost per barrel when y barrels of Whyall are produced is $c(y)$ dollars, where

$$c(y) = 0.5y^2 - 10y + 90.$$

a) What should y be if average cost per barrel is to be minimized?
b) What is the minimum average cost per barrel?
c) How do we know b) is the minimum average cost per barrel?

47. If the total cost of producing y yards of Yardall is

$$C(y) = 0.001y^2 + 2y + 500$$

find the marginal cost at outputs of:
a) 1000 yards.   b) 2000 yards.

48. If the total cost of producing t tons of Tonal, in dollars, is

$$C(t) = 0.0005t^3 - 0.3t^2 + 100t + 30,000,$$

compute marginal cost at outputs of:
a) 100 tons.   b) 200 tons.   c) 300 tons.

49. Prove from the definition of derivatives and limit theorems that

$$\frac{d}{dx}[f(x) + g(x)] = f'(x) + g'(x).$$

9.15 TESTING STATIONARY POINTS FOR MAXIMA AND MINIMA

We know how to find the maximum or minimum of a quadratic and, by looking at its x^2 term, tell whether the point at hand is a maximum or a minimum. In this section we show how to test whether a point of zero slope is a maximum or minimum (or neither) for functions other than quadratics. We lead up to the tests by discussing stationary points, concavity, and inflection points.

The slope of a line tangent to a curve at a point is, by definition, the value of the first derivative at that point. Hence, when the first derivative is zero, the tangent line is horizontal and points where this occurs are called *stationary* points.

Definition: A stationary point on $f(x)$ is a point where $f'(x) = 0$.

Figure 9–10 shows

$$f(x) = x^3 - 6x^2 + 37.$$

To find the stationary points, we write

$$f'(x) = 0 \quad \text{where} \quad 3x^2 - 12x = 0$$
$$x(3x - 12) = 0.$$

FIGURE 9-10   (not to scale)

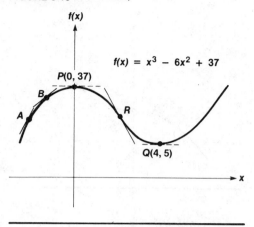

The values of x are found by setting each factor of the last expression equal to zero. Thus,

$$x = 0; \; 3x - 12 = 0$$
$$3x = 12$$
$$x = 4.$$

The function values for $x = 0$ and $x = 4$ are

$$f(0) = 0^3 - 6(0)^2 + 37 = 37$$
$$f(4) = 4^3 - 6(4)^2 + 37 = 5,$$

and the stationary points $P(0, 37)$ and $Q(4, 5)$ are shown on Figure 9–10. P is called a *local* maximum because it is the highest point in its neighborhood, but not the highest point on the curve. Similarly, Q is a *local* minimum. By way of contrast, the stationary point on a vertical parabola is the highest or lowest point on the entire curve, so it is called a *global* as well as a local maximum or minimum. The term *extreme* point means a local or global maximum or minimum. Observe also the point R on Figure 9–10. To the left of R, the curve is shaped like a cup that will spill water, and this section is said to be *concave downward*. To the right of R, the shape is like a cup that holds water and is said to be *concave upward*. The point R, which separates downward and upward concavity, is called an *inflection point*.

Next, consider the function shown in Figure 9–11,

$$g(x) = x^3 + 8.$$

We find

$$g'(x) = 3x^2$$

and

438

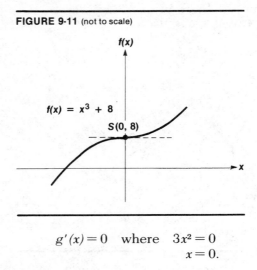

FIGURE 9-11 (not to scale)

$$g'(x) = 0 \quad \text{where} \quad 3x^2 = 0$$
$$x = 0.$$

The corresponding function value is

$$g(0) = (0)^3 + 8 = 8,$$

and the stationary point, $S(0, 8)$ is shown on Figure 9–11. By definition the tangent line is horizontal at S, as shown by the dotted line. We see that the calculus definition of a tangent leads to a situation where the tangent line crosses the curve at the point of tangency. Observe also that S is an inflection point as well as a stationary point and therefore is called a *stationary inflection point*. By way of contrast, point R on Figure 9–10 is just an inflection point because the tangent crossing the curve at R is not horizontal.

Now we consider

$$h(x) = x^4 + 5 \quad \text{and} \quad k(x) = 6 - x^4,$$

which are shown in Figures 9–12A and 9–12B. We find

$$h'(x) = 4x^3 \quad \text{and} \quad k'(x) = -4x^3,$$

so

$$h'(x) = 0 \text{ where } x = 0 \quad \text{and} \quad k'(x) = 0 \text{ where } x = 0.$$

This leads to the stationary points $T(0, 5)$ on Figure 9–12A and $V(0, 6)$ on Figure 9–12B. There is nothing unusual about these points, except that T is a global minimum and V is a global maximum, but we want to have these points and the points on Figures 9–10 and 9–11 for reference in the discussion that follows.

At a point such as A on Figure 9–10 where the slope, $f'(x)$, is positive, the tangent line slopes upward to the right, and we say the graph is rising or that $f(x)$ is increasing. For contrast, $f'(x)$ at point R is negative,

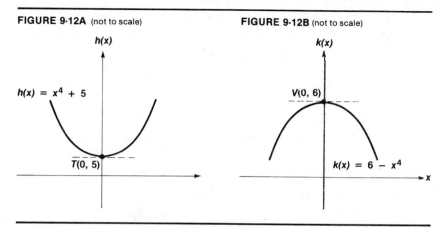

FIGURE 9-12A (not to scale)

h(x)

h(x) = x⁴ + 5

T(0, 5)

FIGURE 9-12B (not to scale)

k(x)

V(0, 6)

k(x) = 6 − x⁴

and $f(x)$ is decreasing. Of course, a point where $f'(x)$ is neither positive nor negative, that is $f'(x) = 0$, is a stationary point.

Definition: *A function is increasing (decreasing) at a point where its derivative is positive (negative).*

Now the derivative of a function, $f'(x)$, is itself a function—the slope function. Its derivative is

$$\frac{d}{dx}f'(x) = f''(x),$$

where $f''(x)$ is the *second derivative*.[3] If the derivative of $f'(x)$, which is $f''(x)$, is positive (negative), the slope function is increasing (decreasing). Now look at points A and B on Figure 9–10. The slope at B is less than the slope at A because the curve is concave downward and decreasing slopes mean the slope function is *decreasing* so its derivative, $f''(x)$, must be *negative*. In general, a curve is concave downward at a point where the second derivative is negative. This will be true, for example, at $x = -1$, $x = 0$, and $x = +1$ for the function plotted on Figure 9–10, that is,

$$f(x) = x^3 - 6x^2 + 37$$
$$f'(x) = 3x^2 - 12x$$
$$f''(x) = 6x - 12$$

---

[3] The second derivative may also be symbolized as

$$\frac{d^2f(x)}{dx^2},$$

which is read as "*d* second *f* of *x*, *dx* second."

and

$$f''(-1) = -18$$
$$f''(0) = -12$$
$$f''(1) = -6.$$

Moreover, we have

$f'(0) = \quad 0$; stationary point where $x = 0$.
$f''(0) = -12$; curve concave downward at $x = 0$.
Conclusion: There is a local maximum where $x = 0$.

Turning next to the point $Q(4, 5)$ on Figure 9–10, we find

$f'(4) = 0 \quad$; stationary point where $x = 4$.
$f''(4) = +12$; curve concave upward where $x = 4$.
Conclusion: There is a local minimum where $x = 4$.

In summary, if x^* is a value such that $f(x^*) = 0$ (stationary point), the point is a local maximum (minimum) if $f''(x^*)$ is negative (positive). We now formulate a procedure for finding possible local extreme points and testing to determine whether they are local maxima or minima.

Second derivative test

1. Find $f'(x)$, set it equal to zero, and solve for candidate values, x^*.
2. Find $f''(x)$ and evaluate $f''(x^*)$.
   a) If $f''(x^*)$ is $-$, a local maximum occurs where $x = x^*$.
   b) If $f''(x^*)$ is $+$, a local minimum occurs where $x = x^*$.
   c) If $f''(x) = 0$, the test fails to determine what happens at $x = x^*$.

The reason the test fails when $f''(x^*) = 0$ can be seen by examining Figures 9–11, 9–12A, and 9–12B in the foregoing. All of the stationary points in these figures, $S(0, 8)$, $T(0, 5)$, and $V(0, 6)$ have $x^* = 0$ and, as shown next, the second derivative is also zero at these points.

> *Exercise.* Write the expression for the second derivatives of the functions under consideration and compute the values of these derivatives at $x^* = 0$. The functions are: a) $g(x) = x^3 + 8$. b) $h(x) = x^4 + 5$. c) $k(x) = 6 - x^4$. Answer: a) $g''(x) = 6x$; $g''(0) = 0$. b) $h''(x) = 12x^2$; $h''(0) = 0$. c) $k''(x) = -12x^2$; $k''(0) = 0$.

Thus, we see that if both first and second derivatives are zero, the point at hand could be a stationary inflection point (Figure 9–11), a minimum (Figure 9–12A), or a maximum (Figure 9–12B). The second derivative test gives us no help in determining the type of point at hand if both derivatives are zero at the point. We shall discuss what to do in this case after practicing cases where the second derivative test works. Happily, the test does work in a majority of the problems we will encounter.

Example.   Find local extreme points, if any exist, and state which type of point has been found. Prove by testing.

$$f(x) = 2x^2 - 12x + 50.$$

We start with this problem because we know the answer. The curve is a vertical parabola opening upward, and its extreme point is the vertex, which is a local (and global) minimum. Formally, we write

$f'(x) = 4x - 12$ and $f'(x) = 0$ when $4x - 12 = 0$, so $x^* = 3$.
$f''(x) = 4$ and $f''(x^*) = f''(3) = 4$, which is positive.

Hence, a local minimum occurs where $x^* = 3$. The minimum value of the function is

$$f_{min} = f(3) = 2(3)^2 - 12(3) + 50 = 32.$$

Note that

$$f''(x) = 4$$

is a positive constant not depending on x. This means the curve is concave upward at all points which, of course, is the case for an upward-opening parabola.

Example.   Find local extreme points, if any exist. State the nature of points found and prove by testing.

$$f(x) = 20x^{1/2} - 2x.$$

The first derivative is

$$f'(x) = 20\left(\frac{1}{2}\right)x^{-1/2} - 2$$

$$= 10x^{-1/2} - 2.$$

Setting $f'(x) = 0$, we have

$$10x^{-1/2} - 2 = 0 \quad \text{or} \quad \frac{10}{x^{1/2}} - 2 = 0.$$

Now multiply both sides by $x^{1/2}$ to obtain

$$\left(\frac{10}{x^{1/2}}\right)x^{1/2} - 2(x^{1/2}) = 0(x^{1/2})$$

$$10 - 2x^{1/2} = 0$$
$$10 = 2x^{1/2}$$
$$5 = x^{1/2}.$$

To obtain x we must *square* both sides.

$$(5)^2 = (x^{1/2})^2$$
$$25 = x.$$
$$x^* = 25.$$

To test x^*, we start with

$$f'(x) = 10x^{-1/2} - 2$$

and obtain the second derivative,

$$f''(x) = 10\left(-\frac{1}{2}\right)x^{-3/2} = -5x^{-3/2}$$

$$f''(x^*) = f''(25) = -5(25)^{-3/2} = -\frac{5}{(25)^{3/2}}.$$

We need not complete the calculation because it is sufficient to note that the result is negative, showing we have a local maximum at the point where $x^* = 25$, and

$$f_{max} = f(25) = 20(25)^{1/2} - 2(25)$$
$$= 20(5) - 2(25)$$
$$= 50.$$

Exercise. Find the local extreme point of the function and prove which type it is by testing. *a)* $f(x) = 30x - 3x^2 + 25$. *b)* $g(x) = x - 6x^{1/2} + 20$. Answer: *a)* $x^* = 5$, $f(5) = 100$ is a local maximum because $f''(5) = -6$. *b)* $x^* = 9$, $g(9) = 11$ is a local minimum because $g''(9) = 3/[2(9)^{3/2}]$ is positive.

End point considerations. In applied problems there may be a restriction on the *range* of permissible values the independent variable may have. Typically, x must not be negative $(x \geq 0)$, or x must be strictly positive $(x > 0)$. Again, if x is the number of hours a machine is operated during a day, then

$$\text{range: } 0 \leq x \leq 24.$$

When we find a local extreme point by setting the derivative equal to zero, the x^* obtained may be outside the range. For example, if

$$f(x) = 0.01x^2 - 1.2x + 100; \ 0 \leq x \leq 50$$

it may be verified that $x^* = 60$, $f(x^*) = 64$ would be a local minimum, but this is outside the range, and the minimum (global) value of the function occurs at the end point, 50, and is $f(50) = 65$.

Secondly, the value of a function at the end point of the range may be greater (less) than the value found at a local maximum (minimum) and, finally, a function may have no *local* extreme points, in which case its global maximum and minimum occur at the range end points.

It follows that in working with an unfamiliar function it would be wise to keep end point considerations in mind. However, in this book, problems have been designed so that the reader need not be concerned with end points, and we shall call attention to these considerations on the rare occasion where they are relevant.

9.16 ALTERNATIVE TESTS

We need an alternative procedure when the first derivative test fails; that is, when both $f'(x^*)$ and $f''(x^*)$ equal zero. Moreover, in some problems encountered later, the expression for $f''(x)$ can be quite complicated and in such cases it is simpler to use an alternative even though the second derivative test would work. One alternative is the *original function test*. Here we evaluate the original $f(x)$ a *little* to the left and right of the candidate value, x^*, where "a little" means closer to x^* than the x-coordinate of any other stationary point. Recall that a stationary point can be a local maximum (Figure 9–13A), a local minimum (Figure 9–13B), or a stationary inflection point (Figure 9–13C). The stationary point occurs where $x = x^*$, with l and r being, respectively, points a little to the left and to the right of x^*. It is clear that:

At a local maximum (13A): both $f(l)$ and $f(r)$ are less than $f(x^*)$.
At a local minimum (13B): both $f(l)$ and $f(r)$ are greater than $f(x^*)$.
At an inflection (13C): $f(l) > f(x^*)$ and $f(r) < f(x^*)$.
   or (not shown): $f(l) < f(x^*)$ and $f(r) > f(x^*)$.

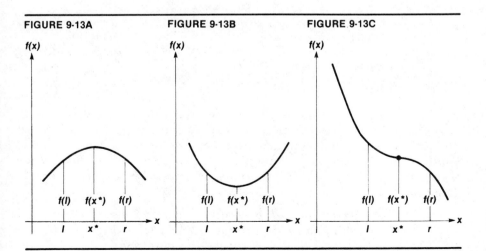

FIGURE 9-13A   FIGURE 9-13B   FIGURE 9-13C

In brief, both less means a maximum, both greater means a minimum, and neither of the foregoing means a stationary inflection.

Example. Find local maxima or minima, if any, and state which has been found. Prove by testing.

$$f(x) = 2x^3 - 18x^2 + 54x - 45.$$

We have

$$f'(x) = 6x^2 - 36x + 54$$
$$f'(x) = 0 \text{ where } 6x^2 - 36x + 54 = 0$$
$$6(x^2 - 6x + 9) = 0$$
$$6(x - 3)(x - 3) = 0$$
$$x^* = 3.$$
$$f''(x) = 12x - 36; \quad f''(x^*) = f''(3) = 12(3) - 36 = 0.$$

The second derivative test fails, so we select a point $l = 2$ (to the left of $x^* = 3$) and a point $r = 4$ (to the right) and calculate:

$f(x) = 2x^3 - 18x^2 + 54x - 45$		
$f(l) = f(2)$	$f(x^*) = f(3)$	$f(r) = f(4)$
$2(2)^3 - 18(2)^2 + 54(2) - 45$ $= 7$	$2(3)^3 - 18(3)^2 + 54(3) - 45$ $= 9$	$2(4)^3 - 18(4)^2 + 54(4) - 45$ $= 11$

Conclusion: There is a stationary inflection point where $x^* = 3$.

Exercise. Find the local maxima or minima, if any, and state which has been found. Prove by testing.

$$f(x) = x^3 - 3x^2 + 3x + 5.$$

Answer: No local extreme points exist. There is a stationary inflection where $x^* = 1$, $f(1) = 6$. $f(l)$ is less than $f(1)$ and $f(r)$ is greater than $f(1)$.

The *first derivative test* is a second alternate. From Figure 13, we can see (13A) that the slope is positive to the left and negative to the right of a local maximum, and the opposite is true at a local minimum (13B). However, the slope has the *same* sign on both sides of a stationary inflection (13C). The slope, of course, is given by $f'(x)$. Hence:

At a local maximum: $f'(l)$ is positive and $f'(r)$ is negative.
At a local minimum: $f'(l)$ is negative and $f'(r)$ is positive.
At an inflection: $f'(l)$ and $f'(r)$ have the same sign.

Example. Apply the first derivative test to the last example. Copying from the last example,

$$f(x) = 2x^3 - 18x^2 + 54x - 45$$
$$f'(x) = 6x^2 - 36x + 54$$
$$f'(x) = 0 \text{ at } x^* = 3.$$

$f'(l) = f'(2)$	$f'(x^*) = f'(3)$	$f'(r) = f'(4)$
$6(2)^2 - 36(2) + 54$ $= 6$	$6(3)^2 - 36(3) + 54$ $= 0$	$6(4)^2 - 36(4) + 54$ $= 6$

Conclusion: $x^* = 3$ is at stationary inflection point because $f'(l)$ and $f'(r)$ have the same sign.

The second derivative test works in the majority of the problems in this book and is the easiest to use. The alternative tests will be required in the questions in the next problem set, for practice, but alternative tests will seldom be needed as we proceed.

9.17 PROBLEM SET 9–5

1. When x gallons of Exall are produced, the average cost per gallon is $A(x)$ where

$$A(x) = \frac{100}{x} + 0.04x + 1.$$

   a) How many gallons should be produced if average cost per gallon is to be minimized?
   b) Compute the minimum average cost per gallon.
   c) Prove b) is a minimum.

2. If we use 100 linear feet of fence to enclose a rectangular plot of land, we need $2x$ feet for one pair of sides (or x feet for one of these sides), leaving $100 - 2x$ feet for the other pair of sides (or $50 - x$ feet for one of these sides). The rectangle is then x feet by $(50 - x)$ feet and its area, $A(x)$, is

$$A(x) = x(50 - x) = 50x - x^2.$$

   a) Find the value of x that maximizes the area of the plot.
   b) Prove a) yields a maximum.
   c) What are the dimensions of the maximum area rectangle?
   d) What is the maximum area?

3. In many sampling surveys, only a small number (n) of a large population of people are interviewed. Published survey results state what proportion (p) of the n people interviewed answered yes to a particular question. Inherently, because only a fraction of the population was interviewed, the proportion p is expected to be in error. Statisticians measure this error by $V(p)$, where V stands for *variance*, and

$$V(p) = \frac{p}{n} - \frac{p^2}{n}.$$

Thus variance or error is a function of p for a *given* n.

a) Find the value of p that maximizes $V(p)$; that is, the proportion that leads to the largest expected error.

b) Prove *a)* is a maximum.

c) Taking $n = 10$, find $V(0.5)$ and $V(0.1)$.

4. The profit realized when y gallons of Whyall are made and sold is

$$P(y) = 20y - 0.005y^2.$$

a) Find the number of gallons that should be made to maximize profit.

b) Prove *a)* yields a maximum.

c) Compute the maximum profit.

For each of the following, find the coordinates of local extreme points, if any exist. In each case, state the type of point which has been found and prove the statement by showing a test.

5. $f(x) = 5x - x^2$.

6. $f(x) = 2x^2 - 12x + 20$.

7. $f(x) = 3 - x^2$.

8. $f(x) = 2x^2 + 20$.

9. $f(x) = 30x - 3x^2 + 10$.

10. $f(x) = x^2 - x + 1$.

11. $f(x) = x^3 - 12x^2 + 12$.

12. $f(x) = x^3 - 3x^2 + 2$.

13. $f(x) = 3x^3 + 27$.

14. $f(x) = 14 - 2x^3$.

15. $f(x) = 21x^2 - 2x^3 - 60x + 20$.

16. $f(x) = 6x^2 - x^3 - 9x$.

17. $f(x) = x^3 - 9x^2 + 27x$.

18. $f(x) = x^3 + 15x^2 + 75x$.

19. $f(x) = x^4 - 32x + 100$.

20. $f(x) = 108x - x^4$.

21. $f(x) = 4x^3 - 3x^4 + 4$.

22. $f(x) = x^4 - 4x^3 + 30$.

23. $f(x) = 2x - 8x^{1/2}$.

24. $f(x) = 9x^{2/3} - 2x$.

25. $f(x) = x^{3/2} - 12x^{1/2}$.

26. $f(x) = 10x^{1/3} - x^{2/3}$.

27. $f(x) = 4x + \dfrac{64}{x}$.

28. $f(x) = 4x + \dfrac{250}{x^2}$.

9.18 MAXIMA AND MINIMA: APPLICATIONS

In this section we present a series of examples that involve the application of calculus rules and procedures developed in earlier sections.

Example 1. A rectangular warehouse with a flat roof is to have a floor area of 9600 square feet. The interior is to be divided into store-room and office space by an interior wall parallel to one pair of the sides of the building. The roof and floor areas will be 9600 square feet for any building, but the total wall length will vary for different dimensions. For example, a 96 by 100 foot building could have a 96-foot interior wall, two 96-foot and two 100-foot exterior walls for a total length of 3(96) + 2(100) = 488 feet.

> **Exercise.**   If the building dimensions are 40 × 240 = 9600 square feet, what will be the total length of wall if the interior wall is 40 feet long? Answer: 3(40) + 2(240) = 600 feet.

The problem is to find the dimensions that minimize the total amount of wall. Letting x and y be the dimensions as in Figure 9–14, we see that the total amount of wall, w, is

$$w = 3x + 2y,$$

subject to the constraint that

$$xy = 9600 \quad \text{or} \quad y = \frac{9600}{x}.$$

FIGURE 9-14

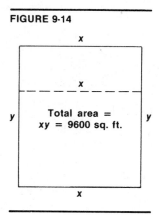

As first stated, w depends upon *two* variables, x and y, and our calculus to this point deals with only *one* independent variable. However, since $xy = 9600$, we can replace y by $9600/x$ and write w as a function of the single variable, x. Thus,

$$w(x) = 3x + 2\frac{(9600)}{x}.$$

We next find

$$w'(x) = 3 - \frac{2(9600)}{x^2}$$

and determine stationary points by setting $w'(x) = 0$, thus:

$$3 - \frac{2(9600)}{x^2} = 0 \quad \text{so} \quad x^2 = \frac{2(9600)}{3} = 6400.$$

$$x = \pm 80.$$

We discard $x = -80$ because x must be positive in this problem. To determine the type of stationary point at $x = 80$, we find the second derivative,

$$w''(x) = \frac{4(9600)}{x^3}$$

and note that $w''(80)$ is positive, so we have a local minimum. Recalling that $y = 9600/x = 9600/80 = 120$, we find that the building dimensions should be 80×120, with the interior wall being 80 feet long. The minimum total wall length is

$$w = 3x + 2y = 3(80) + 2(120) = 480 \text{ feet.}$$

Exercise. Suppose the floor area is to be 10,000 square feet, with no interior wall. *a)* What is the function $w(x)$? *b)* What dimensions minimize $w(x)$? Answer: *a)* $w(x) = 2x + 20,000/x$. *b)* A square building, 100 by 100 feet.

Example 2. A rectangular plot of land is to be enclosed by a fence. Fence for the horizontal sides costs $10 per running foot, while that for the vertical sides costs $5 per running foot. What is the maximum area that can be enclosed if $1500 is available for purchasing the fence?

We shall carry through the solution of this problem by listing questions and answers in a step-by-step sequence in order to provide a reference framework for solving optimization problems. We shall use x and y, respectively, for the horizontal and vertical dimensions, and A for the area.

1) What quantity is to be maximized? Answer: The area, A.
2) What is the formula for this quantity? Answer: $A = xy$.
3) How can this quantity be made a function of a single variable? Answer: By expressing y in terms of x, or vice versa and substituting into (2).
4) What information is available to accomplish (3)? Answer: We know the total amount to be spent is $1500 and this must equal the cost of the two horizontal sides, which will be $2(\$10x)$ plus the cost of the vertical sides which will be $2(\$5y)$; hence,

$$2(10x) + 2(5y) = 1500 \quad \text{or} \quad 20x + 10y = 1500.$$

Therefore,

$$y = \frac{1500 - 20x}{10} = 150 - 2x.$$

5) Express the quantity to be optimized as a function of a single variable. Answer: $A(x) = x(150 - 2x) = 150x - 2x^2$.
6) Find the first derivative, set it equal to zero, and solve. Answer: $A'(x) = 150 - 4x$ is zero when $x = 150/4 = 37.5$.
7) Test the result in (6) by the second derivative. Answer: $A''(x) = -4$ is always negative, so we have a local maximum.
8) Evaluate the remaining variable, and find the optimum. Answer: $y = 150 - 2(37.5) = 75$. The dimensions of the maximum area enclosure are 37.5 by 75 feet, and the maximum area is 2,812.5 square feet.

Example 3. *Parameterizing a model.* When we write $y = mx + b$ to represent the equation of a straight line, the arbitrary letters m and b represent constants (slope and y-intercept) that distinguish one line from another. Such *arbitrary constants* (that is, letters standing for constants) are called *parameters.* Now return to Example 2 and let $\$h$ (instead of 10) be the cost per horizontal foot, and $\$v$ (instead of 5) be the cost per vertical foot.

Similarly, let $\$C$ (instead of 1500) represent the total cost. We would then have

$$C = 2x(\$h) + 2y(\$v)$$

from which

$$y = \frac{C - 2xh}{2v} \tag{1}$$

and

$$A(x) = x\left(\frac{C - 2xh}{2v}\right). \tag{2}$$

The model for $A(x)$ has now been parameterized, and we seek the optimal solution in terms of the parameters C, h, and v. We have from (2)

$$A(x) = \frac{xC}{2v} - \frac{x^2h}{v}.$$

Remembering that the parameters represent constants, we find

$$A'(x) = \frac{C}{2v} - \frac{2xh}{v} \tag{3}$$

and $A'(x) = 0$ when

$$\frac{C}{2v} - \frac{2xh}{v} = 0.$$

Multiplying by $2v$, we have

$$C - 4xh = 0 \quad \text{or} \quad x = \frac{C}{4h}. \tag{4}$$

From (1) we obtain

$$y = \frac{C - 2xh}{2v} = \frac{C - \left(\frac{2C}{4h}\right)h}{2v} = \frac{C - \frac{C}{2}}{2v}$$

and, multiplying numerator and denominator of the last by 2, we find

$$y = \frac{2C - C}{4v} = \frac{C}{4v}. \tag{5}$$

The parameterized expression for the area is, from (4) and (5),

$$A = xy = \left(\frac{C}{4h}\right)\left(\frac{C}{4v}\right) = \frac{C^2}{16hv}.$$

To show that we have a local maximum, we start with the first derivative (3), and take the second derivative, which is

$$A''(x) = \frac{-2h}{v}.$$

$A''(x)$ is negative because h and v, which are costs, are positive, so we have a local maximum.

The obvious advantage of using parameters instead of specific numbers in solving a problem of a given type is that the solution of the parameterized model is a *general* solution that covers all specific cases. For the problem type under discussion, we see that if $C = \$1500$ is to be spent, with $h = \$10$ and $v = \$5$, then from (4) and (5), the optimal dimensions are

$$x = \frac{C}{4h} = \frac{1500}{4(10)} = 37.5 \text{ feet}$$

and

$$y = \frac{C}{4v} = \frac{1500}{4(5)} = 75 \text{ feet}.$$

Exercise. If, in Example 2, the horizontal and vertical cost per foot are $2 and $3, and $2400 is to be spent, what dimensions will maximize the area? Answer: 300 by 200 feet.

Example 4. A rectangular manufacturing plant with a floor area of 16,875 square feet is to be built on a straight road. The front of the plant must be set back 60 feet from the road, and buffer strips (grass and trees) 30 feet on each side and 20 feet at the back must be provided, as shown in Figure 9–15. Because of the high cost of

FIGURE 9-15

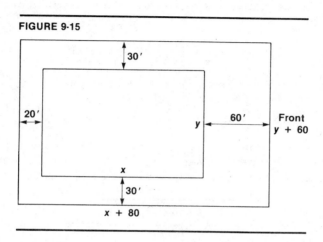

land, builders seek to minimize the total area, plant plus buffers. This area is

$$A = (x + 80)(y + 60).$$

Since the *plant* area is 16,875, we have

$$xy = 16875 \quad \text{so} \quad y = \frac{16875}{x}.$$

Hence,

$$A(x) = (x + 80)\left(\frac{16875}{x} + 60\right)$$

$$= 16875 + 60x + \frac{80(16875)}{x} + 4800.$$

We find

$$A'(x) = 60 - \frac{80(16875)}{x^2},$$

and $A'(x)$ is zero when

$$60 - \frac{80(16875)}{x^2} = 0$$

$$x^2 = \frac{80(16875)}{60} = 22500,$$

$$x = 150 \text{ feet.}$$

The second derivative test shows this dimension will minimize $A(x)$. It follows that

$$y = \frac{16875}{x} = \frac{16875}{150} = 112.5 \text{ feet.}$$

The plant dimensions should be 112.5 feet at the front by 150 feet deep. The total land dimensions will then be

$$x + 80 = 230 \text{ feet by } y + 60 = 172.5 \text{ feet.}$$

The minimal total land area is $(230)(172.5) = 39{,}675$ square feet.

Example 5. Figure 9–16 shows a box with square top and bottom and rectangular sides. The total surface area of this box is $2x^2$ for

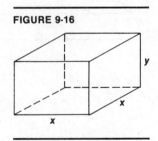

FIGURE 9-16

top and bottom, plus $4xy$ for the four sides. Thus,

$$A = 2x^2 + 4xy.$$

The volume of the box is the area of the base times the altitude. Thus,

$$V = x^2 y.$$

Suppose that we require a box of volume 2592 cubic inches, and we seek to minimize the cost of materials for the sides, top, and bottom. Side material costs 6 cents per square inch, and top and bottom material 9 cents per square inch. Total cost, C, will be

$C = $ (top and bottom area) at 9 cents + (four side areas) at 6 cents

$C = (2x^2)(9) + (4xy)(6).$

From

$$V = x^2y$$

or

$$2592 = x^2y$$

we have

$$y = \frac{2592}{x^2}.$$

Substituting for y in the cost equation yields

$$C(x) = 18x^2 + \frac{24(2592)}{x}.$$

The derivative is

$$C'(x) = 36x - \frac{(2592)(24)}{x^2},$$

and $C'(x) = 0$ when

$$36x = \frac{(2592)(24)}{x^2}.$$

Multiplying by x^2 we find

$$36x^3 = (2592)(24)$$

or

$$x^3 = 1728.$$

Hence,

$$x = (1728)^{1/3} = 12 \text{ inches.}$$

It follows that

$$y = \frac{2592}{x^2} = \frac{2592}{144} = 18 \text{ inches.}$$

As the following exercise shows, the cost function has a local minimum at $x = 12$ inches and $y = 18$ inches.

Exercise. For the foregoing: *a)* Find the cost of a 12 by 12 by 18 inch box. *b)* Prove this cost is a minimum by the second derivative test. Answer: *a)* 7776 cents, or \$77.76. *b)* $C''(x) = 36 + (48)(2592)/x^3$ is positive when $x = 12$.

Example 6. According to United Parcel Service requirements stated at the opening of the chapter, the length plus girth of a package must not exceed 108 inches. Suppose packages are to be cylinders, as shown in Figure 9–17, and we seek the dimensions (length L and

FIGURE 9-17

radius r) that will yield maximum volume. The volume of a cylinder is the circular base area, πr^2, times the height L, so

$$V = \pi r^2 L.$$

The girth of the cylinder is the perimeter of the circular cross-section, $2\pi r$, so length plus girth is $L + 2\pi r$ and

$$L + 2\pi r = 108 \quad \text{or} \quad L = 108 - 2\pi r.$$

Hence, substituting

$$V(r) = \pi r^2 (108 - 2\pi r) = 108\pi r^2 - 2\pi^2 r^3.$$

The derivative is

$$V'(r) = 216\pi r - 6\pi^2 r^2.$$

Setting $V'(r) = 0$,

$$6\pi r(36 - \pi r) = 0$$

from which

$$r = 0 \quad \text{and} \quad r = \frac{36}{\pi}.$$

The second derivative is

$$V''(r) = 216\pi - 12\pi^2 r$$

and

$$V''\left(\frac{36}{\pi}\right) = 216\pi - 12\pi^2\left(\frac{36}{\pi}\right) = -216\pi,$$

so $r = 36/\pi$ yields a local maximum. Next,

$$L = 108 - 2\pi\left(\frac{36}{\pi}\right) = 36$$

so the optimal dimensions for maximum volume are a radius of $36/\pi$ (about 11.5 inches) and a height of 36 inches. The maximum volume will be

$$V = \pi \left(\frac{36}{\pi}\right)^2 (36) = \frac{(36)^3}{\pi} = \frac{46656}{\pi} = 14{,}851 \text{ cubic inches.}$$

Exercise. In the last example, suppose the limitation on length plus girth is the parameter, M inches. Express the dimensions r and L for maximum volume in terms of M. Answer: $r = M/3\pi$ and $L = M/3$.

Example 7. In normal operations, a plant employs 100 workers working 8 hours a day for a total of $100(8) = 800$ labor-hours of work per day. In normal operations, productivity averages 30 units per labor-hour worked. Thus, in a normal day, production is

$$P = \text{(labor-hours worked)(output per labor-hour)}$$
$$= (800)(30) = 24{,}000 \text{ units.}$$

If the work level (labor-hours) is raised above 800, management estimates that average output per labor-hour falls off at the rate of 2.5 units for each extra 100 labor-hours, or by 0.025 units for each labor-hour in excess of 800. For example, if 840 labor-hours are worked, the excess would be $840 - 800 = 40$ and average productivity would be $30 - 40(0.025) = 29$ units per labor-hour. Total output would then be

$$P = 840(29) = 24{,}360 \text{ units.}$$

The problem is to find the work level, x labor-hours, that maximizes output. Reviewing the illustrative calculations, we find that for $x = 840$ labor-hours,

$$P(840) = 840[30 - (840 - 800)(0.025)].$$

In general,

$$P(x) = x[30 - (x - 800)(0.025)]$$
$$= 30x - 0.025x^2 + 20x$$
$$= 50x - 0.025x^2.$$

We find

$$P'(x) = 50 - 0.05x$$

and

$$P'(x) = 0 \quad \text{when} \quad x = 1000 \text{ labor-hours.}$$

Also

$$P''(x) = -0.05$$

so we have a local maximum, which is

$$P(1000) = 50(1000) - 0.025(1000)^2 = 25,000 \text{ units.}$$

9.19 EVALUATION OF PARAMETERS

Students often ask where the values for parameters used in mathematical models are obtained. Even a partial answer to this question would require more pages than are appropriate in this text, so we shall confine ourselves to a few comments on matters suggested by the last example. Here it was assumed that a single product was involved, so reasonably accurate data for labor-hours worked and output would be available in regularly kept payroll and production records. These and other regularly kept records may be direct sources for parameter values. However, we often have to adapt available data to our purposes, or arrange for the collection of data that is not part of the regular information system. Thus, in the last example, it is not likely that the parameter 0.025, representing productivity decline in response to increasing work level, would be available in the information system. We could, however, secure available data on labor-hours and output for periods of different work (labor-hour) levels and try to learn what happens to output per labor-hour as the work level changes. A standard method for doing this is to graph the data as illustrated in Figure 9–18.

As is the case in most applied problems, the plotted points are scattered and we try to represent the central trend of the scatter using,

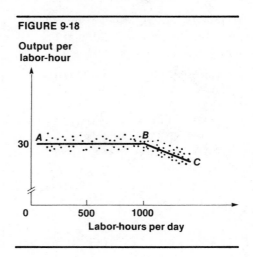

FIGURE 9-18

in this case, two straight line segments.[4] The trend of *AB* on Figure 9–18 is horizontal at 30 units per labor-hour, and the slope of the trend segment *BC* would be the estimate of productivity decline, which was the value −0.025 in Example 7.

The thumbnail sketch just given has in it a number of elements on parameter evaluation, but it oversimplifies what can be a difficult problem. The parameter values we are able to obtain range from very rough to highly accurate approximations and, at times, properly could be described as guesses. It is axiomatic that good information is needed for good decision making, and one of the challenges that face managers of the future as computer technology moves forward is to provide the data base from which reliable estimates of key parameters can be obtained efficiently.

9.20 AN INVENTORY MODEL

Most readers probably have seen advertisements featuring inventory clearance sales. Such sales serve to emphasize that it costs money to carry a stock in inventory. *Carrying* costs include the cost of warehouse space, record keeping, insurance, damage losses, and obsolescence. Additionally, inventory often is acquired with borrowed money, and interest charges on this money can be a significant cost. In this model, we consider the case of a manufacturer who produces a product in batches, or lots, periodically, places the produced lot in inventory, then sells from this inventory until it is exhausted and a new lot is produced. On the production side, the manufacturer can achieve the cost economies of mass production if large lots are produced, but large lots will incur higher inventory carrying costs than small lots. The problem at hand, then, is to determine what lot size, *L*, should be produced to obtain the optimal (minimum-cost) balance between production costs and inventory carrying costs.

To make the problem more specific, suppose that a manufacturer plans to produce 98,000 units of a product during a year. A lot, of size *L*, to be determined, is to be made periodically, and every time a lot is made it is necessary to set up the appropriate machinery and other production facilities before production starts. The *setup cost* is then a fixed cost incurred for each lot produced. Let us suppose this setup cost is $500. When production commences, the cost of making a unit is constant at $5 per unit. Inventory cost is to be determined on the basis that it costs $0.50 per year to carry one unit in inventory. However, when a lot is made and placed in inventory, there are *L*

---

[4] Generally, in practice, lines and curves are fitted to actual data by the least-squares method (see Index for reference). The procedure shows how to manipulate the data and obtain the desired parameters which, in the case of a straight line, are the slope and *y*-intercept.

units in inventory, but as the product is sold during the interval until production of the next lot, the number of units in inventory decreases to zero. Thus the largest and smallest numbers of units in inventory are, respectively, L and 0, and we shall assume that on the average,

$$\frac{L+0}{2} = \frac{L}{2}$$

units are carried in inventory during the year. We assign parameters as follows:

Number of units to be made in a year: $N = 98,000$
Number of units to be made in each lot: L
Fixed setup cost per lot: $F = \$500$
Variable cost per unit made: $v = \$5$
Annual inventory carrying cost per unit
in average inventory: $i = \$0.50$
Average inventory during a year: $\dfrac{L}{2}$.

We now determine the expressions for costs incurred. First we note that if, for example, each lot contains 14,000 units, then to make 98,000 units in a year, the number of lots required would be

$$\text{Lots per year} = \frac{\text{Units per year}}{\text{Units per lot}} = \frac{98,000}{14,000} = 7.$$

Consequently, if the optimal lot size (to be determined) is L, we would have

$$\text{Lots per year} = \frac{98,000}{L} = \frac{N}{L}.$$

Each time a lot is made, the setup cost is $F = \$500$, so the total setup cost for a year will be

Total setup cost per year $=$ (Lots per year)(Setup cost per lot)

$$= \frac{N}{L}\,(F) = \frac{98,000(500)}{L}. \tag{1}$$

Next we consider total annual inventory cost, which is the number of units carried in inventory during the year, on the average, times the carrying cost per unit.

$$\text{Inventory carrying cost for year} = \left(\frac{L}{2}\right)i = \frac{0.50L}{2}. \tag{2}$$

Finally, no matter what the lot size, 98,000 units will be made after setups during the year, and add a cost of 98,000 units times $v = \$5$ per unit to total cost. Thus:

$$\text{Total variable cost for year} = vN = 5(98{,}000). \tag{3}$$

The sum of the costs (1), (2), (3), is the total cost for the year, and this, a function of L, is

$$C(L) = \frac{98{,}000(500)}{L} + \frac{0.50L}{2} + 5(98{,}000) \tag{4}$$

or, in parameterized form,

$$C(L) = \frac{NF}{L} + \frac{iL}{2} + vN. \tag{5}$$

Seeking to minimize $C(L)$ in (5), we find

$$C'(L) = -\frac{NF}{L^2} + \frac{i}{2}$$

and $C'(L) = 0$ when

$$-\frac{NF}{L^2} + \frac{i}{2} = 0 \quad \text{so} \quad L = \sqrt{\frac{2NF}{i}}.$$

Inasmuch as

$$C''(L) = \frac{2NF}{L^3}$$

is positive for any applied problem (that is, N, F, and L are positive), we have a local minimum. For the illustrative parameter values,

$$L = \sqrt{\frac{2(98{,}000)(500)}{0.5}} = \sqrt{196{,}000{,}000} = 14{,}000 \text{ units/lot.}$$

It follows that the firm would make $98{,}000/14{,}000 = 7$ lots each year or a lot every $365/7$ days, that is, a lot every 52 days.

   Exercise. The annual requirement for another product made by the foregoing firm is 7,200 units. The setup cost per batch is \$100 and the inventory cost is \$1 per unit in average inventory. *a)* What batch size will minimize total annual cost? *b)* How often should a lot be made? Answer: *a)* 1,200 per lot. *b)* Six lots will be made in a year, or a lot every two months.

In the model

$$C(L) = \frac{NF}{L} + \frac{iL}{2} + vN$$

note that vN is constant, so its derivative is zero. Consequently, the parameter v does not appear in the optimal solution. Notice also that when L is small, the setup cost term NF/L is large but the inventory cost $iL/2$ is small, and the reverse holds when L is large. The optimal balance of the cost terms occur when

$$L = \sqrt{\frac{2NF}{i}}.$$

Finally, observe that the optimal L is not proportional to N. That is, if the annual requirement was reduced by 19 percent to 81 percent of its old value, so that the new requirement is $0.81N$, then

$$L = \sqrt{\frac{2(0.81N)F}{i}} = 0.9\sqrt{\frac{2NF}{i}}$$

so the optimal lot size is now 0.9 or 90 percent of (or 10 percent below) its old value.

Exercise. If the annual requirement is cut to ¼ its old value, how would this affect the optimal lot size? Answer: The optimal lot size would now be ½ the old value.

9.21 PROBLEM SET 9–6

1. A rectangular warehouse is to have 3,300 square feet of floor area and is to be divided into two rectangular rooms by an interior wall. Cost per running foot is $125 for exterior walls and $80 for the interior wall.
   a) What dimensions will minimize total wall cost?
   b) What is the minimum cost?

2. (Similar to Problem 1, but the area is to be found.) If $49,500 has been allocated for walls,
   a) What are the dimensions of the largest warehouse that can be built?
   b) What is the floor area of this warehouse?

3. (Problem 1 in parameterized form.) The floor area is to be A square feet and the cost per running foot is e for exterior walls and i for the interior wall. Let x and y be the warehouse dimensions, with x being the length of the interior wall.
   a) Write the expression for total wall cost, $C(x)$.
   b) Find the expression for x that minimizes $C(x)$.

4. (Problem 2 in parameterized form.) If D are allocated for wall construction then, using the parameters in Problem 3,
   a) Write the expression for the enclosed area, $A(x)$.
   b) Find the expression for x that will maximize $A(x)$.

5. Both interior and exterior walls of a 13,500-square-foot rectangular warehouse cost $100 per running foot. The warehouse is to be divided into

eight rooms by three interior walls running in the x direction and one running in the y direction.

    a) What dimension will lead to minimal total wall cost?

    b) What is this minimal cost?

6. A rectangular area of 1050 square feet is to be enclosed by a fence, then divided down the middle by another piece of fence. The fence down the middle costs $0.50 per running foot, and the other fence costs $1.50 per running foot. Find the minimum cost for the required fence.

7. Fence is required on three sides of a rectangular plot. Fence for the two ends costs $1.25 per running foot; fence for the third side costs $2 per running foot. Find the maximum area that can be enclosed with $100 worth of fence.

8. A rectangular cardboard poster is to contain a 96-square-inch rectangular section of printed material, have a 2-inch border top and bottom, and a 3-inch border on each side. Find the dimensions and area of the smallest poster that meets these specifications. (Note: Let x and y be the dimensions of the 96-square-inch area).

9. A rectangular-shaped manufacturing plant with a floor area of 600,000 square feet is to be built in a location where zoning regulations require buffer strips 50 feet wide front and back, and 30 feet wide at either end. (A buffer strip is a grass and tree belt that must not be built upon.) What plot dimensions will lead to minimum total area for plant and buffer strips? What is this minimum total area? (Note: Let x and y be the dimensions of the 600,000 square foot area.) If the plant dimensions were made 1500 by 400 feet rather than the dimensions leading to minimum area, by how much would the total plot area exceed the minimum area?

10. In the United Parcel Service example at the beginning of the chapter, the length plus girth of a package was restricted to 108 inches. Suppose a shipper uses rectangular boxes with square ends made of a perforated material to provide ventilation to the box contents. To secure maximum ventilation, the shipper wants the total surface area to be as large as possible. Use x as the side of the square base and L as the length.

    a) Write the expression for the area, A, in terms of L and x.

    b) Express $A(x)$ as a function of x alone.

    c) What value of x maximizes $A(x)$? Prove this is a maximum.

    d) What are the dimensions of the maximum-area box?

11. (See Problem 10.) Suppose the maximum length plus girth is M. Express the optimal x and L in terms of M.

12. A box with a square top and bottom is to be made to contain a volume of 64 cubic inches. What should be the dimensions of the box if its surface area is to be a minimum? What is this minimum surface area?

13. A box with a square bottom and no top is to be made to contain a volume of 500 cubic inches. What should be the dimensions of the box if its surface area is to be a minimum? What is this minimum surface area?

14. A box with a square top and bottom is to be made to contain 250 cubic inches. Material for top and bottom costs $2 per square inch and material for the sides costs $1 per square inch. What should be the dimensions of the box if its cost is to be a minimum? What is the minimum cost?

15. A box with a square bottom and no top is to be made to contain 100 cubic inches. Bottom material costs five cents per square inch and side materials costs two cents per square inch. Find the cost of the least expensive box that can be made.

16. A box with a square bottom and no top is to be made from a 6 by 6 inch piece of material by cutting equal-sized squares from the corners, then turning up the sides. What should the dimensions of the squares be if the box is to have maximum volume? (Note: The quadratic expression encountered in the solution is factorable.)

17. A box with a rectangular bottom and no top is to be made from a rectangular piece of material with dimensions 16 by 30 inches by cutting equal-sized squares from the corners, then turning up the sides. What should the dimensions of the squares be if the box is to have maximum volume? (Note: The quadratic expression encountered in the solution is factorable.)

18. A cylindrical storage tank is to contain $V = 16,000\pi$ cubic feet (about 400,000 gallons). The cost of the tank is proportional to its area, so the minimal-cost tank will be the one with minimum area. The volume of a cylinder of radius r and height h is $V = \pi r^2 h$. Its surface area is the area of top and bottom, $2\pi r^2$, plus the side area, $2\pi rh$. Find the dimensions, r and h, of the minimal-area tank. (Note: We have used $V = 16,000\pi$ so that π will cancel out.)

19. Suppose the tank in Problem 18 is to be built into the ground to catch runoff water, so it needs no top. Suppose, further, that the cost of the base of the tank is $10 per square foot and the sides $8.64 per square foot. What dimensions will lead to the minimal-cost tank?

20. A consulting firm conducts training sessions for employees of various companies. The charge to a company sending employees to a session is $50 per employee, less $0.50 for each employee in excess of 10. That is, for example, if 12 employees are sent, the charge per employee would be $49.00 and the total prorated charge to the company would be $12(49.00) = \$588.00$. The consulting firm further has a fixed total charge for groups of x or more, where x is the number that maximizes the prorated group charge. What should x be, and what is the maximum total group charge to a company?

21. A household appliance service organization has a parts stockroom and a garage at its central office location. Its trucks and drivers service customers in a roughly circular area of radius r around the central office. The number of customers per square mile is approximately $80/\pi$ (about 25) in any circular area around the office. Therefore the number of calls in a month is found by multiplying the number of customers per square mile by the number of square miles in the service area.

   a) What is the expression for the number of calls in a month?.

   b) The company figures travel cost at $2 per mile and computes mileage per call at $r/2$ miles out from the garage plus $r/2$ miles back in, for a total of r miles per call, on the average. The travel charge per call, excluding parts and labor, is fixed at $24. What is the expression for the net travel income per call? ("Net" means after deducting mileage cost.)

   c) What is the expression for the total net monthly travel income?

d) What service area radius will maximize total net monthly travel income?

e) What is the maximum net monthly travel income?

22. (This is a variation on the inventory model discussed in an earlier section.) A retail firm orders a product from a supplier Q units at a time. During a year, the firm will order $N = 2400$ units. The cost per unit ordered is $u = \$4$, and the cost of preparing and handling is figured at $c = \$12$ per order. The annual cost of carrying the average inventory of $Q/2$ units is $p = 0.25$ times (or 25 percent of) the purchase cost of $Q/2$ units. What are the parameterized expressions for:

a) The number of orders placed in a year?

b) Total ordering cost per year?

c) The cost of Q units?

d) Inventory cost per year?

e) $S(Q)$, the sum of ordering and inventory cost per year?

f) The order quantity, Q, which minimizes $S(Q)$?

g) What is the optimal order quantity for the parameter values given in the problem statement?

23. A manufacturing process generates $W = 4096$ cubic feet of waste per year. The waste is accumulated in a cubical container of side x feet (area $= 6x^2$, volume $= x^3$), which lasts one year. The container costs $K = \$2$ per square foot. When full, the container is emptied and the interior surface (also assumed to be $6x^2$ square feet) is decontaminated at a cost of $d = \$0.50$ per square foot. What are the parameterized expressions for:

a) The yearly container cost?

b) The number of decontaminations per year?

c) The yearly decontamination cost?

d) The yearly sum, $C(x)$, of container plus decontamination cost?

e) The container dimension, x, which minimizes $C(x)$?

Finally,

f) What is the optimum container dimension for the parameter values given in the problem statement?

24. A company operates a fleet of Beta model trucks. Study shows that gallons of fuel consumed per mile of driving, $F(x)$, is related to the speed at which a truck is driven, x miles per hour, by the function

$$F(x) = \frac{k_1}{x} + k_2 x; \ 10 \le x \le 80,$$

where k_1 and k_2 are parameters that vary somewhat from truck to truck.

a) Find and write the expression for the speed, x^*, which will lead to minimal fuel consumption per mile of driving.

b) What speed will provide minimal fuel consumption per mile of driving for a truck having $k_1 = 4.9$ and $k_2 = 0.004$?

25. (See 24). Fuel cost is d dollars per gallon and truck drivers are paid p dollars per hour of driving.

a) Write the expression for the value of x^* that will minimize the combined cost of fuel and driver, per mile driven, in terms of the parameters d, p, k_1 and k_2.

b) With $k_1 = 4.9$ and $k_2 = 0.004$, what speed will minimize the cost in (a) if fuel is \$0.52 per gallon and drivers are paid \$9 per hour?

26. The United Parcel Service (UPS) will pick up and deliver packages whose length plus girth does not exceed 108 inches. Zeeall Corporation ships a granular grinding abrasive in rectangular boxes with square ends. What is the largest volume of abrasive Zeeall can ship in a box? (Neglect the volume of the materials of which the box is made.)

9.22 THE DERIVATIVE OF $[f(x)]^n$

We have made extensive use of the *simple* power rule,

$$\frac{d}{dx}(x^n) = nx^{n-1},$$

and now wish to learn how to expand this rule to take the derivative of expressions such as

$$\frac{d}{dx}(2x-7)^{10}$$

$$\frac{d}{dx}(3x^2 - 2x + 5)^{3/2},$$

where the base of the power is something other than simply x; that is, we want a rule for finding

$$\frac{d}{dx}[f(x)]^n.$$

The rule is easily learned and applied. We shall state it, give illustrations, and then show its source.

Function power rule

$$\frac{d}{dx}[f(x)]^n = n[f(x)]^{n-1}\frac{df(x)}{dx}$$
$$= n[f(x)]^{n-1}f'(x).$$

Thus, the new rule starts out as does the simple rule with the (power) times the (function to the power minus one), then is completed by multiplying by the derivative of the function.

Example. Find the derivative of $g(x) = (2x-7)^{10}$.

Here we could expand $g(x)$ by raising $(2x-7)$ to the tenth power and take the derivative of the result term by term using the simple power rule—but this is almost unthinkable. By the function power rule,

$$\frac{d}{dx}(2x-7)^{10} = 10(2x-7)^9 \frac{d}{dx}(2x-7)$$

$$= 10(2x-7)^9 (2)$$

$$= 20(2x-7)^9.$$

Exercise.   Find $h'(x)$ if $h(x) = (3x^2 + 5)^{100}$. Answer: $600x(3x^2 + 5)^{99}$.

Example.   Find $f'(x)$ if $f(x) = (3x^2 - 2x + 5)^{3/2}$.
We have

$$\frac{d}{dx}(3x^2 - 2x + 5)^{3/2} = \frac{3}{2}(3x^2 - 2x + 5)^{1/2} \frac{d}{dx}(3x^2 - 2x + 5)$$

$$= \frac{3}{2}(3x^2 - 2x + 5)^{1/2}(6x - 2)$$

$$= \frac{3}{2}(3x^2 - 2x + 5)^{1/2}(2)(3x - 1)$$

$$= 3(3x^2 - 2x + 5)^{1/2}(3x - 1).$$

Exercise.   Find $f'(x)$ if $f(x) = (2x^3 - 3x^2 - 10)^{4/3}$. Answer: $8(2x^3 - 3x^2 - 10)^{1/3}(x)(x - 1)$.

Example.   Find $f'(x)$ if $f(x) = \dfrac{50}{0.2x + 5}$.
To obtain proper form, $[f(x)]^n$, which is not a fractional form, we write

$$f(x) = \frac{50}{0.2x + 5} = 50(0.2x + 5)^{-1}.$$

Then

$$f'(x) = 50 \frac{d}{dx}(0.2x + 5)^{-1}$$

$$= 50(-1)(0.2x + 5)^{-2} \frac{d}{dx}(0.2x + 5)$$

$$= -50(0.2x + 5)^{-2}(0.2)$$

$$= -10(0.2x + 5)^{-2}$$

$$= -\frac{10}{(0.2x + 5)^2}.$$

Example. Find $f'(1)$ if $f(x) = 10x - \dfrac{54}{(5x^2 + 4)^{1/2}}$. Changing to proper form, we have

$$f(x) = 10x - 54(5x^2 + 4)^{-1/2}$$

$$f'(x) = 10 - 54\left(-\frac{1}{2}\right)(5x^2 + 4)^{-3/2}\frac{d}{dx}(5x^2 + 4)$$

$$= 10 + 27(5x^2 + 4)^{-3/2}(10x)$$

$$= 10 + \frac{270x}{(5x^2 + 4)^{3/2}}.$$

Then,

$$f'(1) = 10 + \frac{270}{(5 + 4)^{3/2}} = 10 + \frac{270}{9^{3/2}} = 10 + \frac{270}{(9^{1/2})^3} = 10 + \frac{270}{3^3}$$

$$= 10 + \frac{270}{27} = 20.$$

Source of the power function rule. This, like all derivative rules, starts with the limit definition of a derivative. For a function $g(x)$ this is

$$\frac{d}{dx}g(x) = g'(x) = \lim_{\Delta x \to 0}\frac{\Delta g(x)}{\Delta x} = \lim_{\Delta x \to 0}\frac{g(x + \Delta x) - g(x)}{\Delta x}.$$

Here we shall use

$$[f(x)]^n \quad \text{as} \quad g(x)$$

so

$$\frac{d}{dx}[f(x)]^n = \lim_{\Delta x \to 0}\frac{\Delta[f(x)]^n}{\Delta x}. \tag{1}$$

To show the results more clearly, we shall for the present write f in place of $f(x)$, leave off the limit symbol until the end, and manipulate the difference quotient

$$\frac{[\Delta f(x)]^n}{\Delta x} = \frac{\Delta f^n}{\Delta x}.$$

Next, we apply a device often used by mathematicians to change a form at hand into a different, but equivalent, form that is wanted. The device is to multiply and divide the given form by the *same* quantity, which, in this case, is Δf. Thus,

$$\frac{\Delta f^n}{\Delta x} = \frac{\Delta f^n}{\Delta x}\left(\frac{\Delta f}{\Delta f}\right) = \frac{\Delta f^n}{\Delta f}\left(\frac{\Delta f}{\Delta x}\right),$$

where, at the right, the factors have been rearranged as permitted by the commutative property for multiplication. Then we take the limit as $\Delta x \to 0$ to obtain the derivative:

$$\lim_{\Delta x \to 0} \frac{\Delta f^n}{\Delta x} = \lim_{\Delta x \to 0} \frac{\Delta f^n}{\Delta f}\left(\frac{\Delta f}{\Delta x}\right).$$

A limit theorem says the limit of a product is the product of the limits, so

$$\lim_{\Delta x \to 0} \frac{\Delta f^n}{\Delta x} = \left[\lim_{\Delta x \to 0} \frac{\Delta f^n}{\Delta f}\right]\left[\lim_{\Delta x \to 0} \frac{\Delta f}{\Delta x}\right].$$

Now we must remember that as Δx approaches zero, $\Delta f(x)$, which is

$$\Delta f(x) = \Delta f = f(x + \Delta x) - f(x),$$

also approaches zero. Hence, changing $\Delta x \to 0$ to $\Delta f \to 0$ in the first factor on the right of the foregoing limit expression, we have

$$\lim_{\Delta x \to 0} \frac{\Delta f^n}{\Delta x} = \left[\lim_{\Delta f \to 0} \frac{\Delta f^n}{\Delta f}\right]\left[\lim_{\Delta x \to 0} \frac{\Delta f}{\Delta x}\right]. \tag{2}$$

We may now write

$$\lim_{\Delta x \to 0} \frac{\Delta f^n}{\Delta x} = [nf^{n-1}][f'(x)] \tag{3}$$

because, by definition, in (2),

$$\lim_{\Delta f \to 0} \frac{\Delta f^n}{\Delta f} = \frac{d}{df}f^n = nf^{n-1}$$

and

$$\lim_{\Delta x \to 0} \frac{\Delta f}{\Delta x} = f'(x).$$

Finally, recalling that f is $f(x)$, (3) becomes

$$\lim_{\Delta x \to 0} \frac{\Delta[f(x)]^n}{\Delta x} = n[f(x)]^{n-1} f'(x),$$

which is the power function rule.

We pause now for a practice problem set and follow this with a section on applications that require use of the power function rule.

9.23 PROBLEM SET 9–7

Find $f'(x)$. Simplify where possible. Leave no negative exponent in answers.

1. $f(x) = (6x - 5)^5$.   2. $f(x) = (2x + 6)^5$.

3. $f(x) = (2x)^3$.   4. $f(x) = (6x)^{1/3}$.

5. $f(x) = (4x)^{1/2}$.

6. $f(x) = (9x)^{4/3}$.

7. $f(x) = (8x - 3)^{3/2}$.

8. $f(x) = (12x - 9)^{5/3}$.

9. $f(x) = (3x^2 - 6x + 2)^{5/2}$.

10. $f(x) = (x^3 - 3x^2 + 6x)^{4/3}$.

11. $f(x) = (2x - 3)^{1/2}$.

12. $f(x) = (3x^2 + 5)^{2/3}$.

13. $f(x) = \dfrac{4}{2x - 3}$.

14. $f(x) = \dfrac{6}{3x - 5}$.

15. $f(x) = \left(\dfrac{1}{x} - 2\right)^2$.

16. $f(x) = \left(5 - \dfrac{1}{x^2}\right)^3$.

17. $f(x) = \dfrac{9}{(3x - 5)^2}$.

18. $f(x) = \dfrac{12}{(2x + 10)^3}$.

19. $f(x) = 5x + \dfrac{10}{3x + 2}$.

20. $f(x) = 0.1x + \dfrac{5}{5 - 0.2x}$.

21. $f(x) = 3x + \dfrac{1}{(5 + 2x)^{1/2}}$.

22. $f(x) = \dfrac{1}{(3x - 7)^{1/2}} - 2x$.

23. Find $g'(2)$ if $g(x) = 10x + \dfrac{18}{(5 + 2x)^{1/2}}$.

24. Find $h'(3)$ if $h(x) = 7x + \dfrac{4}{(x^2 - 1)^{1/3}}$.

9.24 APPLICATIONS: POWER FUNCTION RULE

In this section we present four examples. Practice problems relating to the examples in this section may be found in the next problem set.

Example 1. When x gallons of Exall are produced, the average cost per gallon is $A(x)$ dollars, where

$$A(x) = \frac{200}{0.1x + 5} + 0.05x, \quad x > 0.$$

a) Find the value of x^* where $A(x)$ is stationary.

b) Prove that this value of x occurs at a local minimum of $A(x)$.

c) Compute the minimum average cost per gallon.

a) We start by rewriting $A(x)$ as

$$A(x) = 200(0.1x + 5)^{-1} + 0.05x$$

then

$$A'(x) = 200(-1)(0.1x + 5)^{-2}(0.1)^* + 0.05,$$

where * calls attention to application of the power function rule. Continuing, we set $A'(x)$ equal to zero and solve for x.

$$A'(x) = -20(0.1x + 5)^{-2} + 0.05$$

$$= \frac{-20}{(0.1x + 5)^2} + 0.05$$

and $A'(x) = 0$ when

$$\frac{-20}{(0.1x + 5)^2} + 0.05 = 0$$

$$0.05 = \frac{20}{(0.1x + 5)^2}.$$

Multiplying both sides by the denominator,

$$0.05(0.1x + 5)^2 = 20$$

$$(0.1x + 5)^2 = \frac{20}{0.05} = 400.$$

Taking the square root of both sides (the $1/2$ power)

$$[(0.1x + 5)^2]^{1/2} = (400)^{1/2}$$
$$0.1x + 5 = \pm 20$$

$$\begin{array}{ll} 0.1x + 5 = 20 & \qquad 0.1x + 5 = -20 \\ 0.1x = 15 & \qquad 0.1x = -25 \\ x^* = 150. & \qquad x^* = -250. \end{array}$$

We discard $x^* = -250$ because it is negative and the problem statement requires that x be greater than zero.

 b) To show $x^* = 150$ yields a local minimum, we start with

$$A'(x) = -20(0.1x + 5)^{-2} + 0.05$$

and find the second derivative,

$$A''(x) = -20(-2)(0.1x + 5)^{-3}(0.1)$$

$$= \frac{4}{(0.1x + 5)^3}$$

then

$$A''(x^*) = A''(150) = \frac{4}{(15 + 5)^3},$$

which is positive, so there is a local minimum where $x^* = 150$.

 c) To find the minimum average cost per gallon of Exall, we write

$$A(x) = \frac{200}{0.1x + 5} + 0.05x.$$

Then,

$$A(150) = \frac{200}{0.1(150) + 5} + 0.05(150)$$

$$= \frac{200}{20} + 7.5$$

$$= \$17.5 \text{ per gallon.}$$

Example 2. *a)* Find the slope of the line tangent to the curve representing

$$f(x) = (169 - x^2)^{1/2}$$

at the point where $x = 12$.

b) Find the vertical intercept of the tangent line in *a)*.

A meaningful application of the method of solving this problem appears in the next problem set.

a) We first find the slope function, which is the derivative

$$f'(x) = \frac{1}{2}(169 - x^2)^{-1/2}(-2x)*,$$

where * calls attention to application of the power function rule.

$$f'(x) = \frac{-x}{(169 - x^2)^{1/2}}.$$

Hence the tangent slope where $x = 12$ is

$$f'(12) = \frac{-12}{(169 - 12^2)^{1/2}} = \frac{-12}{(169 - 144)^{1/2}} = \frac{-12}{(25)^{1/2}} = \frac{-12}{5} = -2.4.$$

b) The function value at $x = 12$ is

$$f(12) = (169 - 12^2)^{1/2} = (25)^{1/2} = 5,$$

so the point at hand has coordinates $(12, 5)$. The intercept of the tangent line we seek has coordinates $(0, b)$. If we now set the slope of the line through $(12, 5)$ and $(0, b)$ equal to -2.4, from *(a)*, we have

$$\frac{b-5}{0-12} = -2.4$$
$$b - 5 = (-12)(-2.4)$$
$$b - 5 = 28.8$$
$$b = 33.8,$$

which is the desired intercept.

Example 3. A bus company uses the function

$$P(x) = (3 + 0.6x)^{1/2} - 0.1x, \quad x > 0$$

to estimate the net weekly profit, in hundreds of dollars, if a particular bus route is x miles long. How long should the route be to maximize the net profit, and what is the maximum net profit?

First, we obtain

$$P'(x) = \frac{1}{2}(3 + 0.6x)^{-1/2}(0.6)* - 0.1 = 0.3(3 + 0.6x)^{-1/2} - 0.1,$$

where * calls attention to application of the power function rule. Continuing,

$$P'(x) = \frac{0.3}{(3 + 0.6x)^{1/2}} - 0.1$$

and $P'(x) = 0$ where

$$\frac{0.3}{(3 + 0.6x)^{1/2}} - 0.1 = 0$$

$$\frac{0.3}{(3 + 0.6x)^{1/2}} = 0.1.$$

Multiplying by the denominator leads to

$$0.3 = (0.1)(3 + 0.6x)^{1/2}.$$

Then, *squaring* both sides, we have

$$(0.3)^2 = (0.1)^2[(3 + 0.6x)^{1/2}]^2$$

$$0.09 = (0.01)(3 + 0.6x)$$

$$\frac{0.09}{0.01} = 3 + 0.6x$$

$$9 = 3 + 0.6x$$

$$6 = 0.6x$$

$$\frac{6}{0.6} = x$$

$$10 = x^*,$$

so a stationary point exists at $x^* = 10$ miles. To test this point, we return to

$$P'(x) = 0.3(3 + 0.6x)^{-1/2} - 0.1$$

and find the second derivative,

$$P''(x) = (0.3)\left(-\frac{1}{2}\right)(3 + 0.6x)^{-3/2}(0.6)$$

$$= -0.09(3 + 0.6x)^{-3/2}$$

$$= -\frac{0.09}{(3 + 0.6x)^{3/2}}$$

and

$$P''(x^*) = P''(10) = -\frac{0.09}{(3 + 6)^{3/2}},$$

which is negative, proving there is a maximum at $x^* = 10$. From

$$P(x) = (3 + 0.6x)^{1/2} - 0.1x,$$

the maximum is

$$\begin{aligned} P_{max} = P(10) &= [3 + 0.6(10)]^{1/2} - (0.1)(10) \\ &= (9)^{1/2} - 1 \\ &= \$2 \text{ hundred per week.} \end{aligned}$$

Example 4. Points A and D on Figure 9–19 are to be connected by highways. Construction cost above BD is \$2 hundred thousand per

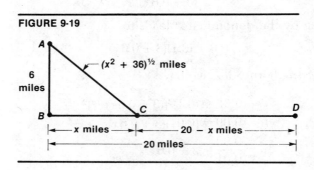

FIGURE 9-19

mile and cost along BD is \$1.6 hundred thousand per mile. Consequently, it would be more costly to run a highway directly from A to D, the shortest distance, than to run a section from A to a point C, and another section from C to D. The problem is to find where the intersection, C, should be if cost is to be minimized. As shown in Figure 9–19, AB is 6 miles and BD is 20 miles. If we let BC be x miles, then CD is $20 - x$ miles. Also, because ABC is a right triangle,

$$AC = \sqrt{x^2 + 6^2} = (x^2 + 36)^{1/2} \text{ miles.}$$

Total highway cost then will be, in hundreds of thousands of dollars,

$$\$2(AC) + 1.6(CD)$$
$$C(x) = 2(x^2 + 36)^{1/2} + 1.6(20 - x).$$

Proceeding to the derivative, and indicating the application of the power function rule by *, we have,

$$C'(x) = 2\left(\frac{1}{2}\right)(x^2 + 36)^{-1/2}(2x^*) - 1.6$$

$$= \frac{2x}{(x^2 + 36)^{1/2}} - 1.6.$$

Setting $C'(x)$ equal to zero yields

$$\frac{2x}{(x^2 + 36)^{1/2}} - 1.6 = 0$$

$$\frac{2x}{(x^2 + 36)^{1/2}} = 1.6.$$

Now multiply both sides by the denominator, then square both sides to obtain

$$2x = 1.6(x^2 + 36)^{1/2}$$

$$(2x)^2 = (1.6)^2[(x^2 + 36)^{1/2}]^2$$

$$4x^2 = 2.56(x^2 + 36)$$

$$4x^2 = 2.56x^2 + 92.16$$

$$4x^2 - 2.56x^2 = 92.16$$

$$1.44x^2 = 92.16$$

$$x^2 = \frac{92.16}{1.44}$$

$$x^2 = 64$$

$$x^* = 8, \ x^* = -8 \text{ miles.}$$

We discard $x^* = -8$ because x must be in the interval $0 \leq x \leq 20$ in this problem. To test $x^* = 8$ by the second derivative, we would have to find the derivative of

$$C'(x) = \frac{2x}{(x^2 + 36)^{1/2}} - 1.6,$$

and we do not yet have a rule for taking the derivative of a *quotient* of two functions. Consequently, we shall apply the *original function* left-right test, with the left side of $x^* = 8$ being $l = 7$ and the right side being $r = 9$. Thus, for

$C(x) = 2(x^2 + 36)^{1/2} + 1.6(20 - x)$
$C(l) = C(7) = 2(49 + 36)^{1/2} + 1.6(20 - 7) = 2(85)^{1/2} + 1.6(13) = 39.24$
$C(x^*) = C(8) = 2(64 + 36)^{1/2} + 1.6(20 - 8) = 2(100)^{1/2} + 1.6(12) = 39.20$
$C(r) = C(9) = 2(81 + 36)^{1/2} + 1.6(20 - 9) = 2(117)^{1/2} + 1.6(11) = 39.23.$

We see that both $C(l)$ and $C(r)$ are greater than $C(x^*)$ so a minimum occurs at the stationary point $x^* = 8$ miles. The minimum cost is

$$C_{min} = C(8) = \$39.2 \text{ hundred thousand}$$
$$= \$3,920,000.$$

By way of contrast, a direct road from A to D on Figure 9–19 would have a length of

$$[(20)^2 + (6)^2]^{1/2} = (436)^{1/2} = 20.880613$$

and the cost would be

$$\$2(20.880613) = \$41.76123 \text{ hundred thousand}$$
$$= \$4,176,123.$$

This exceeds the minimal amount, $3,920,000, found in the foregoing by

$$4,176,123 - 3,920,000 = \$256,123,$$

or about $256 thousand.

9.25 PROBLEM SET 9–8

1. When y gallons of Whyall are produced, the average cost per barrel is $A(y)$, where

$$A(y) = \frac{2500}{0.04y + 9} + 0.16y, \quad y > 0.$$

   a) Find the value, y^*, that minimizes average cost per barrel.
   b) Compute the minimum average cost per barrel.

2. When x gallons of Exall are produced, the average cost per barrel is $A(x)$, where

$$A(x) = \frac{4000}{0.1x + 20} + 0.25x, \quad x > 0.$$

   a) Find the value, x^*, which minimizes average cost per barrel.
   b) Compute the minimum average cost per barrel.

3. Profit realized when x thousand gallons of Exall are produced and sold is $P(x)$ thousand dollars, where

$$P(x) = (100 + 10x)^{1/2} - 0.2x.$$

   a) Find the value, x^*, which leads to maximum profit.
   b) Compute the maximum profit.

4. The output of a chemical process that is applied for t hours is $k(t)$ hundreds of pounds, where

$$k(t) = (6 + 0.3t)^{1/2} - 0.05t.$$

   a) Find the value, t^* hours, which leads to maximum output.
   b) Compute the maximum output.

5. The section of circular roadway on Figure A is part of the graph of the function

$$f(x) = (10,000 - x^2)^{1/2}.$$

An exit is planned at $P(60, 80)$, and the straight exit path is to be tangent to the circular roadway at point P. Find the vertical intercept, b, of the intersection point, Q.

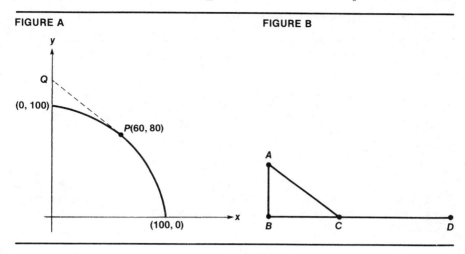

FIGURE A   FIGURE B

6. (See Problem 5). Find the vertical intercept if the circular section is

$$f(x) = (225 - x^2)^{1/2}$$

and the point on this section is $P(12, 9)$.

7. In Figure B, points B, C, and D are on a horizontal line and A is 3 miles above B on a perpendicular to BD. Straight line sections of road are to be constructed from A to C, then from C to D. Construction cost along BD is $2 hundred thousand per mile, but the cost above BD is $2.5 hundred thousand per mile. The distance BD is 10 miles.

   a) How many miles from B should the intersection C be located if cost is to be minimized?

   b) Compute the minimum cost.

   c) How much more than the minimum would be the cost of a single segment from A to D?

8. Answer a), b), and c) of Problem 7 if A is 8.4 miles above B, BD is 15 miles, cost along BD is $2 hundred thousand per mile and cost above BD is $2.9 hundred thousand per mile.

9. a) Find the expression for x^* at the stationary point on

$$f(x) = \frac{a}{bx + c} + kx.$$

   b) If a, b, c, k, and x are all positive, determine by the second derivative test whether x^* is at a local maximum or a local minimum.

10. a) Find the expression for x^* at the stationary point on

$$f(x) = (ax + b)^{1/2} - kx.$$

   b) If a, b, and k are all positive, determine by the second derivative test whether x^* is at a local maximum or a local minimum.

9.26 PRODUCT AND QUOTIENT RULES

We now extend our list of rules to include the derivative of the product of two functions and the quotient of two functions; that is, for example, to find

$$\frac{d}{dx}(x-1)(x^2+2)^{4/3} \quad \text{and} \quad \frac{d}{dx}\left(\frac{2x^2+3}{x+1}\right),$$

or, more generally,

$$\frac{d}{dx}[f(x)g(x)] \quad \text{and} \quad \frac{d}{dx}\left[\frac{f(x)}{g(x)}\right].$$

The product rule. To find the derivative of any function, we start with the definition that the derivative is the limit of the difference quotient as $\Delta x \to 0$. For

$$f(x)g(x)$$

the difference quotient is

$$\frac{f(x+\Delta x)g(x+\Delta x)-f(x)g(x)}{\Delta x}. \tag{1}$$

Now

$$\Delta f(x)=f(x+\Delta x)-f(x)$$

so

$$f(x+\Delta x)=f(x)+\Delta f(x).$$

For clarity, we shall abbreviate the last as

$$f(x+\Delta x)=f+\Delta f.$$

Similarly,

$$g(x+\Delta x)=g+\Delta g.$$

The difference quotient, (1), then is

$$\frac{(f+\Delta f)(g+\Delta g)-fg}{\Delta x}=\frac{fg+f(\Delta g)+g(\Delta f)+(\Delta f)(\Delta g)-fg}{\Delta x}$$

$$=\frac{f(\Delta g)+g(\Delta f)+(\Delta f)(\Delta g)}{\Delta x}$$

$$=f\left(\frac{\Delta g}{\Delta x}\right)+g\left(\frac{\Delta f}{\Delta x}\right)+\Delta f\left(\frac{\Delta g}{\Delta x}\right).$$

The derivative is the limit of the last expression as $\Delta x \to 0$, and the limit may be taken term by term. Thus,

$$\frac{d}{dx}[f(x)g(x)] = \frac{d}{dx}[fg]$$

$$= \lim_{\Delta x \to 0} f\left(\frac{\Delta g}{\Delta x}\right) + \lim_{\Delta x \to 0} g\left(\frac{\Delta f}{\Delta x}\right) + \lim_{\Delta x \to 0} \Delta f\left(\frac{\Delta g}{\Delta x}\right). \qquad (2)$$

Now f and g are to be thought of as function values (at a *given* point x) that do not vary as we approach the point by letting $\Delta x \to 0$, so f and g may be placed outside the limit symbols. And, of course, as $\Delta x \to 0$, so also do Δf and Δg approach zero. Finally, the limit of the product at the right end of (2) is the product of the limits. Hence, (2) may be written as

$$\frac{d}{dx}[fg] = f\lim_{\Delta x \to 0}\left(\frac{\Delta g}{\Delta x}\right) + g\lim_{\Delta x \to 0}\left(\frac{\Delta f}{\Delta x}\right) + \left[\lim_{\Delta f \to 0}\Delta f\right]\left[\lim_{\Delta x \to 0}\left(\frac{\Delta g}{\Delta x}\right)\right]. \qquad (3)$$

Now, because

$$\lim_{\Delta x \to 0}\left(\frac{\Delta g}{\Delta x}\right) = g' = g'(x),$$

$$\lim_{\Delta x \to 0}\left(\frac{\Delta f}{\Delta x}\right) = f' = f'(x),$$

and

$$\left[\lim_{\Delta f \to 0}\Delta f\right]\left[\lim_{\Delta x \to 0}\left(\frac{\Delta g}{\Delta x}\right)\right] = 0\,(g') = 0,$$

the expression (3) becomes

$$\frac{d}{dx}[fg] = fg' + gf'.$$

Replacing the independent variable, we have:

Product rule:$\dfrac{d}{dx}[f(x)g(x)] = f(x)g'(x) + g(x)f'(x).$

In words, *the derivative of the product of two functions is the first times the derivative of the second, plus the second times the derivative of the first.*

To show that the product rule works, note that

$$\frac{d}{dx}(x^3)(x^5) = \frac{d}{dx}(x^8) = 8x^7$$

by the power rule. If we now think of

$$f(x) = x^3, \quad \text{the first function}$$
$$g(x) = x^5, \quad \text{the second function,}$$

then

$$\frac{d}{dx}(x^3)(x^5) = x^3 \frac{d}{dx}(x^5) + x^5 \frac{d}{dx}(x^3)$$

$$= x^3 (5x^4) + x^5 (3x^2)$$

$$= 5x^7 + 3x^7$$

$$= 8x^7,$$

as it should be.

Example. Find

$$\frac{d}{dx}(x-1)(x^3+2)^{4/3}.$$

Here, the first function is $(x-1)$ and the second is $(x^3+2)^{4/3}$.

$$\frac{d}{dx}(x-1)(x^3+2)^{4/3} = (x-1)\frac{d}{dx}(x^3+2)^{4/3} + (x^3+2)^{4/3}\frac{d}{dx}(x-1)$$

$$= (x-1)\left(\frac{4}{3}\right)(x^3+2)^{1/3}(3x^2)* + (x^3+2)^{4/3}(1)$$

$$= 4x^2(x-1)(x^3+2)^{1/3} + (x^3+2)^{4/3},$$

where * (two lines above) calls attention to application of the power function rule. The expression last written has $(x^3+2)^{1/3}$ as a common factor, and to show this we write $(x^3+2)^{4/3}$ as $(x^3+2)^1 (x^3+2)^{1/3}$, giving

$$4x^2(x-1)(x^3+2)^{1/3} + (x^3+2)^1(x^3+2)^{1/3}$$
$$= (x^3+2)^{1/3}[4x^2(x-1) + (x^3+2)]$$
$$= (x^3+2)^{1/3}[4x^3 - 4x^2 + x^3 + 2]$$
$$= (x^3+2)^{1/3}(5x^3 - 4x^2 + 2).$$

Exercise. *a)* By the simple power rule, find
$$d/dx(2x+1)(x-1) = d/dx(2x^2 - x - 1).$$
b) Do *(a)* by the product rule. *c)* Find $d/dx(x^2+3)(2x+5)^{3/2}$.
Answer: *a)* $4x - 1$. *b)* $(2x+1)(1) + (x-1)(2) = 4x - 1$.
c) $3(x^2+3)(2x+5)^{1/2} + 2x(2x+5)^{3/2}$. Factoring the last yields
$(2x+5)^{1/2}(7x^2 + 10x + 9)$.

The quotient rule. This rule can be derived easily from the product rule if we remember that by the power function rule

$$\frac{d}{dx}\left[\frac{1}{g(x)}\right] = \frac{d}{dx}[g(x)]^{-1} = -1[g(x)]^{-2}g'(x).$$

We start by changing the quotient to a product, then apply the product rule.

$$\frac{d}{dx}\left[\frac{f(x)}{g(x)}\right] = \frac{d}{dx}\{f(x)[g(x)]^{-1}\}$$

$$= f(x)\frac{d}{dx}[g(x)]^{-1} + [g(x)]^{-1}\frac{d}{dx}f(x)$$

$$= f(x)(-1)[g(x)]^{-2}g'(x) + [g(x)]^{-1}f'(x)$$

$$= \frac{-f(x)g'(x)}{[g(x)]^2} + \frac{f'(x)}{g(x)}$$

$$= \frac{-f(x)g'(x)}{[g(x)]^2} + \frac{f'(x)}{g(x)}\left[\frac{g(x)}{g(x)}\right]$$

$$= \frac{-f(x)g'(x) + f'(x)g(x)}{[g(x)]^2},$$

where in the next to the last step we multiplied the rightmost term by $g(x)/g(x)$ to obtain a common denominator. Rearranging the last expression, we have:

Quotient Rule: $\dfrac{d}{dx}\left[\dfrac{f(x)}{g(x)}\right] = \dfrac{g(x)f'(x) - f(x)g'(x)}{[g(x)]^2}.$

In words, *the derivative of a quotient is the denominator times the derivative of the numerator, minus the numerator times the derivative of the denominator, all over the denominator squared.* For example,

$$\frac{d}{dx}\left(\frac{2x+5}{3x-7}\right) = \frac{(3x-7)\dfrac{d}{dx}(2x+5) - (2x+5)\dfrac{d}{dx}(3x-7)}{(3x-7)^2}$$

$$= \frac{(3x-7)(2) - (2x+5)(3)}{(3x-7)^2}$$

$$= \frac{6x - 14 - (6x+15)}{(3x-7)^2}$$

$$= -\frac{29}{(3x-7)^2}.$$

Exercise. Find $\dfrac{d}{dx}\left(\dfrac{5x}{3-4x}\right).$ Answer: $\dfrac{15}{(3-4x)^2}.$

Example. Find the derivative of $x^2/(3x+2)^{1/2}$.
We have

$$\frac{d}{dx}\left[\frac{x^2}{(3x+2)^{1/2}}\right] = \frac{(3x+2)^{1/2}(2x) - x^2\left(\frac{1}{2}\right)(3x+2)^{-1/2}(3)}{[(3x+2)^{1/2}]^2}$$

$$= \frac{(3x+2)^{1/2}(2x) - \dfrac{3x^2(3x+2)^{-1/2}}{2}}{(3x+2)}.$$

To remove the divisor, 2, and the negative exponent, we multiply the numerator and denominator by

$$2(3x+2)^{1/2}$$

as follows:

$$\left[\frac{(3x+2)^{1/2}(2x) - \dfrac{3x^2(3x+2)^{-1/2}}{2}}{(3x+2)}\right]\left[\frac{2(3x+2)^{1/2}}{2(3x+2)^{1/2}}\right]$$

$$= \frac{(3x+2)^{1/2}(2x)(2)(3x+2)^{1/2} - \dfrac{3x^2(3x+2)^{-1/2}}{2}(2)(3x+2)^{1/2}}{(3x+2)(2)(3x+2)^{1/2}}$$

$$= \frac{4x(3x+2) - 3x^2}{2(3x+2)^{3/2}} = \frac{9x^2 + 8x}{2(3x+2)^{3/2}} = \frac{x(9x+8)}{2(3x+2)^{3/2}}.$$

The reader may wish to refer to the foregoing examples when working on the next set of problems.

9.27 PROBLEM SET 9–9

Find the first derivative of each of the following. Simplify results and factor where possible. Do not leave negative exponents or complex fractions (fractions containing fractions) in answers.

1. $f(x) = (3x-2)(2x+5)$.

2. $f(x) = (7x+3)(4-3x)$.

3. $f(x) = (x^2+2)(3x-5)$.

4. $f(x) = (3-x^2)(5x+6)$.

5. $f(x) = x(x-1)^4$.

6. $f(x) = x^2(x+5)^3$.

7. $f(x) = x^2(x+3)^{3/2}$.

8. $f(x) = x^3(6x-1)^{2/3}$.

9. $f(x) = 2x(3x^2+7)^{1/3}$.

10. $f(x) = 3x(2x^3+5)^{1/2}$.

11. $f(x) = \dfrac{x}{x-1}$.

12. $f(x) = \dfrac{x-2}{x+1}$.

13. $f(x) = \dfrac{x^2}{2x+3}$.

14. $f(x) = \dfrac{x^3}{3x+5}$.

15. $f(x) = \dfrac{x}{3+2x^2}$.

16. $f(x) = \dfrac{3x}{1-2x^2}$.

17. $f(x) = \dfrac{x}{(3x+2)^{1/2}}$.

18. $f(x) = \dfrac{2x}{(2x+3)^{1/2}}$.

19. $f(x) = \dfrac{2x+1}{(x^2+5)^{1/3}}$.

20. $f(x) = \dfrac{3-5x}{(x^3+2)^{1/3}}$.

21. When a Jack truck is driven at a speed of x miles per hour, it travels $m(x)$ miles per gallon of fuel consumed, where

$$m(x) = \frac{x}{5.76 + 0.0036x^2} .$$

At what speed should a truck be driven if $m(x)$ is to be maximized?

22. Answer Problem 21 for a Jill truck for which

$$m(x) = \frac{x}{4.5 + 0.005x^2} .$$

23. Answer Problem 21 in terms of the parameter k_1 and k_2 if

$$m(x) = \frac{x}{k_1 + k_2x^2} .$$

24. If the proportion of defective transistors in a very large stock of transistors is p, then the proportion of good transistors is $1 - p$. For example, if 5 percent (0.05 as a proportion) are defective, then 95 percent (0.95) are good. Now suppose that p is unknown. One transistor is selected at random and inspected; then a second is selected and inspected, and so on until the *first* defective is found, and this is the *tenth* one. It is shown in probability that the chance that this will happen if the proportion defective is p, is $C(p)$, where

$$C(p) = p(1-p)^9.$$

What value of p will maximize $C(p)$, the chance that the first defective will be the tenth one inspected?

25. (See Problem 24.) The chance that the first defective found is the nth transistor inspected is

$$C(p) = p(1-p)^{n-1},$$

where, of course, n is a parameter. In terms of this parameter, what value of p maximizes $C(p)$?

26. In one run of a process, the number of pounds of Hypop that can be produced is x, where $0 < x \le 300$. The production cost *per pound* is, in dollars

$$\frac{100}{40 - 0.1x} .$$

Hypop sells for $10 *per pound*.

   a) Find the number of pounds that should be made in a run of the process if profit (revenue from sales, minus cost) is to be maximized.

   b) Compute the maximum profit.

27. (See Problem 26.) A new process has been designed to make an improved product, Hypop-II, which sells at $15 *per pound*. In one run of this process, x pounds are produced at a *per-pound* cost of

$$\frac{100}{60 - 0.05x}; \quad 0 < x \le 750.$$

a) Find the number of pounds to be made in a run if profit is to be maximized.

b) Compute the maximum profit.

28. A rectangular manufacturing plant with a floor area of 150,000 square feet is to be built in a town where zoning regulations require buffer strips 50 feet wide front and back, and 30 feet at each end. (A buffer strip is a grass and tree belt that must not be built upon.) Make a sketch of the overall rectangular plot and let its dimensions be x and y, so that the total area is $A = xy$. Then sketch the smaller rectangular plant of area 150,000 square feet inside the larger rectangle, and mark the dimensions of the buffer strips.

a) Write the expression for the 150,000 square foot area in terms of x and y.

b) Solve (a) for y.

c) Substitute the expression for y found in (a) into $A = xy$ to get $A(x)$.

d) Find the value for x^* that minimizes the total area, A, and then compute the corresponding y value.

e) Compute the minimal total area.

29. Sam, a modern artist, has submitted a sketch of a proposed piece to a client who wants a simple, clean-lined decoration to place in the lobby of a new office building. Sam's sketch shows an 8-foot column whose cross section is a right triangle, so the column has three plane surfaces. The client commissions Sam to do the work, but stipulates that the hypotenuse of the triangle must be 40 inches *and* the triangular cross sectional area is to be maximized. What dimensions should be used for the two sides of the right triangle if these stipulations are to be met? Recall that the area of a right triangle is one-half the product of the lengths of the sides, and the sum of the squares of the sides equals the square of the hypotenuse. State answer to the nearest one-hundredth of an inch.

9.28 SKETCHING GRAPHS OF POLYNOMIALS

A function of a single variable, such as

$$f(x) = 0.1x^3 - 1.8x^2 + 8.1x + 2,$$

in which each term is either a constant or a constant times the variable to a power that is a positive whole number, is called a polynomial. The *degree* of a polynomial is its highest power, so $f(x)$ just written is of degree three. Third-degree polynomials are called *cubics*. Quadratics, which graph as vertical parabolas, are second-degree polynomials, and we need not repeat our earlier discussion of these nor of first-degree polynomials, which are straight lines.

If a polynomial is set equal to zero and solved for x, it can be proved that the maximum possible number of roots is the degree of the polynomial or less than the degree by a multiple of two. For example, if we write

$$f(x) = 0.1x^3 - 1.8x^2 + 8.1x + 2$$
$$f'(x) = 0.3x^2 - 3.6x + 8.1$$

and set the resultant *second*-degree polynomial $f'(x)$ equal to zero to solve for extreme points, we have

$$0.3x^2 - 3.6x + 8.1 = 0,$$

which has either *two* roots or two minus two (equals zero) roots. Hence, a cubic (whose derivative is second-degree) can have two local extreme points or none. In the case at hand, we factor the last expression to obtain

$$0.3(x^2 - 12x + 27) = 0$$
$$0.3(x - 9)(x - 3) = 0$$
$$x - 9 = 0, \quad \text{so } x^* = 9$$
$$x - 3 = 0, \quad \text{so } x^* = 3.$$

The second derivative is

$$f''(x) = 0.6x - 3.6$$
$$f''(9) = 5.4 - 3.6 = \quad 1.8, \quad \text{so } x^* = 9 \text{ is at a local minimum,}$$
$$f''(3) = 1.8 - 3.6 = -1.8, \quad \text{so } x^* = 3 \text{ is at a local maximum.}$$

Evaluating $f(x)$ at these extreme points we find

$$f(9) = 0.1(9^3) - 1.8(9^2) + 8.1(9) + 2 = 2,$$

so (9, 2) is the minimum, and

$$f(3) = 0.1(3^3) - 1.8(3^2) + 8.1(3) + 2 = 12.8,$$

so (3, 12.8) is the local maximum. A *sketch* showing the important characteristics of $f(x)$ can be made from these two points. See Figure 9–20. We have included the intercept, $x = 0$, $f(0) = 2$ in the figure

FIGURE 9-20 (not to scale)

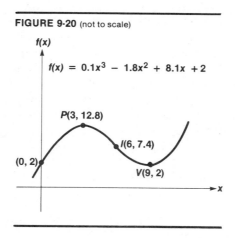

because it is easily obtained. Note that to the right of the peak, P, the curve is concave downward, so the second derivative is negative. However, beyond point I, which is an ordinary (nonstationary) inflection point, the curve becomes concave upward and the second derivative becomes positive. Thus, $f''(x)$ is negative to the left of I and positive to the right of I, and this suggests that the second derivative is zero at the inflection point. In general, it is true that *for a polynomial,*[5] *the second derivative must equal zero at an inflection point.* However, this does not mean that $f''(x) = 0$ *only* at an inflection point, as shown by reference to Figure 9–12A, where $h''(x) = 0$ at a local minimum, and Figure 9–12B, where $k''(x) = 0$ at a local maximum. Thus, for a polynomial, $f''(x)$ must equal zero at an inflection point, but $f''(x) = 0$ does not necessarily mean we have an inflection point. In the case at hand,

$$f'(x) = 0.3x^2 - 3.6x + 8.1$$
$$f''(x) = 0.6x - 3.6$$
$$f''(x) = 0 \quad \text{when} \quad 0.6x - 3.6 = 0; \quad x = 6.$$

From $f(x)$ we find $f(6) = 7.4$, so we have $I(6, 7.4)$ as shown on Figure 9–20. Note that the coordinates of I are the averages of the respective coordinates of P and V, the local maxima and minima. This will always be the case for a cubic that has two local extreme points.

Smooth curves and continuity. When we sketched the cubic shown in Figure 9–20, we assumed that the curve had to rise continually to the right of $V(9, 2)$ because if it were ever to turn around and start falling it would have to have another extreme point—and it has only two. By the same argument, it follows that the curve falls forever as we move to the left of point P. The assumption that leads to these conclusions is, to use a suggestive term, that the curve is everywhere a *smooth* curve. Technically, this means that $f(x)$ is *differentiable* (has a unique value for the derivative) for every value of x and, interpreted graphically, the function has a *unique* tangent line at every point. Figures 9–21 and 9–22 show two functions whose graphs are not smooth curves. The definition of $f(x)$ in Figure 9–21 may seem strange, but it is proper because for every value of x, there is one and only one value for $f(x)$. However, $f(x)$ does not have a derivative at $x = 2$ and therefore does not have a tangent line at $x = 2$. The reason this is so is that the derivative is the limit of the difference quotient, and for this to exist, the limit must be the same whether we approach 2 from the left (as in 1.9, 1.99, 1.999, etc.) or from the right (as in 2.1, 2.01, 2.001). In Figure 9–21 this requirement for a limit is not met at $x = 2$. The idea of a smooth curve, then, is that it progresses smoothly and does not suddenly change direction at a point, as is the case in Figure 9–21.

---

[5] It may not be true for nonpolynomials. For example, $f(x) = x^{1/3}$ has an inflection point at $(0, 0)$, but $f''(x)$ is undefined at $(0, 0)$.

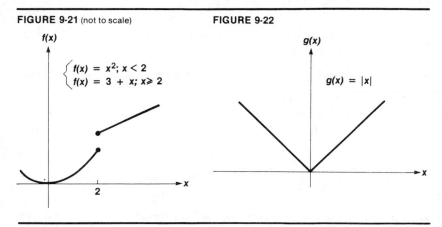

FIGURE 9-21 (not to scale)   **FIGURE 9-22**

Figure 9–22 is a graph of the absolute value of x, as indicated by the vertical bars in

$$g(x) = |x|.$$

If

$$x = 0, 1, 2, 3, \cdots ; \; g(x) = 0, 1, 2, 3, \cdots$$

and if

$$x = -1, -2, -3, \cdots ; \; g(x) = |-1|, |-2|, |-3|, \cdots = 1, 2, 3 \cdots .$$

Thus, $g(x)$ has a graph consisting of a half-line that bisects the first quadrant and another half-line that bisects the second quadrant. These half-lines meet at the origin forming the angle shown in Figure 9–22 at the point $(0, 0)$. For the reason cited when discussing $f(x)$, $g(x)$ is not differentiable (does not have a derivative) at $x = 0$. The "curve" in Figure 9–22 is smooth everywhere but at $x = 0$.

Both Figures 9–21 and 9–22 have a point where the function does not have a derivative, but they have an important difference. Figure 9–21 has a break or *discontinuity*, but Figure 9–22 does not. In suggestive terms, a *continuous* curve can be drawn without lifting pencil from paper. This is true for all of Figure 9–22 and true for all of Figure 9–21 *except* for the discontinuity at $x = 2$. Thus, we see that smooth curves, such as the one in Figure 9–20, automatically are continuous, but that a continuous curve, such as the one in 9–22, may not be smooth. We shall have occasion to be concerned with discontinuities later. For the present, it is sufficient to know that *polynomials are differentiable at all points. Hence, they are smooth curves and, therefore, continuous at all points.*

Example.   Sketch the graph of

$$f(x) = x^3 - 12x^2 + 50x - 60.$$

We find

$$f'(x) = 3x^2 - 24x + 50$$
$$f'(x) = 0 \quad \text{when} \quad 3x^2 - 24x + 50 = 0.$$

The quadratic last written cannot be factored, so we shall apply the quadratic formula

$$x = \frac{-b \pm \sqrt{b^2 - 4ac}}{2a}$$

with $a = 3$, $b = -24$, $c = 50$ to obtain

$$x = \frac{24 \pm \sqrt{576 - 600}}{6} = \frac{24 \pm \sqrt{-24}}{6}.$$

The square root of a negative is not a real number, so we conclude that $f(x)$ has no local extreme points. Turning to the second derivative, we find

$$f''(x) = 6x - 24$$
$$f''(x) = 0 \quad \text{when} \quad 6x - 24 = 0; \quad x = 4.$$

Now, $f''(x) = 0$ *may* mean we have an inflection point. If so, concavity must be of different sign to the left and right of $x = 4$. Using $l = 3$ as a value to left of 4, and $r = 5$ as a value to the right, we find

$$f''(l) = f''(3) = 6(3) - 24 = -6 \quad \text{(concave down)},$$
$$f''(r) = f''(5) = 6(5) - 24 = +6 \quad \text{(concave up)}.$$

Hence there is an ordinary (nonstationary) inflection at

$$x = 4$$
$$f(4) = (4)^3 - 12(4)^2 + 50(4) - 60 = 12.$$

The sketch of $f(x)$ is shown in Figure 9–23. The curve is concave

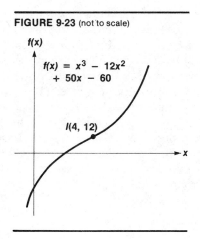

FIGURE 9-23 (not to scale)

f(x)

f(x) = x³ − 12x² + 50x − 60

I(4, 12)

x

downward at all points to the left of $I(4, 12)$, and concave upward at all points to the right of $I(4, 12)$.

Summary. A cubic always has one, and only one, inflection point. Depending upon its coefficients, it may also have one pair, and only one pair, of local extreme points, a maximum and a minimum, with the inflection point halfway between. Another fact, which we shall not pause to prove, is that as in Figure 9–23, a cubic having no local extreme points that rises to the right is concave downward to the left of its inflection point and concave upward to the right of the inflection, whereas if the cubic falls to the right it is concave upward (downward) to the left (right) of the inflection point.

Cubics with no local extreme points play an important role in economic cost analyses. (See Index reference to Cost analysis.)

Example. Sketch the graph of

$$f(x) = x^4 - 8x^2 + 26.$$

This is a fourth degree polynomial, a *quartic*. We find

$$f'(x) = 4x^3 - 16x,$$

which, when set equal to zero, is a cubic and may have 3 roots or $3 - 2 = 1$ root. Hence, a quartic may have three or one local extreme points. Continuing,

$$f'(x) = 0 \quad \text{where} \quad 4x^3 - 16x = 0$$
$$4x(x^2 - 4) = 0$$
$$4x(x - 2)(x + 2) = 0$$
$$x^* = 0, 2, -2$$

are candidate extreme points.

$$f''(x) = 12x^2 - 16$$
$$f''(0) = -16 \qquad \text{(local maximum)},$$
$$f''(2) = 32 \qquad \text{(local minimum)},$$
$$f''(-2) = 32 \qquad \text{(local minimum)}.$$

FIGURE 9-24 (not to scale)

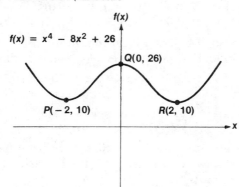

$f(x)$

$f(x) = x^4 - 8x^2 + 26$

$Q(0, 26)$

$P(-2, 10)$

$R(2, 10)$

x

The extreme points are

$x^* = 0; f(0) = 0^4 - 8(0)^2 + 26 = 26;$ $\qquad$ $Q(0, 26)$, maximum.
$x^* = 2; f(2) = 2^4 - 8(2)^2 + 26 = 10;$ $\qquad$ $R(2, 10)$, minimum.
$x^* = -2; f(-2) = (-2)^4 - 8(-2)^2 + 26 = 10;$ $\quad$ $P(-2, 10)$, minimum.

We plot these three local extreme points on Figure 9–24, then sketch the curve that makes Q a maximum and P and R minima.

Exercise. *a)* How many inflection points are there in Figure 9–24? *b)* What are the x-coordinates of these points? Answer: *a)* 2. *b)* $x = \pm (4/3)^{1/2}$ or about ± 1.155.

Example. Sketch the graph of

$$f(x) = 0.1(x - 10)^4 - 25.6x + 340.8.$$

We find

$$f'(x) = 0.4(x - 10)^3 - 25.6$$

and

$$f'(x) = 0 \quad \text{when} \quad 0.4(x - 10)^3 - 25.6 = 0$$

$$(x - 10)^3 = \frac{25.6}{0.4}$$

$$(x - 10)^3 = 64.$$

Taking the cube root of both sides,

$$[(x - 10)^3]^{1/3} = (64)^{1/3}$$
$$x - 10 = 4$$
$$x^* = 14$$

and

$$f(x^*) = f(14) = 0.1(14 - 10)^4 - 25.6(14) + 340.8 = 8$$

so we have the point $(14, 8)$. Then

$$f''(x) = 0.4(3)(x - 10)^2 = 1.2(x - 10)^2$$
$$f''(x^*) = f''(14) = 1.2(14 - 10)^2 = 1.2(16) = 19.2,$$

which is positive, so $(14, 8)$ is a local minimum.

To check for inflection points, we set $f''(x) = 0$. Thus,

$$f''(x) = 1.2(x - 10)^2 \text{ is zero where } 1.2(x - 10)^2 = 0$$
$$x = 10.$$

However,

$$f''(x) = 1.2(x - 10)^2$$

is positive to the left and right of $x = 10$, so we do not have an inflection point here and, as shown on Figure 9–25, the curve has a local minimum and no inflection points.

FIGURE 9-25 (not to scale)

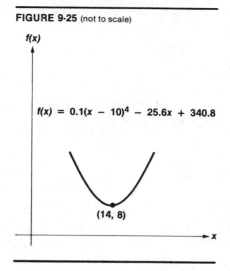

$f(x) = 0.1(x - 10)^4 - 25.6x + 340.8$

(14, 8)

Example.   Sketch the graph of

$$f(x) = x^4 - 8x^3 + 18x^2 - 27.$$

We find

$$f'(x) = 4x^3 - 24x^2 + 36x$$
$$f(x) = 0 \quad \text{where} \quad 4x^3 - 24x^2 + 36x = 0$$
$$4x(x^2 - 6x + 9) = 0$$
$$4x(x - 3)(x - 3) = 0$$
$$x^* = 0, \, x^* = 3.$$
$$f(0) = (0)^4 - 8(0)^3 + 18(0)^2 - 27 = -27.$$
$$f(3) = (3)^4 - 8(3)^3 + 18(3)^2 - 27 = 0,$$

so $(0, -27)$ and $(3, 0)$ are candidate extreme points. Taking the second derivative, we have

$$f''(x) = 12x^2 - 48x + 36$$
$$f''(0) = +36, \quad \text{so } (0, -27) \text{ is a local minimum.}$$
$$f''(3) = 12(3)^2 - 48(3) + 36 = 0, \quad \text{so the test fails.}$$

To test $x = 3$, we can use the left-right test with the original function, the first derivative, or the second derivative. We shall use the last, choosing $l = 2$ and $r = 4$ as points to the left and right, respectively, of $x = 3$. We find:

$$f''(l) = f''(2) = 12(2)^2 - 48(2) + 36 = -12 \quad \text{(concave down)}$$
$$f''(r) = f''(4) = 12(4)^2 - 48(4) + 36 = 36 \quad \text{(concave up).}$$

The stationary point at $x = 3$, $f(3) = 0$, is a *stationary inflection* point because, as just shown, concavity changes sign as we go from the left to the right of $x = 3$. Checking further for concavity, we set $f''(x) = 0$. Thus,

$$f''(x) = 12x^2 - 48x + 36$$

is zero where

$$12x^2 - 48x + 36 = 0$$
$$12(x^2 - 4x + 3) = 0$$
$$12(x - 3)(x - 1) = 0$$
$$x = 3 \quad \text{(already checked)}$$
$$x = 1.$$

We have already found a stationary inflection at $x = 3$. Right-left testing of $x = 1$ using $l = 0.5$ (being careful that a *little* to the left does not go beyond the stationary point at $x = 0$), and $r = 2$.

$$f''(l) = f''(0.5) = 12(0.5)^2 - 48(0.5) + 36 = 15 \quad \text{(concave up)}$$
$$f''(r) = f''(2) = 12(2)^2 - 48(2) + 36 = -12 \quad \text{(concave down)}.$$

Hence, concavity changes and there is an ordinary inflection point at $x = 1$, $f(1) = -16$. The sketch of the function is shown in Figure 9–26.

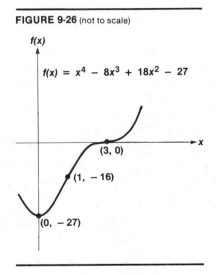

FIGURE 9-26 (not to scale)

f(x)

$f(x) = x^4 - 8x^3 + 18x^2 - 27$

(3, 0)

(1, −16)

(0, −27)

x

Summary. The graph of a quartic may have a single local extreme point, one local extreme point and two inflection points, or three local extreme points and two inflection points.

The procedures we have discussed can, in principle, be applied to sketch any polynomial. However, as the degree of the polynomial rises above four, problems arise in solving the equations $f'(x) = 0$ and $f''(x) = 0$, so we shall not discuss higher degree polynomials. Because our discussion has been quite detailed, it is worth noting that the procedures presented involved only finding and interpreting the first and second derivatives.

9.29 SKETCHING RATIONAL FUNCTIONS

A rational function is a function that is the quotient, or ratio, of two polynomials. Examples are

$$f(x) = \frac{2x^2 - 3}{x + 5}$$

$$f(x) = \frac{x^3 - 3x^2 + 2x - 7}{x^2 + 3x}$$

$$f(x) = \frac{x}{x - 5}.$$

We start with the last function

$$f(x) = \frac{x}{x - 5}$$

and our first concern is the discontinuity at $x = 5$. Here the denominator becomes zero, and $f(5) = 5/0$ is undefined. However, if x approaches 5 from the right through a sequence such as 5.1, 5.01, 5.001, and so on, we find

$$f(5.1) = \frac{5.1}{0.1} = 51$$

$$f(5.01) = \frac{5.01}{0.01} = 501$$

$$f(5.001) = \frac{5.001}{0.001} = 5001.$$

The closer x is to 5, the larger is $f(x)$. Thus the curve rises higher and higher as x becomes closer to 5, but x cannot equal 5. Similarly, as x approaches 5 from the left through a sequence such as 4.9, 4.99, 4.999, and so on,

$$f(4.9) = \frac{4.9}{-0.1} = -49$$

$$f(4.99) = \frac{4.99}{-0.01} = -499$$

$$f(4.999) = \frac{4.999}{-0.001} = -4999,$$

so as $x \to 5$ from the left, the curve falls further and further downward, but, again, x cannot equal 5. These characteristics are shown in the

492

FIGURE 9-27 (not to scale)

x = 5

partial sketch of $f(x)$ in Figure 9–27, where the vertical line is $x = 5$. Coming in from the left of $x = 5$ the curve falls forever, *becoming closer to but never touching* $x = 5$. We describe this by saying $x = 5$ is an *asymptote* of $f(x)$ or that $f(x)$ falls, approaching $x = 5$ *asymptotically* from the left. Similarly, $f(x)$ rises and approaches $x = 5$ asymptotically as the curve comes in from the right.

The next questions concern what happens to $f(x)$ as x moves off indefinitely to the left and to the right. We shall use the symbols

$x \rightarrow -\infty$ means x decreases (moves off to the left) without limit.
$x \rightarrow \infty$ means x increases (moves off to the right) without limit.

The symbol ∞ (infinity) is not a number, and to investigate $x \rightarrow -\infty$, we use a sequence such as $x = -100, -1000, -10,000$ and so on, and similarly for $x \rightarrow \infty$. In

$$f(x) = \frac{x}{x-5}$$

$$f(-1000) = \frac{-1000}{-1000-5} = \frac{1000}{1005} = 0.995$$

$$f(-10,000) = \frac{-10,000}{-10,000-5} = \frac{10,000}{10,005} = 0.9995$$

and it does not take many such trials before it becomes clear that for very large values of x, the constant -5 in the denominator of

$$f(x) = \frac{x}{x-5}$$

becomes almost, but not quite, inconsequential, so $f(x)$ becomes almost, but not quite, $x/x = 1$. However, as the foregoing sequence shows, when $x \to -\infty$, $f(x)$ increases, but is always a bit less than 1. Hence, $f(x)$ approaches the horizontal line one unit above the x-axis asymptotically as $x \to -\infty$. Figure 9–28 shows the asymptote as the constant function $A(x) = 1$. The reader may verify by using a sequence such as 100, 1000, 10,000, $\cdots$ that as $x \to \infty$, $f(x)$ approaches $A(x) = 1$ asymptotically from above, as shown on Figure 9–28. Separately, the two

FIGURE 9-28

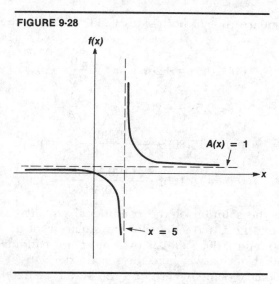

branches are smooth curves, but the function $f(x)$ itself has a discontinuity at $x = 5$. The function has no local extreme points. The reader may wish to verify this by the usual first derivative method, and also verify from the derivative that $f'(x)$ is always negative (except at $x = 5$). This is consistent with Figure 9–28, which shows that tangents to the curve always slant downward to the right. The concavity (downward for $x < 5$ and upward for $x > 5$) can be verified by the second derivative, which is negative for $x < 5$ and positive for $x > 5$.

Example. Sketch the graph of

$$f(x) = 15x - \frac{100x}{60 - 0.5x}.$$

In the last example, we went immediately to the new idea of asymptotes. However, it is safer first to examine functions for extreme points. To do so, we apply the quotient rule to find

$$f'(x) = 15 - \frac{(60 - 0.5x)(100) - (100x)(-0.5)}{(60 - 0.5x)^2}$$

$$= 15 - \frac{6000}{(60 - 0.5x)^2}$$

$$f'(x) = 0 \quad \text{where} \quad 15 - \frac{6000}{(60 - 0.5x)^2} = 0$$

$$15(60 - 0.5x)^2 - 6000 = 0$$

$$(60 - 0.5x)^2 = \frac{6000}{15} = 400.$$

Taking the square root of both sides yields ± 20 for the square root of 400.

$$60 - 0.5x = +20, \quad x^* = \frac{40}{0.5} = 80$$

$$60 - 0.5x = -20, \quad x^* = \frac{80}{0.5} = 160$$

$$f(80) = 15(80) - \frac{100(80)}{60 - 0.5(80)} = 800$$

$$f(160) = 15(160) - \frac{100(160)}{60 - 0.5(160)} = 3200.$$

Note that in the solution of the conditional equality, we took both the positive and negative values for the square root of 400. We now have (80, 800) and (160, 3200) as extreme point candidates. Next we rewrite the first derivative, take the second derivative, and evaluate it at the stationary points where $x^* = 80$ and $x^* = 3200$.

$$f'(x) = 15 - \frac{6000}{(60 - 0.5x)^2} = 15 - 6000(60 - 0.5x)^{-2}$$

$$f''(x) = -6000(-2)(60 - 0.5x)^{-3}(-0.5)$$

$$= \frac{-6000}{(60 - 0.5x)^3}$$

$$f''(80) = \frac{-6000}{(60 - 40)^3} \quad \text{is negative (maximum)}$$

$$f''(160) = \frac{-6000}{(60 - 80)^3} \quad \text{is positive (minimum).}$$

Returning to

$$f(x) = 15x - \frac{100x}{60 - 0.5x},$$

we note that a discontinuity occurs where the denominator is zero; that is, where

$$60 - 0.5x = 0$$

$$x = \frac{60}{0.5} = 120.$$

As in the previous example, $f(x)$ will approach $x = 120$ asymptotically. We now have sufficient information to sketch the graph of $f(x)$ shown in Figure 9–29. It is not necessary to examine $f(x)$ as $x \to \pm\infty$ because

FIGURE 9-29 (not to scale)

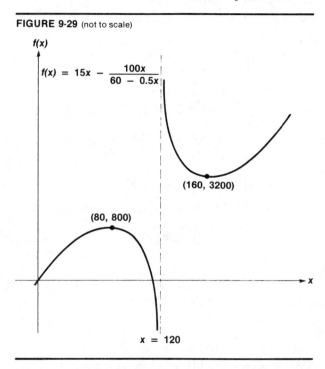

$$f(x) = 15x - \frac{100x}{60 - 0.5x}$$

(160, 3200)

(80, 800)

$x = 120$

it is clear that the left branch falls forever and the right branch rises forever.

It is now time for the reader to practice curve sketching. We shall show how to sketch curves of additional functions as they arise in the next chapter.

9.30 PROBLEM SET 9–10

Sketch the graphs of the following functions. Place coordinates of extreme points and inflection points on the sketch. If a curve has an asymptote, put it and its equation on the graph. Graphs need not be to scale, and coordinate axes need not be shown.

1. $f(x) = x^3 - 18x^2 + 96x - 100$.
2. $f(x) = 9x^2 - x^3 - 15x$.

3. $f(x) = 5 - x^3$.

4. $f(x) = 0.2x^3 + 4$.

5. $f(x) = x^3 - 15x^2 + 80x - 70$.

6. $f(x) = -0.5x^3 + 6x^2 - 25x + 42$.

7. $f(x) = 50x^2 - x^4 - 100$.

8. $f(x) = x^4 - 18x^2 + 90$.

9. $f(x) = (0.001)(x - 10)^4 + 20$.

10. $f(x) = 0.01(6 - x)^4 + 0.32x + 0.56$.

11. $f(x) = x^4 - 8x^3 + 18x^2 - 27$.

12. $f(x) = 10 - 24x^2 - x^4$.

13. $f(x) = \dfrac{1}{x}$.

14. $f(x) = \dfrac{100}{x^2}$.

15. $f(x) = \dfrac{2x}{x - 4}$.

16. $f(x) = \dfrac{x}{0.1x - 1}$.

17. $f(x) = 9x - \dfrac{48x}{3 - x}$.

18. $f(x) = \dfrac{2x}{x - 2} + x$.

9.31 COST ANALYSIS

An important model used in economic analysis of cost assumes producers have control over the quantity (number of units) they will produce, and that these units will be sold. However, the price received per unit is not under the producers' control but is, instead, established by competitive market forces, and this price is the same for all units. Hence, if the selling price is established at, say, \$425 per unit, the producer's problem is to determine what number of units should be produced and sold at this price in order to maximize profit or minimize loss. A key factor involved in this decision is the concept of *marginal cost*, which was defined earlier in the chapter as the derivative of the total cost function. In this section we shall adopt cubics having an inflection point, but no local extreme points, to represent total cost functions. Consider, for example,

$$\text{Total cost} = \text{TC} = C(q) = 0.0015q^3 - 0.9q^2 + 200q + 60{,}000,$$

where $C(q)$ is the total cost of producing q units of product. The graph of this function, sketched by procedures presented earlier in the chapter, is shown on Figure 9–30. Observe that as output, q, increases from zero, the curve is at first *concave downward* which means *marginal cost is decreasing*. Beyond the inflection point I, where $q = 200$, the curve becomes and remains *concave upward*, so *marginal cost is increasing*. These characteristics of a cubic that rises to the right and has no local extreme points correspond to the cost behavior postulated in economics. That is, as output increases from zero, cost advantages of large scale production are realized (marginal cost de-

FIGURE 9-30 (not to scale)

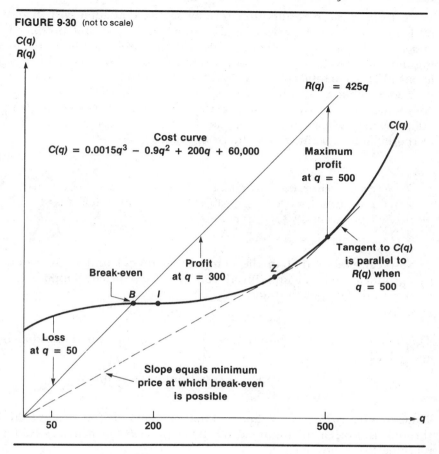

C(q)
R(q)

$R(q) = 425q$

$C(q)$

Cost curve
$C(q) = 0.0015q^3 - 0.9q^2 + 200q + 60,000$

Maximum
profit
at $q = 500$

Break-even

B   *I*

Profit
at $q = 300$

Z

Tangent to $C(q)$
is parallel to
$R(q)$ when
$q = 500$

Loss
at $q = 50$

Slope equals minimum
price at which break-even
is possible

50   200   500   *q*

creases). However, there comes a point *(I)*, where production increases require additional costs, such as overtime pay for workers, and marginal cost begins to increase.

Turning now to

$$\text{Marginal cost} = \text{MC} = C'(q) \text{ at } q \text{ units output,}$$

we have, for the foregoing cost function,

$$\text{MC} = C'(q) = 0.0045q^2 - 1.8q + 200.$$

Thus,

at $q = 50$, $\text{MC} = C'(50) = 0.0045(50)^2 - 1.8(50) + 200$
$$= \$121.25 \text{ per unit.}$$
at $q = 100$, $\text{MC} = C'(100) = 0.0045(100)^2 - 1.8(100) + 200$
$$= \$65 \text{ per unit.}$$

We interpret these results precisely by saying that total cost *at q =* 50 is increasing at the rate of $121.25 per additional unit produced, and similarly for MC = $65 *at q =* 100. Less precisely, but more meaningfully, we say that at $q = 50$, producing an additional unit adds about $121. 25 to total cost, whereas at $q = 100$, producing an additional unit adds about $65 to total cost. Observe that the derivative values computed, which are marginal costs, decrease as q goes from 50 to 100, which is a consequence of the downward concavity to the left of the inflection point on the curve in Figure 9–30.

Exercise. For the foregoing cost function, compute marginal cost at output levels: *a)* 300 units. *b)* 400 units. Answer: *a)* $65 per unit. *b)* $200 per unit.

Now remember that in the competitive model under discussion, the producer sells all units produced at a fixed price. Suppose this price is p dollars per unit and

$$p = \$425 \text{ per unit.}$$

The total revenue, TR, the producer receives for the q units produced and sold, is

$$\text{TR} = R(q) = (\text{price per unit})(\text{number of units})$$
$$\text{TR} = R(q) = 425q.$$

Total revenue, $R(q)$, is the linear function shown as the straight line through the origin in Figure 9–30. Its slope is $R'(q)$, which is called marginal revenue.

$$\text{Marginal revenue} = \text{MR} = R'(q) = \$425,$$

where, of course, the derivative is the *constant* slope of the straight line and means an additional unit produced and sold adds $425 to total revenue. The producer's profit, $P(q)$, is the difference between total revenue and total cost.

$$P(q) = R(q) - C(q).$$

Observe the vertical line segment at $q = 50$ on Figure 9–30. Here cost is above revenue, meaning that a loss is incurred at this level of output. Further to the right, the two curves intersect at B. Here revenue equals cost and we are at the *break-even point*. As we move to the right of break-even, the vertical separation of the revenue line and the cost curve represents profit. Profit (the separation) first increases, then decreases, so that at some value of q, profit attains its maximum value, and it is this value of q that we seek. Remembering that

$$P(q) = R(q) - C(q),$$

we can determine the optimal value of q by setting the derivative of $P(q)$ equal to zero. Thus,

$$\frac{d}{dq} P(q) = \frac{d}{dq} R(q) - \frac{d}{dq} C(q)$$
$$P'(q) = R'(q) - C'(q)$$
$$P'(q) = 0 \quad \text{where} \quad R'(q) - C'(q) = 0$$
$$R'(q) = C'(q) \quad \text{or} \quad C'(q) = R'(q).$$

The last says that for profit to be maximized, $C'(q)$, which is marginal cost, must equal $R'(q)$, which is marginal revenue.

Principle: *For profit to be maximized, marginal cost must equal marginal revenue.*

Inasmuch as $C'(q)$ and $R'(q)$ are slopes, this principle means that profit is maximized where the slope of a line tangent to the cost curve on Figure 9–30 equals the slope of the revenue line; that is, where these two lines are parallel as shown in the figure at $q^* = 500$. At this point we have the maximum separation between cost and revenue. All this means is that to the left of $q^* = 500$, where the cost slope is less than the revenue slope, an additional unit produced will add less to total cost than the $425 additional revenue obtained when it is sold. For example,

at $q = 450$, $\quad MC = C'(q) = 0.0045(450)^2 - 1.8(450) + 200 = \$301.25$
$$R'(q) = \$425,$$

and an additional unit brings in more in revenue than it adds to cost, so profit increases.

> **Exercise.** Compute and interpret the comparison of $R'(550)$ and $C'(550)$. Answer: $R'(550) = \$425$, $C'(550) = \$571.25$, so an additional unit adds more to cost than it returns in revenue and profit decreases.

The marginal-cost-equals-marginal-revenue principle says, in effect, to increase production as long as an additional unit returns more in revenue than it adds to cost because the difference then adds to profit. We now return to

$$R'(q) = \$425$$
$$C'(q) = 0.0045q^2 - 1.8q + 200.$$

Then, to maximize profits, we equate marginal cost and marginal revenue to obtain

$$0.0045q^2 - 1.8q + 200 = 425$$
$$0.0045q^2 - 1.8q - 225 = 0.$$

We recognize this as a quadratic equation of the form

$$aq^2 + bq + c = 0$$

with $a = 0.0045$, $b = -1.8$, $c = -225$. The solutions are given by the quadratic formula

$$q = \frac{-b \pm \sqrt{b^2 - 4ac}}{2a}$$

$$= \frac{-(-.18) \pm \sqrt{(-1.8)^2 - 4(0.0045)(-225)}}{2(0.0045)}$$

$$= \frac{1.8 \pm \sqrt{7.29}}{0.009}$$

$$= \frac{1.8 \pm 2.7}{0.009}$$

from which

$$q^* = \frac{1.8 + 2.7}{0.009} = 500 \text{ units}$$

$$q^* = \frac{1.8 - 2.7}{0.009} = -100 \text{ units.}$$

We discard the negative q^* because output cannot be negative, but point out in passing that it is possible for both q^* values to be positive and, if this occurs, the larger one is selected because the smaller one would occur where profit is a local minimum rather than a local maximum.

To compute the maximum profit, we write

$$P(q) = R(q) - C(q)$$
$$= 425q - [0.0015q^3 - 0.9q^2 + 200q + 60{,}000]$$
$$P_{max} = P(q^*) = P(500) = 425(500) - [0.0015(500)^3 - 0.9(500)^2$$
$$+ 200(500) + 60{,}000]$$
$$= 212{,}500 - 122{,}500$$
$$= \$90{,}000,$$

so production and sale of 500 units yields the maximum profit of $90,000.

Exercise. If the producer's cost function is $C(q) = 0.002q^3 - 0.06q^2 + 50q + 1600$ and selling price is $266 per unit, a) Find the optimal level of production, q^*. b) Compute maximum profit. Answer: a) $q^* = 200$. b) $28,000. [Note: $\sqrt{5.1984} = 2.28$.]

Shutdown price level.   The producer decides how many units to make and, if the selling price is low, the decision may be to produce $q = 0$ units; that is, to halt or shut down production. To see what is involved in this decision, recall the cost function introduced earlier,

$$\text{TC} = C(q) = \underbrace{0.0015q^3 - 0.9q^2 + 200q}_{\text{Total variable cost} = \text{TVC} = V(q)} + \underbrace{60,000}_{\text{Fixed cost}}.$$

Observe that total cost consists of a part that varies with q, the level of output, and a fixed component that is present whatever the level of output. As shown in the last expression, the variable component is called

$$\text{Total variable cost} = \text{TVC} = V(q)$$

and

$$\text{TVC} = V(q) = 0.0015q^3 - 0.9q^2 + 200q.$$

If the producer shuts down, so that $q = 0$, then

$$\text{TVC} = V(0) = 0$$

but

$$\text{TC} = C(0) = 60,000$$

and, of course, revenue is zero, so a loss of $60,000 is incurred. However, it should not be inferred that a producer should shut down just because a loss might be incurred because if it is possible to operate at a loss of, say, $30,000, operation would be preferable to shutting down and incurring a loss of $60,000. In general, the idea is that it is worthwhile to operate if total revenue exceeds total variable cost because then the difference will pay some of the $60,000 of fixed cost and the loss will be less than the $60,000 incurred by shutting down. The principle involved is to:

Operate when total revenue exceeds total variable cost.
Shut down when total revenue is less than total variable cost.

For our purposes, we shall want to express the principle in terms of averages. Average revenue per unit sold is, of course, the constant selling price. Average variable cost when q units are produced is total variable cost of q units divided by q. Thus,

$$\text{Average variable cost per unit} = \text{AVC} = \frac{\text{Total variable cost of } q \text{ units}}{q}$$

$$\text{AVC} = \frac{V(q)}{q}.$$

Inasmuch as total revenue exceeds total variable cost when average revenue (price) exceeds average variable cost, we may restate the last principle as:

Operate when price exceeds average variable cost.
Shut down when price is less than average variable cost.

Finally, remembering that the producer controls the level of output and therefore the average variable cost per unit, it follows that the producer can determine the level of output at which average variable cost is a *minimum*. If the selling price exceeds this minimum, the producer should operate. If the selling price is short of the minimum, the producer should shut down. To determine the minimum average variable cost in our example, we write

$$\text{AVC} = \frac{V(q)}{q} = \frac{0.0015q^3 - 0.9q^2 + 200q}{q} = 0.0015q^2 - 0.9q + 200.$$

The derivative is

$$\frac{d}{dq}(\text{AVC}) = 0.003q - 0.9$$

and

$$\frac{d}{dq}(\text{AVC}) = 0 \quad \text{where} \quad 0.003q - 0.9 = 0$$

$$0.003q = 0.9$$
$$q^* = 300 \text{ units.}$$

The second derivative is positive so AVC is a minimum at $q^* = 300$ units and the minimum is

$$\text{AVC}(300) = 0.0015(300)^2 - 0.9(300) + 200 = \$65 \text{ per unit.}$$

Consequently, the producer should shut down if the selling price is less than this minimum average variable cost per unit, $65.

Principle: The shutdown price level equals the minimum value of average variable cost.

In summary, to find the shutdown price:

1. Set up the expression for $AVC(q) = \dfrac{V(q)}{q}$.
2. Find the derivative of $AVC(q)$, set it equal to zero and solve for q^*.
3. $AVC(q^*)$ is shutdown price level.

Exercise. Find the shutdown price level for the cost function of the last exercise, $C(q) = 0.002q^3 - 0.06q^2 + 50q + 1600$. Answer: $49.55 per unit.

Break-even analysis. At the break-even point B shown on Figure 9–30, total cost equals total revenue. Thus,

$$C(q) = R(q)$$
$$0.0015q^3 - 0.9q^2 + 200q + 60{,}000 = 425q$$
$$0.0015q^3 - 0.9q^2 - 225q + 60{,}000 = 0.$$

The last is a cubic equation that must be solved for q. While a general method for solving cubics does exist, presentation of the method here would carry us too far afield, so we shall simply note that by trial and error, using a hand calculator as we did in Section 8.23 (Chapter 8), the point B occurs where q is about 178 (more closely, 177.74), so the break-even level of operation at a price of $425 per unit is about 178 units. We also note in passing the straight line $R(q) = 425q$ also intersects the cost curve (extended) at two points other than B, but these are not of interest because q will be negative for the intersection to the left of B, and the intersection to the right of B is beyond the point of maximum profit, $q^* = 500$.

A matter of interest to a producer is the *minimum* price at which it is possible to operate and break even. To see what is involved, we must remember that each different price level leads to a different slope of the revenue line on Figure 9–30, and that if the price is sufficiently low, the revenue line will not intersect the cost curve, but will always be below it. The *minimum* price that still makes it possible to break even is the slope of a revenue line that is tangent to the cost curve as shown by the dotted line on Figure 9–30. To find this unknown price level or slope, we observe that total cost must equal total revenue *and* marginal cost (the slope of the cost curve) must equal marginal revenue (price, or the slope of the revenue line). Thus,

$$C(q) = R(q) = pq \tag{1}$$

and

$$C'(q) = R'(q) = p. \tag{2}$$

If we substitute $p = C'(q)$ from (2) into (1), we have

$$C(q) = pq = C'(q)q. \tag{3}$$

For our cost function

$$C(q) = 0.0015q^3 - 0.9q^2 + 200q + 60{,}000$$
$$C'(q) = 0.0045q^2 - 1.8q + 200.$$

Substituting the last two expressions into (3), we have

$$0.0015q^3 - 0.9q^2 + 200q + 60{,}000 = q(0.0045q^2 - 1.8q + 200)$$

which, after expanding and collecting terms, becomes

$$0.003q^3 - 0.9q^2 - 60{,}000 = 0.$$

Again, the equation to be solved is a cubic. Through trial and error, $q = 416$ (more closely, 415.7) is a solution of the equation and, from (2),

$$P = C'(q)$$
$$= C'(415.7)$$
$$= 0.0045(415.7)^2 - 1.8(415.7) + 200 = \$229.4.$$

Thus, rounding, if the price level is \$229 per unit, the producer can break even by making and selling 416 units, and it is impossible to break even at any price level below \$229 per unit.

Economists usually present the foregoing ideas in terms of averages where, again, average revenue per unit is simply the price per unit, and average *total* cost per unit, ATC, is total cost of producing q units divided by q. That is,

$$\text{Average total cost per unit} = \text{ATC} = \frac{C(q)}{q}.$$

We may now define break-even as the point where price per unit equals average total cost per unit. That is, if ATC and price are both \$229.4 per unit, the producer receives from his sales precisely what it cost to produce the product. To find the minimum price level at which break-even is possible, it is now necessary to find where ATC is a minimum, as follows:

$$\text{ATC} = \frac{C(q)}{q}.$$

Taking the derivative by the quotient rule leads to

$$\frac{d}{dq}(\text{ATC}) = \frac{q\dfrac{d}{dq}C(q) - C(q)\dfrac{d}{dq}q}{q^2}$$
$$= \frac{qC'(q) - C(q)(1)}{q^2}.$$

The condition for a local minimum is that this derivative be zero. Thus,

$$\frac{qC'(q) - C(q)}{q^2} = 0.$$

Multiplying both sides by q^2 gives

$$qC'(q) - C(q) = 0$$
$$qC'(q) = C(q)$$

or, by rearrangement,

$$C(q) = C'(q)q \tag{3}$$

and we have used the number (3) again to call attention to the fact that the present method of analysis yields the same condition as (3) earlier in the discussion. We have not proved by testing that (3) provides a minimum, but this can be done and it is left for the reader to do in the next set of problems.

In summary:

To find the lowest price at which break-even is possible,

1. Set up $\text{ATC} = \dfrac{C(q)}{q}$.
2. Find the derivative of ATC, set it equal to zero, and solve for q^*.
3. The desired price is $\text{ATC}(q^*)$.

To provide practice in using the foregoing procedure, we shall use simplified cost functions in order to avoid the need to solve cubic equations by trial and error.

Example. Find the lowest price level at which break even is possible if

$$C(q) = 0.004q^3 + 120q + 8000.$$

1. Set up the expression for the average total cost as a function of q:

$$\text{ATC} = \frac{C(q)}{q} = \frac{0.004q^3 + 120q + 8000}{q} = 0.004q^2 + 120 + \frac{8000}{q}.$$

2. Find the derivative of the foregoing expression with respect to q, set it equal to zero, and solve the equation for q^*:

$$\frac{d(\text{ATC})}{dq} = 0.008q - \frac{8000}{q^2}$$

$$\frac{d(\text{ATC})}{dq} = 0 \quad \text{where} \quad 0.008q - \frac{8000}{q^2} = 0$$

$$0.008q^3 - 8000 = 0$$

$$q^3 = \frac{8000}{0.008}$$

$$q^3 = 1,000,000$$

$$q^* = (1,000,000)^{1/3}$$

$$q^* = 100.$$

3. Substitute the value of $q*$ into the expression for average total cost to find the desired price:

$$ATC(100) = 0.004(100)^2 + 120 + \frac{8000}{100} = \$240.$$

Hence, it is possible to break even at a selling price of $240 per unit if $q = 100$ units are made and sold, but it is not possible to break even at any output level if the selling price is less than $240 per unit.

9.32 PROBLEM SET 9–11

1. The cost of producing q units of a product is

$$C(q) = 0.001q^3 - 0.15q^2 + 10q + 100$$

and each unit is sold for $10.
a) What level of output leads to maximum profit?
b) Compute maximum profit.
c) Find the shutdown price level.

2. The cost of producing q units is

$$C(q) = 0.0015q^3 - 0.9q^2 + 200q + 28,000$$

and each unit is sold for $200 per unit.
a) What level of output leads to maximum profit?
b) Compute maximum profit.
c) Find the shutdown price level.

3. The cost of producing q units is

$$C(q) = 0.001q^3 - 0.15q^2 + 10q + 100$$

and each unit is sold for $3.70.
a) What level of output leads to maximum profit?
b) Compute maximum profit and interpret the outcome.
c) Would the producer shut down at this price level? Why or why not?

4. The cost of producing q units is

$$C(q) = 0.0008q^3 - 0.6q^2 + 300q + 60,000$$

and each unit is sold for $300.
a) What level of output leads to maximum profit?
b) Compute maximum profit and interpret the outcome.
c) Would the producer shut down at this price level? Why or why not?

5. Find the lowest selling price at which break-even is possible if

$$C(q) = 0.01q^2 + 11q + 400.$$

6. Find the lowest selling price at which break-even is possible if

$$C(q) = 0.05q^2 + 12q + 320.$$

7. Find the lowest selling price at which break-even is possible if

$$C(q) = 0.0004q^3 + 10q + 100.$$

8. (Optional: Trial-and-error solution by hand calculator.) Find the lowest selling price at which break-even is possible if

$$C(q) = 0.0015q^3 - 0.9q^2 + 200q + 28{,}000.$$

9. If $C(q)$ is the total cost of producing q units of a product, and $R(q)$ is total revenue received from the sale of these q units
   a) Write the expression for profit when q units are produced and sold.
   b) Prove that marginal cost must equal marginal revenue if profit is to be maximized.
   c) If $R(q)$ is a linear function and $C(q)$ is a cubic where maximum profit occurs to the right of the inflection point, prove that b) yields a maximum by the second derivative test. That is, write the expression for $P''(q)$ and state why it must be negative.

10. Given a cost function, $C(q)$,
    a) Write the expression for average total cost, ATC(q), when q units are produced.
    b) Prove that if ATC(q) is to be a minimum, average total cost must equal marginal cost.
    c) Find the second derivative of ATC(q), substitute the expression for $C'(q)$ from (b) into this derivative and simplify the resulting expression as much as possible. Now if $C''(q)$ is positive, as it is for a cubic cost curve to the right of its inflection, state why the condition in (b) yields a minimum.

9.33 REVIEW PROBLEMS

1. If $f(x) = 3x^2 + 2x + 5$, find the value of, or the algebraic expression for:

   a) $f(0)$. b) $f(1)$. c) $f(-1)$. d) $f(5)$. e) $f(2a)$. f) $f\left(\dfrac{1}{x+1}\right)$.
   g) $f(x+1) - f(x)$.   h) $f(x) - f(x - a)$.

2. a) If $f(x) = 2xy$, write the expression for $[f(a)]^2$.
   b) If $g(y) = 2xy$, write the expression for $g(a - 1)$.

3. If $f(x) = 32x^{-1} - 2x^{-2/3} + 24x^{-1/2}$, find $f(64)$.

4. Find the expression for $\Delta f(x)$ if
   a) $f(x) = 7 - 2x$.   b) $f(x) = x^2 + 3x - 6$.

- 5. If $f(x) = 3x^2 - 2x + 4$;
    a) Write the expression for $\Delta f(x)$.
    b) Compute $\Delta f(x)$ if x changes from 2 by the amount $\Delta x = 0.1$.

6. The total cost of making p pounds of Poundal is $C(p)$ dollars, where

$$C(p) = 50 + 1.5p + 0.02p^2.$$

   a) Write the expression for the marginal cost of the pth gallon.
   b) Find the marginal cost of the fifth gallon.
   c) Find the marginal cost of the fortieth gallon.

Find each of the following limits if it exists.

7. $\lim\limits_{x \to 0} (3x^2 - 2x + 5)$.

8. $\lim\limits_{b \to a} (a^3 + 3a^2b + 3ab^2 + b^3)$.

9. $\lim\limits_{x \to 1} \left(\dfrac{x^2 + 1}{x - 1} \right)$.

10. $\lim\limits_{x \to 4} \left(\dfrac{3x}{0.5x - 2} \right)$.

11. $\lim\limits_{x \to 8} (2x^{-2/3})$.

12. $\lim\limits_{x \to 0} \left(\dfrac{x^{3/2}}{x} \right)$.

13. $\lim\limits_{a \to b} (a^2 + 2ab + b^2)$.

14. $\lim\limits_{x \to 1.25} \left(\dfrac{16x^2 - 25}{4x - 5} \right)$.

15. $\lim\limits_{x \to 0} \left(\dfrac{x}{x^{1/3}} \right)$.

16. $\lim\limits_{\Delta x \to 0} \left[\dfrac{\Delta x(x - 1)}{\Delta x} \right]$.

17. $\lim\limits_{x \to 0} \left(8 + \dfrac{x^2}{x} \right)^{1/3}$.

18. $\lim\limits_{a \to 0} \dfrac{\dfrac{1}{(x + a)^2} - \dfrac{1}{x^2}}{a} ; x \neq 0$.

19. Write the delta limit definition for the derivative of $f(x)$.

Apply the delta definition of Problem 19 to find the derivatives for Problems 20–25.

20. $f(x) = 2x^2 + 3$.

21. $f(x) = x^2 + 3x - 2$.

22. $f(x) = \dfrac{1}{1 - x}$.

23. $f(x) = \dfrac{2}{3x - 5}$.

24. $f(x) = x^{1/2}$. [Hint: After the difference quotient has been set up, multiply its numerator and denominator by $(x + \Delta x)^{1/2} + x^{1/2}$ and proceed.]

25. $f(x) = x^{3/2}$. [Hint: After the difference quotient has been set up, multiply its numerator and denominator by $(x + \Delta x)^{3/2} + x^{3/2}$ and proceed.]

26. Find the derivative by the delta limit definition, then find the slope of the curve at the point where $x = 2$ if $f(x) = x^2 - 5x$.

27. Find the derivative by the delta limit definition, then find the slope of the curve at the point where $x = 2$ if $f(x) = 12/x$.

In Problems 28–39, find $f'(x)$. Do not leave negative exponents in answers.

28. $f(x) = 2x - 3$.

29. $f(x) = 5 - 4x$.

30. $f(x) = 1.5x^2 - 2x^3 - 4x + 5$.

31. $f(x) = 0.25x^4 - 2.5x^2 + x - 6$.

32. $f(x) = ax^3 - bx + 1$.

33. $f(x) = xy - ax^2$.

34. $f(x) = \dfrac{3}{2x^2} - \dfrac{5}{x}$.

35. $f(x) = \dfrac{1}{3x^3} - \dfrac{1}{x}$.

36. $f(x) = 3x^{1/2} + 2x^{1/3} + 1$.

37. $f(x) = 6x^{4/3} - 3x^{2/3} - 2x + 6$.

38. $f(x) = \dfrac{12}{x^{2/3}} - \dfrac{16}{x^{3/4}} + x$.

39. $f(x) = x^{-1/2} - x^{-1/3} + 4x - 25$.

Find each of the following:

40. $\dfrac{d}{dx} (ax^3 - bx^2 + cx - d)$.

41. $\dfrac{d}{dy} (xy^2 - 2y + 3x^2y - 2x + 3)$.

Find the slope of the line tangent to each curve at the indicated value of x:

42. $f(x) = 2 - 0.5x;\ x = 2.$

43. $f(x) = \dfrac{x^3}{6} - \dfrac{18}{x} + 2x - 1;\ x = 2.$

44. $f(x) = 18x^{1/3};\ x = 8.$

45. $f(x) = x + \dfrac{12}{x^{1/2}};\ x = 4.$

Find the value(s) of x for which the slope is zero.

46. $f(x) = 0.25x^2 - x + 4.$

47. $f(x) = 2x + \dfrac{72}{x}.$

48. $f(x) = 4x^3 - 3x^2.$

49. $f(x) = x^3 - 12x.$

50. $f(x) = x - 3x^{1/2}.$

51. $f(x) = 0.1x + \dfrac{12.8}{x^{1/2}}.$

For Problems 52 and 53, find the maximum or minimum value of $f(x)$, state which has been found, and describe the graph of $f(x)$.

52. $f(x) = 0.5x^2 - 50x + 2500.$ 53. $f(x) = 10x - 0.2x^2 - 5.$

54. If the total cost of producing y yards of Yardall is, in dollars,

$$C(y) = 0.002y^2 + 5y + 100,$$

find marginal cost at outputs of:

a) 2000 yards.   b) 2500 yards.

55. If the total cost of producing t tons of Tonall is, in dollars,

$$C(t) = 0.001t^3 - 0.15t^2 + 10t + 100,$$

find marginal cost at outputs of:

a) 10 tons.   b) 50 tons.   c) 60 tons.

For Problems 56–65, find the coordinates of local extreme points, if any exist. In each case, state the type of point which has been found, and prove the statement by testing.

56. $f(x) = 0.1x^2 - 4x + 50.$

57. $f(x) = 20x - x^2.$

58. $f(x) = 2x^3 - 3x^2 - 12x.$

59. $f(x) = 8x - x^3 - 5x^2 + 50.$

60. $f(x) = x^3 + x^2 + x - 4.$

61. $f(x) = 2x^4 - 216x + 500.$

62. $f(x) = 3x^4 - 16x^3 + 24x^2 + 10.$

63. $f(x) = 2x^3 - 9x^2 + 12x.$

64. $f(x) = 2x + \dfrac{98}{x}.$

65. $f(x) = 96x^{1/2} - 6x.$

66. Interior and exterior walls of a rectangular 80,000-square-foot warehouse cost \$90 per running foot. The warehouse is to be divided into ten rooms by four interior walls in the x direction and one interior wall in the y direction. What should be the warehouse dimensions if wall cost is to be minimized?

67. A rectangular area is to be enclosed, then divided into thirds by two fences across the area parallel to one pair of the sides. If the area to be enclosed is 1250 square feet, what dimensions will lead to the use of a minimum amount of fence?

68. (See Problem 67.) If the fence on the two ends costs \$0.64 per running foot and the dividers and the sides cost \$1 per running foot, what should

be the dimensions if the cost of the fence is to be a minimum, and what is this minimum cost?

69. If a rectangular area is to be fenced in the manner of Problem 67, what is the maximum area that could be enclosed with 1000 feet of fence?

70. A rectangular manufacturing plant with a floor area of 5400 square feet is to be built in a location where zoning regulations require buffer strips 30 feet wide front and back, and 20 feet wide at either end. (A buffer strip is a grass and tree belt that must not be built upon.) What plot dimensions will lead to minimum total area for plant and buffer strips? What is this minimum total area?

71. A box with square bottom and no top is to contain 32 cubic inches. Find the dimensions that will lead to a box of minimum area. What is this minimum area?

72. If the bottom material for the box in Problem 71 costs eight cents per square inch, and the side material costs one cent per square inch, what dimensions will lead to minimum cost? What is this minimum cost?

73. A box with no top is to be made from an 8 by 15 inch piece of cardboard by cutting equal sized squares from the corners, then turning up the sides. What should be the dimensions of the squares if the box is to have maximum volume? (Note: The quadratic expression encountered in the solution is factorable.)

74. A parcel delivery service accepts cylindrical packages whose length, L, plus girth, $2\pi r$, does not exceed 120 inches. A shipper who uses cylindrical cartons, perforated, wishes to design a carton with maximum ventilation (area). What should be the length and radius of the carton?

75. An appliance service company is located centrally in a roughly square area x miles on a side. It charges $27 per call, not including parts and labor, and travel cost is figured at $1.50 per mile. The average distance traveled per call is $1.2x$ miles. In a month, the average number of calls per square mile of service area is 30.
    a) What should x be if net travel income (which excludes parts and labor) is to be maximized?
    b) What is this maximum?

76. Following the inventory model of this chapter (Section 9.20), suppose it costs $250 to set up a plant to make a batch (lot) of a product and $2 for each unit made after set-up. Inventory cost is $1.20 per year per unit in average inventory. The annual requirement for the product is 9600 units.
    a) Write the expression for total annual cost as a function of the lot size, L.
    b) What lot size yields minimum cost?

77. A parcel delivery service picks up and delivers packages whose length plus girth does not exceed M inches. A shipper uses rectangular cartons with square ends. In terms of the parameter M, find the dimensions the carton should have if its volume is to be maximized.

In Problems 78–85, find $f'(x)$. Simplify where possible. Leave no negative exponents in answers.

78. $f(x) = (3x - 2)^{15}$.

79. $f(x) = (2x^3 + 3x^2 + 4x - 50)^{3/2}$.

80. $f(x) = (x^3 - 3x^2 + 5)^{1/3}$.

81. $f(x) = (10x)^{1/2}$.

82. $f(x) = \dfrac{1}{4(8x - 6)}$.

83. $f(x) = \left(5 - \dfrac{2}{x}\right)^3$.

84. $f(x) = \dfrac{2}{27(10 - 3x)^{3/2}}$.

85. $f(x) = x - \dfrac{1}{(5 - 0.3x)^{1/3}}$.

86. Find the slope of the curve at the point where $x = 1$.

$$f(x) = \left(5 + \frac{3}{x}\right)^{1/3}.$$

87. Find the value of x at the points on the curve where the tangent line is horizontal.

$$f(x) = x + \frac{50}{(2 + 0.5x)}.$$

88. Given the function

$$f(x) = 25(x^2 + 9)^{1/2} + 20(10 - x), \quad x \geq 0,$$

   a) Find the x-coordinate of the local extreme point.
   b) By the left-right original function test determine whether the extreme point is a local maximum or minimum.

89. When x tons of Tonall are produced, the average cost per gallon is $A(x)$, where

$$A(x) = \frac{4000}{0.1x + 20} + 0.25x, \quad x > 0.$$

   a) Find the value of x at the local extreme point of $A(x)$.
   b) Prove that *a)* is a local minimum.
   c) Compute the minimum average cost per ton.

90. Profit per tree grown and sold by a tree grower depends upon the height of a tree at the time of sale. Taking h as tree height in inches, the profit per tree, in dollars, is approximated by

$$p(h) = (10 + 2h)^{1/2} - 0.1h.$$

   a) What tree height provides maximum profit per tree?
   b) What is the maximum profit per tree?

91. Points B, C, and D are on a horizontal line with C between B and D. Point A is 4.2 miles vertically above B, and BD is 14 miles. Straight road sections are to be constructed from A to C, then from C to D. Along BD, construction cost is $200 thousand per mile and above BD, cost is $290 thousand per mile.
   a) How far from B should the intersection at C be if construction cost is to be minimized?

    b) Compute the minimum construction cost.

    c) How much is saved by the two-section construction choice rather than a single road from A to D?

92. (Optional.) This question has the material on summation (see Index for reference) as a prerequisite.

Given any set of n numbers, x_1, x_2, x_3, $\cdots$ x_n:

    a) What is the summation expression for the average of the n numbers?

    b) Let a be an arbitrary number, subtract a from each x_i, square this, and sum for all x_i to obtain

$$S(a) = \sum_{i=1}^{n} (x_i - a)^2.$$

Find the expression for a which will minimize the sum of squares, $S(a)$.

    c) How is this expression related to the set of numbers?

93. (Note: This problem is algebraically difficult, so we shall give the answer, $x = 5/3$ miles.) Point $R(x, 0)$ is on the x-axis. To its left, and above, is point $P(1, 1)$ and to its right, also above, is $Q(3, 2)$. A plant is to be built at R and the sum of its distances from points P and Q is to be minimized. Find the x-coordinate of $R(x, 0)$. Note that x is between 1 and 3, so x must be positive.

94. Given

$$f(x) = h(a^2 + x^2)^{1/2} + g(b - x),$$

where a, b, h, and g are all positive and $h < g$, find the expression for x if it is to mark a point where $f(x)$ has a local extreme point.

Find the first derivative for Problems 95–102. Simplify results where possible. Do not leave negative exponents or complex fractions (fractions containing fractions) in answers.

95. $f(x) = x(2x - 1)^6$.

96. $f(x) = 2x(x^3 - 5x)^{1/2}$.

97. $f(x) = 2x^2(3x^2 + 5)^{1/2}$.

98. $f(x) = (2x + 1)(4x - 5)^{1/2}$.

99. $f(x) = \dfrac{3x + 2}{2 - 3x}$.

100. $f(x) = \dfrac{x}{1 - 0.5x^2}$.

101. $f(x) = \dfrac{3x + 2}{(6x^2 + 5)^{1/3}}$.

102. $f(x) = \dfrac{2x - 1}{(2x + 3)^{1/2}}$.

Find the value of x at local extreme points for Problems 103–105. State whether each extreme point is a local maximum or minimum.

103. $f(x) = \dfrac{3x}{8 + 0.5x^2}$, $x > 0$.

104. $f(x) = x(1 - x)^{1/2}$, $x \leq 1$.

105. $f(x) = 2x + \dfrac{8x}{0.5x - 1}$.

For Problems 106–115, sketch graphs of the functions. Place coordinates of extreme points and inflection points on the sketch. If a curve has an asymptote, put it and its equation on the graph. Graphs need not be to scale, and coordinate axes need not be shown.

106. $f(x) = 8x - 0.5x^2 - 20.$

107. $f(x) = x^2 - 10x + 35.$

108. $f(x) = x^3 - 6x^2 + 15x - 4.$

109. $f(x) = x^3 - 12x^2 + 21x + 100.$

110. $f(x) = 32x - (x + 7)^4 + 200.$

111. $f(x) = 0.04x^4 - 2x^2 + 30.$

112. $f(x) = x^4 - 16x^3 + 72x^2 - 128.$

113. $f(x) = 2x - \dfrac{50x}{1 - 0.2x} + 15.$

114. $f(x) = \dfrac{2x}{0.5x - 4}.$

115. $f(x) = \dfrac{36}{x}.$

116. The cost of producing q units of a product is

$$C(q) = 0.0005q^3 - 0.1q^2 + 10q + 30{,}000$$

and each unit is sold at a price of \$285.
a) What output leads to maximum profit?
b) Compute the maximum profit.
c) Find the shutdown price level.

117. Find the lowest selling price per unit at which break-even is possible if the cost of producing q units is

$$C(q) = 0.004q^3 + 10q + 1000$$

and all units made are sold.

118. (Optional—Trial-and-error solution by hand calculator.) Find the lowest selling price per unit at which break-even is possible if the cost of producing q units is

$$C(q) = 0.001q^3 - 0.15q^2 + 10q + 100$$

and all units made are sold.

119. The cost of producing q units of a product is

$$C(q) = 0.0008q^3 - 0.6q^2 + 300q + 60{,}000.$$

a) What output leads to maximum profit?
b) If selling price is \$300 per unit, compute maximum profit and interpret the outcome.
c) Should the producer operate at this price level? Why or why not?

Further topics in differential calculus

10

10.1 INTRODUCTION

In Chapter 9, we learned the basic ideas of differential calculus and applied them to solve numerous optimization (maximum-minimum) problems that could be formulated in terms of one class of functions known as power functions. Optimization problems arise in many situations where the appropriate formulation includes functions other than power functions, so in the first part of this chapter, problem-solving ability is expanded by developing derivative rules for exponential and logarithmic functions and applying them to optimization problems. Next we provide a rule (the chain rule) that makes it possible to find the derivative when a functional form is implied, but not stated explicitly, and apply this rule to develop the formula for the *multiplier*, which is a fundamental determinant of the behavior of the economy. As we shall see, a derivative involved in the multiplier is a *rate*, and rate interpretations are a new (non-optimization) application of the derivative that we proceed to explore and illustrate in a variety of situations. In the last part of the chapter we again expand our set of calculus tools by showing how optimization problems can be solved, and rate interpretations can be made, when a function has *two* independent variables rather than one, as has been the case up until this point. This introduction to multivariate calculus and its applications concludes our work in differential calculus.

The presentation in this chapter assumes that the reader is familiar with the basic ideas presented in Chapter 9. Additionally, it is assumed that the reader is familiar with natural logarithms and the rules of logarithms presented in the first part of Chapter 7.

10.2 DERIVATIVES OF EXPONENTIAL FUNCTIONS

If $1000 is deposited in a bank account that earns interest at the rate of 8 percent compounded annually, the amount in the account after t years is

$$A(t) = 1000(1.08)^t.$$

Observe that the independent variable, t, in the last expression is the exponent of the power of the constant base, 1.08. Functions that have a constant base and a variable exponent are called *exponential functions*. Other examples are

$$f(x) = 2^x, \quad g(x) = e^{-0.1x}, \quad h(x) = 3e^{2x-5},$$

where, in $g(x)$ and $h(x)$,

$$e = 2.718282$$

to six places of decimals. This important constant is the base of the natural (ln) system of logarithms presented in Chapter 7. Calculus applications involving exponential functions, of which there are many, typically are expressed with e as the base because this choice leads to a remarkably simple derivative, which is

$$\frac{d}{dx}(e^x) = e^x;$$

that is, the derivative of the function e^x is the function itself. To show how the last statement comes about, we start as always with the definition of the derivative,

$$\frac{df(x)}{dx} = \lim_{\Delta x \to 0} \frac{f(x + \Delta x) - f(x)}{\Delta x}.$$

Hence,

$$\frac{d(e^x)}{dx} = \lim_{\Delta x \to 0} \frac{e^{x+\Delta x} - e^x}{\Delta x}. \tag{1}$$

Now, by a rule of exponents, we can write

$$e^{x+\Delta x} = e^x e^{\Delta x}$$

and use this to re-write (1) as

$$\frac{d(e^x)}{dx} = \lim_{\Delta x \to 0} \frac{e^x e^{\Delta x} - e^x}{\Delta x}$$

and, by factoring,

$$\frac{d(e^x)}{dx} = \lim_{\Delta x \to 0} e^x \left[\frac{e^{\Delta x} - 1}{\Delta x} \right]$$

$$= e^x \lim_{\Delta x \to 0} \left[\frac{e^{\Delta x} - 1}{\Delta x} \right] \tag{2}$$

where, in the last line, we have placed e^x outside the limit symbol. This is permitted because x is to be thought of as being a coordinate of a fixed point that does not vary as Δx changes and approaches zero.

We cannot evaluate the limit needed in the foregoing by any elementary procedure so, instead, we state and illustrate what happens to

$$\frac{e^{\Delta x} - 1}{\Delta x}$$

as Δx approaches zero. The important point, which can be proved rigorously, is that as Δx becomes smaller and smaller, approaching zero, $e^{\Delta x}$ gets closer and closer to the value $(1 + \Delta x)$. To illustrate, we find in Table IX at the back of the book that for $\Delta x = 0.02$

$$e^{\Delta x} = e^{0.02} = 1.02\ 02 \text{ compared to } (1 + \Delta x) = 1.02,$$

and with a smaller Δx, $\Delta x = 0.01$

$$e^{\Delta x} = e^{0.01} = 1.01\ 01 \text{ compared to } (1 + \Delta x) = 1.01.$$

Readers having appropriate hand calculators may verify that with $\Delta x = 0.001$

$$e^{\Delta x} = e^{0.001} = 1.001\ 000\ 5, \text{ compared to } (1 + \Delta x) = 1.001.$$

Note in the last three arithmetic expressions that as Δx becomes smaller, $e^{\Delta x}$ becomes closer to $(1 + \Delta x)$. Accepting as a fact that

$$e^{\Delta x} \text{ approaches } (1 + \Delta x) \text{ as } \Delta x \to 0,$$

the ratio in (2) approaches 1 as a limit. That is,

$$\frac{e^{\Delta x} - 1}{\Delta x} \text{ approaches } \frac{(1 + \Delta x) - 1}{\Delta x} = \frac{\Delta x}{\Delta x} = 1.$$

Hence, (2) becomes

$$\frac{d(e^x)}{dx} = e^x \lim_{\Delta x \to 0} \left(\frac{e^{\Delta x} - 1}{\Delta x}\right) = e^x(1) = e^x.$$

Simple exponential rule, base e: $\dfrac{d(e^x)}{dx} = e^x.$

The simple function, e^x, is not nearly so common as expressions such as

$$e^{-0.1x}, \quad e^{-0.5x^2}, \quad e^{2x-3},$$

or, in general,

$$e^{f(x)},$$

where $f(x)$ is some function other than simply x itself. The reader has had much practice with the power function rule, which states

$$\frac{d[f(x)]^n}{dx} = n[f(x)]^{n-1}f'(x).$$

The procedure that led to this rule, when applied to

$$\frac{de^{f(x)}}{dx},$$

yields a corresponding result.

Exponential function rule, base e: $\dfrac{d}{dx}[e^{f(x)}] = e^{f(x)}f'(x).$

Example. Find the first and second derivatives of

$$f(x) = e^{-0.5x}.$$

We have

$$f'(x) = \frac{d}{dx}(e^{-0.5x}) = e^{-0.5x}\frac{d}{dx}(-0.5x)$$

$$= e^{-0.5x}(-0.5)$$

$$f''(x) = \frac{d}{dx}[e^{-0.5x}(-0.5)]$$

$$= -0.5\frac{d}{dx}e^{-0.5x}$$

$$= -0.5\left[e^{-0.5x}\frac{d}{dx}(-0.5x)\right]$$

$$= -0.5(e^{-0.5x})(-0.5)$$

$$= 0.25e^{-0.5x}.$$

> **Exercise.** Find the first and second derivatives of
>
> $$f(x) = e^{0.1x+2}.$$
>
> Answer: $f'(x) = 0.1e^{0.1x+2}$. $f''(x) = 0.01e^{0.1x+2}$.

The important idea to be kept in mind is that the derivative of an exponential expression with base e consists of two factors. The first is the expression itself, and the second is the derivative of the function in the exponent.

Example. Find the local extreme point of $f(x)$, and prove it is a local minimum.

$$f(x) = e^{0.2x} - 3x + 50.$$

We have

$$f'(x) = e^{0.2x}(0.2) - 3$$
$$f'(x) = 0 \quad \text{where} \quad e^{0.2x}(0.2) - 3 = 0$$
$$0.2e^{0.2x} = 3$$
$$e^{0.2x} = \frac{3}{0.2}$$
$$e^{0.2x} = 15.$$

We next take the natural logarithm of both sides of the last equation and write

$$\ln (e^{0.2x}) = \ln 15.$$

By a rule of logarithms (see Chapter 7), the last may be written as

$$0.2x \ln e = \ln 15$$
$$0.2x(1) = \ln 15$$

because $\ln e = 1$. Hence,

$$x^* = \frac{\ln 15}{0.2} = \frac{2.70805}{0.2} = 13.54025$$

and x^* is a candidate extreme point. To evaluate the function

$$f(x) = e^{0.2x} - 3x + 50$$

at x^*, remember that x^* came from solving

$$e^{0.2x} = 15.$$

Hence,

$$f(x^*) = e^{0.2x} - 3(13.54025) + 50$$
$$= 15 - 3(13.54025) + 50$$
$$= 24.37925.$$

To prove we have a local minimum, we return to

$$f'(x) = 0.2e^{0.2x} - 3$$

and find

$$f''(x) = 0.2e^{0.2x}(0.2)$$
$$= 0.04e^{0.2x}$$
$$f''(x^*) = 0.04e^{0.2(13.54025)}.$$

There is no need to evaluate $f''(x^*)$ because *e to any power is positive,* so $f''(x^*)$ is positive, proving x^* is at a local minimum point. We conclude that

$$f_{min} = f(13.54025) = 24.37925.$$

> **Exercise.** Given $f(x) = e^{2x} - 7x + 5.88466$: *a)* Find the local minimum value of $f(x)$. *b)* Prove *(a)* is a local minimum. Answer: *a)* $f_{min} = 5$. *b)* $f''(x) = 4e^{2x}$, which is always positive.

The derivative of $f(x) = a^x$. The exponential rule we have used requires that the base of the exponential be e. To generalize the rule to cover all bases, a, where $a > 0$ but not equal to 1, we consider the function a^x. By a rule of logarithms,

$$\ln a^x = x \ln a.$$

By the definition of a logarithm (see Chapter 7), the last means

$$a^x = e^{x \ln a}. \tag{3}$$

Here we have converted a^x into a form whose derivative we can take by applying the exponential function rule. Thus:

$$\frac{d}{dx}(a^x) = \frac{d}{dx} e^{x \ln a}$$

$$= e^{x \ln a} \frac{d}{dx}(x \ln a)$$

$$\frac{d}{dx}(a^x) = e^{x \ln a}(\ln a). \tag{4}$$

Substituting from (3) into (4) we have

$$\frac{d}{dx} a^x = e^{x \ln a}(\ln a) = a^x \ln a.$$

The last says that for a base a, the exponential rule contains a conversion factor, $\ln a$. The same factor appears in the exponential function rule. Thus:

520

Exponential rules, base _a:_

$$\frac{d(a^x)}{dx} = a^x \ln a$$

$$\frac{d}{dx}[a^{f(x)}] = a^{f(x)} \cdot f'(x) \ln a.$$

Example. Find the derivative of _a)_ $f(x) = 2(3)^x$. _b)_ $g(x) = 10^{-0.3x}$. We have:

_a)_ $f'(x) = 2\dfrac{d}{dx}(3)^x$

$= 2(3)^x \ln 3$

$= 2(3^x)(1.09861)$

$= (2.19722)(3^x).$

_b)_ $g'(x) = \dfrac{d}{dx}(10^{-0.3x})$

$= 10^{-0.3x}\dfrac{d}{dx}(-0.3\,x) \ln 10$

$= 10^{-0.3x}(-0.3) \ln 10$

$= 10^{-0.3x}(-0.3)(2.30259)$

$= -(0.69078)(10^{-0.3x}).$

> **_Exercise._** Find the derivative of: _a)_ $f(x) = 2^x$. _b)_ $g(x) = 5 - (4^{1-2x})$. Answer: _a)_ $f'(x) = (0.69315)(2^x)$. _b)_ $g'(x) = (2.77258)$ (4^{1-2x}), from Table X; or more accurately by hand calculator, $(2.77259)(4^{1-2x})$.

Graphs of exponential functions. The characteristics of exponential functions are shown in Figures 10–1 and 10–2 by the sketches of

$$f(x) = 2^x \quad \text{and} \quad g(x) = e^{-0.1x}.$$

FIGURE 10-1 (not to scale)

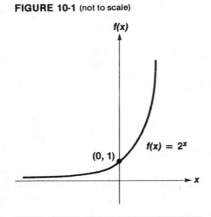

x	$f(x) = 2^x$
0	1
1	2
2	4
3	8
4	16
5	32
−1	1/2
−2	1/4
−3	1/8
−4	1/16
−5	1/32
−6	1/64

FIGURE 10-2 (not to scale)

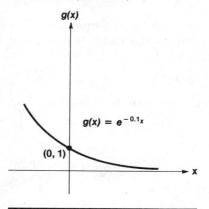

x	$g(x) = e^{-0.1x}$
0	1
10	0.37
20	0.14
30	0.05
40	0.02
−10	2.72
−20	7.39
−30	20.09
−40	54.60
−50	148.41

Observe that the value of x making the exponent zero, which in the case of both $f(x)$ and $g(x)$ is $x = 0$, yields the value 1 for the exponential. Both functions approach the x-axis asymptotically. In the case of $f(x) = 2^x$, we see that

$$f(-4) = 2^{-4} = \frac{1}{16}, \quad f(-5) = 2^{-5} = \frac{1}{32},$$

so this function approaches 0 (the x-axis) as we move to the left. The opposite holds for $g(x)$, as shown in the table accompanying Figure 10–2. It is important to remember that the base of an exponential must be positive, but not 1, and such a base to *any* power is always positive—never zero or negative.

It follows from the foregoing that an exponential can be sketched if we know what its asymptote is, in which direction (to the left or right) it approaches the asymptote, and have one point, the *zero exponent point,* as a starting point. The functions we will want to sketch consist of a constant term, which may be zero, and an exponential term as in

$$h(x) = 10 + 3(2^{1-x}),$$

where the constant term is 10 and the exponential term is

$$3(2^{1-x}).$$

Observe that the base of the exponential, 2, is greater than one and that the coefficient of x is negative. If we move to the right on the graph to larger values of x, such as $x = 100$, then

$$3(2^{1-x}) = 3(2^{-99}) = \frac{3}{2^{99}}$$

is a very small number, and becomes ever smaller as x increases. This means the curve approaches its asymptote to the right. Inasmuch as the exponential term approaches zero,

$$h(x) = 10 + 3(2^{1-x})$$

approaches 10, so the asymptote, which we shall symbolize as $A(x)$, is the horizontal line

$$A(x) = 10.$$

To obtain the zero exponent starting point, we note that the exponent

$$1 - x = 0 \quad \text{where} \quad x = 1$$

and

$$h(1) = 10 + 3(2)^0 = 10 + 3(1) = 13.$$

Hence $(1, 13)$ is the starting point. The curve may now be sketched as shown in Figure 10–3.

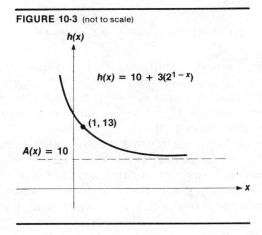

FIGURE 10-3 (not to scale)

$h(x)$

$h(x) = 10 + 3(2^{1-x})$

$(1, 13)$

$A(x) = 10$

x

If the base of the exponential is greater than one and the coefficient of x is positive, as in Figure 10–1, the curve approaches its asymptote to the left. Hence for bases greater than one, the curve approaches its asymptote to the right (left) if the coefficient of x is negative (positive). If the base is between zero and one, its *reciprocal* is greater than one, and we can change to the reciprocal by changing the sign of the exponent. For example,

$$f(x)=(0.4)^x=\left(\frac{1}{0.4}\right)^{-x}=(2.5)^{-x},$$

and, by the right-left principle just stated for bases greater than 1, $f(x)$ approaches its asymptote to the right.

Procedure for sketching exponentials

1. Draw the horizontal asymptote, which is $A(x) =$ constant term.
2. Find the zero exponent point.
3. Find the *direction of approach indicator* by taking the sign of x for bases greater than 1 and the opposite sign for bases between 0 and one. A positive (negative) indicator means approach is to the left (right).

Example.   Describe the characteristics of the graph of

$$f(x)=20-9(2^{0.5x-3}).$$

1. The asymptote is the horizontal line $A(x)=20$.
2. The zero exponent occurs where

$$0.5x-3=0$$
$$0.5x=3$$
$$x=\frac{3}{0.5}=6.$$

$$f(6)=20-9(2^0)=11.$$

Hence, $(6, 11)$ is a starting point.

3. The base is greater than 1, so the positive coefficient of x means the curve approaches its asymptote to the left.

FIGURE 10-4 (not to scale)

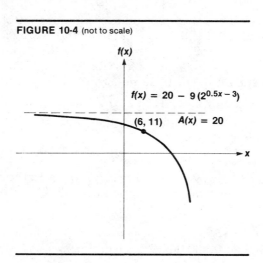

$f(x) = 20 - 9\,(2^{0.5x - 3})$

$(6, 11)$ $A(x) = 20$

As shown in Figure 10–4, the curve rises to the left of (6, 11) toward the asymptote and falls ever more steeply to the right of (6, 11).

Exercise. Describe the graph of $f(x) = 20 + 10e^{3-0.5x}$. Answer: Starting at (6, 30) the curve falls to the right toward its asymptote, $A(x) = 20$. To the left of (6, 30) it rises ever more steeply.

10.3 RESPONSE FUNCTIONS

In the planning stages of a promotional effort to sell a new product, the Newprod Company uses the exponential function

$$r(t) = 0.40 - 0.40e^{-0.02t}$$

as an indicator of the proportion of all potential customers who will have responded to the promotion in its first t days of operation. Thus, for example, during the first 50 days

$$\begin{aligned}
r(50) &= 0.40 - 0.40e^{-0.02(50)} \\
&= 0.40 - 0.40e^{-1} \\
&= 0.40 - 0.40(0.3679) \\
&= 0.253
\end{aligned}$$

or 25.3 percent of potential customers are expected to respond. It follows from the discussion in the last section that the exponential response function has $A(x) = 0.40$ as its asymptote and that the proportion responding rises as time goes on (t increases), but never reaches 0.40.

Newprod market research personnel estimate the total number of potential customers at 5,000,000 and that, on the average, one response will yield a revenue of $2. The cost of running the promotion consists of a fixed cost of $155,200 plus a variable cost of $24,000 each day the promotion is continued. In summary:

Proportion responding in t days: $r(t) = 0.40 - 0.40e^{-0.02t}$.
Number of potential customers: 5,000,000.
Average revenue per response: $2.
Cost in t days: $C(t) = (155,200 + 24,000t)$ dollars.

Total revenue in t days, symbolized as $R(t)$, will be $2 times the number of responses, and the number of responses will be $r(t)$, the proportion responding, times 5,000,000. Thus:

$$R(t) = \$2(\text{proportion responding})(5,000,000)$$
$$R(t) = 10,000,000(0.4 - 0.4e^{-0.02t}).$$

Profit at t days, $P(t)$, will be total revenue minus cost

$$P(t) = R(t) - C(t)$$
$$= 10,000,000(0.4 - 0.4e^{-0.02t}) - (155,200 + 24,000t)$$
$$P(t) = 4,000,000 - 4,000,000e^{-0.02t} - 155,200 - 24,000t.$$

Newprod now wants to know how many days the promotion should be continued to maximize profit. Taking the derivative,

$$P'(t) = -4,000,000e^{-0.02t}(-0.02) - 24,000$$
$$= 80,000e^{-0.02t} - 24,000.$$

To have an extreme point $P'(t)$ must be zero. Hence,

$$80,000e^{-0.02t} - 24,000 = 0$$
$$80,000e^{-0.02t} = 24,000$$
$$e^{-0.02t} = \frac{24,000}{80,000}$$
$$e^{-0.02t} = 0.3.$$

Taking the natural logarithm of both sides yields

$$\ln (e^{-0.02t}) = \ln(0.3)$$
$$-0.02t \ln e = \ln(0.3)$$
$$-0.02t = \ln(0.3)$$
$$t = \frac{\ln(0.3)}{-0.02}$$
$$= \frac{-1.20397}{-0.02}$$
$$t^* = 60.2 \text{ days.}$$

It may be verified that $P''(t)$ is negative for all values of t, so we have maximum profit if the promotion is continued for about 60 days. To compute the maximum, we find $P(t^*)$, remembering that $e^{-0.02t} = 0.3$. Thus,

$$P_{max} = P(60.2) = 4,000,000 - 4,000,000(0.3) - 155,200 - 24,000(60.2)$$
$$= \$1,200,000.$$

Practice with response functions will be provided in the next set of problems.

10.4 PROBLEM SET 10–1

Find the first and second derivatives of each of the following functions.

1. $f(x) = 2e^x$.
2. $f(x) = 5^x$.
3. $f(x) = 7^x$.
4. $f(x) = 3e^x$.
5. $f(x) = 5e^{-0.2x}$.
6. $f(x) = 2.5e^{-0.4x}$.

7. $f(x) = 20e^{3-0.1x}$.

8. $f(x) = 10e^{0.3x-6}$.

9. $f(x) = e^{x^2}$.

10. $f(x) = e^{-x^2}$.

11. $f(x) = (1/2)^{-3x}$.

12. $f(x) = (0.4)^{-0.3x}$.

13. $f(x) = 1000(1.06)^x$.

14. $f(x) = 500(1.08)^{-x}$.

In Problems 15–22, sketch the curve. Show the starting point coordinates and the asymptote with its equation. The sketches need not be to scale, and the coordinate axes need not be shown.

15. $f(x) = 5^x$.

16. $f(x) = 5^{-x}$.

17. $f(x) = 2^{3-0.5x}$.

18. $f(x) = 3^{0.5x-2}$.

19. $f(x) = (0.5)^x$.

20. $f(x) = (0.25)^{-x}$.

21. $f(x) = 15 - 10e^{3-0.1x}$.

22. $f(x) = 20 + 15e^{2-0.1x}$.

23. Find the slope of the line tangent to the following at the point where $x = 2$.

$$f(x) = e^{-0.1x^2+2x-3}.$$

24. Newprod Company estimates the total potential number of customers for a new product is 1,000,000. It plans to operate a promotional campaign to sell the product and uses the response function

$$r(t) = 0.25 - 0.25e^{-0.01t}$$

as a measure of the proportion of total customer potential responding to the promotion after it has been in operation for t days. On the average, one response generates $5 in revenue. Campaign costs consist of a fixed cost of $15,000 plus a variable cost of $1000 per day of operation.

    a) How long should the campaign continue if profit (revenue minus cost) is to be maximized?

    b) Compute the maximum profit.

25. Solve Problem 24 if the response function is

$$r(t) = 0.25 - 0.25e^{-0.02t}$$

and other facts remain as given.

26. An oil deposit contains 1,000,000 barrels of oil, which, after being pumped from the deposit, yields a revenue of $12 per barrel. The proportion of the deposit that will have been pumped out after t years of pumping is

$$0.9 - 0.9e^{-0.16t}.$$

Operating costs are $345,600 per year.

    a) How long should pumping be continued to maximize profit?

    b) Compute the maximum profit.

27. Answer Problem 26 if revenue per barrel is $15.

28. The revenue from, and cost of, operating an undertaking for t years are, respectively,

$$R(t) = 4e^{0.3t} \quad \text{and} \quad C(t) = 1.5e^{0.4t}$$

millions of dollars. How long should the undertaking be continued if profit is to be maximized?

In Problems 29–34, find the value of the function at its local extreme point and state whether this is a local maximum or minimum.

29. $f(x) = 25x - e^x$.

30. $f(x) = 0.1x + e^{-0.1x}$.

31. $f(x) = 0.02x + e^{-0.1x}$.

32. $f(x) = e^{(x - 0.1x^2)}$.

33. $f(x) = 10x - e^{0.2x}$.

34. $f(x) = 0.05x + e^{-0.1x} + 4$.

10.5 DERIVATIVES OF LOGARITHMIC FUNCTIONS

To derive the simple logarithmic derivative rule, we start with

$$g(x) = \ln_e x. \tag{1}$$

Applying the definition of a logarithm, the last means

$$e^{g(x)} = x. \tag{2}$$

Taking the derivative with respect to x by application of the exponential function rule, we have

$$\frac{d}{dx}[e^{g(x)}] = \frac{d}{dx}(x)$$

$$e^{g(x)}g'(x) = 1$$

$$g'(x) = \frac{1}{e^{g(x)}}.$$

But from (2), the right side of the last expression is $1/x$. Hence,

$$g'(x) = \frac{1}{x}.$$

Thus,

$$\frac{d}{dx}[g(x)] = g'(x) = \frac{1}{x},$$

and from (1), $g(x)$ is $\ln x$, so

$$\frac{d}{dx}(\ln x) = \frac{1}{x}.$$

Thus, the simple logarithmic function, $\ln x$, has the reciprocal of x as its derivative. Moreover, as might by now be expected, the logarithm of $f(x)$, $\ln f(x)$, has as its derivative the reciprocal of $f(x)$ multiplied by $f'(x)$.

Derivative rule: Natural logarithmic functions

$$\frac{d}{dx}(\ln x) = \frac{1}{x}$$

$$\frac{d}{dx}\ln[f(x)] = \frac{f'(x)}{f(x)}.$$

Because of the simplicity of the derivative rules when the natural logarithm system is used (which is the reason for the name *natural*), we shall not use base 10 (common) logarithms in our work with calculus. But, to be complete, we note here that the foregoing rules can be converted to base 10 by dividing the derivative by ln 10.

Example. Find $f'(2)$ if

$$f(x) = \ln(2x + 1) + \ln x.$$

We have

$$f'(x) = \frac{1}{2x+1} \frac{d}{dx}(2x+1) + \frac{1}{x}$$

$$= \frac{2}{2x+1} + \frac{1}{x}.$$

$$f'(2) = \frac{2}{5} + \frac{1}{2} = 0.4 + 0.5 = 0.9.$$

Exercise. Find $h'(1)$ if $h(x) = \ln(x^2 + 3x + 1) + 2 \ln x$. Answer: 3.

Sketching the graph of $g(x) = \ln x$. Inasmuch as $\ln 1 = 0$, the graph of this function passes through $(1, 0)$ as shown on Figure 10–5. The table accompanying the graph shows that to the right of $(1, 0)$ the curve rises, but not very rapidly, because ln x increases slowly as x increases. To the left of $(1, 0)$, where x is less than one, ln x is negative and the closer x is to zero, the further the curve drops below the

x	$g(x) = \ln x$
1	0
10	2.30
100	4.61
1000	6.91
0.1	−2.30
0.01	−4.61
0.001	−6.91
0.0001	−9.21

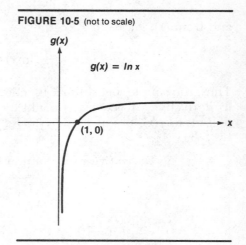

FIGURE 10-5 (not to scale)

x-axis. But note that *x* cannot be zero or negative because only positive numbers have logarithms. It follows that as *x* approaches zero, the curve falls and approaches the vertical axis asymptotically.

Exercise. Prove that $f(x) = \ln x$ is concave downward at all permissible values of *x*. Answer: This is true because $f''(x) = -1/x^2$, which is negative for all values of *x*.

10.6 PROBLEM SET 10–2

Find the derivatives of the following functions.

1. $f(x) = \ln x$.
2. $f(x) = 3 \ln(2x)$.
3. $f(x) = \ln(2x + 3)$.
4. $f(x) = \ln\left(\dfrac{1}{x}\right)$.
5. $f(x) = \ln (e^x)$.
6. $f(x) = \ln(2x + 5)^{1/3}$.
7. $f(x) = \ln(x^2 + 2x)$.
8. $f(x) = \ln(2x^3 - 6x)$.
9. $f(x) = \ln(3x + 2)^{1/2}$.

In Problems 10–15, find the value of $f(x)$ at its local extreme point and state whether this is a local maximum or minimum.

10. $f(x) = 20x - 10 \ln x$.
11. $f(x) = 100 \ln x - 0.5x^2$.
12. $f(x) = 3x - 12 \ln x$.
13. $f(x) = \ln(x^2 - 10x + 35)$.
14. $f(x) = 100x^2 - 72 \ln x$.
15. $f(x) = 0.5x^2 - 4x - 5 \ln x + 25$.
16. When *x* ounces of seed costing $2 per ounce are sown on a plot of land, the crop yield is $\ln(2x + 1)$ bushels worth $25 per bushel. How many ounces should be sown if the worth of the crop minus the cost of the seed is to be maximized?
17. Solve Problem 16 if seed cost changes to $2.50 per ounce and the crop is worth $30 per bushel.

10.7 THE CHAIN RULE AND IMPLICIT DIFFERENTIATION

The function rules we have developed and worked with are special cases of a general rule called the chain rule. Consider, for example, the exponential function rule

$$\frac{d}{dx} e^{f(x)} = e^{f(x)} \cdot \frac{df(x)}{dx}.$$

If we write *f* rather than *f(x)*, the last appears as

$$\frac{d}{dx} e^f = e^f \cdot \frac{df}{dx}.$$ (Note: This implies *f* is a function of *x*.)

As the note at the right states, because the derivative is specified as being with respect to *x*, the statement *implies* that *f* is a function of *x*. Note carefully that the last expression asks for the derivative of e^f,

530

which is a function of *f, not with respect to f as we would expect, but with respect to x.* Similarly,

$$\frac{d}{dx}(z^5)$$

asks for the derivative of z^5, which is a function of z, not with respect to z as we would expect, but with respect to x. The chain rule states how the derivative of a function of one variable can be taken with respect to a different variable where, by implication, the two variables are functionally related.

Chain Rule:

$$\frac{d}{dx}g[f(x)] = \frac{d}{df}g[f(x)] \cdot \frac{d}{dx}f(x) \quad \text{(Complete form.)}$$

$$\frac{dg(f)}{dx} = \frac{dg(f)}{df} \cdot \frac{df}{dx} \qquad \text{(Abbreviated form.)}$$

Applied to the problem of finding

$$\frac{d}{dx}(z^5),$$

the chain rule says to take the derivative of the function of z, which is z^5, with respect to z (which is what we would expect), then multiply this by the derivative of z with respect to x. Thus,

$$\frac{d}{dx}(z^5) = \frac{d}{dz}(z^5)\frac{dz}{dx} = 5z^4 \cdot \frac{dz}{dx}.$$

Again, to determine

$$\frac{d}{dw}e^{2y}$$

we must use the chain rule because the derivative is with respect to w, and the expression whose derivative is sought, e^{2y}, is a function not of w, but of y. Hence,

$$\frac{d}{dw}e^{2y} = \frac{d}{dy}e^{2y} \cdot \frac{dy}{dw}$$

$$= e^{2y}\frac{d}{dy}(2y) \cdot \frac{dy}{dw}$$

$$= e^{2y}(2) \cdot \frac{dy}{dw}.$$

To be sure the chain rule has been applied correctly, refer to the first of the three lines of the foregoing and, on the right, cover the denominator of the first factor with one finger and the numerator of the second factor with another finger. What remains in view should be the same as the expression on the left. Note that we avoided saying cancel dy on the right because d/dy does not mean d divided by dy but rather is an instruction meaning to take the derivative with respect to y, and we cannot cancel part of an instruction. Note also that we wrote

$$\frac{dy}{dw}$$

rather than the more formal statement for the derivative of y with respect to w,

$$\frac{d}{dw}\,(y),$$

which is what dy/dw means.

Definition: $\dfrac{dy}{dx}$ means $\dfrac{d}{dx}\,(y).$

Example. Write the expression for the derivative with respect to w for the following.

a) $2y^3$. b) $e^{x^2} + 5w$. c) $\ln(3z + 2)$.

We have:

a) $\dfrac{d}{dw}\,(2y^3) = 2\dfrac{d}{dw}\,(y^3) = 2\dfrac{d}{dy}\,(y^3) \cdot \dfrac{dy}{dw} = 2(3y^2) \cdot \dfrac{dy}{dw} = 6y^2 \cdot \dfrac{dy}{dw}.$

b) $\dfrac{d}{dw}\,(e^{x^2} + 5w) = \dfrac{d}{dw}\,(e^{x^2}) + \dfrac{d}{dw}\,(5w) = \dfrac{d}{dx}\,(e^{x^2}) \cdot \dfrac{dx}{dw} + 5$

$$= e^{x^2}(2x) \cdot \dfrac{dx}{dw} + 5.$$

c) $\dfrac{d}{dw}\,\ln(3z + 2) = \dfrac{d}{dz}\,\ln(3z + 2) \cdot \dfrac{dz}{dw} = \left(\dfrac{1}{3z + 2}\right)(3) \cdot \dfrac{dz}{dw} = \dfrac{3}{3z + 2} \cdot \dfrac{dz}{dw}.$

Exercise. Write the expressions for the derivative with respect to x of: a) $\ln y$. b) $2z^{3/2} + x^2$. c) e^{3w}. Answer: a) $\left(\dfrac{1}{y}\right) \cdot \dfrac{dy}{dx}$. b) $3z^{1/2} \cdot \dfrac{dz}{dx} + 2x$. c) $3e^{3w} \cdot \dfrac{dw}{dx}$.

   The three function rules (power, exponential, and logarithmic) we have used are, as already noted, special cases of the chain rule, and no purpose would be served by applying the chain rule to problems we can solve with the function rules. The importance of the chain rule is its application to problems where a function cannot be stated explicitly. To see what is meant here, recall that prior to this section all expressions were of the form

$$f(x) = \quad \text{or} \quad g(x) =$$

and so on, where the expression on the right involved x and constants, but *not* the function. These expressions are said to define the function *explicitly*. By way of contrast, consider

$$2[y(x)]^5 + 7[y(x)] = 3x^2 - 2.$$

Here it is impossible to solve this expression explicitly to obtain $y(x) =$ an expression not involving the function. For convenience, and in keeping with convention, we shall write y instead of $y(x)$ and say that

$$2y^5 + 7y = 3x^2 - 2$$

implies y is a function of x. The method of finding

$$\frac{d}{dx} y(x) = \frac{dy}{dx}$$

is then called *implicit* differentiation, and the chain rule must be invoked. Thus, taking the derivative with respect to x, we have

$$\frac{d}{dx}(2y^5 + 7y) = \frac{d}{dx}(3x^2 - 2)$$

$$2\frac{d}{dx}(y^5) + 7\frac{d}{dx}(y) = 6x$$

$$2 \cdot \frac{d}{dy}(y^5)\frac{dy}{dx} + 7\frac{dy}{dx} = 6x$$

$$2(5y^4)\frac{dy}{dx} + 7\frac{dy}{dx} = 6x.$$

Factoring dy/dx on the left yields

$$\frac{dy}{dx}(10y^4 + 7) = 6x$$

$$\frac{dy}{dx} = \frac{6x}{10y^4 + 7}.$$

Exercise. Find dy/dx if: a) $y^4 = x^3$. b) $y^3 - 3y^2 - 2x^3 + 3x = 0$. c) In (a) and (b), y is an abbreviation of what symbol?

Answer: a) $\dfrac{dy}{dx} = \dfrac{3x^2}{4y^3}$. b) $\dfrac{dy}{dx} = \dfrac{6x^2 - 3}{3y^2 - 6y} = \dfrac{2x^2 - 1}{y^2 - 2y}$. c) $y(x)$.

It is instructive to review part (a) of the exercise because

$$y^4 = x^3 \tag{1}$$

can be written in *explicit* form by taking the fourth root of both sides to give

$$(y^4)^{1/4} = (x^3)^{1/4}$$
$$y = x^{3/4}$$

or

$$y(x) = x^{3/4}.$$

We now can find $y'(x)$ by the simple power rule as,

$$y'(x) = \frac{3}{4} x^{-1/4} = \frac{3}{4x^{1/4}}.$$

Hence, we now have dy/dx, or $y'(x)$ as

$$y'(x) = \frac{dy}{dx} = \frac{3}{4x^{1/4}} \tag{2}$$

and, from the exercise answer,

$$\frac{dy}{dx} = \frac{3x^2}{4y^3}. \tag{3}$$

The two results certainly look different, but they are, in fact, equivalent. To see that this is true, let us replace the y^3 in (3) by an equivalent expression from (1),

$$y^4 = x^3.$$

To change y^4 to y^3 we must raise both sides to the power $3/4$, thus:

$$(y^4)^{3/4} = (x^3)^{3/4}$$
$$y^3 = x^{9/4}.$$

Hence, (3) becomes

$$\frac{dy}{dx} = \frac{3x^2}{4y^3} = \frac{3x^2}{4x^{9/4}} = \frac{3x^{2-9/4}}{4} = \frac{3x^{-1/4}}{4} = \frac{3}{4x^{1/4}},$$

so (3) is equivalent to (2).

There is no need to invoke the chain rule for an expression that can be stated as an explicit function but, as the last discussion illustrates, we may choose to use the chain rule if we wish and the outcome of implicit differentiation will be equivalent to the derivative of the explicit function, although the expression for the two may appear to be different. Our immediate interest in the chain rule arises in the next section, where we encounter expressions such as

$$f(y) = y - x^2$$

and we want to find the expression for the derivative of y with respect to x, dy/dx. We start by writing the expression for the derivative with respect to x.

$$\frac{d}{dx} f(y) = \frac{d}{dx} (y) - \frac{d}{dx} (x^2),$$

and, on the left, see the need for the chain rule because we have to take the derivative of a function of y not with respect to y, but with respect to x. The chain rule says this is carried out by doing what is natural; that is, take the derivative of $f(y)$ with respect to y, then multiply this by dy/dx. Thus,

$$\frac{df(y)}{dy} \cdot \frac{dy}{dx} = \frac{dy}{dx} - 2x.$$

To solve for the desired dy/dx, we isolate terms containing dy/dx on one side, factor, and proceed as follows:

$$\frac{df(y)}{dy} \frac{dy}{dx} - \frac{dy}{dx} = -2x$$

$$\frac{dy}{dx} \left[\frac{df(y)}{dy} - 1 \right] = -2x$$

$$\frac{dy}{dx} = \frac{-2x}{\dfrac{df(y)}{dy} - 1} = \frac{2x}{1 - \dfrac{df(y)}{dy}}.$$

The significance of what seems to be symbol manipulation in the last example will be seen in the next section when we attach interpretations to the derivatives in the result.

Exercise. Find the expression for dy/dx if $y = C(y) + x$.
Answer:

$$\frac{dy}{dx} = \frac{1}{1 - \dfrac{dC(y)}{dy}}.$$

10.8 MARGINAL PROPENSITY TO CONSUME AND THE MULTIPLIER

The major part of the income people have available for their use is spent on food, clothing, shelter, medical care, transportation, recreation, and so on, and the remainder is saved in one form or another. The amount spent is called consumption expenditure or, briefly, consumption. Consumption, of course, is a function of income. If we symbolize income by Y, as is conventional in economics, and consumption as the function $C(Y)$, $C(Y)$ is called a *consumption function*. Suppose, for example, that

$$C(Y) = 30 + 0.6\,Y$$

and that Y and $C(Y)$ are in billions of dollars. At income level $Y = \$100$ billion,

$$C(Y) = C(100) = 30 + 0.6(100) = \$90 \text{ billion,}$$

so that people having $100 billion of income available for use (referred to as *disposable* income) spend $90 billion on consumption, and the remaining $10 billion represents savings. While aggregate numbers for consumption and savings are important numbers, the topic of this section is concerned with the *rate at which consumption changes per additional $1 of income received.* This is the derivative of $C(Y)$ with respect to Y and is called the *marginal propensity to consume,* MPC.

Definitions:

Marginal Propensity to Consume $= \text{MPC} = \dfrac{dC(y)}{dy}$.

Marginal Propensity to Save $= \text{MPS} = 1 - \text{MPC} = 1 - \dfrac{dC(y)}{dy}$.

The second part of the last definition means simply that if MPC is the part of an extra $1 of income going to consumption, the remaining part of the $1, $1 - \text{MPC}$, goes to saving.

For the foregoing consumption function, we have

$$C(Y) = 30 + 0.6\,Y$$
$$\text{MPC} = \frac{dC(Y)}{dY} = 0.6$$
$$\text{MPS} = 1 - \text{MPC} = 1 - 0.6 = 0.4.$$

Thus, of an additional $1 of income, $0.60 is spent on consumption and $0.40 is saved.

> **Exercise.** For the consumption function $C(Y) = 20 + 0.75\,Y$:
> *a)* Find MPC and MPS. *b)* Interpret the answer to *(a)*. *c)* What

proportion (or percentage) of an income of $100 billion would
be spent? Answer: *a)* MPC = 0.75, MPS = 0.25. *b)* Seventy-
five cents of an *additional* $1 of income would be spent and
twenty-five cents would be saved. *c)* 0.95 or 95 percent.

Note in part *(c)* of the last exercise that

$$C(100) = 20 + 0.75(100) = 95,$$

so $95 billion of the $100 billion, or 95 percent would be spent. The
proportion spent is not the same as the marginal propensity to consume
because in economics marginals always refer to the rate at which a
function changes per increase of 1 in the independent variable. That
is, marginals are slopes of lines (derivatives).

The Multiplier. At income level Y, $C(Y)$ is spent on consumption,
and the difference

$$\text{Income} - \text{Consumption expenditure} = Y - C(Y)$$

is the amount saved. In our elementary analysis, we shall suppose that
business investment opportunities exist, and all savings are invested
so saving equals investment. Letting I represent investment, we have

$$I = Y - C(Y).$$

Now, of course, investment by someone generates income for others.
For example, a businessman who invests $1000 to pay the wages for
two workers to build an addition on a building, using material the
businessman owns, generates a first round of a matching $1000 of
income to the workers. However, this is only the beginning round,
and the important point we wish to demonstrate is that the $1000
investment generates a multiple of $1000 of income, and the multiplier
is greater than one. To see this, we start with the $1000 investment,
which generates $1000 of income to the workers. However, the work-
ers now spend part of the $1000 income (remember MPC) on consump-
tion, and this part becomes income to the suppliers of the consumption
items. In turn, these suppliers spend part of their income on consumption, and so on, and on. Just how much income received at one link
in this chain is spent on consumption and becomes income at the
next link depends, of course, on the marginal propensity to consume,
MPC. We now analyze the situation, starting with the last algebraic
expression.

$$I = Y - C(Y).$$

What we want to know is by how much does income, Y, change per
additional dollar of investment; that is, we want dY/dI. Taking the
derivative of the last *with respect to I*, we have

$$\frac{d}{dI}(I) = \frac{d}{dI}(Y) - \frac{d}{dI}[C(Y)].$$

We must invoke the chain rule for the rightmost term. We have

$$1 = \frac{dY}{dI} - \frac{dC(Y)}{dY} \cdot \frac{dY}{dI}.$$

Remembering that dY/dI is sought, we proceed as follows:

$$1 = \frac{dY}{dI}\left[1 - \frac{dC(Y)}{dY}\right]$$

$$\frac{dY}{dI} = \frac{1}{1 - \dfrac{dC(Y)}{dY}}.$$

Next, remembering that $dC(Y)/dY$ is the marginal propensity to consume, MPC, we may write

$$\frac{dY}{dI} = \frac{1}{1 - \text{MPC}} = \frac{1}{\text{MPS}}.$$

The rate at which income increases per additional \$1 of investment (investment = saving), dY/dI, is called the *multiplier*. Thus,

$$\boldsymbol{Multiplier} = \frac{dY}{dI} = \frac{1}{1 - \text{MPC}} = \frac{1}{\text{MPS}}.$$

Returning to the consumption function at the beginning of this section,

$$C(Y) = 30 + 0.6Y$$

$$\text{MPC} = \frac{dC(Y)}{dY} = 0.6.$$

$$\text{Multiplier} = \frac{1}{1 - \text{MPC}} = \frac{1}{1 - 0.6} = \frac{1}{0.4} = 2.5.$$

The multiplier here means that 2.5 dollars of income is generated by an additional \$1 of investment. It is clear that MPC is a critical factor in income generation. When investment opportunities exist, a high MPC generates new income in an amount that is a large multiple of the sum invested.

> *Exercise.* a) Find the multiplier if the consumption function is $C(Y) = 10 + 0.9Y$. b) Interpret the answer to (a). Answer: a) 10. b) One additional dollar of investment generates \$10 of additional income.

We must leave more extensive analysis of the topics introduced in this section for courses in economics. However, our introduction serves to demonstrate the important contribution calculus can make to this analysis and also serves our purpose of illustrating a situation where the chain rule arises. We close the section by an example involving a consumption function that is not linear.

Example. *a)* Suppose Y is family income in tens of thousands of dollars so that, for example, $Y = 2$ means \$20,000, and

$$C(Y) = 6 + 0.6Y - e^{-0.4Y}$$

represents the family consumption function. Find MPC for families with \$1 and \$2 of income.

b) Suppose Y is in trillions of dollars and the $C(Y)$ in part *(a)* represents the national income of a country. Find the multiplier at an income level of \$0.8 trillion.

Here $C(Y)$ is not linear so MPC is a function of income that we shall symbolize as MPC (Y).

a) For

$$C(Y) = 6 + 0.6Y - e^{-0.4Y}$$

$$\text{MPC}(Y) = \frac{dC(Y)}{dY} = 0.6 - \frac{d}{dY}(e^{-0.4Y}) = 0.6 + 0.4e^{-0.4Y}.$$

$$\text{MPC}(1) = 0.6 + 0.4e^{-0.4(1)} = 0.6 + 0.4e^{-0.4}$$
$$= 0.6 + 0.4(0.6703) = 0.87.$$

$$\text{MPC}(2) = 0.6 + 0.4e^{-0.4(2)} = 0.6 + 0.4e^{-0.8}$$
$$= 0.6 + 0.4(0.4493) = 0.78.$$

Part *(a)* is illustrative of a common observation that the higher the income level of a family, the smaller MPC tends to be. That is, at a higher income, a smaller fraction of an additional dollar is spent on consumption and a larger fraction is saved.

b) Here we have the same MPC function as in *(a)*,

$$\text{MPC}(Y) = 0.6 + 0.4e^{-0.4Y}.$$
$$\text{MPC}(0.8) = 0.6 + 0.4e^{-0.4(0.8)}$$
$$= 0.6 + 0.4e^{-0.32}$$
$$= 0.6 + 0.4(0.7261)$$
$$= 0.89.$$

Then, from our earlier derivation,

$$\text{Multiplier} = \frac{1}{1 - \text{MPC}} = \frac{1}{1 - 0.89} = \frac{1}{0.11} = 9.1.$$

10.9 PROBLEM SET 10–3

Employ the chain rule to carry out each of the following:

1. $\dfrac{d}{dx} f(y)$.

2. $\dfrac{d}{dz} f(x)$.

3. $\dfrac{d}{dh} p(q)$.

4. $\dfrac{d}{dw} g(f)$.

5. $\dfrac{d}{dx} (y^4)$.

6. $\dfrac{d}{dy} (x^3)$.

7. $\dfrac{d}{dz} (e^{2w})$.

8. $\dfrac{d}{dq} (2e^{-0.5 p})$.

9. $\dfrac{d}{dx} \ln(2y + 3)$.

10. $\dfrac{d}{dw} \ln(z^2 - 3z)$.

11. $\dfrac{d}{dy} (xy)$.

12. $\dfrac{d}{dx} \left(\dfrac{x}{y}\right)$.

Find $\dfrac{dy}{dx}$ by implicit differentiation:

13. $y^3 - x^2 = 0$.

14. $y^5 - 3x^3 = 0$.

15. $2y^3 - 3y^2 - x^6 - 6x + 10 = 0$.

16. $4y^5 - 2y^2 + y - 2x - 50 = 0$.

17. $x = y + e^{-y}$.

18. $x = 2y - e^{y^2}$.

19. $e^x \ln y = 5$.

20. $e^{-y} \ln x = 1$.

21. $xy^3 = x^2 + 5$.

22. $x^2 y^3 = 2x + 10$.

23. Find the expression for $\dfrac{dy}{dx}$ if y is a function of x and

$$x = y - g(y).$$

24. Find the expression for $\dfrac{dw}{dz}$ if w is a function of z and

$$w^2 = z^2 + f(w).$$

25. At income level Y, consumption expenditures are $C(Y)$ billion dollars, where

$$C(Y) = 34 + 0.68 Y.$$

   a) Find the marginal propensity to consume.
   b) Interpret (a).
   c) How much income is generated by an additional dollar of investment?
   d) At income level 200, what proportion of total income is spent?

26. At income level Y, consumption expenditures are $C(Y)$ billion dollars, where

$$C(Y) = 54 + 0.63 Y.$$

   a) Find the marginal propensity to consume.
   b) Interpret (a).

    c) How much income is generated by an additional dollar of investment?

    d) At income level 200, what proportion of total income is spent?

27. For the consumption function

$$C(Y) = 9 + 0.7\,Y - e^{-0.3\,Y},$$

find the multiplier at income level

   a) 3.   *b)* 10.

28. For the consumption function

$$C(Y) = 6 + 0.8\,Y - e^{-0.2\,Y},$$

find the multiplier at income level

   a) 2.   *b)* 15.

10.10 THE DERIVATIVE AS A RATE

The word *rate* has many meanings, so it is important at the outset to understand what it means in the calculus context. Refer to Figure 10–6, which shows $P(t)$ as the total profit from a venture at time

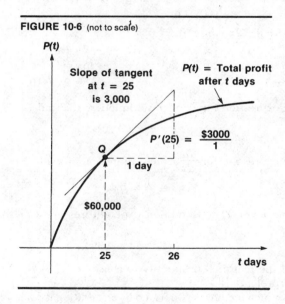

FIGURE 10-6 (not to scale)

t days. Point Q is at 25 days and $P(25) = \$60,000$ is total profit at 25 days. The derivative at $t = 25$,

$$P'(25) = 3000,$$

is the slope of the tangent line at point Q. As the figure shows, this value is

$$\frac{\$3000}{1 \text{ day}}.$$

To describe this ratio by saying profit is $3000 per day does not bring out the full meaning of what is shown on Figure 10–6, and, furthermore, it is easy to misinterpret the statement and assume it means that profit from the venture is $3000 per day *every* day. The value $3000 does not mean profit per day every day, nor does it mean average profit per day. It does mean that the *tangent line at point Q* rises $3000 per day every day, and we describe this by saying that *at $t = 25$ days* profit is increasing at the rate of $3000 per additional day the venture continues. Thus, in calculus, rates are tangent slopes at points, so we call them *point* rates or, when time is the independent variable, *instantaneous* rates, and whenever we describe a derivative in rate terminology it is to be understood that this means a point rate.

As examples of point rates already introduced, recall that marginal cost *at* output level q units is the change (increase) in total cost per additional unit made, and that marginal propensity to consume *at* income level Y is the change (increase) in total amount spent per additional dollar of income received.

Example. The cost of renting a car for one day consists of a fixed charge of $15, plus $0.20 per mile the car is driven. Letting m be the number of miles driven, one day's cost, $C(m)$, is

$$C(m) = 15 + 0.20m.$$

Interpret $C'(m)$ in rate terminology.

We have

$$C'(m) = 0.20,$$

which is a constant because $C(m)$ is a linear function. The rate interpretation is that the day's cost increases at the rate of $0.20 (or 20 cents) per additional mile driven. Note that it would be incorrect to state that the $0.20 means the cost is 20 cents per mile, because, for example, if the car is driven 100 miles, the cost is $35 and cost per mile is $0.35. This $0.35 is the *average* cost per mile if the car is driven 100 miles.

> **Exercise.** The cost of printing and binding a paperback book is $C(n)$ dollars, where n is the number of copies made. If $C(n) = 1,500 + 1.5n$, find $C'(n)$ and interpret it in rate terminology. Answer: $C'(n) = 1.5$ means that total cost increases at the rate of $1.50 per additional book made.

Example. In an attempt to attract renters who drive a car many miles in one day, an enterprising car rental manager decided to offer a one-day rental cost computed from

$$C(m) = 15 + 0.20m - 0.00004m^2.$$

Find $C'(500)$ and $C'(1000)$ and interpret these values in rate terminology.

Observe the negative term in $C(m)$. This represents a reduction in cost that, though modest when the car is driven a small number of miles, becomes significant if the car is driven, say, 1000 miles. We find

$$C'(m) = 0.20 - 0.00008m$$
$$C'(500) = 0.16$$
$$C'(1000) = 0.12.$$

Here, the cost function is not linear, so $C'(m)$ varies as m changes. $C'(500) = 0.16$ means that at 500 miles, cost is increasing at the rate of 16 cents per additional mile driven, and $C'(1000) = 0.12$ means that at 1000 miles, cost is increasing at the rate of 12 cents per additional mile driven.

10.11 RELATIVE RATE OF CHANGE

The change from a beginning value divided by the beginning value is referred to as a *relative* rate of change and may be expressed as a proportion or as a percent. On Figure 10–6, for example, the $3,000 is a change from $P(25) = $60,000$, so the relative rate is

$$\frac{3000}{60000} = 0.05 \quad \text{or} \quad 5 \text{ percent.}$$

We describe the last by saying that *at $t = 25$ days* profit is increasing at the rate of 5 percent per additional day. This rate is

$$\frac{P'(25)}{P(25)}$$

and, in general,

$$\textit{Relative rate of change of } f(x) \textit{ at a point} = \frac{f'(x)}{f(x)}.$$

In passing, we note that the logarithmic function rule states that

$$\frac{d}{dx} \ln[f(x)] = \frac{1}{f(x)}[f'(x)] = \frac{f'(x)}{f(x)},$$

so that the relative rate of change of a function is the derivative of the logarithm of the function. However, we prefer to think of the definition as stated because it brings out the concept of dividing the change by the beginning value.

Example.   If $1000 is deposited in a bank account that pays 6 percent interest compounded annually, the amount in the account at time *t* years is *A(t)* where

$$A(t) = 1000(1.06)^t.$$

Find the relative rate of change at $t = 5$ years and describe this number in rate terminology.

First we find

$$A'(t) = 1000(1.06)^t \ln(1.06).$$

The relative rate expression then is

$$\frac{A'(t)}{A(t)} = \frac{1000(1.06)^t \ln(1.06)}{1000(1.06)^t}$$
$$= \ln 1.06$$
$$= 0.05827$$
$$= 5.827\%.$$

Noting that the rate is independent of *t*, we state that at any point in time the amount in the account is increasing at the rate of 5.827 percent per additional year. This result may seem to be incorrect because the example states that interest is 6 percent per year, not 5.827 percent. The reason for this is that the procedure we have used translates the rate of 6 percent compounded *once* a year to an equivalent rate of 5.827 percent compounded *continuously* (or instantaneously). A complete discussion of this matter can be found in Chapter 8. The next exercise shows that if interest is compounded continuously, the difference just noted does not occur.

Exercise.   If $1000 is deposited in an account paying 6 percent interest compounded continuously, the amount in the account at time *t* years is

$$A(t) = 1000e^{0.06t}.$$

Find the percent rate of change at time $t = 5$ years and describe this in rate terminology. Answer: At any time, *t* years, the amount is increasing at the rate of 6 percent per additional year.

Example. Total sales volume of a new product, in thousands of gallons, is $G(t)$ at time t years, where

$$G(t) = 12t + 4(e^{-0.5t} - 1).$$

Find the percent rate of change at $t = 4$ years and describe this in rate terminology.

First we find

$$G'(t) = 12 + 4e^{-0.5t}(-0.5)$$
$$= 12 - 2e^{-0.5t}.$$

. The relative rate is

$$\frac{G'(t)}{G(t)} = \frac{12 - 2e^{-0.5t}}{12t + 4(e^{-0.5t} - 1)}$$

and at $t = 4$

$$\frac{G'(4)}{G(4)} = \frac{12 - 2e^{-2}}{48 + 4(e^{-2} - 1)} = \frac{11.7293}{44.5413} = 0.263,$$

so at $t = 4$ years sales volume is increasing at the rate of 26.3 percent per additional year.

Sensitivity analysis. In Chapter 9 we developed an inventory model to describe the costs associated with acquiring units of a product and carrying some of these units in inventory. The objective was to determine the lot size (the number of units to be acquired periodically during the year) so that cost would be minimized. In parameterized form the solution for L, the lot size, was

$$L = \left[\frac{2NF}{i} \right]^{1/2}$$

where

$N =$ Number of units to be acquired in one year.
$F =$ Fixed cost when a lot is acquired.
$i =$ Annual inventory cost per unit in average inventory.

If i is subject to error, or to change, we would like to know what effect a variation in i would have on the optimum lot size, L. If this effect is small, we say the lot size formula is not very sensitive to changes in i. We now consider L to be a function of i and write

$$L(i) = \left[\frac{2NF}{i} \right]^{1/2} = (2NF)^{1/2} i^{-1/2}$$

and find

$$\frac{dL(i)}{di} = L'(i) = (2NF)^{1/2} \left(-\frac{1}{2} \right) i^{-3/2}$$

$$= \underbrace{(2NF)^{1/2}(i^{-1/2})}_{L(i)}\left(-\frac{1}{2}\right)i^{-1}$$

$$= L(i)\left(-\frac{1}{2}\right)i^{-1}$$

$$= -\frac{L(i)}{2i}. \tag{1}$$

Now suppose the parameter values used to compute the optimum L were

$$N = 500, \quad F = \$100, \quad i = \$10$$

so

$$L = \left[\frac{2(500)(100)}{10}\right]^{1/2} = 100 \text{ units.}$$

If $L = 100$ and $i = 10$, then (1) is

$$\frac{dL(i)}{di} = -\frac{L}{2i} = -\frac{100}{2(10)} = -5,$$

which says that the optimum lot size decreases at the rate of 5 units per additional \$1 of inventory cost per unit. Consequently, if the unit inventory cost changes from \$10 to, say, \$10.20, the optimal lot size would decrease by *approximately* 1 from 100 to 99; that is, the change from \$10 to \$10.20 is an additional \$0.20 and the decrease at the rate of 5 units per \$1 is, proportionately, 1 unit for \$0.20. Note that the word approximately in the last sentence was italicized. This was done to call attention to the fact that the derivative applies precisely only at a point, and only approximately if we move 1 unit from the point. In effect, the approximation means we are moving along the tangent line rather than along the curve. The exact change would be found by computing $L(10.20) - L(10)$, and this turns out to be a decrease of 0.985 units rather than 1 unit.

Exercise. The optimal (minimal) value of a function *g(h)* occurs at $h = 600/k$, where k is a parameter. *a)* Write the expression for the rate of change of h with respect to k. *b)* Evaluate *(a)* if $k = 4$. *c)* Using *(b)*, by how much, approximately, would h change if k increased by 0.2? Answer: *a)* $-600/k^2$. *b)* -37.5. *c)* h would decrease by about 7.5.

10.12 PROBLEM SET 10–4

1. If $C(x)$ is the cost of making (printing and binding) x books, and

    $$C(x) = 2000 + 2.50x \text{ dollars,}$$

    find $C'(1000)$ and interpret the number in rate terminology.

2. If the total cost of renting a car for one day and driving it x miles is

    $$C(x) = 20 + 0.2x - 0.00005x^2 \text{ dollars,}$$

    find $C'(500)$ and $C'(1000)$ and interpret them in rate terminology.

3. Two thousand dollars deposited in a bank account paying 8 percent interest compounded annually will grow to the amount

    $$A(t) = 2000(1.08)^t$$

    in t years. Find $A'(5)$ and $A'(10)$ and interpret the results in rate terminology.

4. Two thousand dollars deposited in an account paying 8 percent interest compounded continuously will grow to the amount

    $$A(t) = 2000e^{0.08t}$$

    in t years. Find $A'(5)$ and $A'(10)$ and interpret the results in rate terminology.

5. The cost of renting a car for one day and driving it x miles is

    $$C(x) = 16 + 0.20x \text{ dollars.}$$

    a) Write the expression for $a(x)$, the *average* cost per mile if the car is driven x miles.
    b) Find $a'(40)$ and interpret it in rate terminology.

6. The total cost of producing x tons of a product is

    $$C(x) = 0.01x^3 - 3x^2 + 300x + 10,000 \text{ dollars.}$$

    a) Write the expression for $a(x)$, the average cost per ton if x tons are produced.
    b) Find $a'(100)$ and interpret it in rate terminology.
    c) Find $a'(500)$ and interpret it in rate terminology.

7. [Note: If a group of 10 miners work a "shift" of 8 hours each, the product (10 laborers) times (8 hours) = 80 is called 80 labor-hours.]

    A shift of miners produce $T(m)$ tons of ore when m labor-hours are worked, where

    $$T(m) = 1.2m - 0.15m(\ln m).$$

    Find $T'(100)$ and interpret it in rate terminology.

8. Total sales of a new product in thousands of dollars when the product has been sold for t years is

    $$G(t) = 20t + 18(e^{-0.8t} - 1).$$

    Find $G'(2)$ and interpret it in rate terminology.

9. The potential number of customers in a store's trading area t years from the time the store opens is $N(t)$, where

$$N(t) = \frac{200,000}{1 + 50e^{-0.8t}}.$$

Find $N'(5)$ and interpret it in rate terminology.

10. Find the relative rate of change in the function

$$f(x) = x^{1/2}$$

at $x = 5$ and interpret it in percent terminology.

11. The potential number of customers in a store's trading area t years from the time the store opens is

$$N(t) = 50,000e^{0.1t}.$$

Find the relative rate of change at $t = 5$ years and interpret it in percent terminology.

12. Total sales volume of a new product, in thousands of gallons, when the product has been sold for t years is

$$G(t) = 20t + 18(e^{-0.8t} - 1).$$

Find the relative rate of change at $t = 2$ years and interpret it in percent terminology.

13. A deposit of \$5000 in a bank account paying 7 percent interest compounded annually will grow in t years to the amount

$$A(t) = 5,000(1.07)^t.$$

Find the relative rate of change at $t = 10$ years and interpret it in percent terminology.

14. A formula for the optimum lot size, L, which minimizes cost, is

$$L = \left(\frac{2NF}{i}\right)^{1/2}$$

where N, F, and i are parameters.
   a) What is the optimum lot size if $N = 500$, $F = 100$, and $i = 10$?
   b) How sensitive is the optimum lot size to a change in the parameter F? To answer this, write the expression for the rate at which L changes per unit change in F.
   c) Using the expression in *(b)* and the numbers from *(a)*, estimate how much the optimum lot size would change if F increased from its value of 100 to the value 104.

15. (See Problem 14.)
   a) What is the optimal lot size if $N = 1000$, $F = 200$, and $i = 10$?
   b) How sensitive is optimal lot size to changes in the parameter N?
   c) Using the expression in *(b)* and the numbers from *(a)*, estimate how much the optimal lot size would change if N decreased from its value of 1000 to the value 900.

16. A revenue function is maximized if its independent variable takes the value

$$x = \frac{50}{h} + \frac{k}{2}$$

where h and k are parameters.
   a) What is the optimal value of x if $h = 10$ and $k = 20$?
   b) How sensitive is the optimal value of x to changes in the parameter h? To answer this, write the expression for the rate at which x changes per unit change in h.
   c) Using the expression in (b) and the numbers from (a), estimate how much the optimal value of x would change if h increased from its value of 10 to the value 10.5.

17. a) Find $f'(5)$ and interpret it in rate terminology if

$$f(x) = \ln x.$$

   b) Table X at the back of the book shows

$$f(5) = \ln 5 = 1.60944,$$

   but the table does not give ln 5.001. Estimate how much ln x will change if x changes from 5 to 5.001.
   c) Determine from (b) the approximate value of ln 5.001.

18. a) Find $f'(4)$ and interpret it in rate terminology if

$$f(x) = \sqrt{x}.$$

   b) We know $f(4) = 2$ and wish to estimate $f(4.01) = \sqrt{4.01}$. Estimate how much the square root of x will change if x increases from 4 to 4.01.
   c) Determine from (b) the approximate value of $\sqrt{4.01}$.

10.13 ELASTICITY OF DEMAND

The number of units of a product that consumers are willing and able to purchase at a given price per unit is called the *demand* for the product at that price. Demand changes if price per unit changes, and typically these changes are in opposite directions so that an increase (decrease) in price is accompanied by a decrease (increase) in demand. To obtain a measure of the relationship between price and demand changes that is comparable from product to product, economists convert absolute changes to percentage changes. Thus, for example, if a price increase from \$2.00 per unit to \$2.20 per unit (a 10 percent *increase*) is accompanied by a decline in demand from 1,000,000 units to 800,000 units (a 20 percent *decrease*), the ratio

$$\frac{-20 \text{ percent change in demand}}{10 \text{ percent change in price}} = -2$$

would be computed. The minus sign arises because of the opposite directions of price and demand movements. However, it is easier to interpret the number if the minus sign is omitted, so economists usually define the price *elasticity of demand* as a *positive number* computed from the ratio

$$\frac{\text{percent change in demand}}{\text{percent change in price}},$$

leaving it to the user of the number to understand that price and demand move in opposite directions. Thus, the elasticity measure, 2, computed in the foregoing, would be interpreted by saying a 2 percent demand decrease accompanies a 1 percent price increase. The implication of an elasticity of 2 to the seller of the product is that a price increase will be accompanied by a decline in total revenue received. Thus, in the example at hand,

$$(1{,}000{,}000 \text{ units})(\$2.00 \text{ per unit}) = \$2{,}000{,}000$$
$$(800{,}000 \text{ units})(\$2.20 \text{ per unit}) = \$1{,}760{,}000$$

so total revenue at \$2.20 per unit is less than total revenue at \$2.00 per unit. This behavior will be true whenever the percent decrease in demand is greater than the percent increase in price; that is, whenever elasticity is greater than 1 and, of course, if elasticity is less than 1, a price increase is accompanied by an increase in total revenue. If elasticity equals 1, a state called *unitary* elasticity of demand, the percent changes in demand and price nullify each other and total revenue is the same before and after a price change.

> **Exercise.** At a given price level, a product has an elasticity of 0.5. *a)* Interpret the elasticity measure. *b)* What implication does this number have to sellers of the product? Answer: *a)* A demand decrease of 1/2 percent accompanies a price increase of 1 percent. *b)* Total revenue will increase if price per unit is increased.

Elasticity of demand for a given product generally is not constant but is a function of the demand and the associated price. If we let q units represent the demand, then price per unit can be expressed as a function of q. We shall call this function the *demand function* and symbolize it as $p(q)$. The graph of a demand function is called a *demand curve*. As illustrated in Figure 10–7, demand curves typically fall to the right and are concave upward; that is, at all points, $p'(q)$ is negative and $p''(q)$ is positive. Our purpose in this section is to develop the

550

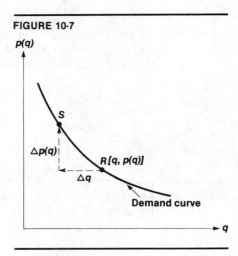

FIGURE 10-7

calculus definition of elasticity of demand and we start by considering the demand curve in Figure 10–7. At point R on the curve, demand is q units and price per unit is $p(q)$. If unit price changes (increases) by $\Delta p(q)$ from its level at R, demand changes (decreases) by Δq units. Remembering that a percent change calculation requires that we divide the amount of change by the beginning value, and the beginning value is at point R, we write

$$\text{percent change in } p = \frac{\Delta p(q)}{p(q)} (100\%)$$

$$\text{percent change in } q = \frac{\Delta q}{q} (100\%).$$

We now divide the percent change in demand by the percent change in price, inserting a minus sign to convert the negative ratio to a positive number, and write

$$-\frac{\frac{\Delta q}{q}(100\%)}{\frac{\Delta p(q)}{p(q)}(100\%)} = -\frac{\Delta q}{q}\left[\frac{p(q)}{\Delta p(q)}\right] = -\frac{p(q)}{q}\left[\frac{\Delta q}{\Delta p(q)}\right] = -\frac{p(q)}{q}\left[\frac{1}{\frac{\Delta p(q)}{\Delta q}}\right].$$

Note that in the rightmost part of the last line we converted

$$\frac{\Delta q}{\Delta p(q)} \quad \text{to} \quad \frac{1}{\frac{\Delta p(q)}{\Delta q}}.$$

Our purpose in so doing was to prepare for taking the limit of the expression as Δq approaches zero, which we show next. The limit is called the *point* elasticity of demand and we shall designate this by $e(q)$. Thus,

$$e(q) = \lim_{\Delta q \to 0} \left[-\frac{p(q)}{q} \cdot \frac{1}{\dfrac{\Delta p(q)}{\Delta q}} \right]$$

$$= -\frac{p(q)}{q} \lim_{\Delta q \to 0} \left[\frac{1}{\dfrac{\Delta p(q)}{\Delta q}} \right]$$

where, in the last line, q and $p(q)$ are the constant coordinates of point R and may be placed outside the limit symbol. Now, of course,

$$\lim_{\Delta q \to 0} \frac{\Delta p(q)}{\Delta q} = p'(q)$$

so,

$$e(q) = -\frac{p(q)}{q} \lim_{\Delta q \to 0} \left[\frac{1}{\dfrac{\Delta p(q)}{\Delta q}} \right] = -\frac{p(q)}{q} \left[\frac{1}{p'(q)} \right].$$

$$\textbf{\textit{Definition:}} \; e(q) = -\frac{p(q)}{q} \left[\frac{1}{p'(q)} \right].$$

This calculus definition assigns an elasticity measure to each point on a demand curve and resolves a problem that arises when two points on the curve are used to compute elasticity. The nature of the two-point problem can be understood by reference to Figure 10–7, where we started from R and went to S, so the *beginning* values used in computing percent changes were those at R. It would be just as reasonable to start from S and go to R. Then the changes in q and $p(q)$ would be of the same size, but the beginning values would be those at S, so the percent changes would not be the same as those computed with R as a starting point, and we could obtain different elasticity measures using the same two points on the demand curve. The calculus point definition of elasticity avoids the ambiguity inherent in the two-point definition.

   Example.   Compute elasticity of demand at $q = 100$ units for the demand function

$$p(q) = \frac{100}{0.01q + 1} = 100(0.01q + 1)^{-1}.$$

552

First we find the derivative of

$$p'(q) = \frac{d}{dq}[100(0.01q+1)^{-1}]$$
$$= 100(-1)(0.01q+1)^{-2}(0.01)$$
$$= -\frac{1}{(0.01q+1)^2}.$$

At $q = 100$,

$$p'(100) = -\frac{1}{[(0.01)(100)+1]^2} = -\frac{1}{(1+1)^2} = -\frac{1}{4} = -0.25.$$

Next we need the price at demand $q = 100$,

$$p(q) = p(100) = \frac{100}{0.01(100)+1} = \frac{100}{1+1} = \$50 \text{ per unit.}$$

Finally,

$$e(100) = -\frac{p(q)}{q}\left[\frac{1}{p'(q)}\right] = -\frac{50}{100}\cdot\left[\frac{1}{-0.25}\right] = \frac{50}{25} = 2.$$

The result, $e(100) = 2$, is interpreted by saying that, *approximately,* a demand decrease of 2 percent accompanies a price increase of 1 percent. The word approximately is italicized in the last sentence to call attention to the fact that $e(q)$ is computed *at* a point and applies only approximately if a change is made from the point. However, as a practical matter, a demand level is not a precise constant, so it is reasonable to speak of an elasticity of about 2 when the demand level is about 100 units.

Exercise. a) Compute elasticity of demand at demand 100 units for a product whose demand function is

$$p(q) = 60e^{-0.025\,q}.$$

b) Interpret the answer to part *(a)*. Answer: *a)* $e(100) = 0.4$. *b)* Approximately, a demand decrease of 0.4 percent (4 tenths of one percent) accompanies a price increase of 1 percent.

A property of $e(q)$ that is worthy of note is that if the revenue function

Revenue = (Number of units demanded)(Price per unit)
$$R(q) = q[p(q)]$$

has a maximum value, the maximum occurs where $e(q) = 1$. To see why this is so, we first find the condition that must be true if $R(q)$ is to have a local extreme point, and this condition is that $R'(q) = 0$. We find by the product rule

$$R'(q) = q\frac{d}{dq}p(q) + p(q)\frac{d}{dq}(q)$$

$$R'(q) = qp'(q) + p(q)(1). \tag{1}$$

$$R'(q) = 0 \quad \text{where} \quad qp'(q) + p(q) = 0$$

$$qp'(q) = -p(q)$$

$$p'(q) = -\frac{p(q)}{q}. \tag{2}$$

Now we write the definition of elasticity

$$e(q) = -\frac{p(q)}{q}\left[\frac{1}{p'(q)}\right]$$

and substitute the expression for $p'(q)$ from (2) into the definition, obtaining

$$e(q) = -\frac{p(q)}{q}\left[\frac{1}{-\dfrac{p(q)}{q}}\right] = \frac{p(q)}{q}\left[\frac{q}{p(q)}\right] = 1,$$

showing that $e(q)$ must equal 1 if $R'(q)$ is to be zero, as required for a local extreme point. We have not shown that the extreme point is a local maximum for revenue, but only that if there is a maximum, it must occur where $e(q) = 1$. Consequently, it will be necessary to determine by a test whether a local maximum has been found.

Example. a) If the demand function for a product is

$$p(q) = 60e^{-0.025\,q},$$

write the expression for the revenue function, find q^* at its local extreme point, and show that this q^* yields a local maximum. b) Compute $e(q^*)$.

a) The revenue expression is

$$\text{Revenue} = (\text{price per unit})(\text{number of units})$$
$$R(q) = p(q)q$$
$$R(q) = (60e^{-0.025\,q})\,(q).$$

Applying the product rule, we have

$$R'(q) = 60\left[e^{-0.025\,q}\frac{d}{dq}(q) + q\frac{d}{dq}(e^{-0.025\,q})\right]$$

$$= 60[e^{-0.025\,q}(1) + q(e^{-0.025\,q})(-0.025)]$$

$$R'(q) = 60e^{-0.025\,q}[1 + q(-0.025)]$$

$R'(q) = 0$ where

$$60e^{-0.025}(1 - 0.025q) = 0.$$

Because e to a power is never zero, we may divide both sides of the last equation by the first two factors to obtain

$$1 - 0.025q = 0$$
$$-0.025q = -1$$
$$q = \frac{-1}{-0.025}$$
$$q^* = 40 \text{ units.}$$

To show $q^* = 40$ yields a local maximum we need to show that $R''(40)$ is negative. Starting with

$$R'(q) = 60e^{-0.025q}(1 - 0.025q)$$

we again apply the product rule to find

$$R''(q) = 60\left[e^{-0.025q}\frac{d}{dq}(1 - 0.025q) + (1 - 0.025q)\frac{d}{dq}e^{-0.025q}\right]$$
$$R''(q) = 60[e^{-0.025q}(-0.025) + (1 - 0.025q)(e^{-0.025q})(-0.025)].$$

Now remember that at $q^* = 40$, $(-0.025)q^* = -(0.025)(40) = -1$. Hence,

$$R''(q^*) = R''(40) = 60[e^{-1}(-0.025) + (1 + 1)e^{-1}(-0.025)]$$
$$= 60e^{-1}[-0.025 + 2(-0.025)]$$
$$= 60e^{-1}(-0.075)$$

and because e^{-1} is positive, $R''(40)$ is negative and we have a local maximum for revenue at 40 units demand.

 b) To compute $e(q^*) = e(40)$, we need $p(40)$ and $p'(40)$.

$$p(40) = 60e^{-0.025(40)} = 60e^{-1}$$

which we will leave as is to avoid introducing an approximation.

$$p'(q) = 60e^{-0.025q}(-0.025)$$
$$= -1.5e^{-0.025q}$$
$$p'(40) = -1.5e^{-1}.$$

Finally,

$$e(40) = -\frac{p(40)}{40}\left[\frac{1}{p'(40)}\right]$$
$$= -\frac{60e^{-1}}{40}\left(\frac{1}{-1.5e^{-1}}\right) = \frac{60}{40(1.5)} = \frac{60}{60} = 1,$$

verifying that $e(q^*) = 1$ if revenue has a local maximum at demand $q^* = 40$.

> **Exercise.** *a)* If the demand function for a product is
>
> $$p(q) = 100 - 0.1q,$$
>
> write the expression for the total revenue function, find q^* at its local extreme point, and show that this q^* yields a local minimum. *b)* Compute $e(q^*)$. Answer: *a)* $R(q) = 100q - 0.1q^2$ and $q^* = 500$. $R''(q) = -0.2$, so q^* is at a local maximum. *b)* $e(500) = 1$.

In closing this section we note that the term elasticity we have been using is only one of several elasticity concepts treated in economics and, to be more descriptive, the measure developed in this section should be referred to as the *price elasticity of demand.*

10.14 PROBLEM SET 10–5

1. The demand function for a product is $p(q) = 220 - 0.5q$.
   a) Compute elasticity of demand at 200 units demand.
   b) Interpret the answer to part *(a)*.
2. The demand function for a product is $p(q) = 75 - 0.1q$.
   a) Compute elasticity of demand at 500 units demand.
   b) Interpret the answer to part *(a)*.
3. The demand function for a product is $75 - 0.1q$.
   a) Write the expression for the revenue function, find the q^* where revenue has a local extreme point, and show that this point is a local maximum.
   b) Compute $e(q^*)$.
4. The demand function for a product is $p(q) = 220 - 0.5q$.
   a) Write the expression for the revenue function, find the q^* where revenue has a local extreme point and show that this point is a local maximum.
   b) Compute $e(q^*)$.
5. The demand function for a product is $p(q) = 100e^{-0.0008q}$.
   a) Compute elasticity of demand at 1000 units demand.
   b) What level of demand would lead to maximum revenue?
6. The demand function for a product is $p(q) = 20e^{-0.0025q}$.
   a) Compute elasticity of demand at 500 units demand.
   b) What level of demand would lead to maximum revenue?
7. The demand function for a product is $p(q) = \dfrac{1000}{q}$.
   a) Find the expression for $e(q)$ and simplify it as much as possible.
   b) Write the expression for the revenue function.
   c) Interpret the outcomes of parts *(a)* and *(b)*.

8. The demand function for a product is $p(q) = \dfrac{1000}{q^{1/2}}$. Find the expression for $e(q)$ and simplify it as much as possible.

9. The demand function for a product is $p(q) = \dfrac{k}{q^{1/m}}$ where k and m are positive constants. Find the expression for $e(q)$ and simplify it as much as possible.

10. Do the following demand functions have the properties of negative slope and upward concavity? State reasons for answers.

   a) $p(q) = \dfrac{1000}{0.5q + 1}$.

   b) $p(q) = 100 - 2 \ln q$.

10.15 CALCULUS OF TWO INDEPENDENT VARIABLES

All of our work in differential calculus thus far has involved functions having one independent variable. Thus, for example, if a producer makes only the product Exall, the cost of producting x gallons of Exall is $C(x)$, a function of the single variable x. However, if the producer makes two products, x gallons of Exall and y gallons of Whyall, production cost is a function of both x and y, which we symbolize as

$$C(x, y),$$

and C is now a function of two independent variables. Clearly, the function concept can be extended to more than two independent variables and we refer to calculus of two or more independent variables as *multivariate* calculus. This subject area is too extensive to treat in any detail in this book, but we can learn some of its important aspects and applications by considering the case of two independent variables. Consider the function

$$f(x, y) = 3x^2 - 2xy - 8y + y^2 + 44.$$

A pair of numbers, one for x and the other for y, yield a value for $f(x, y)$. Thus, at $x = 1$, $y = 2$, we find

$$f(1, 2) = 3(1)^2 - 2(1)(2) - 8(2) + (2)^2 + 44 = 31.$$

Exercise. Find $f(2, 6)$ for the foregoing function. Answer: 20.

If we treat y in $f(x, y)$ as a constant and take the derivative of $f(x, y)$ with respect to x, the procedure is called taking the *partial derivative* with respect to x or, more briefly, the partial with respect to x. Among the symbols used to designate this procedure are

$$f_x; \quad \frac{\partial f(x, y)}{\partial x}; \quad \frac{\partial f}{\partial x}$$

where ∂ is called a "round delta." Actually, the large and small Greek deltas are Δ and δ; ∂ is really not a letter but is a symbol devised to represent taking a partial derivative. For

$$f(x, y) = 3x^2 - 2xy - 8y + y^2 + 44$$
$$f_x = 6x - 2y;$$

that is, if y is treated as a constant then in the second term, $-2xy$, $-2y$ is constant and the derivative of $(-2y)x$ with respect to x is the constant, $-2y$. Similarly, the derivatives with respect to x of $-8y + y^2 + 44$ are all zero. In round delta notation we would write

$$\frac{\partial}{\partial x} f(x, y) = \frac{\partial}{\partial x}(3x^2 - 2xy - 8y + y^2 + 44)$$
$$= 6x - 2y.$$

Next we find f_y, the partial of f with respect to y. This means to take the derivative of $f(x, y)$ treating x as a constant. We find

$$f_y = -2x - 8 + 2y.$$

The *second partial of $f(x, y)$ taken twice with respect to x is symbolized* as

$$f_{xx}; \quad \frac{\partial^2 f(x, y)}{\partial x^2}; \quad \frac{\partial^2 f}{\partial x^2},$$

and it means to take the partial with respect to x, then take the partial of this result again with respect to x. For our example,

$$f_x = 6x - 2y,$$

so

$$f_{xx} = 6.$$

The last, in round delta notation, is

$$\frac{\partial^2 f}{\partial x^2} = \frac{\partial}{\partial x}\left(\frac{\partial f}{\partial x}\right) = \frac{\partial}{\partial x}(6x - 2y) = 6.$$

Similarly, we have

$$f_y = -2x - 8 + 2y$$
$$f_{yy} = 2.$$

Finally we may take the partial with respect to x,

$$f_x = 6x - 2y,$$

then take the partial of f_x *with respect to* y. This is symbolized as

$$f_{xy} \quad \text{or} \quad \frac{\partial f}{\partial y \partial x}$$

and is called the second partial of $f(x, y)$, first with respect to x, then with respect to y. Thus, with $f_x = 6x - 2y$,

$$f_{xy} = -2.$$

We may obtain the same result by finding

$$f_{yx};$$

that is, find the partial of $f(x, y)$ first with respect to y, then take the partial of the result with respect to x. In our example,

$$f_y = -2x - 8 + 2y$$
$$f_{yx} = -2,$$

which is the same as f_{xy}. The equality of f_{xy} and f_{yx} is true wherever these derivatives are continuous, and will be true for all functions we shall deal with.

Exercise. $f(x, y) = 2x + 3y + x^2y + 2xy^3 - 20$. Find: *a)* f_x. *b)* f_y. *c)* f_{xx}. *d)* f_{yy}. *e)* f_{xy}. *f)* f_{yx}. Answer: *a)* $2 + 2xy + 2y^3$. *b)* $3 + x^2 + 6xy^2$. *c)* $2y$. *d)* $12xy$. *e)* $2x + 6y^2$. *f)* $2x + 6y^2$.

Partial derivatives, like derivatives of a function of one variable, can be interpreted as slopes or rates of change. For example, if the cost of making x gallons of Exall and y gallons of Whyall is

$$C(x, y) = 3x + 2y + 10 \text{ dollars,}$$

then

$$C_x = 3 \quad \text{and} \quad C_y = 2.$$

We can say that if production of Whyall is held constant, cost increases at the rate of $C_x = 3$ dollars for each additional gallon of Exall made, whereas if production of Exall is held constant, cost increases at the rate of $C_y = 2$ dollars for each additional gallon of Whyall made.

If, unlike the example just given, the partials are not constant, then the rate interpretation is made at a point; that is, a pair of values, (x, y).

Example. Total profit when x gallons of Exall and y gallons of Whyall are produced and sold is given by

$$P(x, y) = 100x - x^2 - 2xy + 200y - 3y^2 \text{ dollars.}$$

Find $P(15, 20)$, $P_x(15, 20)$, and $P_y(15, 20)$. Interpret the partials in rate terminology.

First we find

$$P(15, 20) = 100(15) - (15)^2 - 2(15)(20) + 200(20) - 3(20)^2 = \$3475.$$

Next,

$$P_x(x, y) = 100 - 2x - 2y$$
$$P_x(15, 20) = 100 - 2(15) - 2(20) = 30.$$

Thus, with Whyall production held constant (at $y = 20$) profit is increasing at the rate of \$30 per additional gallon of Exall made and sold. Similarly

$$P_y(x, y) = -2x + 200 - 6y$$
$$P_y(15, 20) = -2(15) + 200 - 6(20) = 50$$

and with production of Exall constant (at $x = 15$), profit is increasing at the rate of 50 dollars per additional gallon of Whyall made and sold.

Note in the foregoing example that profit at the point where $x = 15$, $y = 20$, is $P(15, 20) = \$3475$ and both partials indicate that profit can be increased by making and selling additional amounts of Exall and Whyall. This suggests that production should be increased and raises the question of whether profit increases indefinitely or reaches a local maximum at some combination of outputs of Exall and Whyall. We shall answer this question after the next set of problems.

> **Exercise.** For the profit function of the last example, find: *a)* $P(26, 27)$. *b)* $P_x(26, 27)$. *c)* $P_y(26, 27)$. *d)* Interpret the answer to parts *(b)* and *(c)*. Answer: *a)* \$3733. *b)* $P_x = -6$. *c)* $P_y = -14$. *d)* At 26 units of Exall and 27 units of Whyall, profit is *decreasing* at the rate of \$6 per additional unit of Exall and 14 dollars per additional unit of Whyall.

10.16 PROBLEM SET 10–6

For the following find: *a)* f_x. *b)* f_{xx}. *c)* f_y. *d)* f_{yy}. *e)* f_{xy}.

1. $f(x, y) = 3x - 2y + 6$.
2. $f(x, y) = 2x + 5y - 10$.
3. $f(x, y) = x^2 + y^2 + 3x - 2y - 9$.
4. $f(x, y) = x^2 - y^2 - 5x + 4y + 7$.
5. $f(x, y) = 3x^2 - 2xy + y^2 + 4$.
6. $f(x, y) = 2x^3 + 3xy + 4y - 6$.
7. $f(x, y) = x^{1/2}y^{1/2}$.
8. $f(x, y) = x^{2/3}y^{1/3}$.
9. $f(x, y) = 2xy^{1/2} - 3x^{1/3}y$.
10. $f(x, y) = x^{3/2}y^2$.
11. $f(x, y) = xye^x$.
12. $f(x, y) = xy \ln y$.

13. $f(x, y) = \dfrac{x-y}{x+y}$.

14. $f(x, y) = \dfrac{x+y}{x-y}$.

15. $f(x, y) = \ln(3x + 2y)$.

16. $f(x, y) = e^{2x+3y}$.

Perform the following:

17. $\dfrac{\partial}{\partial z}(3z^2 - 2xz)$.

18. $\dfrac{\partial}{\partial w}(w^2 - 3z^2 w)$.

19. $\dfrac{\partial^2}{\partial w^2}(w^2 - 3z^2 w)$.

20. $\dfrac{\partial^2}{\partial z^2}(3z^2 - 2xz)$.

21. $\dfrac{\partial^2}{\partial w \partial z}(w^3 - 3zw)$.

22. $\dfrac{\partial^2}{\partial z \partial x}(z^3 + 3xz)$.

23. If $f(x, y) = 3x + 2y$, find:
    a) $f(1, 2)$.
    b) $f_x(1, 2)$.
    c) $f_{xx}(1, 2)$.
    d) $f_y(1, 2)$.
    e) $f_{yy}(1, 2)$.
    f) $f_{xy}(1, 2)$.

24. If $f(x, y) = 2x + 3y - 6$, find:
    a) $f(2, 3)$.
    b) $f_x(2, 3)$.
    c) $f_{xx}(2, 3)$.
    d) $f_y(2, 3)$.
    e) $f_{yy}(2, 3)$.
    f) $f_{xy}(2, 3)$.

25. If $f(x, y) = x^2 + xy - 3y^2 + 5$, compute:
    a) $f(2, 1)$.
    b) $f_x(2, 1)$.
    c) $f_{xx}(2, 1)$.
    d) $f_y(2, 1)$.
    e) $f_{yy}(2, 1)$.
    f) $f_{xy}(2, 1)$.

26. If $f(x, y) = 2x^2 - 5xy + y^2 - 6$, compute:
    a) $f(3, 2)$.
    b) $f_x(3, 2)$.
    c) $f_{xx}(3, 2)$.
    d) $f_y(3, 2)$.
    e) $f_{yy}(3, 2)$.
    f) $f_{xy}(3, 2)$.

27. During a period of operation a producer makes and sells x gallons of Exall and y gallons of Whyall at a cost of

$$C(x, y) = 3x^2 - 2xy + y^2 - 12x - 4y + 61 \text{ dollars.}$$

Find $C_x(3, 5)$ and $C_y(3, 5)$ and interpret these numbers in rate terminology.

28. (See the cost function in Problem 27.) Find $C_x(3, 7)$ and $C_y(3, 7)$ and interpret these numbers in rate terminology.

29. A producer makes and sells x gallons of Exall and y gallons of Whyall during a period of operation. Profit is

$$P(x, y) = 50x - 0.05x^2 + 110y - 0.10y^2 \text{ dollars.}$$

Compute each of the following and interpret the result in rate terminology.

    a) $P_x(400, 500)$.
    b) $P_y(400, 500)$.
    c) $P_x(600, 500)$.
    d) $P_y(400, 600)$.

30. The total output of an industry, V million dollars, is a function of L (expenditure for labor) and C (dollars of capital invested in the industry). Suppose

$$V(L, C) = 8L^{1/3}C^{2/3}.$$

Compute the following and interpret the result in rate terminology.
    a) $V_L(64, 27)$.
    b) $V_C(64, 27)$

10.17 MAXIMA AND MINIMA: TWO INDEPENDENT VARIABLES

Reviewing the example of the last section where $P(x, y)$ was the profit function for a producer who made and sold x gallons of Exall and y gallons of Whyall,

$$P(x, y) = 100x - x^2 - 2xy + 200y - 3y^2,$$

we found

$$P_x = 100 - 2x - 2y \quad \text{and} \quad P_y = -2x + 200 - 6y.$$

At $x = 15$, $y = 20$

$$P_x(15, 20) = 30 \quad \text{and} \quad P_y(15, 20) = 50,$$

so profit is increasing per additional unit of Exall if production of Whyall is constant, and profit also is increasing per additional unit of Whyall if Exall production is held constant. However, the opposite (profit rates decreasing) occurs when $x = 26$ and $y = 27$ because

$$P_x(26, 27) = -6 \quad \text{and} \quad P_y(26, 27) = -14$$

are both negative. The change from increasing to decreasing rates suggest that a maximum profit exists. If such is the case at some pair of (x, y) values, then for this pair of values profit must be neither increasing nor decreasing, so *both* P_x and P_y must be zero. This is similar to the one-variable case condition that the first derivative must be zero at a local extreme point, except that with two independent variables, both first partial derivatives must be zero.

Condition: Both first partial derivatives must be zero at a local extreme point.

Consequently, in seeking to maximize the profit function $P(x, y)$ at hand, we set $P_x = 0$ and $P_y = 0$ to give equations e_1 and e_2:

$$e_1: 100 - 2x - 2y = 0$$
$$e_2: 200 - 2x - 6y = 0.$$

These equations are now solved simultaneously. The simplest procedure is to subtract, as follows:

$$e_1 - e_2: (100 - 2x - 2y) - (200 - 2x - 6y) = 0$$
$$-100 + 4y = 0$$
$$4y = 100$$
$$y = 25.$$

Substituting $y = 25$ into e_1, we find

$$100 - 2x - 2(25) = 0$$
$$-2x = -50$$
$$x = 25$$

so that $x^* = 25$, $y^* = 25$ marks a stationary point that may (or may not) be a local extreme point. In any event,

$$P(x^*, y^*) = P(25, 25)$$
$$= 100(25) - (25)^2 - 2(25)(25) + 200(25) - 3(25)^2$$
$$= \$3750.$$

To determine whether we have a local extreme point and, if so, whether it is a maximum or minimum, we use a test which, as might be anticipated, involves the second partial derivatives. We shall not develop the source of this test because to do so would carry us too far afield.

Test for local extreme points: Two independent variables

1. Find f_x and f_y, set both equal to zero, and solve the resultant equations simultaneously to obtain the candidate values x^* and y^*.
2. Compute $f_{xx}(x^*, y^*)$, $f_{yy}(x^*, y^*)$, and $f_{xy}(x^*, y^*)$.
3. Compute $D = (f_{xx})(f_{yy}) - (f_{xy})^2$ from the values in (2).
4. If $D > 0$: $\begin{cases} \text{Local maximum if both } f_{xx} \text{ and } f_{yy} \text{ are negative.} \\ \text{Local minimum if both } f_{xx} \text{ and } f_{yy} \text{ are positive.} \end{cases}$

    If $D < 0$:   Candidate is neither a maximum nor minimum.

    If $D = 0$:   The test fails.

In the example at hand, with $x^* = 25$, $y^* = 25$

$$P_x = 100 - 2x - 2y, \quad \text{so} \quad P_{xx} = -2$$
$$P_y = -2x + 200 - 6y, \quad \text{so} \quad P_{yy} = -6$$
$$P_{xy} = -2.$$

Hence, Step (3) in the foregoing test provides

$$D = (-2)(-6) - (-2)^2 = 8$$

so in Step (4) we have $D > 0$, and both f_{xx} and f_{yy} are negative. This proves we have a maximum, and

$$\text{Maximum profit} = P_{max} = P(25, 25) = \$3750$$

as previously computed. The producer should make and sell 25 gallons of Exall and 25 gallons of Whyall to achieve this maximum profit.

> **Exercise.** a) Find (x^*, y^*) at the candidate extreme point for
>
> $$f(x, y) = 3x^2 - 2xy + y^2 - 12x - 4y + 61.$$
>
> b) Find $f(x^*, y^*)$. c) Show by test whether (b) is a local maximum or minimum. Answer: a) $x^* = 4$, $y^* = 6$. b) $f(4, 6) = 25$. c) $D = 8$, $f_{xx} = 6$, $f_{yy} = 2$. Local minimum.

If D, defined in the foregoing extreme point test, is greater or less than zero, the test works. However, if $D = 0$, the test fails and the point in question may not be a local extreme point. Situations where $D = 0$ fortunately do not occur for the functions we shall deal with, but in passing we note that when $D = 0$ it is possible to determine whether or not an extreme point occurs at (x^*, y^*) by moving to nearby points which we may symbolize as $[(x^* + h), (y^* + k)]$, where h and k are arbitrary small numbers, determining either algebraically or by computation, whether the function values at nearby points are greater than (less than) $f(x^*, y^*)$, making $f(x^*, y^*)$ a local minimum (maximum).

Example. To meet customer demand for a product, the producer makes a batch, or *lot*, of L units of the product periodically. The producer's strategy is to establish the period of time between lot productions so that a new lot is not made until some time after current inventory has been exhausted. Thus, there is an out-of-stock or stock-out interval. During this interval, orders received from customers are placed in "back-order" status and are filled immediately when a new lot is made. Thus, the producer makes L units, uses some to fill back-orders, and places the remainder, I units, in inventory. Taking into account various costs, including a cost associated with being out of stock, the producer has developed the following cost-function model.

$$C(L, I) = \frac{1{,}350{,}000}{L} + \frac{20I^2}{L} + 15L - 30I.$$

What values of L and I will minimize this cost function?

We proceed by taking the first partial derivatives:

$$C_L = -\frac{1{,}350{,}000}{L^2} - 20\frac{I^2}{L^2} + 15$$

$$C_I = \frac{40I}{L} - 30.$$

If there is an extreme point, the first partials must equal zero. Hence,

$$-\frac{1{,}350{,}000}{L^2} - 20\frac{I^2}{L^2} + 15 = 0 \tag{1}$$

$$\frac{40I}{L} - 30 = 0. \tag{2}$$

The common solution of the pair of equations can be found by solving (2) for I in terms of L and substituting this into (1). Thus, from (2)

$$\frac{40I}{L} - 30 = 0$$

$$40I - 30L = 0$$

$$40I = 30L$$

$$I = \frac{30L}{40} = 0.75L. \tag{3}$$

Substituting (3) into (1), we have

$$-\frac{1,350,000}{L^2} - \frac{20(0.75L)^2}{L^2} + 15 = 0$$

$$-\frac{1,350,000}{L^2} - \frac{20(0.75)^2L^2}{L^2} + 15 = 0$$

$$-\frac{1,350,000}{L^2} - 11.25 + 15 = 0$$

$$3.75 = \frac{1,350,000}{L^2}$$

$$3.75L^2 = 1,350,000$$

$$L^2 = \frac{1,350,000}{3.75}$$

$$L^2 = 360,000$$
$$L^* = 600 \text{ units.}$$

With $L = 600$, we find from (3) that

$$I = 0.75L$$
$$= 0.75(600)$$
$$I^* = 450 \text{ units.}$$

To prove the values found do minimize cost, $C(L, I)$, we find

$$C_{LL} = \frac{2(1,350,000)}{L^3} + 40\frac{I^2}{L^3}$$

$$C_{LL}(600, 450) = \frac{2(1,350,000)}{(600)^3} + \frac{40(450)^2}{(600)^3} = 0.05.$$

$$C_{II} = \frac{40}{L} \text{ and } C_{II}(600,450) = \frac{40}{600} = 0.06667.$$

$$C_{LI} = -\frac{40I}{L^2} \text{ and } C_{LI}(600,450) = -\frac{40(450)}{(600)^2} = -0.05.$$

Both C_{LL} and C_{II} are positive and

$$D = (C_{LL})(C_{II}) - (C_{LI})^2 = (0.05)(0.06667) - (-0.05)^2 = 0.0008$$

which is greater than zero, so we have a minimum, which is

$$C(600, 450) = \frac{1,350,000}{600} + \frac{20(450)^2}{600} + 15(600) - 30(450) = \$4500.$$

Hence, the producer should make $L^* = 600$ units in each lot and place $I^* = 450$ units in inventory. This means that the remaining 150 units are used to fill back orders. The advantage the producer achieves by his out-of-stock strategy arises from the cost saving achieved by not having to carry these 150 units in inventory.

The last example serves not only to illustrate a current application of multivariate calculus but also to bring out the point that the set of simultaneous equations encountered when first partials are set equal to zero may be very difficult to solve. The task was relatively easy in our example because we had only two variables and a relatively simple cost function. Typically, when the function is more complicated and there are more than two variables, solutions of the set of simultaneous equations are approximated by specially prepared computer programs that also deal with the situation, earlier mentioned, where $D = 0$. Procedure for approximating solutions are treated in texts on *numerical analysis*.

10.18 PROBLEM SET 10–7

Find the value of the function at its extreme point, if one exists. State whether an extreme point is a local maximum or minimum.

1. $f(x, y) = 2x^2 + 3y^2 + 10.$ 2. $f(x, y) = 4x + 6y - x^2 - y^2 + 10.$
3. $f(x, y) = 10xy - 5x^2 - 6y^2 + 20x.$
4. $f(x, y) = 3x^2 + y^2 - 2xy - 12y + 104.$
5. $f(x, y) = x + y + \dfrac{9}{x} + \dfrac{4}{y}$ (x and y positive).
6. $f(x, y) = 4xy + 8x + 20y - 4x^2 - 4y^2.$
7. $f(x, y) = xy - x - 2y + 2.$
8. $f(x, y) = 2x^2 - 3y^2.$
9. $f(x, y) = 3xy - 3x^2 - y^2 + 6x + 10y + 3.$
10. $f(x, y) = x^2 + 3y^2 - 4xy + 8x - 6y + 5.$
11. $f(x, y) = 10e^{-x^2} - y^2.$
12. $f(x, y) = 3x^2 - 5e^{-y^2}.$
13. Profit earned by making and selling x gallons of Exall and y gallons of Whyall is

$$P(x, y) = 50x - 0.05x^2 + 110y - 0.10y^2 \text{ dollars.}$$

   a) What number of units of each product will maximize profit?
   b) Prove (a) is a maximum.
   c) Compute maximum profit.

14. One unit of product can be made by using the Extron machine for x hours and the Whytron machine for y hours. The cost of making one unit is

$$C(x, y) = 3x^2 - 2xy + y^2 - 12x - 4y + 50 \text{ dollars.}$$

   a) What numbers of hours on each machine will minimize the cost of making one unit?
   b) Prove (a) is a minimum.
   c) Compute the minimum cost.

15. The inventory cost model discussed in the example near the end of the last section is to be applied to another product and the cost function is

$$C(L, I) = \frac{750,000}{L} + \frac{20I^2}{L} + \frac{25L}{2} - 25I,$$

where L is the number of units to be made in each lot and I is the number to be placed in inventory each time a lot is made. The remaining part of the lot is used to fill back orders.

a) Find the values of L and I that minimize $C(L, I)$.
b) Compute the minimum cost.

16. Another company follows the strategy of the company in Problem 15, but uses the cost model

$$C(L, p) = \frac{108,160}{L} + \frac{Lp^2}{2} + 2L(1 - p)^2,$$

where $C(L, p)$ is in thousands of dollars, L is the lot size, and p is the proportion of a lot which is to be placed in inventory.

a) Find the values of L and p that minimize $C(L, p)$.
b) Prove (a) is a minimum.
c) Compute the minimum cost.

17. (This problem is somewhat difficult, algebraically.) The cost model in Problem 15 is a special case of the general model

$$C(L, I) = \frac{cd}{L} + \frac{(a + b)I^2}{2L} + \frac{bL}{2} - bI,$$

where a, b, c, and d are parameters. Find the expressions for L and I that minimize cost in terms of these parameters.

10.19 LEAST-SQUARES CURVE FITTING

The reader may wish to review the meaning and properties of the summation operation indicated by

$$\sum_{i=1}^{n} x_i$$

before studying this section (see Index for reference). To aid in this review, we start the section with a simple exercise in minimization. First, suppose that we have selected a sample of $n = 3$ people and found from them that their weekly incomes, in hundreds of dollars, are

$$x_1 = 3, \quad x_2 = 5, \quad x_3 = 10$$

where the subscripts refer, respectively, to the first, second, and third persons. The *average* of this sample of $n = 3$, designated by $\bar{x}$ (read as "x bar"), is

$$\bar{x} = \frac{3 + 5 + 10}{3} = \frac{18}{3} = 6.$$

This is

$$\bar{x} = \frac{x_1 + x_2 + x_3}{3}$$

or, in summation symbols,

$$\bar{x} = \frac{\sum\limits_{i=1}^{n} x_i}{n}.$$

Next, we show the three numbers in our sample as points on a line segment, Figure, 10–8. We have placed an arbitrary point, *a*, on this

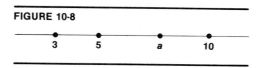

FIGURE 10-8

line. Now we express the distance between each of the sample numbers and *a* as

$$a - 3, \qquad a - 5, \quad \text{and} \quad a - 10.$$

These distances we call the *deviations* of the sample numbers from the point *a*. The object of this warm-up exercise is to find the value *a* must have if the sum of the squares of the deviations is to be minimized. This sum is

$$S = (a - 3)^2 + (a - 5)^2 + (a - 10)^2.$$

Taking the derivative with respect to *a*, we find

$$\frac{dS}{da} = 2(a - 3) + 2(a - 5) + 2(a - 10),$$

and

$$\frac{dS}{da} = 0 \quad \text{when} \quad 2(a - 3) + 2(a - 5) + 2(a - 10) = 0$$

$$a - 3 + a - 5 + a - 10 = 0$$

$$3a = 3 + 5 + 10$$

$$a = \frac{3 + 5 + 10}{3} = 6$$

$$a = \bar{x}.$$

The second derivative is positive, so we have a minimum and have proved that a must be the average of the x's. That is, the sum of the squares of the deviations of this group of numbers from their average, $\bar{x}$, is less than the sum of the squares of the deviations from any other number.

To complete our warm-up, we next show that the last result is true for any set of n numbers $\{x_i\}$. The expression for the sum of squares of the deviations of the x_i from a number a is

$$S = \sum_{i=1}^{n} (a - x_i)^2.$$

Taking the derivative, we write

$$\frac{dS}{da} = \frac{d}{da} \sum_{i=1}^{n} (a - x_i)^2,$$

but the derivative of a sum is the sum of the derivatives, so we may place d/da inside the Σ symbol, thus:

$$\frac{dS}{da} = \sum_{i=1}^{n} \frac{d}{da} (a - x_i)^2$$

$$= \sum_{i=1}^{n} [2(a - x_i)]$$

$$= \sum_{i=1}^{n} [2a - 2x_i]$$

$$= \sum_{i=1}^{n} 2a - \sum_{i=1}^{n} 2x_i.$$

Now the sum of a constant with respect to i as i goes from 1 to n is n times the constant, and $2a$ is a constant, so

$$\frac{dS}{da} = n(2a) - 2 \sum_{i=1}^{n} x_i$$

$$\frac{dS}{da} = 0 \quad \text{where} \quad n(2a) - 2 \sum_{i=1}^{n} x_i = 0$$

$$n(2a) = 2 \sum_{i=1}^{n} x_i$$

$$a = 2 \frac{\displaystyle\sum_{i=1}^{n} x_i}{2n}$$

$$= \frac{\displaystyle\sum_{i=1}^{n} x_i}{n}$$

$$= \bar{x}.$$

The second derivative is

$$\frac{d}{da}\left[n(2a) - 2 \sum_{i=1}^{n} x_i \right] = 2n,$$

which is positive because n, the count of the numbers in a set, is positive. Hence, we have a minimum, and we have proved that for any group of numbers, the sum of the squares of their deviations from their average is less than the sum of their squared deviations from any other number.

So much for the review warm-up. We now turn to the main objective of this section, which is to develop the procedure most generally used to fit mathematical models to actual observed data. This procedure is called *least-squares curve fitting*. In applications, the user determines by examination of the data, or from knowledge of applicable functions, the type of curve that is to be used. The unknown quantities are the constants or *parameters* that the selected curve function should have if it is to match actual observations as closely as possible. For example, if the choice of curve is a straight line, the functional form is

$$y(x) = mx + b,$$

and the problem is to take actual observed data and apply the least-squares method to them to get values for m and b so that the line with this slope and intercept fits the observed data more closely than any other line.

To see what is involved, suppose the observed data in Table 10–1 are x_i, the number of units made, and y_i, the corresponding cost of making the x_i units.

Table 10–1

Number of units made x_i	Total cost y_i
1	4
2	3
3	8
4	11
5	10

The (x_i, y_i) points, $(1, 4)$, $(2, 3)$, and so on, are plotted in Figure 10–9. Notice that the points scatter somewhat and do not all fall on one straight line. The figure is called, aptly, a scatter diagram. To begin with, a straight line has been drawn *freehand* to describe the tendency for cost to rise as output increases. We do not yet have an equation

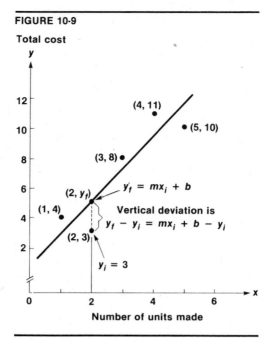

FIGURE 10-9

for this line and, clearly, freehand drawing by different people would lead to varying lines. To say which of all possible lines is the line that "best" fits the actual observed data, it is necessary to specify the criterion by which a line is to be judged to be the best-fitting line. In descriptive, but imprecise, terms, we want the line that comes *closest* to the points taken as a group. To make this more precise, look at Figure 10–9 and notice that for $x_i = 2$, the actual observed value, which is 3, has been labeled y_i. Now, denoting the as yet unknown value for y on the line when $x_i = 2$ as y_f (where y_f means the fitted value for y), the equation of the line is

$$y_f = mx_i + b,$$

and the *vertical* deviation of the observed point from the line is the difference of the vertical coordinates; that is,

$$\text{deviation} = y_f - y_i$$
$$= mx_i + b - y_i.$$

Moreover,

$$\text{squared deviation} = (mx_i + b - y_i)^2.$$

The smaller this squared deviation, the closer is the line to the point, and the smaller the sum of the squared deviations for *all* observed points, the closer is the line to the group of points. Hence,

Least-Squares best-fitting line criterion

The line that best fits a set of observed points is the one whose slope and intercept, m *and* b, *are such that the sum of the squares of the vertical deviations of the points from the line is a minimum.*

If we write the sum of squares as

$$S(m, b) = \sum_{i=1}^{n} (mx_i + b - y_i)^2, \tag{1}$$

the problem then becomes that of finding the expressions for m and b that will minimize S. Consequently, m and b are independent variables. We must set the first partials with respect to m and to b equal to zero and solve the resultant pair of equations for m and b. We have

$$\frac{\partial S}{\partial m} = \sum_{i=1}^{n} \frac{\partial}{\partial m} (mx_i + b - y_i)^2$$

$$= \sum_{i=1}^{n} 2(mx_i + b - y_i)(x_i)$$

$$= 2 \sum_{i=1}^{n} (mx_i^2 + bx_i - x_i y_i).$$

Now, although m and b are variables in taking the derivative, i is the index (or variable) in the summations, so in these summations, m and b are treated as constants. Hence,

$$\frac{\partial S}{\partial m} = 2 \left[m \sum_{i=1}^{n} x_i^2 + b \sum_{i=1}^{n} x_i - \sum_{i=1}^{n} x_i y_i \right]. \tag{2}$$

Taking the partial of (1) with respect to b yields

$$\frac{\partial S}{\partial b} = \frac{\partial}{\partial b} \sum (mx_i + b - y_i)^2$$

$$= \sum_{i=1}^{n} 2(mx_i + b - y_i)(1)$$

$$= 2 \left[m \sum_{i=1}^{n} x_i + \sum_{i=1}^{n} b - \sum_{i=1}^{n} y_i \right].$$

The middle term in the last expression is the summation of a constant with i going from 1 to n and equals nb. Hence,

$$\frac{\partial S}{\partial b} = 2\left[m\sum_{i=1}^{n} x_i + nb - \sum_{i=1}^{n} y_i \right]. \tag{3}$$

After setting (2) and (3) equal to zero and dividing both sides of each equation by 2, we have

$$e_1: \quad m \sum_{i=1}^{n} x_i^2 + b \sum_{i=1}^{n} x_i - \sum_{i=1}^{n} x_i y_i = 0$$

$$e_2: \quad m \sum_{i=1}^{n} x_i + nb \quad - \sum_{i=1}^{n} y_i = 0,$$

which are to be solved. First we solve e_2 to get b in terms of m. Thus,

$$e_2: \quad m \sum_{i=1}^{n} x_i + nb - \sum_{i=1}^{n} y_i = 0$$

$$nb = \sum_{i=1}^{n} y_i - m \sum_{i=1}^{n} x_i$$

$$b = \frac{\sum_{i=1}^{n} y_i - m \sum_{i=1}^{n} x_i}{n}. \tag{4}$$

We now substitute (4) into e_1 and solve for m.

$$e_1: \quad m\sum_{i=1}^{n} x_i^2 + b \sum_{i=1}^{n} x_i - \sum_{i=1}^{n} x_i y_i = 0$$

$$m \sum_{i=1}^{n} x_i^2 + \left[\frac{\sum_{i=1}^{n} y_i - m \sum_{i=1}^{n} x_i}{n} \right] \sum_{i=1}^{n} x_i - \sum_{i=1}^{n} x_i y_i = 0.$$

Next, expanding the center term and multiplying both sides by n,

$$nm \sum_{i=1}^{n} x_i^2 + \sum_{i=1}^{n} x_i \sum_{i=1}^{n} y_i - m\left(\sum_{i=1}^{n} x_i \right)^2 - n \sum_{i=1}^{n} x_i y_i = 0$$

$$m\left[n \sum_{i=1}^{n} x_i^2 - \left(\sum_{i=1}^{n} x_i \right)^2 \right] + \sum_{i=1}^{n} x_i \sum_{i=1}^{n} y_i - n \sum_{i=1}^{n} x_i y_i = 0.$$

Placing the two terms at the right on the other side of the equation and then dividing both sides by the coefficient of m, we obtain

$$m = \frac{n \sum_{i=1}^{n} x_i y_i - \left(\sum_{i=1}^{n} x_i \right)\left(\sum_{i=1}^{n} y_i \right)}{n \sum_{i=1}^{n} x_i^2 - \left(\sum_{i=1}^{n} x_i \right)^2}. \tag{5}$$

Expression (5) tells us how to compute the slope of the best fitting line from the observed data. Having done this, we can then compute the intercept, b, from expression (4).

Formulas for slope and intercept of the best-fitting (least-squares) straight line

(Note: The summations are over all n values of x_i and y_i.)

$$m = \frac{n\Sigma x_i y_i - (\Sigma x_i)(\Sigma y_i)}{n\Sigma x_i^2 - (\Sigma x_i)^2}$$

$$b = \frac{\Sigma y_i - m\Sigma x_i}{n}.$$

To compute m and b according to the formulas we need

n = Number of points
Σx_i = Sum of the x observations
Σy_i = Sum of the y observations
$\Sigma x_i y_i$ = Sum of the products: (x observation)(y observation)
$(\Sigma x_i)^2$ = Square of the sum of the x observations.

We carry out the necessary calculation for the data of Table 10–1 in Table 10–2.

TABLE 10–2

x_i	y_i	$x_i y_i$	x_i^2
1	4	4	1
2	3	6	4
3	8	24	9
4	11	44	16
5	10	50	25
15	36	128	55
Σx_i	Σy_i	$\Sigma x_i y_i$	Σx_i^2

For the $n = 5$ points

$$m = \frac{5(128) - (15)(36)}{5(55) - (15)^2} = \frac{100}{50} = 2;$$

$$b = \frac{36 - 2(15)}{5} = \frac{6}{5} = 1.2;$$

$$y_f = 2x + 1.2.$$

Thus, the sum of the squares of the vertical deviations of the five points on Figure 10–9 is at the smallest possible value if the line has the equation

$$y_f = 2x + 1.2.$$

The line can be plotted using its intercept $(0, 1.2)$ and another point, say

$$x = 5, \quad y_f = 2(5) + 1.2 = 11.2,$$

or (5, 11.2). This is, in fact, the line shown on Figure 10–9, so examination of that figure will show how closely the line fits the points.

> **Exercise.** Given the three observed points (1, 2); (2, 4); (3, 3), find: a) Σx_i. b) Σy_i. c) $\Sigma x_i y_i$. d) Σx_i^2. e) $(\Sigma x_i)^2$. Now compute: f) m. g) b. Then, h) Write the equation of the best fitting least-squares straight line. Answer: a) 6. b) 9. c) 19. d) 14. e) 36. f) 0.5. g) 2. h) $y_f = 0.5x + 2$.

Least-squares methodology is widely used to fit functions of various types to observed data. It is interesting to note that some relatively inexpensive hand calculators have been designed so that the user need only enter the observed data values and push the appropriate buttons to obtain the slope and intercept of the best-fitting least-squares line. We should remark in passing that the methodology can be extended to fit functions other than straight lines to observed data, and also to fit functions that have more than one independent variable. These topics are treated in statistics under the heading of *regression analysis*.

10.20 PROBLEM SET 10–8

For each of the following data sets: Find m and b for the best-fitting least-squares line and write the equation of the line. [Note: The reader will find it instructive in each case to plot the given data and the fitted line to see graphically how well the line fits the data.]

1.

x_i	y_i
3	4
1	3
11	11

2.

x_i	y_i
7	12
2	10
3	11

3.

x_i	y_i
2	7
2	6
1	8
3	5

4.

x_i	y_i
1	8
1	9
4	1
2	7

5. (For practice on a hand calculator.) The following data are values of x_i, hours worked, and y_i, number of units produced. Find and write the equation of the best-fitting least-squares line.

x_i:	10	20	14	35	40	25	15	31	50	58	50	62
y_i:	25	40	30	45	50	45	35	47	52	54	55	56

6. (For practice on a hand calculator.) The following are data for 12 new different model automobiles. (x_i = Automobile engine size in cubic inches, obtained from stickers on the automobiles; y_i = Miles per gallon for highway driving as provided by the Environmental Protection Agency.)

x_i:	85	140	151	200	231	232	250	301	302	305	400	425
y_i:	28	26	26	21	18	20	18	15	16	16	13	11

Find and write the equation of the best-fitting least-squares straight line.

10.21 LAGRANGE MULTIPLIERS

In Chapter 9, we solved numerous maximum-minimum problems by a process we shall now refer to as the *substitution* method. The problems as originally set up involved functions of two variables, but additional information was provided so that it was possible to express one variable in terms of the other and, by substitution, state the original problem as a function of one variable.

Example. By the substitution method, find the local extreme point on

$$f(x, y) = x^2 + y^2 - 8x - 6y + 30 \tag{1}$$

subject to

$$2x + y = 21. \tag{2}$$

The statement (2) restricts the values x and y can assume, and is called a *constraint*. We proceed to solve (2) for y, obtaining

$$y = 21 - 2x \tag{3}$$

and substitute (3) into (1), making it a function of x only. Thus,

$$f(x) = x^2 + (21 - 2x)^2 - 8x - 6(21 - 2x) + 30.$$

We now proceed to find the stationary point of $f(x)$, as follows:

$$f'(x) = 2x + 2(21 - 2x)(-2) - 8 + 12$$
$$f'(x) = 10x - 80$$

and

$$f'(x) = 0 \quad \text{where} \quad 10x - 80 = 0$$
$$x^* = 8.$$

From (3) we find

$$y = 21 - 2(8)$$
$$y^* = 5.$$

Inasmuch as

$$f''(x) = 10$$

is positive, we have a local minimum, which is, from (1),

$$f(x^*, y^*) = f(8, 5) = (8)^2 + (5)^2 - 8(8) - 6(5) + 30$$
$$f_{min} = 25.$$

It is worth noting that $f(x, y)$ in expression (1) itself has a minimum without regard to the constraint (2). By the methods presented earlier in the chapter, it can be verified that this *unconstrained* minimum is 5 and occurs at $x = 4$, $y = 3$. The higher minimum we obtained, 25, by conforming to the constraint (2) is called a *constrained* minimum and, in general, problems of finding maxima and minima subject to one or more constraints are referred to as problems in *constrained optimization*.

The objective of this section is to show another method that is often applied to problems in constrained optimization, the method of *Lagrangian multipliers*. The Lagrangian method starts by making one side of the constraint zero; thus, (2) is re-written as

$$2x + y - 21 = 0.$$

Now a name is given to the left side, say, $g(x, y)$. Thus,

$$g(x, y) = 2x + y - 21.$$

Next, we form the

Lagrangian expression: $h(x, y, \lambda) = f(x, y) - \lambda g(x, y)$.

The quantity λ (lambda) is called a *Lagrangian multiplier*. It can be shown, although we shall not do so here, that in order for $f(x, y)$ to have a maximum or minimum subject to the constraint, all first partial derivatives of $h(x, y, \lambda)$ must equal zero. That is,

$$\frac{\partial h}{\partial x} = 0; \quad \frac{\partial h}{\partial y} = 0; \quad \frac{\partial h}{\partial \lambda} = 0.$$

In our example,

$$h(x, y, \lambda) = x^2 + y^2 - 8x - 6y + 30 - \lambda(2x + y - 21)$$

$$\frac{\partial h}{\partial x} = h_x = 2x - 8 - 2\lambda$$

$$\frac{\partial h}{\partial y} = h_y = 2y - 6 - \lambda$$

$$\frac{\partial h}{\partial \lambda} = h_\lambda = -(2x + y - 21).$$

Setting the three partials equal to zero, we have

$$\begin{array}{llll} e_1: & 2x & - 8 - 2\lambda = 0 \\ e_2: & 2y - 6 - \lambda = 0 \\ e_3: & -2x - y + 21 & = 0. \end{array}$$

We now proceed to eliminate λ by multiplying e_2 by 2 and subtracting the result from e_1, as indicated next by $e_1 - 2(e_2)$:

$$e_4: \quad 2x - 4y + \ \ 4 = 0 \quad (e_1 - 2e_2)$$
$$e_3: -2x - \ \ y + 21 = 0.$$

Then, adding e_4 and e_3,

$$e_5: -5y + 25 = 0 \qquad (e_4 + e_3)$$
$$-5y = -25$$
$$y^* = 5.$$

Substituting $y^* = 5$ into e_3 yields

$$-2x - (5) + 21 = 0$$
$$-2x = -16$$
$$x^* = 8,$$

and we have $x^* = 8$, $y^* = 5$, the same values as those obtained by the substitution method at the beginning of the section. We can also determine the value for λ. Thus, inserting $x^* = 8$, $y^* = 5$ into e_2 yields

$$2(5) - 6 - \lambda = 0$$
$$-\lambda = -4$$
$$\lambda = 4.$$

The *sign* of λ, here positive, is of significance when the so-called Kuhn-Tucker conditions are applied to try to determine whether or not an extreme point exists.[1] However, discussion of these conditions goes beyond our objective of introducing the Lagrangian method, so we shall not need the value of λ for the problems in this section. In all the problems that will appear, the quantity

$$D = (h_{xx})(h_{yy}) - (h_{xy})^2$$

will be positive, and when this is true at (x^*, y^*) we have a local minimum (maximum) if h_{xx} and h_{yy} are both positive (negative). In the present example

$$h_x = 2x - 8 - 2\lambda; \quad h_y = 2y - 6 - \lambda;$$
$$h_{xx} = 2 \qquad\qquad ; \quad h_{yy} = 2 \qquad\qquad ; \quad h_{xy} = 0$$

so

$$D = (2)(2) - (0)^2 = 4$$

is positive, both second partials are positive, and we have a local minimum, which we determined earlier to be

---

[1] See, for example, Daniel Teichroew, *An Introduction to Management Science:* John Wiley and Sons, New York, 1964.

$$f_{min} = f(8, 5) = 25.$$

In summary:

Method of Lagrangian multipliers

1. Write the constraint as $g(x, y) = 0$.
2. Form the Lagrangian expression

$$h(x, y, \lambda) = f(x, y) - \lambda\, g(x, y).$$

3. Find h_x, h_y, h_λ and set them equal to zero. Eliminate λ and solve for x^* and y^*.
4. Check to be sure that D is positive, and determine whether the point at hand is a local maximum or minimum.
5. Compute the value of the local maximum or minimum.

Exercise. By the Lagrangian method, find the local extreme point for $f(x, y) = xy - x^2 - y^2 + 100$, subject to the constraint $x + y = 10$. *a)* State the Lagrangian expression. *b)* Write the values of x^* and y^*. *c)* Compute $f(x^*, y^*)$ and state whether it is a local maximum or minimum. Answer: *a)* $h(x, y, \lambda) = xy - x^2 - y^2 + 100 - \lambda(x + y - 10)$. *b)* $x^* = 5$, $y^* = 5$. *c)* $f(5, 5) = 75$ is a local maximum.

In the illustrations we have used, the Lagrangian method clearly is more cumbersome than the substitution method. In more complicated problems, it may be difficult, or impossible, to solve the constraint equation for one of the variables and in such problems the Lagrange procedure is selected. However, the major reason for using the Lagrange procedure turns out to be that the sign of λ can be very helpful in the difficult problem of determining whether an extreme point exists and, if so, whether it is a local maximum or minimum.

10.22 PROBLEM SET 10–9

Solve the following constrained optima problems by the method of Lagrangian multipliers. For each problem:

a) Write the Lagrangian expression.
b) Find and state the values of x^* and y^*.
c) Compute the value of $f(x^*, y^*)$ and state whether it is a local maximum or minimum.

1. $f(x, y) = 20x + 10y - x^2 - y^2$, subject to $x + 2y = 10$.
2. $f(x, y) = x^2 + 2y^2 - 3x - 10y + 20.25$, subject to $2x + 3y = 19$.
3. $f(x, y) = 3x^2 + 4y^2 - xy - 300$, subject to $x + y = 16$.

4. $f(x, y) = 2xy - x^2 - 2y^2 + 200$, subject to $2x - y = 30$.
5. $f(x, y) = 4xy - y^2 - 5x^2 + 16x + 10y$, subject to $2x + y = 60$.
6. $f(x, y) = 2x^2 + y^2 + 2x + 4y$, subject to $x + y = 8$.

10.23 REVIEW PROBLEMS

Find the first and second derivatives of each of the following functions.

1. $f(x) = 5e^{0.4x}$.
2. $f(x) = (1.09)^x$.
3. $f(x) = 4e^{0.5x-5}$.
4. $f(x) = e^{-0.5x^2}$.
5. $f(x) = 3e^{2x-x^2}$.
6. $f(x) = \left(\dfrac{1}{4}\right)^x$.

Find local extreme values, if any exist, and state which type of value has been found.

7. $f(x) = e^{2x} - 10x + 4$.
8. $f(x) = e^{-0.1x} + 0.9x$.
9. $f(x) = xe^{-0.2x} + 3$.
10. $f(x) = 11 - x^2 - e^{x^2}$.

Sketch graphs of the following. Show the starting point coordinates and the asymptote with its equation. Sketches need not be to scale, and the coordinate axes need not be shown.

11. $f(x) = 30 + 20e^{3+0.2x}$.
12. $f(x) = 30 - 20(0.5)^{x-10}$.
13. The revenue from, and the cost of, operating an undertaking for t years are, respectively, in millions of dollars,

$$R(t) = 3e^{0.05t} \quad \text{and} \quad C(t) = 1.5e^{0.08t}.$$

    a) How long should operations continue if profit is to be maximized?
    b) Compute maximum profit.

14. The total potential audience for a promotional campaign is 10,000 customers. Revenue averages $3 per response to the campaign. Campaign costs are a fixed amount of $500, plus $300 per day the campaign continues. The proportion of the total audience responding by time t days is

$$1 - e^{-0.25t}.$$

    a) How long should the campaign continue if profit is to be maximized?
    b) Compute maximum profit.

15. The total potential audience for a promotional campaign is 2000 customers. Revenue averages $5 per response to the campaign. Costs are $105.36 per day, plus a fixed cost of $100. The proportion of potential audience responding by time t days is

$$1 - (0.9)^t.$$

    a) How long should the campaign continue if profit is to be maximized?
    b) Compute maximum profit.

Find the derivatives of the following functions.

16. $f(x) = 2 \ln 3x$.
17. $f(x) = \ln(5x - 4)$.

18. $f(x) = \ln\left(\dfrac{1}{2x^2}\right).$

19. $f(x) = \ln(x^3 + x^2 + x - 5).$

20. $f(x) = \ln(2x + 3)^{1/2}.$

21. $f(x) = \ln(xe^x).$

Find the value of $f(x)$ at its local extreme point, if any exists, and state whether this value is a local maximum or minimum.

22. $f(x) = 50 \ln x - x^2,\ x > 0.$

23. $f(x) = x^2 - 4x - 16 \ln x + 30,\ x > 0.$

24. $f(x) = 100 \left[\dfrac{\ln(0.5x)}{x}\right],\ x > 0.$

25. $f(x) = -0.5x^2 + 5x + 50 \ln x,\ x > 0.$

26. $f(x) = 3 \ln(2x^2 - 16x + 40).$

By the chain rule, carry out each of the following.

27. $\dfrac{d}{dh}\, p(q).$

28. $\dfrac{d}{dx}\, (3y^5).$

29. $\dfrac{d}{dx}\, \ln(y^2 - 2y).$

30. $\dfrac{d}{dx}\, (e^{y^2}).$

31. $\dfrac{d}{dx}\, (2y^{3/2} - x^2).$

32. $\dfrac{d}{dy}\, (x^3 - 3y).$

Find dy/dx by implicit differentiation.

33. $y^4 - x^2 = 0.$

34. $2y^3 - 3y^2 - x^3 + 2x^2 + 10 = 0.$

35. $e^y \ln x = x^2.$

36. $xy^3 = x^3 + 10.$

37. $y + e^y = x.$

38. $y = h(y) + x^2.$

39. At income level Y, consumption expenditures are $C(Y)$ billion dollars, where

$$C(Y) = 25 + 0.875\, Y.$$

   a) Find the marginal propensity to consume.
   b) Interpret (a).
   c) How much income is generated by an additional dollar of investment?
   d) At income level $500 billion, what proportion of total income is spent?

40. For the consumption function

$$C(Y) = 10 + 0.75\, Y - e^{-0.25\, Y}$$

find the multiplier at income levels
   a) 1.   b) 4.

41. Find the expression for the multiplier if the consumption function is

$$C(Y) = a + bY.$$

42. Find the expression for the multiplier if the consumption function is

$$C(Y) = a + bY + e^{kY}.$$

43. If $P(x)$ is total profit when x gallons of Exall are made and sold and

$$P(x) = 4x - 200,$$

compute $P'(500)$ and interpret the result in rate terminology.

44. A car rental agency rents a car for one day at a cost figured from

$$C(x) = 15 + 0.10x - 0.00005x^2 \text{ dollars,}$$

where x is the number of miles the car is driven. Compute $C'(500)$ and interpret the result in rate terminology.

45. Five thousand dollars deposited in a bank account paying 7 percent interest compounded annually will grow to the amount

$$A(t) = 5000(1.07)^t$$

in t years. Compute $A'(20)$ and interpret the result in rate terminology.

46. (See Problem 45.) If interest is 7 percent compounded continuously,

$$A(t) = 5000e^{0.07t}.$$

Compute $A'(20)$ and interpret the result in rate terminology.

47. The cost of making (printing and binding) x books is

$$C(x) = 2500 + 2.5x \text{ dollars.}$$

a) Write the expression for $a(x)$, the average cost per book if x books are made.
b) Compute $a'(100)$ and interpret the result in rate terminology.

48. The cost of producing x tons of a product is

$$C(x) = 0.0002x^3 - 0.06x^2 + 8x + 500 \text{ dollars.}$$

a) Write the expression for $a(x)$, the average cost per ton if x tons are produced.
b) Compute $a'(50)$ and interpret the result in rate terminology.

49. If $1500 is deposited in a bank account paying 9 percent interest compounded annually, the amount in the account at time t years is

$$A(t) = 1500(1.09)^t.$$

Find the percent rate of change of $A(t)$ at time $t = 10$ years.

50. (See Problem 49.) If the account pays 9 percent compounded continuously, then

$$A(t) = 1500e^{0.09t}.$$

Find the percent rate of change of the amount in the account at $t = 10$ years.

51. The number of potential customers in a store's trading area t years from the time the store opens is

$$N(t) = \frac{200{,}000}{1 + 50e^{-0.1t}}.$$

Find the percent rate of change in number of potential customers at $t = 5$ years.

52. Find the percent rate of change in $f(t)$ at time $t = 7$ years, if

$$f(t) = 5t + 2te^{-0.4t}.$$

53. A cost function is minimized when its independent variable, y, has the optimal value

$$y = \frac{10}{k} + 2a,$$

where a and k are parameters.
a) What is the optimal value of y if $k = 5$ and $a = 10$?
b) How sensitive is y to changes in the parameter k? To answer this, write the expression for the rate at which y changes per unit change in k.
c) Using the expression in (b) and the numbers from (a), estimate by how much the optimal value of y would change if k changes from 5 to 5.5.

54. a) Find $f'(0)$ and interpret it in rate terminology if

$$f(x) = e^x.$$

b) What is the value of $f(0) = e^0$?
c) Using (a), estimate by how much e^x will change if x increases from 0 to 0.001.
d) From (b) and (c), estimate the value of $e^{0.001}$.

55. The demand function for a product is

$$p(q) = 250 - 0.1q.$$

a) Compute elasticity of demand at 1000 units demand.
b) Interpret the answer to part (a).

56. The demand function for a product is

$$p(q) = 120 - 20 \ln(q + 1).$$

a) Compute elasticity of demand at 100 units demand.
b) Interpret the answer to part (a).

57. The demand function for a product is

$$p(q) = 50e^{-0.002q}.$$

a) Compute elasticity of demand at demand 400 units.
b) Interpret the answer to part (a).

58. The demand function for a product is

$$p(q) = 50e^{-0.002q}.$$

a) Write the expression for the revenue function, find the q^* where revenue has a local extreme point, and show that this is a local maximum.
b) Compute elasticity of demand at the demand q^* found in part (a).

For each of the following find:

a) f_x. b) f_{xx}. c) f_y. d) f_{yy}. e) f_{xy}.

59. $f(x, y) = 2x^3 - 3xy + 4y^2 - 6.$
60. $f(x, y) = (x^2 + y^2)^{1/2}.$

61. $f(x, y) = 2y^5 - 6x^2y^3 + 3x - 2y.$

62. $f(x, y) = y \ln x.$

63. $f(x, y) = xe^y.$

64. $f(x, y) = \dfrac{x^2}{2y + 1}.$

Carry out the following:

65. $\dfrac{\partial}{\partial y}(x^2 - 3y^2).$

66. $\dfrac{\partial^2}{\partial x^2}(x^3y^2).$

67. $\dfrac{\partial^2}{\partial x \partial y}(x^3y^2).$

68. $\dfrac{\partial^2}{\partial x^2}(xe^y).$

69. If $f(x, y) = 5x - 4y$, compute
   a) $f_x(1, 2).$ b) $f_{xx}(1, 2).$ c) $f_y(1, 2).$ d) $f_{yy}(1, 2).$
   e) $f_{xy}(1, 2).$

70. During a period of operations, a producer makes and sells x gallons of Exall and y gallons of Whyall at a cost, in cents, of

$$C(x, y) = 2x^2 - 4xy + 3y^2 - 6x + 10y.$$

Find $C_x(10, 15)$ and $C_y(10, 15)$ and interpret these numbers in rate terminology.

For the following, find the value of the function at its local extreme point, if one exists, and state whether a value found is a local maximum or minimum.

71. $f(x, y) = xy - x^2 - y^2 + 15x.$

72. $f(x, y) = x + y + \dfrac{25}{x} + \dfrac{16}{y}$; x and y positive.

73. $f(x, y) = 3x^2 - 3xy + y^2 - 6x + 32.$

74. $f(x, y) = \ln x + \ln y - 0.2x - 0.5y + 5.$

75. $f(x, y) = \dfrac{1,600,000}{x} + 2xy^2 + \dfrac{x(1 - y)^2}{2}$, $x > 0.$

76. (This problem is difficult, algebraically.) Given

$$f(x, y) = \dfrac{ab}{x} + \dfrac{cxy^2}{2} + \dfrac{dx(1 - y)^2}{2}$$

where x, y, and all parameters (a, b, c, and d) are positive.
   a) Find the expressions for x^* and y^* that minimize $f(x, y)$, in terms of the parameters a, b, c, and d.
   b) Prove (a) yields a minimum.

For the following data sets: a) Find the equation of the best-fitting least squares straight line. b) Plot the data and the line on a graph.

77. x_i: 2 4 8 10 16
    y_i: 2 5 4 10 9

78. x_i: 5 6 3 2 6 4 6 2 7 4
    y_i: 4 3 2 5 5 3 6 3 5 5

Solve the following problems in constrained optimization by the method of Lagrangian multipliers.

a) Write the Lagrangian expression.

b) Find and state the values of x^* and y^*.

c) Compute $f(x^*, y^*)$ and state whether it is a local maximum or minimum.

79. $f(x, y) = 2x^2 + 3y^2 - 20x - 60y + 300$, subject to $2x + 3y = 60$.

80. $f(x, y) = 6xy - 4x^2 - 3y^2 + 26x + 3y$, subject to $2x + y = 30$.

Integral calculus

11.1 INTRODUCTION

We have learned how to find the derivatives of functions and that numerous applications of the derivative stem from its interpretation as the slope of the curve representing the function. It is quite amazing to learn that if we start with a function, $f(x)$, carry out the process that is the inverse of taking the derivative, the result provides an area that has $f(x)$ as part of its boundary. The inverse process is first called taking the *anti-derivative of $f(x)$* then, later, we shall refer to it as *integrating $f(x)$*. Many of the applications of the inverse process stem from interpreting the result as an area that represents quantities such as dollars of profit or pounds of output.

Students generally find integral calculus easier than differential calculus because, having had extensive practice in finding derivatives, the idea of doing the process in reverse, so to speak, is not hard to grasp. However, experience indicates that, at the beginning, students tend to get the two procedures mixed up. The need, therefore, is to concentrate on practicing integration until it is firmly fixed in mind. To that end, we shall postpone introduction of applications for a while so that the reader can develop confidence in using the rules and symbols of integration.

In the first part of the chapter, only power functions are considered, and emphasis is placed on practicing integrating such functions; then attention turns to areas having power functions as part of their boundaries. Then, again using only power functions, applications involving area interpretations are presented. At this point, the reader will have learned most of the new ideas covered in the chapter. The middle part of the chapter simply expands the list of integration rules to include exponential and logarithmic forms. The final section of the chapter introduces integration out of the context of area as the method of solving equations containing derivatives, known as differential equations, and illustrates the application of this important branch of mathematics.

11.2 ANTIDERIVATIVES: THE INDEFINITE INTEGRAL

Addition and subtraction are examples of *inverse* operations where one operation annuls the effect of the other. Thus, if we start with the number 50, add 10 to it, then subtract 10, we have the original number 50. Similarly, multiplication and division are inverse operations and, as a last example, cubing a number is annulled by the inverse operation of taking the cube root. Thus, starting with 2,

$$2^3 = 8 \quad \text{and} \quad \sqrt[3]{8} = 2.$$

Now recall the rule for the operation of taking the derivative of x to a power, which is

$$\frac{d}{dx}(x^{\text{power}}) = (\text{power})x^{(\text{power}-1)}.$$

Thus, the operation is carried out *subtracting* one from the power and *multiplying* by the power. The operations we might *expect* to annul the operation of taking this derivative are to take the result

$$(\text{power})x^{(\text{power}-1)}$$

add one to the power and *divide* by the power, but this is not correct, for two reasons. The first reason is that the divisor is wrong, as may be seen by noting that

$$\frac{d}{dx}(x^3) = 3x^2$$

and if we take the result, $3x^2$, add one to its power giving 3, and divide by *its* power, 2, we would obtain

$$\frac{3x^{2+1}}{2} = \frac{3x^3}{2},$$

which is not the x^3 we started with. Clearly, what must be done is to add one to the power in $3x^2$ and *divide by this new power*, which is 3. Then we have

$$\frac{3x^{2+1}}{2+1} = \frac{3x^3}{3} = x^3,$$

as desired. However, the correct procedure just shown is still incomplete, because if we have

$$\frac{d}{dx}(x^3 + 10) = 3x^2,$$

and apply the procedure to $3x^2$, we cannot recover the constant 10. What we can say is that

$$3x^2$$

is the derivative of x^3 plus an arbitrary constant, C, and write

$$\text{antiderivative } (3x^2) = x^3 + C.$$

The word *antiderivative* means the operation which is the inverse of taking the derivative. Thus,

$$\text{antiderivative } (3x^2)$$

means all expressions whose derivative is $3x^2$. Again,

$$\text{antiderivative } (x^3) = \frac{x^{3+1}}{3+1} + C = \frac{x^4}{4} + C$$

and means that $(x^4)/4$ plus an arbitrary constant constitutes all expressions which have x^3 as their derivative.

> ***Exercise.*** *a)* Find antiderivative *(x)*. *b)* What does the answer to *(a)* mean? Answer: *a)* $(x^2/2) + C$. *b)* $(x^2/2) + C$ constitutes all expressions that have x as their derivative.

For reasons that will become clear when we introduce interpretations of the antiderivative operation, we shall represent the operation by an elongated S; thus,

$$\int$$

is called the *integral symbol* and means the antiderivative of the expression following, as in

$$\int x^2 \, dx = \frac{x^3}{3} + C.$$

The dx is a *single* symbol and is called the *differential of x*. That is, just as Δx means the change in x and not Δ times x, so dx does not mean d times x. For the moment, we shall think of dx in the same sense as we did when we wrote

$$\frac{d}{dx}$$

to mean the derivative with respect to x. That is, the dx means the independent variable is x and we are to integrate with respect to x. Continuing with terminology, the function to be integrated is called the *integrand*, the outcome of the integration is called the *integral*, and the arbitrary constant is called the *constant* of *integration*. Thus, in

$$\int x^2 \, dx = \frac{x^3}{3} + C,$$

the integrand is x^2, the constant of integration is C, and

$$\frac{x^3}{3} + C$$

is the *indefinite* integral, where the italicized word is inserted because of the presence of the arbitrary constant, C.

The general rule for the indefinite integral of x to a constant power is:

Simple power rule: $\int x^n \, dx = \dfrac{x^{n+1}}{n+1} + C;$ if $n \neq -1$.

Observe that the rule is inapplicable if the power is $n = -1$ for then the divisor, $(n+1)$, in the integral would be zero. We shall see later that if $n = -1$, the integral is the natural logarithm of x.

Example. Integrate the following functions:

$$a)\ 1/y^2. \qquad\qquad b)\ x^{1/2}.$$

We write

$$a)\ \int \frac{1}{y^2} \, dy = \int y^{-2} \, dy = \frac{y^{-2+1}}{-2+1} + C = \frac{y^{-1}}{-1} + C = -\frac{1}{y} + C.$$

Then,

$$b)\ \int x^{1/2} \, dx = \frac{x^{1/2+1}}{\frac{1}{2}+1} + C = \frac{x^{3/2}}{\frac{3}{2}} + C = \frac{2}{3} x^{3/2} + C.$$

Exercise.   Integrate the following: *a) z⁵. b)* $1/(w^{1/2})$. Answer: *a)* $(z^6/6) + C$. *b)* $2w^{1/2} + C$.

Properties of the integration operation.   These properties parallel those of the derivative operation. First, we call attention to the fact that the derivative of x is 1, so that

$$\int 1 \, dx = x + C,$$

which is correct because

$$\frac{d}{dx} (x + C) = 1,$$

which is the integrand of

$$\int 1 \, dx.$$

Conventionally, the factor 1 is not written, as is the case for example when we write x rather than one times x. That is,

$$\int dx = \int 1 \, dx = x + C.$$

Similarly, we know that

$$\frac{d}{dx}(3x) = 3,$$

so that

$$\int 3 \, dx = 3x + C$$

and, in general, for any constant k,

$$\int k \, dx = kx + C.$$

Exercise.   Find the following: *a)* $\int (-2 \, dx)$. *b)* $\int dy/2$. *c)* $\int dz$. Answer: *a)* $-2x + C$. *b)* $(1/2)y + C$. *c)* $z + C$.

As was the case in taking derivatives, a constant factor may be placed inside or outside the operation symbol. For example,

$$\int 3x \, dx = 3 \int x \, dx = 3 \cdot \frac{x^2}{2} + C.$$

We could have written the last integral using K to be a constant, as

$$3 \left(\frac{x^2}{2} + K \right) = \frac{3x^2}{2} + 3K,$$

but, inasmuch as K is an arbitrary constant, so is $3K$ an arbitrary constant, and only one symbol is required for such a constant. We shall follow convention and write the constant of integration as a single letter, frequently C.

Again as in the case of taking derivatives, integration may be performed on an expression term by term. Thus,

$$\int (3z^3 + 2z + 5 + m)\, dz = \int 3z^3\, dz + \int 2z\, dz + \int 5\, dz + \int m\, dz$$

$$= 3\int z^3\, dz + 2\int z\, dz + 5\int dz + m\int dz$$

$$= 3\left(\frac{z^4}{4}\right) + 2\left(\frac{z^2}{2}\right) + 5z + mz + C$$

$$= \frac{3z^4}{4} + z^2 + 5z + mz + C.$$

In the last example, note that m is to be considered as a constant because the symbol dz specifies the variable is z.

We summarize the last two properties as follows:

The integral may be taken term by term, and constant factors may be placed inside of or outside of the integral sign.

Exercise. *a)* Write the symbols for integrating $(6y - 10y^4 + b - 1)$ with respect to y. *b)* Write the integral. Answer: *a)* $\int (6y - 10y^4 + b - 1)\, dy$. *b)* $3y^2 - 2y^5 + by - y + C$.

We shall have occasion from time to time to question whether an integral is correct. The answer, of course, is

An indefinite integral is correct if the derivative of the integral is the integrand.

Example. Is the following correct?

$$\underbrace{\int (7x + 5)^2\, dx}_{\text{integrand}} \overset{?}{=} \underbrace{\frac{(7x + 5)^3}{3} + K}_{\text{integral}}. \tag{1}$$

To check, we find the derivative of the integral,

$$\frac{d}{dx}\left[\frac{(7x + 5)^3}{3} + K\right] = \frac{d}{dx}\left[\frac{(7x + 5)^3}{3}\right] + \frac{d}{dx}\, (K)$$

$$= \frac{3(7x + 5)^2(7)^*}{3}$$

$$= (7x + 5)^2(7),$$

where * calls attention to application of the power function derivative rule. The integral is *incorrect* because its derivative contains the factor 7, which does not appear in the integrand. However, this tells us that we can obtain the correct result by dividing the integral in (1) by 7. That is, the correct integral is

$$\int (7x + 5)^2 \, dx = \frac{(7x + 5)^3}{3(7)} + C$$

where, again, we have chosen to use the symbol C rather than $K/7$, which would arise in dividing (1) by 7.

Exercise. Is the following correct? Why or why not?

$$\int (2x + 9)^{-1/2} \, dx \overset{?}{=} (2x + 9)^{1/2} + C.$$

Answer: It is correct because the derivative of $[(2x + 9)^{1/2} + C]$, by application of the power function derivative rule, is the integrand, $[(2x + 9)]^{-1/2}$.

Integration rule for powers of a linear function. Returning to the correct integral in the foregoing example,

$$\int (7x + 5)^2 \, dx = \frac{(7x + 5)^3}{3(7)} + C,$$

the integrand is a *linear* function, $(7x + 5)$, to a constant power. The integral can be obtained by proceeding first as we do for a simple power function, that is, adding one to the power and dividing by the new power, to give

$$\frac{(7x + 5)^3}{3}$$

provided that we also divide by the coefficient of x, which is 7, and write

$$\frac{(7x + 5)^3}{3(7)} + C.$$

Observe that the constant 5 in the linear expression $(7x + 5)$ has no effect on the result. If the linear function is expressed in general form as

$$mx + b,$$

we may write the following general rule:

Power rule for linear functions:

$$\int (mx+b)^n \, dx = \frac{(mx+b)^{n+1}}{m(n+1)} + C, \ n \neq -1.$$

Example. Integrate the following: $\int \frac{dx}{(5x-6)^3}$.

To obtain the proper form, we write

$$\int \frac{dx}{(5x-6)^3} = \int (5x-6)^{-3} \, dx$$

$$= \frac{(5x-6)^{-3+1}}{5(-3+1)} + C$$

$$= \frac{(5x-6)^{-2}}{-10} + C$$

$$= -\frac{1}{10(5x-6)^2} + C.$$

Exercise. For $\int (5-2x)^{-3/2} dx$: a) What are m and n? b) Write the integral, simplified where possible, with a positive exponent. Answer: a) $m = -2$, $n = -3/2$. b) $[1/(5-2x)^{1/2}] + C$.

Before turning to a practice problem set, it is worth noting that the product, quotient, and chain rules that applied in taking derivatives do not have general counterparts in integration. As a consequence, it is necessary to pay careful attention to the *form* of the expression to be integrated and be sure that it matches the form of the integration rule being applied. At the moment, we have only the two forms specified in

$$\int x^n \, dx = \frac{x^{n+1}}{n+1} + C; \ n \neq -1$$

$$\int (mx+b)^n \, dx = \frac{(mx+b)^{n+1}}{m(n+1)} + C; \ n \neq -1.$$

11.3 PROBLEM SET 11–1

Write the expression for the indefinite integral. Simplify results where possible. Express answers with positive exponents.

1. $\int dx$.
2. $\int dz$.
3. $\int 5 \, dy$.
4. $\int -7 \, dw$.
5. $\int (1+x) \, dx$.
6. $\int (3-y) \, dy$.

7. $\int (2x^2 - 3x + 4)\, dx.$ 8. $\int (3y^3 + 4y - 1)\, dy.$

9. $\int p\, dq.$ 10. $\int q\, dp.$

11. $\int (pq)\, dp.$ 12. $\int (pq)\, dq.$

13. $\int (x^3 + x^4 - 1)\, dx.$ 14. $\int (y^2 + y^5 - 1)\, dy.$

15. $\int (3y^2 + 5y^4 + 1)\, dy.$ 16. $\int (4x^3 - 6x^5 + 1)\, dx.$

17. $\int x^{-2}\, dx.$ 18. $\int 7^{-3}\, dy.$

19. $\int \dfrac{dy}{y^3}.$ 20. $\int \dfrac{dx}{x^4}.$

21. $\int (5 - 2y^{-3})\, dy.$ 22. $\int (7 - 3x^{-4})\, dx.$

23. $\int \left(2x - \dfrac{1}{x^2} + 1\right) dx.$ 24. $\int \left(y + \dfrac{2}{y^3} + 1\right) dy.$

25. $\int p^{1/2}\, dp.$ 26. $\int q^{1/3}\, dq.$

27. $\int 3x^{-4/3}\, dx.$ 28. $\int 5y^{-3/2}\, dy.$

29. $\int \left(2 - \dfrac{1}{x^2} - \dfrac{2}{3x^{5/3}}\right) dx.$ 30. $\int \left(2x - \dfrac{2}{x^3} - \dfrac{1}{4x^{4/3}}\right) dx.$

31. $\int 12(x^2 + x^3)\, dx.$ 32. $\int 3(x^2 + 2x)\, dx.$

33. $\int 16(2x - 9)^3\, dx.$ 34. $\int 30(3x + 5)^5\, dx.$

35. $\int (3x - 9)^{-2}\, dx.$ 36. $\int (2x + 3)^{-4}\, dx.$

37. $\int \dfrac{dx}{(5 - 3x)^{1/2}}.$ 38. $\int \dfrac{dw}{(7 - 2w)^3}.$

39. $\int \dfrac{8dx}{(2x + 5)^{1/3}}.$ 40. $\int \dfrac{12dx}{(3x - 7)^{1/2}}.$

11.4 THE DEFINITE INTEGRAL

If we write the indefinite integral

$$\int 2x\, dx = x^2 + C$$

and subtract the *value* of the integral at $x = 1$ from the value of $x = 3$, we obtain

$$[(3)^2 + C] - [(1)^2 + C]$$
$$= 9 + C - 1 - C$$
$$= 8,$$

a *definite* number. To symbolize the operation just carried out, we write

$$\int_1^3 2x\, dx = x^2 \Big|_1^3 = (3)^2 - (1)^2 = 8$$

where the vertical bar is an evaluation symbol and means to evaluate the expression to its left at the *upper limit*, 3, and from this subtract the value of the expression at the *lower limit*, 1. In general,

$$\int_a^b f(x)\, dx$$

is called the *definite* integral, with a as the lower limit and b as the upper limit. We read the symbols as 'the integral from a to b of $f(x)\,dx$.' The integral is a definite quantity that depends on the limits, and contains no arbitrary constant of integration.

Example. Find the value of $\int_9^{36} x^{1/2}\,dx$.

First, we write the definite integral as

$$\int_9^{36} x^{1/2}\,dx = \frac{2}{3} x^{3/2} \Big|_9^{36},$$

then insert the limits and subtract to obtain

$$\frac{2}{3}(x)^{3/2} \Big|_9^{36} = \frac{2}{3}[(36)^{3/2} - (9)^{3/2}]$$

$$= \frac{2}{3}[(36^{1/2})^3 - (9^{1/2})^3]$$

$$= \frac{2}{3}[(6)^3 - (3)^3]$$

$$= \frac{2}{3}(216 - 27)$$

$$= \frac{2}{3}(189)$$

$$= 126.$$

Exercise. Evaluate: *a)* $\int_1^3 x\,dx$. *b)* $\int_0^8 x^{-1/3}\,dx$. Answer: *a)* 4. *b)* 6.

We point out again that the value of a definite integral depends upon the values of its limits or, we may say, is a function of its limits. For example,

$$\int_a^b 4x^3\,dx = x^4 \Big|_a^b = b^4 - a^4,$$

and the resulting expression is a function of the limits.

Exercise. Find $\int_0^k 2x\,dx$. Answer: k^2.

Following the practice problem set, we will return to the definite integral and explain what it represents.

11.5 PROBLEM SET 11–2

Evaluate the following definite integrals.

1. $\displaystyle\int_{2}^{5} 2\,dx.$

2. $\displaystyle\int_{1}^{3} 3\,dx.$

3. $\displaystyle\int_{-1}^{4} 2x\,dx.$

4. $\displaystyle\int_{-3}^{9} 3x\,dx.$

5. $\displaystyle\int_{2}^{6} (x+1)\,dx.$

6. $\displaystyle\int_{-1}^{3} (2x-1)\,dx.$

7. $\displaystyle\int_{0}^{8} x^{2/3}\,dx.$

8. $\displaystyle\int_{0}^{9} x^{1/2}\,dx.$

9. $\displaystyle\int_{1}^{2} (x^2 - 3x + 5)\,dx.$

10. $\displaystyle\int_{0}^{3} (x^2 + 5x - 2)\,dx.$

11. $\displaystyle\int_{1}^{9} (5 + y^{-1/2})\,dy.$

12. $\displaystyle\int_{1}^{8} (1 + 2y^{-1/3})\,dy.$

13. $\displaystyle\int_{2}^{6} (2x - 3)^{1/2}\,dx.$

14. $\displaystyle\int_{1}^{8} (5x - 4)^{-1/2}\,dx.$

15. $\displaystyle\int_{1}^{6} \frac{60\,dx}{(3x+2)^2}.$

16. $\displaystyle\int_{0}^{6} \frac{20}{(4x+1)^{3/2}}\,dx.$

Write the following definite integrals in terms of constants and the limits.

17. $\displaystyle\int_{a}^{b} 3x^2\,dx.$

18. $\displaystyle\int_{c}^{d} 4x^3\,dx.$

19. $\displaystyle\int_{1}^{n} (x+1)\,dx.$

20. $\displaystyle\int_{n}^{1} (2x-1)\,dx.$

11.6 THE DEFINITE INTEGRAL AS AN AREA

In differential calculus, we learned that, in geometrical terms, the derivative of a function evaluated at a point is the slope of the line tangent to the curve at the point. In this section we shall show that the definite integral, in geometrical terms, represents an *area* having the integrand function as one of its boundaries. To verify this statement in a simple case, suppose the integrand is the constant function

$$f(x) = 10$$

shown in Figure 11–1. The function $f(x) = 10$ forms the upper boundary of a rectangular area whose other boundaries are the x-axis and verticals at $x = 2$ and $x = 5$. We know that this area is

Area of rectangle: (base)(height) = 3(10) = 30.

FIGURE 11-1

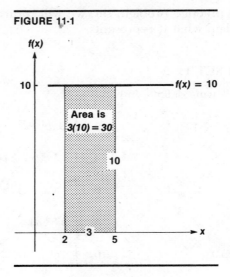

This area we describe as the area *under f(x) over the interval 2 to 5.*
Now observe that the definite integral of $f(x) = 10$, from 2 to 5, is

$$\int_2^5 f(x)\,dx = \int_2^5 10\,dx$$

$$= 10x \Big|_2^5$$

$$= 10(5-2)$$

$$= 30,$$

and we see in this simple case that the definite integral gives the
area under $f(x)$ over the interval.

Next consider

$$f(x) = x$$

over the interval 0 to b, shown in Figure 11–2. From geometry,

$$\text{Area of a right triangle} = \frac{1}{2}(\text{base})(\text{height})$$

$$= \frac{1}{2}(b)(b)$$

$$= \frac{b^2}{2},$$

as shown in the figure. Turning now to the definite integral of
$f(x) = x$ from 0 to b, we have

FIGURE 11-2

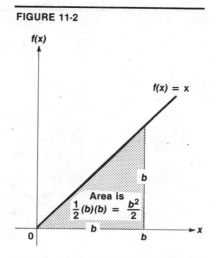

$$\int_{0}^{b} f(x)\, dx = \int_{0}^{b} x\, dx = \frac{x^2}{2}\bigg|_{0}^{b}$$

$$= \frac{b^2}{2} - 0$$

$$= \frac{b^2}{2},$$

and, again, the definite integral gives the area under $f(x)$.

Thus far, formulas of plane geometry have been applied to verify that the definite integral gives the area. The challenge at hand is to demonstrate that the definite integral provides areas when the function part of the boundary is a curve of some sort rather than a straight line. The simplest function to work with is

$$f(x) = x^2,$$

shown in Figure 11–3. Ordinary plane geometry does not provide a formula for the shaded area shown in this figure. However, if we sketch in a horizontal (dashed) line to form a rectangle with base b and height $f(b) = b^2$, clearly, b^3, the rectangular area, is greater than the shaded area, and the error which would exist if we used b^3 to *approximate* the shaded area is labeled 'error' on Figure 11–3. Equally clear is the fact that the shaded area is only a *part* of the rectangular area, b^3. Our objective is to show that this part is 1/3. That is, we want to show that the shaded area is

$$\int_{0}^{b} x^2\, dx = \frac{x^3}{3}\bigg|_{0}^{b} = \frac{b^3}{3} - 0 = \frac{1}{3}b^3,$$

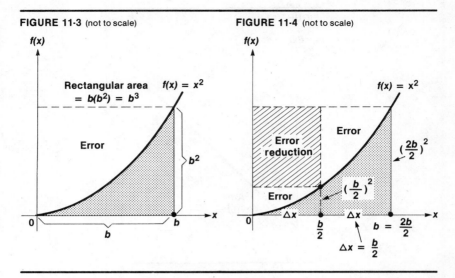

FIGURE 11-3 (not to scale)

FIGURE 11-4 (not to scale)

and this is not easy to do. However, we shall pursue the matter not just for the sake of proving the point, but also to develop a fundamental definition that states that the definite integral is the *limit* of a sum, and it is the sum concept which accounts for adopting an elongated S as the integral symbol.

To proceed, note in Figure 11–4 that we have divided the interval from 0 to b into two equal subdivisions each of length

$$\Delta x = \frac{b}{2}.$$

We now form the two rectangles shown, and take the sum of the two rectangular areas to approximate the shaded area under the curve. This new approximation will still be in error, but the error will be less than that in Figure 11–3 by the amount shown as 'error reduction' in Figure 11–4. For the first rectangle,

$$\text{base} = \frac{b}{2}, \ \text{height} = f\left(\frac{b}{2}\right) = \left(\frac{b}{2}\right)^2; \ \text{area} = \left(\frac{b}{2}\right)\left(\frac{b}{2}\right)^2.$$

For the second rectangle,

$$\text{base} = \frac{b}{2}, \ \text{height} = f(b) = f\left(\frac{2b}{2}\right) = \left(\frac{2b}{2}\right)^2; \ \text{area} = \left(\frac{b}{2}\right)\left(\frac{2b}{2}\right)^2.$$

The approximation using two rectangles is the sum of the areas,

$$\frac{b}{2}\left(\frac{b}{2}\right)^2 + \left(\frac{b}{2}\right)\left(\frac{2b}{2}\right)^2 = \frac{b^3}{2^3}(1 + 2^2).$$

The reason for writing b as $(2b/2)$ was to get the form at the right end of the last line, because this form can be generalized. Thus, to obtain a closer approximation by dividing b into three equal parts of length $\Delta x = b/3$, it may be verified that the sum of the *three* rectangular areas will be

$$\frac{b^3}{3^3}(1^2 + 2^2 + 3^2),$$

and for any number of subdivisions, n, of length $\Delta x = \dfrac{b}{n}$,

$$\text{Sum of the } n \text{ rectangular areas} = \frac{b^3}{n^3}(1^2 + 2^2 + 3^2 + \cdots + n^2). \quad (1)$$

There is a formula, which we shall not derive, stating that

$$1^2 + 2^2 + 3^2 + \cdots + n^2 = \frac{n^3}{3} + \frac{n^2}{2} + \frac{n}{6}. \quad^1 \quad (2)$$

Substituting (2) into (1) we have the approximate area under $f(x) = x^2$ as

$$\text{Sum of } n \text{ rectangular area} = \frac{b^3}{n^3}\left(\frac{n^3}{3} + \frac{n^2}{2} + \frac{n}{6}\right).$$

Expanding the last, we have:

$$\text{Approximate area under } f(x) = x^2 \text{ is } \frac{b^3}{3} + \frac{b^3}{2n} + \frac{b^3}{6n^2}. \quad (3)$$

The more rectangles we use (the larger n is), the closer the approximation. Of course, as n increases, the base length of each rectangle, Δx, decreases because

$$\Delta x = \frac{b}{n}. \quad (4)$$

That is, as n becomes ever larger, Δx approaches zero. We now express (3) in terms of Δx. From (4)

$$n = \frac{b}{\Delta x}$$

and, substituting the last into (3), we find

$$\text{Approximate area under } f(x) = x^2 \text{ is } \frac{b^3}{3} + \frac{b^3}{2\left(\dfrac{b}{\Delta x}\right)} + \frac{b^3}{6\left[\dfrac{b^2}{(\Delta x)^2}\right]}$$

$$= \frac{b^3}{3} + \frac{b^2(\Delta x)}{2} + \frac{b(\Delta x)^2}{6}. \quad (5)$$

---

[1] For example, $1^2 + 2^2 + 3^2 + 4^2 + 5^2 = 1 + 4 + 9 + 16 + 25 = 55$. By the formula with $n = 5$, we have $(5^3/3) + (5^2/2) + (5/6) = 330/6 = 55$.

Now it should be clear that as n becomes very large, Δx becomes very close to zero, and the value of the last expression becomes very close to $b^3/3$. Thus, as Δx approaches zero, the expression (5) approaches a limiting value, which is $b^3/3$, so that

$$\lim_{\Delta x \to 0} \left[\frac{b^3}{3} + \frac{b^2(\Delta x)}{2} + \frac{b(\Delta x)^2}{6} \right] = \frac{b^3}{3}.$$

Thus, we have shown that the area under $f(x) = x^2$ over the interval 0 to b is

$$\int_0^b x^2 \, dx = \frac{x^3}{3} \bigg|_0^b = \frac{b^3}{3},$$

but to do this we had to apply the limit concept.

To see the origin of integral symbolism, let us write

$$\underset{0}{\overset{b}{S}} \, (n \text{ rectangular areas})$$

to mean the sum of n rectangular areas over the interval 0 to b. Each of these rectangles has a base of Δx and a height that is a value of $f(x)$, so we may write the sum in the form

$$\underset{0}{\overset{b}{S}} f(x) \, \Delta x.$$

We now symbolize the limit of this sum by replacing S by $\int$, and replacing Δx by dx. Thus,

$$\lim_{\Delta x \to 0} \underset{0}{\overset{b}{S}} f(x) \Delta x = \int_0^b f(x) dx.$$

Our demonstration shows why, and in what sense (the limit sense), the area under a curve is defined as the definite integral of a function. Of course, the demonstration applies only to a special simple function, $f(x) = x^2$. The demonstration that it applies *generally* is so important that proof of this fact is called the *fundamental theorem of calculus*. Readers who wish to see the nature of this proof, which is a landmark in the progress of mathematics, should read the addendum at the end of the chapter. The points we wish to stress here are, first, that the basic definitions of both differential and integral calculus involve taking a limit as $\Delta x \to 0$ and, second, that the fundamental theorem shows how these two limits are related. That is,

$$\int_0^b f(x) \, dx,$$

which is the limit of the sum of the rectangular areas we have just discussed, can be determined by finding the function that has $f(x)$ as a derivative, the antiderivative of $f(x)$, and evaluating the antiderivative from 0 to b. Readers who do not see the significance of this second point should review the detail of the previous effort to establish the limit of the sum of the areas in the case of the simple function x^2 and then contemplate the unthinkable complexity of problems that would arise if the fundamental theorem was not true and we had to attempt to find a separate limit each time a new function was encountered.

In the last discussion, zero was used as the lower limit. If some

---

FIGURE 11-5 (not to scale)

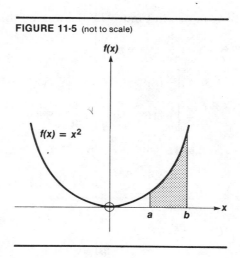

$f(x)$

$f(x) = x^2$

a b x

---

other value, a, was used, the area under the curve over the interval a to b, shown in Figure 11–5, can be found by taking the area over 0 to b and subtracting from this the area over 0 to a, thus:

$$\int_0^b x^2 \, dx - \int_0^a x^2 \, dx = \left(\frac{x^3}{3}\right)\Big|_0^b - \left(\frac{x^3}{3}\right)\Big|_0^a$$

$$= \frac{b^3}{3} - \frac{a^3}{3}.$$

This, of course, is the same as

$$\int_a^b x^2 \, dx = \frac{x^3}{3}\Big|_a^b = \frac{b^3}{3} - \frac{a^3}{3}.$$

We have, in general:

> *Definition: The area under a curve means the area bounded by a section of a function, $f(x)$, the x-axis, and verticals at $x = a$ and $x = b$, where b is greater than a. It is computed by evaluating*

$$\int_a^b f(x)\, dx.\,^2$$

Exercise. Find the area under $f(x) = x^2$ over the interval $x = 1$ to $x = 4$. Answer: 21.

Example. Find the area under $f(x) = x^{1/3} + 5$ over the interval $x = 1$ to $x = 8$.

The area sought is shown in Figure 11–6 and is computed as follows:

$$\int_1^8 (x^{1/3} + 5)\, dx = \left(\frac{3}{4} x^{4/3} + 5x\right)\Big|_1^8$$

$$= \left[\frac{3}{4}(8)^{4/3} + 5(8)\right] - \left[\frac{3}{4}(1)^{4/3} + 5(1)\right]$$

$$= \left[\frac{3}{4}(16) + 40\right] - \left[\frac{3}{4} + 5\right]$$

$$= 12 + 40 - \frac{3}{4} - 5$$

$$= 46.25.$$

FIGURE 11-6 (not to scale)

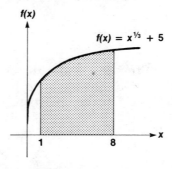

$f(x) = x^{1/3} + 5$

Exercise. Compute the area under $f(x) = 3x^{1/2} - 2$ over the interval $x = 4$ to $x = 16$. Answer: 88.

---

[2] If the section of the curve lies below the axis so $f(x)$ is negative, the definite integral will be negative and its sign must be changed to represent area, which, by definition, is positive. If part of $f(x)$ is below and part above the axis, the definite integral is computed for the segment where $f(x)$ is negative. This value, with its sign changed, is added to the value of the definite integral computed over the segment where $f(x)$ is positive. However, we shall not be concerned with this procedure because all applications in this text deal with functions that are positive over the interval of interest.

Example. Sketch the function $f(x) = 10x - x^2$, then find the area bounded by the function and the x-axis.

From our work in Chapter 9, we know

$$f(x) = 10x - x^2$$

is a vertical parabola opening downward. The vertex is the local maximum that occurs where $f'(x) = 0$. Hence,

$$f'(x) = 10 - 2x, \quad \text{and} \quad f'(x) \text{ is zero where } x = 5, \, f(5) = 25,$$

so $(5, 25)$ is the vertex, as shown on Figure 11–7. The curve opens

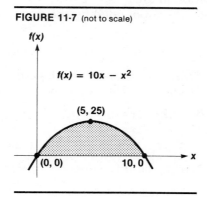

FIGURE 11-7 (not to scale)

downward from the vertex and so must intersect the x-axis at two points. These points occur where $f(x) = 0$, and we find

$$f(x) = 0 \quad \text{where} \quad 10x - x^2 = 0$$
$$x(10 - x) = 0$$
$$x = 0$$
$$x = 10;$$

so the intercepts are $(0, 0)$ and $(10, 0)$ as shown on Figure 11–7. The area sought is computed as follows:

$$\int_0^{10} (10x - x^2) \, dx = \left(5x^2 - \frac{x^3}{3} \right) \Big|_0^{10}$$
$$= \left(500 - \frac{1000}{3} \right) - (0 - 0)$$
$$= \frac{500}{3}.$$

When the endpoints of the interval over which an area is to be computed are given, the computation involves only integrating and evaluating the integral. But, as the last example shows, the problem specification may require that the interval endpoints be found and, in this case, a sketch of the function may be helpful.

Exercise. Sketch the function $f(x) = 16 - x^2$ and compute the area bounded by the curve and the x-axis. Answer: The curve is a parabola with vertex at $(0, 16)$, opening downward, with intercepts $(-4, 0)$ and $(4, 0)$. The area is $256/3$.

Areas with two functions as boundaries. This heading refers to cases where two functions in addition, possibly, to the axes and vertical lines, are boundaries of an area.

Example. Find the area bounded by the functions

$$f(x) = 15 - 2x - x^2$$
$$g(x) = 9 - x.$$

Here, $f(x)$ is again a parabola opening downward. Its vertex is found in the usual manner to be $(-1, 16)$, as follows:

$$f'(x) = 0: \quad -2 - 2x = 0$$
$$x = -1$$
$$f(-1) = 15 - 2(-1) - (-1)^2 = 16.$$

As shown next, the horizontal intercepts are at $x = -5$ and $x = 3$.

x-intercepts: $f(x) = 0$:

$$15 - 2x - x^2 = 0$$
$$(-x + 3)(x + 5) = 0$$
$$x = 3; x = -5.$$

A sketch of $f(x)$ is shown on Figure 11–8. The function

$$g(x) = 9 - x$$

is a straight line that can be plotted from two points. Frequently, the intercepts are the simplest points to find:

Horizontal (x) intercept: $\quad g(x) = 0$
$$0 = 9 - x$$
$$x = 9$$
Point is $(9, 0)$.
Vertical intercept $(x = 0)$: $\quad g(x) = 9 - 0$
$$g(x) = 9$$
Point is $(0, 9)$.

After drawing $g(x)$ on Figure 11–8, we see the specified (shaded) area extends over the interval $x = a$ to $x = b$, where dashed vertical lines have been drawn. We can find this area by taking the area under $f(x)$ over the interval from a to b, which would include also the area under the line $g(x)$, and subtracting from this the area under the line. Thus,

$$A = \int_a^b f(x)\, dx - \int_a^b g(x)\, dx,$$

which is the same as

$$A = \int_a^b [f(x) - g(x)]\, dx.$$

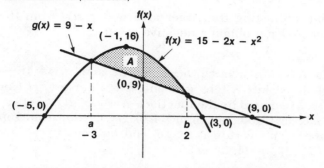

FIGURE 11-8 (not to scale)

The values of a and b are the x-coordinates of the points of intersection of $f(x)$ and $g(x)$. At these points, $g(x) = f(x)$.

Intersections:
$$g(x) = f(x)$$
$$9 - x = 15 - 2x - x^2$$
$$x^2 + x - 6 = 0.$$
$$(x + 3)(x - 2) = 0$$
$$x = -3;\ x = 2.$$

Hence, $a = -3$ and $b = 2$, so the area is

$$A = \int_{-3}^{2} [f(x) - g(x)]\, dx$$

$$= \int_{-3}^{2} [15 - 2x - x^2 - (9 - x)]\, dx$$

$$= \int_{-3}^{2} [6 - x - x^2]\, dx$$

$$= \left(6x - \frac{x^2}{2} - \frac{x^3}{3} \right) \Big|_{-3}^{2}$$

$$= \left[6(2) - \frac{(2)^2}{2} - \frac{(2)^3}{3} \right] - \left[6(-3) - \frac{(-3)^2}{2} - \frac{(-3)^3}{3} \right]$$

$$= \left[12 - 2 - \frac{8}{3} \right] - \left[-18 - \frac{9}{2} + 9 \right]$$

$$= 19 - \frac{8}{3} + \frac{9}{2} = \frac{114 - 16 + 27}{6} = \frac{125}{6}.$$

In passing, we call attention to the fact that area is defined to be a positive number, so if an area calculation gives a negative result, the work should be checked. In the last example, a negative result could mean the arithmetic is faulty, or it could mean that the functions in the integrand are in the wrong order; that is, if we set up the area on Figure 11–8 as

$$\int_{-3}^{2} [g(x) - f(x)] \, dx$$

we will be subtracting the larger area under $f(x)$ from the smaller area under $g(x)$ and obtain a negative result.

Exercise. *a)* Sketch $h(x) = 8 - 2x$ and $g(x) = 16 - x^2$. *b)* Find the points where $g(x)$ and $h(x)$ intersect. *c)* Compute the area bounded by the functions. Answer: *a)* $h(x)$ is a straight line passing through $(0, 8)$ and $(4, 0)$; $g(x)$ is a downward opening parabola with vertex at $(0, 16)$ and intercepts $(-4, 0)$; $(4, 0)$. *b)* $(-2, 12)$ and $(4, 0)$. *c)* 36.

11.7 PROBLEM SET 11–3

Find the area under the curve of the following functions over the given x-intervals.

1. $f(x) = 2x$; $x = 1$ to $x = 2$.
2. $f(x) = 3x$; $x = 1$ to $x = 5$.
3. $f(x) = 3x + 2$; $x = 1$ to $x = 2$.
4. $f(x) = 2x + 3$; $x = -1$ to $x = 1$.
5. $f(x) = \dfrac{6}{x^2}$; $x = 1$ to $x = 3$.
6. $f(x) = \dfrac{4}{x^3}$; $x = 1$ to $x = 2$.
7. $f(x) = \dfrac{40}{(2x + 1)^2}$; $x = 0$ to $x = 2$.
8. $f(x) = \dfrac{50}{x^{3/2}}$; $x = 1$ to $x = 25$.

Find the areas described in each of the following problems. (Sketches will be helpful.)

9. Find the area bounded by the axes and

$$f(x) = 10 - 0.5x.$$

10. Find the area bounded by the axes and

$$f(x) = 5 + x.$$

11. Find the area bounded by the x-axis and

$$f(x) = 30x - 3x^2.$$

12. Find the area bounded by the x-axis and

$$f(x) = 4x - x^2 + 21.$$

13. For the functions
$$f(x) = 1 + x \quad \text{and} \quad g(x) = 10 - 2x,$$
   a) Find the first-quadrant area bounded by the functions and the y-axis.
   b) Find the first-quadrant area bounded by the functions and the axes.

14. Find the first-quadrant area bounded by the axes and the functions
$$f(x) = 0.5x + 2; \ g(x) = 2x - 4.$$

15. Find the area bounded by the functions
$$f(x) = x^2 + 1 \quad \text{and} \quad g(x) = 10.$$

16. Find the area bounded by the functions
$$f(x) = 34 - x^2 \quad \text{and} \quad g(x) = 9.$$

17. Find the area bounded by the functions
$$f(x) = x^2 - 8x + 20 \quad \text{and} \quad g(x) = 14 - x.$$

18. Find the area bounded by the functions
$$f(x) = 20 - 2x \quad \text{and} \quad g(x) = 12x - 2x^2.$$

19. [Note: $f(x)$ is the upper half of a *horizontal* parabola that opens to the right and has the origin as its vertex.] Find the area bounded by
$$f(x) = 8x^{1/2} \quad \text{and} \quad g(x) = x^2.$$

20. (See Note, Problem 19.) Find the area bounded by
$$f(x) = 6x^{1/2} \quad \text{and} \quad g(x) = 0.4x.$$

11.8 INTERPRETIVE APPLICATIONS OF AREA

The area of the rectangle shown in Figure 11–9 is
$$30(40) = 1200.$$
If asked what this area represents, our reply depends upon what the 30 and 40 represent. Thus, if the rectangle is a piece of level land with dimensions 30 feet by 40 feet, then

$$(30 \text{ feet})(40 \text{ feet}) = 1200 \ (\text{feet})^2,$$

or 1200 square feet. Here, the *unit of measure* is feet for both numbers, so the product of the numbers is square feet. However, if 30 means 30 gallons of gasoline and 40 means 40 miles per gallon of gasoline, then the first unit is *gallon* and the second is *miles per (divided by) gallon,* and we have

$$30(\text{gallon})40 \left(\frac{\text{mile}}{\text{gallon}} \right) = 1200(\text{mile}),$$

so the gallon unit cancels and the area of the rectangle represents the 1200 miles traveled by a car that consumes 30 gallons of gasoline and travels 40 miles per gallon consumed. Once again, if a person rents a motel room for 30 days (unit is *day*) and pays $40 per day (unit is dollar/day), we find

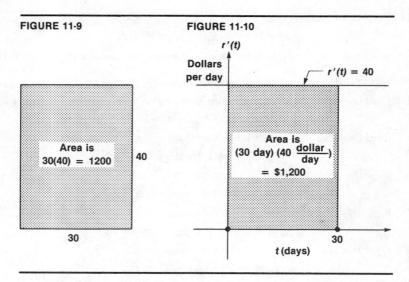

FIGURE 11-9   FIGURE 11-10

$$30(\text{day})40\left(\frac{\text{dollar}}{\text{day}}\right) = 1200(\text{dollar}),$$

and the area represents the \$1200 the person pays for the 30-day rental at \$40 per day.

In the foregoing, observe that miles per gallon and dollars per day are *rates*. The *derivative* of a function also can be interpreted as a rate. Thus, the *total* rent function in the last example is

$$r(t) = t(\text{day})40\left(\frac{\text{dollar}}{\text{day}}\right)$$

$$r(t) = 40t$$

and

$$r'(t) = 40$$

is the rate at which rent changes per additional day's rental. We picture this as in Figure 11–10, with the constant rate function plotted on the vertical. Then we can compute the area as

$$\int_0^{30} r'(t)\, dt.$$

Observe that the integrand, $r'(t)$, has the unit (dollar/day), whereas dt, which refers to the horizontal axis has the unit (day). Hence,

$$r'(t)\, dt$$

or

$$\left(\frac{\text{dollar}}{\text{day}}\right)(\text{day}) = (\text{dollar})$$

and

$$\int_0^{30} r'(t)\, dt = \int_0^{30} 40\, dt = 40t \Big|_0^{30} = 1200 \text{(dollar)}.$$

The point we wish to emphasize is that for

$$\int_a^b f(x)\, dx$$

an area interpretation is in terms of the product of the unit on the vertical, f(x), axis times the unit of dx, which is the unit on the horizontal axis. This point is important because it arises in numerous applications of integrals.

Example 1. The total amount of coal a country will consume in a period of years depends upon the rate of consumption, and this rate increases as time, *t* years, increases. Suppose it is estimated that the consumption rate, *r' (t)*, *t* years from now, will be

$$r'(t) = (20 + 1.2t) \text{million tons per year.}$$

Compute the total amount of coal the country will consume in the next ten years.

Figure 11–11 shows the rate function, *r'(t)*, as a rising straight line. The shaded area, marked A, is

$$\int_0^{10} (20 + 1.2t)\, dt = (20t + 0.6t^2) \Big|_0^{10}$$
$$= (200 + 60) - (0)$$
$$= 260.$$

The vertical has the unit (ton/year) and the horizontal has the unit (year). Hence,

$$260 \text{ is } \left(\frac{\text{ton}}{\text{year}}\right)(\text{year}) = (\text{ton})$$

and means that the area 260 represents 260 million tons consumed in 10 years.

Example 2. (See Example 1.) How much coal will be consumed in the following ten years, that is, during the second decade from now?

The answer here is the area marked as B on Figure 11–11. It is

$$\int_{10}^{20} (20 + 1.2t)\, dt = (20t + 0.6t^2) \Big|_{10}^{20}$$
$$= (400 + 240) - (200 + 60)$$
$$= 380 \text{ million tons.}$$

Example 3. (See Example 1.) If the total supply of coal available to the country now and in the future is 2500 million tons, how long will it be until the total supply is exhausted?

610

FIGURE 11-11 (not to scale)

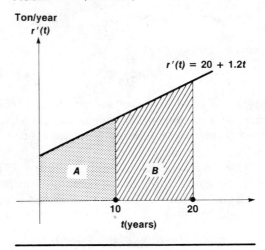

Here, the total supply will be exhausted when the amount consumed, which is the integral, equals 2500. We write

$$\int_0^t (20 + 1.2t)\, dt = 2500$$

$$(20t + 0.6t^2)\Big|_0^t = 2500$$

$$(20t + 0.6t^2) - (0 + 0) = 2500$$

$$0.6t^2 + 20t - 2500 = 0.$$

We must now solve the quadratic equation just written which has coefficients $a = 0.6$, $b = 20$, $c = -2500$, and according to the quadratic formula,

$$t = \frac{-b \pm \sqrt{b^2 - 4ac}}{2a} = \frac{-20 \pm \sqrt{(20)^2 - 4(0.6)(-2500)}}{2(0.6)}$$

$$= \frac{-20 \pm \sqrt{400 + 6000}}{1.2}$$

$$= \frac{-20 \pm \sqrt{6400}}{1.2}$$

$$= \frac{-20 \pm 80}{1.2}.$$

Using the negative sign of $\pm$ in the last gives a negative value for t, which is not in the permissible range. Hence,

$$t = \frac{-20 + 80}{1.2} = \frac{60}{1.2} = 50$$

so the supply will be exhausted in 50 years.

Exercise. Maintenance cost on a new machine is at the rate of $100t$ dollars per year at time t years. *a)* Compute total maintenance cost for the first four years. *b)* How many years will it take for total maintenance cost to amount to $5000? Answer: *a)* $800. *b)* 10 years.

Accounting for a fixed component. In the last exercise, it was assumed, reasonably, that maintenance cost at time $t = 0$ was zero and therefore it was not necessary to take this zero cost into account. Suppose, however, that a company spends $2000 to have an advertising campaign prepared and then plans to run the campaign at a cost rate of $C'(t) = 900$ dollars per week at time t weeks. If we compute the total cost of running the campaign for 10 weeks as

$$\int_0^{10} C'(t)\, dt = \int_0^{10} 900\, dt = 900t \Big|_0^{10} = \$9000,$$

the $9000 does not include the fixed preparation cost of $2000. The total cost would be

$$\text{Total cost} = \text{Fixed cost} + \int_0^{10} 900\, dt$$
$$= 2000 + 9000$$
$$= \$11{,}000.$$

Thus it follows that if a total is obtained by integrating a rate, and a fixed element is present, this element must be added to the value of the definite integral.

Example. The fixed cost incurred when g gallons of Exall are produced is $2500 and marginal cost at g gallons of output is

$$C'(g) = 0.00045g^2 - 0.18g + 20.$$

Find the total cost of producing 100 gallons.

Here we must recall that marginal cost at output level g gallons is the *rate* at which total cost is changing per additional gallon made. That is, marginal cost has the unit (dollar/gallon) and dg has the unit (gallon). Hence,

$$\int C'(g)\, dg$$

has the unit

$$\left(\frac{\text{dollar}}{\text{gallon}} \right) (\text{gallon}) = (\text{dollar}).$$

The total cost of making 100 gallons, fixed cost included, is

$$\$2500 + \int_0^{100} (0.00045g^2 - 0.18g + 20) \, dg$$

$$= 2500 + (0.00015g^3 - 0.09g^2 + 20g) \Big|_0^{100}$$

$$= 2500 + [0.00015(100)^3 - (0.09)(100)^2 + 20(100)] - [0]$$
$$= \$2500 + \$1250$$
$$= \$3750.$$

Exercise. Marginal cost at output level p pounds of Poundall is

$$1 + (p + 1)^{-1/2}$$

and fixed cost is \$400. Find the total cost of making 399 pounds of the product. Answer: \$837.

11.9 INTERPRETING THE AREA BOUNDED BY TWO FUNCTIONS

Suppose that an operation provided a company income at the rate of $I'(t)$ dollars per day at time t days, where

$$I'(t) = 110 + 4t^{1/2}.$$

The operation was started at an initial fixed expense of \$2000. Variable expense at the rate of $E'(t)$ per day is incurred at time t days, where

$$E'(t) = 20 + 7t^{1/2}.$$

The income and expense rate functions are shown on Figure 11–12. From a profit maximization viewpoint, the operation should be continued as long as income per day exceeds expense per day; that is, as long as $I'(t)$ is above $E'(t)$ on Figure 11–12. This situation exists up until the intersection point, b, which we find by equating $E'(t)$ and $I'(t)$, thus:

$$20 + 7t^{1/2} = 110 + 4t^{1/2}$$
$$3t^{1/2} = 90$$
$$t^{1/2} = 30$$
$$t = (30)^2$$
$$t = 900 \text{ days,}$$

which is the value shown for b on Figure 11–12. If we write

$$\text{Net profit} = \int_0^{900} I'(t) \, dt - \int_0^{900} E'(t) \, dt - 2000,$$

FIGURE 11-12 (not to scale)

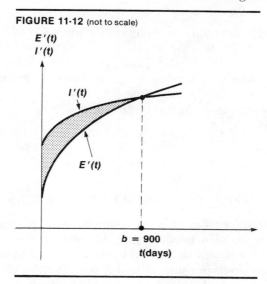

then, on the right, the first term is total income in 900 days, the middle term is total *variable* expense, and the difference in these two is the shaded area on Figure 11–12 and represents profit before the fixed expense of $2000. The three terms therefore represent the net profit obtained, and this is the maximum profit, possible only if the operation is terminated at $t = 900$ days. We have

$$\text{Net profit} = \int_0^{900} I'(t)\, dt - \int_0^{900} E'(t)\, dt - 2000$$

$$= \int_0^{900} [I'(t)\, dt - E'(t)]\, dt - 2000$$

$$= \int_0^{900} [(110 + 4t^{1/2}) - (20 + 7t^{1/2})]\, dt - 2000$$

$$= \int_0^{900} (90 - 3t^{1/2})\, dt - 2000$$

$$= (90t - 2t^{3/2}) \Big|_0^{900} - 2000$$

$$= [90(900) - 2(900)^{3/2}] - [0] - 2000$$

$$= 81{,}000 - 2(30)^3 - 2000$$

$$= \$25{,}000.$$

614

Exercise. Income from an operation at time t months from its initiation is at the rate of $(10 + 0.2t)$ thousand dollars per month and variable expense is at the rate of $(5 + 0.3t)$ thousand dollars per month. If fixed start-up cost was $5 thousand, find *a)* The optimum time to terminate the operation. *b)* The maximum net profit that can be achieved. Answer: *a)* 50 months. *b)* $120 thousand.

11.10 CONSUMERS' AND PRODUCERS' SURPLUS

In the economic model of pure competition it is assumed that all consumers (buyers) of a product pay the *same* price per unit for a product. This price comes about by the interplay of competitive market forces and is the price per unit at which the quantity of product consumers are willing and able to buy (called consumer *demand*) is matched by the quantity producers (sellers) are willing and able to supply. The purpose of this section is to illustrate that this competitive situation benefits both consumer and supplier, and to develop a measure of these benefits.

Looking first at the consumer (buyer) side of the situation, it is a matter of observation that even though a product, say gasoline, sells competitively at 99 cents per gallon, there are buyers who would be willing to pay $3 or more per gallon and, clearly, the lower price established by competition benefits these buyers. Economists use downward sloping curves, called *demand* curves, to represent the observed relationship between the number of units, q, demanded by consumers and $p_d(q)$, the selling price per unit. The subscript d on $p_d(q)$ signifies that this is a demand function to distinguish it from the supply function, which will be introduced later and named $p_s(q)$. Figure 11–13 shows an illustrative demand curve. As the figure shows, low price is accompanied by high demand, and high price is accompanied by low demand. The point $E(q_m, p_m)$ is the *equilibrium* point; that is, the market price that matches the quantity consumers demand with the quantity producers will supply. The area under the curve,

$$\int_0^{q_m} p_d(q)\, dq,$$

represents dollars because it is the product of price per unit times number of units. This area may be taken as a measure of what consumers might have had to pay for the q_m units if competition did not lead to the constant price, p_m, for all units; that is, if various consumers paid the higher prices on the demand curve to the left of (q_m, p_m).

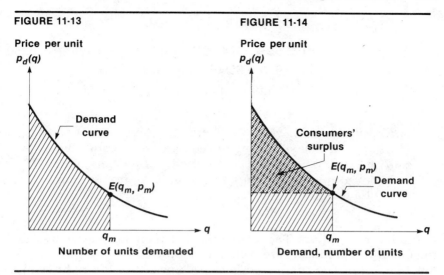

FIGURE 11-13

Price per unit
$p_d(q)$

Demand curve

$E(q_m, p_m)$

q_m
Number of units demanded

FIGURE 11-14

Price per unit
$p_d(q)$

Consumers' surplus

$E(q_m, p_m)$

Demand curve

q_m
Demand, number of units

However, at equilibrium, consumers pay p_m dollars per unit for each of the q_m units demanded, so the total paid is

$$p_m q_m \text{ dollars,}$$

which is the area of the rectangle shown in Figure 11–14. The amount paid, the rectangular area, is less than the whole area under the curve over the interval 0 to q_m, and the *difference* may be taken as a measure of the benefit to consumers of competitive forces that lead to the same unit price for all units sold. Hence, it is called *consumers' surplus* and, as shown by the shaded area on Figure 11–14, it is the total area under the curve over the inverval 0 to q_m, minus the area of the rectangle.

$$\textbf{Consumers' surplus} = \left[\int_0^{q_m} p_d(q)\, dq \right] - (p_m q_m).$$

Example. At market equilibrium, consumers demand 625 thousand gallons of Gallall, which has the demand function

$$p_d(q) = 25 - 0.6q^{1/2},$$

where q is in thousands of gallons and $p_d(q)$ is in dollars per gallon. Compute consumers' surplus.

First we compute p_m as

$$
\begin{aligned}
p_m = p_d(625) &= 25 - 0.6(625)^{1/2} \\
&= 25 - 0.6(25) \\
&= \$10 \text{ per gallon.}
\end{aligned}
$$

Then,

$$\text{Consumer surplus} = \left[\int_0^{625} (25 - 0.6q^{1/2})\, dq \right] - (10)(625)$$

$$= [25q - 0.4q^{3/2}] \Big|_0^{625} - 6250$$

$$= [25(625) - 0.4(625)^{3/2}] - [25(0) - 0.4(0)^{3/2}] - 6250$$

$$= [15{,}625 - 6250] - [0] - 6250$$

$$= \$3125 \text{ thousand.}$$

Exercise. At market equilibrium, consumers demand 100 thousand tons of Tonall, which has the demand function $p_d(q) = 110 - 0.5q$, where q is in thousands of tons and $p_d(q)$ is in dollars per ton. Compute consumers' surplus. Answer: $2500 thousand.

Producers' surplus. The relationship between the market price and the quantities producers are willing to supply is expressed as a supply function, $p_s(q)$. The supply curve slopes upward to the right as illustrated in Figure 11–15 because producers are willing to supply more at higher prices than at lower prices. Again, the equilibrium point $E(q_m, p_m)$ is at the price that matches the amount consumers demand with the amount producers are willing to supply. However, producers receive the same unit price, p_m, for all q_m units they produce and sell, so the total received by producers is

$$(p_m)(q_m)$$

and this is the area of the rectangle shown on Figure 11–15. However, the supply curve implies that producers would have supplied some product at prices lower than p_m, and the area under the supply curve may be taken as a measure of what producers might have received

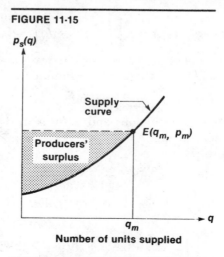

FIGURE 11-15

Number of units supplied

in the absence of an equilibrium price, that is, if various producers sold some product at prices to the left (below) the equilibrium price on the curve. The *difference* between the area of the rectangle and the area under the curve then is a measure of the benefit to producers of market competition that leads to the equilibrium price, p_m, and is called *producers' surplus*. This is the shaded area on Figure 11–15, which is

$$\textbf{Producers' Surplus} = (p_m q_m) - \left[\int_0^{q_m} p_s(q)\, dq \right].$$

Example. At market equilibrium, consumers demand 625 thousand gallons of Gallall, whose supply function is

$$p_s(q) = 2.5 + 0.3q^{1/2},$$

where q is in thousands of gallons and $p_s(q)$ is in dollars per gallon. Compute producers' surplus.

First we compute the equilibrium price,

$$p_s(625) = 2.5 + 0.3(625)^{1/2}$$
$$= 2.5 + 0.3(25)$$
$$= \$10 \text{ per gallon.}$$

Then, producers' surplus is

$$(10)(625) - \left[\int_0^{625} (2.5 + 0.3q^{1/2})\, dq \right]$$

$$= 6250 - [2.5q + 0.2q^{3/2}] \Big|_0^{625}$$

$$= 6250 - [(2.5)(625) + 0.2(625)^{3/2} - (0 + 0)]$$

$$= 6250 - [1562.5 + 0.2(25)^3]$$

$$= 6250 - [1562.5 + 3125]$$

$$= \$1562.5 \text{ thousand.}$$

Exercise. At market equilibrium, consumers demand 100 thousand tons of Tonall, whose supply function is

$$p_s(q) = 10 + 0.5q,$$

where q is in thousands of tons and $p_s(q)$ is in dollars per ton. Compute producers' surplus. Answer: \$2500 thousand.

Now that we have discussed both the demand function and the supply function for a product, we can see that the equilibrium point (q_m, p_m) occurs at a price where the quantity consumers demand equals the amount producers are willing to supply. That is, $E(q_m, p_m)$ is the intersection point of the supply and demand functions for the product.

Thus, in our previous two examples, the product Gallall had supply and demand functions

$$p_s(q) = 2.5 + 0.3q^{1/2}$$
$$p_d(q) = 25 - 0.6q^{1/2}.$$

The equilibrium point occurs where $p_s(q) = p_d(q)$; that is, where

$$2.5 + 0.3q^{1/2} = 25 - 0.6q^{1/2}$$
$$0.9q^{1/2} = 22.5$$
$$q^{1/2} = \frac{22.5}{0.9}$$
$$q^{1/2} = 25$$
$$(q^{1/2})^2 = (25)^2$$
$$q_m = 625.$$

Then, with $q_m = 625$

$$p_s(q) = 2.5 + 0.3(625)^{1/2}$$
$$= 2.5 + 0.3(25)$$
$$p_m = 10.$$

Thus, in the previous two examples the equilibrium point at demand $q_m = 625$ thousand gallons and price $p_m = \$10$ per gallon were the same because they applied to the supply and demand functions for the same product, Gallall. It follows that if we have the supply and demand functions for a product, both producers' and consumers' surplus can be computed by first determining the equilibrium intersection point, then applying the formulas for producers' and consumers' surplus.

Exercise. The supply and demand functions for Tonall are

$$p_s(q) = 10 + 0.5q$$
$$p_d(q) = 110 - 0.5q,$$

where q is in thousands of tons and price is in dollars per ton. By review of the previous two exercises, or by starting anew, find: *a)* The equilibrium point, (q_m, p_m). *b)* Producers' surplus. *c)* Consumers' surplus. Answer: *a)* (100, 60). *b)* \$2500 thousand. *c)* \$2500 thousand.

11.11 PROBLEM SET 11–4

1. Maintenance cost on newly purchased equipment is expected to be at the rate of $(2 + 0.1t)$ thousand dollars per year at time t years.

    a)   Compute total maintenance cost during the first six years.
    b)   Make a sketch showing what has been computed in *(a)*.
    c)   Compute maintenance cost during the second six years.
    d)   At what time, *t*, will the total spent on maintenance reach $60 thousand?

2.   Sales of a currently new product are expected to be at the rate of $(20 - 0.4t)$ thousand dollars per year at time *t* years, $t < 50$.
    a)   Compute total sales during the first 10 years.
    b)   Make a sketch showing what has been computed in *(a)*.
    c)   Compute total sales during the second 10 years.
    d)   At what time, *t*, will total sales reach $375 thousand?

3.   An industry consumes fuel at a rate of $(2 + 0.6t^{1/2})$ million barrels per year at time *t* years. How much fuel will the industry consume in 25 years?

4.   An oil rig pumps oil from a well at the rate of $(360 - 72t^{1/2})$ barrels per year at time *t* years. How much oil will be pumped in the next 9 years?

5.   At time *t* years, an industry consumes fuel at the rate of $(2t + 9)^{1/2}$ million barrels per year. If the total supply of fuel available to the industry now and in the future is 63 million barrels, how many years will the supply last?

6.   At time *t* years, sales of a currently new product are expected to be at the rate of

$$\frac{10}{(0.5t + 16)^{1/2}}$$

million dollars per year. How many years will it take for total sales to amount to $40 million?

7.   The population of a trading area is currently 100 thousand. At time *t* years from now population will be growing at the rate of

$$\frac{20}{(0.5t + 9)^{1/2}}$$

thousand per year. What will total population be 14 years from now?

8.   A fixed cost of $2 thousand has been incurred in setting up an advertising campaign. Variable cost is expected to be at the rate of $(3 + 0.06t^{1/2})$ thousand dollars per month at time *t* months. Estimate total cost if the campaign runs 25 months.

9.   When *t* tons of Tonall are produced, marginal cost in dollars per ton is

$$0.006t^2 - 1.2t + 50.$$

If fixed cost is $1,600, find the total cost of producing 100 tons.

10.   When *b* barrels of Barall are produced, marginal cost in dollars per barrel is

$$0.0045b^2 - 1.8b + 200.$$

If fixed cost is $4000, find the total cost of producing 200 barrels.

11.   A fixed cost of $50 thousand was incurred in setting up an operation. At time *t* months thereafter, the operation yields income at the rate of

$(20 - 0.3t)$ and incurs expense at the rate of $(10 - 0.1t)$, where both rates are in thousands of dollars per month.

a) What is the optimal time to terminate the operation?

b) What will total profit be at the optimal time of termination?

12. A fixed cost of $4100 was incurred in setting up an operation. At time t months thereafter, the operation yields income at the rate of $(2000 - 100t^{1/2})$ and incurs expense at the rate of $(200 + 200t^{1/2})$ where both rates are in dollars per month.

a) What is the optimal time to terminate the operation?

b) What will total profit be at the optimal time of termination?

13. The demand function for a product is

$$p_d(q) = 75 - 0.6q,$$

where q is in millions of barrels and $p_d(q)$ is in dollars per barrel. Market equilibrium occurs at a demand of 100 million barrels.

a) Compute consumers' surplus.

b) Make a sketch showing what was computed in (a).

14. The supply function for a product is

$$p_s(q) = 5 + 0.2q,$$

where q is in millions of tons and $p_s(q)$ is in dollars per ton. Market equilibrium occurs at a demand of 50 million tons.

a) Compute producers' surplus.

b) Make a sketch showing what was computed in (a).

15. The supply function for a product is

$$p_s(q) = (4 + 0.2q)^{3/2}$$

where q is in thousands of truckloads and $p_s(q)$ is in dollars per truckload. Market equilibrium occurs at a demand of 60 thousand truckloads. Compute producers' surplus.

16. The demand function for a product is

$$p_d(q) = \frac{80}{(0.1q + 0.2)^2}$$

where q is in millions of tons and $p_d(q)$ is in dollars per ton. Market equilibrium occurs at a demand for 18 million tons. Compute consumers' surplus.

17. The supply and demand functions for a product are

$$p_s(q) = 10 + 0.1q \quad \text{and} \quad p_d(q) = 100 - 0.2q,$$

where q is in thousands of tons and price is in dollars per ton.

a) Compute consumers' surplus.

b) Compute producers' surplus.

18. The supply and demand functions for a product are

$$p_s(q) = 1 + 0.02q \quad \text{and} \quad p_d(q) = 6 - 0.08q,$$

where q is in millions of pounds and price is in dollars per pound.

a) Compute consumers' surplus.

b) Compute producers' surplus.

11.12 THE INTEGRAL OF $(mx + b)^{-1}$

We first consider the special case of

$$\int (mx + b)^{-1}\, dx = \int \frac{dx}{mx + b}$$

where $m = 1$ and $b = 0$; that is,

$$\int \frac{dx}{x} = \int \left(\frac{1}{x}\right) dx.$$

We know that

$$\frac{d}{dx}(\ln x) = \frac{1}{x},$$

and $1/x$ is the integrand of the integral. Hence,

$$\int \left(\frac{1}{x}\right) dx = \ln x + C.$$

Turning next to the general form,

$$\int \frac{1}{mx + b}\, dx, \tag{1}$$

we would expect the integral to contain $\ln(mx + b)$. However, by the chain rule,

$$\frac{d}{dx}\ln(mx + b) = \frac{1}{mx + b}\frac{d}{dx}(mx + b)$$

$$= \frac{1}{mx + b}(m),$$

which has a factor m that does not appear in the integrand of (1). To remove the unwanted factor, we divide by m. Thus:

$$\int \frac{dx}{mx + b} = \frac{\ln (mx + b)}{m} + C.$$

In words, the integral of one over a linear function to the first power is the natural logarithm of the function divided by the coefficient of the variable.

Example. Evaluate $\displaystyle\int_{-1}^{3} \frac{4\, dx}{2x + 3}$.

The constant factor 4 remains in the integral, which is

$$4\left[\frac{\ln(2x+3)}{2}\right]\Bigg|_{-1}^{3}$$

$$= 2[\ln(2x+3)]\Bigg|_{-1}^{3}$$

$$= 2[\ln 9 - \ln 1]$$

$$= 2[2.19722 - 0]$$

$$= 4.39444.$$

Exercise. Evaluate $\int_{1}^{3} \frac{6\,dx}{3x-2}$. Answer: 3.89182.

11.13 PROBLEM SET 11–5

Carry out the following; simplify results where possible.

1. $\int x^{-1}\,dx.$

2. $\int \frac{dx}{x}.$

3. $\int \frac{2\,dx}{x}.$

4. $\int 3x^{-1}\,dx.$

5. $\int x^{-2}\,dx.$

6. $\int \frac{dx}{x^2}.$

7. $\int \frac{dx}{5x+4}.$

8. $\int \left(\frac{1}{3-x}\right) dx.$

9. $\int \left(\frac{1}{3-0.2x}\right) dx.$

10. $\int 2(0.5x+1)^{-1}\,dx.$

11. $\int \frac{dx}{(2x-1)^2}.$

12. $\int \frac{10\,dx}{(5x+3)^2}.$

13. $\int_{1}^{10} \frac{dx}{x}.$

14. $\int_{1}^{e} x^{-1}\,dx.$

15. $\int_{0}^{2} \frac{dx}{0.5x+4}.$

16. $\int_{5}^{10} \left(\frac{1}{0.6x+1}\right) dx.$

17. If new reserves of a fuel are discovered at the rate of

$$\frac{100}{0.2t+1}$$

million barrels per year at time t years, find the total amount of fuel that will be discovered in the next 25 years.

18. The demand function for a product is

$$p_d(q) = \frac{50}{0.5q+1}$$

where q is millions of pounds and $p_d(q)$ is dollars per pound. Compute consumers' surplus if market equilibrium occurs at a demand of 18 million pounds.

11.14 INTEGRALS OF EXPONENTIAL FUNCTIONS

Inasmuch as the derivative of e^x is e^x, it follows that

$$\int e^x \, dx = e^x + C.$$

If the exponent is the linear function $(mx + b)$, then, as in the case in the last section, the integral has m as a divisor. Thus,

Exponential rule, base e: $\displaystyle \int e^{mx+b} \, dx = \frac{e^{mx+b}}{m} + C.$

Example. Evaluate $\displaystyle \int_0^{10} 30e^{0.06x} \, dx.$

Here the exponent has $m = 0.06$ and $b = 0$. Hence,

$$\int_0^{10} 30e^{0.06x} \, dx = 30\left(\frac{e^{0.06x}}{0.06}\right)\Big|_0^{10}$$
$$= 500[e^{0.6} - e^0]$$
$$= 500[1.8221 - 1]$$
$$= 411.$$

Exercise. Evaluate $\displaystyle \int_1^2 e^{2x-1} \, dx.$ Note that b is not zero. Answer: $0.5(e^3 - e^1) = 8.684.$

Recall that the derivative of an exponential that has a base other than e requires a factor that is the natural logarithm of the base. That is,

$$\frac{d}{dx} a^x = a^x \ln a.$$

It follows that in the inverse process we must divide by $\ln a$. That is,

$$\int a^x \, dx = \frac{a^x}{\ln a} + C.$$

Again, if the exponent is the linear function $(mx + b)$, we must also divide by m. Hence,

Exponential rule, base a: $\displaystyle \int a^{mx+b} \, dx = \frac{a^{mx+b}}{m \ln a} + C.$

Example. Evaluate: $\displaystyle \int_1^2 (0.9)^{2x-1} \, dx.$

Matching the constants of the example with the foregoing rule, we see that $a = 0.9$, $m = 2$, $b = -1$. Hence,

$$\int_1^2 (0.9)^{2x-1}\, dx = \frac{(0.9)^{2x-1}}{2(\ln 0.9)}\Big|_1^2$$

$$= \frac{(0.9)^3 - (0.9)^1}{2(\ln 0.9)}$$

$$= \frac{0.729 - 0.9}{2(-0.10536)}$$

$$= 0.8115.$$

Exercise. Evaluate $\int_2^4 (6)^{0.5x}\, dx$. Answer: 33.49.

Exponentials often arise in applied problems dealing with rates, as illustrated in the following:

Example. A company projects its cost of providing medical care to workers to be at the rate of

$$15e^{0.03t}$$

thousand dollars per year at time t years. $a)$ Compute total medical care cost for the next 10 years. $b)$ How long will it be until total cost amounts to $250 thousand?

The example makes no reference to fixed cost, so we shall assume there is none, in which case the total sought in (a) is

$$\int_0^{10} 15e^{0.03t}\, dt = 15\left(\frac{e^{0.03t}}{0.03}\right)\Big|_0^{10}$$

$$= 500e^{0.03t}\Big|_0^{10}$$

$$= 500[1.3499 - 1]$$

$$= \$175 \text{ thousand.}$$

Part (b) of the example asks us to find the time, t years from now, such that

$$\int_0^t 15e^{0.03t}\, dt = 250.$$

Integrating, we find as before

$$500e^{0.03t}\Big|_0^t = 250$$

$$500[e^{0.03t} - e^0] = 250$$

$$e^{0.03t} - 1 = \frac{250}{500}$$

$$e^{0.03t} = \frac{250}{500} + 1$$

$$e^{0.03t} = 1.5.$$

We solve for t by first taking the natural logarithm of both sides of the last expression. Thus,

$$\ln e^{0.03t} = \ln 1.5$$

$$0.03t \ln e = \ln 1.5$$

$$0.03t(1) = \ln 1.5$$

$$t = \frac{\ln 1.5}{0.03}$$

$$= \frac{0.40547}{0.03}$$

$$= 13.5 \text{ years.}$$

To understand the significance of the foregoing example, note the cost rate

$$15e^{0.03t}$$

now, at $t = 0$, is

$$15e^0 = 15,$$

or at a rate of $15 thousand per year. If this rate were *constant* over time, then total cost would accumulate to $250 thousand in $250/15 = 16.7$ years. However, the exponential factor in

$$15e^{0.03t}$$

indicates that the cost rate per year is increasing and this accounts for the fact that cost will accumulate to $250 thousand in 13.5 years rather than 16.7 years. The reader will find it useful to refer to the last example while working the next set of problems.

11.15 PROBLEM SET 11–6

Carry out the following. Simplify results where possible.

1. $\int e^x \, dx.$

2. $\int 2^x \, dx.$

3. $\int 3^{-x} \, dx.$

4. $\int e^{-x} \, dx.$

5. $\int e^{0.5x} \, dx.$

6. $\int 50^{0.2x} \, dx.$

7. $\int (0.5)^{1-0.4x} \, dx.$

8. $\int e^{2-0.5x} \, dx.$

9. $\int 2e^{3-0.1x} \, dx.$

10. $\int 4e^{5-0.2x} \, dx.$

11. $\int \frac{dx}{(0.8)^x}.$

12. $\int \frac{1}{e^{0.5x}} \, dx.$

Evaluate the following:

13. $\displaystyle\int_0^5 2e^{1-0.2x}\,dx.$ 14. $\displaystyle\int_1^2 4e^{2-0.5x}\,dx.$

15. $\displaystyle\int_1^2 10(0.5)^x\,dx.$ 16. $\displaystyle\int_0^2 5(0.9)^x\,dx.$

17. The total supply of a fuel available now and in the future is 1000 million barrels. At time t years from now, fuel will be consumed at the rate of

$$10e^{0.05t}$$

million barrels per year.
a) How much fuel will be consumed in the next 20 years?
b) How long will the supply of fuel last?

18. At time t years, the cost of maintaining a facility is at the rate of $12e^{0.08t}$ thousands of dollars per year. Assuming there is no fixed cost involved,
a) Find total maintenance cost for the next 10 years.
b) How long will it take for total maintenance cost to reach $300 thousand?

19. At time t years, interest on a bank account is at the rate of $600e^{0.06t}$ dollars per year.
a) What will be total interest accumulation in 12 years?
b) How long will it take for total interest accumulation to reach $5000?

20. Sales of a product are projected to be at the rate of $15e^{-0.2t}$ million pounds per year at time t years.
a) Find total sales in the next 5 years.
b) How long will it take for total sales to reach 60 million pounds?

21. Sales of Exall at time t years are projected to be at the rate of $5 + 15e^{-0.2t}$ million pounds per year. Find total sales in the next 5 years.

22. Sales of Whyall at time t years are projected to be at the rate of $10 + 20e^{-0.4t}$ million gallons per year. Find total sales in the next 5 years.

23. The supply function for a product is

$$p_s(q) = 5 + e^{0.02q},$$

where q is in thousands of pounds and $p_s(q)$ is in dollars per pound. Market equilibrium occurs at a demand of 40 thousand pounds. Compute producers' surplus.

24. The supply function for a product is

$$p_s(q) = 10 + 2e^{0.05q},$$

where q is in thousands of gallons and $p_s(q)$ is in dollars per gallon. Market equilibrium occurs at a demand of 20 thousand gallons. Compute producers' surplus.

11.16 INTEGRALS OF LOGARITHMIC FUNCTIONS

In a later optional section of this chapter, we shall present a formal procedure that proves the rule that

$$\int \ln x\,dx = x(\ln x - 1) + C.$$

For the present, we shall simply demonstrate that the rule is correct by showing that the integrand, ln x, is equal to

$$\frac{d}{dx}[x(\ln x - 1) + C].$$

To do this, we apply the product rule, obtaining

$$x\frac{d}{dx}(\ln x - 1) + (\ln x - 1)\frac{d}{dx}(x) + \frac{d}{dx}(C)$$

$$= x\left(\frac{1}{x} - 0\right) + (\ln x - 1)(1) + 0$$

$$= 1 + \ln x - 1$$

$$= \ln x,$$

proving that the rule above is correct.

The rule for integrating the logarithm of a linear function to the first power is

$$\int \ln(mx + b)\, dx = \frac{(mx + b)[\ln (mx + b) - 1]}{m} + C.$$

Example. Evaluate $\int_0^2 \ln (5x + 3)\, dx$.

Following the foregoing rule, we write

$$\int_0^2 \ln (5x + 3)\, dx = \frac{(5x + 3)[\ln (5x + 3) - 1]}{5} \Big|_0^2$$

$$= \frac{1}{5}[13(\ln 13 - 1) - (3)(\ln 3 - 1)]$$

$$= \frac{1}{5}[13(2.56495 - 1) - 3(1.09861 - 1)]$$

$$= \frac{20.0485}{5}$$

$$= 4.0097.$$

Exercise. Evaluate $\int_1^3 \ln (2x + 1)\, dx$. Answer: 3.16277.

We shall not present the rule for integrating the common (base 10) logarithmic function because common logarithms are practically never used in calculus. However, the rule may be found in Table XII-B at the end of the book.

11.17 PROBLEM SET 11–7

Carry out the following. Simplify where possible.

1. $\int 3 \ln x \, dx$.
2. $\int 2 \ln x \, dx$.
3. $\int (\ln 2x) \, dx$.
4. $\int (\ln 3x) \, dx$.
5. $\int \ln(2x + 1) \, dx$.
6. $\int \ln(3x + 5) \, dx$.
7. $\int 6 \ln(3x - 2) \, dx$.
8. $\int 4 \ln(2x - 1) \, dx$.
9. At time t years, the cost of electric power used by a plant is projected to be at the rate of $4 \ln (2t + 6)$ thousand dollars per year. Find total electric power cost for the next seven years.
10. At time t years, production of a food is projected to be at the rate of $9 \ln (3t + 5)$ million bushels per year. Find total production in the next five years.

11.18 TABLES OF INTEGRALS

In order to fix integration in mind, the reader should memorize, or be able to develop quickly, the integration rules presented up to this point; that is, the rules for integrating the forms

$$(mx + b)^n, \quad \frac{1}{mx + b}, \quad e^{mx+b}, \quad a^{mx+b}, \quad \text{and} \quad \ln (mx + b).$$

These include the simple forms,

$$x^n, \quad \frac{1}{x}, \quad e^x, \quad a^x, \quad \text{and} \quad \ln x,$$

which are the special cases when $m = 1$ and $b = 0$. Rules for other forms used in our work are given in Table XII-B at the end of the book, and a more extensive list of rules can be found by consulting a recent edition of *Standard Mathematical Tables* published by the Chemical Rubber Co., Cleveland, Ohio. Before illustrating the use of a table of integrals, we should point out that the very general rules for taking derivatives (the chain, product, and quotient rules) do not have counterparts in integration. Consequently, rules presented in tables of integrals were developed by specialized procedures, some quite advanced. Even so, we would look in vain in a table to find a rule for

$$\int e^{x^2} \, dx.$$

There is no rule for this integral and, moreover, there are no integration rules for numerous functions that arise in practice. As a consequence, procedures for approximating values of definite integrals are important in solving applied problems. We shall introduce approximation procedures later under the heading *numerical integration*.

To use a table of integral rules, it is necessary to study the *form* of the function to be integrated and, having determined this, the table is searched for the corresponding form. Then constants are matched and the integration is carried out, as illustrated next.

Example. Find

$$\int xe^{0.1x}\,dx.$$

Here the integrand has the form *x times e to a linear function of x*. Running through Table XII-B, we find the corresponding form in Rule 17, which states

$$\int xe^{mx+b}\,dx = \frac{e^{mx+b}(mx-1)}{m^2} + C. \tag{1}$$

To match

$$xe^{0.1x} \quad \text{with} \quad xe^{mx+b},$$

it is necessary that

$$m = 0.1 \quad \text{and} \quad b = 0.$$

Therefore, with these values for *m* and *b*, we have

$$\int xe^{0.1x}\,dx = \frac{e^{0.1x}(0.1x-1)}{(0.1)^2} + C$$

$$= \frac{e^{0.1x}(0.1x-1)}{0.01} + C$$

$$= 100e^{0.1x}(0.1x-1) + C.$$

Example. Find

$$\int \frac{8x\,dx}{(2x-3)^2}.$$

Here, the factor 8 remains in the integral and is not considered when determining the form of the function. The form of

$$\frac{x}{(2x-3)^2}$$

is *x* over the square of a linear function of *x*, and Rule 11 of Table XII-B states that

$$\int \frac{x\,dx}{(mx+b)^2} = \frac{b}{m^2(mx+b)} + \frac{1}{m^2}\ln\,(mx+b) + C.$$

Matching constants in

$$\frac{x}{(2x-3)^2} \quad \text{and} \quad \frac{x}{(mx+b)^2}$$

we see that

$$m = 2 \quad \text{and} \quad b = -3.$$

Consequently,

$$\int \frac{8x\,dx}{(2x-3)^2} = 8\left[\frac{-3}{(2)^2(2x-3)} + \frac{1}{(2)^2}\ln(2x-3)\right] + C$$

$$= \frac{-6}{2x-3} + 2\ln(2x-3) + C,$$

where the last step was determined by multiplying the terms in the bracket of the preceding step by 8.

11.19 ASYMPTOTIC AREAS: IMPROPER INTEGRALS

If we write

$$\int_0^5 \left(\frac{1}{x}\right) dx,$$

the integrand is *undefined* at $x = 0$. This is one type of *improper* integral, and in this case the definite integral is undefined. The expression

$$\int_1^\infty \frac{1}{x^2}\,dx$$

is also an improper integral because its upper limit, denoted by the infinity symbol, ∞, is not a number. We shall define the last to mean the *limit* of

$$\int_1^x \left(\frac{1}{x^2}\right) dx$$

as the upper limit approaches ∞; that is, as x becomes indefinitely large. With this definition, the improper integral may or may not have a value (be defined). In the case at hand, if we write

$$\int_1^\infty \left(\frac{1}{x^2}\right) dx = -\left(\frac{1}{x}\right)\Big|_1^\infty$$

and translate this to mean

$$-\left[\lim_{x\to\infty}\left(\frac{1}{x}\right) - \frac{1}{1}\right], \tag{1}$$

we see that x becomes larger and larger $(x \to \infty)$,

$$\frac{1}{x}$$

becomes smaller and smaller, approaching zero as a limit, so

$$\lim_{x \to \infty} \left(\frac{1}{x} \right) = 0$$

and (1) becomes

$$-\left[\lim_{x \to \infty} \left(\frac{1}{x} \right) - 1 \right] = -[0 - 1] = 1$$

and

$$\int_1^\infty \left(\frac{1}{x^2} \right) dx = 1.$$

Hence, this improper integral is defined and has the value 1. On the other hand,

$$\int_1^\infty \left(\frac{1}{x} \right) dx = \ln x \,\Big|_1^\infty$$

$$= \lim_{x \to \infty} (\ln x) - \ln 1.$$

Here, as x becomes larger and larger, $\ln x$ also becomes larger and larger so does not approach a limit, and

$$\int_1^\infty \left(\frac{1}{x} \right) dx$$

is not defined (does not exist).

To see the area implication of the integral

$$\int_1^\infty \left(\frac{1}{x^2} \right) dx,$$

we sketch the graph of the integrand,

$$f(x) = \frac{1}{x^2}.$$

The important facts to note are, first, that when x is close to zero, $f(x)$ has a large value. For example, with $x = 0.01$ (close to zero)

$$f(0.01) = \frac{1}{(0.01)^2} = \frac{1}{0.0001} = 10{,}000,$$

and the closer x is to zero the larger $f(x)$ becomes, *but x cannot equal zero.* Second, if x becomes very large, $f(x)$ becomes very small. For example, at $x = 100$,

$$f(100) = \frac{1}{(100)^2} = \frac{1}{10{,}000} = 0.0001,$$

and the larger x is, the closer $f(x)$ gets to zero, *but $f(x)$ can never equal zero* no matter how large a number x is. Now refer to Figure 11–16 and note the right branch of the curve. As we move off to the right ($x \to \infty$), $f(x)$ approaches zero (the x-axis), but does not touch it, and we describe this behavior by saying $f(x)$ approaches the x-axis *asymptotically,* or that the x-axis is an *asymptote* of $f(x)$. Similarly, as we move to the left in the first quadrant, $f(x)$ rises higher and higher as x approaches zero, but x can never be zero, so the curve does not touch the vertical axis, but approaches it asymptotically as $x \to 0$. The behavior exhibited by the left branch of Figure 11–16 follows from the fact that

$$f(x) = \frac{1}{x^2}$$

has the same value when x is a given number, whether the number be positive or negative.

FIGURE 11-16

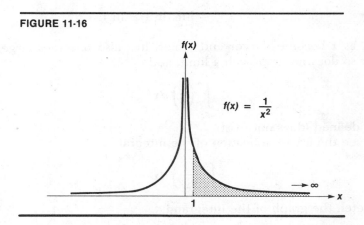

As a consequence of the foregoing, when we find

$$\int_1^\infty \left(\frac{1}{x^2}\right) dx = 1$$

we call this result, 1, the area under $f(x)$ over the interval 1 to ∞, even though the area is not completely enclosed by a finite boundary at the right. This idea of an *asymptotic area* is more than an exercise in applying the limit concept because there are numerous important applications of functions having areas in the asymptotic sense. Indeed, almost every reader of this book will at some point work with the *normal curve* in the study of probability and statistics, and the normal curve, as shown in Figure 11–17, has the property under discussion.

FIGURE 11-17

The normal curve

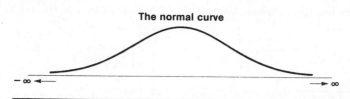

$-\infty \longleftarrow$ $\longrightarrow \infty$

Example.   Find the value, if one exists, for

$$\int_0^\infty \frac{30\ dx}{(2x+3)^2}.$$

We proceed to write

$$\int_0^\infty \frac{30\ dx}{(2x+3)^2} = 30 \int_0^\infty (2x+3)^{-2}\ dx$$

$$= 30 \left[\frac{(2x+3)^{-1}}{2(-1)} \right] \Big|_0^\infty$$

$$= -15 \left(\frac{1}{2x+3} \right) \Big|_0^\infty.$$

Now we reason that at the upper limit,

$$\lim_{x \to \infty} \left(\frac{1}{2x+3} \right) = 0$$

because as x becomes larger and larger, $1/(2x+3)$ has an increasingly large denominator, but a fixed numerator, so its value becomes ever smaller, approaching zero as a limit. Hence,

$$-15 \left(\frac{1}{2x+3} \right) \Big|_0^\infty = -15 \left[0 - \frac{1}{2(0)+3} \right]$$

$$= -15 \left(-\frac{1}{3} \right)$$

$$= 5.$$

Exercise.   Find the value of the following, if a value exists.

$$\int_6^\infty \frac{3\ dx}{(2x-10)^2}.$$

Answer: 3/4.

11.20 PROBLEM SET 11–8

Carry out the following using Table XII-B at the end of the book. Simplify where possible.

1. $\int x(2^x)\,dx.$

2. $\int \dfrac{6x}{2x+1}\,dx.$

3. $\int \dfrac{2x}{(0.5x+1)^2}\,dx.$

4. $\int xe^x\,dx.$

5. $\int xe^{2-0.5x}\,dx.$

6. $\int \dfrac{9x\,dx}{(3x+4)^2}.$

7. $\int \dfrac{2\,dx}{x(0.5x+1)}.$

8. $\int 2xe^{0.4x-1}\,dx.$

9. $\int \dfrac{2x\,dx}{5x-3}.$

10. $\int \dfrac{6\,dx}{x(3x+2)}.$

11. $\int xe^{0.1x^2-2}\,dx.$

12. $\int \dfrac{dx}{1+e^{0.2x}}.$

13. $\int \dfrac{dx}{1+2e^{0.5x}}.$

14. $\int 2xe^{2-0.5x^2}\,dx.$

Find the value of each of the following, if a value exists.

15. $\displaystyle\int_1^\infty 4x^{-3/2}\,dx.$

16. $\displaystyle\int_2^\infty \dfrac{6\,dx}{x^2}.$

17. $\displaystyle\int_{-3}^\infty \dfrac{dx}{(0.5x+2.5)^2}.$

18. $\displaystyle\int_3^\infty 3(0.5x+2.5)^{-3/2}\,dx.$

19. $\displaystyle\int_1^\infty x^{-1/2}\,dx.$

20. $\displaystyle\int_1^\infty x^{-2/3}\,dx.$

21. $\displaystyle\int_0^\infty 5e^{-0.2x}\,dx.$

22. $\displaystyle\int_0^\infty e^{-0.5x}\,dx.$

11.21 NUMERICAL INTEGRATION

Although extensive lists of integration rules are available in published tables, it frequently happens that a table does not have the counterpart of a form that arises in an applied problem. Sometimes it is possible to derive the integration rule for the form at hand either by applying a formal procedure such as the one discussed in the next section, or by some ingenious technique, but there are cases where formal procedures and ingenuity are to no avail. Thus, we note that while

$$\int e^x\,dx = e^x + C$$

is perhaps the simplest of all integration rules, no rule can be found if we write

$$\int e^{x^2}\,dx.$$

Lack of an integration rule poses no particular problem in applications where the value of a definite integral is sought because such values can be approximated quickly and with high accuracy on a computer that has been programmed to carry out numerical (approximate) integration. After the computer has been supplied with the program for numerical integration, we need only supply it with a statement specifying the function to be integrated and the values of the upper and lower limits of integration. The computer will then determine the desired value in less time than we spent supplying it with the problem. Lacking access to a computer, we can also approximate definite integrals easily on hand calculators. The procedure we shall develop for this purpose is called the *trapezoidal rule*. Having done this, we shall state *Simpson's rule*, which generally achieves a given degree of accuracy with fewer calculations than would be required by applying the trapezoidal rule, and we note in passing that there are other rules that outperform Simpson's rule.

The trapezoidal rule approximates the definite integral (which is the area under a curve) by computing the areas of n trapezoids as illustrated in Figure 11–18. To obtain the trapezoids, the interval from a to b, which are the integration limits, is divided into n equal parts, each of width w. That is,

$$w = \frac{b-a}{n}.$$

For illustration, Figure 11–18 uses $n = 5$ parts, leading to 5 trapezoids. The base of the first trapezoid extends from a to $a + w$ and is of

FIGURE 11-18

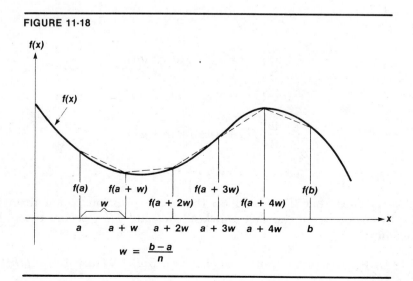

width w; the base of the second trapezoid extends from $a + w$ to $a + 2w$ and is of width w, and so on. The x-values of the points delineating the bases are

$$a, a + w, a + 2w, a + 3w, a + 4w, b.$$

Note for later reference that for $n = 5$, the next to the last point in the sequence is $(a + 4w)$, which in general will become $a + (n - 1)w$.

The (parallel) sides of the trapezoids are the function values for the points on the x-axis; that is,

$$f(a), f(a + w), f(a + 2w), f(a + 3w), f(a + 4w), f(b),$$

where, again, the next to the last function value will in general be $f[a + (n - 1)w]$.

The area of a trapezoid is computed as one-half the base times the sum of the (parallel) sides. Thus, the area of the first trapezoid is

$$\frac{w}{2}[f(a) + f(a + w)].$$

Exercise. What is the expression for the area of the second trapezoid? Answer $(w/2)[f(a + w) + f(a + 2w)]$.

From the foregoing, we see that the sum of the areas of the five trapezoids of Figure 11–18 is:

$$\frac{w}{2}[f(a) + f(a + w)]$$

$$+ \frac{w}{2}[f(a + w) + f(a + 2w)]$$

$$+ \frac{w}{2}[f(a + 2w) + f(a + 3w)]$$

$$+ \frac{w}{2}[f(a + 3w) + f(a + 4w)]$$

$$+ \frac{w}{2}[f(a + 4w) + f(b)].$$

Observe next that if we add up the last expressions, each term in the brackets *except* the first and last, $f(a)$ and $f(b)$, occurs twice, so the sum is

$$\frac{w}{2}[f(a) + 2f(a + w) + 2f(a + 2w) + 2f(a + 3w) + 2f(a + 4w) + f(b)]$$

$$= w\left[\frac{f(a)}{2} + f(a+w) + f(a+2w) + f(a+3w) + f(a+4w) + \frac{f(b)}{2}\right],$$

where the last follows by bringing the divisor 2 into the bracket. Now, remembering that $a + 4w$ in the next to the last term arose because we had $n = 5$ divisions, it follows that in general, for n divisions, the *next to the last term will be*

$$f[a + (n-1)w].$$

Using $\doteq$ to mean equals approximately, we have the

Trapezoidal rule

$$\int_a^b f(x)\, dx \doteq w\left[\frac{f(a)}{2} + f(a+w) + f(a+2w) + \cdots + f[a+(n-1)w] + \frac{f(b)}{2}\right],$$

$$\text{where } w = \frac{b-a}{n}.$$

This rule may be stated more compactly in summation symbol form (see Chapter 6) as

$$\int_a^b f(x)\, dx \doteq w\left\{\frac{f(a)}{2} + \left[\sum_{i=1}^{n-1} f(a+iw)\right] + \frac{f(b)}{2}\right\}.$$

To apply the rule, we first decide the number of subdivisions we wish to use, n, then compute

$$w = \frac{b-a}{n}.$$

With this value of w, we then write the values of

$$a,\ a+w,\ a+2w,\ \cdots,\ a+(n-1)w,\ b.$$

Next we compute half the function value at a, all of the following function values prior to b, half the function value at b, and sum the results. This sum, multiplied by w, is the approximate value of the definite integral. The greater the number of divisions (the larger n is) the more accurate will be the approximation. The following illustration shows the procedure in a simple case where we can check the outcome.

Example. Approximate the following integral by the trapezoidal rule using 5 subdivisions.

$$\int_1^3 x^2\, dx.$$

Since $n = 5$, the interval width is

$$w = \frac{b-a}{5} = \frac{3-1}{5} = \frac{2}{5} = 0.4.$$

The points at which the function x^2 is to be evaluated are $a = 1$, $a + w = 1.4$, and so on as shown in organized fashion in Table 11–1.

TABLE 11–1

Point, x	Function value, x^2		Numbers to be summed
$a\quad = 1$	1	(times 1/2)	0.50
$a + w\ = 1.4$	1.96		1.96
$a + 2w = 1.8$	3.24		3.24
$a + 3w = 2.2$	4.84		4.84
$a + 4w = 2.6$	6.76		6.76
$b\qquad = 3$	9	(times 1/2)	4.50
			Sum = 21.80

Answer: $w(\text{Sum}) = 0.4(21.80) = 8.72.$

Hence, from Table 11–1,

$$\int_1^3 x^2\, dx \doteq 8.72.$$

The exact value of this integral is

$$\int_1^3 x^2\, dx = \frac{x^3}{3}\ \Big|_1^3 = \frac{27}{3} - \frac{1}{3} = \frac{26}{3},$$

which, to two decimals, is 8.67, so our approximation is in error by $8.72 - 8.67$ or about 0.05 when $n = 5$ trapezoids are used. Calculations (not shown) provide 8.68 as the approximation when $n = 10$ trapezoids are used, and the error is $8.68 - 8.67$, or 0.01 and, of course, larger values of n can be used to reduce the error as much as we wish.

Example. Approximate the following by the area of 10 trapezoids.

$$\int_0^1 e^{x^2}\, dx.$$

In this case, no rule can be applied to determine the correct value of the integral. The approximation with $n = 10$ leads to a subdivision width of

$$w = \frac{b-a}{n} = \frac{1-0}{10} = 0.1.$$

The calculations for the approximation are organized in Table 11–2, where the exponential values were taken from Table IX at the end of the book.

TABLE 11–2

x	x^2	$f(x) = e^{x^2}$	Numbers to be summed
0.0	0.00	1.0000	(times $1/2$) = 0.5000
0.1	0.01	1.0101	1.0101
0.2	0.04	1.0408	1.0408
0.3	0.09	1.0942	1.0942
0.4	0.16	1.1735	1.1735
0.5	0.25	1.2840	1.2840
0.6	0.36	1.4333	1.4333
0.7	0.49	1.6323	1.6323
0.8	0.64	1.8965	1.8965
0.9	0.81	2.2479	2.2479
1.0	1.00	2.7183	(times $1/2$) = 1.3591
			Sum = 14.6717

Answer: w(sum) = 0.1(14.6717) = 1.46717.

The answer just obtained,

$$\int_0^1 e^{x^2}\, dx \doteq 1.46717,$$

almost certainly contains some meaningless digits at the right, and we cannot say how many. To shed light on this, we carried out the approximation (calculations not shown) with $n = 20$ and obtained

$$\int_0^1 e^{x^2}\, dx \doteq 1.46378$$

which has the same first three digits as the approximation with $n = 10$, and suggests that the first three digits, 1.46, will remain if n is increased.

Exercise.   *a)* Approximate the following by the areas of 5 trapezoids. Retain 3 decimal places in the answer.

$$\int_1^2 \ln x\, dx.$$

b) Compute the correct value (for comparison) to 3 decimal places. Answer: *a)* 0.3846 or 0.385. *b)* 0.386.

The foregoing approximation procedure in effect replaced segments of the curve of $f(x)$ by straight line segments forming the tops of trapezoids (see Figure 11–18). For a given number of subdivisions, a closer approximation can be obtained by replacing sections of the curve by pieces of *parabolas* because segments of parabolic curves will match the curve of a function more closely than will straight line segments. The parabolic approximation rule, generally called Simpson's rule, requires that we divide the interval from a to b into an *even* number of segments (4 or more) of width

$$w = \frac{b-a}{n}, \quad n \text{ even.}$$

The resulting approximation, which we shall not develop, is

Simpson's rule

$$\int_a^b f(x)dx$$

$$\doteq \frac{w}{3}[f(a) + 4f(a+w) + 2f(a+2w) + 4f(a+3w) + \cdots + f(b)].$$

The multiplier sequence in the foregoing bracket will be 1, 4, 2, 4, 2, 4, $\cdots$ 2, 4, 1. We shall offer exercises in the next set of problems for readers who wish to practice Simpson's rule.

11.22 INTEGRATION BY PARTS

The purpose of this section is to illustrate one formal procedure that can be applied to integrate some functions whose integrals are not immediately obvious. There are other formal procedures that the reader can find in more extensive treatments of calculus under the headings of *substitution, rational fractions,* and *reduction formulas.* But, as we have already mentioned, these procedures work only for functions of particular forms.

Integration by parts separates the factors of the integrand into two parts. One part, containing the differential, is named dv. The remaining part is named u. For example, in

$$\int \ln x \, dx,$$

if we let

$$u = \ln x \quad \text{and} \quad dv = dx,$$

then, taking the derivative with respect to x of $u = \ln x$, we have

$$u = \ln x$$

$$\frac{du}{dx} = \frac{1}{x}$$

which is written as

$$du = \frac{dx}{x}.$$

Next, integrating

$$dv = dx$$

we write

$$\int dv = \int dx$$
$$v = x.$$

Hence, we have

$$u = \ln x; \; dv = dx$$
$$du = \frac{dx}{x}; \; v = x.$$

We now apply the

Rule for Integration by Parts
$$\int u \, dv = uv - \int v \, du. \; [3]$$

Thus,

$$\int \ln x \, dx = \int u \, dv$$
$$= uv - \int v \, du$$
$$= (\ln x)(x) - \int x \cdot \frac{dx}{x}$$
$$= (\ln x)(x) - \int dx$$
$$= (\ln x)(x) - x + C$$
$$= x(\ln x - 1) + C,$$

which is the rule stated without derivation earlier in the chapter. Integration by parts is a useful procedure, but its applicability is limited.

---

[3] The product rule applied to *differentials* states that $d(uv)$ is the first times the differential of the second plus the second times the differential of the first. That is,

$$d(uv) = u \, dv + v \, du.$$

Integrating both sides of the last gives

$$\int d(uv) = \int u \, dv + \int v \, du \quad \text{or} \quad uv = \int u \, dv + \int v \, du,$$

which, by rearrangement, yields

$$\int u \, dv = uv - \int v \, du.$$

Even when it works, it may take trial and error to get started correctly. The trick is to make the proper selection of the parts.

Example. Integrate the following by parts.

$$\int (e^x) x \, dx.$$

Suppose we try taking $u = e^x$, so that

$$\frac{du}{dx} = \frac{d(e^x)}{dx} = e^x$$

and $du = e^x \, dx$.

The remaining part is $dv = x \, dx$, and

$$\int dv = \int x \, dx$$

$$v = \frac{x^2}{2}.$$

We now have

$$u = e^x; \; dv = x \, dx$$

$$du = e^x \, dx; \; v = \frac{x^2}{2}.$$

Then, applying the rule,

$$\int u \, dv = \int (e^x) x \, dx$$

$$= uv - \int v \, du$$

$$= e^x \left(\frac{x^2}{2} \right) - \int \frac{x^2}{2} e^x \, dx,$$

and the integral on the right is more difficult than the original problem. Consequently we start again and choose the parts the other way around. That is, let $u = x$ and $du = e^x dx$. This leads to

$$u = x; \; dv = e^x dx$$

$$du = dx; \; v = \int e^x dx = e^x.$$

Hence,

$$\int u \, dv = \int x e^x dx$$

$$= uv - \int v \, du$$

$$= x e^x - \int e^x dx$$

$$= x e^x - e^x + C$$

$$= e^x (x - 1) + C,$$

which is Rule 17 in Table XII–B at the end of the book, with $m = 1$ and $b = 0$.

Exercise. Integrate by parts: $\int x^2 \ln x \, dx$.

Answer: $\dfrac{x^3}{3} \ln x - \dfrac{x^3}{9} + C.$

11.23 PROBLEM SET 11–9

Approximate the following by the sum of the areas of n trapezoids. (Trapezoidal rule.)

1. $\displaystyle\int_1^4 \ln x \, dx; \ n = 6.$

2. $\displaystyle\int_1^4 e^{-x} dx; \ n = 6.$

3. $\displaystyle\int_0^{0.4} e^{-x^2} dx; \ n = 4.$

4. $\displaystyle\int_0^{0.8} e^{x^2} dx; \ n = 4.$

5. $\displaystyle\int_1^2 \frac{x-1}{x+1} \, dx; \ n = 5.$

6. $\displaystyle\int_2^3 \frac{x+1}{x-1} \, dx; \ n = 5.$

7. $\displaystyle\int_1^3 \frac{\ln x}{x} \, dx; \ n = 5.$

8. $\displaystyle\int_0^{0.5} x e^{-x^2} dx; \ n = 5.$

(Problems 9 and 10 are for readers with appropriate hand calculators.) The *normal curve* or, more precisely, the standard normal probability density function, is without question the most important function in probability and statistics. It is

$$f(x) = \frac{1}{\sqrt{2\pi}} \, e^{-\frac{z^2}{2}} \doteq 0.398942 e^{-z^2/2}; \ -\infty \le z \le \infty.$$

There is no rule for integrating this function to obtain areas, which, in applications, are probabilities. Consequently, to assist the millions of people who use normal probabilities (which includes nearly every student of management and economics), values have been computed and tabulated. Table VIII at the end of this book is a typical presentation. The entry next to $z = 1$ in Table VIII is 0.3413. This means that

$$\frac{1}{\sqrt{2\pi}} \int_0^1 e^{-\frac{z^2}{2}} dz = 0.3413$$

to four decimal places. Inasmuch as very few of the millions of users of tables such as VIII have any understanding of the source of the numbers, a reader who does problem 9 or 10 can properly claim to be "one in a million."

9. (See the foregoing introduction.) Approximate the following integral by the sum of 10 trapezoids.

$$0.398942 \int_0^1 e^{-\frac{z^2}{2}} dz.$$

10. Do Problem 9 using 20 trapezoids.

11. Approximate the following by Simpson's rule using $n = 4$.

$$\int_1^3 \frac{x}{e^x} \, dx.$$

12. Approximate the following by Simpson's rule using $n = 6$.

$$\int_1^4 \frac{\ln x}{x} \, dx.$$

Apply integration by parts and carry out the following.

13. $\int x \ln x \, dx.$

14. $\int x^3 \ln x \, dx.$

15. $\int \frac{\ln x}{x^2} \, dx.$

16. $\int \frac{x + \ln x}{x^2} \, dx.$

11.24 DIFFERENTIAL EQUATIONS

Numerous problems earlier in this chapter start with given information about a *rate* and proceed to determine a desired quantity by integrating the rate. The methodology of differential equations, which we introduce in this section, is widely applied in solving such rate problems and, we should remark, such problems occur very frequently in practice.

A differential equation is one that contains a differential or a derivative. As examples, we write

$$y \, dy - x \, dx = 0$$

$$\frac{dy}{dx} = 2x + 1$$

$$\frac{d^2y}{dx^2} + 3\frac{dy}{dx} + x - 1 = 0.$$

The equation last written contains a second derivative and serves to illustrate the fact that differential equations may involve second and higher order derivatives and differentials. However, in this introduction we shall deal only with first order equations. Before proceeding, we call attention to the use of differentials such as dy and dx as symbols having individual meanings. When these were introduced in differential calculus, it was emphasized that if

$$y = f(x)$$

then

$$\frac{dy}{dx} = f'(x)$$

and dy/dx was not a fraction with dy as numerator and dx as denominator, but a single symbol meaning the same as $f'(x)$. However, when we write

$$\int f'(x) dx,$$

dx appears alone. To achieve consistency, we define dy to be $f'(x)dx$.

That is,

$$\text{if } \frac{dy}{dx} = f'(x),$$

$$\text{then } dy = f'(x)dx.$$

In terms of symbol manipulation, the last can be thought of as being obtained by multiplying both sides of the preceding equation by dx, and this is how we shall describe the procedure. Thus, for example, if

$$\frac{dy}{dx} = 2x$$

then

$$dy = 2x\,dx$$

is a differential equation that is solved by integrating the left with respect to y and the right with respect to x. That is,

$$\int dy = \int 2x\,dx$$
$$y + C_1 = x^2 + C_2$$
$$y = x^2 + C_2 - C_1.$$

Inasmuch as C_2 and C_1 are arbitrary constants, $C_2 - C_1$ is another arbitrary constant, which we may call C. *We shall follow the practice of writing only one constant for one integration step*, so the solution of the foregoing is

$$y = x^2 + C.$$

A solution such as the last, which contains the constant of integration, is called the *general* solution of the differential equation, and each value for C yields a *particular* solution of the differential equation. Thus,

$$y = x^2 + 10$$

is one particular solution, and there are infinitely many particular solutions.

Initial conditions. If we are presented with a differential equation and with a pair of values that must satisfy the solution, the given values are called *initial conditions* or boundary values.

Example. Solve the differential equation

$$dy = x^{1/2}dx$$

subject to the initial condition that $y = 10$ when $x = 9$.

. We write

$$\int dy = \int x^{1/2}\, dx$$

$$y = \frac{2}{3}\, x^{3/2} + C.$$

The point $x = 9$, $y = 10$ must satisfy the last equation. Hence,

$$10 = \frac{2}{3}\, (9)^{3/2} + C$$

$$10 = \frac{2}{3}\, (3)^3 + C$$

$$10 = 18 + C$$

$$-8 = C,$$

so C must be -8, and the *particular* solution required is

$$y = \frac{2}{3}\, x^{3/2} - 8.$$

> *Exercise.* Solve the differential equation
>
> $$dy = 2x\, dx$$
>
> subject to the initial condition that $y = 12$ when $x = 3$. Answer:
> $y = x^2 + 3$.

11.25 SEPARABLE DIFFERENTIAL EQUATIONS

A differential equation that can be arranged in the form

$$g(y)\, dy = f(x)\, dx,$$

where one side is dy times a function of y and the other side is dx times a function of x, is called a separable differential equation. For example,

$$x\frac{dy}{dx} - 1 = 0,$$

after multiplying both sides by dx, becomes

$$x\, dy - dx = 0$$

or

$$x\, dy = dx$$

$$dy = \frac{dx}{x}.$$

The general solution of the last separated form is

$$\int dy = \int \frac{dx}{x}$$

$$y = \ln x + C.$$

Exercise. Write the general solution of $\dfrac{dy}{dx} - 2x = 1$. Answer: $y = x^2 + x + C$.

11.26 FORMS OF THE CONSTANT

The definition and rules of logarithms are applied frequently to obtain *explicit* solutions of differential equations. For reference, these are:

1. Definition: If $\ln a = b$, then $a = e^b$.
2. Product: $\ln (ab) = \ln a + \ln b$.
3. Quotient: $\ln \left(\dfrac{a}{b}\right) = \ln a - \ln b$.
4. Power: $\ln (a^b) = b \ln a$.

Example. Find the general solution of the following differential equation in *explicit* form.

$$\frac{dy}{dx} = \frac{y}{x}; \ x \neq 0.$$

First, multiplying by dx gives

$$dy = \frac{y}{x}\, dx.$$

Then dividing by y provides the separated form

$$\frac{dy}{y} = \frac{dx}{x}.$$

Integrating, we have

$$\int \frac{dy}{y} = \int \frac{dx}{x}$$

$$\ln y = \ln x + C.$$

The last solves the equation in terms of $\ln y$, not explicitly in terms of y. Proceeding, we write

$$\ln y - \ln x = C$$

$$\ln \left(\frac{y}{x}\right) = C,$$

where the last applies to a rule of logarithms. We next apply the definition of the natural logarithm and write

$$\frac{y}{x} = e^C.$$

Inasmuch as e is a constant and C is a constant,

$$e^C \text{ is a constant} = K.$$

That is, there is no need for two constants in the solution. Hence, we have

$$\frac{y}{x} = K$$

$$y = Kx$$

as the *explicit* general solution. This surprisingly simple result is a straight line of slope K that passes through the origin, and a line through the origin is the only function whose slope can be found by dividing the y-coordinate of any one of its points by the x-coordinate; that is,

$$\text{slope} = \frac{dy}{dx} = \frac{y}{x},$$

which is the initial differential equation. Observe, however, that x cannot equal zero, so the differential equation, and therefore the general solution, is not defined for $x = 0$. As a matter of convenience, we shall assume it to be understood that solutions we write exclude values of the independent variable where the differential equation and/or its solution are not defined.

Exercise. a) Write the following equation in separated form. b) Find the explicit general solution.

$$x\frac{dy}{dx} + y = 0.$$

Answer: a) $dy/y = -dx/x$. b) $y = K/x$.

As another example, consider the equation

$$x \, dy = 2y \, dx.$$

Dividing both sides by xy, we have

$$\frac{x\,dy}{xy} = \frac{2y\,dx}{xy}$$

$$\frac{dy}{y} = \frac{2dx}{x}.$$

Integrating leads to

$$\int \frac{dy}{y} = 2 \int \frac{dx}{x}$$

$$\ln y = 2 \ln x + C.$$

Now, applying logarithm rules,

$$\ln y = \ln x^2 + C$$

$$\ln y - \ln x^2 = C$$

$$\ln \left(\frac{y}{x^2}\right) = C$$

$$\frac{y}{x^2} = e^C = K$$

and we have

$$y = Kx^2$$

as the explicit general solution.

Finally, we remind the reader that the rule of exponents that says

$$e^x \cdot e^C = e^{x+C}$$

can be applied in reverse if an expression similar to the one on the right is encountered. That is,

$$e^{x+C} = e^x(e^C) = Ke^x,$$

because if C is an arbitrary constant, so also is e^C an arbitrary constant.

Example   Find the explicit general solution of

$$2x(0.5y + 1)\,dx + dy = 0.$$

First we write

$$dy = -2x(0.5y + 1)\,dx.$$

Division by $(0.5y + 1)$ separates the variables, giving

$$\frac{dy}{0.5y + 1} = -2x\,dx,$$

$$\int \frac{dy}{0.5y + 1} = -2 \int x\,dx.$$

On the left we see a linear form whose integral we recall, or look up in Table XII-B, and have

$$\frac{\ln{(0.5y+1)}}{0.5} = -2\left(\frac{x^2}{2}\right) + C_1$$

$$\ln{(0.5y+1)} = 0.5(-2)\left(\frac{x^2}{2}\right) + 0.5C_1$$

$$= -0.5x^2 + C,$$

where, in the last statement, $0.5C_1$ has been written as another arbitrary constant, C. Application of the logarithm definition leads to

$$0.5y + 1 = e^{-0.5x^2 + C}$$

$$= (e^{-0.5x^2})(e^C)$$

$$= (e^{-0.5x^2})K_1,$$

where, in the last, e^C has been replaced by another arbitrary constant, K_1. Continuing,

$$0.5y = (e^{-0.5x^2})K_1 - 1$$

$$y = \frac{K_1 e^{-0.5x^2} - 1}{0.5}$$

$$= \frac{K_1}{0.5} e^{-0.5x^2} - \frac{1}{0.5},$$

but $K_1/0.5$ is an arbitrary constant which we shall call K, and then write the explicit general solution,

$$y = Ke^{-0.5x^2} - 2$$

where, in the last, -2 is not an *arbitrary* constant.

Exercise. Find the explicit general solution of

$$dy - (2y + 6)\, dx = 0.$$

Answer: $y = Ke^{2x} - 3$.

11.27 APPLICATIONS OF DIFFERENTIAL EQUATIONS

In many applications, including the examples in this section, differential equations arise because the information available consists of an expression for a rate of change, together with initial conditions. When a rate expression is given, it will be helpful in setting up the associated differential equation to keep in mind the *units* associated with the differential and the rate.

Example. Smith owes Brown $100 now, at time $t = 0$, and the amount owed, y, is increasing at the rate of $10 per year at time t years. How much will Smith owe at time $t = 3$ years?

The answer is, of course, $100 + 3(10) = $130 because the rate of change is constant. However, we wish to use the example to illustrate the mode of thinking involved in setting up differential equations. To this end, we note that the rate of increase, $10 per year, has (dollar/year) as its unit of measurement. If this rate continued for dt years, where dt therefore has (year) as its unit of measurement, then

$$10 \left(\frac{\text{dollar}}{\text{year}}\right) \cdot dt(\text{year}) = 10 \, dt(\text{dollar}),$$

so $10dt$ dollars represents the change in the amount owed in time dt years. Calling this change in amount owed dy, we write the differential equation

$$dy = 10 \, dt$$

and its general solution is

$$\int dy = \int 10 \, dt$$
$$y = 10t + C$$
$$y(t) = 10t + C.$$

From the initial conditions, $y = 100$ at $t = 0$, we have

$$y(0) = 100 = 10(0) + C$$
$$100 = C$$

and the particular solution is

$$y(t) = 10t + 100.$$

Hence, at time $t = 3$ years, Smith will owe

$$y(3) = 10(3) + 100 = \$130.$$

Although the mode of thinking just illustrated is precise only when the rate function is linear, it will, nevertheless, lead to correct statements of differential equations, and we shall apply it in coming illustrations.

Example. A bank account contains $5000 now, at time $t = 0$, and yields interest at the rate of 6% (0.06) per year, compounded *continuously*. How much will the account contain at $t = 5$ years?

The consequence of continuous compounding is that at any point t in time, the amount in the account at that time $A(t)$ or A, whatever it is, is increasing at the rate of 6 percent of that amount, or $0.06A$.

At a rate of

$$0.06A$$

per year, the change in the amount in the account, in dt years, is $0.06A\ dt$, and this change in A is dA. Thus,

$$dA = 0.06A\ dt.$$

We proceed to separate the variables and integrate.

$$\frac{dA}{A} = 0.06\ dt;$$

$$\int \frac{dA}{A} = 0.06 \int dt;$$

$$\ln A = 0.06t + C$$

$$A = e^{0.06t + C}$$

$$A = e^{0.06t}(e^C)$$

$$A = Ke^{0.06t}.$$

The general solution of the differential equation is

$$A(t) = Ke^{0.06t}.$$

To evaluate K, we apply the initial condition $A(0) = \$5000$.

$$A(0) = 5000 = Ke^{0.06(0)}$$

$$5000 = Ke^0$$

$$5000 = K.$$

The particular solution is

$$A(t) = 5000e^{0.06t},$$

so at $t = 5$, we have

$$A(5) = 5000e^{0.06(5)}$$

$$= 5000(e^{0.3})$$

$$= 5000(1.349859)$$

$$= \$6749.29.$$

Exercise. If the account in the foregoing started with P dollars rather than $\$5000$, and interest rate i instead of 0.06, what would be the expression for the particular solution? Answer: $A(t) = Pe^{it}$.

Readers who have studied the material on mathematics of finance in Chapter 8 will recognize the formula in the answer to the last exercise was derived by more laborious means in that earlier material.

Example. An oil company and a farmer upon whose land a well has been drilled agree that the farmer shall receive a royalty of $2 per barrel of oil produced. The company wishes to pay for a year's output in a lump sum at the year's end. However, the farmer argues that a year's production does not occur all at once at the end of the year, but is spread out uniformly during the year. Consequently, asserts the farmer, his royalty income is generated *continuously* throughout the year but held by the company until the end of the year, and that he should receive interest. Because the flow is continuous, the company agrees to pay interest at the rate of 6 percent per year, compounded continuously, and consider the flow of royalty as being continuous. If the well produces 10,000 barrels in a year, what should the farmer's royalty be?

At the outset, we state that this example contains an idea that is hard to grasp, and that is the idea of a continuous flow of royalties being compounded continuously. To sort the pieces out, we recall from the last example that whatever the amount A or $A(t)$ is in the account at time t years, that amount will grow at the rate of $0.06A$ dollars per year and in dt years will increase by $0.06A\,dt$ dollars. Next we note that 10,000 barrels per year at $2 royalty per barrel is a royalty inflow at the rate of $20,000 per year, which is $20,000dt$ dollars in dt years. Hence, there are two components of increase; $0.06A\,dt$ and $20,000dt$. The change in the royalty account in dt years is, therefore,

$$dA = 0.06A\,dt + 20,000\,dt.$$

We proceed to solve the last differential equation.

$$dA = (0.06A + 20000)\,dt$$

$$\frac{dA}{0.06A + 20000} = dt$$

$$\int \frac{dA}{0.06A + 20000} = \int dt$$

$$\frac{\ln(0.06A + 20000)}{0.06} = t + C_1$$

$$\ln(0.06A + 20000) = 0.06t + 0.06C_1$$

$$= 0.06t + C$$

$$0.06A + 20000 = e^{0.06t + C}$$

$$= e^{(0.06\,t)}(e^C)$$

$$= Ke^{0.06t}$$

$$0.06A = Ke^{0.06t} - 20000$$

$$A = \frac{Ke^{0.06t} - 20000}{0.06}.$$

The last is an explicit general solution. As to initial conditions, we recall that royalties accumulate from 0 at the start of the year, $t = 0$, to the amount at the end of the year. Hence, $A = 0$ at $t = 0$, so

$$A(0) = 0 = \frac{Ke^0 - 20000}{0.06}$$

$$0 = K(1) - 20000$$

$$20,000 = K.$$

The particular solution is

$$A = A(t) = \frac{20,000e^{0.06\,t} - 20000}{0.06}$$

$$= \frac{20,000(e^{0.06\,t} - 1)}{0.06}.$$

Consequently, at the end of the year, $t = 1$, the farmer should receive

$$A(1) = \frac{20,000(e^{0.06} - 1)}{0.06}$$

$$= \frac{20,000(1.0618365 - 1)}{0.06}$$

$$= \$20,612.18,$$

which is about \$612 more than a lump sum end-of-the-year payment of \$20,000.

If, in the last example, money flowed in at the rate of R dollars per year rather than \$20,000, and the interest rate was i rather than 0.06, the expression

$$A(t) = \frac{20,000(e^{0.06\,t} - 1)}{0.06}$$

would turn out to be

$$A(t) = \frac{R(e^{it} - 1)}{i},$$

which is the *formula for a continuous flow compounded continuously.*

Example. Suppose that a country has 200 million dollars worth of its old one-dollar bills in circulation and that an average of 2 million dollars worth of one-dollar bills flow through the country's banks each banking day. Starting today, $t = 0$ days, old bills appearing at banks are destroyed and replaced by new one-dollar bills. *a)* Find the expression for the amount of new currency in circulation at time t banking days. *b)* How much new currency will be in circulation after 50 banking days?

The key point to note in solving this problem is that even though 2 million one-dollar bills appear at banks each day, on the average, at any point in time except $t = 0$, not all of these bills will be old bills. For example, at the time when half the old bills have been replaced, only one million of the two million bills appearing will be old bills. Thus, the rate at which old bills appears is a function of t, and is a fraction of \$2 million per day. To construct this fraction, we let $N(t)$, or N, be the number of *new* bills in circulation at time t. Then, since we started with 200 million old bills, the number of old bills in circulation is 200 − (amount of new bills),

$$\text{Number of old bills in circulation} = 200 - N.$$

The total number in circulation remains at 200 million. Hence,

$$\text{Proportion of old bills in circulation} = \frac{200 - N}{200}.$$

Taking this proportion of the total 2 million per day appearing at banks, we have for the rate of increase in the number of new bills, per day:

$$\frac{200 - N}{200} (2) \text{ million per day}$$

$$= \frac{200 - N}{100} \text{ million per day.}$$

Consequently, the change dN in new bills in dt days will be

$$dN = \frac{200 - N}{100} dt.$$

We now solve the last differential equation, as follows:

$$\frac{dN}{200 - N} = \frac{dt}{100} = 0.01 \, dt$$

$$\int \frac{dN}{200 - N} = \int 0.01 \, dt$$

$$-\ln (200 - N) = 0.01 t + C$$

$$\ln (200 - N) = -0.01 t - C$$

$$200 - N = e^{-0.01 t - C}$$

$$= (e^{-C})(e^{-0.01 t})$$

$$200 - N = K e^{-0.01 t}$$

$$-N = K e^{-0.01 t} - 200$$

$$N = 200 - K e^{-0.01 t}$$

or

$$N(t) = 200 - Ke^{-0.01t}$$

becomes the explicit general solution. At time $t = 0$, no new bills are in circulation, so the initial condition is $N(0) = 0$. Thus,

$$N(0) = 0 = 200 - Ke^0$$
$$0 = 200 - K$$
$$K = 200.$$

Hence,

$$N(t) = 200 - 200e^{-0.01t}$$
$$N(t) = 200(1 - e^{-0.01t})$$

is the explicit particular solution. To find the number of new bills in circulation at time $t = 50$ banking days, we compute

$$N(50) = 200[1 - e^{-0.01(50)}]$$
$$= 200(1 - e^{-0.5})$$
$$= 200(1 - 0.60653)$$
$$= \$78.69 \text{ million.}$$

All of the examples in this section have involved *time* rates of change, and while time rates are very common in practice and the coming problems will all involve time rates, we should not leave the impression that applications of differential equations are restricted to such rates. Thus, all of the marginals in economics (marginal cost, marginal propensity to consume, and so on) are rates not involving time and we could, if we wished to expand our examples, pose problems in terms of these marginals.

11.28 PROBLEM SET 11–10

Find the explicit general solution of each of the following.

1. $dy - dx = 0$.
2. $dy - 3x^2\, dx = 0$.
3. $x\, dy = dx$.
4. $dy = y\, dx$.
5. $\dfrac{dy}{dx} = \dfrac{y}{x}$.
6. $\dfrac{dy}{dx} = \dfrac{y}{x^2}$.
7. $2\, dy - \dfrac{y\, dx}{x^{1/2}} = 0$.
8. $x\, dy + y\, dx = 0$.
9. $(0.2x + 3)\, dy = dx$.
10. $(0.5x + 2)\, dy = 2\, dx$.
11. $dy - (0.5y + 2)\, dx = 0$.
12. $dy = (0.2y + 3)\, dx$.

Find the explicit particular solution using the stated initial conditions.

13. $dy - (x + 1)\, dx$; $y = 26$ when $x = 6$.
14. $\dfrac{dy}{dx} = 2x$; $y = 9$ when $x = 3$.

15. $\frac{dy}{dx} = y;\ y = 1.0874$ when $x = 1$.

16. $x\,dy + y\,dx = 0;\ y = 2$ when $x = 3$.

17. $dy = (0.2y + 3)\,dx;\ y = 4$ when $x = 0$.

18. $dy - (0.5y + 2)\,dx = 0;\ y = 1$ when $x = 0$.

19. $x\,dy - y\,dx - dx = 0;\ y = 2$ when $x = 1$.

20. $x\,dy + y\,dx + dx = 0;\ y = 3$ when $x = 2$.

21. Chris has just opened a sandwich shop and at time t days from opening expects to sell hamburgers at the rate of $150 + 6t$ per day.
   a) Represent total sales after t days by S or $S(t)$ and set up the relevant differential equation.
   b) What are the initial conditions?
   c) Find the particular solution of the differential equation.
   d) How many hamburgers will Chris sell in the first 50 days?
   e) How long will it take for Chris to sell 45,000 hamburgers?

22. (See Problem 21.) At time t days, Chris expects to sell hot dogs at the rate of $100 + 2t$ per day.
   a) Set up the relevant differential equation.
   b) What are the initial conditions?
   c) Find the particular solution of the differential equation.
   d) How many hot dogs will Chris sell in the first 50 days?
   e) How long will it take for Chris to sell 60,000 hot dogs?

23. Population of a town now, at $t = 0$ years, is 100 thousand. At any time t years, population grows at a per-year rate which is 10 percent of the population at that time.
   a) Set up the relevant differential equation, using $P(t)$ or P as population at time t years.
   b) Find the particular solution of the differential equation.
   c) Compute population 10 years from now.

24. Now, at time $t = 0$, \$2000 is deposited in a bank account that yields interest at 5 percent per year. Interest is compounded continuously, so that at any point t years in time, the amount in the account grows at an annual rate that is 5 percent of the amount in the account at that time. Let $S(t)$, or S, be the amount in the account at time t years.
   a) Set up the relevant differential equation.
   b) Find the particular solution of the equation.
   c) Compute the amount in the account 20 years from now.

25. A bank account now, at $t = 0$ years, contains \$2000. Interest is at an annual rate of 8 percent per year but is compounded continuously so that at any time t, the account will increase at the annual rate of 8 percent of the amount in the account at that time. Moreover, money is to be added to the account at a rate of \$500 per year. The addition is to be thought of as a continuous flow so that \$500 per year is \$500 dt in dt years.
   a) Let A or $A(t)$ be the amount in the account at time t years and set up the relevant differential equation.
   b) Find the particular solution of the differential equation.

c) How much will the account contain 10 years from now?

d) How long will it take for the account balance to grow to $18,500?

26. (See Problem 25.) A bank account now, at $t = 0$ years, contains $5000. Interest is at an annual rate of 8 percent, compounded continuously. However, money is to be withdrawn from the account continuously at the rate of $500 per year.

a) Set up the relevant differential equation.

b) Find the particular solution of the equation.

c) How much will the account contain 10 years from now?

d) How many years will it take to reduce the account balance to zero?

27. The total supply of a fuel available now and in the future is 200 million barrels. At the present moment, $t = 0$, none of this total supply has been consumed. At time t years, the fuel is being consumed at the rate of

$$5e^{0.01t}$$

billion barrels per year. Let C, or $C(t)$, be the amount consumed at time t years.

a) Set up the relevant differential equation.

b) Find the particular solution of the equation.

c) Compute the total amount of fuel which will be consumed in the next 10 years.

d) How long will the fuel supply last?

28. Solve Problem 27 if annual consumption of the fuel at time t is at the rate of $5e^{0.02t}$ million barrels per year.

11.29 ADDENDUM: THE FUNDAMENTAL THEOREM OF CALCULUS

Early in this chapter we showed how the area under the curve

$$f(x) = x^2$$

over an interval can be approximated by breaking the interval into n segments of width Δx, forming rectangles having the segments as bases, and summing the areas of the rectangles. The limit of this sum as Δx approaches zero was defined to be the value of the definite integral and to be the area under the curve. Applying this procedure to determine integrals of a variety of functions would be a formidable task and, fortunately, this task turns out to be unnecessary. The theorem that shows the task is unnecessary is called the *fundamental theorem of calculus*. This theorem says that instead of applying the summing process to a function over an interval, and then determining a limit, all that need be done is to find the antiderivative of the function and evaluate it over the interval in the usual manner. Thus, to establish the fundamental theorem, we need to show that the area under $f(x)$ over the interval 0 to x equals the

$$[\text{antiderivative of } f(x)] \Big|_0^x.$$

FIGURE 11-19

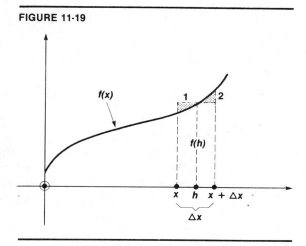

We begin by referring to Figure 11–19 and noting that the amount of area under $f(x)$ over the interval 0 to x depends on where the point x is; that is, the area in question is a function of x that we shall call the area function, $A(x)$. Next observe that the interval has been extended by an amount Δx to the point $x + \Delta x$. The point h has been selected so that when the rectangle shown is constructed, the little areas shown as 1 and 2 on Figure 11–19 are equal and, consequently, the area *under the curve over Δx equals the area of the rectangle.* We shall rely on intuition to assure ourselves that such a point, h, exists, but it is a matter of importance in theoretical mathematics to prove it exists and to state the conditions that insure that it exists. Readers wishing to pursue this matter should refer to *mean value* theorems in a text on calculus theory.

What we plan to do is write two expressions for the area of the rectangle in Figure 11–19; that is, two expressions for the same quantity. These expressions must, of course, be equal, and we shall deduce the fundamental theorem from this equality. The first expression for the area is simply the base of the rectangle, Δx, times its height, $f(h)$; that is,

$$\Delta x f(h). \tag{1}$$

To obtain the second expression, recall that $A(x)$ is the function specifying the area under the curve over the interval 0 to x. Thus, $A(x + \Delta x)$ is the area over the interval 0 to $x + \Delta x$. If we now subtract $A(x)$ from this to obtain $A(x + \Delta x) - A(x)$, this difference is the area under the curve over the interval x to $x + \Delta x$ and *because of the way h was chosen,* this area equals the area of the rectangle. Hence, our second expression for the area of the rectangle is

$$A(x + \Delta x) - A(x). \tag{2}$$

Equating (1) and (2), we have

$$A(x + \Delta x) - A(x) = \Delta x f(x)$$

or,

$$\frac{A(x + \Delta x) - A(x)}{\Delta x} = f(x). \tag{3}$$

Our strategy should be clear now. The foregoing steps have provided, on the left side of (3), the *difference quotient* for the area function. The limit of this difference quotient as Δx approaches zero is the derivative of the area function, which is $A'(x)$. Moreover, by reference to Figure 11–19, it is clear that as Δx approaches zero, h moves toward x and $f(h)$ approaches $f(x)$ as a limit. Thus,

$$\lim_{\Delta x \to 0} \frac{A(x + \Delta x) - A(x)}{\Delta x} = \lim_{\Delta x \to 0} f(h)$$

$$A'(x) = f(x). \tag{4}$$

The area function we seek, $A(x)$, can be obtained by taking the antiderivative of $A'(x)$. From (4) we find

$$\text{antiderivative}[A'(x)] = \text{antiderivative}[f(x)]$$
$$A(x) = \text{antiderivative}[f(x)].$$

$A(x)$ was defined as the area over the interval 0 to x, so $A(0) = 0$ and

$$A(x) - A(0) = \text{antiderivative}[f(x)] \Big|_0^x$$

$$A(x) = \text{antiderivative}[f(x)] \Big|_0^x.$$

This establishes what we set out to show; namely, that the area over the interval 0 to x equals the antiderivative of $f(x)$ evaluated between the limits 0 and x. Now, recalling that

$$\int_0^x f(x)\, dx$$

is *defined* to be the limit of a sum (and thus makes no reference to derivatives), we see that the fundamental theorem establishes the relationship between the integral and the derivative because it states that the definite integral can be evaluated by finding the function that has $f(x)$ as its derivative and evaluating that function between the limits.

The mechanical procedure of finding an antiderivative is, by itself, a barren exercise. The area interpretation of the definite integral as the limit of a sum, on the other hand, has many fruitful interpretations, but would be difficult to apply if the limit procedure had to be carried out for every new function encountered. What the fundamental theorem shows is that the otherwise barren mechanics of antiderivatives indeed are important because they serve to replace the cumbersome limit process required in the definition of the very useful definite integral. It is easy to fall into the habit of thinking that integrals are the same as antiderivatives. In fact, the rules that we call integration rules and designate by the integral sign were developed, and are proved, by antiderivative procedures. We need not try to break this habit, of course, because the fundamental theorem assures that it will not lead us astray. However, readers interested in the intellectual development of mankind may find satisfaction in knowing why it is that from a myriad of theorems, mathematicians have selected one as worthy of being called the *fundamental* theorem of calculus.

11.30 REVIEW PROBLEMS

Write the expression for the indefinite integral. Simplify results where possible. Express answers with positive exponents.

1. $\int dq$.
2. $\int dp$.
3. $\int k\,dx$.
4. $\int 5\,dz$.
5. $\int (1 + y^2)\,dy$.
6. $\int (2 + 3p^2)\,dp$.
7. $\int (a + bx)\,dx$.
8. $\int (1 + cx)\,dx$.
9. $\int 12(2x^2 - x^3 + 3x + 2)\,dx$.
10. $\int 6(10x^4 - 8x^3 + 6x^2 + 4x + 3)\,dx$.
11. $\int (x^{2/3} + 2x^{-3/2})\,dx$.
12. $\int \left(x^2 - \dfrac{1}{x^2} \right) dx$.
13. $\int \left(1 + \dfrac{2}{x^3} + \dfrac{4}{x^{1/2}} \right) dx$.
14. $\int \left(1 - \dfrac{2}{x^{2/3}} \right) dx$.
15. $\int 30(3x - 5)^4\,dx$.
16. $\int \dfrac{15\,dx}{(7 - 5x)^2}$.
17. $\int \dfrac{6\,dx}{(8 - 3x)^{4/3}}$.
18. $\int 12(4x - 9)^{2/3}\,dx$.

Evaluate the following definite integrals.

19. $\displaystyle\int_1^5 dx$.
20. $\displaystyle\int_2^5 (1 + x)\,dx$.
21. $\displaystyle\int_1^{16} x^{3/4}\,dx$.
22. $\displaystyle\int_1^2 \left(3 + 4x + \dfrac{10}{x^2} \right) dx$.
23. $\displaystyle\int_2^3 \dfrac{12\,dx}{(3x - 5)^3}$.
24. $\displaystyle\int_0^6 \dfrac{dx}{(0.5x + 1)^2}$.

Express the definite integral in terms of its limits, simplifying where possible.

25. $\displaystyle\int_a^{2a} \left(\dfrac{x}{a} + \dfrac{a}{x^2} \right) dx$.
26. $\displaystyle\int_0^b \dfrac{ab}{(ax + b)^2}\,dx$.

Find the area under the curve of the following functions over the given x-intervals.

27. $f(x) = (8x + 4)^{1/2}$, $x = 0$ to $x = 4$.

28. $f(x) = \dfrac{20}{(0.5x + 4)^2}$, $x = 2$ to $x = 12$.

Find the areas described in each of the following problems. (Sketches will be helpful.)

29. Find area bounded by the axes and

$$f(x) = 10 - 0.2x.$$

30. Find the area bounded by an axis and

$$f(x) = 60x - 12x^2 - 48.$$

31. Find the area bounded by the x-axis and

$$f(x) = 3x - x^2 + 18.$$

32. Find the area bounded by

$$f(x) = 19 - x \quad \text{and} \quad g(x) = 25 - x^2.$$

33. Find the first-quadrant area bounded by an axis and

$$f(x) = x^2 \quad \text{and} \quad g(x) = \frac{x^2}{2} + 2.$$

34. Maintenance cost on newly purchased equipment is expected to be at the rate of $(1 + 0.2t)$ thousand dollars per year at time t years.
    a) Compute total maintenance cost during the first five years.
    b) Make a sketch showing what was computed in (a).
    c) Compute maintenance cost during the second five years.
    d) At what time, t, will the total spent on maintenance reach $20 thousand?

35. At time t years, an industry consumes fuel at the rate of $(1 + 0.3t)$ million barrels per year. How much fuel will the industry consume in the next 9 years?

36. At time t years, sales of a currently new product are expected to be at the rate of

$$\frac{5}{(0.4t + 1)^{1/2}}$$

million dollars per year. How many years will it take for total sales to amount to $20 million?

37. Selling expense has a fixed component of $5000 per year, *each year*, and a variable component estimated to be at the rate of $(10 + 0.1t)$ thousand dollars per year at time t years. Estimate total selling expense for the next four years.

38. Sales of a product are projected to be at the rate of $(81 - 3t)^{1/2}$ thousand tons per year at time t years. Find total sales during the next 24 years.

39. Sales of a product are projected to be at the rate of $(0.9t + 9)^{1/2}$ thousand tons per year at time t years. Find total sales during the next 30 years.

40. The population of a trading area is currently 100 thousand. At time t years from now, population will be growing at the rate of

$$\frac{20}{(0.5t + 9)^{1/2}}$$

thousand per year. How long will it take for population to reach 260 thousand?

41. When t tons of Tonall are produced, marginal cost in dollars per ton is

$$0.003t^2 - 0.4t + 25.$$

If fixed cost is $1000, find the total cost of producing 200 tons.

42. A fixed cost of $20,000 was incurred in setting up an operation. At time t months thereafter, income and expense are at the rates, respectively, of $(4100 - 100t^{1/2})$ and $(500 + 260t^{1/2})$ dollars per month.
   a) What is the optimal time to terminate the operation?
   b) What will total profit be at the optimal time of termination?

43. The demand function for a product is

$$P_d(q) = \frac{100}{(0.2q + 1)^2},$$

where q is in millions of gallons and $P_d(q)$ is in dollars per gallon. Market equilibrium occurs at a demand of 20 million gallons. Compute consumers' surplus.

44. The supply and demand functions for a product are

$$P_s(q) = 2 + 0.06q \quad \text{and} \quad P_d(q) = 10 - 0.02q.$$

   a) Find the equilibrium market price.
   b) Compute producers' surplus.
   c) Compute consumers' surplus.
   d) Sketch a graph containing both the supply and demand functions. Show by two different shadings which area represents producers' surplus and which represents consumers' surplus.

Carry out the following, simplifying where possible.

45. $\int \frac{3}{x}\, dx.$

46. $\int \frac{dx}{2x + 3}.$

47. $\int 2(5x + 4)^{-1}\, dx.$

48. $\int \left(\frac{1}{0.5x + 10}\right) dx.$

49. $\int_0^{10} 6(2x + 3)^{-1}\, dx.$

50. $\int_5^{50} \frac{dx}{0.2x + 1}.$

51. If new fuel reserves are discovered at the rate of

$$\frac{40}{0.25t + 1}$$

million tons per year at time t years, find the amount of fuel that will be discovered in the next 20 years.

52. The demand function for a product is

$$p_d(q) = \frac{40}{0.2q + 3},$$

where q is in millions of gallons and $p_d(q)$ is in dollars per gallon. Market equilibrium occurs at a demand of 10 million gallons. Compute consumers' surplus.

Carry out the following, simplifying where possible.

53. $\int 6e^{2x+5} \, dx.$

54. $\int e^{-0.5x+1} \, dx.$

55. $\int 2^x \, dx.$

56. $\int 3^{0.5x} \, dx.$

57. $\int_5^{10} e^{0.2x-1} \, dx.$

58. $\int \frac{2 \, dx}{e^{0.5x-1}}.$

59. $\int 10(0.5)^x \, dx.$

60. $\int_1^2 10^x(\ln 10) \, dx.$

61. The total supply of a fuel available now and in the future is 200 billion barrels. At time t years from now, fuel will be consumed at the rate of

$$2e^{0.08t}$$

billion barrels per year.
a) How much fuel will be consumed in the next 10 years?
b) How long will the fuel supply last?

62. At time t years, interest on a bank account is accumulating at the rate of

$$800e^{0.08t}$$

dollars per year.
a) What will total interest accumulation be in five years?
b) How long will it take for interest accumulation to reach $10,000?

63. The supply function for a product is

$$p_s(q) = 2 + e^{0.01q}$$

where q is in millions of pounds and $p_s(q)$ is in dollars per pound. Market equilibrium occurs at a demand of 100 million pounds. Compute producers' surplus.

Carry out the following, simplifying where possible.

64. $\int 5 \ln x \, dx.$

65. $\int \ln (5x) \, dx.$

66. $\int \ln (0.3x + 5) \, dx.$

67. $\int \ln (10x + 3) \, dx.$

68. $\int_1^5 10 \ln x \, dx.$

69. $\int_0^4 \ln (5x + 7) \, dx.$

70. At time t years from now, production of a food is projected to be at a rate of $8 \ln (0.5t + 4)$ million bushels per year. Find total production during the next 12 years.

Carry out the following, simplifying results where possible.

71. $\int 4xe^{1-0.25\,x^2}\,dx.$

72. $\int \dfrac{6\,dx}{3+5e^{2x}}.$

73. $\int \dfrac{4\,dx}{x(0.4x+3)}.$

74. $\int \dfrac{2x\,dx}{0.5x+9}.$

75. $\int 3xe^{5-0.1x}\,dx.$

76. $\int 3x(5^x)\,dx.$

77. $\int \dfrac{9x\,dx}{(3x+2)^2}.$

78. $\int 5xe^{-0.4x}\,dx.$

79. $\displaystyle\int_0^\infty 10e^{-0.5x}\,dx.$

80. $\displaystyle\int_1^\infty 20x^{-5/4}\,dx.$

81. $\displaystyle\int_0^\infty \dfrac{2\,dx}{(0.2x+1)^2}.$

82. $\displaystyle\int_1^\infty \dfrac{30\,dx}{x^3}.$

Approximate the following by the sum of the areas of n trapezoids. (Trapezoidal rule.)

83. $\displaystyle\int_0^3 \ln\,(x^2+1)\,dx;\ n=6.$

84. $\displaystyle\int_0^2 e^{0.5x^2-0.3}\,dx;\ n=5.$

85. Approximate the following by Simpson's rule using n = 6 subdivisions.

$$\int_0^3 \ln\,(x^2+1)\,dx.$$

Apply integration by parts and carry out the following:

86. $\int xe^x\,dx.$

87. $\int x^2e^x\,dx.$ (Use answer to Problem 86.)

88. $\displaystyle\int \dfrac{xe^x\,dx}{(1+x)^2}.$

Find the explicit general solution of each of the following differential equations.

89. $dy+dx=0.$

90. $3y^2\,dy-2x\,dx=0.$

91. $x^2\,dy-dx=0.$

92. $x^2\,dy-y^2\,dx=0.$

93. $x\,dy+2\,dx=0.$

94. $(2x+3)\,dy=dx.$

95. $dy-0.4x(5y+4)\,dx=0.$

Find the explicit particular solution for the following using the stated initial conditions.

96. $dy-2x\,dx=0;\ y=25$ when $x=5.$

97. $x^2\,dy-y^2\,dx;\ y=1.8$ when $x=18.$

98. $dy-y\,dx=0;\ y=10$ when $x=0.$

99. $dy-(0.4y+3)\,dx=0;\ y=2.5$ when $x=0$

100. Chris has just opened a pizza shop and at time t days from opening expects to sell pizzas at the rate of $50+2t$ per day.

a) Represent total sales after t days by S or $S(t)$ and set up the relevant differential equation for $dS(t).$

b) What are the initial conditions?

c) Find the particular solution of the differential equation.

d) How many pizzas will Chris sell in the first 20 days?

e) How long will it take for Chris to sell 3600 pizzas?

101. A bank account now, at $t=0$ years, contains $1000. Interest accrues at the annual rate of 6 percent per year, but is compounded continuously so that at any time t, the account will increase at the annual rate of 6 percent of the amount in the account at that time. Moreover, money is to be added to the account at the rate of $300 per year. The addition is to be thought of as a continuous flow, so that $300 per year is ($300dt$) in dt years.

   a) Let A or $A(t)$ be the amount in the account at time t years and set up the relevant differential equation.

   b) Find the particular solution of the differential equation.

   c) How much will be in the account 10 years from now?

102. A trading area has 20,000 potential customers for a new product and an advertising promotion is planned to sell the product. $R(t)$ is the total number of potential customers who will have responded to the promotion at time t days. At time t days, responses occur at the daily rate of 2 percent of those who have *not* responded.

   a) Write the relevant differential equation for $dR(t)$.

   b) What are the initial conditions?

   c) Find the explicit particular solution of the differential equation.

   d) How many of the potential customers will have responded at $t=10$ days?

   e) How many days will it take before 10,000 have responded?

Probability

12

12.1 INTRODUCTION

Almost all students of management and economics study statistics at some point during their undergraduate years. Typically, statistics courses devote only a small amount of time to probability, even though probability forms the foundation upon which statistical inference rests. Consequently, one objective of this chapter is to lay the foundation for the study of statistical inference. A second objective is to encourage readers to think in the probabilistic mode that is characteristic of decision-making problems. Finally, we should add, probability has its own interesting and useful applications and is worthy of study in its own right.

12.2 PROBABILITY AND ODDS

A weather forecaster may state in his radio broadcast that the probability of rain tomorrow is 60% and later, while playing poker, tell friends that the odds on rain tomorrow are 3 to 2. Other methods exist for expressing the degree of certainty, or uncertainty. In mathematics, we express probabilities as numbers on a scale from 0 to 1, inclusive. Thus, letting $P(R)$ represent the probability of rain, we convert 60% to its decimal equivalent and write

$$P(R) = 0.60.$$

The odds statement, 3 to 2, is viewed as meaning there are $3 + 2 = 5$ "chances" on tomorrow's weather, of which 3 are for rain, so

$$P(R) = \frac{3}{3 + 2} = \frac{3}{5} = 0.60$$

as before.

> **Exercise.** If the odds are 13 to 7 that the stock market will rise tomorrow, what is the decimal value of the probability of a rise? Answer: 0.65.

Odds are in common use, but suffer lack of comparability because they do not have a fixed base. Thus, if the odds for a stock price rise are 11 to 7 while those for a rise in bond yields are 3 to 2, a conversion is needed to express the likelihood of rises in comparable form, and probabilities in decimal form are an appropriate conversion.

> **Exercise.** Convert the last two odds statements as decimal probabilities. Answer: 0.611 and 0.600.

12.3 SOURCES OF PROBABILITIES

If we plan to draw a card from a well-shuffled standard deck of 52 cards, we would find it natural to assess the probability that the card will be a spade as

$$P(S) = (13 \text{ spades})/(52 \text{ cards}) = \frac{13}{52} = \frac{1}{4} = 0.25.$$

What this implies is that if we repeat the experiment of drawing a card from a well-shuffled deck of 52 cards many times, we would expect the proportion or *relative* frequency of drawing a spade to approach 0.25 as a limit as the number of drawings increases. *Relative frequency over the long run* under constant conditions (*e.g.*, drawing from a well-shuffled deck) is one source of probability measures. Most people feel confident in such probability assignments and refer to them as *objective* probabilities that can be verified by experiment. A novice in a poker game may feel that a pair of kings and a pair of aces should beat three deuces, but such is not the case. In poker, the more unusual (less probable) hand wins, and three of a kind is more unusual than two pairs—not because of intuition or any subjective factor, but because of mathematical calculation that can be verified by experiment.

We should point out at once that objective probabilities, which arise as relative frequencies over the long run under constant conditions, are not as prevalent in management decision-making as they are in games of chance. To see why, consider a firm that is deciding on whether or not to bid on a $5 million project, at a point in time, and in the competitive environment of that time. The firm knows it will cost about $100,000 to prepare the bid so, naturally, it will not prepare a bid unless the probability of winning the contract is sufficiently high to justify the cost. How does the firm assess the probability of winning? Past experience will be of some help, but it is unlikely that the circumstances surrounding this particular situation have occurred very often, if at all, in the past, so that a relative-frequency objective probability is not at hand. Nevertheless, a decision based upon a probability assessment must be made. We may assume that the appropriate officials of the firm, drawing upon both quantified and nonquantified experience and projection, will make the assessment. We refer to the results as a *subjective* probability.

In practice, probability assessments range from highly objective to highly subjective. In this chapter, we shall not distinguish between the two because probability theory is not concerned with the source of the probabilities we assign. The theory requires only that we consider *all* of the *different* events that may occur in a given situation and assign to each event a *nonnegative* number in a manner such that the sum of all the probabilities is one. For example, in drawing a card from a deck, the events red card, spade, and club are one way to specify *all* the *different* events that can occur, and we may assign probabilities 0.5, 0.25, and 0.25, respectively, to these events.

Certain and impossible events. We assign a probability of 1 to an event that is certain to occur, such as the event of getting a red or a black card when drawing from a deck. The probability zero is assigned to an impossible event, such as drawing a red spade from the deck. Again, a weatherman may assign a probability of 0 for rain tomorrow if he believes rain is impossible, or 1 if he believes rain is certain. In summary, if $P(E)$ is the probability of event E,

$$0 \le P(E) \le 1$$
$$P(\text{impossible event}) = 0$$
$$P(\text{certain event}) = 1.$$

12.4 PROBABILITY SYMBOLS AND DEFINITIONS

We shall introduce probability terminology by reference to the data in Table 12–1. These data show the results of a poll of *all* 500 employees in a firm on the question of whether to change the work week from five 8-hour days to four 10-hour days.

TABLE 12–1

	Favor (F)	Disfavor (D)	Neutral (N)	Totals
Men *(M)*	70	10	20	100
Women *(W)*	170	150	80	400
Total	240	160	100	500

The right margin shows there are 100 men in the group of 500 workers. If we were to draw an individual at random[1] from the 500, it would be natural to assign the probability of selecting a man, *P(M)*, as

$$P(M) = \frac{100}{500} = 0.20.$$

Similarly, the probability that the individual disfavors, *P(D)*, is

$$P(D) = \frac{160}{500} = 0.32.$$

> ***Exercise.*** From Table 12–1, find *a)* P(W). *b)* P(F). Answer: *a)* 0.80. *b)* 0.48.

Joint events: Intersection. In Table 12–1, sketch a horizontal line across the *M* row and a vertical line down the *F* column. At the *intersection* of the two lines there are 70 people who are both men *and* in favor. In set terminology, the symbol ∩ represents intersection, so the event at hand is *M* intersection *F*, or

$$M \cap F. \text{ [2]}$$

---

[1] The concept of random selection underlies all probability rules. In the case of card playing, randomness is sought by shuffling the cards before dealing a hand. In the case of Table 12–1, we could shuffle the 500 employee time cards or use other methods to insure random selection.

[2] Later in the chapter we will, for brevity, omit ∩ and express the intersection as *MF*. See also Appendix 1, Section A1.7.

An alternative description refers to the intersection as the *joint* event, *M* and *F*. Clearly,

$$P(M \cap F) = \frac{70}{500} = 0.14.$$

Similarly,

$$P(W \cap N) = \frac{80}{500} = 0.16,$$

and is the probability that the selected individual is a woman *and* is neutral. Remember always that the intersection means *both* occur and is characterized by the word *and*.

> *Exercise.* Write the symbols for the probability that a man who disfavors is selected, and compute this probability. Answer: $P(M \cap D) = {}^{10}\!/_{500} = 0.02.$

Mutually exclusive (disjoint) events. In Table 12–1 the intersection

$$F \cap D$$

is empty (has no elements) because an individual cannot both favor and disfavor the proposal. In set terminology

$$F \cap D = \emptyset$$

where $\emptyset$ is the *empty* or *null* set. We shall say *F* and *D* are disjoint or, more commonly, that they are *mutually exclusive*. Clearly,

$$P(F \cap D) = 0$$

and, in general,

A and B are mutually exclusive if $P(A \cap B) = 0$

> *Exercise.* If a card is drawn from a deck, what is the probability that it will be red and be a spade? Why? Answer: $P(R \cap S) = 0$ because red and spade are mutually exclusive.

Conditional probability. Suppose next that an individual has been selected. We are told that the individual is a man, but not told his attitude toward the proposal. What now is the probability that this individual is in favor? Here it is *given* that we have a man, so in Table

12–1 we consider only the 100 men and the 70 of these who are in favor. We find that

$$P(\text{in favor, given man}) = P(F \mid M) = \frac{70}{100} = 0.70,$$

where the vertical line segment in $P(F \mid M)$ is read as *given*, and the whole symbol as the probability of F, given M.

Similarly, if we are told the person selected is neutral and asked to find the probability that the person is a man, we compute

$$P(M \mid N) = \frac{20}{100} = 0.20.$$

Exercise. If the person selected is in favor, write the symbols for, and compute the probability that, the person is a woman. Answer: $P(W \mid F) = {}^{170}\!\!\diagup\!\!_{240} = 0.708$.

Dependent and independent events. Table 12–1 shows

$$P(M) = \frac{100}{500} = 0.200 \quad \text{but} \quad P(M \mid F) = \frac{70}{240} = 0.292.$$

We note that

$$P(M \mid F) \neq P(M)$$

and say that M and F are *dependent* in the probability sense. The idea is that the probability that the selected person is a male is related to, or depends on, whether the person is in favor. In this case, the *unconditional* probability that the selected person is a man is 0.200, but it is more likely (0.292) that the person is a man if we are given that the person is in favor.

On the other hand, we note that

$$P(M) = \frac{100}{500} = 0.20 \quad \text{and} \quad P(M \mid N) = \frac{20}{100} = 0.20$$

so that

$$P(M \mid N) = P(M)$$

which we translate by saying that the probability that the selected person is a male does not depend on (is independent of) whether the person is neutral. In general, we state that *in the probability sense,*

If $P(A \mid B) = P(A)$, A and B are independent.
If $P(A \mid B) \neq P(A)$, A and B are dependent.

Exercise. See Table 12–1. Are the following independent or dependent? Why? *a) W* and *D. b) W* and *N.* Answer: *a)* $P(W| D) = {}^{150}\!/_{160}$ which does not equal $P(W) = 0.8$, so *W* and *D* are dependent. *b)* $P(W| N) = 0.80 = P(W)$, so *W* and *N* are independent.

In the foregoing definition of independence, we emphasized that independence as a mathematical concept is independence *in the probability sense.* We should note that events that we think or feel (intuitively) are independent or dependent may or may not prove to be so in the probability sense. Thus, in the poll of workers under discussion, we might feel intuitively that women would tend not to be in favor of changing from an 8-hour to a 10-hour day because the longer workday would interfere with the home responsibilities of some women. That is, intuitively we might feel that women would be more likely not to favor a longer day than workers in general and, therefore, that Disfavor and Women are dependent. The test in the probability sense may or may not bear out this intuitive feeling, and the test is:

If $P(D| W)= P(D)$ *D* and *W* are independent.
If $P(D| W)\neq P(D)$ *D* and *W* are dependent.

From Table 12–1 we find

$$P(D| W) = \frac{150}{400} = 0.375.$$

$$P(D) = \frac{160}{500} = 0.32$$

so *D* and *W* are dependent. Moreover, the last two numbers confirm the intuitive feeling that women are more likely to disfavor a longer workday than workers in general. By way of contrast, we might feel intuitively that men would be less likely to be neutral on the question than workers in general because men would prefer to work longer days in order to have a three-day weekend. Here intuition is again suggesting dependence. However, if we compute from Table 12–1

$$P(N| M) = \frac{20}{100} = 0.2$$

$$P(N) = \frac{100}{500} = 0.2,$$

we see that *N* and *M* are independent *in the probability sense,* and the data do not support intuition in this case.

Another matter worthy of special attention is the distinction between mutual exclusiveness and independence. The first point to note is that *if each of two events, A and B, has a nonzero probability and the events are mutually exclusive they are necessarily dependent;* that is, they cannot be independent. To see why this is so, suppose that

$$P(A) = 0.3 \qquad P(B) = 0.7.$$

Now, if A occurs, B *cannot* occur because A and B are mutually exclusive. Hence,

$$P(B \mid A) = 0,$$

but

$$P(B) = 0.7,$$

so

$$P(B \mid A) \neq P(B),$$

and the events are dependent. The question of independence for mutually exclusive events thus always has the answer that the events are dependent in the probability sense. On the other hand, *if A and B are not mutually exclusive, they may be independent or they may be dependent.* The last statement will be made clear in the next illustration.

Example. An urn (jar) contains 200 small glass balls. Each ball has a left and a right half of different colors. The left half may be R, W, or B (for Red, White, or Blue) and the right side may be G or Y (for Green or Yellow). A ball with red and yellow can be said to have some red (or some yellow). There are 80 balls with some red; 70 with white and green; 30 with white and yellow; 20 with blue and green; 140 with some green. Answer the following:

1. Are blue and green mutually exclusive? Why?
2. Are blue and green independent? Why?
3. Are white and yellow mutually exclusive? Why?
4. Are white and yellow independent? Why?
5. Are blue and yellow mutually exclusive? Why?
6. Are blue and yellow independent? Why?

The numbers given in this problem are summarized in the left table, and the table at the right provides the remaining numbers by using the given data and the requirements of row and column sums; that is, for example, the row sum for white must be 100, and to make the column sum for green be 140, the red and green element must be 50.

	Given data				*Completed table*		
	Green	Yellow	Totals		Green	Yellow	Totals
Red			80	Red	50	30	80
White	70	30		White	70	30	100
Blue	20			Blue	20	0	20
Totals	140		200	Totals	140	60	200

We turn now to questions 1–6 of the example statement, using the completed table at the right.

1. There are 20 balls that are both blue and green, so blue and green are not mutually exclusive; that is, $P(B \cap G) \neq 0$.
2. If we compute

$$P(B \mid G) = \frac{20}{140} = \frac{1}{7}; \; P(B) = \frac{20}{200} = \frac{1}{10}$$

we see that $P(B \mid G) \neq P(G)$, so blue and green are dependent, not independent. Or, we could have computed

$$P(G \mid B) = \frac{20}{20} = 1; \; P(G) = \frac{140}{200} = \frac{7}{10}$$

and $P(G \mid B) \neq P(G)$, showing again that blue and green are not independent.

3. There are 30 balls that are white and yellow, so white and yellow are not mutually exclusive; that is, $P(W \cap Y) \neq 0$.
4. If we compute

$$P(W \mid Y) = \frac{30}{60} = 0.50, \; P(W) = \frac{100}{200} = 0.50,$$

we note that $P(W \mid Y) = P(W)$, so white and yellow are independent. Or, we could have computed

$$P(Y \mid W) = \frac{30}{100} = 0.3; \; P(Y) = \frac{60}{200} = 0.3$$

and $P(Y \mid W) = P(Y)$ showing again that yellow and white are independent.

    Note: Observe in the answers to 1–4 that if two events are not mutually exclusive, they may be either dependent (as in 1 and 2) or independent (as in 3 and 4).

5. There is no ball with blue and yellow, so blue and yellow are mutually exclusive; that is, $P(B \cap Y) = 0$.

6. Mutually exclusive events with nonzero probabilities automatically are dependent, so blue and yellow are dependent. Numerically, the dependence is exhibited by showing that $P(B|Y) \neq P(B)$. These probabilities are

$$P(B|Y) = \frac{0}{60} = 0; \; P(B) = \frac{20}{200} = 0.10.$$

Summarizing the last example, we observe in the answer to 1 that if a ball has blue, it also can have green, so blue and green are not mutually exclusive; moreover, the two demonstrations of the answer to 2 show that *if we know* one color is green (blue), this knowledge does affect the probability that the ball also has blue (green) because green and blue are dependent. Then, in 3, we determine yellow and white also are not mutually exclusive, but 4 shows that *if we know* one color is yellow (white) this knowledge does not affect the probability that the other color is white (yellow) because yellow and white are independent. Finally, 5 shows blue and yellow are mutually exclusive, so of necessity they are dependent; that is, as 6 shows, *if we know* one color is blue (yellow) this does affect the probability that the other color is yellow (blue); that is, this latter probability must be zero.

Exercise. If we draw two cards from a deck and let A_1 and A_2 represent first card ace and second card ace, respectively, are A_1 and A_2 mutually exclusive? Independent? Explain. Answer: The events A_1 and A_2 (e.g., first card ace of spades, second card ace of clubs) both can occur, so $P(A_1 \cap A_2) \neq 0$ and the events are not mutually exclusive. However, the probability that the second card is an ace *does* depend upon whether the first card was an ace. That is, because there are four aces in the deck of 52 cards, the probability that the first card is an ace is 4/52. If we know the first card drawn was an ace, 3 aces remain in a deck of 51 cards and the probability of the second card being an ace, *given* that the first card is an ace, is 3/51.

Union of events. Refer to Table 12–1. The expression

$$M \cup F,$$

where $\cup$ is read as *union* and $M \cup F$ as M *union* F, means the set of individuals who are either men *or* in favor *or* both. The table shows there are 100 men and an additional uncounted 170 in favor, for a total of 270, so

$$P(M \cup F) = \frac{270}{500} = 0.54.$$

Another counting method we shall find useful soon is to add the number of men to the number in favor, then subtract the *doubly* counted number of men in favor. This yields $100 + 240 - 70$, so

$$P(M \cup F) = \frac{100 + 240 - 70}{500} = \frac{270}{500} = 0.54,$$

as before.

On the other hand, *F* and *D* are mutually exclusive (the intersection is empty), so

$$P(F \cup D) = \frac{240 + 160 - 0}{500} = 0.80.$$

Exercise. From Table 12—1, find: *a)* $P(M \cup W)$. *b)* $P(F \cup N)$. *c)* $P(W \cup D)$. *d)* $P(W \cup F)$. Answer: *a)* 1. *b)* 0.68. *c)* 0.82. *d)* 0.94.

Complementary events. Two events are said to be complementary if the sum of their probabilities is one. That is, letting E' (*E* prime) be the complement of *E*, then

$$P(E) + P(E') = 1$$

$$P(E') = 1 - P(E).$$

E' may be read as "not *E*." Similarly, if *R* means "rain," R' means "not rain" and if $P(R) = 0.40$, then

$$P(R') = 1 - P(R) = 1 - 0.40 = 0.60.$$

Again, if *W* means woman, W' is not woman (therefore man). In Table 12–1

$$P(W') = 1 - P(W) = 1 - 400/500 = 0.2.$$

Exercise. From Table 12–1, *a)* What would constitute F'? *b)* What is $P(F')$? Answer: *a)* F' (not favor) would include Disfavor and Neutral. *b)* $P(F') = 1 - P(F) = 1 - 240/500 = 0.52.$

For further practice with complements, note that $F \cup D$ includes all joint events containing an *F* or a *D* (or both). The only remaining event is not F and not D; that is, $F' \cap D'$, so $F' \cap D'$ is the complement of $F \cup D$ and we may write

$$P(F \cup D) = 1 - P(F' \cap D') \quad \text{or} \quad P(F' \cap D') = 1 - P(F \cup D).$$

Similarly,

$$P(M \cup F) = 1 - P(M' \cap F') \quad \text{or} \quad P(M' \cap F') = 1 - P(M \cup F).$$

Further practice with these symbols is provided in the following set of problems.

12.5 PROBLEM SET 12–1

1. The table shows, for example, that area A has 30 large stores and area C has 150 small stores.

	Store size ($ volume)		
Geographic area	Large L	Medium M	Small S
A	30	45	75
B	150	125	275
C	20	130	150

Find the following probabilities:

a) $P(M)$. b) $P(B)$. c) $P(M \cap S)$.
d) $P(B \cap M)$. e) $P(A \cap C)$. f) $P(A \cap L)$.
g) $P(L \cap A)$. h) $P(A \mid L)$. i) $P(P \mid A)$.
j) $P(A \mid M)$. k) $P(S \mid B)$. l) $P(L \cup M)$.
m) $P(B \cup S)$. n) $P(M \cup C)$. o) $P(B \cup C)$.
p) $P(A')$. q) $P(M')$. r) $P(L' \cap M')$.

What event, joint event, or union of events is the complement of

s) $M' \cap S'$? t) $B' \cap C'$? u) $A \cup C$?
v) $A \cup M$? w) $B' \cap M'$? x) $L' \cap C'$?

2. See table of Problem 1.
   a) Does $P(M \mid A) = P(M)$?
   b) What does the answer to (a) mean?
   c) Are A and L independent? Explain.
   d) Are B and L independent? Explain.
   e) What is $P(A \cap B)$? What does this mean?

3. a) Voters in an area are classified as Democrat, Independent, or Republican. If 48 percent are Democrats and 10 percent are Independents, what is the probability that a voter selected at random is a Republican?
   b) If the probability of rain tomorrow is 0.4, what is the probability that it will not rain?

4. An urn contains 600 glass balls. Each ball has a left half and a right half of different colors. The left may be G or Y (for Green or Yellow) and the right half may be R, W, or B (for Red, White, or Blue). The numbers of balls of various colors are shown in the table.

	R	*W*	*B*
G	30	42	138
Y	270	78	42

a) Are white and blue mutually exclusive? Why?
b) Are white and blue independent? Why?
c) Are yellow and white mutually exclusive? Why?
d) Are yellow and white independent? Why?
e) Are yellow and red mutually exclusive? Why?
f) Are yellow and red independent? Why?

5. An urn contains 400 small glass balls. Each ball has a left half and a right half of different colors. The left half may be R, W, or B (for Red, White, or Blue), and the right half may be Y, G, or T (for Yellow, Green or Tan). Each ball has R, W, or B on one side and Y, G, or T on the other, except that none of the balls has red on one side and green on the other. A ball that has, say, red and yellow, can be said to have some red (or some yellow). There are 200 balls having some red; 40 are red and yellow; 10 are white and yellow; 50 are blue and yellow; 50 have some green and 20 are white and green; 120 have some blue. Construct a tabular form having three rows labeled R, W, and B and three columns headed Y, G, and T. Place the given numbers in their proper position in the table, then fill in the remaining elements and totals.

A ball is selected at random. What is the probability that it has the following colors on it?

a) Some red.   *b)* Some green.   *c)* White or green.
d) Red and blue.   *e)* Red and green.   *f)* Both sides red.
g) Some yellow.   *h)* Red or green.   *i)* Blue and tan.

A ball (which we cannot see) has been selected. Answer the following according to the information provided.

j) If the ball drawn has some blue, what is the probability that the other color is green?
k) If the ball drawn has some yellow, what is the probability that the other color is red?
l) If the ball drawn is green, what is the probability that the other color is red?
Answer the following:
m) Are white and yellow mutually exclusive? Why?
n) Are white and yellow independent? Why?
o) Are white and tan mutually exclusive? Why?
p) Are white and tan independent? Why?

q) Are red and green mutually exclusive? Why?

r) Are red and green independent? Why?

6. In the table, *H, M, L* stand for "high absenteeism," "medium absenteeism," and "low absenteeism," respectively. *S* stands for "salaried" and *P* for "hourly paid." The table shows, for example, that there were 40 salaried workers who had high absentee records. Assigning probabilities as relative frequencies, show that method of pay and absenteeism (for these data) are independent in the probability sense.

	H	*M*	*L*
S	40	60	100
P	60	90	150

7. A plant has 300 workers, 200 of whom are females. Classified by *H, M, L* absenteeism, there are 30 workers in the *H* class and 30 workers in the *L* class. Assuming independence of sex and absenteeism, how many workers should there be in each of the six possible classes?

12.6 PROBABILITY RULES

Table 12–1 showed the numbers of people in each category, the marginal totals, and the grand total, 500. We now divide each number in Table 12–1 by 500 to yield the *probabilities* shown in Table 12–2.

TABLE 12–2

	Favor (F)	Disfavor (D)	Neutral (N)	Total
Men *(M)*	0.14	0.02	0.04	0.20
Women *(W)*	0.34	0.30	0.16	0.80
Total	0.48	0.32	0.20	1.00

Note: Intersections, such as $M \cap F$, are very often of interest, and for brevity we shall henceforth omit the intersection symbol and express a joint event such as $M \cap F$ as simply *MF*. We shall refer to *MF* as *M and F*, keeping in mind that the word *and* means *both M and F*. The union symbol, as in $M \cup F$ shall be retained, and we shall refer to $M \cup F$ as *M or F*, keeping in mind that this includes *M*, or *F*, or *MF*.

From Table 12–1 we computed

$$P(M \cup F) = \frac{100 + 240 - 70}{500}$$

$$= \frac{100}{500} + \frac{240}{500} - \frac{70}{500}$$
$$= 0.20 + 0.48 - 0.14 = 0.54.$$

The corresponding calculation can be made directly from Table 12–2 as

$$P(M \cup F) = P(M) + P(F) - P(MF)$$
$$= 0.20 + 0.48 - 0.14 = 0.54.$$

On the other hand, from Table 12–2,

$$P(F \cup D) = P(F) + P(D) - P(FD)$$
$$= 0.48 + 0.32 - 0$$
$$= 0.80$$

where, here, $P(FD) = 0$ because F and D are mutually exclusive. We have the following rule.

Addition rule.   The probability that A or B occurs is the probability of A plus the probability of B, minus the probability of A *and* B. Thus,

$$P(A \cup B) = P(A) + P(B) - P(AB).$$

If A and B are mutually exclusive, then

$$P(A \cup B) = P(A) + P(B).$$

Exercise.   From Table 12–2 find: *a)* $P(M \cup N)$. *b)* $P(M \cup W)$. *c)* $P(D \cup N)$. Answer: *a)* $0.20 + 0.20 - 0.04 = 0.36$. *b)* $0.20 + 0.80 - 0 = 1$. *c)* $0.32 + 0.20 - 0 = 0.52$.

As another example, suppose a political candidate runs for two offices, A and B. She assesses her probabilities of winning at 0.30 and 0.20 for A and B, respectively, and thinks she has an outside chance, probability 0.05, of winning both offices. The probability of winning A or B would then be

$$P(A \cup B) = P(A) + P(B) - P(AB)$$
$$= 0.30 + 0.20 - 0.05$$
$$= 0.45.$$

Exercise.   An investor thinks the probability that stock P will rise tomorrow is 0.70, and that Q will rise is 0.80. He thinks there is a 50–50 chance that both will rise. What is his probability that P or Q will rise? Answer: $0.70 + 0.80 - 0.50 = 1.00$.

It is worth noting in the last exercise that the investor's subjective probability assignments lead logically to the conclusion that he is certain (probability 1) that P or Q will rise. Note also that if the investor had assessed the probability that both will rise at 0.4 (rather than 0.5), then $0.7 + 0.8 - 0.4 = 1.1$. This would lead the investor to reassess probabilities because the probability of an event cannot exceed 1. The last sentence illustrates a fundamental reason for understanding probability rules; namely, to monitor probability assessments and insure the internal logical consistency of such assessments.

Returning to Table 12–1, recall that

$$P(F \mid M) = \frac{70}{100} = 0.70.$$

If we express the latter in the equivalent manner,

$$P(F \mid M) = \frac{\dfrac{70}{500}}{\dfrac{100}{500}} = \frac{0.14}{0.20} = 0.70$$

and relate the 0.14 and 0.20 to Table 12–2, we observe that

$$P(MF) = 0.14 \quad \text{and} \quad P(M) = 0.20.$$

It follows that

$$P(F \mid M) = \frac{P(FM)}{P(M)}.$$

Thus, the probability of F given M is the probability of the joint event, FM, divided by the probability of M. Correspondingly,

$$P(M \mid F) = \frac{P(MF)}{P(F)}$$

which, from Table 12–2, is

$$P(M \mid F) = \frac{0.14}{0.48} = 0.292.$$

Exercise. Complete the following: a) $P(N \mid W) =$ ___ . b) $P(W \mid N) =$ ___ . Compute a) and b) from Table 12–2. Answer: a) $P(N \mid W) = P(NW)/P(W)$. b) $P(W \mid N) = P(WN)/P(N)$. The probabilities are a) 0.20 and b) 0.80.

Conditional probability rule. The probability that B will occur, given that A has occurred, is the probability of AB divided by the probability of A. Thus,

$$P(B|A) = \frac{P(AB)}{P(A)}.$$

The rule just stated can be solved for $P(AB)$ after multiplying both sides by $P(A)$. The result is:

Joint probability rule.   The probability of AB is the probability of A times the probability of B, given A. Thus,

$$P(AB) = P(A)P(B|A).$$

As an example, suppose a box contains two defective and three good items, $DDGG$. Two items are to be selected and we seek the probability, $P(GD)$, that the first is good and the second defective.

$$P(GD) = P(G)P(D|G).$$

The probability, $P(G)$, that the first is good is ⅗. We reason that if the first selected is G, the four remaining are $DDGG$, so the probability, $P(D \mid G)$, that the second is defective, given the first is good, is ¾. Hence,

$$P(GD) = \frac{3}{5}\left(\frac{2}{4}\right) = 0.30.$$

Observe that *order* is significant in the context of the last example. Thus, GD means *first* good, *second* defective. Similarly, GGD means first good, second good, third defective. Using a continuation of the joint probability rule,

$$P(GGD) = P(G)P(G|G)P(D|GG),$$

which means that $P(GGD)$ is the probability that the first is good times the probability that the second is good, given that the first is good, times the probability that the third is defective, given that the first two are good. Arithmetically,

$$P(GGD) = \frac{3}{5} \cdot \frac{2}{4} \cdot \frac{2}{3} = 0.20.$$

Exercise.   For the foregoing example, compute: a) $P(DD)$. b) $P(DDG)$. c) $P(DDD)$. Answer: a) 0.1. b) 0.1. c) 0.

Recalling that independence means

$$P(B|A) = P(B),$$

we may substitute $P(B)$ for $P(B \mid A)$ in the joint probability rule and obtain $P(AB) = P(A)P(B)$.

Joint probability rule, independent events. If A and B are independent in the probability sense, then the probability of the joint event AB is the probability of A times the probability of B. Thus,

$$P(AB) = P(A)P(B).$$

We may use this rule as a test for independence. For example, in Table 12–2, we note that

$$P(M) = 0.20, \quad P(N) = 0.20, \quad \text{and} \quad P(MN) = 0.04.$$

Hence,

$$P(MN) = P(M)P(N) = 0.04,$$

so M and N are independent.

> ***Exercise.*** Refer to Table 12–2. Are W and D independent? Explain. Answer: $P(W) = 0.80$, $P(D) = 0.32$, whereas $P(WD) = 0.30$. $P(WD) \neq P(W)P(D)$, so W and D are not independent.

The *assumption* of independence underlies a number of probability applications. For example, suppose that, by test, a fire alarm functions successfully, S, in the presence of fire 90 percent of the time and fails, F, 10 percent of the time. For added protection, a store installs two alarms that operate independently. If there is a fire, what is the probability that both will fail? Here $P(F) = 0.1$ and because of independence,

$$P(FF) = P(F)P(F) = 0.1(0.1) = 0.01.$$

> ***Exercise.*** For the foregoing example, find the probability that: *a)* Both will function successfully. *b)* The first will function and the second fail. Answer: *a)* $P(SS) = 0.9(0.9) = 0.81$. *b)* $(0.9)(0.1) = 0.09$.

12.7 EXPERIMENT, EVENT, SAMPLE SPACE

The word *experiment* usually is associated with the natural sciences. However, it is also used in a very broad sense in probability to specify what we plan to do. For example, one experiment may be to select 10 items from a continuous production line, examine them, and record the number of defectives. Another experiment would be to select a sample of 100 families and record the date at which they last purchased an automobile.

An outcome of an experiment is called an *event,* and a totality of all possible mutually exclusive events is called a *sample space* for the experiment. For example, consider the experiment of testing two fire alarms by subjecting them to fire and recording whether they worked or failed. *One* way to construct the sample space is to use ordered pairs such as *SF,* which means the first was successful and the second failed. We have

Events, sample space 1
SS
SF
FS
FF

.

Observe that the occurrence of one of the events (say *SS*) excludes the possibility of occurrence of any of the others, so the events are mutually exclusive. Moreover, every possible event is listed, so this is a proper sample space.

Often, more than one sample space can be constructed for an experiment. In the case at hand, we could describe all possible events by listing the three mutually exclusive events 0 fail, 1 fails, 2 fail. Thus,

Events, sample space 2 (Number failing)
0
1
2

.

Returning to Sample Space 1 and recalling the assumption of independence with $P(S) = 0.9$ and $P(F) = 0.1$, we can compute $P(SS) = (0.9)(0.9) = 0.81$, and so on leading to the following listing of the sample space and probabilities.

Sample space 1, with probabilities	
Event	Probability
SS	0.81
SF	0.09
FS	0.09
FF	0.01
	1.00

.

The sum of the probabilities for the entire sample space (which includes all possible mutually exclusive events) must be 1, as shown.

Turning now to probabilities for Sample Space 2, the event 0 fail is the event *SS*, so $P(0) = 0.81$. Similarly, the event 2 fail is the event *FF*, so $P(2) = 0.01$. However, the event 1 fails occurs if *FS* or *SF* occurs. Keeping in mind that the latter two are mutually exclusive, we have

$$P(FS \cup SF) = P(FS) + P(SF) = 0.09 + 0.09 = 0.18.$$

We thus have:

Sample space 2, with probabilities	
Number failing	*Probability*
0	0.81
1	0.18
2	0.01
	1.00

A special way of constructing a sample space is to place each event in one or the other of two categories. In the last table, for example, we could say *E* is the event zero fail and *E'* is the event one or more fail. We have

$$P(E) = 0.81$$
$$P(E') = 0.19$$
$$P(E) + P(E') = 1.00.$$

According to our earlier definition, *E'* is the complement of *E*. Clearly, complementary events constitute a sample space because they are mutually exclusive and the sum of their probabilities is one. The rule for complementary events,

$$P(E') = 1 - P(E),$$

finds frequent application. Returning to the fire alarm example, the most important consideration is the probability that at least one of the two alarms functions in case of fire. This could be computed as

$$P(SS) + P(SF) + P(FS),$$

but the computation is simplified if we note that the event *FF* is the complement of the three just mentioned. Hence, the desired probability is

$$1 - P(FF) = 1 - 0.01 = 0.99.$$

Exercise. If the store had three independent alarms with $P(S) = 0.9$ and $P(F) = 0.1$, what is the probability that at least one functions successfully? Answer: $1 - P(0) = 1 - P(FFF) = 1 - (0.1)(0.1)(0.1) = 0.999$.

Note on terminology. Readers are sometimes puzzled by the term *at least*, and similar terms. Some equivalences are shown in the following to help make clear the meaning of terms that appear quite frequently. In reading the equivalences, assume x can be 0 or any positive whole number; that is, 0, 1, 2, 3, · · · and so on.

Equivalences

a) x is at least 5; x is 5 or more; $x \geq 5$; $x > 4$.
b) x is at most 4; x is 4 or less; $x \leq 4$; $x < 5$.
c) x is 3; x is exactly 3; $x = 3$.
d) x is less than 5; x is 4 or less; $x < 5$; $x \leq 4$.
e) x exceeds 6; x is greater than 6; $x > 6$; $x \geq 7$.

12.8 PROBLEM SET 12–2

1.   Convert the following to a probability table.

| | Store size ($ Volume) | | |
Geographic area	Large L	Medium M	Small S
A	30	45	75
B	150	125	275
C	20	130	150

2.   *a)*   Complete the following probability table. (Note C' and F' are complements of C and F, respectively.)

	F	F'	Total
C	0.24		0.60
C'			
Total	0.40		

Find the following probabilities:

b) $P(CF')$. c) $P(C \cup F)$. d) $P(C \cap C')$.
e) $P(F \mid C)$. f) $P(C \mid F)$. g) $P(F \mid F')$.
h) Are C and F independent? Why?
i) Why is $P(F \cup F') = 1$?

3. A political candidate runs for two offices, A and B. She assesses her probabilities of winning at 0.6 and 0.2, respectively, and thinks she has only a probability of 0.01 of winning both.
   a) What is the probability that she wins one office *or* the other?
   b) If the candidate sets $P(A) = 0.7$, $P(B) = 0.4$, and $P(AB) = 0.01$, what advice should she be given? Why?

4. A box contains seven good and three defective items. If a sample of two items is selected, what is the probability that:
   a) Both will be good?
   b) Both will be defective?
   c) *Exactly one* will be defective. (Note: Two events are involved.)
   d) If a sample of 3 items is selected, what is the probability that *at least one* will be defective? (Hint: What is the complement of at least one defective?)

5. A weatherman states that if it is colder tomorrow, the probability of snow is 0.7. He also states the probability that it will be colder is 0.5. What is the probability it will be colder *and* snow tomorrow? Why?

6. The probability that a student will graduate with honors and get a good job is 0.09, whereas the probability that a student will graduate with honors is 0.1. What is the probability that a student will get a good job if he graduates with honors?

7. In the manufacture of an expensive product, each item is inspected independently by two women. The probability that either woman will correctly classify an item as defective is 0.95. If a defective item is inspected, what is the probability that:
   a) Both will correctly classify the item?
   b) Neither will correctly classify the item?
   c) At least one will correctly classify the item?
   d) *Exactly* one will correctly classify the item? (Two events.)
   e) The second will correctly classify the item if the first did not? Why?

8. Three coins are to be tossed.
   a) Write the sample space in terms of *HHT*, and so on, where *HHT* means first coin heads, second heads, and third tails. (There are 8 events.)
   b) Write the sample space in terms of the number of heads.
   c) Construct a probability table for *a)*.
   d) Construct a probability table for *b)*.

   What is the probability of:
   e) At least one head? f) At least two heads?
   g) Exactly one head? h) Exactly two heads?

9. An election results in a tie among two women (W_1 and W_2) and one man *(M)*. It is decided to select two people at random from the three and appoint the first selected as president and the second selected as vice president.

   a) Construct a sample space using, for example, W_1M to mean woman one is president, and the man is vice president. Assign probabilities to each event in the sample space.

   Using the results of *a)*, what is the probability that:

   b) The man will be president?
   c) The man will be either president or vice president?
   d) What is the probability women will occupy both offices?

10. See Problem 9, but do not refer to the sample space.
   a) Using *W* to represent either woman, write the symbol for the joint event which specifies the man is vice president.
   b) Compute the probability for the event in *a)* using the conditional probability rule.
   c) Write the symbols for the event that a man is selected for neither office.
   d) Compute the probability for the event in *c)* using the conditional probability rule.

12.9 PRACTICE WITH PROBABILITY RULES

In this section we present a series of examples to provide further practice with probability rules.

Example 1. A job applicant assigns probabilities as follows: The probability, *P(A)*, of being offered a job at company *A* is 0.6; the probability, *P(B)*, of being offered a job at company *B* is 0.5; the probability of being offered a job at both companies is 0.4. What, consequently, is the probability of being offered a job with at least one of the two companies?

Here we shall apply a tabular approach to the problem. The given probabilities are entered in the table at the left, and the entries in the table at the right follow as logical consequences.

	Given:					*Completed Table:*		
	B	*B'*	*Total*			*B*	*B'*	*Total*
A	0.4	..	0.6		*A*	0.4	0.2	0.6
A'	..	..	..		*A'*	0.1	0.3	0.4
Total	0.5	..	..		Total	0.5	0.5	1.0

The event in question consists of the mutually exclusive events *AB'*; *A'B;* and *AB*, the sum of whose probabilities is 0.7. The complement of the event in question is $A'B'$, so that the desired probability could have been found as $1 - 0.3 = 0.7$.

When only two events are at hand, problems often can be solved quite easily by construction of a two-by-two table as in Example 1.

Exercise. The probability of good weather, G, is 0.6 and the probability of accident, A, is 0.014. The probability of the joint event, accident and good weather, is 0.006. Find the probability of accident if the weather is not good. (Hint: Make a two-by-two table.) Answer: 0.02.

The completion of a two-by-two table may require application of the probability rule for joint events, as shown next.

Example 2. Ten percent of the workers in an area are accountants; 60 percent of accountants read the area paper Goodnews, and 30 percent of those who are not accountants read Goodnews. If a worker is selected at random, what is the probability that the worker reads Goodnews?

If we make a tabular format with A meaning accountant and G meaning a reader of Goodnews, the only direct entry that can be made from the given information is $P(A) = 0.10$, which follows from the fact that 10 percent are accountants. From the latter, we find $P(A') = 1 - 0.10 = 0.90$ and start the table as shown next.

	G	G'	Total
A............			0.10
A'...........			0.90
Total	——	——	1.00

In order to complete the table we need joint probabilities. *The key matter to think about is the meaning of a statement such as* "60 percent of the accountants read Goodnews." The proper interpretation starts by observing that the statement is limited to accountants; that is, *accountant is given,* and the 60 percent is the probability of reading Goodnews given accountant,

$$P(G \mid A) = 0.60.$$

By similar reasoning, the meaning of "30 percent of those who are not accountants read Goodnews" is

$$P(G \mid A') = 0.30.$$

We can now determine the probabilities for the joint events AG and $A'G$. These are

$$P(AG) = P(A)P(G \mid A) = 0.10(0.6) = 0.06$$
$$P(A'G) = P(A')P(G \mid A') = 0.90(0.30) = 0.27.$$

Completing the table, we have

	G	G'	Total
A...........	0.06	0.04	0.10
A'	0.27	0.63	0.90
Total	0.33	0.67	1.00

and the answer to the question is the probability of G, which is

$$P(G) = 0.33.$$

Tree diagrams. A tree diagram such as that used in the next example often is helpful in solving probability problems, especially when more than two events are involved.

Example 3. Solve the problem of Example 2 by a tree diagram.

We start by showing an initial *fork* (the oblong at the left of Figure 12–1A) and drawing two *branches* from the fork. One is A, for "accoun-

FIGURE 12-1A

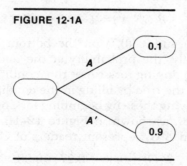

tant," and the other A', for "not accountant." The respective probabilities for A and A' (0.1 and 0.9) are shown at the right ends of the two branches that end at the righthand forks. Next, Figure 12–1B shows two branches from each of the righthand forks of Figure 12–1A. Each of the new pair of branches contains G and G', for those who are and are not readers of Goodnews. The new branch probabilities, indicated in parentheses, are *conditional* probabilities. Thus, $G(0.6)$ on the top right branch of Figure 12–1B is the probability of G *given* A, as stated in Example 2. Inasmuch as

692

FIGURE 12-1B

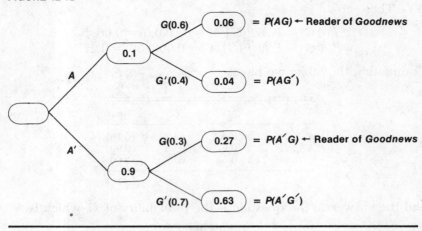

$$P(G \mid A) + P(G' \mid A) = 1$$

we have the conditional probability

$$P(G' \mid A) = 1 - P(G \mid A) = 1 - 0.6 = 0.4,$$

which is shown as $G'(0.4)$ on Figure 12–1B. Similarly, from Example 2, $P(G \mid A') = 0.3$, so

$$P(G' \mid A') = 1 - 0.3 = 0.7,$$

as shown by $G(0.3)$ and $G'(0.7)$ on the bottom pair of branches of Figure 12–1B. Finally, the probability at the end of a branch is the probability at the beginning fork times the (conditional) branch probability. We complete the tree by filling in the conditional branch probabilities, then multiplying these by beginning fork probabilities to obtain the end probabilities, as shown in Figure 12–1B. The first and third joint events, AG and $A'G$, represent readers of Goodnews, so

$$P(G) = 0.06 + 0.27 = 0.33,$$

as before.

Defining a *fork* to mean the point from which *branches* emerge, we may state the following:

Rules for trees

1. The sum of the probabilities on all branches from a fork is one.
2. The probability at the end of a branch is the probability at the beginning fork multiplied by the (conditional) branch probability.

3. The sum of the probabilities at the ends of the branches from a fork equals the probability at the fork.
4. The probability at a fork is the joint probability of all events leading to that fork.
5. The probability of an event X is the sum of the probabilities of all joint events in which X appears.

Example 4. To advertise the opening of a new store, management plans to hold one large outdoor display of fireworks on Thursday, Friday, or Saturday of the opening week if it does not rain. A meteorologist states that the probability of rain on Thursday is 0.6, but if it rains on Thursday, the probability of rain on Friday is 0.7, and if it rains on both Thursday and Friday, the probability of rain on Saturday is 0.10. What is the probability that the firework display will occur?

The tree in Figure 12–2 contains the given information, using R for rain and R' for not rain. We want only events containing R' (not rain) and, furthermore, if R' appears on a branch the fireworks display is held on the corresponding day and that part of the tree ceases at

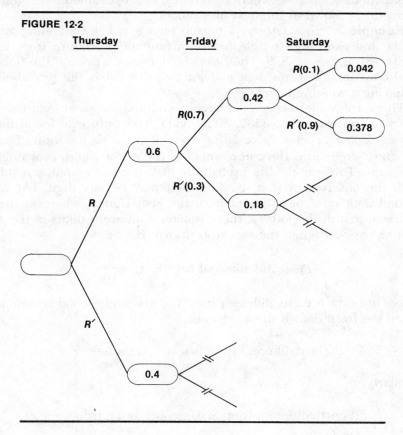

FIGURE 12-2

the end of that branch. Figure 12–2 presents the relevant branches of the tree and the given probabilities.

Exercise. Verify the probabilities in Figure 12–2. What is the probability that the fireworks display will be held? Answer: $0.378 + 0.18 + 0.4 = 0.958$.

Now that we have done the problem the "hard" way, we can see that the only joint event that would result in the fireworks display *not* being held is rain on all three days. Hence, the top path, *RRR*, with probability 0.042, is the *complement* of the desired event, so the answer to our problem is

$$1 - 0.042 = 0.958,$$

as shown in the exercise.

Situations often arise where events are, or are assumed to be, independent, as shown in the next illustration.

Example 5. In a lottery, a person buys a ticket containing four digits, and each digit is selected by a random procedure from the digits $0, 1, 2, \cdots, 8, 9$. The ticket holder wins a prize if the ticket held contains the same digit at least twice. What is the probability of having a winning ticket?

There are numerous ways of having a number occur at least twice. We have, as examples, 3537, 3339, 3333. The *only* way for a digit not to appear at least twice is for every digit to be different. Thus, *all digits different* is the complement of the event whose probability is sought. To compute the probability of this complement, we start with the observation that the first digit may be any digit, but the second digit must be different from the first. Hence, whatever digit is drawn first in the lottery, there remain 9 different digits in the 10 that can appear when the second is drawn. Hence,

$$P(\text{Second different from first}) = \frac{9}{10}.$$

There are now 8 digits different from the first and second remaining when the third digit is drawn. Hence,

$$P(\text{Third different from first and second}) = \frac{8}{10}.$$

Similarly,

$$P(\text{Fourth different from first, second, and third}) = \frac{7}{10}.$$

Because the outcome of the drawing of a digit does not depend on what happened in previous drawings, we have independent events, and the probability that all three occur is the product of the three probabilities.

$$P(\text{All digits different}) = \frac{9}{10} \cdot \frac{8}{10} \cdot \frac{7}{10} = \frac{504}{1000} = 0.504.$$

Finally,

$$P(\text{At least two digits same}) = 1 - P(\text{All digits different})$$
$$= 1 - 0.504 = 0.496.$$

12.10 PROBLEM SET 12–3

1. A salesman has two prospective customers, A and B, to call on one day. He assigns the probability of making a sale to A at 0.4, of making a sale to B at 0.3, and the probability of making a sale to both at 0.1. What is the probability of making a sale to at least one of the customers?

2. In order to insure that weather will not prevent making an air trip, an executive makes plane reservations for two successive days and assesses the probability of being able to fly either day at 0.95. What is the probability that the executive will be able to make the trip? (Assume independence.)

3. If $P(X \mid Y) = 0.7$, what is $P(X' \mid Y)$?

4. Given $P(X) = 0.6$, $P(Y) = 0.4$, and $P(XY) = 0.1$. Find: $P(X \mid Y)$ and $P(Y \mid X)$.

5. If $P(Y \mid X) = 0.4$, $P(X \mid Y) = 0.5$, and $P(XY) = 0.2$, find $P(X)$ and $P(Y)$.

6. If $P(Y \mid X) = 0.72$, $P(XY) = 0.18$, and $P(Y') = 0.4$, find $P(X \cup Y)$.

7. A candidate runs for two political offices, A and B. He assigns 0.4 as the probability of being elected to both, 0.7 as the probability of being elected to A if he is elected to B, and 0.8 as the probability of being elected to B if he is elected to A.

   a) What is the probability of being elected to A?
   b) What is the probability of being elected to B?
   c) What is the joint probability of being elected to neither?
   d) What is the probability of being elected to at least one of the offices?

8. The probability of snow is 0.4, of colder 0.5; and the conditional probability of snow if colder is 0.7. Find the probability of:

   a) Colder and snow.
   b) No snow.
   c) Either colder or snow.
   d) Neither colder nor snow.

9. Job candidates are screened by means of a preliminary interview. The probability is 0.6 that a screened candidate will be a good worker. Screened candidates are given a test. If the candidate is one who will prove to be a good worker, the probability of his passing the test is 0.8. If the candidate is one who will prove to be a poor worker, the probability of his passing the test is 0.4.

696

a)   What is the probability that a screened candidate will be a good worker and pass the test?

b)   What is the probability that a screened candidate will not be a good worker and will pass the test?

10.   Fran plans to offer a new product for sale and, in assessing the chances that the product will be successful, Fran has to take her competitor, Judy, into account because Judy may offer a competing new product for sale. Fran thinks that the chance of Judy competing is 0.4. Fran assesses the probability that she will be successful to be 0.85 if Judy does not compete, but only 0.25 if Judy does compete. Compute Fran's chance of being successful.

11.   The first time an insurance salesman calls on a new client, he has a probability of 0.1 of selling a policy. *If* he does not sell the policy on the first call, he makes a second call and has a probability of 0.3 of making the sale. *If* he does not make the sale on either of the first two calls, he makes one final call and has a probability of 0.05 of making the sale. What is the probability that the insurance man will sell the policy to a client?

12.   A company assesses the probability that its product will fail during the first month after sale as 0.01. The probability it will fail during the next 11 months if it did not fail during the first month is 0.001. The company guarantees the product for the first year. What is the probability that the product will fail in the first year?

13.   See Problem 12. Suppose the probability of failure during the first month is 0.05, the probability of failure during the next 5 months (if the product did not fail in the first month) is 0.02, and the probability of failure for the remainder of the year if failure did not occur during the first noted time intervals is 0.01. What is the probability of failure during the first year?

14.   If snow, colder are the events with $P(S) = 0.8$, $P(C) = 0.6$, and $P(S'C) = 0.1$:

a)   Show that the events are not independent in the probability sense.

b)   What probability would $(S'C)$ have if the events were to be independent in the probability sense?

15.   The probability that machine A will break down on a particular day is

$$P(A) = 1/50.$$

Similarly, for machine B:

$$P(B) = 1/80.$$

Assuming independence, on a particular day:
a)   What is the probability that both will break down?
b)   What is the probability that neither will break down?
c)   What is the probability that one *or* the other will break down?
d)   What is the probability that exactly one machine will break down?

16.   A pair of dice is rolled, and a coin is tossed. Assuming independence, what is the probability of

   a) Heads on the coin and a seven on the dice?
   b) Heads on the coin or a seven on the dice?
   c) Heads on the coin and an even number on the dice?
   d) Heads on the coin and a number greater than eight on the dice?

17. Five parts go into the assembly of item *X*. The assembly is defective if any one of the parts is defective, and each part has a probability of 0.03 of being defective. Assuming independence, what is the probability that an assembly is defective?

18. Suppose an arena has two events, hockey and basketball, scheduled on Thursday and another two, hockey and track, on Friday. If two are selected to be played and two cancelled, calculate by probability rules the probability that there will be a hockey game on Thursday or on Friday.

19. Company *A* plans to bid on a contract. It does not know whether or not a competitor, Company *B*, will bid, but assesses the probability that it will bid at 0.6. *A* judges that it has a probability of 0.8 of winning if *B* does not bid, but only a probability of 0.4 if *B* does bid. What is the probability that *A* wins?

20. An arena has scheduled two events on Thursday and three on Friday of a particular week. To obtain time to carry out repairs, two of the events, selected at random, are to be postponed. Find the probability that there will be an event on Thursday and an event on Friday.

21. Five people at a party were born in the month of December. What is the probability that at least two of the five have the same birthday?

22. (For those with hand calculators.) A mathematics class has 30 students. What is the probability that at least two have the same birthday? (The same day of the same month.) Assume independence and assume a year has 365 days.

12.11 BAYES' RULE

As a result of past hiring procedure, a company finds that 60% of its employees are good workers, *G*, and 40% are poor workers, which we shall designate by the complement, *G'*. The woman in charge of hiring believes that the proportion of good workers can be increased by designing a test to be administered to job applicants, and hiring only those who pass, *P*, the test. A consulting firm supplies the test and offers to administer it for a fee to applicants. Because of the cost, it is decided to determine first how well the test discriminates between good and poor workers by trying it on current employees. It is found that 80% of the good workers and 40% of the poor workers pass the test. It is important to understand that these last two numbers are *conditional* probabilities because the first applies only to good workers and the second applies to poor workers. That is, *if* a worker is a good worker, the probability of passing is 0.80, so

$$P(P \mid G) = 0.80.$$

Similarly,

$$P(P \mid G') = 0.40.$$

Note carefully that in general

$$P(P \mid G) \neq 1 - P(P \mid G')$$

because there is no *necessary* relationship between a good worker's passing the test and a poor worker's passing the test.[3]

At first glance it may appear that the test functions well because it is twice as likely (0.80 vs 0.40) that an employee will pass if he is a good worker as it is that he will pass if he is a poor worker. The important point to note, however, is that the real question at hand is whether the test should be used in selecting employees from job applicants. Thus, the issue is not whether an employee will pass if he is a good worker, $P(P \mid G)$, but rather $P(G \mid P)$, which is the probability that an employee will be a good worker if he passes the test. Thus, we know $P(P \mid G)$ and we seek the *inverse* probability, $P(G \mid P)$. We have given:

$$P(G) = 0.60 \qquad P(G') = 1 - P(G) = 0.40.$$
$$P(P \mid G) = 0.80 \qquad P(P \mid G') = 0.40.$$

From the foregoing, we can compute:

$$P(PG) = P(G)P(P \mid G) = (0.60)(0.80) = 0.48.$$
$$P(PG') = P(G')P(P \mid G') = (0.40)(0.40) = 0.16.$$

We now have

	P	P'	Total
G	0.48	. .	0.6
G'	0.16	. .	. .
Total	0.64	. .	. .

It follows from the table that

$$P(G \mid P) = \frac{0.48}{0.64} = 0.75.$$

To see the significance of the last result, recall that 60% of current employees are good workers. This means that past or prior employment

---

[3] Similarly, the probability that it will snow tomorrow if it is colder, $P(S \mid C)$ tells us nothing about $P(S \mid C')$, the probability that it will snow if it does not get colder. However, if the *given* is the *same* as in $P(S \mid C)$ and $P(S' \mid C)$, then $P(S' \mid C) = 1 - P(S \mid C)$, which means simply that if it is colder it *must* snow or not snow, so $P(S \mid C) + P(S \mid C') = 1$.

procedures, without the test, had a probability of $P(G) = 0.6$, which we shall call the *prior* probability, of selecting a good worker. If, now, we change the selection procedure and employ only those who pass the test, we *revise* the probability of hiring a good worker to $P(G|P) = 0.75$, and this revised probability is called the *posterior* probability. Management must decide whether the additional information provided by a test which results in an increase in the probability of hiring good workers from 0.60 to 0.75, is worth the cost of administering the test.

To develop the symbolic procedure (Bayes' Rule) for determining $P(G|P)$ from $P(P|G)$, refer to the foregoing calculation of $P(G|P) = 0.75$,

$$P(G|P) = 0.75 = \frac{0.48}{0.64}.$$

From the foregoing table, we see that the number 0.48 at the right is

$$P(PG) = 0.48$$

and the number 0.64 is

$$P(PG) + P(PG') = 0.48 + 0.16.$$

Hence,

$$P(G|P) = 0.75 = \frac{0.48}{0.48 + 0.16}$$

$$P(G|P) = \frac{P(PG)}{P(PG) + P(PG')}. \tag{1}$$

In the last, at the right, we have probabilities for the joint events PG and PG'. The rule for joint probabilities allows us to write

$$P(PG) = P(P|G)P(G)$$
$$P(PG') = P(P|G')P(G').$$

Substituting the last into (1) yields:

Bayes' Rule: $P(G|P) = \dfrac{P(P|G)P(G)}{P(P|G)P(G) + P(P|G')P(G')}.$

The rule can be remembered easily if we note that in the numerator on the right we start with $P(P|G)$ which is simply the left side, $P(G|P)$, inverted. The numerator is completed in the manner of any joint probability by multiplying by $P(G)$ to give $P(P|G)P(G)$. The first term in the denominator repeats the numerator, and the second term follows by replacing the G in the first term by G'. As another example,

700

$$P(B \mid A) = \frac{P(A \mid B)P(B)}{P(A \mid B)P(B) + P(A \mid B')P(B')}.$$

Exercise. Write Bayes' Rule for $P(X \mid Y)$. Answer: $P(X \mid Y) = [P(Y \mid X)P(X)]/[P(Y \mid X)P(X) + P(Y \mid X')P(X')]$.

To help fix in mind the inverse probability concept expressed by Bayes' Rule, we shall review the last example by constructing tree diagrams. Recall that the given information was

$$P(G) = 0.60 \quad \text{so} \quad P(G') = 0.40.$$
$$P(P \mid G) = 0.80 \quad \text{and} \quad P(P \mid G') = 0.40.$$

From the foregoing, we compute

$$P(GP) = P(G)P(P \mid G) = 0.6(0.8) = 0.48.$$
$$P(G'P) = P(G')P(P \mid G') = 0.4(0.4) = 0.16.$$

Figure 12–3A shows the probabilities now at hand. Note that this figure has *good* as given and shows the branch probability of passing the

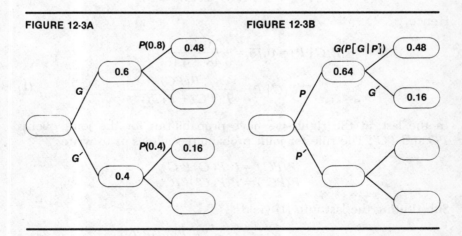

FIGURE 12-3A   **FIGURE 12-3B**

test, if good, is 0.8. Figure 12–3B is what we want because it shows *pass* as given, and the branch (conditional) probability $P(G \mid P)$ is the desired probability of good, given pass. To determine this, we first observe that the probabilities for the joint events GP and $G'P$ on Figure 12–3A are the same as the probabilities for PG and PG', respectively, on Figure 12–3B, as shown. Moreover, the sum of these branch probabilities is the fork probability, 0.64, shown on Figure 12–3B, and, finally, following a tree rule,

$$0.64P(G\,|\,P) = 0.48.$$

From the last we have

$$P(G\,|\,P) = \frac{0.48}{0.64} = 0.75,$$

as before.

The tree presentation just given serves well to show the inverse probability concept inherent in Bayes' Rule. For problem solving, one may use the symbolic formula, a table, or the tree approach.

We now offer a second application of Bayes' Rule.

Example. Suppose the probability that a person has disease X is

$$P(X) = 0.09.$$

The probability that medical examination will indicate the disease if a person has it is

$$P(I\,|\,X) = 0.06.$$

The probability that medical examination will indicate the disease if a person does not have it is

$$P(I\,|\,X') = 0.05.$$

What is the probability that a person has the disease if medical examination so indicates?

By formula:

$$\begin{aligned}
P(X\,|\,I) &= \frac{P(I\,|\,X)P(X)}{P(I\,|\,X)P(X) + P(I\,|\,X')P(X')} \\
&= \frac{(0.6)(0.09)}{(0.6)(0.09) + (0.05)(0.91)} \\
&= \frac{(0.054)}{(0.0995)} = 0.543.
\end{aligned}$$

The same result could be obtained by applying the tabular analysis used earlier in this section. We see that the prior probability of the person having the disease, 0.09, has been revised to the posterior probability 0.543 as a consequence of the additional information that medical examination indicated the disease.

12.12 PROBLEM SET 12–4

1. Supervisors rate 87.5 percent of the workers as good workers. On a work-aptitude test, 64 percent of the good workers and 24 percent of the other (not good) workers obtained passing grades. If a worker passes the test, what is the probability that this is a good worker?

2. The probability that a person has disease X is 0.05. The probability that a medical test will indicate the disease is present is 0.80 if the person has the disease and 0.02 if the person does not have the disease. What is the probability that a person has the disease if the medical test so indicates?

3. A fellow notes that his girl friend is happy on 60 percent of his visits to her home and that 40 percent of the times when she is happy she makes a drink for him. She makes drinks on 10 percent of the visits when she is not happy. If the fellow arrives on a visit and finds his friend making drinks, what is the probability that she is happy?

4. The probability that a customer will be a bad debt is 0.01. The probability that he will make a large down payment if he is a bad debt is 0.20, and the probability that he will make a large down payment if he is not a bad debt is 0.60.

   a) Suppose that a customer makes a large down payment. Find the posterior probability that he will be a bad debt.

   b) What is the probability that a customer who does not make a large down payment will be a bad debt?

5. The probability that a machine is running properly is 0.95. From time to time, samples of output are selected and measured, and the sample average is computed. If the machine is running properly, the probability that the sample average will be in a certain range is 0.9. If the machine is not running properly, the probability that the sample average will be in this range is 0.04.

   a) A sample is selected, and its average is in the range. What is the probability that the machine is running correctly?

   b) What is the probability that the machine is running correctly if the sample average is not in the range?

12.13 THE BINOMIAL PROBABILITY DISTRIBUTION

Let us assume that a basketball player is successful (hits) on 40% of his free-throw trials. Letting H represent *hit* and M represent *miss*, we assign

$$P(\text{Hit}) = P(H) = p = 0.40$$
$$P(\text{Miss}) = P(M) = q = 1 - p = 0.60.$$

We assume further that the outcome of one trial is *independent* of the outcomes of other trials and that p and q are the same on every trial.[4] If we consider the probability of the five hits in a row, we find

$$P(5) = P(HHHHH) = (0.4)(0.4)(0.4)(0.4)(0.4) = (0.4)^5$$
$$= 0.01024.$$

Next, consider $P(4)$, the probability of four hits in the next five trials. One way this can be done is $HHHHM$; that is, hit on the first

---

[4] This assumption, which is *required* in the development of the binomial distribution, may be subject to question in this case because players do have their ups and downs.

four trials and miss on the fifth. Another way is *HHHMH*. Writing all the events which constitute four hits and a miss, together with their probabilities, we find:

Components of event four hits	Probability
HHHHM	$(0.4)(0.4)(0.4)(0.4)(0.6) = (0.4)^4(0.6)^1$
HHHMH	$(0.4)(0.4)(0.4)(0.6)(0.4) = (0.4)^4(0.6)^1$
HHMHH	$(0.4)(0.4)(0.6)(0.4)(0.4) = (0.4)^4(0.6)^1$
HMHHH	$(0.4)(0.6)(0.4)(0.4)(0.4) = (0.4)^4(0.6)^1$
MHHHH	$(0.6)(0.4)(0.4)(0.4)(0.4) = (0.4)^4(0.6)^1$

The five events just written are mutually exclusive, so the total probability, $P(4)$, is the sum of the five probabilities. Moreover, the five probabilities are equal, so that

$$P(4) = 5(0.4)^4(0.6)^1 = 5(0.01536) = 0.0768.$$

In the last statement, note that

$$P(4) = \text{(number of ways of getting four hits and a miss)} \, p^4 q^1.$$

If we next ask for the probability of three hits in the next five trials, we would have

$$P(3) = \text{(number of ways of getting three hits and two misses)} \, p^3 q^2.$$

One way to get three hits is *HHHMM*, another way is *HHMHM*, and if we persist we will find there are 10 different sequences in the event (three hits, two misses) in five trials. Fortunately, we can compute this number by the formula

$$C_3^5 = \frac{5!}{3!(5-3)!} = \frac{5!}{3!(2!)} = \frac{5(4)3!}{3!(2)(1)} = \frac{20}{2} = 10.$$

The symbol at the left in the foregoing is read as *the number of combinations of five taken three at a time*. In the present application, it is the number of different ways of arranging five things of which three are of one kind and the remaining $5 - 3 = 2$ are of another kind.

We now have

$$P(3) = C_3^5 p^3 q^2$$
$$= 10(0.4)^3(0.6)^2 = 10(0.02304) = 0.2304.$$

> **Exercise.** Find the probability of two hits and three misses in the next five trials. Answer: $C_2^5 p^2 q^3 = 10(0.4)^2(0.6)^3 = 10(0.03456) = 0.3456.$

We may now generalize our demonstrations and state that if n independent trials are made, where p is the probability of outcome A and $q = 1 - p$ is the probability of the complementary outcome A' on any trial, then the probability that A will occur x times in the n trials is:

Binomial probability rule: $P(x) = C_x^n \, p^x q^{n-x}$.

As an example, let us compute the probability that the basketball player hits exactly three times in his next 10 tries. We have

$$n = 10, \quad p = 0.4, \quad q = 0.6, \quad x = 3,$$

and

$$P(3) = C_3^{10} \, p^3 q^{10-3} = \frac{10!}{3!7!}(0.4)^3(0.6)^7$$

$$= \frac{10(9)(8)}{3!}(0.4)^3(0.6)^7$$

$$= 120(0.064)(0.0279936)$$

$$= 0.215.$$

Exercise. Compute the probability that the player will hit exactly twice in his next six trials. Answer: $15(0.16)(0.1296) = 0.31104$.

If we seek the probability that $x = 0$ in $n = 7$ trials, it is clear that there is only *one* way, *MMMMMMM*, that this can occur. The binomial rule states

$$P(0) = C_0^7(0.4)^0(0.6)^7$$

$$= \frac{7!}{0!7!}(0.4)^0(0.6)^7.$$

The count,

$$\frac{7!}{0!7!},$$

must be *one* so, for consistency, we state:

Definition: $0! = 1$.

Then,

$$P(0) = \frac{7!}{0!7!}(0.4)^0(0.6)^7 = 1(1)(0.6)^7$$

$$= 0.0280.$$

Binomial situations arise in many applications. For example, we may record an inspected item as good or defective, check male or female on a questionnaire, classify accounts receivable as active or bad debts, and so on. In some situations we apply the binomial rule even though the requisite constancy of probability from trial to trial is only approximately correct. For example, if we draw items from a batch of 100,000 items, of which 100 are defective, we would have

$$P\text{(first is defective)} = P(D_1) = \frac{100}{100,000} = 0.001.$$

The probability that the second is also defective, correctly computed, is not 0.001 but is

$$P(D_2 \mid D_1) = \frac{99}{99,999} = 0.00099.$$

The difference noted between 0.001 and 0.00099 would be lessened if we were drawing from a batch larger than 100,000. To call attention to this approximate use of the binomial, we shall indicate that selection is made from a (very) *large* group and *not* state the number in the group. For example, suppose that 35 percent of the people in a *large* area are independent voters. We select 10 people and compute the probability that exactly six are independents as, approximately,

$$P(6) = C_6^{10}(0.35)^6(0.65)^4$$
$$= \frac{10(9)(8)(7)}{(4)(3)(2)(1)}(0.35)^6(0.65)^4$$
$$= 210(0.001838)(0.1785) = 0.0689.$$

Exercise. In a large batch of items, 10 percent are defective. If a sample of five is selected, what is the probability that exactly three will be defective? Answer: $C_3^5(0.1)^3(0.9)^2 = 10(0.001)(0.81) = 0.0081$.

At least one occurrence. Some applications require the computation of the probability of at least one occurrence. For example, suppose the assembly of a machine requires five independent operations, and the probability that any operation will result in a defect is 0.01. Further, the assembly is defective if one or more operations is defective; that is, if the number of defective operations is 1, 2, 3, 4, or 5. Hence, we seek the probability of at least one defective operation, which is the complement of zero defective.

706

$$P(\text{at least one defective}) = 1 - P(0 \text{ defective})$$
$$= 1 - C_0^5 (0.01)^0 (0.99)^5$$
$$= 1 - 1(1)(0.99)^5$$
$$= 1 - 0.951$$
$$= 0.049.$$

Exercise. A town has three ambulances for emergency transportation to the hospital. The probability that any one of the ambulances will be available at a point in time is 0.90. If a person calls for an ambulance, what is the probability that at least one will be available? Answer: $1 - (0.1)^3 = 0.999$.

As a variation on the last example and exercise, consider a car salesman who makes telephone calls during the day to obtain prospective car buyers. He assesses the probability of obtaining a prospect on a given call as 0.10. How many calls should he make if his goal is to have a probability of 0.90 of obtaining at least one prospect? Here we have:

$$P(\text{at least one prospect}) = 1 - P(0 \text{ prospects}) = 0.90,$$

or

$$1 - C_0^n (0.10)^0 (0.90)^n = 0.90$$
$$1 - (0.90)^n = 0.90$$
$$-(0.90)^n = -0.10$$
$$(0.90)^n = 0.10.$$

We solve for n by taking logarithms of both sides, writing $\ln (0.90)^n$ as $n \ln (0.90)$. Thus,

$$n \ln (0.90) = \ln (0.10)$$
$$n = \frac{\ln (0.10)}{\ln (0.90)}$$
$$= \frac{-2.30259}{-0.10536} = 21.9,[5]$$

or about 22 calls.

Exercise. A complicated computer program has a flaw or "bug" in it, and will not execute properly. The program is to be sent to n experts, each of whom has a probability of 0.4

[5] Natural logarithms (ln) may be found in Table X.

of finding the bug. What should n be if the probability that at least one expert will find the flaw is to be 0.99? Answer: 9.

12.14 CUMULATIVE BINOMIAL PROBABILITIES

The binomial rule as applied in the foregoing computes the probability of *exactly* x occurrences in n trials. Thus, in examples where we drew, say, five items from a large lot and recorded the number of defectives, the binomial rule calculates $P(2)$, the probability of exactly two defectives, or $P(1)$, and so on. In its real-world application, the purpose of drawing the sample is to make a decision on whether or not to *accept the whole lot* on the basis of the sample evidence. Thus, the quality control department may have a *sampling plan* that specifies that five items are to be drawn and inspected, and the *whole lot* is to be accepted if no more than one defective is found in the sample. That is, the lot is accepted if the sample contains zero or one defective; otherwise, the lot is rejected. Hence, if the *lot* is 10 percent defective

$$
\begin{aligned}
P(\text{Acceptance}) &= P(0 \text{ defective}) + P(1 \text{ defective}) \\
&= P(0) + P(1) \\
&= C_0^5 (0.1)^0 (0.9)^5 + C_1^5 (0.1)^1 (0.9)^4 \\
&= (0.9)^5 + 5(0.1)(0.9)^4 \\
&= 0.59049 + 0.32805 \\
&= 0.91854.
\end{aligned}
$$

The probability of acceptance is about 0.92, so if lots 10 percent defective are submitted to this sampling plan, 92 percent will be accepted and only 8 percent rejected.

> ***Exercise.*** If, in the foregoing, lots submitted are one percent defective, what is the probability of acceptance and rejection? Answer: $(0.99)^5 + 5(0.01)(0.99)^4 = 0.951 + 0.048 = 0.999$ as the probability of acceptance and 0.001 as the probability of rejection.

If we consider 10 percent defective as poor quality and one percent defective as good quality, then the sampling plan at hand, with a sample of $n = 5$, has a high probability (0.999) of accepting good quality, but also an undesirably high probability (0.92) of accepting poor quality. We can easily see that the probability of accepting poor quality would be reduced by selecting a larger sample, say 20, and accepting the lot only if *none* of the 20 is defective. Now, if lots are 10 percent defective,

$$P(\text{Acceptance}) = P(0 \text{ defective}) = C_0^{20}(0.1)^0(0.9)^{20} = 0.12.$$

This plan offers better protection against the acceptance of poor lots. On the other hand, a good lot (say one percent defective) now has

$$P(\text{Acceptance}) = P(0 \text{ defective}) = C_0^{20}(0.01)^0(0.99)^{20} = 0.82,$$

so in reducing the probability of accepting poor lots (from 0.92 to 0.12) we also lower the probability of accepting good lots (from 0.999 to 0.82). Clearly, the choice of sampling plan (number in the sample to be inspected and the number of defectives permitted in the sample) can be adjusted to achieve a balance between the chances of accepting good and poor lots. For example, we may select 25, and accept the lot if it has no more than three defectives. If p is the proportion defective in a lot, then

$$\begin{aligned}
P(\text{Acceptance}) &= P(0, 1, 2, \text{ or } 3 \text{ defectives}) \\
&= P(0) + P(1) + P(2) + P(3) \\
&= C_0^{25}\, p^0 q^{25} + C_1^{25}\, p^1 q^{24} + C_2^{25}\, p^2 q^{23} + C_3^{25}\, p^3 q^{22} \\
&= \sum_{x=0}^{3} C_x^{25}\, p^x q^{25-x}.
\end{aligned}$$

We refer to the last expression as a *cumulative* binomial probability, and hasten to add that tables are available for these cumulative probabilities. Tables XIII-A ($n = 10$) and XIII-B ($n = 25$) at the end of the book will be used in our examples and exercises.[6] For the problem at hand, we find from Table XIII-B that for $p = 0.01$

$$\sum_{x=0}^{3} C_x^{25}(0.01)^x(0.99)^{25-x} = 1.000.$$

The tabular entries have been rounded to three decimals and the last-written number, 1.000, does not mean exactly 1, but a number less than 1 which rounds to 1.000. We interpret this result by saying that if $n = 25$ items are selected from a lot which is one percent defective and three defectives are allowed in the sample, it is almost certain that the lot will be accepted.

Exercise. For the foregoing find the probability of acceptance if the lot is 10 percent defective. Answer: 0.764.

---

[6] Cumulative binomial probabilities to five decimal places for p in intervals of 0.01 and for selected values of n up to 1,000 may be found in *Tables of the Cumulative Binomial Probability Distribution*, Harvard University Press, 1955.

To demonstrate an understanding of cumulative probabilities, we should be able to write summation expressions similar to the one we wrote in the foregoing. Thus, if we toss a coin 50 times and seek the probability of getting *at most* 10 heads, we have $n = 50$, $p = 0.5$. The probability of 0, 1, 2, $\cdot$ $\cdot$ $\cdot$, 10 heads would be expressed as

$$\sum_{x=0}^{10} C_x^{50}(0.5)^x(0.5)^{50-x}.$$

Exercise. If a salesperson's probability of making a sale on any contact is 0.08, express in summation symbols the probability of making at most 15 sales in 100 contacts. Answer:

$$\sum_{x=0}^{15} C_x^{100}(0.08)^x(0.92)^{100-x}.$$

Now, for brevity, we shall express the cumulative probability for, say, 0, 1, 2, 3, 4, 5, as

$$\sum_{0}^{5}.$$

Thus, from Table XIII-B, with $n = 25$, $p = 0.4$,

$$\sum_{0}^{5} = 0.029$$

is the probability of *at most* five occurrences. If we seek the probability of *at least* five occurrences, which would be 5, 6, 7, . . . , 25, it would be found as the complementary probability

$$1 - \sum_{0}^{4} = 1 - 0.009 = 0.991.$$

Again, the probability of six to 10 occurrences, inclusive, would be

$$\sum_{0}^{10} - \sum_{0}^{5} = 0.586 - 0.029 = 0.557.$$

Finally, if we want the probability of *exactly* five occurrences, we compute

$$\sum_{0}^{5} - \sum_{0}^{4} = 0.029 - 0.009 = 0.020.$$

Exercise. For $n = 25$, $p = 0.3$, find the probability of: *a)* Less than eight occurrences. *b)* At most five occurrences. *c)* At least 10 occurrences. *d)* From five to 11 occurrences, inclusive. *e)* Exactly eight occurrences. Answer: *a)* 0.512. *b)* 0.193. *c)* 0.189. *d)* 0.866. *e)* 0.165.

As an application of Table XIII-B, suppose a student takes a 25-question true-false test and determines the answer to each by flipping a coin, so that the probability of getting the correct answer is $p = 0.5$ for each question. If 60 percent or more correct is passing, what is the probability of passing? We note that 60 percent or more is 0.6(25) = 15 or more correct. Hence,

$$P(\text{Pass}) = \sum_{15}^{25} = 1 - \sum_{0}^{14} = 1 - 0.788 = 0.212.$$

The student's probability of not passing is 0.788, or about 0.8. In other terminology, the odds are about 4 to 1 against passing.

> ***Exercise.*** If a student takes a 25-question four-choice multiple choice examination and judges that his probability of getting any question correct is about 0.6, what is the probability that he will get 80 percent or more correct? Answer: 0.029.

12.15 PROBLEM SET 12–5

Compute the answers to Problems to 1–8 using the binomial rule.

1. A basketball player has a probability of 0.3 of hitting on any shot from the foul line. What is the probability of:
   a) Exactly one hit in three trials?
   b) At least one hit in three trials?
   c) Exactly two hits in five trials?
   d) More than two hits in four trials?

2. The probability that an inspector will properly classify an item is 0.8. If each item is inspected independently by three inspectors, what is the probability that at least one will properly classify the item?

3. A company has bid on five projects, assessing the probability of winning a contract at 0.6. To have a successful year, it must win at least two of the contracts. What is the probability for a successful year?

4. A sample of six items is selected from a large lot. The *lot* is accepted if the *sample* contains no more than one defective item. Find the probabilities of accepting and rejecting a lot if the proportion defective in the lot is: *a)* 0.1. *b)* 0.2.

5. A department store employs four people who take orders over the telephone. Each person is busy taking an order 70 percent of the time. What

is the probability that an operator will be free to take an order at the time of a call:

a)   If one customer calls at a point in time?

b)   If three customers call at the same time?

6.   See Problem 5. How many telephone operators should the store have if the probability that an operator will be free when a customer calls is to be 0.83?

7.   A true-false examination has five questions, and a student guesses the answer for each question, assigning probability of 0.5 of being correct. Assuming independence, what is the probability that he gets

a)   All five correct?   b)   At least four correct?

c)   Exactly three correct?   d)   At least three correct?

e)   At least four incorrect?

8.   A test has five four-choice questions, and a student guesses the answer to each question. Assuming independence, what is the probability of

a)   All five correct?   b)   At least four correct?

c)   Exactly three correct?   d)   At least three correct?

9.   The probability that an item in a large group is defective is 0.05. Express the following by use of the summation symbol, but do not try to calculate the answer.

a)   The probability that at most 15 of 100 items purchased are defective.

b)   The probability that more than 10 of 200 items purchased are defective.

Use Tables XIII-A and B to answer Problems 10–15.

10.   A sample of 25 items is selected from a large lot and the lot is rejected if the sample contains more than four defectives. Find the probability of acceptance if the proportion defective in a lot is: a) 0.05. b) 0.20.

11.   Repeat Problem 10 assuming the lot is rejected if more than one defective is found in the sample.

12.   A test has 25 five-choice questions. A student gives answers at random. What is the probability that he gets:

a)   Less than 40 percent correct?

b)   More than 40 percent correct?

c)   At least 20 percent correct?

d)   Exactly eight correct?

e)   Six to 10, inclusive, correct?

f)   At most five correct?

13.   Repeat Problem 12 if the student assesses his probability of getting a correct answer at 0.5.

14.   Ten percent of the (very large) supply of tires offered for sale around the country have faulty values. If a person buys 10 of these tires (assumed to be a random selection), what is the probability that the buyer will get

a)   No faulty tires?   b)   Exactly one faulty tire?

c)   At least one faulty tire?   d)   Two or three faulty tires?

15. A company has 10 employees who, on the average, are absent from work on 5 percent of the working days. What is the probability that on a given day

a) Exactly 2 are absent?   b) Exactly 9 are present?
c) More than 2 are absent?   d) One or two are absent?

12.16 EXPECTED MONETARY VALUE (EMV)

If we bet $1 that heads will appear on the toss of a coin for which we have assigned

$$P(H) = P(T) = 1/2$$

we win $1 if heads appears, and win −$1 (lose $1) if tails appears. We compute the *expected monetary value* of the *act* of tossing the coin as

EMV = (payoff if *event H* occurs)$P(H)$ + (payoff if *event T* occurs)$P(T)$
= $1(1/2) + (−$1)(1/2) = 0.

More generally, the expected monetary value of an *act* is the sum of the products formed by multiplying the dollar payoff of each *event* by the probability of the event. It is assumed that the events constitute a sample space. For example, if we consider act A as having three events, E_1, E_2, and E_3, with probabilities 0.4, 0.5, and 0.1, respectively, and payoffs $10, −$8, and $2, then

EMV of act A = 0.4(10) + 0.5(−8) + 0.1(2) = $0.20.

Example. An urn contains five red, one white, and four green balls, and we assign probabilities

$$P(R) = 0.5$$
$$P(W) = 0.1$$
$$P(G) = 0.4.$$

A ball is to be drawn, and the payoffs are red ball, lose $1; white ball, win $3; green ball, win nothing. Compute the EMV of the act of drawing a ball.

P	Event	Payoff
0.5	R	−1
0.1	W	3
0.4	G	0

EMV = 0.5(−1) + 0.1(3) + 0.4(0)
= −$0.20.

Exercise. What is the expected monetary value of the act of tossing two coins if the payoffs are $0 for zero heads, $1 for one head, and −$1 (loss of $1) for two heads? Answer: (0) (1/4) + ($1)(1/2) + (−$1)(1/4) = $0.25.

Expected monetary value has been advanced as one criterion to aid decision making. The notion is that we list the various events that might arise in a certain situation and assign a probability to each event. In addition, we consider the payoffs that would occur for each event for each decision we might make, the decisions being the choice of act 1, act 2, and so on. Suppose that we use EMV to choose between act 1 and act 2 in Table 12–3. According to the EMV criterion, the decision maker would choose act 1 rather than act 2 because act 1 has the higher expected monetary value.

TABLE 12–3
Payoff table

		Act	
P	Events	A_1	A_2
0.3	E_1	$2.00	$2.00
0.4	E_2	1.00	3.00
0.3	E_3	8.00	3.00
1.0			

EMV, act $1 = 0.3(2) + 0.4(1) + 0.3(8) = 3.4.$
EMV, act $2 = 0.3(2) + 0.4(3) + 0.3(3) = 2.7.$

Example. Items are manufactured for sale. Each unit made and sold yields a profit of $3; each unit made but not sold yields a loss of $1. It is believed that zero, one, two, or three units might be demanded by customers, but the event *four or more* units demanded is considered

TABLE 12–4

Events (Number of units demanded)	Probability of number of units being demanded
0	0.2
1	0.4
2	0.3
3	0.1
4 or more	0.0
	1.0

impossible and assigned probability zero. Other probabilities are assigned by experience and judgment (Table 12–4). Use the expected monetary value to decide whether to make zero units (act 1), one unit (act 2), two units (act 3) or three units (act 4).

The payoff table can be filled in from the given information. For example, if two units are made and one unit is demanded, one of the two would yield a profit of $3 and the other a loss of $1, for a payoff of $2 net. Again, if one unit is made and two are demanded, the payoff is $3 on the single unit made. These and the remaining payoffs are shown in Table 12–5.

TABLE 12–5
Payoff table

		Acts			
P	Events (Units demanded)	A_1 (Make 0)	A_2 (Make 1)	A_3 (Make 2)	A_4 (Make 3)
0.2	0	0	−1	−2	−3
0.4	1	0	3	2	1
0.3	2	0	3	6	5
0.1	3	0	3	6	9
0.0	4 or more	0	3	6	9
1.0					

EMV of $A_1 = 0.0$
EMV of $A_2 = 2.2$
EMV of $A_3 = 2.8$
EMV of $A_4 = 2.2$

The decision would be to choose A_3, the act with the highest EMV, and so make two units.

EMV can be useful criterion in some decisions. However, it is easy to illustrate that this criterion does not have general applicability. For example, the EMV of A_1 in the following table is $2500, compared to an EMV of $200 for A_2, and yet some persons would prefer A_2 to A_1.

P	Event	A_1	A_2
0.5	E_1	$10,000	$400
0.5	E_2	− 5000	0

EMV of $A_1 = 2500
EMV of $A_2 = 200$

The point here is that even though the EMV of A_1 is much larger than that of A_2, some people would not feel they could afford a loss of \$5000, which would arise if they chose A_1 and event E_2 occurred. Others would prefer A_2 on the ground that they cannot lose if they choose A_2, and have a 0.5 probability of gaining \$400. Of course, a person possessing a large amount of money might well choose A_1 because he can afford to lose \$5000 and thinks a 50–50 gamble of winning \$10,000 or losing \$5000 is sensible.

The last illustration shows that the act chosen depends upon the person making the decision and the amounts involved. EMV may or may not be a proper guide for action. A criterion applicable when EMV is not appropriate is *expected utility value* (EUV), which allows a person to inject his or her own circumstances and inclinations into the analysis. Exploration of the EUV criterion would carry us beyond our immediate goals.[7]

One-time decisions. The probability of heads when a coin is tossed, 0.5, means that as the number of tosses increases (approaches infinity in the limit sense), the proportion or relative frequency of heads approaches 0.5 as a limit. However, if we toss a coin only once, heads or tails will appear and the 0.5 probability does not tell us which will occur. Similarly, the EMV of an act is computed from probabilities and means the average payoff we would expect to arise if the act was performed an increasingly large number of times under constant conditions, but this EMV does not tell us what will occur if the act is performed only once. Consequently, some people contend that it is not correct to apply the EMV criterion to a one-time decision. Other people, while agreeing with the long-run interpretation of EMV, contend that EMV may be applied to one-time decisions, arguing that if a person would choose act A over and over again in a repeated series of decisions, it would not be unreasonable to choose A if the decision circumstances occur only once. The latter group would also point out that management-administrative decisions typically are of the one-time variety because the circumstances under which real-world decisions are made are not "constant over the long run." The controversy between those who would and those who would not use probability considerations in one-time decisions really centers upon the question of what information a decision-maker would choose to consider in making a one-time decision where the outcome is uncertain. Thus, whether or not probabilities would be considered in chancy one-time decisions is a choice left to the decision maker. In the writer's view, it is almost unthinkable that a manager or administrator would not consider proba-

---

[7] The reader is encouraged to investigate Robert Schlaifer, *Introduction to Statistics for Business Decisions* (New York: McGraw-Hill Book Co., 1961) as the next step toward achieving a fuller understanding of the role of probability in business decisions.

bilities when making decisions where the outcome is uncertain, even if a one-time decision is at hand.

12.17 PROBLEM SET 12–6

1. A pair of dice is to be rolled. If the number appearing is even, you win that even number of dollars; if the number appearing is odd, you lose that odd number of dollars. Compute the EMV of the act "rolling the pair of dice."

2. An act is accompanied by three possible events with probabilities 0.2, 0.3, and 0.5, and payoffs $2, $3, and −$1, respectively. Compute the expected monetary value of the act.

3. Urn number one contains four red, nine white, and seven green balls with payoffs $2, −$4, and $2, respectively. Urn number two contains four red and six black balls with payoffs $3 and −$1.80, respectively. If act 1 is selecting a ball from urn number one and act 2 is selecting a ball from urn number two, which act should be chosen according to the criterion of expected monetary value?

4. Which act should be chosen according to EMV?

Payoff table

P	Event	A_1	A_2	A_3	A_4
0.2	E_1	$2	$1	$0	$0
0.1	E_2	2	2	−1	−3
0.4	E_3	2	3	3	3
0.3	E_4	2	2	4	5

5. If you make a unit of product and it is sold (demanded), you gain $5; if you make a unit that is not sold, you lose $2. You assign probabilities as follows:

Number of units demanded	Probability of number of units demanded
0	0.10
1	0.20
2	0.25
3	0.40
4	0.05
5 or more	0.00

According to the EMV criterion, how many units should you make?

6. In setting premiums to charge for protection against various hazards, insurance companies must start with a base figure (exclusive of overhead and

profit), which represents their expected loss. A building is to be insured in the amount of $60,000 for fire damage. The probabilities of total, 75 percent, 50 percent, and 25 percent losses in a year are, respectively, 0.0001, 0.00015, 0.0005, and 0.001. Assuming these are the only losses to be considered,

   a) What base figure should be used in computing the annual premium?
   b) Why do the probabilities given not add up to 1?

7. Think seriously about your present circumstances, and then decide in each case whether you would choose act 1 or act 2. For example, in part *(a)*, would you prefer a 0.6 probability of gaining $3, 0.4 of losing $1, to a gamble which has a 0.6 probability of gaining $1? (There are no correct answers to this question.)

a) P	A_1	A_2
0.6	$3	$1
0.4	-1	0.

b) P	A_1	A_2
0.6	$30	$10
0.4	-10	0.

c) P	A_1	A_2
0.5	$3000	$500
0.5	-1000	0.

12.18 REVIEW PROBLEMS

1. The table shows, for example, that 40 cars of make Y had gear train malfunctions.

	Malfunction		
Make of car	Electrical (E)	Gear train (G)	Carburetor (C)
X	17	60	23
Y	20	40	60
Z	15	48	117

   Find the following probabilities:

   a) $P(Y)$.   *b)* $P(E)$.   *c)* $P(C)$.   *d)* $P(X \cap G)$.
   e) $P(G \cap X)$.   *f)* $P(E \cap G)$.   *g)* $P(Z \cap X)$.   *h)* $P(Y \cap C)$.
   i) $P(X \mid C)$.   *j)* $P(C \mid X)$.   *k)* $P(E \mid Z)$.   *l)* $P(Z \mid G)$.
   m) $P(X \cup Y)$.   *n)* $P(X \cup C)$.   *o)* $P(G \cup C)$.   *p)* $P(Y \cup G)$.

2. See the table of Problem 1.

   a) Are Z and C independent? Explain.
   b) Are Y and C independent? Explain.
   c) What is $P(X \cap Y)$? What does this mean?

3. *a)* In some areas, one may often predict tomorrow's weather correctly by stating it will be the same as today's weather. In such areas, is tomorrow's weather independent of today's weather? Explain.
   b) A card is to be drawn from a deck. Let B represent black card and D represent diamond. Are B and D mutually exclusive? Independent? Explain.

c) Consider the physical traits of brown eyes, *B,* and dark hair, *D.* Are *B* and *D* mutually exclusive? Do *you* think *B* and *D* are independent?

4. *a)* Complete the following table if all event pairs such as *AX* are independent.

	X	Y	Z	Total
A.............				0.40
B.............				0.60
Total	0.30	0.20	0.50	

Find the following probabilities from the completed table.

b) $P(A|X)$. *c)* $P(X|A)$. *d)* $P(X)$. *e)* $P(A)$.
f) $P(A \cup X)$. *g)* $P(X \cup Y)$.

5. *a)* Complete the following probability table. (Note: *A'* and *B'* are complements of *A* and *B*, respectively.)

	B	B'	Total
A.............			0.70*
A'		0.12*	
Total		0.34*	

b) Are *A* and *B* independent? Explain.

Compute:
c) $P(B'|A')$. *d)* $P(A \cup B')$.

6. An investor assesses the probability that the Dow-Jones stock market average will rise tomorrow as 0.65, and the probability that the price of stock *X* will rise if the Dow-Jones rises as 0.90. What is the probability that the Dow-Jones will rise and *X* will rise? Why?

7. A candidate runs for offices *A* and *B,* assessing the probability of winning both at 0.10, and the probability of winning *B* at 0.25. What is the probability of winning *A* if he wins *B?* Why?

8. An investor has funds in banks *A* and *B.* He assesses the probability that *A* will fail at 0.0001, and assigns the same failure probability to *B.* Further, he thinks the probability that both banks will fail is 0.00001.
a) What is the probability that *A* or *B* will fail?
b) Are the events *A* fails, *B* fails independent? Why?
c) What is the probability that *B* fails if *A* fails?

9. A box of eight items contains six good and two defective items. If a sample of two items is selected, what is the probability that:
a) Both will be good?
b) Both will be defective?

c) Exactly one will be defective? (Note: Two events are involved.)

d) If three items are drawn, what is the probability that at least one will be defective?

10. A box contains four items, of which two are good and two defective. A sample of two items is selected.

   a) Letting G and D stand for good and defective, write a sample space for the experiment using four events and enter the probabilities for each event.

   b) Write a sample space using as events the count of the number of defectives in the sample, and enter the probabilities for each event.

   c) Write a sample space using equally likely events. (Hint: Let G_1D_2 mean good number one and defective number two.)

11. Graduation exercises are to be held outdoors on Friday if it does not rain. If it rains on Friday, the exercises will be postponed until Saturday and held outdoors if it does not rain, and indoors if it rains. The probability that it will rain on Friday is 0.3, and the probability it will not rain on Saturday if it rains on Friday is 0.4. Find the probability that the exercises will be held outdoors.

12. Two women and three men are equally qualified for two positions in a firm. The firm decides to select two of the five at random.

   a) Write a sample space for the experiment using equally likely events.

   b) From *(a)* determine the probability that a man and a woman are selected.

   c) Calculate the answer to *(b)* by probability rules.

13. A job applicant assigns probabilities as follows: The probability, *P(A)*, of being offered a job at company A is 0.4, the probability of being offered a job at B is $P(B) = 0.3$, and the probability of being offered jobs at both companies is 0.12. What is the probability of being offered a job at at least one of the two companies? At exactly one of the two companies?

14. Probability of colder is assigned at 0.7, probability of snow at 0.4, and probability of neither colder nor snow at 0.2. What is the probability

   a) That it will get colder and snow?

   b) That it will get colder but not snow?

   c) That it will snow but not get colder?

15. A test has two questions. A student assigns probability 0.6 of getting the first correct, 0.3 of getting the second correct, and 0.25 of getting both wrong.

   a) Show that with this assignment of probabilities, the outcome of the second question is not independent of the outcome of the first in the probability sense.

   b) What would the probability of getting both wrong be if the outcomes are to be independent in the probability sense?

16. The claim is made that whether or not an employee's attendance record is good depends upon the sex of the employee. On the basis of the table, using probability terminology, refute the claim.

	Number of employees with	
Sex	Good attendance records	Poor attendance records
Male	40	10
Female	80	20

17. The probability that machine A will break down on a particular day is $P(A) = 1/100$; similarly, for machine B, $P(B) = 1/200$. Assuming independence, on a particular day,
   a) What is the probability that both will break down?
   b) What is the probability that neither will break down?
   c) What is the probability that one or the other will break down?
   d) What is the probability that exactly one will break down?

18. Five parts go into the assembly of item X. The assembly is defective if any one of the parts is defective, and each part has probability of 0.01 of being defective. Assuming independence, what is the probability that an assembly is defective?

19. There are five intersections between two cities where a driver can bear left or bear right. If, and only if, the proper turn is made at each intersection will a driver starting from one city arrive at the second city. Suppose that the driver flips a coin to choose each turn. What is the probability that he will arrive at the second city?

20. Given $P(X) = 0.5$, $P(Y) = 0.7$, and $P(XY) = 0.30$, find $P(X|Y)$ and $P(Y|X)$.

21. If $P(Y|X) = 0.80$, $P(X|Y) = 0.75$, and $P(XY) = 0.60$, find $P(X)$ and $P(Y)$.

22. If $P(Y|X) = 0.80$, $P(XY) = 0.20$, $P(Y') = 0.30$, find $P(X \cup Y)$.

23. Job candidates are screened by means of a preliminary interview. The probability is 0.6 that a screened candidate will be a good worker. Screened candidates are given a test. If the candidate is one who will prove to be a good worker, the probability of his passing the test is 0.90. If the candidate is one who will prove to be a poor worker, the probability of his passing the test is 0.40.
   a) What is the probability that a screened candidate will be a good worker and pass the test?
   b) What is the probability that a screened candidate will not be a good worker and will pass the test?

24. A candidate runs for two political offices, A and B. He assigns 0.27 as the probability of being elected to both, 0.50 as the probability of being elected to A if he is elected to B, and 0.90 as the probability of being elected to B if he is elected to A.
   a) What is the probability of being elected to A?
   b) What is the probability of being elected to B?

c) What is the probability of being elected to neither?

d) What is the probability of being elected to at least one of the offices?

25. Mr. M thinks his probability of winning an election is 0.9 if Ms. W does not run, but only 0.3 if Ms. W does run. Mr. M judges the probability that Ms. W will run to be 0.4. What is the probability that Mr. M wins the election?

26. Complete the Bayes' Rule formulation that starts with $P(A'|B) =$

27. The probability that a customer will be a bad debt is 0.01. The probability that he will make a large down payment if he is a bad debt is 0.10, and the probability that he will make a large down payment if he is not a bad debt is 0.50. Suppose that a customer makes a large down payment. Find the posterior probability that he will be a bad debt.

28. Complete the tabular analysis for Problem 27, and find the probability that a customer will not be a bad debt if he does not make a large down payment.

29. The probability that a person has disease X is $P(X) = 0.008$. The probability that medical examination will indicate the disease if a person has it is $P(I|X) = 0.75$, and the probability that examination will indicate the disease if a person does not have it is $P(I|X') = 0.01$. What is the probability that a person has the disease if medical examination so indicates?

30. The probability that a machine is running properly is 0.80. From time to time, samples of output are selected and measured, and the sample average is computed. If the machine is running properly, the probability that the sample average will be in a certain range is 0.95. If the machine is not running properly, the probability that the sample average will be in this range is 0.05. A sample is selected, and its average is outside the range. What is the probability that the machine is running properly?

Do Problems 31–41 using the binomial rule.

31. A baseball player assesses his probability of getting on base each time at bat at 0.20. What is the probability that he will get on base:

a) Exactly once in his next three times at bat?

b) At least once in his next three trips?

c) At least twice in his next three trips?

d) Exactly twice in six trips?

32. The four engines of an airplane operate independently and on an overseas flight the probability that an engine will fail is 0.001. On such a flight, what is the probability that:

a) Exactly one will fail?

b) More than one will fail?

33. A true-false examination has five questions, and a student assigns probability of 0.7 of being correct on each of the questions. Assuming independence, what is the probability that he gets:

a) All five correct?   *b)* At least four correct?

c) Exactly three correct?   *d)* At least three correct?

e) At least four incorrect?

34. A test has five five-choice questions and a student guesses the answer to each question. Assuming independence, what is the probability of:
    a) All five correct?   b) At least four correct?
    c) Exactly three correct?   d) At least three correct?

35. A test has five four-choice questions and a student assigns probability 0.8 of getting each one correct. What is the probability of getting:
    a) All five correct?   b) At least three correct?
    c) At least one correct?

36. The probability of success on a single trial of an event is 0.3. In 100 independent trials, write, but do not evaluate, the expression for the probability of:
    a) At least one success.   b) At most one success.
    c) Exactly five successes.   d) At most three successes.

37. A test has 50 five-choice questions and a student guesses the answer to each question. Write, but do not evaluate, the expression for each of the following probabilities:
    a) Ninety percent or more correct.   b) At least one correct.

38. The probability that any particular item is defective is 0.10. Using binomial probabilities, find:
    a) The probability of at most one defective in 10 items.
    b) How many items would have to be selected to have a probability of 0.99 that the group would contain at least one good item.

39. A binomial experiment consists of m trials. Write the summation expression using m, p, q, r, and x for:
    a) The probability of at most r occurrences in the m trials.
    b) The probability of more than r occurrences in the m trials.

40. A department store employs three people who take orders over the telephone. Each person is busy taking an order 80 percent of the time. What is the probability that an operator will be free to take an order at the time of the call:
    a) If one customer calls at a point in time?
    b) If two customers call at a point in time?

41. See Problem 40. How many telephone operators should the store have if the probability that an operator will be free when a customer calls is to be approximately 0.6?

Do Problems 42–44 using Table XIII-B.

42. A sample of 25 items is selected from a large lot, and the lot is rejected if the sample contains any defective items. Find the probability of acceptance if the proportion defective in the lot is:
    a) 0.01.   b) 0.10.   c) 0.20.

43. Repeat Problem 42 assuming the lot is rejected if more than two defectives are found in the sample.

44. From past experience, a market research firm knows that 60 percent of the people contacted by telephone will agree to a telephone interview. If the firm calls 25 people, what is the probability of completing:
    a) At most 10 interviews?   b) At least 10 interviews?

   c) More than 15 interviews?

   d) From 15 to 20 interviews, inclusive?

   e) Exactly 10 interviews?

45. The act is to draw a card. If the card is a face card you win $20; if it is not a face card, you lose $20. Compute the expected monetary value of the act. (Face cards are jack, queen, king, and ace.)

46. The act is rolling a pair of dice. If the dice come up 2, 7, or 11, you win $100; otherwise you lose $16. Compute the expected monetary value of the act.

47. Which act should be chosen according to the EMV criterion?

		Payoffs		
Probability	Event	A_1	A_2	A_3
0.5	E_1	$5	$4	$4
0.3	E_2	5	7	7
0.2	E_3	5	3	5

48. If you make a unit of a product and it is sold (demanded), you gain $10; if you make a unit and it is not sold, you lose $5. You assign probabilities as follows:

Number of units demanded	Probability of number of units demanded
0	0.1
1	0.2
2	0.4
3	0.3
4 or more	0.0

According to the EMV criterion, how many units should you make?

49. One million tickets are sold at $1 each for a lottery. There is a first prize of $100,000, two second prices of $50,000, 10 third prizes of $1000, and 20 fourth prizes of $500. If a person buys a ticket, what is the (expected) value of the ticket?

50. A company offers insurance covering damage of 20 percent, 40 percent, 60 percent, 80 percent, or 100 percent. The owner of a particular property wishes to insure it for $200,000. The company assesses yearly damage probabilities (for the various respective damage percentages) at 0.0010, 0.0008, 0.0006, 0.0004, and 0.0002. What base figure (before overhead and profit) should the company use in establishing the annual premium to charge for insuring the property?

Probability in the continuous case

13

13.1 INTRODUCTION

The binomial rule discussed in Chapter 12 assigned probabilities to the *number of occurrences* of an event in a certain number of trials. As such, the events were counting numbers, 0, 1, 2, 3, and so on, with no event between a successive pair of these numbers. The set, 0, 1, 2, 3, $\cdots$, is called a *discrete* set because its members are separated. If we were to represent 0, 1, 2, 3, $\cdots$ graphically, it would consist of a set of separated points. In this chapter we introduce probability calculations for the *continuous* (as contrasted to discrete) case, and here graphical representation of the set of events consists of a continous line or line segment. We shall first present the basic ideas in the continuous case by means of relatively simple functions, and then discuss the normal probability model and show some applications of this widely used model.

13.2 THE UNIFORM PROBABILITY DENSITY FUNCTION

Fran, an executive in a company, is expecting a friend to arrive by ship one afternoon. Upon inquiry, Fran learns from a dock official that the ship will certainly arrive between 1 and 6 o'clock, but just when is a matter of chance. Fran decides to assign one (certainty) as the probability of arrival from 1 to 6 o'clock and assume that arrival is equally likely at any time in this interval. Fran has a business appointment that will prevent her from being at the dock from 2 to 3:30 o'clock and calculates the chance that she will not be there when her friend arrives is 0.30 because she will not be present for 1.5 hours in the 5 hour interval from 1 to 6 o'clock; that is, 1.5/5 = 0.3.

We shall use the foregoing to introduce the method of measuring probabilities by areas. Figure 13-1 shows a horizontal interval from

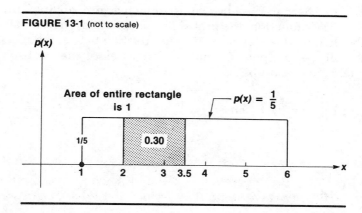

FIGURE 13-1 (not to scale)

1 to 6 representing the time of arrival of Fran's friend. Because time is a *continuous* variable, the interval is a continuous line segment rather than a discrete (separated) set of points. A rectangle of height ⅕ has been drawn over the horizontal base, and its upper boundary is the constant function

$$p(x) = \frac{1}{5}; \quad 1 \le x \le 6.$$

The area of this entire rectangle is

$$A = 5\left(\frac{1}{5}\right) = 1,$$

and the 1 means it is *certain* (probability of 1) that the ship will arrive in this time interval. The shaded area over the horizontal interval from 2 to 3.5 (3:30), which is

$$(3.5 - 2)\left(\frac{1}{5}\right) = 1.5\left(\frac{1}{5}\right) = 0.30,$$

represents the probability that the ship will arrive during this time period. The function $p(x) = \frac{1}{5}$ is called a *probability density function* or, more briefly, a *density function*. Note carefully that the density function does not provide probabilities. Rather, *areas under a density function over a horizontal interval are probabilities assigned to the horizontal interval*. Density functions are also called *probability distributions*, and we shall use the terms density function and probability distribution interchangeably.

The density function of Figure 13–1 is called the *uniform* density function, or uniform distribution, because it is a constant function representing Fran's assumption that arrival of the ship is *equally likely* to occur anywhere in the interval. Inasmuch as probabilities are areas under a density function, the probability assigned to a horizontal interval is the definite integral of the density function between limits which are the interval endpoints. Thus, in Fran's case, the probability of arrival between 2 and 3.5 is

$$\int_2^{3.5} p(x)\, dx = \int_2^{3.5} \left(\frac{1}{5}\right) dx = \frac{1}{5} x \bigg|_2^{3.5} = \frac{1}{5}(3.5 - 2) = \frac{1}{5}(1.5) = 0.3.$$

Exercise. By integration find the probability that the ship will arrive between 3 and 3:30 o'clock if $p(x) = \frac{1}{5}$. Answer: 0.10

One consequence of area assignment of probabilities should be noted; namely, the probability associated with a point is zero. Thus, in Fran's case, the probability that the ship will arrive at *exactly* 2 o'clock is zero, where exactly means an instant (duration zero). This consequence arises because a point (instant) has zero width and the area (probability) over a point necessarily is zero. In applications, of course, we shall be concerned with intervals, so the zero point probability causes no difficulty. However, if we use the symbol

$$P(2 \le x \le 3.5) = 0.3$$

to represent Fran's probability assignment, we should realize that

$$P(2 \le x \le 3.5) = P(2 < x < 3.5)$$

because the probabilities for exactly 2 and exactly 3.5 are zero.

13.3 CONVERTING $f(x)$ TO A DENSITY FUNCTION OVER AN INTERVAL

If we have a function $f(x)$ that is nonnegative over an interval, we can convert $f(x)$ to a density function by dividing $f(x)$ by the total area under $f(x)$ over the interval.

Example. Convert $f(x) = 4x - x^2 - 3$, $1 \le x \le 3$, to a density function.

First we find the total area under $f(x)$ as

$$\int_1^3 (4x - x^2 - 3)\, dx$$

$$= 2x^2 - \frac{x^3}{3} - 3x \Big|_1^3$$

$$= \left[2(9) - \frac{27}{3} - 9 \right] - \left[2(1) - \frac{1}{3} - 3 \right]$$

$$= 18 - 9 - 9 - 2 + \frac{1}{3} + 3$$

$$= \frac{4}{3}.$$

Then

$$p(x) = \frac{f(x)}{\dfrac{4}{3}} = \frac{3}{4} f(x)$$

$$p(x) = \frac{3}{4} [4x - x^2 - 3]; \quad 1 \le x \le 3.$$

The foregoing function, $p(x)$, is a vertical parabola opening downward, with x-intercepts at $x = 1$ and $x = 3$, as shown in Figure 13–2. The

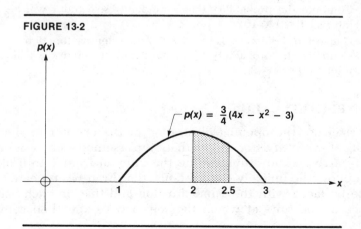

FIGURE 13-2

$p(x) = \frac{3}{4}(4x - x^2 - 3)$

conversion procedure simply changes $f(x)$, which has an area of $\frac{4}{3}$ over the interval, to the density function, $p(x)$, which has an area of 1, as required. Once assured that we have a density function, probabilities for intervals are computed as definite integrals.

Example. For the density function of Figure 13–2, find the probability that a randomly selected value of x will lie in the interval $2 \le x \le 2.5$.

The desired probability is the shaded area in Figure 13–2. We compute it as

$$\int_{2}^{2.5} p(x)\, dx = \int_{2}^{2.5} \frac{3}{4}(4x - x^2 - 3)\, dx$$

$$= \frac{3}{4}\left(2x^2 - \frac{x^3}{3} - 3x\right)\Bigg|_{2}^{2.5}$$

$$= \frac{3}{4}\left[\left(-\frac{0.625}{3}\right) - \left(-\frac{2}{3}\right)\right]$$

$$= \frac{3}{4}\left(\frac{1.375}{3}\right) = \frac{1.375}{4} = 0.34375.$$

13.4 PROBLEM SET 13–1

1. *a)* Convert $f(x) = 20$, $0 \le x \le 10$, to a density function.
    b) Compute the probability that a randomly selected x will lie in the interval $4 \le x \le 7$.

2. *a)* Convert $f(x) = x$, $0 \le x \le 10$, to a density function.
    b) Compute the probability that a randomly selected x will lie in the interval $4 \le x \le 5$.

3. *a)* Convert $f(x) = x^{1/2}$, $0 \le x \le 9$, to a density function.
    b) Compute the probability that a randomly selected x will lie in the interval $1 \le x \le 4$.

4. *a)* Convert $f(x) = 2 - x^{-2}$, $1 \le x \le 20$, to a density function.
    b) Compute the probability that a randomly selected x will lie in the interval $1 < x < 10$.

5. *a)* Convert $f(x) = 12x - 3x^2$, $0 \le x \le 4$, to a density function.
    b) Compute the probability that a randomly selected x will lie in the interval $0 < x < 2$.

13.5 EXPECTED VALUE

We shall use the opening illustration of the chapter to show the meaning of *expected value*. Recall that Fran assumed a ship was equally likely to arrive at any time during the interval from 1 to 6 o'clock. Now suppose the unlikely circumstance that Fran has repeated occasions to be faced with the same situation and that on each occasion she records the time at which the ship arrives and then computes the average of these times. Because times on the interval 1 to 6 are equally likely to occur, we would expect that, on repeated trials, the average time would be the midpoint of this interval, which is at 3.5, or 3:30 o'clock, and this is called the *expected* time of arrival. Thus, expected value is an average value. What we wish to show first is that Fran's expected value for the time of arrival can be computed as

$$\int_1^6 xp(x)\, dx;$$

that is, the definite integral over the entire range of *x*'s of *x* times the density function. For Fran's density function, $p(x) = \frac{1}{5}$, we compute the expected value of *x*, symbolized by $E(x)$, as

$$E(x) = \int_1^6 x\left(\frac{1}{5}\right) dx$$

$$= \frac{1}{5}\left(\frac{x^2}{2}\right)\Big|_1^6$$

$$= \frac{36 - 1}{10} = \frac{35}{10} = 3.5,$$

which is the value anticipated earlier in this section.

Definition: Expected value of $x = E(x) = \int_{\text{all } x} xp(x)\, dx,$

where the "all *x*" on the integral sign means that the limits on the integral are the endpoints of the entire interval of permissible values of *x*.

The expected value of *x*, $E(x)$, is also referred to as the mean (or average) of the *x* values and is designated by μ (the Greek letter mu.) Thus,

$$E(x) = \mu.$$

The mean, μ, has another interpretation that aids understanding; namely, μ is the *x*-coordinate of the *center of gravity* of a graphical representation of the density function. To see what is meant here, suppose Figure 13–1 was drawn on a piece of cardboard, then the rectangle was cut out with scissors and placed vertically across a knife edge, ($\wedge$), as shown in Figure 13–3. The rectangle would just balance, not tipping down at one side, if the knife edge is at the center of gravity, $\mu = 3.5$.

By way of contrast, consider

$$p(x) = \frac{10 - x}{50}; \quad 0 \le x \le 10,$$

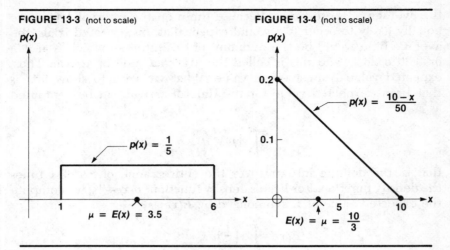

FIGURE 13-3 (not to scale) FIGURE 13-4 (not to scale)

which is shown in Figure 13–4. If we cut this triangle out, it will not balance at the midpoint of the base, which is at $x = 5$, but at a point, μ, which is to the left of the midpoint. To see where μ is, we compute

$$\mu = E(x) = \int_0^{10} xp(x)\, dx$$

$$= \int_0^{10} x\left(\frac{10-x}{50}\right) dx$$

$$= \int_0^{10} \frac{10x - x^2}{50}\, dx$$

$$= \frac{1}{50}\left(5x^2 - \frac{x^3}{3}\right)\Big|_0^{10}$$

$$= \frac{1}{50}\left[\left(500 - \frac{1000}{3}\right) - 0\right]$$

$$= 10 - \frac{20}{3} = \frac{10}{3}.$$

Thus, $\mu = 10/3$, as shown on Figure 13–4.

Exercise. Find the expected value of x, $E(x) = \mu$, if $p(x) = x/18$, $0 \le x \le 6$. Answer: $\mu = 4$.

Example. Jon sells oil in amounts up to 20 gallons, and uses

$$p(x) = 0.004x + 0.01,\ 0 \le x \le 20$$

as the density function for amounts purchased by customers. *a)* Find the average (expected value) of the amounts purchased by customers. *b)* Estimate the amount that will be purchased by 30 customers.

a) We compute the expected value as

$$\mu = \int_0^{20} xp(x)dx = \int_0^{20} x(0.004x + 0.01)dx$$

$$= \int_0^{20} (0.004x^2 + 0.01x)dx$$

$$= \left(\frac{0.004x^3}{3} + \frac{0.01x^2}{2} \right) \Big|_0^{20}$$

$$= \frac{0.004(20)^3}{3} + \frac{0.01(20)^2}{2} - 0$$

$$= \frac{32}{3} + 2$$

$$= \frac{38}{3} \text{ gallons per customer.}$$

b) With an average purchase of 38/3 gallons per customer, total purchases by 30 customers would be estimated at

(number of customers)(average purchased per customer)

$$= 30\left(\frac{38}{3}\right) = 380 \text{ gallons.}$$

Exercise. In the last example, estimate total sales to 90 customers. Answer: 1140 gallons.

13.6 VARIANCE AND STANDARD DEVIATION

The standard deviation is denoted by the small Greek letter σ (sigma) and its square, σ^2, is called the variance. The definition for the variance of x, $V(x)$, is

$$V(x) = \sigma^2 = \int_{\text{all } x} (x - \mu)^2 p(x) \, dx.$$

The term *variance* has been chosen to emphasize that σ^2 is a measure of variability of x about its mean, μ, and this is captured in the definition by the factor

$$(x - \mu)^2$$

because $x - \mu$ represents a deviation or variation of x from its mean. In the beginning, variance has little, if any, intuitive appeal, but as we proceed through the chapter, the significance of this measure will become clear. For the present, we shall consider only computations of the variance.

Example. Compute the variance and standard deviation of x if its density function is $p(x) = x/288$, $0 \le x \le 24$.

We must compute μ first. This is

$$\mu = \int_0^{24} x\left(\frac{x}{288}\right)dx = \frac{1}{288}\int_0^{24} x^2\,dx = \frac{1}{288}\left(\frac{x^3}{3}\right)\Big|_0^{24} = \frac{1}{288}\left(\frac{13824}{3}\right) = 16.$$

Next we find

$$V(x) = \sigma^2 = \int_0^{24} (x - \mu)^2 p(x)\,dx$$

$$= \int_0^{24} (x - 16)^2\left(\frac{x}{288}\right)dx$$

$$= \frac{1}{288}\int_0^{24} (x^2 - 32x + 256)(x)\,dx$$

$$= \frac{1}{288}\int_0^{24} (x^3 - 32x^2 + 256x)\,dx$$

$$= \frac{1}{288}\left(\frac{x^4}{4} - 32\frac{x^3}{3} + 256\frac{x^2}{2}\right)\Big|_0^{24}$$

$$= \frac{1}{288}(82{,}944 - 147{,}456 + 73{,}728)$$

$$= \frac{9216}{288}$$

$$\sigma^2 = 32$$

$$\sigma = \sqrt{32} = 5.66.$$

Thus, the variance is $\sigma^2 = 32$ and the standard deviation is the square root of the variance, which is approximately $\sigma = 5.66$.

> **Exercise.** Find the mean, variance, and standard deviation of x if the density function is $p(x) = x/72$, $0 \le x \le 12$. Answer: $\mu = 8$; $V(x) = \sigma^2 = 8$; $\sigma = 2.828$.

Figures 13–5 and 13–6 show the density functions of the last example and exercise. Note that the density function of Figure 13–5 spreads out over a wider interval than is the case in Figure 13–6, and it is

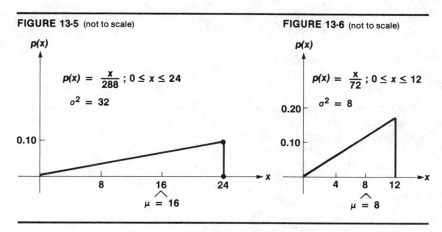

FIGURE 13-5 (not to scale)

$p(x)$

$$p(x) = \frac{x}{288}; \ 0 \le x \le 24$$

$$\sigma^2 = 32$$

0.10

8   16   24

$\mu = 16$

FIGURE 13-6 (not to scale)

$p(x)$

$$p(x) = \frac{x}{72}; \ 0 \le x \le 12$$

0.20 $\sigma^2 = 8$

0.10

4   8   12

$\mu = 8$

the spread or variability that is measured by the variance or its square root, the standard deviation. We shall see the importance of σ in more understandable form a little later in the chapter when we introduce the *normal* probability density function.

13.7 PROBLEM SET 13–2

1. Given $p(x) = 0.05$, $0 \le x \le 20$,
   a) Find μ.   *b)* Find σ^2.   *c)* Find σ.
2. Given $p(x) = 0.01$, $0 \le x \le 100$,
   a) Find μ.   *b)* Find σ^2.   *c)* Find σ.
3. Given $p(x) = \dfrac{x}{18}$, $0 \le x \le 6$,

   a) Find the expected value of x.
   b) Find the variance of x.
   c) Find the standard deviation of x.
4. Given $p(x) = 2x/9$, $0 \le x \le 3$,
   a) Find the expected value of x.
   b) Find the variance of x.
   c) Find the standard deviation of x.
5. Fran sells oil to customers in amounts of x hundred gallons, x going from 0 to 5, with the density function

   $$p(x) = 0.04x + 0.1,   0 \le x \le 5.$$

   a) Find average (expected) sales per customer.
   b) Estimate total sales to 72 customers.
6. Sam sells oil to customers in amounts of x thousand gallons, x going from 0 to 2 with the density function

   $$p(x) = 0.2x + 0.3,   0 \le x \le 2.$$

a) Find average (expected) sales per customer.
b) Estimate total sales to 90 customers.

7. The density function

$$p(x) = \frac{3(1-x^2)}{4}, \qquad -1 \le x \le 1$$

is a vertical parabola opening downward, with vertex at $(0, 3/4)$ and intercepts at $(-1, 0)$ and $(1, 0)$.
a) Sketch the density function.
b) Why must μ equal zero?
c) Verify that $\mu = 0$.
d) Find the standard deviation of x.

8. The density function

$$p(x) = \frac{3}{32}(4 - x^2), \qquad -2 \le x \le 2$$

is a vertical parabola opening downward, with vertex at $(0, 3/8)$ and intercepts at $(-2, 0)$ and $(2, 0)$.
a) Sketch the density function.
b) Why must μ equal zero?
c) Verify that $\mu = 0$.
d) Find the standard deviation of x.

13.8 THE EXPONENTIAL DISTRIBUTION

A commonly occurring applied problem arises in situations where arrival times have to be considered. For example, in a bank where tellers wait upon customers, satisfactory (timely) service depends not only on the number of tellers but the times of arrival of customers. A number of tellers that would be adequate to provide good service to all customers if arrival times were uniformly spaced would not be able to maintain timely service to all if arrivals bunch up from time to time. One model of arrival times that has been useful in some applications is based upon an exponential function of the form

$$p(x) = be^{-bx}, \qquad 0 \le x \le \infty. \tag{1}$$

This is a proper density function because it is nonnegative and has the required area of one, as shown next.

$$\int_0^\infty be^{-bx}\,dx = \frac{be^{-bx}}{-b}\bigg|_0^\infty = -e^{-bx}\bigg|_0^\infty = -\frac{1}{e^{bx}}\bigg|_0^\infty$$

$$= -\left[\lim_{x \to \infty}\frac{1}{e^{bx}} - \frac{1}{e^0}\right]$$

$$= -[0 - 1]$$

$$= 1.$$

The expected value of x is

$$\mu = E(x) = \int_0^\infty x(be^{-bx})dx \tag{2}$$

$$= b\int_0^\infty xe^{-bx}\,dx.$$

Rule 17 from Table XII-B at the back of the book shows the desired integral is

$$\mu = b\left[\frac{e^{-bx}(-bx-1)}{b^2}\right]\Big|_0^\infty = -\frac{1}{b}\left(\frac{bx+1}{e^{bx}}\right)\Big|_0^\infty. \tag{3}$$

The evaluation of (3) at ∞ requires that we find

$$\lim_{x\to\infty}\left(\frac{bx+1}{e^{bx}}\right) \tag{4}$$

and an evaluation of this particular type has not been encountered at a previous point in the text. The problem is that both the numerator, $bx+1$, and the denominator, e^{bx}, grow without limit (approach ∞) as x goes toward ∞. To readers who have developed an understanding of the growth in functions, it will be sufficient to state that the exponential function in the denominator increases much more rapidly than the linear function in the numerator and, as a consequence, the expression in (4) becomes smaller and smaller, approaching zero as a limit as x becomes very large ($x\to\infty$). To *prove* that the limit of (4) is zero, we have to call upon the following application of a procedure known as *L'Hospital's Rule:*

> ***Rule:*** If $f(x)$ and $g(x)$ approach ∞ as $x\to\infty$, then
>
> $$\lim_{x\to\infty}\frac{f(x)}{g(x)} = \lim_{x\to\infty}\frac{f'(x)}{g'(x)},$$
>
> provided $f'(x)$, $g'(x)$, and the limit of their quotient exist.

To apply the rule to (4), we find the derivative of the numerator and denominator to be:

$$\text{Numerator:}\quad \frac{d}{dx}(bx+1) = b$$

$$\text{Denominator:}\quad \frac{d}{dx}(e^{bx}) = be^{bx}.$$

Then, by the rule,

$$\lim_{x\to\infty}\left(\frac{bx+1}{e^{bx}}\right) = \lim_{x\to\infty}\left(\frac{b}{be^{bx}}\right) = \lim_{x\to\infty}\left(\frac{1}{e^{bx}}\right) = 0.$$

736

Hence, returning to (3),

$$\mu = -\frac{1}{b}\left(\frac{bx+1}{e^{bx}}\right)\Bigg|_0^\infty$$

$$= -\frac{1}{b}\left[\lim_{x\to\infty}\left(\frac{bx+1}{e^{bx}}\right) - \frac{b(0)+1}{e^0}\right]$$

$$= -\frac{1}{b}[0-1]$$

$$= \frac{1}{b}.$$

Consequently,

$$\mu = \frac{1}{b} \qquad \text{or} \qquad b = \frac{1}{\mu}$$

so that the density function, (1), $p(x) = be^{-bx}$, becomes the *exponential density function*

$$p(x) = \frac{1}{\mu}\left(e^{-x/\mu}\right) = \frac{e^{-x/\mu}}{\mu}.$$

The cumulative exponential density function. Figure 13–7 shows the shape of exponential density functions. The probability that ran-

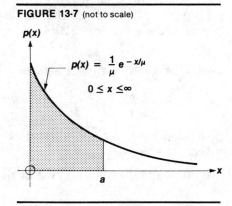

FIGURE 13-7 (not to scale)

$$p(x) = \frac{1}{\mu}e^{-x/\mu}$$

$$0 \le x \le \infty$$

domly occurring x will be in the interval 0 to a is the shaded area in the figure, which is calculated to be

$$\int_0^a p(x)\,dx = \int_0^a \frac{e^{-x/\mu}}{\mu}\,dx$$

$$= \frac{1}{\mu}\int_0^a e^{-x/\mu}\,dx$$

$$= \frac{1}{\mu} \frac{(e^{-x/\mu})}{-\frac{1}{\mu}} \Big|_0^a$$

$$= \frac{1}{\mu} (e^{-x/\mu})(-\mu) \Big|_0^a$$

$$= -(e^{-x/\mu}) \Big|_0^a$$

$$= -[e^{-a/\mu} - e^0]$$

$$= -e^{-a/\mu} + 1$$

$$= 1 - e^{-a/\mu}.$$

The last expression provides the probability over the interval 0 to a specific value of x, namely $x = a$. In general, for any value of x, we may use $P(0 \text{ to } x)$ to represent the probability assigned to the interval from 0 to x and write:

Cumulative exponential density function, Mean μ:

$$P(0 \text{ to } x) = 1 - e^{-x/\mu}.$$

Note carefully that a *cumulative* density function is a formula for direct computation of probabilities, but a density function must be integrated to obtain a probability.

Example. If a variable has the exponential distribution with a mean of 10, find the probability that a randomly selected value of the variable lies in the following intervals:

a)   0 to 5.   *b)*   1 to 4.   *c)*   More than 3.

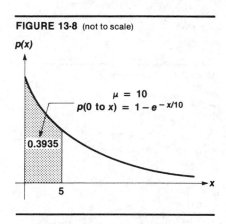

FIGURE 13-8 (not to scale)

a) Figure 13–8 shows the desired probability as a shaded area over the interval 0 to 5. The area is

$$P(0 \text{ to } 5) = 1 - e^{-5/10} = 1 - e^{-0.5} = 1 - 0.6065 = 0.3935.$$

b) Figure 13–9 shows the desired probability as a shaded area over the interval 1 to 4. This area is computed as the area over 0 to 4 minus the area over 0 to 1. Thus,

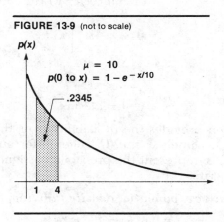

FIGURE 13-9 (not to scale)

$$
\begin{aligned}
P(1 \text{ to } 4) = P(0 \text{ to } 4) - P(0 \text{ to } 1) &= 1 - e^{-4/10} - (1 - e^{-1/10}) \\
&= 1 - e^{-0.4} - 1 + e^{-0.1} \\
&= e^{-0.1} - e^{-0.4} \\
&= 0.9048 - 0.6703 \\
&= 0.2345.
\end{aligned}
$$

c) Figure 13–10 shows the desired probability for "more than 3" as the *tail* area extending over the interval from 3 to ∞. Because the total area over 0 to ∞ is 1, the shaded area is

$$P(3 \text{ to } \infty) = 1 - P(0 \text{ to } 3) = 1 - (1 - e^{-3/10}) = e^{-0.3} = 0.7408.$$

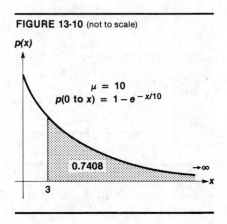

FIGURE 13-10 (not to scale)

> *Exercise.* If x has the exponential density function with mean 5, find the probability that a randomly occurring value of x will lie in the interval *a)* 0 to 5. *b)* 2 to 3. *c)* 4 or more. Answer: *a)* 0.6321. *b)* 0.1215. *c)* 0.4493.

The expected value, μ, in the exponential case is the *average interval between occurrences*. This could be the average *time* interval between calls received at a telephone switchboard or the average *distance* traveled between accidents when driving a truck. In neither case, however, is there any compelling reason that the exponential function is appropriate for determining the probability of an occurrence in a given interval. The question, as always, is whether the exponential model adequately describes a real-world situation. In considering the use of the model it should be kept in mind that an interval starts here and now, and a current probability calculation is not influenced by what has already occurred. For example, if the exponential model is used to compute the probability that the next call at a switchboard will come within the next 5 minutes, the interval starts now, extends 5 minutes, and the probability is not influenced by occurrence of previous calls. This characteristic sometimes is described by saying the exponential is *memoryless*. The same term applies in the binomial model; that is, for example, a coin does not have a memory and what happens on the next toss is not influenced by the previous tossing history of the coin.

Example. On the average, 30 customers per hour arrive at a bank. Assuming exponential density, what is the probability that the next customer arrives within the next 3 minutes?

Here the average of 30 is given *per hour*, but the question has to do with *minutes*. Consequently, the proper value for μ is the average number of minutes between arrivals. We have

$$30 \text{ arrivals in one hour} = 30 \text{ arrivals in 60 minutes}$$

$$\mu = \frac{60 \text{ minutes}}{30 \text{ arrivals}}$$

$$\mu = 2 \text{ minutes between arrivals.}$$

Then,

$$P(0 \text{ to } 3 \text{ minutes}) = 1 - e^{-3/2} = 1 - e^{-1.5} = 1 - 0.2231 = 0.7769.$$

The foregoing example shows the exponential density function in applied context and also calls attention to the need for caution in selecting the proper value for μ.

13.9 PROBLEM SET 13–3

Use the cumulative exponential density function for Problems 1–6.

1. If $\mu = 2$, find the probabilities for the following intervals:
    a) 0 to 3.   b) 2 to 4.   c) More than 5.

2. If $\mu = 4$, find the probabilities for the following intervals.
    a) 0 to 2.   b) 3 to 5.   c) More than 4.

3. The average interval between customer arrivals at a gasoline station is 4 minutes. Find the probability that
    a) The next customer arrives in the next 3 minutes.
    b) No customer will arrive during the next 2 minutes.

4. The average time between breakdowns of a computer is 10 days. Find the probability that
    a) The next breakdown will occur in the next 5 days.
    b) There will be no breakdown in the next 8 days.

5. On the average, a stockbroker receives 40 calls per eight-hour day from clients.
    a) Find the probability that the next call will come during the next hour.
    b) If the broker takes a half-hour off, what is the probability that no calls will come while the broker is away?

6. For new television sets of a certain model, the time until first failure of the picture tube averages five years.
    a) What is the probability of tube failure in the first nine months?
    b) If the manufacturer guarantees to replace the tube if it fails within three months, what is the probability that the manufacturer will *not* have to replace the tube on a newly-purchased set?

7. If the function $f(x) = ke^{-4x}$ is to be a density function over $0 \le x \le \infty$, the value of k must be such that the integral over 0 to ∞ equals 1.
    a) Integrate $f(x)$ over the interval 0 to ∞ and write the expression for the definite integral treating k as an unknown constant.
    b) Set the expression in (a) equal to 1, solve for k, and write the density function, $p(x)$.
    c) Find μ, the expected value of x, by integration.

8. If the function $f(x) = k/x^3$ is to be a density function over $1 \le x \le \infty$, the value of k must be such that the integral over 1 to ∞ equals 1.
    a) Integrate $f(x)$ over the interval 1 to ∞ and write the expression for the definite integral, treating k as an unknown constant.
    b) Set the expression in (a) equal to 1, solve for k, and write the density function, $p(x)$.
    c) Find μ, the expected value of x, by integration.

13.10 THE NORMAL DENSITY FUNCTION

The widely-used normal density function (normal distribution) has the bell shape shown in Figure 13–11. The function that is the source of the normal distribution is

FIGURE 13-11

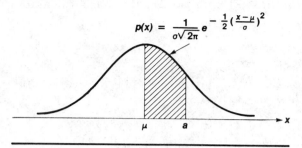

$$p(x) = \frac{1}{\sigma\sqrt{2\pi}}e^{-\frac{1}{2}\left(\frac{x-\mu}{\sigma}\right)^2}$$

$$f(x) = e^{-\frac{1}{2}\left(\frac{x-\mu}{\sigma}\right)^2}, \quad -\infty \leq x \leq \infty.$$

To qualify as a density function it is necessary to determine k so that

$$k\int_{-\infty}^{+\infty} f(x)\,dx = 1.$$

By advanced methods, it can be proved that

$$k = \frac{1}{\sigma\sqrt{2\pi}}.$$

Hence the equation of the normal probability density function is

$$p(x) = \frac{1}{\sigma\sqrt{2\pi}}e^{-\frac{1}{2}\left(\frac{x-\mu}{\sigma}\right)^2}.$$

The parameters in the last expression are μ, the mean or expected value, and σ, the standard deviation. As shown in Figure 13–11, a normal distribution has symmetry with respect to a vertical line at μ; that is, in Figure 13–11, the curve has the same height at a point a units to the left or right of μ. The standard deviation measures the variability of x around the mean so that a distribution with a small value of σ concentrates a major proportion of the probability (area) over a small interval around μ. If σ is large, the major central proportion of the probability is spread over a wide interval around μ. Often, we specify a normal distribution by $N(\mu, \sigma)$, meaning "normal, with mean mu and standard deviation sigma." For example, $N(100, 5)$ specifies a normal distribution with mean 100, standard deviation 5 (variance 25). This specific distribution function is

$$p(x) = \frac{1}{5\sqrt{2\pi}}e^{-\frac{1}{2}\left(\frac{x-100}{5}\right)^2}.$$

Normal curves may be described as being bell-shaped, symmetrical about the mean, and approaching the x-axis asymptotically in both directions.

> **Exercise.** Write the equation of the normal distribution $N(40, 8)$. Geometrically, how would this distribution compare with $N(30, 2)$? Answer:
>
> $$p(x) = [1/8\sqrt{2\pi}]\left[e^{-\frac{1}{2}\left(\frac{x-40}{8}\right)^2}\right].$$
>
> This distribution would have its mean to the right of $N(30, 2)$ and would spread out more than $N(30, 2)$.

13.11 NORMAL PROBABILITY TABLE

Starting from first principles to find the probability that a randomly selected x will lie in a specified interval, we would try to integrate the normal density function between the interval limits. As it happens, this function cannot be integrated exactly, but it is possible to make approximations. It seems natural to suggest that the approximations for various sets of limits be made and the outcomes tabulated, but this raises a problem. There is not just one normal distribution. There is a normal distribution for each of the limitless combinations of mu and sigma. Fortunately, this latter problem can be handled by changing every normal distribution to a *standard* form. To see how this is done, consider the problem of evaluating

$$\frac{1}{\sigma\sqrt{2\pi}}\int_\mu^a e^{-\frac{1}{2}\left(\frac{x-\mu}{\sigma}\right)^2}\,dx.$$

(See Figure 13–11.) We introduce a new variable, z, called the standardized normal deviate, where

$$z = \frac{x-\mu}{\sigma}.$$

We must now make the following conversions:

If

$$z = \frac{x-\mu}{\sigma}$$

then

$$\frac{dz}{dx} = \frac{d}{dx}\left(\frac{x-\mu}{\sigma}\right) = \frac{1}{\sigma}$$

$$\sigma \, dz = dx.$$

The last statement says we must replace dx in the integral by $\sigma \, dz$. Attention turns next to the limits on the integral. The limit $x = \mu$ becomes

$$z = \frac{x-\mu}{\sigma} = \frac{\mu-\mu}{\sigma} = 0.$$

Similarly, the limit $x = a$ becomes

$$z = \frac{a-\mu}{\sigma}.$$

Substituting, we now have the expression

$$\frac{1}{\sqrt{2\pi}} \int_0^{\frac{a-\mu}{\sigma}} e^{-\frac{z^2}{2}} \, dz$$

in place of the original integral. See Figure 13–12.

FIGURE 13-12

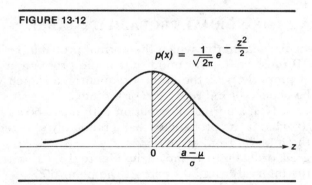

$$p(x) = \frac{1}{\sqrt{2\pi}} e^{-\frac{z^2}{2}}$$

To see what has been accomplished, consider the two normal distributions $N(50, 5)$ and $N(78, 2)$. Asked to integrate $N(50, 5)$ between the limits 50, which is μ, and 55, which is a, we find

$$z = \frac{55-50}{5} = 1$$

for the z corresponding to 55. The z corresponding to μ is, of course, always zero. Our problem here is to integrate the standardized normal distribution from $z = 0$ to $z = 1$.

Suppose, next, we are asked to integrate the second distribution, $N(78, 2)$, between the x limits 78, which is μ, and 80, which is a. We find again that the z limits are 0 and 1. Both problems lead to the evaluation of

$$\frac{1}{\sqrt{2\pi}} \int_0^1 e^{-\frac{z^2}{2}} \, dz.$$

Clearly, the same evaluation arises if the transformation to z yields the interval from $z = 0$ to $z = 1$, no matter what normal distribution is at hand.

Exercise. Given $N(20, 3)$, what z-values correspond to $x = 26$ and $x = 18.5$? Answer: 2, $-1/2$.

Values of the definite integral

$$\frac{1}{\sqrt{2\pi}} \int_0^z e^{-\frac{z^2}{2}} \, dz$$

are given in Table VIII at the end of the book for various values of z. We find, for example, when $z = 1$, the tabulated probability is 0.3413.

13.12 USING THE NORMAL PROBABILITY TABLE

The particular manner in which the normal probabilities are given in Table VIII must be kept in mind so that the problem at hand will be matched properly with the table. Frequently, a sketch will serve to lessen the chance of improper use of the table.

Example. Given a normal distribution with mean 50 and standard deviation 10, what is the probability that a randomly selected number will lie in the interval from 50 to 65?

The desired probability is shown in Figure 13–13 as the shaded area over the interval from 50 to 65. At the point 65:

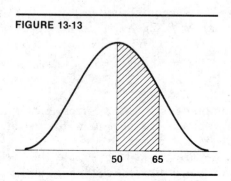

FIGURE 13-13

$$z = \frac{65 - 50}{10} = 1.5.$$

From Table VIII, with $z = 1.5$, we find the desired probability to be 0.4332.

Example. Given $N(50,10)$, as in the last example, what is the probability that x will fall in the interval 42 to 50? As Figure 13–14 shows,

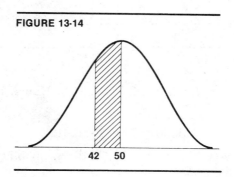

FIGURE 13-14

42 lies to the left of the mean. However, the curve is symmetrical. The probability is the same for a given value of z, whether it be positive or negative. Here

$$z = \frac{42 - 50}{10} = -0.8.$$

The desired probability is found by entering Table VIII with z equal 0.8. It is 0.2881.

Exercise. Given $N(50, 10)$ as in the last example, what is the probability that x will be in the interval *a)* 50 to 62? *b)* 45 to 50? Answer: *a)* 0.3849. *b)* 0.1915.

Example. Given $N(100, 20)$ what is the probability that x will be greater than 145?

The pertinent area is shown in Figure 13–15. It is a *right tail* area. Table VIII provides areas only for intervals that start at the mean (that is, at $z = 0$). We can find the desired area by applying the table, then subtracting from 0.5, inasmuch as the entire area to the right of the mean is 0.5.

$$z = \frac{145 - 100}{20} = 2.25.$$

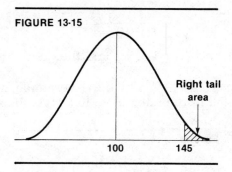

FIGURE 13-15

Right tail area

100   145

The tabulated entry for $z = 2.25$ is 0.4878. Hence the desired area is

$$0.5000 - 0.4878 = 0.0122.$$

Example. Given $N(2, 0.1)$, what is the probability that x will lie in the interval 1.95 to 2.1?

Figure 13–16 shows that we must add two tabulated entries.

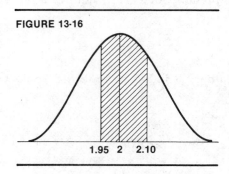

FIGURE 13-16

1.95 2 2.10

At 1.95, z is -0.5, and the area is 0.1915.
At 2.1, z is 1, and the area is 0.3413.

Adding, we find the desired probability as

$$0.1915 + 0.3413 = 0.5328.$$

Example. Given $N(15, 2)$, find the probability that x will lie in the interval from 16 to 17.

Analysis of Figure 13–17 shows that the tabulated areas at $z = 1$ and $z = 0.5$ must be subtracted. Hence:

$$0.3413 - 0.1915 = 0.1498.$$

FIGURE 13-17

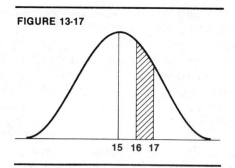

15 16 17

13.13 PROBLEM SET 13–4

1. Given a normal distribution with mean 150 and standard deviation 20, find the area over each of the following intervals:

    a) 150 to 160, inclusive.   b) 150 up to but not including 160.
    c) Beyond 170.   d) 160 to 170.
    e) Below 135.   f) 145 to 165.
    g) 142 to 150.   h) 138 to 146.

2. Given $N(50, 5)$ find the probability that a randomly selected x will lie in each of the following intervals:

    a) 45 to 65.   b) 40 to 60.
    c) 35 to 65.   d) Outside the interval from 40 to 60.
    e) Above 46.   f) 44 to 48.
    g) Below 55.   h) Above 56.

3. Given $N(10, 2)$, find the probability that a randomly selected x will lie in each of the following intervals:

    a) Above 6.   b) 11 to 15.
    c) Below 9.   d) 8 to 10.
    e) 8 to 9, excluding the end   f) 8 to 12.
        points.
    g) Above 11.5.   h) 9 to 13.
    i) Above 16.

13.14 ESTIMATING THE MEAN AND THE STANDARD DEVIATION

The normal probability density function as a mathematical model can be related approximately to numerous real-world situations. Suppose, for example, that an automatic machine is adjusted to produce shafts that have a diameter of 2 inches. As shaft after shaft comes off the machine, it is not expected that each shaft will have a diameter of precisely 2 inches. Shaft diameters will vary somewhat, the amount of variability being dependent upon the precision of the machine. Quality control engineers often assume that the probability of various departures from the nominal figure, 2 inches, can be estimated by application of the normal distribution.

If we are to apply a normal distribution to a real-world problem, we must have numbers for the mean and standard deviation of the distribution. In the mathematical model, of course, these quantities represent the central point and the degree of spread of an infinite set of values of the variate x. In the real world, we work with a finite set of numbers and think of them as a *sample* of x's drawn at random from the infinite set. The numbers in the sample reflect the central tendency and variability of the infinite set. We estimate the mean and the standard deviation for the infinite set from the numbers in the sample. In the remainder of this section, we shall show how the estimates are made.

Suppose that we select a sample of 10 shafts from the output of a machine and measure the diameter of each shaft, obtaining the numbers in the next table:

Diameters of 10 shafts, in inches				
2.00	2.01	1.97	2.02	1.97
1.99	2.03	1.99	2.01	2.01

Because we plan to use the sample to *estimate* μ and σ, we would prefer to have a sample of many more than 10 units. However, our purpose is only to show how the estimates are made, and a small sample will suffice. The mean, μ, for the model, is estimated by computing the mean (the ordinary average) of the sample. The procedure followed in estimating σ starts by estimating the variance, σ^2. Recalling that variance involves the squares of the amounts by which values differ from the mean, we proceed as follows.

1. Find the average of the sample data.
2. Subtract the average from a sample number, and square the difference.
3. Repeat Step 2 for each number in the sample, and then sum all the squares.
4. Divide the sum of squares in Step 3 by $n - 1$, where n is the number of values in the sample. The result of this step is the desired estimate of the variance.

The steps are illustrated in Table 13–1. The difference between each value of x and the sample average, 2, is shown in the second column whose heading is $x - 2$; that is, sample value minus sample average. The squares of the differences are shown in the third column, whose

TABLE 13–1

Sample value x	$x-2$	$(x-2)^2$
2.00	0.00	0.0000
1.99	−0.01	0.0001
2.01	0.01	0.0001
2.03	0.03	0.0009
1.97	−0.03	0.0009
1.99	−0.01	0.0001
2.02	0.02	0.0004
2.01	0.01	0.0001
1.97	−0.03	0.0009
2.01	0.01	0.0001
20.00	0.00	0.0036

Average $=\dfrac{20}{10}=2;\ n=10.$

Variance estimate $=\dfrac{0.0036}{9}=0.0004.$

Standard deviation estimate $=\sqrt{0.0004}=0.02.$

heading is descriptive of the method for obtaining its entries. The sum of squares is 0.0036; dividing this sum of squares by one less than the number of values in the sample, that is, by $n-1$ where n is 10, we obtain the variance estimate, 0.0004. The square root of the variance estimate, 0.02, is the estimated standard deviation.

The summation symbol, Σ, large sigma, can be put to good use in summarizing instructions for computing estimates of mean and standard deviation. Conventionally, if the variate at hand is x and we wish to indicate the average of a sample of x's, we write the variate with a bar over it, thus: $\bar{x}$, which is read as "x bar" or "bar x." If we now let Σx represent the sum of a set of values of x, and n represent the number of values in the set, the average of the set is

$$\bar{x}=\frac{\Sigma x}{n}.$$

Returning to Table 13–1, n is 10, Σx is 20; hence:

$$\bar{x}=\frac{20}{10}=2.$$

Turning to the second column of Table 13–1, we express the operation of subtracting the average from each value of x by the instruction $x-\bar{x}$. In the third column, we square each difference, an operation denoted by $(x-\bar{x})^2$. The sum of the third column is expressed as $\Sigma(x-\bar{x})^2$. Finally, dividing the last expression by $n-1$ provides the variance estimate, which we now label as s^2:

$$s^2 = \frac{\Sigma(x - \bar{x})^2}{n - 1}.$$

The standard deviation estimate, s, is

$$s = \sqrt{\frac{\Sigma(x - \bar{x})^2}{n - 1}}.$$

Example. Compute $\bar{x}$ and s for the values of x given in the first column of Table 13–2.

TABLE 13–2

x	$x - \bar{x}$	$(x - \bar{x})^2$
5	−2.2	4.84
8	0.8	0.64
8	0.8	0.64
9	1.8	3.24
6	−1.2	1.44
36		10.80

$$\bar{x} = \frac{\Sigma x}{n} = \frac{36}{5} = 7.2.$$

$$s^2 = \frac{\Sigma(x - \bar{x})^2}{n - 1} = \frac{10.80}{4} = 2.7.$$

$$s = \sqrt{\frac{\Sigma(x - \bar{x})^2}{n - 1}} = \sqrt{\frac{10.80}{4}} = \sqrt{2.7} = 1.64.$$

13.15 PROBLEM SET 13–5

Compute $\bar{x}$ and s for each given set of x's:

1. 8, 8, 9, 6, 4.
2. 17, 15, 14, 20, 18, 16, 10, 12, 15, 13.
3. 8, 13, 12, 13, 10.
4. 2, 2, 6, 5, 1, 5.
5. 0.64, 0.58, 0.57, 0.57, 0.52, 0.58, 0.60, 0.58.

13.16 $N(\mu, \sigma)$ IN THE REAL WORLD

In applications the areas for the normal distribution are interpreted as percentages and proportions as well as probability, as we shall see in the examples that follow. In each example the mean and the standard deviation are presumed to have been estimated from sample data by the methods of the previous section.

Example. An automatic filling machine can be set to pour a certain number of ounces of fluid into containers. Data collected from runs of the machine show that the *average* amount of fill per container is, for practical purposes, equal to the machine setting. However, the amount of fill per can varies, and the standard deviation of amount of fill per can has been estimated from sample data to be 0.4 ounces.

a) If the machine is set to pour 32 ounces, what is the probability that a container will receive less than 31.5 ounces?

We interpret this problem as one of finding the probability that x will be less than 31.5 in a normal distribution whose mean is 32 and

FIGURE 13-18

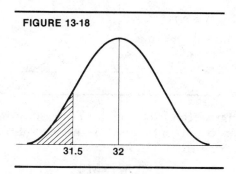

whose standard deviation is 0.4. Figure 13–18 illustrates the probability sought.

$$\text{At } 31.5, \ z = \frac{31.5 - 32}{0.4} = -1.25.$$

From Table VIII, with $z = 1.25$, the tabulated area is 0.3944. The shaded area in the sketch is

$$0.5000 - 0.3944 = 0.1056.$$

We estimate the probability that a can will receive less than 31.5 ounces to be 0.1056.

b) Given the circumstances in part *a)*, what percent of the containers will receive less than 31.5 ounces?

The probability, 0.1056, here is interpreted as the percent equivalent, 10.56 percent; that is, over the long run, about 10.56 percent of the containers will receive less than 31.5 ounces.

c) Given the circumstances in part *a)*, if 1000 containers are filled, approximately how many will contain less than 31.5 ounces?

Following the interpretation of part *b)*, we find 10.56 percent of 1000. This is

$$0.1056(1000) = 105.6$$

or about 106 containers.

Example. Given the same machine as in the last example, suppose that we wish to set the machine so that not more than 5 percent of the containers will receive less than 32 ounces. What should be the setting?

The nature of this problem is seen in Figure 13–19. We wish to

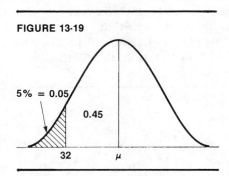

FIGURE 13-19

find the mean (the setting) so that the probability below 32 is 0.05 (that is, the given 5 percent). This means, by subtraction, that the area over the interval from 32 to μ must be

$$0.5 - 0.05 = 0.45.$$

We can find, by *inverse* use of Table VIII, what value of z corresponds to the area 0.45.

Turning to Table VIII, we see in the field that 0.4500 is between the tabulated values 0.4495 and 0.4505. The corresponding marginal values, z, are 1.64 and 1.65, respectively. By interpolation, $z = 1.645$.

Now, it is important to understand that in terms of the units of the original problem, *a value of z means z times the standard deviation.* In the present case, $z = 1.645$ means

$$1.645(0.4 \text{ ounces}) = 0.658 \text{ ounces.}$$

Hence, the interval from 32 up to μ is 0.658 ounces, which tells us the machine should be set at

$$\mu = 32 + 0.658 = 32.658 \text{ ounces}$$

if not more than 5 percent of the containers are to receive less than 32 ounces.

The italicized words in the immediately foregoing refer to the definition

$$z = \frac{x - \mu}{\sigma}$$

from which it follows that

$$z\sigma = x - \mu.$$

The interval $x - \mu$ is the distance from x to the mean. It equals z (for the x in question) times the standard deviation. In passing, we note that the question whether to add $z\sigma$ to x or to subtract $z\sigma$ from x to locate the mean depends upon which side of the mean our sketch shows x to be. In the present case the mean is to the right of 32, so we added $z\sigma$.

Example. If light bulbs of a certain type have an average burning life of 500 hours and a standard deviation of 40 hours, as estimated from experience, within what limits symmetrically located above and below 500 will the burning lives of half of such bulbs lie?

The desired limits are shown by question marks in Figure 13–20.

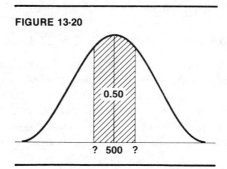

FIGURE 13-20

0.50

? 500 ?

If we want half (50 percent) of the numbers to fall between these limits, which are symmetrically located with respect to the mean, then 25 percent will be in each of the intervals 500 to ?. Entering Table VIII, we find that the z corresponding to an area of 0.25 is 0.675. Hence the length of the interval from 500 to ? is 0.675 times the standard deviation:

$$0.675(40) = 27.$$

The desired limits are

$$500 \pm 27 \quad \text{or} \quad 473 \text{ to } 527 \text{ hours.}$$

Example. A testing service reports the average score on a test as 130 points, and the standard deviation of test scores as 20 points. Jones takes the text and scores 118 points. What is Jones's percentile standing? See Figure 13–21.

We interpret percentile standing as the percent scoring less than Jones, and estimate it as being the percent below 118 in $N(130, 20)$:

FIGURE 13-21

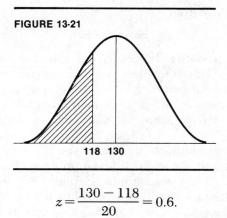

118 130

$$z = \frac{130 - 118}{20} = 0.6.$$

From Table VIII, with $z = 0.6$, we obtain the area 0.2257. Figure 13–21 shows that the area *below* Jones's 118 points is found by the subtraction

$$0.5000 - 0.2257 = 0.2743 = 27.43\%.$$

Hence, we estimate Jones's score as being at the 27th percentile.

Example. Study of records of the number of units of item K3 in inventory day by day shows average inventory to be 50 units, and the standard deviation of the day-by-day numbers to be 10 units. Suppose that, during a day, 25 units are requested. What is the probability that there will not be enough units in inventory to meet this day's requests?

We interpret this problem as one of finding the probability that x is less than 25 in $N(50, 10)$.

$$z = \frac{25 - 50}{10} = -2.5.$$

From Table VIII, with $z = 2.5$, we obtain the area 0.4938. According to Figure 13–22, the desired area is

FIGURE 13-22

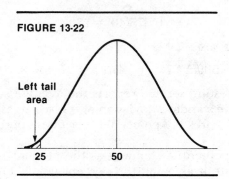

Left tail
area

25   50

$$0.5000 - 0.4938 = 0.0062.$$

The probability of not being able to meet the requests is the small figure 0.0062; that is, the chances are only six in a thousand that such requests could not be satisfied.

13.17 PROBLEM SET 13–6

All problems in this set are to be solved under the assumption that the normal distribution is applicable.

1. Given $N(50, 5)$:
    a) The probability is 0.17 that x will lie in the interval from 50 up to what number?
    b) The probability is 0.10 that x will exceed what number?
    c) The probability is 0.05 that x will be less than what number?
    d) The probability is 0.50 that x will lie in what interval symmetrically located above and below 50?
    e) The probability is 0.95 that x will be less than what number?

2. Assuming that a machine will turn out parts whose average diameter is the figure at which the machine is set, and that the standard deviation has been estimated from sample data to be 0.001 inches:
    a) If the machine is set at 2.00 inches, what percentage of parts made will have diameters exceeding 2.001 inches?
    b) If the machine is set at 0.100 inches, what percentage of parts made will have diameters exceeding 0.098 inches?
    c) If the machine is set at 1.500 inches, and 1000 parts are made, approximately how many parts will have diameters less than 1.498 inches?
    d) Specifications state that parts are to have 2.000-inch diameter, tolerance plus or minus 0.002; that is, parts whose diameters differ by more than 0.002 from 2.000 are to be classified as defectives. What percent of the parts will be defective?
    e) Within what limits symmetrically located about a setting of 1.575 inches will the diameters of 95 percent of the parts lie?
    f) If the machine is set at 2.250 inches, 5 percent of the parts will have diameters less than what number of inches?
    g) A loss of 40 cents is incurred for every part that is scrapped because its diameter is too *small*. Specifications state that the part is to have diameter 1.500 inches, tolerance ± 0.0015 inches. The machine is set at 1.500 inches. If 2000 parts are made, what will be the cost of the scrapped undersized parts?

3. The amounts poured by an automatic can-filling machine average out at the machine setting. Collected data indicate that the standard deviation of amounts of fill per can is 0.2 ounces.
    a) If the machine is set to pour 32 ounces, what is the probability that a can will contain less than 31.9 ounces?
    b) If the machine is set to pour 32 ounces, what percent of the cans will contain at least 31.9 ounces?

c) If the machine is set to pour 32 ounces and 2000 cans are filled, approximately how many cans will contain less than 31.7 ounces?

d) If the can label states the contents to be 32 ounces and the machine is set to pour 32.25 ounces, what percent of the cans will be under-filled?

e) If the can labels state contents to be 16 ounces, at what level should the machine be set if 99 percent of the cans are to contain at least 16 ounces?

4. The average breaking strength of certain connectors is 2000 pounds, $s = 80$ pounds.

a) Within what limits symmetrically located about 2000 will the breaking strength of 75 percent of the connectors lie?

b) What percent of the connectors will have a breaking strength of more than 2100 pounds?

c) The breaking strength of 95 percent of the connectors will exceed what number of pounds?

d) If a connector with breaking strength less than 1800 pounds is classified as defective, what percent of connectors will be defective?

5. A testing service reports the average score on a certain test to be 200 points. $s = 25$ points.

a) What percent of those taking the test score more than 275?

b) What percent of those taking the test score less than 160?

c) Half of those taking the test make scores in what interval symmetrically located above and below 200?

d) Ninety-five percent of those taking the test score above what number of points?

e) What would be the percentile standing of a person who scored 260 points on the test?

13.18 REVIEW PROBLEMS

1. *a)* Convert $f(x) = x + 1$, $0 \leq x \leq 10$, to a probability density function.
   b) What is the probability that a randomly selected x will lie in the interval 0 to 1? The interval 5 to 6?

2. *a)* Convert $f(x) = x^{1/3} - 1$, $1 \leq x \leq 8$, to a probability density function.
   b) Find the probability that a randomly selected x will lie in the interval 1 to 27/8.

3. *a)* Convert $f(x) = 50$, $20 \leq x \leq 45$, to a probability density function.
   b) Find the probability that a randomly selected x will lie in the interval 30 to 40.

4. Convert $f(x) = 10(x + 2)^{-3}$, $0 \leq x \leq \infty$, to a probability density function.

5. Given that

$$p(x) = \frac{12x - 3x^2}{32}$$

is a probability density function over the interval $x = 0$ to $x = 4$, find the probability that x will lie in the intervals 0 to 1; 0 to 2; 2 to 3.

6. If $p(x) = kx^{3/2}$ is a probability density function over the interval 0 to 100, what must be the value for k?

7. Given the probability density function $p(x) = 0.02$, $0 \le x \le 50$:
    a) Find the expected value, μ.
    b) Find the variance, σ^2.
    c) Find the standard deviation, σ.

8. Given the probability density function $p(x) = 0.012x + 0.04$, $0 \le x \le 10$:
    a) Find the expected value, μ.
    b) Find the variance, σ^2.
    c) Find the standard deviation, σ.

9. Find μ, the expected value of x, for the probability density function

$$p(x) = \frac{3}{x^4}, 1 \le x \le \infty.$$

10. The quantity of fish brought to port varies from day to day and has the probability density function

$$p(x) = 0.003(x^2 - 20x + 100), 0 \le x \le 10,$$

where x is in thousands of pounds per day.
    a) Find μ, the expected value of x.
    b) Estimate the total quantity of fish that will be brought to port in a 30-day period.

11. If x has the exponential density function with $\mu = 10$, find the probability that a randomly selected x will be in the interval
    a) 0 to 4.   b) 1 to 5.   c) More than 8.

12. The average interval between arrivals of customers at a bank is two minutes. Assuming the exponential density function applies, what is the probability that
    a) The next customer will arrive within the next five minutes?
    b) No customer will arrive during the next three minutes?

13. On the average, a bank teller provides service to 15 customers per hour. Assuming the exponential density function applies
    a) What is the probability that it will take less than five minutes to service the next customer?
    b) If customer B comes into the bank at the moment the teller starts servicing A, the only other customer present, what is the probability that B will have to wait more than three minutes for service?

14. Given a normal distribution with mean 30 and standard deviation 6, find the area over each of the following intervals:
    a) 30 to 33.   b) 36 to 42.   c) Less than 27.
    d) More than 48.   e) 24 to 33.   f) More than 21.

15. Given $N(100, 5)$, find the probability that a randomly selected x will lie in each of the following intervals:
    a) 90 to 105.   b) 105 to 115.   c) 92 to 96.
    d) Less than 103.   e) More than 88.   f) 112 to 115.

16. Compute the mean, $\bar{x}$, and the standard deviation, s, for each of the following:

a) 3, 7, 7, 9, 14.   b) 3, 5, 7, 7, 8.
c) 15, 15, 15.   d) 1.2, 0.7, 1.2, 1.3.

17. Given $N(20, 4)$:
    a) The probability is 0.20 that x will lie in the interval from 20 up to what number?
    b) The probability is 0.16 that x will exceed what number?
    c) The probability is 0.025 that x will be less than what number?
    d) The probability is 0.50 that x will lie in what interval symmetrically located above and below 20?
    e) The probability is 0.95 that x will be less than what number?

18. Assuming that a machine will turn out parts whose average diameter is the figure at which the machine is set, and that the standard deviation has been estimated from sample data to be 0.002 inches:
    a) If the machine is set at 2.500 inches, what percentage of parts made will have diameters exceeding 2.503 inches?
    b) If the machine is set at 2.000 inches, what percentage of parts made will have diameters exceeding 1.999 inches?
    c) If the machine is set at 2.250 inches, and 1000 parts are made, approximately how many parts will have diameters less than 2.245 inches?
    d) Specifications state that parts are to have 2.500-inch diameter, plus or minus 0.003; that is, parts whose diameters differ by more than 0.003 from 2.500 are to be classified as defectives. If the machine is set at 2.500 inches, what percent of parts will be defective?
    e) Within what limits symmetrically located about a setting of 3.000 inches will the diameters of 95 percent of the parts lie?
    f) If the machine is set at 2.500 inches, 10 percent of the parts will have diameters less than what number of inches?
    g) A loss of 50 cents is incurred for every part that is scrapped because its diameter is too small. Specifications state that the part is to have diameter 2.000 inches, tolerance ± 0.004 inches. If the machine is set at 2.000 inches and 5000 parts are made, what will be the cost of the scrapped undersized parts?

19. If 100 coins are tossed and the number of heads recorded, the number of heads, x will be *approximately* normally distributed with mean 50 and standard deviation 5. Assuming $N(50, 5)$,
    a) What percent of the time would more than 65 heads appear in tossing 100 coins?
    b) What is the probability that the number of heads will differ by more than 15, one way or the other, from 50?
    c) What percent of the time would the number of heads be between 40 and 60?

20. If light bulbs of a certain type have an average burning life of 1000 hours with a standard deviation of 100 hours,
    a) Within what limits symmetrically located above and below 1000 will burning lives of 50 percent of the bulbs fall?
    b) If a large number of these bulbs burn continuously, how long will it be before 40 percent have burned out?
    c) How many bulbs out of 2000 will have burning lives exceeding 1250 hours?

21. The amounts poured by an automatic can-filling machine average out at the machine setting. Collected data indicate that the standard deviation of the amounts of fill per can is 1 percent of the machine setting; that is, for example, if the machine is set at 50 ounces, the standard deviation is 1 percent of $50 = 0.50$ ounces.

    a) If the machine is set to pour 20 ounces, what is the probability that a can will contain less than 19.7 ounces?

    b) If the machine is set to pour 50 ounces, how many cans out of 1000 will contain between 49 and 51 ounces?

    c) If can labels state contents to be 48 ounces and the machine is set to pour 49 ounces, what percent of the cans will be underfilled?

    d) If specifications call for cans to contain at least 100 ounces, and the machine is set to pour 101 ounces, what percent of cans will meet specifications?

    e) In (d), if cans cannot hold any more than 102.3 ounces, what percent of cans will overflow?

    f) If the maximum amount cans hold is 50.8 ounces and the machine is set at 50 ounces, some cans will overflow. Overflows are defectives and are removed from the batch under production. How many cans will have to be filled if the number left after removing defectives is to be 5000?

22. A testing service reports the average score on a test to be 500 points with a standard deviation of 100 points.

    a) What percent of those taking the test score above 600? Above 700?

    b) Seventy-five percent of those taking the test make scores in what interval symmetrically located about the average?

23. Inventory on hand for item X is brought up to 150 at the end of each day. Demand for X averages 100 units a day, with a standard deviation of 20 units. On a particular day, what is the probability that demand will exceed inventory on hand?

Sets

appendix

1

A1.1 INTRODUCTION

It is not necessary to cover all of this appendix systematically before starting the text. The material contained here is referred to at various points in the text, and it will be sufficient to review the relevant sections when the referral is made.

A1.2 SET TERMINOLOGY

Some simple notions about groups or collections or *sets* are core ideas in mathematics. We use braces to indicate a set, and specify the *members* or *elements* of the set within the braces. Thus,

$$\{\text{Boston, Wellesley, Newton}\}$$

is a set whose members (elements) are the cities Boston, Wellesley, and Newton. Again,

$$\{3, 4, 7, 8\}$$

is a set of numbers whose elements (members) are 3, 4, 7, and 8. The symbols ϵ and $\notin$ are membership symbols, the first being read as "is a member of," the second as "is not a member of." For example,

$$3 \in \{3, 4, 7, 8\}$$
$$5 \notin \{3, 4, 7, 8\}$$

say that 3 is a member (or element) of the set $\{3, 4, 7, 8\}$ but 5 is not an element of this set.

Each element of a set must be unique. For example, in set terminology, the numbers 1, 3, 4, 4 would be a set whose elements are 1, 3, and 4. A set may have no elements, a finite number of elements, or an unlimited number of elements. The set with no elements is called the *empty* (or *null*) set and is symbolized by $\emptyset$, without braces. A set with an unlimited number of elements is said to be *infinite* set. Two sets are *equal* if, and only if, they contain exactly the same elements. The expression

$$\{\text{Integers between 5 and 6}\} = \emptyset$$

says the set of integers between 5 and 6 has no elements; it is the empty set.

> **Exercise.**   Read the following aloud:
>
> $\{\text{Red spades in a bridge deck}\} = \emptyset.$

The expression

$$\{\text{Integers between 5 and 7}\} = \{6\}$$

says that the set of integers between 5 and 7 is the set with the single element, 6. The next expression states that the set of integers between 5 and 10 is the set whose elements are 6, 7, 8, 9

$$\{\text{Integers between 5 and 10}\} = \{6, 7, 8, 9\}.$$

> **Exercise.**   Express in set symbols that the numbers whose square is 25 are 5 and −5.

The set specified by

$$\{\text{Integers}\}$$

is an infinite set, so we cannot list all its members.

> **Exercise.**   Using braces and the symbols for set membership, express the facts that ½ is not an integer and 13 is an integer. Answer: ½ $\notin$ {Integers}; 13 $\in$ {Integers}.

It is conventional to use a capital letter, without braces, when an entire set is to be named by one symbol. For example, we might let E represent the (infinite) set of even integers; thus

$$E = \{\text{Even integers}\}.$$

We may then refer to E by statements such as

$$3 \notin E \quad \text{and} \quad 4 \in E,$$

which say 3 is not a member of E and 4 is a member of E.

Set terminology is not limited to collections of numbers. We may talk, for example, about the set of residents of New York, classifying residents as being members of the set and nonresidents as not being members of the set.

Exercise. In set symbols, write the relationship of \$ to L, and Q to L, if $L = \{\text{Capital English letters}\}$.

A1.3 SET SPECIFICATION

A set is specified by identifying its elements. The *roster* method of specification *lists* each member of the set. The *descriptive* method states the rule or condition that distinguishes members of the set from nonmembers. Thus,

$$S = \{1, 3, 5, 7, 9\}$$

specifies a set by the roster method. The same set could be specified by the condition that its members be odd numbers between 0 and 10; a descriptive specification therefore would be

$$S = \{\text{Odd numbers between 0 and 10}\}.$$

In either specification it is clear whether an object is or is not a member of the set, for example,

$$3 \in S \quad \text{and} \quad 4 \notin S.$$

Exercise. Specify $W = \{\text{Days in the week}\}$ by the roster method.

The set of positive integers, like any infinite set, cannot be listed completely. We may specify the set by the descriptive method as $I = \{\text{Positive integers}\}$ or by

$$I = \{1, 2, 3, \ldots\}$$

where the three dots are read "and so on." We shall classify this last expression as being a roster specification, although in fact it is a roster-like descriptive specification.

> ***Exercise.*** Specify the set of positive odd integers by the "three dot" convention.

A1.4 SOLUTION SETS FOR EQUATIONS[1]

The statement, "two plus three equals five," is a *sentence* that can be written as

$$2 + 3 = 5.$$

Similarly, the sentence, "y plus three equals five," can be expressed as

$$y + 3 = 5.$$

Sentences that use the equality symbol $=$ are called equations. If we replace the symbol y by a number that makes the equation become a *true* sentence, the replacement number is a member of the *solution set* of the equation. In the case at hand, the solution set has only one element, 2. We may write

$$\{y : y + 3 = 5\} = \{2\}$$

where, at the left, the symbol y : is read as *the set of y's such that,* and the whole statement says the set of y's such that $y + 3 = 5$ is the set with the element 2.

> ***Exercise.*** *a)* Write the set symbols for $y + 1 = 6$. *b)* Translate the symbols $\{y : y - 2 = 10\} = \{12\}$. Answer: *a)* $\{y : y + 1 = 6\} = \{5\}$. *b)* The set of y's such that y minus two is ten is the set with the element 12.

A1.5 RELATIONS AND FUNCTIONS

A core concept in mathematics is the relationship, or correspondence, between the elements of two sets. A kind of correspondence known as a *function* is of particular interest, and to bring out the special nature of a functional relationship, think of one set of elements

---

[1] Extension of this material appears in Section 26 of Chapter 1.

as the set of mothers and the other set as the set of children. The set of mothers and the set of children are related, obviously, but there is a difference between the relationship of children to mothers and that of mothers to children. To see this difference, let children be the *starting* set and mothers be the *ending* set. If we now take a child from the starting set there is one, *and only one*, corresponding mother in the ending set. This one-to-one correspondence is an example of a *functional* correspondence. If, however, mothers are the starting set and we take a mother from this set, there can be one *or more* corresponding children in the ending set, so we have a relationship that is not one-to-one, and this is not a functional relationship.

One way to specify a function is to write pairs of elements in *order;* that is, the first element is the element from the starting set and the second is the corresponding element from the ending set. Thus, for example, the set of *ordered* pairs

$$\{(\text{child, mother})\}$$

is a function, but

$$\{(\text{mother, child})\}$$

is a relationship, not a function. Similarly, if we let x be any number whatsoever and $2x$ be twice whatever the number x is, the set of ordered pairs

$$\{(x, 2x)\}$$

is a function because for every first number, x, there is one, and only one, second number, $2x$. Clearly, this set of ordered pairs is infinite. Some ordered pairs in the set are

$$(0, 0); (1, 2); (5/2, 5).$$

Usually, we choose to represent functional correspondence by an equation, if it is possible to do so. For example, the function

$$\{(x, 2x)\}$$

can be represented by the equation

$$y = 2x.$$

When it is necessary to specify which element is the starting element, this element is specified in parentheses and is called the *independent* variable. For example, the last equation with functional specification would be

$$y(x) = 2x,$$

meaning x is the starting value *chosen independently* (arbitrarily). For example, taking x as 2, arbitrarily, we find the second element of the ordered pair, symbolized by $y(2)$, to be

$$y(2) = 2(2) = 4$$

and the ordered pair

$$(2, 4)$$

is in the *solution set*. The complete (infinite) solution set is *the set of ordered pairs, (x, y), such that y = 2x* and we symbolize this by

$$\{(x, y): y = 2x\},$$

where the colon is read as "such that."

In the work we shall encounter in this book, it will be helpful to view a function as a *rule* stating how the member of the ending set is obtained for an arbitrarily selected member of the starting set. For example, if we have

$$f(x) = 5x + 1,$$

where $f(x)$ is read as "the f function of x," or "f of x," the rule is

multiply 5 by x, then add 1.

Thus, selecting x arbitrarily to be 20, we find

5 times 20, plus $1 = 100$ plus $1 = 101$.

The last is simple substitution. However, we shall have to work with expressions such as

$$f(a + 3),$$

and in such a case it is important to remember that the rule tells us that $f(a + 3)$ means

multiply 5 by $a + 3$, then add 1.

We shall leave further discussion of this last matter to be developed at a point in the text where it becomes necessary.[2] For the present, we ask only that the reader be able to respond to questions such as those in the next exercise.

> **Exercise.** Given $g(z) = 2z + 3$, *a)* How is $g(z)$ read? *b)* What is $g(4)$? *c)* State (write) the function rule. *d)* How would $g(a + b)$ be obtained? Answer: *a)* "The g function of z" or "g of z." *b)* 11. *c)* Multiply 2 by the value of z, then add 3. *d)* Multiply 2 by $a + b$, then add 3.

---

[2] See Chapter 9, Section 3.

766

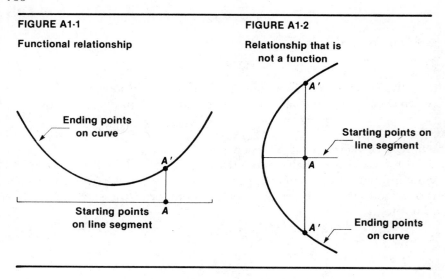

FIGURE A1-1

Functional relationship

Ending points
on curve

Starting points
on line segment

A

A'

FIGURE A1-2

**Relationship that is
not a function**

A'

Starting points on
line segment

A

A'

Ending points
on curve

We shall have frequent occasions to represent functions graphically, taking points on a horizontal line as being members of the starting set and a point on the graph *vertically* above or below the starting point as the corresponding member of the ending set. For example, Figure A1–1 represents a functional correspondence because for each point, such as *A*, in the starting set, there is one and only one point, *A'*, in the ending set. However, the relationship in Figure A1–2 is not a functional correspondence because a vertical line through a member of the starting set intersects the graph (the ending set) in two points.

A1.6 PROBLEM SET A1–1

1. Read the following aloud: for example, $A = \{5, 10, 15, \ldots, 100\}$ is read, "*A* is the set whose elements are 5, 10, 15, and so on, through 100."
   - *a)* $B = \{2, 4, 6\}$.
   - *b)* $C = \{0\}$.
   - *c)* $Q = \{1, 3, 5, \ldots, 29\}$.
   - *d)* $F = \{a, e, i, o, u\}$.
   - *e)* $L = \{\text{George, Charles}\}$.
   - *f)* $K = \{2, 4, 6, \ldots\}$.
2. Specify each of the following sets by the descriptive method:
   - *a)* $F = \{a, e, i, o, u\}$.
   - *b)* $S = \{3, 6, 9, \ldots\}$.
   - *c)* $P = \{1, 2, 3, 4, 5\}$.
   - *d)* $G = \{A, B, C, D, E\}$.
   - *e)* $R = \{100, 101, 102, \ldots\}$.
   - *f)* $J = \{-1, -3, -5, \ldots, -101\}$.
3. Specify each of the following by roster:
   - *a)* {Odd numbers less than 9}.
   - *b)* {First four months of the year}.
   - *c)* {Positive even multiples of 7}.

d) $\left\{\begin{array}{l}\text{Positive numbers, less than 1000,}\\\text{that are exactly divisible by 5}\end{array}\right\}$.

e) $\left\{\begin{array}{l}\text{Numbers representing the ratios}\\\text{formed by dividing integers by 0}\end{array}\right\}$.

f) {Squares of the integers from 1 through 10}.

4. Why is {0} not equal to Ø?

5. Using set membership symbols and the roster specification of the sets, write the following:

   a) 1/2 is not a positive integer.

   b) 64 is a multiple of 4.

   c) *a* is a lower case vowel.

   d) 4 is not a positive odd number.

   e) $ is not a letter in the lower case English alphabet.

   f) This problem does not have a part *g*.

6. Given $A = \{1, 3, 4, 7\}$; $B = \{3, 7, 12\}$; $C = \{1, 5, 8\}$, write the following sets:

   a) The set containing all elements that are members of *A*, or members of *B*, or members of both *A* and *B*.

   b) The set of elements that are members of both *A* and *B*.

   c) The set of elements that are members of both *B* and *C*.

   d) The set of elements that are members of *A* but not members of *B*.

   e) The set of elements that are members of both *A* and *C*.

   f) The set of elements that are members of all three sets.

7. a) Write the set symbols for $y + 2 = 10$.

   b) Translate the symbols $\{y : y + 5 = 8\} = \{3\}$.

8. a) Write the set symbols for $m - 6 = 4$ and its solution set.

   b) Translate the symbols $\{q : 2q = 6\} = \{3\}$.

9. What is the solution set for $\{z : 2z + 5 = 2z + 7\}$?

10. What is the solution set for $\{y : 4y + 1 = 12\}$?

In Questions 11–14, the first mentioned set is the starting set.

11. Each of the United States has two Senators.

    a) Is the relationship between the set of states and the set of Senators a functional correspondence? Why?

    b) Is the relationship between the set of Senators and the set of states a functional correspondence? Why?

12. a) Is the relationship between the set of individual apples in an orchard and the set of individual apple trees in an orchard a functional correspondence? Why?

    b) Is the relationship between the set of apple trees and the set of apples a functional correspondence? Why?

13. Draw a horizontal line and a slant line intersecting the horizontal line, then draw vertical lines intersecting the first two lines. Does the set of points at the intersection of such verticals and the horizontal line have a functional correspondence with the respective set of points at the intersection of verticals and the slant line? Why?

14. Draw a circle and a horizontal line that is a diameter of the circle. If vertical lines are drawn, does the set of points at the intersection of the vertical and the diameter have a functional correspondence with the respective set of points at the intersection of the vertical and the circle? Why?

15. If $h(z) = 4z + 7$
    a) How is $h(z)$ read?
    b) What is $h(2)$?
    c) State the function rule.
    d) How would $h(x + 2)$ be obtained?

16. If $f(x) = 3x + 10$
    a) How is $f(x)$ read?
    b) What is $f(5)$?
    c) State the function rule.
    d) How would $f(x + 1)$ be obtained?

17. If $y = 2x + 7$, write the symbolic expression for the solution set.

18. If $y = 5x$, write the symbolic expression for the solution set.

19. Does $y = 10x + 25$ represent a functional correspondence? Why?

20. If the starting set is 0 or any whole number and the elements of the ending set are obtained by dividing five by the starting element, is there a functional relationship between the sets? Why?

A1.7 SUBSETS, UNIONS, AND INTERSECTIONS[3]

If every element of a set B is also an element of a set A, then B is called a *subset* of A. For example, if

$$A = \{1, 3, 6, 9\} \quad \text{and} \quad B = \{3, 6\}$$

then B is a subset of A. Moreover, in this example, B is a *proper* subset of A because it does not contain all the elements of A. If

$$C = \{8, 9, 10\} \quad \text{and} \quad D = \{8, 9, 10\}$$

then D is an *improper* subset of C, and vice versa, because all the elements of one are elements of the other. It follows that if two sets are equal, they are improper subsets of each other. The empty set, $\emptyset$, is by definition a proper subset of every set except itself. If $S = \{a, b, c\}$, then the proper subsets of S are

$$\{a\}, \{b\}, \{c\}, \{a, b\}, \{a, c\}, \{b, c\}, \text{and } \emptyset.$$

Exercise. List all eight subsets of $K = \{2, 4, 6\}$. [4]

The *intersection* of two sets A and B is a set whose elements are elements of *both* A and B. The intersection symbol is like an inverted U. Thus

$$A \cap B$$

---

[3] This material is applied in Section 4 of Chapter 12.

[4] If a set has n elements, then the set will have 2^n subsets, counting the null set and the improper subset.

is read as "A intersection B," or "the intersection of A and B." If we have

$$A = \{4, 8, 9, 11\} \quad \text{and} \quad B = \{8, 11, 14\} \quad \text{and} \quad C = \{3, 5\}$$

then

$$A \cap B = \{8, 11\} \quad \text{and} \quad A \cap C = \emptyset.$$

The last expression says that the intersection of A and C is the empty set because the two sets have no common members. Sets with no common members also are said to be *disjoint* sets.

> *Exercise.* Given $A = \{a, b, c, d, e\}$ and $B = \{e, f, g, \ldots, z\}$, write $A \cap B$.

The *union* of sets A and B is a set containing those elements that are members of A or members of B, or members of both A and B. A U-like symbol represents the union; thus,

$$A \cup B$$

is read "A union B" or "the union of A and B." If

$$A = \{a, b, c, d, e\} \quad \text{and} \quad B = \{c, e, f, k\}$$

then

$$A \cup B = \{a, b, c, d, e, f, k\}.$$

> *Exercise.* Find $A \cup B$ if $A = \{2, 7, 8, 9, 13\}$ and $B = \{1, 8, 13, 22\}$.

It is helpful to illustrate the ideas of intersection and union diagrammatically as in Figure A1–3. If we think of A as a set containing all

FIGURE A1-3

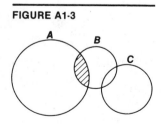

FIGURE A

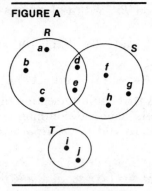

the points in circle A, and similarly for B and C, then $A \cap B$ is the set of points indicated by the shaded area; $A \cup B$ is the set of points containing all the points in both circles A and B; $A \cup C$ contains all the points in circles A and C; and $A \cap C = \emptyset$ because A and C have no points in common, so are disjoint.

A1.8 PROBLEM SET A1–2

1. Figure A shows set R with points a, b, c, d, and e and similarly for sets S and T. Write the following sets by the roster method:
    a) $R \cup S$. b) $R \cup T$. c) $R \cap S$.
    d) $R \cap T$. e) $S \cup T$. f) $S \cap T$.

2. If $A = \{1, 3, 5, \ldots\}$, $B = \{0, 2, 4, 6, \ldots\}$, and $C = \{5, 7, 9\}$ write the following sets by the roster method:
    a) $A \cup B$. b) $A \cap B$. c) $A \cap C$.
    d) $B \cap C$. e) $A \cup C$.

3. If A is the set of face cards in a standard 52-card deck and B is the set of queens, what will be the elements of $A \cup B$ and $A \cap B$?

4. If A is the set of face cards in a standard 52-card deck, B is the set of red cards, and C is the set of nines, what will be the elements in $A \cup B$, $A \cap B$, $A \cap C$, $B \cap C$?

5. If M represents the set of all points on one line in a plane, and N the points on a different line in the plane, what is $M \cap N$?
    a) If the lines are parallel? b) If the lines are not parallel?

6. If R means rain tomorrow and W means warmer tomorrow, what do the following mean?
    a) $R \cup W$. b) $R \cap W$.

A1.9 REVIEW PROBLEMS

1. Read the following aloud: for example, $A = \{5, 10, 15, \ldots, 100\}$ is read, "A is the set whose elements are 5, 10, 15, and so on, through 100."

a) {5, 6, 7, 9}.   b) $B = $ {Amherst, Babson, Colgate, Dartmouth}.

c) $C = $ {1, 3, 5, 7, . . .}.   d) $D = $ {A, B, C, . . . , Z}.

e) $E = \left\{ 1, \dfrac{1}{2}, \dfrac{1}{4}, \dfrac{1}{8}, \cdots \right\}$ f) $F = $ {Tim, Dick, Harry}.

2. Specify each of the following sets by the descriptive method:

    a) $A = $ {1, 3, 5, 7, 9, . . .}.   b) $B = $ {5, 10, 15, . . . }.

    c) $C = $ {101, 102, 103, . . . , 999}.   d) $D = $ {M, A, R, Y}.

3. Specify each of the following by the roster method:

    a) $A = $ {Letters in the word HARVARD}.

    b) $B = $ {First nine positive prime numbers}.

    c) $C = $ {Positive odd multiples of 4}.

    d) $D = $ {The first five positive even multiples of 3}.

4. Using set membership symbols and the roster method of specification of the sets, write the following:

    a)   The letter S is among the letters in the word BABSON.

    b)   7 is not a positive integral multiple of 3.

    c)   3 is not a positive integral power of 2.

    d)   5 is not a lower case English letter.

5. a)   Write the set symbols for $t + 1 = 20$ and its solution set.

    b)   Translate the symbols $\{y : 3y + 5 = 11\} = \{2\}$.

6. Given the relation $y = x$:

    a)   Is the relation a function? Explain.

    b)   What is the function rule?

    c)   What is meant by the "solution set" for the equation?

    d)   How many elements are there in the solution set?

    e)   What are the elements in the solution set corresponding to $x = 1$, 2, 3, 4?

7. What is the solution set for $\{t : t = t + 1\}$?

8. Suppose x is any positive whole number and y is any positive whole number that is an exact divisor of x. Is the relation between y and x a function? Explain.

9. Given $A = $ {5, 10, 12, 13, 15}; $B = $ {2, 10, 13, 14}; $C = $ {12, 16, 17}; write the following sets:

    a)   The set containing all elements that are members of A or members of B, or members of both A and B.

    b)   The set of elements that are members of both A and B.

    c)   The set of elements that are members of both B and C.

    d)   The set of elements that are members of A but not members of B.

    e)   The set of elements that are members of both A and C.

    f)   The set of elements that are members of all three sets.

10. Figure B shows set R with points p, q, r, s, and t, and similarly for sets S and T. Write the following sets by the roster method:

    a) $R \cup S$.   b) $R \cup T$.   c) $R \cap S$.

    d) $R \cap T$.   e) $S \cup T$.   f) $S \cap T$.

11. If $A = $ {5, 10, 12, 15, 19}, $B = $ {3, 10, 15}, and $C = $ {7, 12}, write the following sets by the roster method:

    a) $A \cup B$.   b) $A \cap B$.   c) $A \cap C$.

    d) $B \cap C$.   e) $A \cup C$.   f) $B \cup C$.

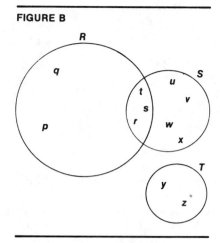

FIGURE B

12. Make a diagram showing sets R, S, and T (like the diagram in Problem 10) corresponding to the following requirements:
   a) $R \cap T = \{x\}$,   b) $R \cap S = \{z\}$.   c) $T \cap S = \emptyset$.
   d) $R \cup S = \{v, x, y, z\}$.   e) $R \cup T = \{w, x, y, z\}$.

13. If sets M and N are, respectively, the sets of points on the perimeters of two concentric circles, what is $M \cap N$?

14. If P is the set of all cars with air conditioners, Q is the set with automatic shifting, and R is the set with manual (or stick) shifting, what are the following?
   a) $P \cup Q$.   b) $P \cap Q$.   c) $Q \cap R$.   d) $P \cap R$.

15. a) Write the set symbols for $y + 5 = 15$ and its solution set.
   b) Translate the symbols $\{x : x - 2 = 10\} = \{12\}$.

16. What is the solution set for $\{z : z + 7 = 30\}$?

17. Is 5 in the solution set of $\{x : x < 10\}$? Why?

18. If x is a number in the starting set, and the corresponding number in the ending set is any number less than x, do the sets have a functional correspondence?

19. Each secretary in a company works for three executives, and always the same three.
   a) Does the starting set of secretaries have a functional correspondence to the set of executives? Why?
   b) Does the starting set of executives have a functional correspondence to the set of secretaries?

20. If $f(x) = 20x + 50$
   a) How is $f(x)$ read?   b) What is $f(10)$?
   c) State the function rule.   d) How would $f(x + 1)$ be obtained?

21. Write the symbolic expression for the solution set of $y = 12x + 6$.

22. Does $z = 3x + 5$ represent a functional correspondence? Why?

23. How many ordered pairs are there in the solution set for $y = x + 3$?

Elements of algebra

A2.1 INTRODUCTION

This appendix presents the basic definitions, conventions, and rules of algebra, together with discussions and illustrations of the fundamental properties of numbers. Each topic is accompanied by examples that will meet the needs of those who use isolated parts of the appendix for reference purposes. However, topics are presented in logical order so that the needs of those seeking a substantial review of elements will be met by starting at the beginning and moving in an orderly fashion through the material.

A2.2 THE REAL NUMBERS

The positive and negative whole numbers, with zero, form the set of *integers*. Combining the integers with the fractions, we have the set of *rational* numbers. Thus, a rational number (ratio number) is a number that can be expressed in the form of a fraction that has integers as numerator and denominator. For reasons to be discussed soon, fractions with denominator zero are excluded from our number system.

773

> *Exercise.* How would the number 1.15 be classified? Answer: This is a rational number because it can be expressed as 115/100, which is the ratio of the two integers 115 and 100.

Numbers such as the square root of 2 or the cube root of 6 cannot be expressed as a ratio of two integers, and serve as examples of *irrational* numbers. The set of all rational and irrational numbers is called the set of *real* numbers. Only real numbers will be used in this book. Numbers with an imaginary component (that is, a component involving the square root of -1) are not in the set of reals and play no role in our discussions.

A2.3 JUSTIFICATION FOR MATHEMATICAL STATEMENTS

Justifications for mathematical statements fall into three rough categories: first, definitions and conventions, which represent common agreement about the meanings of words and symbols; second, fundamental properties of numbers; and third, rules that are derived from fundamental properties. Definitions and conventions are, of course, a part of statements of fundamental properties and rules. As illustrations of the categories, observe that it is *conventional* to write the product of the numbers a and b without a multiplication sign as ab. It is true by *definition* that the sum of a number and its negative is zero. Thus:

$$+a + (-a) = 0.$$

> *Exercise.* Why is a the same as $1a$? Answer: *Conventionally*, if a quantity is written without a coefficient, the coefficient is assumed to be one.

The statement that the sum of two numbers is the same whether found by adding the first to the second or by adding the second to the first is a *fundamental property* of numbers. Finally, the idea that "minus a minus is plus" is a *rule* that can be derived using fundamental properties, definitions, and conventions.

> *Exercise.* Of the operations *multiply, divide, subtract,* which has the property just illustrated for *add?* Answer: *Multiply.*

A2.4 RULES OF SIGN

The numbers $+2$ and -2 are opposites in the sense that each is the negative of the other. The statement that -2 is the negative of $+2$ is true by definition; the statement that the negative of -2 is $+2$ is a rule that can be derived from definitions and axioms.[1] Thus:

$$-(+2) = -2$$

and

$$-(-2) = +2.$$

Another derived rule justifies the statement

$$+(-2) = -(+2).$$

That is, the positive of the negative of a number is the same as the negative of the positive of the number. As a consequence of this rule, the sign of a number is the same no matter what the *order* is of the signs immediately preceding the number. For example:

$$+[-(-5)] = -[+(-5)] = -[-(+5)].$$

The ultimate sign is shown by working with the last expression to be

$$-[-(+5)] = -[-5] = +5.$$

> *Exercise.* Write $-[-(-5)]$ with one sign. Answer: -5.

A2.5 SIGN SEQUENCE RULE

Any sequence of signs immediately preceding a number can be converted to a single ultimate sign, the ultimate sign being *minus* if the sequence has an *odd* number of *minus* signs, and being *plus* if the sequence has an *even* number of minus signs. For example:

$$-[+(-3)] = +3$$

and

$$-\{-[+(-3)]\} = -3.$$

A2.6 ADDITION AND SUBTRACTION OF SIGNED NUMBERS

If two numbers have the same sign, their sum is found by adding the two numbers and affixing the common sign. Thus:

---

[1] The rule is not derived here.

$$(+4) + (+3) = +7$$

and

$$(-3) + (-4) = -7.$$

Exercise. Add $(-10) + (-4)$. Answer: -14.

The sum of a positive and a negative number is found by taking the difference between the two numbers, disregarding sign, and affixing the sign of the larger to the difference. For example:

$$(+7) + (-3) = +4$$

and

$$(-7) + (+3) = -4.$$

Exercise. Add: $(+5) + (-3)$. Answer: $+2$.

In terms of fundamentals, subtraction is developed from addition. Subtraction can be converted to addition by following the rule that states that subtracting a number is the same as adding its negative. This rule, together with a rule already stated, permits us to write

$$+3 - (+4) = +3 + (-4) = -1.$$

As another example:

$$-3 - (-4) = -3 + (+4) = +1.$$

Exercise. Write as an addition problem, and find the sum: $+5 - (+11)$. Answer: $+5 + (-11) = -6$.

Observe that the minus symbol is used both to designate negative and to indicate the operation of subtraction. Again, the plus symbol designates both positive and the operation of addition. For example, we write $-(+2)$ to designate the *negative* of $+2$, but in the expression $+3 - (+2)$, the same symbol, $-(+2)$, means to *subtract* $+2$. Now, let us adopt two conventions. First, we shall assume that an unsigned number is positive; second, we shall indicate the subtraction of a number by a minus sign, rather than by addition of its negative. Thus, instead of $+3 - (+2)$ we shall write the condensed statement, $3 - 2$.

The conventions just stated, taken together with the rule for sign sequence, contribute to simplicity of expression, as illustrated next:

$$+1 + (+2) - (+3) + (-4) - (-5) - [+(-4)]$$

becomes

$$1 + 2 - 3 - 4 + 5 + 4 = 12 - 7 = 5.$$

> **Exercise.** Simplify and evaluate $+(-1) + (-3) - (+2) + (+4) + [-(-6)]$. Answer: 4.

We usually seek to reduce additions and subtractions of signed numbers to simplified form, as just shown. However, in the interest of keeping track of fundamentals, we should be able to convert from the simplified form to an expression in terms of addition of signed numbers. Thus:

$$4 - 3 + 2$$

means

$$+4 + (-3) + (+2).$$

> **Exercise.** What does $2 + 5 - 8$ mean in terms of addition of signed numbers? Answer: $+2 + (+5) + (-8)$.

A2.7 NEGATIVE OF A SUM

The negative of the sum of two numbers is the sum of the negatives of the numbers. Thus:

$$-(7 + 5) = (-7) + (-5) = -7 - 5 = -12.$$

It will be observed that as we go from $-(7 + 5)$ to $-7 - 5$, the effect is to remove the parentheses and change the signs of the numbers inside the parentheses. The methodology applies generally; that is, parentheses preceded by a minus sign alone can be removed by changing the sign of each term inside the parentheses.[2] As examples:

$$-(4 - 6) = -4 + 6$$

and

$$-(4 + 8) = -4 - 8.$$

---

[2] See Section A2.18 for the meaning of *term*.

Parentheses preceded by a plus sign alone (or no sign) serve no necessary purpose and may be omitted. Thus:

$$+(8 - 6) = 8 - 6$$

and

$$(2 - 4 + 3) = 2 - 4 + 3.$$

Exercise. Remove grouping symbols: $(2 + 3 - 5) - (5 - 2 + 4)$. Answer: $2 + 3 - 5 - 5 + 2 - 4$.

We have introduced the foregoing at this point because the sign changes that occur when parentheses with a preceding minus sign are removed need special emphasis. Actually, the rule stated can be derived, and will be derived in Section 12.

A2.8 MULTIPLICATION AND DIVISION OF SIGNED NUMBERS

Multiplication is a fundamental operation with numbers, and division is defined in terms of multiplication. One rule of sign suffices for both operations—namely, if two numbers have the same sign, their product (or quotient) is positive; if two numbers are of opposite sign, their product (or quotient) is negative. In multiplication, for example:

$$(-4)(-2) = 8$$

and

$$2(-4) = -8.$$

Observe the use of parentheses in the last written expression. In the first, the absence of a symbol between the two sets of parentheses means that the numbers inside the parentheses are to be multiplied. We could have written $-4(-2)$ instead of $(-4)(-2)$ without ambiguity, but multiplication would not be in order if the expression read $(-4) - 2$. The last written number is a difference, not a product.

The division symbol, $\div$, is not used often in algebraic statements. The fractional form of expression is preferred. Thus, 8 divided by 4 is written as $8/4$. Moreover, the general practice in mathematics is to speak of the last number as "8 over 4," not "4 into 8," and the student is advised to follow this practice.

According to the rule for division of signed numbers:

$$\frac{8}{-4} = -2, \frac{-8}{4} = -2, \text{ and } \frac{-8}{-4} = 2.$$

Exercise. Evaluate: $-2(4)$; $3(6)$; $-(-3)(-5)$; $-4/-2$; $6/-3$; $-(6/-3)$; $-(-6/3)$; $-(-6/-3)$. Answer: -8; 18; -15; 2; -2; 2; 2; -2.

The sign of a term that involves only multiplications and divisions of numbers can be determined by counting the number of minus signs; if the count is odd, the ultimate sign of the term is minus; if the count is even, the ultimate sign of the term is positive. For example:

$$\frac{-4(2)(-3)}{-2(-1)(4)} = 3$$

and

$$\frac{-(-4)(-5)}{2(-1)(-10)} = -1.$$

Exercise. Write as a single signed number: $\dfrac{-2(3)(6)(-10)}{12(-5)}$.

Answer: -6.

A2.9 PROBLEM SET A2–1

Reduce to a single signed number, or zero:

1. $-2 + (-3) - [-(-4)] + (-5)$.
2. $-(-8) + 8 - (+3)$.
3. $-(+4) + (-4) - [-(-4)]$.
4. $6 - (+3) + (+5) - (-7) + (-2) - (+10)$.
5. $3.15 - (+1.08) - (-3.27)$.
6. $-4 - (3)(-2) + (-5)(-4)(-1) - (3)(2)$.
7. $-4(3) + (-1)(2)(-6)(-1) - (-1)(-1)(-1)$.
8. $-(-4 + 3) + [-1(-2)]$.
9. $14(-2) - (-13) + (-3)(-2) - (-5)$.
10. $(3 - 4) - (5 - 2)$.
11. $3 - (-2) - (7 - 5)$.
12. $-5 + (7 - 3) - (2 - 4)$.
13. $+(3 - 6) - (-5 + 4)$.
14. $-(-5 - 7) + (3 - 2)$.
15. $12 - (-3) + (-2) - (+2)$.
16. $-25 - (30 - 10)$.
17. $\dfrac{-3(20)}{-4}$.
18. $\dfrac{5(-2)(3)}{-4(-2.5)}$.
19. $\dfrac{-(-7)}{-2}$.
20. $\dfrac{6(-2)(-1)(-3)}{-(-12)(-2)}$.

21. $\dfrac{-(-26)(16)}{-[-(-13)]}$.

22. $\dfrac{4(-1)(-2)(3)(-7)}{(-5)(-3)(2)}$.

23. $\dfrac{-3-(5-20)}{5-(-2)}$.

24. $\dfrac{6-(-7+3)}{-[-(-2+7)]}$.

25. $\dfrac{-(8-3)}{-2+3(-1)}$.

26. $\dfrac{40-2(-5)}{2-(15-3)}$.

Rewrite as the sum of signed numbers:

27. $3-4+5-2+6$.

28. $3-2+4-5$.

29. $-1.5-3-2.5$.

30. $-3+7-4+2$.

31. $-10+7-4+3$.

32. $12-6+5-7$.

Reduce to a single signed number:

33. $-3(2)+(3-7)-(2+4)$.

34. $-[-2+(-3)]-2(-3)+(-3)(-4)$.

35. $-(3-5)+(2-6)$.

36. $-[+(-3)+(-5)]+[-2+(-3)]$.

37. $-[+(-5)]+[-(-3)]+(8-4)-(2-6)$.

38. $-(-3)+(-7)+(8)-(-6)$.

A2.10 REPRESENTING NUMBERS BY LETTERS

Mathematics beyond arithmetic is characterized by its use of letter representation of numbers. The sum, difference, and quotient of two numbers in literal form, such as a and b, are written in the usual manner as, respectively, $a+b$, $a-b$, and a/b. The product conventionally is written as ab, the multiplication symbol being omitted. This convention is, of course, in conflict with the place system of decimal notation. For example, 34 means thirty-four, not 3 times 4. We would interpret $34a$ as thirty-four times a, not 3 times 4 times a. It follows that care must be exercised when writing numbers that have literal and numerical factors. Sometimes a dot is employed to indicate multiplication, as in $3 \cdot 4ab$. We shall generally follow the more common practice, already introduced, and use parentheses; thus, $3(4)ab$ or $(3)(4)ab$. The first of the last two expressions is the simpler. The second is equivalent to the first, but uses two sets of parentheses where only one is required.

The first advantage of literal representation of numbers is their utility in stating generalizations. For example, we may illustrate the point that the sum of any two real numbers is a real number by mentioning that the sum of 3 and 4 is 7, and 7 is a real number. To generalize the point and state it as a property of all real numbers, we need a statement that is not restricted to a particular pair of numbers. The statement desired is: *"If a and b are any real numbers, then $a + b$ is a real number."* In this statement the word *any* means that the property applies to *all* real numbers. If someone asks if the statement applies to x and y, we inquire whether x and y are real numbers; if they are, the statement applies, and $x + y$ is a real number. The student will find it helpful to keep in mind the *all* implication of the word *any* in mathematical statements.

The fundamental properties we list for real numbers are formal statements of observations of arithmetic. The very fact that they seem obvious is a primary reason for stating them as fundamental properties. The fact that they seem obvious is not, however, justification for thinking time spent on them is wasted. Many questions that perplex beginning students arise simply because of failure to have in mind the fundamental properties of numbers.

A2.11 CLOSURE

Countless illustrations can be written showing that when two real numbers are added, subtracted, multiplied, or divided (division by zero excluded), the outcome itself is a real number. In other words, the "answers" to problems in addition, subtraction, multiplication, and division of real numbers are part of the real number system. Indeed, our algebra with literal numbers would falter if we did not assume *closure* to be a fundamental property. The closure property is this: *If a and b are any real numbers, then $a + b$, $a - b$, ab, and a/b are real numbers, except that a/b is not defined if b is zero.*

The property is sometimes described by stating that the real numbers are closed with respect to addition, subtraction, multiplication, and division, except division by zero. The property extends to more than two numbers according to the logic that says, for example, that if a, b, and c are real numbers, then $a + b$ is a real number, and this last real number when added to c yields a real number, so that $a + b + c$ is a real number.

The concept of closure can be understood easily by thinking of addition of clock numbers. If we start at 2 and add 3, we go to 5. If we start at 10 and add 4 we go to 2. True, the sums in such additions are contrary to those obtained for real numbers, but the point is that the addition of any two clock numbers has an answer that is a clock number, so that we may say the clock numbers are closed with respect to addition.

As a simple example of lack of closure, it is easy to see that the set of integers is not closed with respect to division; for example, 3 and 7 are integers, but 3/7 is not an integer. Again, the real numbers are not closed with respect to the operation "square root," because negative numbers are real, but the square roots of negatives are not real.

782

> **_Exercise._** Is the set of odd numbers closed with respect to addition? Explain. Answer: No. The sum of two odd numbers is not an odd number.

A2.12 DISTRIBUTIVE PROPERTY

If we wish to write an expression in which the *sum* of the numbers a and b is to be multiplied by 2, a convention must be established to ensure that the expression is interpreted as 2 times the sum, and not 2 times one of the numbers. Parentheses accomplish the purpose. We write $2(a + b)$ and intepret this to mean that the quantity inside the parentheses is a number to be multiplied by 2, the number inside being the sum of a and b. It may be helpful to practice a bit here by reading the following parallel statements:

In words	*In symbols*
Multiply 2 by a	$2a$
Multiply 2 by a, and then add b to the product	$2a + b$
Multiply 3 by the sum of a and c	$3(a+c)$
To a, add b times the sum of c and d	$a + b(c+d)$
Multiply the sum of a and b by the sum of c and d	$(a+b)(c+b)$

The distributive property is illustrated by the observation that

$$2(3 + 4) = 2(7) = 14$$

can be evaluated in the alternative manner:

$$2(3) + 2(4) = 6 + 8 = 14.$$

As another example:

$$-7[+5 + (-6)] = -7[-1] = 7$$

can be evaluated as

$$-7[+5 + (-6)] = -7(+5) + (-7)(-6) = -35 + 42 = 7.$$

These and the countless similar observations one could devise are generalized by stating the distributive property of the real numbers. In brief, we say that multiplication may be distributed over addition (or subtraction). More formally, we state that *if a, b, and c are any real numbers, then*

$$a(b + c) = ab + ac.$$

This property finds extensive application in algebra. As examples:

$$2a(b + c + 3d) = 2ab + 2ac + 6ad$$
$$-3(a - 2b) = -3a + 6b.$$

In the last example, it is instructive to match the procedure with the formal statement of the property. To do so, we remember that $-3(a - 2b)$ is shorthand for $-3[+a + (-2b)]$. Applying the distributive property to the rightmost statement, we obtain

$$-3(+a) + (-3)(-2b)$$

which, reverting to shorthand, is $-3a + 6b$.

Exercise. Apply the distributive property to $-3(-2 + a)$. Answer: $6 - 3a$.

As another example, if c does not equal zero,

$$\frac{a}{c}(x + 2y) = \frac{a}{c}(x) + \frac{a}{c}(2y).$$

The point here is that the distributive property holds for any real numbers, and a/c is a real number if $c \neq 0$.

Exercise. Why is a/c a real number? Answer: In the foregoing it was stated that a and c are real numbers. Inasmuch as the real numbers are closed with respect to division if $c \neq 0$, a/c must be a real number.

In closing this section, we note that

$$-1(2 - 3) = -1(2) - 1(-3) = -2 + 3$$

by the distributive property. This, of course, is justification for the earlier stated rule that the negative of the sum of two numbers is the sum of the negatives of the numbers.

A2.13 COMMUTATIVE PROPERTIES

The observation that we may find the sum of 3 and 4 by starting with 4 and then adding 3, or by starting with 3 and then adding 4, is one of countless illustrations that the sum of two numbers is the same in whatever *order* the numbers are added. Thus:

$$3 + 4 = 4 + 3$$

and

$$-2 + (-7) = -7 + (-2).$$

In shorthand, the last statement would have been

$$-2 - 7 = -7 - 2.$$

Formally, the commutative property for addition states that *if a and b are any real numbers, then*

$$a + b = b + a.$$

The property is not limited to the sum of two numbers, of course. It extends by closure to any series of numbers in addition. For example, if we have $a + b + c$, then $a + b$ is a real number, by closure, and we may change the order of this real number and c to obtain $c + a + b$. Again, the commutative property for addition justifies the following rearrangement:

$$2a + 3b + 6 + b + 7 = 2a + 3b + b + 6 + 7.$$

Finally, we note that

$$a - b = -b + a$$

because $a - b$ is shorthand for $a + (-b)$, which, by the commutative property, is $-b + a$.

> **Exercise.** Rearrange $c - a + b$ so that the letters are in alphabetical order. Answer: $-a + b + c$.

The commutative property for multiplication states that *if a and b are any real numbers, then* $ab = ba$. That is, the *order* of the factors in a product may be changed. As numerical examples, we have:

$$3(4) = 4(3); \quad 2(-5) = -5(2); \quad -3(-2) = -2(-3).$$

The property extends to more than two numbers by closure. For example, in abc, ab is a real number, by closure, and the commutative property says we may change the order of this real number and c; hence:

$$abc = cab.$$

Similarly:

$$abcd = cdab = bcda.$$

A2.14 ASSOCIATIVE PROPERTIES

Associative properties for addition and multiplication have to do with the manner of *grouping* numbers. If we have the addition problem

$$2 + 3 + 4$$

we can find the sum as

$$(2 + 3) + 4$$

or as

$$2 + (3 + 4)$$

that is, as 5 + 4 or as 2 + 7. The same sum results whether we group the first two and then add the third number or whether we group the second two and add to the first number. Formally, the associative property for addition states that *if a, b, and c are any real numbers, then*

$$(a + b) + c = a + (b + c).$$

The parentheses in the last statement are not necessary. They serve only to indicate the different groupings. The number in question is $a + b + c$, no matter what order or grouping is used.

> *Exercise.*   According to the associative property, 5 + 2 + 8 may be evaluated in what ways? Answer: 7 + 8, 5 + 10, and (if the commutative property is applied first), 13 + 2.

The associative property for multiplication asserts that *if a, b, and c are any real numbers, then*

$$a(bc) = (ab)(c).$$

Again, unnecessary parentheses are used to show that the property has to do with the manner of grouping the factors. To illustrate the property, we observe that (3)(4)(2) can be evaluated as (12)(2) or as (3)(8). That is, we can multiply the product of the first pair of numbers by the third number, or multiply the first number by the product of the second pair of numbers.

> *Exercise.*   Illustrate the associative property for multiplication by reference to the product (3)(5)(2). Answer: The product may be evaluated as (15)(2) or as (3)(10).

A2.15 IMPORTANCE OF FUNDAMENTAL PROPERTIES

The distributive, associative, commutative, and closure properties of the real numbers are important because they are the fundamental justifications for many of the steps taken in algebraic procedures. Asked if $(a + b)c$ is the same as $c(a + b)$, we answer yes, and cite the commutative law for multiplication as justification. Now:

$$c(a + b) = ca + cb$$

by the distributive property, and the commutative property assures us that the right side can be altered to obtain

$$c(a + b) = ac + bc.$$

From what has been said, it follows that

$$(a + b)c = ac + bc.$$

The point here is that the distributive law does not tell us directly how to expand an expression such as $(2 + a)b$, but we can easily justify by fundamentals that the expansion may be written as

$$(2 + a)b = 2b + ab.$$

Exercise. In $(2)(a)(3) = (2a)(3) = 2(3)(a) = 6a$, what justifies each of the three expressions following the $=$ sign? Answer: Associative law for multiplication; commutative law for multiplication; associative law for multiplication.

Carrying on one step further, let us learn how to expand

$$(a + b)(c + d).$$

We know by closure that $(a + b)$ is a real number, and the distributive law justifies writing

$$(a + b)(c + d) = (a + b)c + (a + b)d.$$

The right member of the last expression can be expanded, leading to

$$(a + b)(c + d) = (a + b)c + (a + b)d = ac + bc + ad + bd.$$

The distributive law works both ways. Writing it in reverse, we have

$$ab + ac = a(b + c).$$

Converting in this manner from the sum of ab and ac to the product of a by $(b + c)$ is referred to as *factoring*. Hence the distributive property is the fundamental underlying factoring. When applied to obtain

$$2a + 3a = (2 + 3)a = 5a$$

the procedure is called *combining like terms.* As other examples:

$$2ab - 3ab + 5ab - 6 = 4ab - 6$$
$$4a - 2a + 5b - 2b = 2a + 3b.$$

In the last two examples the associative law for addition justifies the separate combinations of like terms. We may say in general that terms having the same literal factor can be added (subtracted) by adding (subtracting) their numerical coefficients. Thus, in

$$3ab + 4ab - 2ab$$

each term has the same literal factor, ab. Combining the numerical coefficients, we have

$$(3 + 4 - 2)ab = 5ab.$$

> ***Exercise.*** Remove symbols of grouping and combine like terms: $a(2 - b) - 3(a - 2b)$. Answer: $a(2) - ab - 3a + 6b = 6b - ab - a$.

The following example shows how the fundamental properties come into play. Starting with the expression at the top left, we change to the expression on the next line, indicating on that next line, at the right, the justification for the step:

$$-2a(3b) + ab$$
$$= -2(3b)a + ab \qquad \text{Commutative law, multiplication}$$
$$= -6ba \quad + ab \qquad \text{Associative law, multiplication}$$
$$= -6ab \quad + ab \qquad \text{Commutative law, multiplication}$$
$$= -5ab. \qquad \text{Distributive law (combining like terms, or factoring)}$$

A similar example follows:

$$2 + ba + 3$$
$$= 2 + \quad 3 + ba \qquad \text{Commutative law, addition}$$
$$= \qquad 5 + ba \qquad \text{Associative law, addition}$$
$$= \qquad 5 + ab. \qquad \text{Commutative law, multiplication}$$

We do not mean to suggest by these examples that the reader cite the fundamental reason for every algebraic step he takes. We do mean to suggest, and strongly, that he become so thoroughly familiar with these fundamentals that he never has to ask his instructor questions such as:

"Is $(2 + 3b)(-c)$ the same as $-c(2 + 3b)$?"

788

Exercise. What fundamental law specifies that the two last written expressions are equal? Answer: The commutative law for multiplication.

A2.16 PROBLEM SET A2–2

1. When we write ab or cxy, in what way does algebraic convention differ from the usual decimal notation?
2. What would abc mean in decimal notation?
3. What is the connotation of the word *any* in mathematical statements?
4. If we add the odd number 3 to the odd number 5, the sum is an even number, 8. State this odd-even relationship in a manner that shows it is a fundamental property of integers.
5. Explain the property of closure by using addition of clock numbers as an example.
6. Why does $3a + c$ not mean three times the sum of a and c?
7. How many specific numerical illustrations of the distributive property do you think you could devise?
8. Explain what is meant by the statement that the distributive property equates a product involving a sum to a sum of products.

What fundamental property or convention justifies each of the following?

9. $a(2) = 2a$.
10. $2 - a = -a + 2$.
11. $a(2 + 3) = a2 + a3$.
12. $a + 2 + b = a + b + 2$.
13. $+a = a$.
14. $a(b + c) = (b + c)a$.
15. $a + (-3) = a - 3$.
16. $a2 + a3 = 2a + 3a$.
17. $2 + 3 + b = 5 + b$.
18. $3a2 = 3(2)a$.
19. $3(2)a = 6a$.
20. $+b + (-a) = b - a$.
21. $5 - 4 + b = 1 + b$.
22. $2a + 2b = 2(a + b)$.
23. $a + 2 + b = a + b + 2$.
24. $a + (b + c) = (b + c) + a$.
25. $a(2)(4) = a(8)$.

Combine like terms:

26. $2b + (-3b)$.
27. $2a - 3b + 5a - 2b$.
28. $3abc - 2d - (-2abc)$.
29. $5 - (-3x) + 2 - (+2x)$.
30. $2abc - (-3abc) + 3abc$.
31. $3 + 2a - 3b + 5 - (-2b)$.

Name the fundamental property justifying each step:

32. $\quad -b + a$
$= \quad a - b$.

33. $\quad 2b(3a)$
$= 2b3(a)$
$= 2(3)ba$
$= 6ba$
$= 6ab$.

34. $\quad 3 + xy + \ 5 + 3(2ab)$
$\quad = 3 + 5 \ + xy + 3(2ab)$
$\quad = \quad \ 8 \ + xy + 3(2ab)$
$\quad = \quad \ 8 \ + xy + 6ab.$

35. $\quad 3 + a + 2$
$\quad = 3 + 2 + a$
$\quad = \quad \ 5 + a.$

36. $\quad a + 3 + c + 2$
$\quad = a + c + 3 + 2$
$\quad = a + c + 5.$

37. $\quad acdb$
$\quad = adcb$
$\quad = adbc.$

Write in algebraic form:

38. The sum of a and b.
39. To a, add the sum of b and c.
40. To the sum of a and b, add twice the product of c and d.
41. From twice the sum of a and b, subtract three times the sum of c and $2d$.
42. Multiply the sum of a, b, and c by the product of 2 and d.

A2.17 REMOVING GROUPING SYMBOLS

Instructions to remove grouping symbols are carried out by systematic application of the distributive property. Thus:

$$a[b - c(d + 2)] = a[b - cd - 2c] = ab - acd - 2ac.$$

Observe that the innermost grouping symbols, the parentheses, were removed first. This order of attack lessens the chance of errors of omission. As another example:

$$a - \{-2 - 3[-4 + (5 - a)]\} = a - \{-2 - 3[-4 + 5 - a]\}$$
$$= a - \{-2 + 12 - 15 + 3a\}$$
$$= a + 2 - 12 + 15 - 3a$$
$$= -2a + 5.$$

Exercise.   Remove symbols of grouping and combine like terms: $3x - 2[y - 4(x - 3y)]$. Answer: $11x - 26y$.

Care must be exercised in the interpretation of grouping symbols. Thus:

$$5 - 3(b + c) = 5 - 3b - 3c.$$

If we had wished to indicate that the difference of 5 and 3 was to be multiplied by the sum of b and c, the expression would have been

$$(5 - 3)(b + c) = 2b + 2c.$$

Extension of the distributive property justifies the expansions

$$(3 - a)(2b + 5) = 6b + 15 - 2ab - 5a$$

and

$$(a - b + 2)(c + d) = ac - bc + 2c + ad - bd + 2d.$$

A2.18 DEFINITIONS: EXPRESSION, TERM, FACTOR

Any statement involving mathematical symbols may be referred to as a mathematical *expression*. When an expression consists of parts separated by plus or minus signs, or by an equal sign, the parts, together with their signs, are called *terms* of the expression. Thus, in the expression

$$a + 2b - 3ac + 4$$

the terms are $+a$, $+2b$, $-3ac$, and $+4$.

Each term in an expression consists of one or more *factors*, a factor being one of the separate multipliers in a *product*. Thus, in the expression $2a - 3bc$, the term $+2a$ consists of the two factors, 2 and a; the term $-3bc$ consists of the three factors, -3, b, and c.

The words we have just defined make possible clear descriptions of mathematical statements. We shall provide illustrations for practice purposes. Consider the expression

$$(2a + b)(x + 2y).$$

As written, the expression is a single term consisting of the factors $(2a + b)$ and $(x + 2y)$. We may go on to say that the first factor is an expression containing the two terms, $+2a$ and $+b$; the second factor is also an expression of two terms, $+x$ and $+2y$.

> *Exercise.* The expression $3ab + 4a - 3$ has how many terms? What is the composition of the first term? Do all terms have a common factor other than 1? Answer: The expression has three terms; the first term is composed of the factors 3, a, and b; the terms do not have a common factor other than 1.

The distributive property assures us that

$$3(a - 2b + c) = 3a - 6b + 3c.$$

If we think of this last statement as multiplying the parenthetical expression by 3, we come to the general statement that to multiply an expression by a number, we must multiply *each term* of the expression by the number. On the other hand, if we think of

$$3(2a) = 6a$$

as multiplying the term $2a$ by 3, we come to the general statement that to multiply a term by a number, we multiply *one factor* of the term by that number.

According to a rule of sign:

$$-(a - b + 2c) = -a + b - 2c.$$

If we think of this as changing the sign of the parenthetical expression, we see that to change the sign of an expression, we change the sign of every term of the expression. On the other hand, the sign of a term is changed by changing the sign of one of its factors. For instance, thinking of $-[a(-b)(c)(-2)]$ as changing the sign of the bracketed expression, the outcome of the sign change can be written as $-a(-b)(c)(-2)$ or $a(b)(c)(-2)$ or $a(-b)(-c)(-2)$ or $a(-b)(c)(2)$.

Finally, we may state that the sign of a term is unchanged if the signs of an even number of its factors are changed. For example:

$$-ab(-c) = a(-b)(-c) = abc$$

and

$$-3(4 - a) = 3(a - 4).$$

A2.19 ELEMENTARY FACTORING

We are assured by the distributive property that

$$ab + ac = a(b + c).$$

Thinking of this expression as changing from the form on the left to that on the right, we see that the sum of two terms has been converted to the product of two factors. Conversion from the sums and differences of terms to a single term with two or more factors is called *factoring*. Observe that an expression in completely factored form has but one term.

When terms have a factor in common, factoring is carried out by writing the product of the common factor times an expression (in grouping symbols) whose terms are the remaining factors of each of the original terms, as shown in the following examples:

$$2xy + axy = xy(2 + a)$$
$$ax - bx = x(a - b)$$
$$6x + 2y - 4a + 8b = 2(3x + y - 2a + 4b).$$

It is conventional to omit the coefficient one when writing a term, and this convention must be kept in mind when factoring. Thus:

$$ab + b = b(a + 1)$$
$$abc - ab + abd = ab(c - 1 + d).$$

Exercise. Factor $xy + ax + x$. Answer: $x(y + a + 1)$.

As another example, we note that each of the terms of

$$2aby - 6abyz - 12xaby$$

has $2aby$ as factors, so we write the factored form as

$$2aby(1 - 3z - 6x).$$

Exercise. Factor $60xyz - 20axy + 5bxy$. Answer: $5xy(12z - 4a + b)$.

Thus far, we have illustrated *monomial* factoring, that is, cases where the common factor has a single term. On rare occasions in this book the need for *binomial* factoring arises. As examples:

$$a(b + 2) - 3(b + 2) = (b + 2)(a - 3)$$
$$1 + i + i(1 + i) = (1 + i)(1 + i).$$

Exercise. Factor $bx + by - x - y$. Answer: $(x + y)(b - 1)$.

Finally, consider the expression

$$(3x + 4)(2x - 1)$$
$$= 3x(2x - 1) + 4(2x - 1)$$
$$= 6xx - 3x + 8x - 4$$
$$= 6x^2 + 5x - 4$$

where, in the last line, xx is written as x^2 (read as x *squared*). In a number of places in the text, we must do problems like the last one, but in reverse; that is, start with a trinomial that contains a term in x^2, a term in x, and a constant, such as

$$6x^2 + 5x - 4$$

and obtain the equivalent pair of binomial factors,

$$(3x + 4)(2x - 1).$$

We may do this by trial and error. As an example, let us factor

$$12x^2 + 7x - 10. \tag{1}$$

First we write

$$(\quad)(\quad),$$

where each set of parentheses contains two terms, a *first* term and a *second* term. The product of the first terms must be $12x^2$, so the first terms could be x and $12x$, $2x$ and $6x$, $3x$ and $4x$, $6x$ and $2x$, $12x$ and x, or any of the last pairs with signs changed. The product of the second terms must be -10, so the second terms could be 1 and -10, 2 and -5, 5 and -2, 10 and -1, or any of the last four pairs with the sign of both numbers changed. Let us try x and $12x$ as first terms, 1 and -10 as second terms. We fill in the parentheses as follows:

$$(x+1)(12x-10).$$

The product of the first terms is $12x^2$, and the product of the second terms is -10, as required. The *test* of our trial is the product of the inner terms $1(12x)$ plus the product of the outer terms $x(-10)$, which is

$$12x - 10x = 2x.$$

Hence,

$$(x+1)(12x-10) = 12x^2 + 2x - 10$$

and this does not have the same middle term as the original expression, (1),

$$12x^2 + 7x - 10,$$

so we try another set of first and second terms from our list. For example,

$$(3x+2)(4x-5).$$

All we need do at each trial is apply the test mentioned several lines back, and determine if the middle term in the expansion of the trial is the required $7x$. The test here yields

$$2(4x) + (3x)(-5) = 8x - 15x = -7x$$

which is the negative of the desired $7x$ so we need only change the signs of the second (or first) terms. Thus, instead of $(3x+2)(4x-5)$ we write $(3x-2)(4x+5)$. The test term is now

$$-2(4x) + (3x)5 = -8x + 15x = 7x,$$

so we have the desired factors of the original expression,

$$12x^2 + 7x - 10 = (3x-2)(4x+5).$$

With a little practice, the proper pairs of terms usually can be found quite rapidly if we take hints from the original expression and the results of a trial. Thus, in

$$2x^2 - 13x + 20,$$

the second terms have the *positive* product 20, so both must be positive or both must be negative. Inasmuch as the middle term is negative, it follows that both second terms are negative. As a trial, we write

$$(2x - 4)(x - 5)$$

and the test term is $-14x$, which is not the desired $-13x$ in the original. However, we will get the desired middle term if we interchange the second terms of the first trial. Thus,

$$2x^2 - 13x + 20 = (2x - 5)(x - 4).$$

Exercise. Factor $10x^2 + 26x + 12$. Answer: $(5x + 3)(2x + 4)$.

The expression

$$x^2 - 9$$

has a square term and a constant, but no middle term. The first term is the square of x and the second term is the square of 3 (that is, 3 times $3 = 3^2 = 9$), so the expression is called the *difference of two squares*. By trial and error we find

$$x^2 - 9 = (x + 3)(x - 3).$$

Thus, the difference of the squares of two numbers is the product of the sum of the numbers times the difference of the numbers. As another example, noting that $4a^2 = (2a)(2a)$ and $16b^2 = (4b)(4b)$, we have

$$4a^2 - 16b^2 = (2a + 4b)(2a - 4b).$$

Exercise. Factor: *a)* $y^2 - 25$. *b)* $y^2 - 25x^2$. Answer: *a)* $(y + 5)(y - 5)$. *b)* $(y - 5x)(y + 5x)$.

A2.20 PROBLEM SET A2–3

Remove grouping symbols and combine like terms, if any:

1. $2ab(c - 2)$.
2. $(a - 2)(b + 1)$.
3. $2 - 3[1 - (+4)]$.
4. $(c + 2)(a - b + 3)$.
5. $-(-2) + 3[a - (1 - b)]$.
6. $10 - 3[4 - 5(-4 + a)]$.
7. $a - 2\{-3 - 2[5a - 2(a - 6)]\}$.
8. $(3x - 2)[a - 2(b + 3)]$.
9. $-(a - x - 2b)$.
10. $(2x + 3y)$.
11. $(a - 2b) - b$.
12. $a + (3x + 2)$.
13. $b - 2b(a - 3)$.
14. $ab[c - 2(x - 5)]$.

15. $ax - 2b(a - 1)$.

16. $(a - bx)(3 - c)$.

17. $(a + b + 1)(x + y)$.

18. $(a - 1)(b + 1)$.

Reduce to a single signed number:

19. $-3(2) - 2(1 - 3)$.

20. $\dfrac{(-3)(2) - (-2)(4)}{-3(5) - 2.5(-4)}$.

21. $\dfrac{0.017(5 - 1.08) + 2.3[0.5 - 6.2(3.1)]}{-4(0.0025)}$.

22. $\dfrac{12 - (6 - 2)(3 - 1)}{(8 - 3)(-1 - 2) + 6}$.

23. $\dfrac{1.7 - [3 - 17(15.4 - 1.6)]}{-0.8(0.245 - 0.37)}$.

24.   Define *expression*, *term*, and *factor*, giving an illustration in each instance.

Factor:

25. $ab - 2b$.

26. $3a + 5a$.

27. $4abc - 2ab + 6a$.

28. $ax - bx + x$.

29. $3ad - 5ac + a$.

30. $4uv - 2xv + 2$.

31. $abx + aby - ab$.

32. $2ax - 6ay + 4az$.

33. $2x + ax + bx$.

34. $-ab - 3ac - a$.

35. $a(x + 1) + b(x + 1)$.

36. $x + 1 + y(x + 1)$.

37. $2(x + y) - a(x + y)$.

38. $ax + bx + ay + by$.

39. $x^2 - x - 2$.

40. $2x^2 - 9x - 5$.

41. $12x^2 - 25x + 12$.

42. $10x^2 + 13x - 3$.

43. $x^2 - x - 6$.

44. $x^2 - 9$.

45. $x^2 - y^2$.

46. $4x^2 - 9y^2$.

Mark (T) for true or (F) for false:

47.   ( ) The expression ab has only one factor.

48.   ( ) The expression $a + 2b - c$ has three terms.

49.   ( ) The expression $a(b - c)$, as written, is a single term.

50.   ( ) Referring to $a(b - c)$, it would be proper to say that the expression has two factors.

51.   ( ) If ab is to be doubled, both a and b must be doubled.

52.   ( ) $(a - b)(-c) = c(b - a)$.

53.   ( ) It is correct to state that "to multiply an expression by a number, every factor of each term in the expression must be multiplied by the number."

54.   ( ) $-a(-b - c)(-d) = ad(b + c)$.

55.   ( ) The sign of a term is changed if the sign of any one of its factors is changed.

56.   ( ) The parentheses in $a + (b + c)$ are unnecessary.

57.   ( ) $3 + 2(a + b) = 5(a + b)$.

58.   ( ) $-a - b = +(-b) + (-a)$.

59.   ( ) $(a + b) - c = -ac - bc$.

60. ( ) To change the sign of a term, it is sufficient to change the sign of one factor of the term.

A2.21 PROPERTIES OF THE NUMBERS ZERO AND ONE

The number *zero* is unique in several respects. First, for any number a

$$0(a) = 0;$$

that is, zero times any number is zero. Second,

$$a + (-a) = 0;$$

that is, the sum of any number and its negative is zero, and this property defines what we mean by the negative of any number. We also say that in addition, a number and its negative *cancel*, meaning their sum is zero. Third,

$$a + 0 = a - 0 = a;$$

that is, a number is not affected by adding zero to it, or subtracting zero from it. Finally,

$$\frac{a}{0} \text{ is not defined.}$$

We shall find the last statement to be of basic importance in our development of calculus. To understand why $a/0$ is not defined, we consider cases such as $6/0$ where $a \neq 0$, and $0/0$ where a is zero. First recall that $6/3$ means to find a number that when multiplied by 3 yields 6. This is the *definition* of division in terms of multiplication, and we prove $6/3$ is 2 by stating $6 = (3)(2)$. Applying the definition to

$$\frac{6}{0},$$

we would seek a number which when multiplied by 0 yields 6. There is no such number, because 0 times any number is zero. Hence, expressions such as $6/0$, $-5/0$, and so on are not defined.

Next consider

$$\frac{0}{0}.$$

The temptation is to say that this expression is 1 because $0(1) = 0$, which satisfies the definition of division. However, we could say also the expression is 2, or 3.17, 0, or any number because

$$0(\text{any number}) = 0.$$

It follows that if $0/0$ were permitted, the results of mathematical operations could be ambiguous or contradictory. To demonstrate the last statement, suppose that a and b both equal 1. Then,

$$a = b,$$

and it is also true if $a = b = 1$ that

$$a^2 = ab$$

and the equality remains if we subtract $b^2 = 1$ from a^2 and from ab, obtaining

$$a^2 - b^2 = ab - b^2.$$

Factoring shows that

$$(a + b)(a - b) = b(a - b).$$

In general, *except for a division of 0*, if two numbers are equal and we divide them by the same number, the results are equal. If we here neglect the exception of division by zero and divide both numbers in the last equality by $a - b$ (which is zero), we have

$$\frac{(a + b)(a - b)}{(a - b)} = \frac{b(a - b)}{(a - b)}.$$

If we again forget the exception and cancel the $(a - b)$'s, we have

$$a + b = b.$$

Now recall that at the beginning we had $a = b = 1$. The last statement then would be

$$1 + 1 = 1 \quad \text{or} \quad 2 = 1,$$

which is the contradictory result we sought to demonstrate.

Remember: Expressions such as 5/0, 0/0, or any number divided by zero are not defined. We shall say alternatively that division by zero is impossible, or that it is not permitted.

Summarizing the properties, we state that *for any real number, a:*

$$(0)\, a = 0$$
$$a + 0 = a - 0 = a$$
$$a + (-a) = 0$$
$$a/0 \text{ is not defined.}$$

The unique properties of one exist in reference to multiplication and its inverse, division. Thus a number is unchanged if it is multiplied or divided by one. We have

$$a = (1)\, a = \frac{a}{1}.$$

The equivalence of a and $1a$ is assumed conventionally in the writing of various algebraic expressions. Any term may be assumed to have a factor of one. Recall the factoring of

$$ab - a = a(b - 1)$$

as an illustration of the point of the last two sentences.

> **Exercise.** A rational number is the quotient of two integers, yet the single number, 4, is rational. Explain. Answer: 4 is the rational number 4/1.

Any number, zero excepted, divided by itself yields a quotient of one. This fact is employed often, as when we write

$$\frac{6}{6} = 1, \text{ or } \frac{ab}{ab} = 1, \text{ or } \frac{x - y}{x - y} = 1.$$

Circumstances arise also where we may wish to multiply an expression by

$$\frac{6}{6} \quad \text{or} \quad \frac{ab}{ab}$$

and this can be done without changing the expression because it is equivalent to multiplication by one.

The words *cancel* and *cancellation* are used with reference to zero and one. In addition, a number and its negative cancel each other, meaning their sum is zero. In multiplication, a number and its reciprocal cancel, meaning their product is one. Thus:

$$a\left(\frac{1}{a}\right) = 1$$

where $1/a$ is called the reciprocal of a. Often, the latter type of cancellation is thought of in terms of division rather than multiplication of reciprocals. Thus, in ab/a, we think of a over a as being one, so that

$$\frac{ab}{a} = 1\,(b) = b$$

and we say that the a's cancel. An important rule to keep in mind when working with fractions is that cancellation (replacement by one) can be performed *only for factors common to numerator and denominator*. We cannot cancel the a's in

$$\frac{a + 2}{a}$$

because *a* is not a factor of the numerator. On the other hand, the numerator of

$$\frac{ax + 2x}{x(b-1)}$$

can be factored to permit cancellation; thus:

$$\frac{x(a+2)}{x(b-1)} = \frac{a+2}{b-1}.$$

Exercise. *x* is a factor of what parts of the expression

$$\frac{ax + 2}{x(b-c)}?$$

Answer: *x* is a factor of the denominator, and of the single term *ax* in the numerator; *x* is not a factor of the numerator.

A2.22 PRODUCT OF FRACTIONS

The product of two fractions is the product of their numerators over (divided by) the product of their denominators. For example:

$$\left(\frac{2}{5}\right)\left(\frac{3}{7}\right) = \frac{6}{35}$$

$$\left(\frac{a}{2}\right)\left(\frac{3}{b}\right) = \frac{a(3)}{2b} = \frac{3a}{2b}$$

$$\left[\frac{a(b+2)}{3}\right]\left(\frac{2}{b}\right) = \frac{a(b+2)(2)}{3b} = \frac{2a(b+2)}{3b}.$$

Exercise. Express as a single fraction without grouping symbols:

$$\left(\frac{3}{x+y}\right)\left(\frac{x-y}{2}\right).$$

Answer: $\dfrac{3x - 3y}{2x + 2y}.$

Generally, cancellation should be performed where it is possible to do so. For example:

$$\left(\frac{6ab}{5c}\right)\left(\frac{c}{3a}\right) = \frac{2b}{5}$$

$$\frac{a(b+2)}{3}\left(\frac{2}{a}\right) = \frac{2(b+2)}{3}$$

$$\left(\frac{2a-2}{b}\right)\left(\frac{1}{2}\right) = \frac{2(a-1)}{b}\left(\frac{1}{2}\right) = \frac{a-1}{b}.$$

As another example, we start with

$$(x+y)\left[3 + \frac{a}{x+y}\right].$$

Any number can be expressed equivalently as the number divided by 1. Thus, the last is the same as

$$\frac{(x+y)}{1}\left[\frac{3}{1} + \frac{a}{x+y}\right] = \frac{(x+y)}{1}\left(\frac{3}{1}\right) + \frac{(x+y)}{1}\left(\frac{a}{x+y}\right)$$

$$= \frac{(x+y)(3)}{1} + \frac{a}{1}$$

$$= 3(x+y) + a.$$

Similarly,

$$3\left(\frac{a}{b}\right) = \left(\frac{3}{1}\right)\left(\frac{a}{b}\right) = \frac{3a}{b}$$

$$2\frac{(a-3)}{b} = \left(\frac{2}{1}\right)\frac{(a-3)}{b} = \frac{2(a-3)}{b}.$$

> ***Exercise.*** Carry out the multiplication, leaving the result as the sum of two fractions:
>
> $$2\frac{a}{b}\left[3 + \frac{x+2}{ax}\right].$$
>
> Answer: $\dfrac{6a}{b} + \dfrac{2x+4}{bx}$.

An equivalent expression is obtained if a given expression is multiplied or divided by -1 an even number of times because the net effect is multiplication or division by $+1$. For example, in

$$\frac{b-a}{-2}$$

we may change the sign of numerator and denominator to give

$$\frac{-(b-a)}{-(-2)} = \frac{-b+a}{2} = \frac{a-b}{2}.$$

Keeping in mind that three signs are associated with a fraction (the signs of the numerator and denominator and the sign of the fraction itself, it is helpful to remember that an equivalent fraction results if any *two* of these signs are changed. Thus, the fraction

$$\frac{b-a}{-2}$$

in the foregoing has the three signs shown in parentheses next:

$$(+)\frac{(+)\,(b-a)}{(-)\,2},$$

and we changed the signs of the numerator and denominator to yield

$$(+)\frac{(-)(b-a)}{+2} = +\frac{-b+a}{2} = \frac{a-b}{2}.$$

Similarly, in

$$-\frac{2y-x}{3} = (-)\frac{(+)(2y-x)}{(+)3}$$

we may change the sign in front of the fraction and the sign of the numerator to give

$$+\frac{(-)(2y-x)}{+3} = +\frac{-2y+x}{+3} = \frac{x-2y}{3}.$$

It is important to remember that numerator and denominator are *expressions*, and to change the sign of an expression it is necessary to change the sign of every *term* in the expression, where the change of a term's sign is accomplished by changing the sign of *one* (or an odd number) of the term's *factors*. For example,

$$\frac{(-b-2)}{3(2a-5xy)} = (+)\frac{(+)(-b-2)}{(+)3(2a-5xy)},$$

where the expression at the right has the three fraction signs indicated in parentheses. Changing the sign of numerator and denominator gives

$$+\frac{(-)(-b-2)}{(-)3(2a-5xy)} = +\frac{b+2}{-6a+15xy} = \frac{b+2}{15xy-6a}.$$

After some practice, readers will be able to use sign changes to simplify expressions or to reduce the number of negative signs that appear in expressions. While the latter use may seem inconsequential, it does occur frequently. To show how sign change can lead to simplification, note that

$$3x - \frac{b-a}{a-b} = 3x - \frac{(+)(b-a)}{+(a-b)}$$

$$= 3x + \frac{(-)(b-a)}{+(a-b)}$$

$$= 3x + \frac{-b+a}{a-b}$$

$$= 3x + \frac{a-b}{a-b}$$

$$= 3x + 1.$$

However, in

$$2x - \frac{a-y}{b} = 2x - \frac{+(a-y)}{b}$$

$$= 2x + \frac{-a+y}{b}$$

$$= 2x + \frac{y-a}{b},$$

the sign changes served only to reduce the original two negative signs in the beginning expression to one in the ending expression.

> ***Exercise.*** *a)* Simplify by sign changes $y + \dfrac{2x-z}{z-2x}$. *b)* In $2x - \dfrac{a+2}{10-a} = 2x + \dfrac{a+2}{a-10}$, what sign changes were made? Answer: *a)* $y - 1$. *b)* The sign of the fraction and the sign of its denominator were changed.

A final point worthy of note in the multiplication of fractions is the use of the word *of* to designate multiplication. Thus, two thirds *of* one half means

$$\left(\frac{2}{3}\right)\left(\frac{1}{2}\right) = \frac{1}{3}.$$

A2.23 ADDITION AND SUBTRACTION OF FRACTIONS

Addition and subtraction of fractions is accomplished by changing each fraction to the same (common) denominator and then placing the sums (differences) of the resultant numerators over the common denominator. A common denominator can always be found by forming the term that has each of the separate denominators as a factor. On the other hand, if all the factors of each denominator are set down and a term is constructed that contains each factor the maximum number of times it appears in any one denominator, this term is called the *lowest common denominator*. Consider

$$\frac{3}{5} + \frac{4}{15} - \frac{2}{3} + \frac{5}{18}.$$

Factors of 5 are 5, 1.
Factors of 15 are 5, 3, 1.
Factors of 3 are 3, 1.
Factors of 18 are 3, 3, 2, 1.
Lowest common denominator is $(5)(3)(3)(2) = 90$.

The mechanical procedure for changing each fraction to the common denominator is illustrated by reference to the fraction 3/5. We divide the lowest common denominator by 5 to obtain the conversion factor 90/5, which is 18, and then multiply the numerator, 3, by the conversion factor to obtain 54. By this procedure, 3/5 is changed to 54/90. We have

$$\frac{3}{5} + \frac{4}{15} - \frac{2}{3} + \frac{5}{18} = \frac{54}{90} + \frac{24}{90} - \frac{60}{90} + \frac{25}{90} = \frac{43}{90}.$$

The mechanical procedure is efficient, but in the interest of emphasizing fundamentals, it should be made clear that the process derives from a fundamental property of the number one; that is, a number is unchanged if it is multiplied by one. For example, when converting 3/5 to a denominator of 90, we observe that 5 must be multiplied by 18 to yield 90, so we multiply 3/5 by 18/18; that is, in this instance the unit multiplier is 18/18. In the case of 4/15 the unit multiplier is 6/6. It follows that the mechanical procedure is a consequence of the more lengthy, but also more fundamental, process shown next:

$$\frac{3}{5} + \frac{4}{15} - \frac{2}{3} + \frac{5}{18} = \frac{3(18)}{5(18)} + \frac{4(6)}{15(6)} - \frac{2(30)}{3(30)} + \frac{5(5)}{18(5)} = \frac{43}{90}.$$

As another example, follow the conversion of each fraction in the next expression to the lowest common denominator, $a(2)(3)$:

$$\frac{2}{a} + \frac{b}{2} + \frac{2c}{3} + \frac{1}{6} = \frac{2}{a}\left(\frac{6}{6}\right) + \frac{b}{2}\left(\frac{3a}{3a}\right) + \frac{2c}{3}\left(\frac{2a}{2a}\right) + \frac{1}{6}\left(\frac{a}{a}\right)$$

$$= \frac{12 + 3ab + 4ac + a}{6a}.$$

Exercise. Add $5 + \dfrac{2}{y} + \dfrac{3}{4} + \dfrac{1}{2x}$.

Answer: $\dfrac{23xy + 8x + 2y}{4xy}$.

As a final example, consider

$$\frac{5}{3} + \frac{c}{2a} - \frac{c}{a(b+2)}.$$

The factors of the denominator are, in turn:

$$3, 1$$
$$2, a, 1$$
$$a, (b+2), 1.$$

The lowest common denominator is $3(2)a(b+2)$. Hence:

$$\frac{5}{3} \frac{(2a)(b+2)}{(2a)(b+2)} + \frac{c}{2a} \frac{(3)(b+2)}{(3)(b+2)} - \frac{c}{a(b+2)} \frac{(3)(2)}{(3)(2)}$$

is equivalent to the original set of fractions. By the associative and commutative properties, all of the denominators in the last may be written as $6a(b+2)$, so placing all the numerators over the common denominator we have

$$\frac{5(2a)(b+2) + c(3)(b+2) - c(3)(2)}{6a(b+2)}$$
$$= \frac{10a(b+2) + 3c(b+2) - 6c}{6a(b+2)}$$
$$= \frac{10ab + 20a + 3bc + 6c - 6c}{6a(b+2)}$$
$$= \frac{10ab + 20a + 3bc}{6a(b+2)}.$$

Exercise. Add: $\dfrac{1}{x} + \dfrac{2}{y+1} - \dfrac{2}{3}$.

Answer: $\dfrac{4x + 3y - 2xy + 3}{3x(y+1)}$.

A2.24 DIVISION OF FRACTIONS

The division of the fraction a/b by the fraction c/d can be written in the form of a third fraction:

$$\frac{\dfrac{a}{b}}{\dfrac{c}{d}} .$$

This last expression can be simplified if it is multiplied by a suitably chosen one. The objective is to convert from the *complex* fraction (that is, a fraction whose numerator or denominator contains a fraction) to a *simple* fraction (which does not have a fraction in its numerator or denominator). Clearly, we can cancel the c/d of the denominator if we multiply it by d/c. We must then also multiply the numerator by d/c, so that the net effect is multiplication by one. Thus:

$$\frac{\dfrac{a}{b}}{\dfrac{c}{d}} = \frac{\dfrac{a}{b}\left(\dfrac{d}{c}\right)}{\dfrac{c}{d}\left(\dfrac{d}{c}\right)} = \frac{ad}{bc} .$$

The process often is described as "inverting the denominator and multiplying"; that is, invert c/d to give d/c, then multiply the numerator by d/c. This description is adequate when numerator and denominator are in completely factored form, but multiplication by a suitably chosen one not only is a more fundamental description, but also is somewhat more direct when numerator and denominator are not in factored form. Consider the problem of reducing the following to a simple fraction:

$$\frac{\dfrac{a}{2}+\dfrac{1}{3}}{\dfrac{1}{2}+b} .$$

We observe that the lowest common denominator of the terms in the numerator and denominator is 6, so we multiply the fraction by 6/6. Thus:

$$\frac{6\left(\dfrac{a}{2}+\dfrac{1}{3}\right)}{6\left(\dfrac{1}{2}+b\right)} = \frac{3a+2}{3+6b} .$$

Exercise. Convert to a simple fraction by multiplying numerator and denominator by 12:

$$\frac{\dfrac{1}{2}+\dfrac{1}{3}}{\dfrac{3}{4}+\dfrac{1}{3}}.$$

Answer: 10/13.

In the next example the lowest common denominator of all terms in numerator and denominator is $3ab$. Hence, we choose our one to be $3ab/3ab$:

$$\frac{\dfrac{2}{a}+\dfrac{1}{b}}{\dfrac{1}{3}+\dfrac{2}{b}}=\frac{\left(\dfrac{2}{a}+\dfrac{1}{b}\right)(3ab)}{\left(\dfrac{1}{3}+\dfrac{2}{b}\right)(3ab)}=\frac{6b+3a}{ab+6a}.$$

A2.25 PROBLEM SET A2–4

Mark (T) for true or (F) for false:

1. ( ) 0/0 equals one.
2. ( ) 0/0 equals zero.
3. ( ) No matter what number a is, $a/0$ is meaningless.
4. ( ) If a is not zero, then $0/a$ equals zero.
5. ( ) The product of any number and zero is zero.
6. ( ) The reciprocal of 4 equals 0.25.
7. ( ) The reciprocal of 3 equals 0.3.
8. ( ) In addition, it is said that a number and its reciprocal cancel.
9. ( ) In division, cancellation is the equivalent of substituting the factor one in place of the product of a number and its reciprocal.
10. ( ) Multiplying a number by its reciprocal gives the same result as dividing the number by itself.
11. ( ) "Inverting and multiplying" is equivalent to multiplying by a reciprocal.
12. What does it mean to say a number and its reciprocal cancel?

Simplify by cancellation where possible:

13. $\dfrac{-3a(-6)}{12c}.$

14. $\dfrac{2a-3}{3}.$

15. $\dfrac{2a-3a}{-a}$.

16. $\dfrac{2a+6}{2}$.

17. $\dfrac{24acd}{4ad}$.

18. $\dfrac{x+y}{y}$.

19. $\dfrac{2xy+6ax+x}{4xy}$.

20. $4(a+2)\left(\dfrac{2x}{a+2}\right)$.

21. State the rule for multiplication of fractions.

Multiply, leaving no grouping symbols in the answer:

22. $\dfrac{ab}{2}\left(\dfrac{3}{4}\right)$.

23. $\dfrac{a+b}{3}\left(\dfrac{2}{5}\right)$.

24. $\dfrac{-2}{3}\left(\dfrac{9a}{8}\right)$.

25. $(-2)\left(\dfrac{a}{3}\right)\left(\dfrac{b+2}{-7}\right)$.

26. $(2)\left(-\dfrac{1}{3}\right)\left(\dfrac{1}{a+b}\right)$.

27. $3(a+2)\left(\dfrac{1}{3}+\dfrac{2b}{a+2}\right)$.

28. $6ab\left(\dfrac{2}{3b}-\dfrac{1}{a}\right)$.

Express with a single minus sign:

29. $\dfrac{-b(c-d)}{-2}$.

30. $-\dfrac{b-2}{-3-a}$.

31. $\dfrac{-2+(b-c)}{-2a}$.

32. $-\dfrac{-2+(b-c)-a}{2x(a+b)}$.

33. $-\dfrac{x+y}{x-y}$.

Reduce to one simple fraction:

34. $\dfrac{2}{3}-\dfrac{1}{2}+\dfrac{1}{6}$.

35. $\dfrac{a}{2}-\dfrac{3}{5}$.

36. $\dfrac{3}{2a}-\dfrac{1}{6}+\dfrac{2}{5b}$.

37. $\dfrac{2}{5}-\dfrac{2(a-10)}{5a}+\dfrac{1}{6}$.

38. $3\frac{1}{2}-2\frac{1}{3}$.

39. $\dfrac{x}{a-2}+\dfrac{1}{b}-2$.

40. $\dfrac{x}{2a}-b+\dfrac{3}{a}$.

41. $3x-\dfrac{1}{2}+\dfrac{2}{12ab}$.

42. $\dfrac{2a}{3(b-1)}-\dfrac{a-1}{4}+\dfrac{1}{6}$.

43. $\dfrac{7}{2(x+3)}-3+\dfrac{5}{4(x+3)}$.

44. Multiply $2\frac{1}{3}$ by $3\frac{1}{4}$, stating the product as a simple fraction.

45. Divide $1\frac{1}{8}$ by $7\frac{1}{3}$, stating the quotient as a simple fraction.

Reduce the following complex fractions to simple fractions by multiplying by a suitably chosen one:

46. $\dfrac{\dfrac{1}{2}+\dfrac{1}{3}-\dfrac{1}{4}}{\dfrac{2}{3}-\dfrac{1}{6}}$.

47. $\dfrac{\dfrac{6}{a}+2}{-\dfrac{3}{b}+\dfrac{5}{a}}$.

48. $\dfrac{\dfrac{ab}{2} - \dfrac{b-3}{c}}{\dfrac{b}{3} - 1}.$ . .

49. $\dfrac{\dfrac{2a}{3b} - \dfrac{1}{c} + 2}{\dfrac{1}{6} - \dfrac{2}{bc}}.$

50. $\dfrac{\dfrac{a}{2} + \dfrac{b}{3} - \dfrac{c}{6}}{b - \dfrac{a}{4}}.$

A2.26 EXPONENTS

The product $(a)(a)(a)(a)(a)$ is denoted by writing a with a superscript of 5; thus, a^5. It is called the fifth power of a. The number a is the *base*, and 5 is the *exponent* of the power. More generally, if n is a positive integer, a^n is read as "a to the nth" and means the term that has a as a factor n times. By convention, we interpret absence of an exponent to mean the exponent is one. We have

$$(a)(a) = a^2$$
$$(a)(a)(a) = a^3$$
$$a = a^1.$$

It is clear that

$$(a^2)(a^3) = (a)(a)[(a)(a)(a)] = a^5.$$

We see that

$$(a^2)(a^3) = a^{2+3} = a^5$$

and it follows that if two powers have the same base, their product is found by writing the common base with the sum of the exponents as its power. For example:

$$x^5(x)x^2 = x^{5+1+2} = x^8$$
$$3a^2(2a^3) = 6a^5$$
$$3^2(3) = 3^3 = 27$$
$$(-2)^3(-2) = (-2)^4 = 16$$
$$(-3)(-3)^2 = (-3)^3 = -27.$$

Exercise. Write $2x^2(3x^5)$ with a single exponent. Answer: $6x^7$.

Turning to division, we have, for example:

$$\frac{a^5}{a^2} = \frac{(a)(a)(a)(a)(a)}{(a)(a)} = (a)(a)(a) = a^3$$

by cancellation. Alternatively, the final exponent, 3, could have been obtained by the subtraction, $5 - 2$, that is, the numerator exponent minus the denominator exponent. In general, if two powers have the same base, their quotient is the common base with an exponent found by the subtraction procedure just mentioned. As examples:

$$\frac{a^4}{a^2} = a^{4-2} = a^2$$

$$\frac{a^2 b^3}{ab} = ab^2$$

$$\frac{(a+b)^3}{a+b} = (a+b)^2$$

$$\frac{5^{12}}{5^{10}} = 5^2 = 25$$

$$\frac{3^4(2a^6)}{3a^4} = 54a^2$$

$$\frac{(-2)^5}{(-2)^2} = (-2)^3 = -8$$

$$\frac{(-3)^4(2x^5)}{-3x^2} = (-3)^3(2x^3) = -54x^3.$$

Exercise. Write $(5x^5)/3x^2$ with a single exponent. Answer: $5x^3/3$.

A2.27 ZERO EXPONENT

Following the procedure of the last discussion, we see that for any number a, not zero:

$$\frac{a}{a} = a^{1-1} = a^0.$$

Inasmuch as the beginning expression, a/a, equals one, we conclude that any nonzero number to the zero power equals one. Thus:

$$1^0 = 1, \quad (ab^3c^2)^0 = 1, \quad (14.6)^0 = 1, \quad (-x)^0 = 1.$$

Exercise. Evaluate $3^0 + (x + 2y)^0$. Answer: 2.

A2.28 NEGATIVE EXPONENTS

When exponents are subtracted, the difference may be negative. For example:

$$\frac{2^3}{2^6} = 2^{3-6} = 2^{-3}.$$

Alternatively, we may evaluate the expression as

$$\frac{2^3}{2^6} = \frac{(2)(2)(2)}{(2)(2)(2)(2)(2)(2)} = \frac{1}{(2)(2)(2)} = \frac{1}{2^3}.$$

We see that

$$(2)^{-3} = \frac{1}{2^3}.$$

The general definition applying to a negative exponent is

$$a^{-n} = \frac{1}{a^n}.$$

As illustrations of the definition, we see that

$$2^{-1} = \frac{1}{2^1} = \frac{1}{2}$$

$$3^{-2} = \frac{1}{3^2} = \frac{1}{9}$$

$$ax^{-2} = \frac{a}{1}\left(\frac{1}{x^2}\right) = \frac{a}{x^2}$$

$$\left(\frac{2}{3}\right)^{-1} = \frac{1}{\left(\frac{2}{3}\right)} = \frac{1}{\frac{2}{3}} = 1\left(\frac{3}{2}\right) = \frac{3}{2}$$

$$\left(\frac{1}{3}\right)^{-2} = \frac{1}{\left(\frac{1}{3}\right)^2} = \frac{1}{\frac{1}{9}} = 1\left(\frac{9}{1}\right) = 9.$$

Exercise. Evaluate $(1/2)^{-2} + 5(2^{-3})$. Answer: $37/8$.

The following examples are self-explanatory and show how in some expressions we may avoid negative exponents by choice of procedure.

$$\frac{x^3}{x^5} = \frac{1}{x^{5-3}} = \frac{1}{x^2}$$

$$\frac{2x}{x^7} = \frac{2x^1}{x^7} = \frac{2}{x^{7-1}} = \frac{2}{x^6}$$

$$\frac{a^2 b^3}{a^4 b} = \frac{b^{3-1}}{a^{4-2}} = \frac{b^2}{a^2}.$$

Exercise. Apply the laws of exponents to simplify the following expression and write the result with positive exponents.

$$\frac{3x(x^{-2})y^5}{4x^4 y^2}.$$

Answer: $\dfrac{3y^3}{4x^5}.$

Finally, we note that the expression

$$\frac{1 + x^{-n}}{2 + a}$$

is, in effect, a complex fraction because of the fractional nature of x^{-n}. Remembering that

$$x^{-n}(x^n) = x^0 = 1,$$

we may obtain a simple fraction by multiplying by x^n / x^n, as follows:

$$\frac{1 + x^{-n}}{2 + a} = \frac{(1 + x^{-n})x^n}{(2 + a)x^n} = \frac{x^n + 1}{(2 + a)x^n}.$$

As another example of the same procedure:

$$\frac{2 + 3^{-2}}{1 + 3^{-1}} = \left(\frac{2 + 3^{-2}}{1 + 3^{-1}}\right)\left(\frac{3^2}{3^2}\right) = \frac{18 + 1}{9 + 3} = \frac{19}{12}.$$

Exercise. Remove negative exponents by multiplying numerator and denominator by x^n: $\dfrac{1 + x^{-n}}{x^{-n} + 2}.$ Answer: $\dfrac{x^n + 1}{1 + 2x^n}.$

A2.29 POWER TO A POWER

The expression $(a^2)^3$ is an example of a power raised to a power; that is, the second power of a is indicated as being raised to the third power. According to definition:

$$(a^2)^3 = (a^2)(a^2)(a^2)$$

which is a^6 or $a^{(3)(2)}$. The procedure is generalized by stating that in raising a power to a power, exponents are multiplied. As other examples:

$$(2^2)^4 = 2^{(2)(4)} = 2^8 = 256.$$

$$(x^3)^2 = x^{(3)(2)} = x^6.$$

$$(x^{-1})^4 = x^{(-1)(4)} = x^{-4} = \frac{1}{x^4}.$$

$$(5^{-2})^{-1} = 5^{(-2)(-1)} = 5^2 = 25.$$

Exercise. Evaluate $[(2)^{-2}]^{-3}$. Answer: 64.

A2.30 FRACTIONAL EXPONENTS

The number

$$8^{1/3}$$

may be read as *eight to the one-third power*. If we apply the rules discussed earlier for integral exponents to this rational fractional exponent, it follows that

$$(8^{1/3})(8^{1/3})(8^{1/3}) = 8^1 = 8,$$

so that the number symbolized as $(8^{1/3})$ must be 2. That is,

$$8^{1/3} = 2.$$

Moreover,

$$8^{2/3} = (8^{1/3})(8^{1/3}) = (2)(2) = 4.$$

Exercise. Express $8^{4/3}$ as an integer. Answer: 16.

The number $8^{1/3}$ is also called the *cube root* of 8 and expressed by the *radical* symbol,

$$\sqrt[3]{8}.$$

The number appearing in the opening of the radical symbol is called the *index* of the root, and the number under the symbol, here 8, is called the *radicand*. We note that the index of the root is the denominator of the fractional exponent. In similar fashion,

$$16^{1/2} = \sqrt[2]{16} = \sqrt{16}$$

is called the square root of 16 and, conventionally, the index 2 is not written. That is, if no index number appears on the radical, the index is assumed to be 2.

Both 4 and −4 are square roots of 16 because

$$(4)(4) = (-4)(-4) = 16.$$

Thus,

$$\sqrt{16} = \pm 4.$$

Exercise. What are the fourth roots of 16? Answer: ±2.

We point out in passing that if we write

$$y(x) = x^{1/2}$$

as a function, then (See Appendix A1) we restrict the square root to its *positive* value because if y is a function of x, there must be one and only one value of y for each value of x.

Even roots of positive numbers have both a positive and a negative value. However, *we shall generally follow the practice of indicating only the positive value for even roots of positive numbers.* Even roots of negative numbers are not real numbers. For example, in

$$(-4)^{1/2}(-4)^{1/2} = (-4)^1 = -4,$$

there is no real number for $(-4)^{1/2}$ that will make the statement true. Clearly, $(0)(0) \neq -4$ and, moreover, the product of numbers of like sign cannot be negative, which would be required if $(-4)^{1/2}$ was a real number as required by the statement.

Odd roots of negative numbers can be found, as in

$$(-8)^{1/3} = -2,$$

but we shall have no occasion for their use. Thus, *when we use a fractional power of a number, x, it will be assumed that x is not negative.*

If we write radical expressions at random, the desired root often is irrational and must be approximated. For example,

$$\sqrt{2} = 2^{1/2} = 1.4142$$

to four decimal places. We can approximate irrational roots by logarithms. (See Index for reference.) In this appendix, we consider only examples where roots are rational and can be determined by inspection. For example, inspecting

$$4^{3/2}$$

we first write the equivalent numbers

$$4^{3/2} = 4^{(1/2)(3)} = (4^{1/2})^3 = 2^3 = 8.$$

When expressions with fractional exponents are to be evaluated, we strongly urge that the fraction be written first. Thus, in evaluating

$$64^{2/3}$$

we write

$$(64)^{2/3} = (64^{1/3})^2 = 4^2 = 16$$

rather than

$$64^{2/3} = (64^2)^{1/3} = (4096)^{1/3} = 16.$$

The point is simply that the latter procedure, although correct, leads to $(4096)^{1/3}$ and it is not easy to tell at a glance what this cube root is.

Exercise. Evaluate $(27)^{2/3}$. Answer: 9.

Previous rules apply when exponents are fractional or negative, as shown by the following examples:

$$(3^{-2})(3^{-3}) = 3^{-2+(-3)} = 3^{-5} = \frac{1}{3^5} = \frac{1}{243}.$$

$$\frac{5^{-6}}{5^{-8}} = 5^{-6-(-8)} = 5^{-6+8} = 5^2 = 25.$$

$$(2^{1/3})(2^{1/2}) = 2^{1/3+1/2} = 2^{2/6+3/6} = 2^{5/6}.$$

$$\frac{a^{2/3}b^2}{ab} = \frac{a^{2/3}b^2}{a^1 b^1} = \frac{b^{2-1}}{a^{1-2/3}} = \frac{b}{a^{1/3}}.$$

$$(a^{1/3})^2 = a^{(1/3)(2)} = a^{2/3}.$$

Fractional powers can be written with radical signs. Thus,

$$16^{3/4} = (\sqrt[4]{16})^3 = 2^3 = 8.$$

In general, we prefer to use fractional exponents rather than radical signs.

A2.31 SUMMARY OF EXPONENT RULES

All the rules for exponents may now be stated in brief form:

$$a^m a^n = a^{m+n}$$

$$\frac{a^m}{a^n} = a^{m-n} = \frac{1}{a^{n-m}}$$

$$(a^m)^n = a^{mn}$$

$$a^0 = 1.$$

The rules of exponents are not restricted to terms having a single factor. We may raise a term to a power by raising each factor of the term to the power. Thus:

$$(2a)^3 = 8a^3$$

$$\left(\frac{a}{2}\right)^2 = \frac{a^2}{4}$$

$$(3x^2y^{1/2})^3 = 27x^6y^{3/2}.$$

Exercise. Express $(4y^4x^3)^{1/2}$ without parentheses. Answer: $2y^2x^{3/2}$.

Observe, however, that in $3a^2$ the absence of parentheses means the exponent applies only to a, not to 3. Again:

$$3(2a^2)^3 = 3(8a^6) = 24a^6.$$

Exponents cannot be applied separately to terms of an expression. In the case of, say, $(a - b)^2$, we cannot simply raise each term to the second power. By definition:

$$(a - b)^2 = (a - b)(a - b).$$

According to the distributive property:

$$(a - b)(a - b) = a^2 - ab - ba + b^2 = a^2 - 2ab + b^2.$$

By way of numerical illustration:

$$(1 - 0.2)^2 = (0.8)^2 = 0.64$$

could be evaluated as

$$(1 - 0.2)(1 - 0.2) = 1^2 - 1(0.2) - 0.2(1) + 0.2^2 = 0.64.$$

This type of exercise is referred to as squaring a binomial, that is, raising the sum of two terms to the second power. We see in general that

$$(a + b)^2 = a^2 + 2ab + b^2$$

and describe the outcome as the "square of the first, plus twice the product of the two, plus the square of the second." The reader may verify that the cube of a binomial is given by

$$(a + b)^3 = a^3 + 3a^2b + 3ab^2 + b^3.$$

Before working the problem set, the reader may wish to practice by verifying the answer stated for the following practice problem set.

A2.32 PRACTICE PROBLEM SET

The reader should look at the following and be able to justify each step by reference to the appropriate rule.

1. $a^{-3}a^5 = a^{-3+5} = a^2$.

2. $(4^{-1})(4^3) = 4^{-1+3} = 4^2 = 16$.

3. $(2^{-3})(2^{-2}) = 2^{-3+(-2)} = 2^{-5} = \dfrac{1}{2^5} = \dfrac{1}{32}$.

4. $\dfrac{5^2}{5^{-3}} = 5^{2-(-3)} = 5^{2+3} = 5^5 = 3125$.

5. $(125)^{4/3} = (125^{1/3})^4 = 5^4 = 625$.

6. $16^{7/4} = (16^{1/4})^7 = 2^7 = 128$.

7. $27^{-1/3} = \dfrac{1}{27^{1/3}} = \dfrac{1}{3}$.

8. $(125)^{-2/3} = \dfrac{1}{(125)^{2/3}} = \dfrac{1}{(125^{1/3})^2} = \dfrac{1}{5^2} = \dfrac{1}{25} = 0.04$.

9. $27^{2/3}: = (\sqrt[3]{27})^2 = (3)^2 = 9$.

10. $9^{-3/2}: = \dfrac{1}{9^{3/2}} = \dfrac{1}{3^3} = \dfrac{1}{27}$.

11. $8^{4/3}: = 16$.

12. $\sqrt{-4}$: not a real number.

13. $(2^{-2})(3^{-2}): = \dfrac{1}{4}\left(\dfrac{1}{9}\right) = \dfrac{1}{36}$.

14. $(2^{1/3})(2^{1/2})(2^{1/6}): = 2^{1/3+1/2+1/6} = 2$.

15. $(1 + 0.01)^2: = (1.01)^2 = 1.0201$.

16. $2^{-2} + 3^{-1}: = \dfrac{1}{4} + \dfrac{1}{3} = \dfrac{7}{12}$.

17. $2^3(1 + 2^{-3}): = 2^3 + 2^0 = 8 + 1 = 9$.

18. $(15)^{12}(15)^{-10}: = 15^2 = 225$.

19. $\dfrac{10^{-4}}{10^{-2}}: = \dfrac{1}{10^{-2+4}} = \dfrac{1}{100}$.

20. $\dfrac{1}{3}(4^{-3/2}): = \dfrac{1}{3}\left(\dfrac{1}{4^{3/2}}\right) = \left(\dfrac{1}{3}\right)\left(\dfrac{1}{8}\right) = \dfrac{1}{24}$.

21. $\dfrac{5 + 2^{-2}}{3}: = \dfrac{(5 + 2^{-2})2^2}{(3)(2^2)} = \dfrac{(5)2^2 + 2^0}{3(2^2)} = \dfrac{21}{12} = \dfrac{7}{4}$.

Combine exponents where possible, and simplify. If possible, do not leave grouping symbols, radical signs, or negative exponents in the result:

22. $a^2xa^3x^2: = a^5x^3$.

23. $2x^0 + (2x)^0: = 2 + 1 = 3$.

24. $(a^2b)(a^{-1})b^3(ab)^{-1}: = a^{2-1-1}b^{1+3-1} = b^3$.

25. $\dfrac{4x^2b}{(2b)^2}: = \dfrac{4x^2b}{4b^2} = \dfrac{x^2}{b}$.

26. $\dfrac{a^3}{(a^x)^2}: = \dfrac{a^3}{a^{2x}} = a^{3-2x}$ or $\dfrac{1}{a^{2x-3}}$.

27. $\dfrac{(2^3)(2^x)}{2^{1+x}}: = 2^{3+x-(1+x)} = 2^2 = 4$.

28. $\dfrac{(x^{1/3})(y^{2/3})^2}{2(xy)^{1/2}} := \dfrac{x^{1/3}y^{1/3}}{2x^{1/2}y^{1/2}} = \dfrac{y^{5/6}}{2x^{1/6}}$.

29. $\dfrac{x^{1/3}\sqrt{b}}{b\sqrt{x}} := \dfrac{x^{1/3}b^{1/2}}{bx^{1/2}} = \dfrac{1}{b^{1/2}x^{1/6}}$.

30. $(a-2b)^2 := a^2 - 4ab + 4b^2$.

31. $\dfrac{1}{3}(3x)^{-2/3} := \dfrac{1}{3(3)^{2/3}(x)^{2/3}} = \dfrac{1}{3^{5/3}x^{2/3}}$.

32. $\dfrac{1+(1+x)^{-n}}{x} := \dfrac{[1+(1+x)^{-n}](1+x)^n}{x(1+x)^n} = \dfrac{(1+x)^n+1}{x(1+x)^n}$.

A2.33 PROBLEM SET A2–5

Evaluate:

1. 2^4.
2. $7(2)^0$.
3. $(1-0.02)^{-2}$ to three decimal places.
4. $\left(\dfrac{3}{4}\right)^{-1}$.

5. $(-3)^{-2}(2)^{-3}$.
6. $\dfrac{(10^{-3})(10^5)}{(10^3)(10^{-4})}$.

7. $\sqrt{-16}$.
8. $3^{-1}+\left(\dfrac{2}{3}\right)^{-2}$.

9. 3^{-2}.
10. $(x-3)^0$.

11. $16^{3/4}$.
12. $\left(\dfrac{1}{8}\right)^{1/3}$.

13. $\dfrac{10^{-4}}{10^{-5}}$.
14. $\dfrac{2}{3}(16)^{-3/4}$.

15. $\sqrt{\dfrac{25}{16}}$.
16. $5x^0+(ax)^0$.

17. $(25)^4(25)^{-3}$.
18. $(1+0.05)^2$.

19. $(32)^{-3/5}$.
20. $\left(\dfrac{2}{3}\right)^{-1}+2^{-3}$.

21. $\sqrt[3]{125}$.
22. $\left(\dfrac{8}{27}\right)^{-2/3}$.

23. $\dfrac{1+3^{-2}}{5}$.
24. $\dfrac{2^{-3}+2^{-1}}{2^{-1}}$.

25. $3^{-2}(1+3^{-1})$.

Combine exponents where possible, and simplify. If possible, do not leave grouping symbols, radical signs, or negative exponents in the result:

26. a^2a.
27. $(ab)(ac)(bc)$.
28. $a^2b^3(ab^4)$.

29. $(abc^2)(a^2cb)$.
30. $(x^2a)(x^2b)$.
31. $\dfrac{a^2b^3}{ab}$.

32. $\dfrac{a^4b^2c}{a^2b^3c^2}$.
33. $\dfrac{xy^3b^3}{x^3yb}$.
34. $\dfrac{a^3(bxy)}{abx^2y}$.

35. $\dfrac{x^3y^2}{x^2y^3}$.

36. $\dfrac{a(bc)^3}{ab^2}$.

37. $\dfrac{(ab)^2(ab)}{a}$.

38. $\dfrac{a(bc)^3}{b(ac)^4}$.

39. $\dfrac{(-3b)^3(2c)^2}{12b^2c}$.

40. $\dfrac{-2b_i^2(-3c)^2}{-(-bc)^3}$.

41. $(xy^2)(ay^{-2})$.

42. $(ab^2c^{-1})(a^2b^{-1}c^2)$.

43. $a(ax)^{-2}$.

44. $\dfrac{2x^{-1}a^3}{(-ax)^2}$.

45. $\dfrac{x^{-2}}{ax}$.

46. $3a^{1/3}b^{1/2}a^2b$.

47. $\dfrac{\sqrt{x}\,\sqrt[3]{y}}{x^{-1}y^{1/2}}$.

48. $\dfrac{a^{2/3}(b^3)^{1/2}}{\sqrt{a}\,\sqrt[3]{b}}$.

49. $\dfrac{2}{3}(3x)^{-1/2}$.

50. $\dfrac{2x^{-1/2}\sqrt{y^3}}{3y^{-1/3}\sqrt{x^3}}$.

51. $\dfrac{2+x^{-2}}{x+3}$.

52. $\dfrac{(1-x)^{-1}+1}{x}$.

53. $\dfrac{x^{-1}+x^{-2}}{3}$.

54. $\dfrac{(ax)^2}{(a^x)^2}$.

55. $\dfrac{x^5+4}{x^2+2}$.

56. $a(a+b)$.

57. $(a-3b)^2$.

58. $a(a-b)^2$.

59. $(x-y)(x+y)$.

60. $a(a^{-1}+1)$.

61. $(a^{1/2}-1)^2$.

62. $\dfrac{a^{-1}+2}{a}$.

63. $\dfrac{(3^2)(3^a)}{3^{a-1}}$.

Mark (T) for true or (F) for false:

64. (   ) $(2^3)(3^2)=6^5$.

65. (   ) $ab^2=a^2b^2$.

66. (   ) $(2ab)^2=4a^2b^2$.

67. (   ) $\dfrac{a^2}{b^2}=\left(\dfrac{a}{b}\right)^2$.

68. (   ) $(a-b)^2=a^2-b^2$.

69. (   ) $(a+1)^{1/2}=\sqrt{a}+1$.

70. (   ) $\dfrac{\sqrt[3]{a}}{\sqrt[3]{b}}=\sqrt[3]{\dfrac{a}{b}}$.

71. (   ) $(1+0.1)^{-2}=\dfrac{1}{1.21}$.

72. (   ) An expression is raised to a power by raising each of its terms to the power.

73. (   ) The sum of the squares of two numbers is the same as the square of the sum of the two numbers.

74. (   ) $a^{-1}+b^{-1}$ is equivalent to the sum of the reciprocals of a and b.

75. (   ) $0^0=1$.

76. (   ) $5x^0=1$.

77. (   ) $(1-0.02)^{-5}=\dfrac{1}{(0.98)^5}$.

78. (   ) $2x^{-1/2}=\dfrac{2}{x^2}$.

79. (   ) $a^{2/3}$ is the cube root of a^2.

80. (   ) $\sqrt{x^2+y^2+z^2}=x+y+z$.

81. (   ) $\sqrt{2x}=2^{1/2}x^{1/2}$.

82. (   ) $\sqrt{-12}$ is not a real number.

83. (   ) $\dfrac{2a^{1/6}}{b^{1/6}}=2\left(\dfrac{a}{b}\right)^{1/6}$.

84. ( ) $a^{1/2}$ is called the square of a.
85. ( ) $(a - 2b)^2 = a^2 - 4ab + 4b^2$.

A2.34 REVIEW PROBLEMS

Reduce to a single signed number, or zero:

1. $-4 + (-2) - [+(-3)]$.
2. $-(-8) + (-4) - (+4) + 2$.
3. $-2(3) + (-3)(-4) - (1)(-5)$.
4. $-(-3) + 2\{-[-3(2)]\} - (+10)$.
5. $-5(-2)(-3) + (3)(-2)(-5)$.
6. $\dfrac{-3(-2)(-4)}{8(-5)}$.

7. $\dfrac{-(5)(3)(-4)}{2(-6)}$.

8. $\dfrac{-(-5)(20)}{-\{-[-4]\}}$.

9. $\dfrac{3(2)(5)(-6)}{(-10)(-18)}$.

10. $\dfrac{(-1)(-2)(-3)(-4)}{(-5)(6)}$.

11. Rewrite as the sum of signed numbers:
    a) $3 - 4$.   b) $-1 + 2 - 3$.
    c) $-1 - 2 - 4 + 5$.   d) $2 + 3$.

12. Reduce to a single signed number:
    a) $-(2 - 5) + 3(-4) - (2 + 1)$.
    b) $-[-2 - (+3)] + (-1)(1 - 3)$.
    c) $-[-(a - b) + (-b + a)]$.

13. What fundamental property or convention justifies each of the following?
    a) $x2y = 2xy$.   b) $4(3z) = 12z$.
    c) $+2 + (-a) = 2 - a$.   d) $3x + 6y = 3(x + 2y)$.
    e) $3a + 2 + y = 3a + y + 2$.   f) $(a + 2) + b = a + (2 + b)$.
    g) $1 + 2 + x = 3 + x$.   h) $x2 + y3 = 2x + 3y$.
    i) $3 + (-x) = 3 - x$.   j) $1a = a$.
    k) $2(a + b) = 2a + 2b$.   l) $5 - xy = -xy + 5$.
    m) $3(x + y) = 3x + 3y$.   n) $5a + 5b = 5(a + b)$.

14. Combine like terms:
    a) $2 + 5x - 4b + 7 - (-3b)$.   b) $ab - a(2 - b) + 3a + 5$.
    c) $x(ay - 3) + 2x + 4axy - 7$.   d) $xy - 2(2 - xy) + 5xy$.

15. Starting with $3x(2y) + 3 + x + 3x$, name the fundamental property that justifies each step:
    a) $3x(2y) + 3 + x + 3x = 3x(2y) + 3 + 4x$.
    b) $= 3(2)xy + 3 + 4x$.
    c) $= 6xy + 3 + 4x$.
    d) $= 6xy + 4x + 3$.
    e) $= (6xy + 4x) + 3$.
    f) $= 2x(3y + 2) + 3$.

16. Starting with $zy + 5 + 2(3 + 4y)$, name the fundamental property that justifies each step:

a) $zy + 5 + 2(3 + 4y) = zy + 5 + 2(3) + 2(4y).$
b) $\qquad\qquad\qquad = zy + 5 + 6 + 2(4y).$
c) $\qquad\qquad\qquad = zy + 5 + 6 + 8y.$
d) $\qquad\qquad\qquad = zy + 11 + 8y.$
e) $\qquad\qquad\qquad = zy + 8y + 11.$
f) $\qquad\qquad\qquad = yz + 8y + 11.$
g) $\qquad\qquad\qquad = (yz + 8y) + 11.$
h) $\qquad\qquad\qquad = (z + 8)y + 11.$
i) $\qquad\qquad\qquad = y(z + 8) + 11.$

17. Remove grouping symbols and combine like terms:
   a) $x - a\{3 - 2(1 - x)\}.$
   b) $5 - 3[2 - 4(-3 + z)].$
   c) $3 - 2[a - x + 3(2x - a)].$
   d) $(3 - a)(x - b + 2).$
   e) $ax - a\{-2x + 3(2x - 1)\}.$

18. Reduce to a single signed number:

$$-\frac{-3 + 2[-5 - (2 - 7)]}{5 + [2 - 3(2 - 3)]}.$$

19. Factor the following:
   a) $2x + ax.$
   b) $xy + 3xy + axy.$
   c) $2 + 4a + 6b.$
   d) $5xy + 4x.$
   e) $3ax + 6ay + 9a.$
   f) $2(x - 1) + a(x - 1).$
   g) $a + b + ax + bx.$
   h) $x^2 + x - 2.$
   i) $10x^2 + 3x - 1.$
   j) $6x^2 + 7x - 20.$
   k) $8x^2 + 16x + 6.$

20. Simplify by cancellation where possible:
   a) $\dfrac{3(3 - a) + x}{2x}.$
   b) $\dfrac{3xy}{2x}.$
   c) $\dfrac{x - 2xy}{3x}.$
   d) $\dfrac{3}{x - y}[2(x + y)].$

21. Multiply, leaving no grouping symbols in the answer:
   a) $\left(\dfrac{2x}{3}\right)\left(\dfrac{6a}{b}\right).$
   b) $-2\left(\dfrac{x - 3}{a}\right).$
   c) $2(x - y)\left(\dfrac{3a}{x - y} - \dfrac{1}{2}\right).$
   d) $10xy\left(\dfrac{1}{5y} - \dfrac{a}{2xy}\right).$

22. Express with a single minus sign:
   a) $\dfrac{-3x - 2y}{5 - a}.$
   b) $-\dfrac{2a - b}{x + b}.$
   c) $-\dfrac{a - b}{-5 - x}.$
   d) $\dfrac{a + b}{-5 - x}.$

23. Reduce to one simple fraction:
   a) $\dfrac{1}{3} - \dfrac{1}{6} + \dfrac{3}{4}.$
   b) $\dfrac{a}{b} + 1.$
   c) $\dfrac{3}{2x} - \dfrac{2}{x} + \dfrac{5}{4x}.$
   d) $\dfrac{y}{x} - 1 + \dfrac{y}{2a}.$
   e) $\dfrac{1}{12xy} + \dfrac{2}{3} - \dfrac{5}{x}.$
   f) $\dfrac{1}{2x - 3} + \dfrac{b}{3}.$

g) $\dfrac{a}{x+y} - \dfrac{b}{z} + 5.$　　　　　　　　*h)* $\dfrac{3}{a} - \dfrac{b}{2} + \dfrac{c}{x+y} - d.$

i) Three fourths of $2\frac{1}{2}$.　　　　　　　　*j)* $5\frac{1}{4} - 2\frac{1}{3}$.

k) $1\frac{7}{8}$ divided by $2\frac{1}{2}$.

24. Reduce the following complex fractions to simple fractions by multiplication by a suitably chosen 1:

a) $\dfrac{\dfrac{2}{3} - \dfrac{1}{4}}{\dfrac{1}{6} + 1}.$　　　　　　　　*b)* $\dfrac{\dfrac{x}{y} - 1}{\dfrac{2}{3} + \dfrac{3}{y}}.$

c) $\dfrac{x - \dfrac{y-4}{a}}{\dfrac{b}{2} + \dfrac{3}{a}}.$　　　　　　　　*d)* $\dfrac{\dfrac{x}{x+y} - 1}{\dfrac{2}{x+y}}.$

25. Evaluate:

a) 3^3.　　　　　　*b)* $2x^0$.　　　　　　*c)* $(2x)^0$.

d) $(1 - 0.1)^{-2}$ (to three decimals).　　*e)* $\left(\dfrac{2}{3}\right)^{-2}$.　　*f)* $(2)^{-2}(-2)^2$.

g) $(75)^{100}(75)^{-98}$.　　*h)* $3^0 - (2x+1)^0$.　　*i)* $\left(\dfrac{1}{16}\right)^{1/4}$.

j) $\left(\dfrac{1}{16}\right)^{-1/4}$.　　*k)* $\dfrac{2}{5}(32)^{1/5}$.　　*l)* $(27)^{2/3}$.

m) $(125)^{4/3}$.　　*n)* $(8)^{-2/3}$.　　*o)* $\left(\dfrac{2}{3}\right)^{-1}$.

p) $\dfrac{(10)^{-5}(10)^2}{(10)^3(10)^{-8}}.$　　*q)* $2^{-3}(2^2 + 2^{-1})$.　　*r)* $\dfrac{3^{-1} - 3^{-2}}{3^{-3}}.$

s) $\sqrt[3]{64}$.　　*t)* $(1 + 0.03)^3$.　　*u)* $\sqrt{-1}$.

v) $\left(\dfrac{2}{3}\right)^{-2}\left(\dfrac{3}{2}\right)$.　　*w)* $\sqrt{\dfrac{4}{25}}$.　　*x)* $(\sqrt[3]{27})^2$.

y) $\sqrt{16^3}$.　　*z)* $\sqrt{9^{-3}}$.

26. Combine exponents, where possible, and simplify. If possible, do not leave grouping symbols, radical signs, or negative exponents in the final result:

a) $a(a^2)(a^3)$.　　　　　　*b)* $xyz(x^2y)$.

d) $(abc)^2(ab)$.　　*e)* $\dfrac{a^2b^3}{abc}$.　　*f)* $\dfrac{a^2yz^3}{ayz^2}$.

g) $\dfrac{(3ab)^2(2c)^3}{12ac^2}$.　　*h)* $\dfrac{(-2x)^3y^2}{4xy}$.　　*i)* $\dfrac{(ab^{-1}c^2)^3}{2b^2c^2}$.

j) $\dfrac{(3x^{-2}y)^2}{2x^3y^{-3}}$.　　*k)* $\dfrac{(-xy^2)^{-3}(2z^2)}{(-2yz)^4}$.　　*l)* $[(x\sqrt{y})^{3/2}]^2$.

m) $\dfrac{ax^{2/3}y^{-1/2}}{bx^{1/2}y^{5/3}}$.　　*n)* $(y^{1/2})^{-3/2}$.　　*o)* $\dfrac{3}{4}(4y)^{-3/2}$.

p) $\dfrac{\sqrt[3]{ab}\,\sqrt{xy}}{9\,(ab)^{-1}(xy)^{3/2}}$.

q) $\dfrac{(2a^{1/2}b^{1/3}c)^3}{4abc}$.

r) $\dfrac{1+a^{-1}}{1-a^{-1}}$.

s) $\dfrac{(a-b)^{-1}-1}{b^2}$.

t) $\dfrac{2a^{x-3}a^5}{a^2}$.

u) $\dfrac{x^{2a}}{x^a}$.

v) $a^2b^3(ab)^{-5}$.

w) $x(2-x)^2$.

x) $a^2(a^{-2}+a^{-3})$.

y) $a(a-b)(a+2b)$.

z) $(5^{x+1})(5a)^{-3}(5a^3)$.

Formulas, equations, and graphs

appendix

3

A3.1 INTRODUCTION

The distributive property asserts that

$$a(b+c) = ab + ac.$$

Inasmuch as the assertion applies for any numbers, a, b, and c, it is called an identity. On the other hand, the statement

$$x + 5 = 7$$

is true only under the condition that x is 2, and for this reason is called a conditional equality.

A three-lined symbol, $\equiv$, can be used to denote an identity. In this book, however, the two-lined symbol, $=$, is used to denote both identities and conditional equalities because the context makes clear which interpretation of the symbol is relevant. Thus, in Appendix 2 the symbol $=$ means identically equal, whereas in most of the remainder of the book the symbol denotes conditional equality. We shall refer to conditional equalities briefly as equations.

In the equation $x + 5 = 7$, the letter x can be called the unknown, and the value of the unknown that makes the statement true, the number 2, can be called the root of the equation. We may also say that a number that makes the statement of equality true satisfies the equation.

The statement $y = x + 2$ is true for various pairs of values for x and y, and we shall call the letters x and y variables. A particular set of numbers, such as

$$x = -1$$
$$y = 1$$

which satisfy the equation, is a solution of the equation. In this book, we use the *variable, solution* terminology (rather than unknown, root terminology) in the discussion of equations. Thus:

$$y = x + 2$$

will be called an equation in two variables. It has an unlimited number of solutions. The equation

$$x + 5 = 7$$

is an equation in one variable. It has the unique solution $x = 2$.

To solve an equation for a variable means to perform whatever operations are necessary to put the equation in a form in which the stated variable appears by itself (with coefficient one and exponent one) on one side of the equality sign, and the expression on the other side of the equality sign does not contain the stated variable. For example:

$$x - y = 3$$

is not solved for either x or y, but

$$x = y + 3$$

is solved for x, and the equation

$$y = x - 3$$

is solved for y.

Exercise. Fill in the blanks: The _____ $x + 10 = 4$ has a single _____, x. The _____ of the _____ is $x = -6$. On the other hand, the _____ $y - x = 7$ has two _____ and has an _____ number of solutions: if we write $y - x = 7$ in the alternate form $y = x + 7$, it is said to be _____ for _____. Answer: Equation, variable, solution, equation. Equation, variables, unlimited, solved, y.

We now turn to some elementary procedures employed when solving an equation for a variable.

A3.2 SOME AXIOMS

Axioms are statements we assume to be true. For example, we assume that if equals are added to equals, the sums will be equal. Inasmuch as an equation is a statement of equality of the numbers on either side of the equal sign, the axiom says that if we add the same number to both sides of an equation, we shall obtain another equation. For example, if we have the statement

$$x - 3 = 7$$

and we add 3 to each side, we obtain

$$x - 3 + 3 = 7 + 3$$

from which we find $x = 10$, and we have solved the equation for x.

In similar fashion, we assume that if equals are subtracted from equals, the differences are equal; if equals are multiplied by equals, the products are equal; and if equals are divided by equals (division by zero excluded), the quotients are equal. In the context of equations, we say that the solutions of an equation are not altered if the same number is added to both sides, or if the same number is subtracted from both sides, or if both sides are multiplied or divided by the same number.

Examples.

$$x - 2 = 3$$
$$x - 2 + 2 = 3 + 2$$
$$x = 5.$$

Add 2 to both sides:

$$x + 2 = 3$$
$$x + 2 - 2 = 3 - 2$$
$$x = 1.$$

Subtract 2 from both sides:

$$\frac{x}{3} = 2$$

Multiply both sides by 3:

$$3\left(\frac{x}{3}\right) = (3)(2)$$

$$x = 6.$$

$$2x = 7$$

Divide both sides by 2:

$$\frac{2x}{2} = \frac{7}{2}$$

$$x = \frac{7}{2}.$$

Exercise. The solution $x = 4$ is obtained by applying which operation to each of the following: $6x = 24$; $x/2 = 2$; $x - 4 =$

0; $x + 2 = 6$? Answer: Divide both sides by 6; multiply both sides by 2; add 4 to both sides; subtract 2 from both sides.

A3.3 SOLUTIONS BY ADDITION AND MULTIPLICATION, WITH INVERSES

The objective in solving an equation for a certain variable is to derive an expression that has that variable alone (coefficient one and exponent one) on one side of the equal sign, and an expression not involving this variable on the other side of the equal sign. Axioms are applied to accomplish this objective. If we are asked to solve

$$ax + b = c$$

for x, we may proceed by subtracting b from both sides to obtain

$$ax = c - b.$$

Next, we divide both sides by a to obtain the desired solution:

$$x = \frac{c - b}{a}.$$

The steps are justified by the axioms. However, it is helpful to understand what steps are required, and this understanding is enhanced if we keep in mind the notion of *inverse* operation. Thus, in solving

$$ax + b = c$$

for x, we wish to remove b from the left side. Inasmuch as b is *added* on the left, we apply the inverse operation and *subtract b* from both sides to obtain

$$ax = c - b.$$

We now observe that x is *multiplied* by a, so we apply the inverse operation and *divide* both sides by a to obtain

$$x = \frac{c - b}{a}.$$

Inverse operations are the tools we need to manipulate equations into desired form.

The student should be prepared to justify each step taken in the solution of an equation. For brevity, he may omit the phrase "both sides of the equation" when citing justification for an operation. Thus, for example, "add 3" will be assumed to mean to add 3 to both sides of the equation. For uniformity, it is to be understood that the outcome of each operation performed is shown on the line following the description of the operation. For example:

$$x - 3 = 5 \qquad \text{Add 3:}$$
$$x = 8.$$

Example.   Solve for x, citing operations performed:

$$4x - 2 = 2x + 5 \qquad \text{Subtract } 2x:$$
$$2x - 2 = 5 \qquad \text{Add 2:}$$
$$2x = 7 \qquad \text{Divide by 2:}$$
$$x = \frac{7}{2}.$$

Exercise.   Solve for x, citing operations performed: $3 - 2x = -5x + 7$. Answer:

$$3 - 2x = -5x + 7 \qquad \text{Add } 5x:$$
$$3 + 3x = 7 \qquad \text{Subtract 3:}$$
$$3x = 4 \qquad \text{Divide by 3:}$$
$$x = \frac{4}{3}.$$

Generally, equations containing fractions can be handled most efficiently by finding the lowest common denominator of all the fractions in the equation, and then multiplying both sides by the l.c.d. (lowest common denominator).

Example.   Solve for x, citing operations performed:

$$2x - \frac{1}{3} = \frac{x}{2} \qquad \text{Multiply by 6 (the l.c.d.):}$$

$$6(2x) - 6\left(\frac{1}{3}\right) = 6\left(\frac{x}{2}\right) \qquad \text{Carry out multiplications:}$$

$$12x - 2 = 3x \qquad \text{Add 2:}$$
$$12x = 3x + 2 \qquad \text{Subtract } 3x:$$
$$9x = 2 \qquad \text{Divide by 9:}$$
$$x = \frac{2}{9}.$$

Careful attention should be paid to the fact that each side of an equation is an expression; and to multiply an expression by a number, we must multiply every *term* of the expression by that number.

It is instructive to follow through the tactics of the last example. We observe fractions in the equation, and we know that the denominator of a fraction is canceled (replaced by one, which need not be written) if the fraction is multiplied by a term having the fraction's denominator as a factor. Rather than treat each fraction separately, we make up a term (lowest) that has the denominators of all the fractions as factors. Then, when both sides are multiplied by this term, all denominators are canceled, and we are led to the statement

$$12x - 2 = 3x.$$

We decide to gather terms involving x on the left, other terms on the right. The term -2 is removed from the left by the inverse operation of adding 2. Similarly, the term $+3x$ on the right is removed by the inverse operation of subtracting $3x$ from both sides. We now have $9x = 2$. The 9 must be removed to obtain the desired solution. Inasmuch as 9 is multiplied by x, we apply the inverse operation and divide both sides by 9 to obtain

$$x = 2/9.$$

Exercise. Solve for x, citing operations performed:

$$\frac{3x}{4} - 5 = \frac{x}{3}.$$

Answer:

$$\frac{3x}{4} - 5 = \frac{x}{3} \qquad \text{Multiply by 12:}$$
$$9x - 60 = 4x \qquad \text{Subtract } 4x:$$
$$5x - 60 = 0 \qquad \text{Add 60:}$$
$$5x = 60 \qquad \text{Divide by 5:}$$
$$x = 12.$$

If the variable to be solved for is inside a grouping symbol, we remove the grouping symbol, citing the distributive property as justification, and then proceed with the solution. For example:

$$3x - \frac{1}{4} = 2 - \frac{1}{6}[x - (2 - x)] \qquad \text{Multiply by 12:}$$
$$36x - 3 = 24 - 2[x - (2 - x)] \qquad \text{Distributive property:}$$
$$36x - 3 = 24 - 2[x - 2 + x] \qquad \text{Distributive property:}$$
$$36x - 3 = 24 - 2x + 4 - 2x \qquad \text{Add } 4x:$$
$$40x - 3 = 24 + 4 \qquad \text{Add 3:}$$
$$40x = 31 \qquad \text{Divide by 40:}$$
$$x = \frac{31}{40}.$$

In the next example, grouping symbols are introduced to emphasize proper procedure, and then are removed in due course.

$$\frac{2}{3(x-1)} - \frac{1}{2} = \frac{1}{4} \qquad \text{Multiply by l.c.d. } 3(4)(x-1):$$

$$3(4)(x-1)\left[\frac{2}{3(x-1)}\right] - 3(4)(x-1)\left(\frac{1}{2}\right) = 3(4)(x-1)\left(\frac{1}{4}\right).$$

The last equation simplifies to:

$$8 - 6(x-1) = 3(x-1) \qquad \text{Distributive property:}$$
$$8 - 6x + 6 = 3x - 3 \qquad \text{Add } 6x; \text{ add } 3:$$
$$17 = 9x \qquad \text{Divide by } 9:$$
$$\frac{17}{9} = x.$$

The same procedures are followed when numbers are in literal form, but the next to the last step often requires factoring. Factoring derives from the distributive property, but we shall follow convention here and cite factoring rather than the distributive property when the need for justification arises.

Example. Solve for x, citing operations performed:

$$\frac{x}{a} - b = 2a(x - b) \qquad \text{Multiply by } a:$$

$$x - ab = 2a^2(x - b) \qquad \text{Distributive property:}$$
$$x - ab = 2a^2x - 2a^2b \qquad \text{Subtract } 2a^2x; \text{ add } ab:$$
$$x - 2a^2x = ab - 2a^2b \qquad \text{Factor:}$$
$$x(1 - 2a^2) = ab - 2a^2b \qquad \text{Divide by } (1 - 2a^2):$$
$$x = \frac{ab - 2a^2b}{1 - 2a^2}.$$

Exercise. Solve for x, citing operations performed:

$$\frac{2x}{a} + \frac{3}{b} = x + 4.$$

Answer:

$$\frac{2x}{a} + \frac{3}{b} = x + 4 \qquad \text{Multiply by } ab:$$

$$2xb + 3a = abx + 4ab \qquad \text{Subtract } abx:$$
$$2xb - abx + 3a = 4ab \qquad \text{Subtract } 3a:$$
$$2xb - abx = 4ab - 3a \qquad \text{Factor:}$$

$$x(2b - ab) = 4ab - 3a \qquad \text{Divide by } (2b - ab):$$

$$x = \frac{4ab - 3a}{2b - ab}.$$

Example. Solve for x, citing operations performed:

$$a = \frac{x}{1 - nx} \qquad \text{Multiply by } (1 - nx):$$

$$a(1 - nx) = x \qquad \text{Distributive property:}$$

$$a - anx = x \qquad \text{Add } anx:$$

$$a = x + anx \qquad \text{Factor:}$$

$$a = x(1 + an) \qquad \text{Divide by } (1 + an):$$

$$\frac{a}{1 + an} = x.$$

Example. Solve for y, citing operations performed:

$$\frac{a}{y} - \frac{1}{b} = 2 \qquad \text{Multiply by } yb:$$

$$ab - y = 2by \qquad \text{Add } y:$$

$$ab = 2by + y \qquad \textit{Factor:}$$

$$ab = y(2b + 1) \qquad \text{Divide by } (2b + 1):$$

$$\frac{ab}{2b + 1} = y.$$

A3.4 PROBLEM SET A3–1

Solve each of the following for x, citing operations performed:

1. $2x - 3 = x + 4$.

2. $4x + 5 = 2x + 12$.

3. $3x - 7 = 2x + 4$.

4. $5 - 2x = 6$.

5. $7x - 5 = 3 - 4x$.

6. $3 - 2x = x - 4$.

7. $\dfrac{x}{3} + \dfrac{1}{2} = 3x$.

8. $\dfrac{x}{4} + \dfrac{x}{2} = 2 - \dfrac{5x}{8}$.

9. $\dfrac{2x}{5} - \dfrac{3x}{2} = 4$.

10. $1 - \dfrac{x}{3} + \dfrac{x}{2} = x - 4$.

11. $\dfrac{1}{7} - \dfrac{x}{3} = x$.

12. $1 - \dfrac{3x}{7} = 0$.

13. $\dfrac{x + 3}{2} = x - \dfrac{1}{4}$.

14. $\dfrac{5 - 2x}{3} + \dfrac{x}{2} = 1$.

15. $\dfrac{2x-1}{3} - \dfrac{1-x}{5} = 0.$

16. $\dfrac{x+1}{2} - \dfrac{x-1}{3} = 5.$

17. $bx + 2 = c.$

18. $ax + 2 - x = 0.$

19. $ax + b = cx.$

20. $ax + b = x - b.$

21. $a(x - a) = 2x.$

22. $\dfrac{x}{a} - \dfrac{1}{2} = 2x.$

23. $\dfrac{2}{a-x} + \dfrac{1}{3} = 4.$

24. $y = \dfrac{x}{b-cx}.$

25. $\dfrac{3}{4} - \dfrac{2x}{3} = 2x(a-1).$

26. $3(x - 2) = 2 - a(x + 2).$

27. $\dfrac{2}{3(x-2)} + \dfrac{3}{a} - \dfrac{1}{2} = 0.$

28. $ax - \dfrac{b}{2} = c + \dfrac{5[a - 2(b-x)]}{6}.$

29. $\dfrac{b}{a} - x = 2a(b - x).$

30. $x - a(b - x) = 2x - 3.$

31. $3(b - x) = 2 + b[x - (3 - x)].$

32. $b(a + x) = a(b + x),$ where $a \neq b.$

A3.5 TRANSPOSITION

Problem Set A3–1 has afforded practice in citing fundamentals as justification for steps followed in the solutions of equations. From now on, we shall lessen the writing burden by omitting step-by-step justification. The burden can be lightened further by applying the rule of *transposition*, which states that a *term* may be moved from one side of an equation to the other by changing its sign. For example:

$$x - 2a = b$$

becomes

$$x = b + 2a$$

by changing the sign of $-2a$ to $+2a$ and placing $+2a$ on the other side. Similarly:

$$x + 7 = y$$

becomes

$$x = y - 7.$$

The transposition rule is a consequence of the axiom that states that the same number may be added to (subtracted from) both sides of an equation. That is, the change from

$$x + 7 = y$$

to

$$x = y - 7$$

is, in effect, subtracting 7 from both sides of the first equation. Transposition is a handy procedure, but we must keep in mind that it applies to *terms*. It would not be correct to say that any quantity can be moved to the other side by transposition. For example, the 2 in $2x = 6$ is not subject to the rule for transposition because 2 is not a term of the left member of the equation.

> ***Exercise.*** The change from $ax = b$ to $x = b/a$ is not accomplished by transposition. How is it accomplished? Answer: By dividing both sides of the equation by a.

A3.6 FORMULAS

Computational procedures are described efficiently by formulas employing the symbolism of algebra. Thus, if x is the length and y the width of a rectangle, the area, A, of the rectangle is expressed by the formula $A = xy$. On the one hand, if we are told a rectangle has a length of 8 inches and a width of 5 inches, we compute the area

$$A = (8)(5) = 40 \text{ square inches.}$$

On the other hand, if we are asked how wide a rectangle of 12 inches length should be if its area is to be 84 square inches, we substitute into the equation, obtaining $84 = 12y$. Solving this equation, we find y, the desired width, is 7 inches. The point here is that we may wish to use a formula to evaluate a variable other than the one for which the formula is solved, and to do so, we call upon the usual operations for solving equations. Actually, the formula $A = xy$ leads easily to two other formulas:

$$x = \frac{A}{y}$$

$$y = \frac{A}{x}.$$

If we have the formula $A = xy$ and are given values for A and x, we can compute y by substituting the given values directly into the equation and solving for y, or we can solve the literal equation, obtaining

$$y = \frac{A}{x}$$

and then evaluate y by substituting the given values into this equation. Generally, it is more efficient to substitute the given numbers directly into the formula if a single evaluation is to be made. If, however, several evaluations are to be made, the formula should be solved for the desired variable before substituting numbers. For example, suppose that we are given

$$y = 5$$
$$a = 2$$

and are asked to evaluate x from the formula

$$y = 2(ax - 3).$$

Substituting the given numbers directly into the formula:

$$5 = 2(2x - 3)$$
$$5 = 4x - 6$$
$$11 = 4x$$
$$\frac{11}{4} = x.$$

On the other hand, if we are asked to carry out a whole series of evaluations of x for various values of y and a, it would be more efficient to solve the formula for x; thus:

$$x = \frac{y + 6}{2a}.$$

The last formula permits rapid evaluation of x for given values of y and a.

> **Exercise.** If $y = (1 + n)/2$, what would be an efficient way to compute n for many different values of y? Answer: Substitute the values for y into the equation $n = 2y - 1$.

A3.7 EXACT EVALUATIONS

Evaluation by formulas often leads to approximations. Unless care is exercised, these approximations may not be accurate enough for the purpose at hand. Suppose, for example, that we wish to compute 3⅓ percent of $1 million. If we change the stated percent to 3.33 percent, and then to the decimal 0.0333, the computation is

$$0.0333(1,000,000) = 33,300.$$

For most purposes the answer obtained would not be satisfactory. We can express the answer to any desired degree of accuracy by changing the stated percent to its exact equivalent, as follows:

$$3\tfrac{1}{3}\% = \frac{3\tfrac{1}{3}}{100} = \frac{\dfrac{10}{3}}{100} = \frac{10}{300} = \frac{1}{30}.$$

The computation with exact numbers would be

$$(1/30)(1,000,000) = 33,333.333\cdots$$

The point of this example is that if exact numbers are used up to the last step, the final answer can be carried accurately to as many places as desired.

> **Exercise.** Compute x exactly from $3x + 2 = 1/b$ if b is $2\tfrac{1}{4}$. Answer: $-23/45$.

As another example, let us evaluate y, given

$$y = \frac{a + b}{1 - r}$$

$$a = \frac{1}{3}$$

$$b = \frac{1}{2}$$

$$r = 1\tfrac{1}{3}\%.$$

To express all numbers in the same form, r is changed to its exact fractional equivalent, $1/75$. Substituting:

$$y = \frac{\dfrac{1}{3} + \dfrac{1}{2}}{1 - \dfrac{1}{75}}.$$

The last is an exact expression for y. If numerator and denominator are multiplied by 150, we obtain

$$y = \frac{125}{148},$$

which again is an exact expression for y. This result could now be converted to an approximate decimal, by long division, accurate to as many places as desired.

A3.8 PROBLEM SET A3–2

Given that $k = 12$, $m = 5$, $n = 4$, compute y from the following formulas:

1. $y = \dfrac{km}{n}$.

2. $y = \dfrac{k^2 m - n^3}{n^2}$.

3. $y = \dfrac{m}{\dfrac{n}{k^2}}$.

4. $y = \sqrt[3]{mn - k}$.

5. $y = (n - m)[n - k(m + n)]$.

6. $y = \dfrac{8m}{\dfrac{n^2}{k}}$.

7. $y = \left(\dfrac{kn}{3}\right)^{3/2}$.

8. $y = \left(\dfrac{4}{9k}\right)^{-2/3}$.

9. $y = \left(m^2 - \dfrac{3k}{4}\right)^{5/4}$.

10. $y = m^2 n + 3[4k - n(m^3 - k^2)]$.

In the following, find the exact value of y in the form of a fraction in lowest terms, given that

$$a = \frac{5}{3},\ b = \frac{1}{7},\ c = \frac{2}{9},\ d = \frac{5}{12},\ x = \frac{3}{8},\ z = \frac{1}{5}.$$

11. $y = \dfrac{x}{a}$.

12. $y = \dfrac{ab}{2}$.

13. $y = \dfrac{a(b + c)}{3}$.

14. $y = d - \dfrac{1}{a}$.

15. $y = ax + bz$.

16. $y = ax^2 - \dfrac{2a}{b}$.

17. $y = \left(1 - \dfrac{a}{d}\right)^3$.

18. $y = \sqrt[3]{\dfrac{c}{2x}}$.

19. $y = \dfrac{a}{2} - b\sqrt{\dfrac{3d}{5}}$.

20. $y = \dfrac{a + 2c - d}{x - 4a}$.

21. Compute y to the nearest cent:

$$y = at + b(t - c)$$
$$a = 16\tfrac{2}{3}\%,\ b = 15\%,\ c = \$75{,}000,\ \text{and } t = \$140{,}000.$$

22. Given that $a = 0.52$, $x = 4$, compute y to two decimal places:

$$y = \frac{a}{1 + \dfrac{x}{100}}.$$

23. The equation relating the cost, C, of an item to its retail price, R, is $C + pR = R$, where p percent is markup on the retail price.
    a) What markup percent on retail is achieved if an item costs $5 and sells at a retail price of $8?
    b) How much can a merchant pay for an item if he wishes to sell it

for $7.50 and achieve a markup which is 20 percent of the retail price?

    c) If a merchant seeks to achieve a markup of 40 percent on retail, at what price should he sell an item which cost him $2.69?

24. The amount of interest, I, earned on P dollars at simple interest of r percent per year for t years is $I = Prt$. Assuming a year to have 360 days:

    a) How long will it take for $500 to earn interest in the amount of $20 if the interest rate per year is $4\frac{1}{2}$ percent?

    b) How many dollars must be invested at $3\frac{1}{3}$ percent per year if interest in the amount of $75 is to be earned in 50 days?

25. Temperature in degrees Celsius, C, is related to temperature in degrees Fahrenheit, F, by the formula

$$C = \tfrac{5}{9}\,(F - 32).$$

Find the Fahrenheit temperature corresponding to the following Celsius readings:

a) 100.	*b)* 0.	*c)* −10.6.
d) 28.	*e)* −10.	*f)* 50.

26. A salesman's earnings, E, amount to 20 percent of his total sales, T, plus a bonus of $12\frac{1}{2}$ percent of any amount he sells in excess of $50,000.

    a) Write the equation relating earnings to sales if sales exceed $50,000.

    b) What must the man's sales be if he is to earn $10,000? $15,000? $20,000?

    c) How much must he sell if his earnings are to be 25 percent of total sales?

27. An executive is to receive compensation, B, amounting to x percent of his company's net profit after taxes. Taxes amount to y percent of net profits after deduction of the executive's compensation.

    a) Letting P represent net profits before taxes, complete the following formula:

$$B = x[P - ?(? - ?)].$$

    b) If the executive receives 25 percent of net profits after taxes, and taxes amount to 52 percent of net profits after deduction of the executive's compensation, how much must net profits before taxes be if the executive is to receive $50,000?

28. A customer buys an article on the installment plan. The price of the article (or the amount he still owes after making a down payment) is B dollars. The seller adds a carrying charge of C dollars and requires that the debt be paid off by n equal payments. Letting y be the number of payments which would be made in a year (that is, if payments are weekly, y is 52; and if payments are monthly, y is 12), and letting r be the equivalent simple interest rate being paid by the customer, a formula states

$$r = \frac{2yC}{B(n+1)}.$$

    a) An article priced at $500 is purchased. The customer pays $100 down (so B is $400). The seller adds $42 as a carrying charge and requires that the debt be repaid in ten equal monthly installments of $44.20 each. What rate of interest is the customer paying? (Find r, and convert it to percent form.)

b) If the simple interest equivalent of the installment rate is 15 percent, what should be the carrying charge on an article priced at $100 (no down payment) if the debt is to be repaid in 20 equal weekly installments?

29. For each dollar of increase in sales, selling expense increases by $$p$ where, it is hoped, p is less than 1. The sales level B at which the company will break even if its fixed expenses are $$F$ is:

$$B = \frac{F}{1-p}.$$

a) Find B if $F = \$12,000$ and $p = 0.4$.
b) Fixed expenses are $10,000 and management wishes to control p so that breakeven will occur at $12,500. Find p.
c) Repeat *(b)* if fixed expenses are $20,000 and breakeven is to occur at $32,000.

30. The average cost per unit is $$A$ when x units are made, and

$$A = \frac{100}{x} + 0.01.$$

Find the average cost per unit if:
a) 10 units are made.
b) If 20 units are made.

How many units should be made if average cost per unit is to be
c) $1.01?   *d)* $0.41?   *e)* $0.21?

A3.9 COORDINATE AXES

An axiom of mathematics states that a correspondence exists between the real numbers and the points on a straight line; that is, for each real number, there is one and only one corresponding point on a line, and each point on a line corresponds to one and only one real number. Hence, if we draw a straight line and mark upon it an arbitrary point, zero, all negative real numbers can be represented by points to the left of zero, all positive numbers by points to the right of zero. According to the axiom, each number corresponds to a unique point on the line, and the totality of real numbers corresponds to the line itself.

A pair of rectangular coordinate axes is obtained by drawing two intersecting perpendicular lines, one horizontal, the other vertical. The intersection of the horizontal and vertical axes so obtained is taken as the zero point on each line and is called the *origin*. On the vertical, points above the origin are used to represent positive numbers, points

838

below to represent negative numbers. On the horizontal axis, positive numbers are to the right of the origin, negative numbers to the left.

Any point in the plane of the axes can be located by means of a number pair called the *coordinates* of the point. Conventionally, coordinates are written in the form *(a, b)*, where the first number in the parentheses is the horizontal coordinate, or *abscissa*, and the second number is the vertical coordinate, or *ordinate*. Thus, the point (2, 3) is located by moving two units to the right of the origin, and then three units vertically upward. The procedure is analogous to the method of directing a person from a particular spot (origin) in a city by telling him the point he wishes to reach can be found by going two blocks east, and then three blocks north.

> *Exercise.* What are the abscissa and the ordinate of a point? Answer: The abscissa is the horizontal coordinate and the ordinate is the vertical coordinate of the point.

Figure A3–1 shows the point (2, 3) just mentioned, together with other aspects of the coordinate system. The axes divide the plane into

FIGURE A3-1

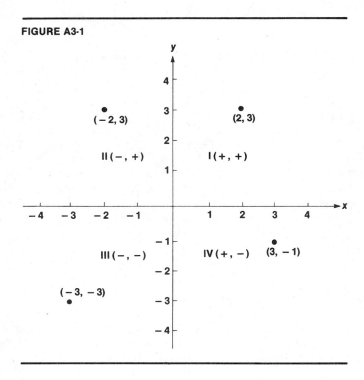

quadrants, which we number counterclockwise as I, II, III, and IV. In quadrant I, both abscissa and ordinate are positive. In II the abscissa is negative, the ordinate positive. In III both coordinates are negative. In IV the abscissa is positive, the ordinate negative.

> *Exercise.* In what quadrants are the points $(-2, 3)$, $(3, 5)$, $(2, -3)$, $(-5, -4)$? Answer: II, I, IV, III.

A3.10 PLOTTING OBSERVATIONAL DATA

By observational data we mean numbers collected in real-world situations—numbers such as costs, sales, units produced, prices, and so on. Frequently, we wish to display such data in a graphical manner, which helps to show relationships which may exist between two variables. Consider the data in Table A3–1.

TABLE A3–1
Production cost per unit of product

Lot number	Number of units in lot (x)	Cost per unit (y)
1	8	$11
2	20	4
3	5	13
4	10	9
5	15	8
6	25	5

Figure A3–2 is a graphic representation (or, more simply, a graph) of the data of Table A3–1. The numbers of units produced have been assigned the general designation x, the costs per unit the general designation y. Following convention, the numbers x are assigned to the horizontal axis, the numbers y to the vertical. By definition, the vertical variable is called the *dependent* variable, and the horizontal variable is called the *independent* variable. In some cases, these definitions make real-world sense. In the present illustration, for example, it is natural to think of production cost as depending on number of units produced, so that cost would be the dependent variable and units produced the independent variable. In many circumstances, however, reference to the vertical axis as the axis of the dependent variable is purely a matter of convention.

The interpretation of Figure A3–2 is simply that unit cost *tends* to decrease as the number of units made increases. It is a graphical illustration of the economies of mass production. We refer to this set of isolated

840

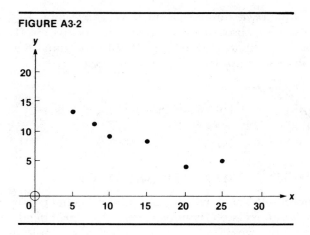

FIGURE A3-2

points as a *discrete* set. If we wish to illustrate the relationship by means of a smooth curve, we may sketch in freehand a curve that comes close to the points, although it does not pass through all of the points. The points on the curve are not, of course, records of observations, because real-world observations are limited in number, whereas a segment of a smooth curve has an unlimited number of points. However, we might wish to use the smooth curve to estimate costs per unit for numbers of units other than those observed.

A3.11 PLOTTING EQUATIONS IN TWO VARIABLES: STRAIGHT LINES

Given the equation

$$y = x + 2,$$

we know that for every x, there is a number y such that the number pair (x, y) satisfies the equation. The entire solution set is an infinite set of ordered pairs. We can generate some members (obviously not all) of the solution set by arbitrarily assigning a value for x, then computing the corresponding value for y. For example,

if $x = 1$ then $y = 1 + 2 = 3$

and $(1, 3)$ is a member of the solution set. Similarly, if $x = 2$, $y = 4$ and $(2, 4)$ is another member of the solution set.

> *Exercise.* Find the points in the solution set corresponding to $x = 0$ and $x = 5$. Answer: $(0, 2)$; $(5, 7)$.

We *plot the graph* of an equation by finding and plotting points that satisfy the equation, then sketching in a smooth curve as suggested by the plotted points. In the case at hand, we may tabulate some of the points as in Table A3–2 where the x values are chosen arbitrarily.

TABLE A3–2

$y = x + 2$	
x	y
0	2
1	3
2	4
3	5
4	6
5	7

The points of Table A3–2 are shown in Figure A3–3, where it is clear that all fall on the same straight line. We would also obtain a straight line if we made the graph of

FIGURE A3-3

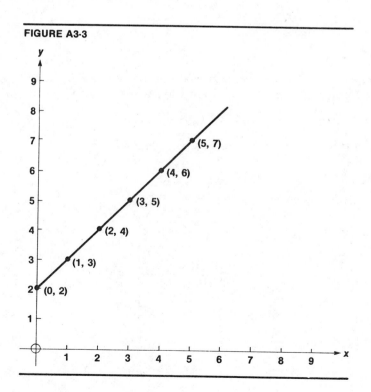

$$y = 3x + 5 \quad \text{or} \quad y = -2x + 7.$$

In general, an equation that can be made into the form

$$y = mx + b,$$

where m and b are constant numbers, has a straight line as its graph and is called a *linear equation*. The graph of such an equation is easily sketched because *two points* suffice to fix a straight line.

Exercise. How would you plot the graph of $y = 2x - 3$? Answer: Pick two arbitrary values of x and find the corresponding values of y to obtain two points. Plot the points and use a straightedge to draw a line through them.

A3.12 VERTICAL PARABOLAS

In the equation

$$y = x^2 - 5$$

if we let

$$x = 0, 1, 2, 3, 4, 5$$

in succession, we find the corresponding values of y to be -5, -4, -1, 4, 11, and 20. In this equation the succession of negatives

$$x = -1, -2, -3, -4, -5$$

leads to the same sequence of y's. In tabular form:

TABLE A3–3

$y = x^2 - 5$	
x	y
-5	20
-4	11
-3	4
-2	-1
-1	-4
0	-5
1	-4
2	-1
3	4
4	11
5	20

FIGURE A3-4

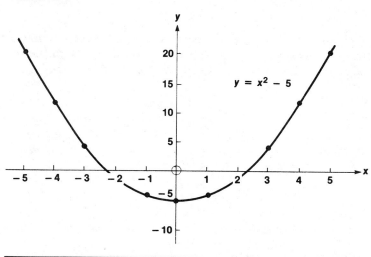

The general shape of the curve is shown in Figure A3–4. It is clear that the curve will continue to rise at both ends because if we substitute x values of 6, 7, and so on (or -6, -7, and so on), the term x^2 in

$$y = x^2 - 5$$

will overpower the number -5 and make y become larger.

> ***Exercise.*** Find the coordinates of the points on $y = x^2 - 5$ where $x = 2.5$ and $x = -2.5$. Check the graph in Figure A3–4 by plotting these points. Answer: The points are (2.5, 1.25) and (-2.5, 1.25).

The graph shown in Figure A3–4 is known as a *vertical parabola.* If we plotted points for

$$y = -2x^2 + 5x - 6 \quad \text{or} \quad y = 3x^2 + 5x - 4,$$

we would again obtain vertical parabolas. In general, any equation that can be made into the form

$$y = ax^2 + bx + c,$$

where a, b, c are constant numbers and $a \neq 0$, is a vertical parabola. The "nose" of the parabola is called the *vertex.* In the case of Figure A3–4, the vertex lies on the vertical axis. This is always the case if

$b = 0$ in $y = ax^2 + bx + c$, as in the plotted equation

$$y = x^2 - 5.$$

Proper positioning of the vertex of a parabola is of key importance in drawing an accurate sketch. This can be done by use of the formula[1] that states that the x-coordinate of the vertex of a vertical parabola occurs at

$$x = -\frac{b}{2a}.$$

Thus, comparing the parabola $y = x^2 - 5$ with $y = ax^2 + bx + c$, we have $a = 1$, $b = 0$, $c = -5$, so

$$x = -\frac{0}{2} = 0$$

is the x-coordinate of the vertex. The corresponding y is -5. The vertex is at $(0, -5)$, as shown in Figure A3–4.

As another example, in

$$y = -2x^2 + 12x - 10$$

we have $a = -2$, $b = 12$. We find

$$x = -\frac{12}{-4} = 3$$

and

$$y = -2(9) + 12(3) - 10 = 8,$$

so $(3, 8)$ is the vertex in this case.

Exercise. Find the coordinates of the vertex of $y = 3x^2 - 12x + 5$. Answer: $(2, -7)$.

In general, peaks and valleys in curves are points of critical importance in sketching accurate graphs. We shall learn more about methods of determining such points and about important interpretations attached to them in Chapter 9. We shall now sketch the graph of the equation of the last exercise.

$$y = 3x^2 - 12x + 5.$$

---

[1] Because a vertical parabola is symmetrical with respect to a vertical line through its vertex at $x = v$ (where v stands for vertex) it follows that the y value is the same at $x + v$ and $x - v$. That is, $a(x + v)^2 + b(x + v) + c = a(x - v)^2 + b(x - v) + c$. If we square, collect terms, factor, and solve, we find $x = -b/2a$.

We found the vertex to be $(2, -7)$, and we may compute some other points as shown in Table A3–4.

TABLE A3–4

\multicolumn{2}{c}{$y = 3x^2 - 12x + 5$}
x	y
-1	20
0	5
1	-4
2	-7
3	-4
4	5
5	20

The desired graph is shown in Figure A3–5.

FIGURE A3-5

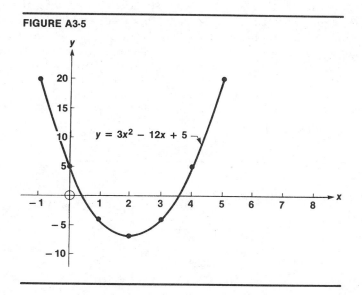

$y = 3x^2 - 12x + 5$

Exercise. Plot $y = x^2 - 8x + 15$ on Figure A3–5. Answer: This parabola rises on either side of the vertex at $(4, -1)$, cutting the x-axis at 3 and 5. Other points are $(2, 3)$, $(6, 3)$, $(0, 15)$, and $(8, 15)$.

As another example, the parabola $y = -4x^2 + 7x - 3$ is found to have its vertex at $(7/8, 1/16)$ and to fall on either side of the vertex. Again, $y = x^2$ is a parabola with vertex at the origin which rises on either side of the vertex.

A3.13 QUADRATIC EQUATIONS

A quadratic equation is of the form

$$ax^2 + bx + c = 0,$$

where a, b, and c are constants, with $a \neq 0$. Thus,

$$3x^2 - 12x + 5 = 0 \qquad (1)$$

is a quadratic equation with $a = 3$, $b = -12$, and $c = 5$. The solution set consists of the values for x which make the equation a true statement. If we relate this equation to

$$y = 3x^2 - 12x + 5, \qquad (2)$$

we see that the solution set of (1) consists of the values of x that make y, in (2), equal to zero. By reference to the graph in Figure A3–5, we see that the graph intersects the x-axis at about $x = 0.5$ and $x = 3.5$, and at these points y, or $3x^2 - 12x + 5$, is zero. An exact determination of the solution set can be found by the *quadratic formula*, which states:

$$x = \frac{-b \pm \sqrt{b^2 - 4ac}}{2a}.$$

In the present case,

$$x = \frac{-(-12) \pm \sqrt{144 - 4(3)(5)}}{2(3)} = \frac{12 \pm \sqrt{84}}{6}.$$

By logarithms, $\sqrt{84}$ is about 9.17 so

$$x = \frac{12 \pm 9.17}{6}$$

$$= \frac{21.17}{6} \quad \text{or} \quad \frac{2.83}{6}$$

$$= 3.53 \quad \text{or} \quad 0.47$$

so the solution set may be written as $x_1 = 0.47$, $x_2 = 3.53$, or $\{0.47, 3.53\}$.

Three cases can arise in the solution of a quadratic. The graphical nature of these cases can be understood by reference to Figure A3–5. In its present state, the graph shows *two different real solutions*. If now we hold the axes fixed and move the curve upward until the vertex is tangent to the *x*-axis (touches it at a single point), we have the case of *two equal real* solutions. Finally, if we again move the parabola upward, it will not intersect the *x*-axis, and we have the case of *no real solutions*. Algebraically, these cases relate to the quantity $b^2 - 4ac$ in the quadratic formula,

$$x = \frac{-b \pm \sqrt{b^2 - 4ac}}{2a}.$$

If $b^2 - 4ac$ is *positive* (greater than zero), we have a positive and a negative value for the square root, and there are *two different real* solutions. If $b^2 - 4ac$ is *zero*, we have only one value for *x* and there are *two equal real* solutions. Finally, if $b^2 - 4ac$ is *negative*, its square root is not a real number, and we have *no real solutions*.

> **Exercise.** Apply the quadratic formula to find the solutions for $x^2 - 9 = 0$. Answer: With $a = 1$, $b = 0$, $c = -9$, we find the two different roots, $x = 3$, and $x = -3$. The same result can be obtained by writing $x^2 = 9$ and noting that the square of +3 and −3 equal 9.

If we apply the formula to

$$3x^2 - 5x + 10 = 0,$$

we have,

$$x = \frac{5 \pm \sqrt{25 - 120}}{6} = \frac{5 \pm \sqrt{-95}}{6}$$

and find that this equation has no real solutions.

> **Exercise.** Solve $x^2 - 4x + 4 = 0$ by the quadratic formula. Answer: $x_1 = x_2 = 2$. We have two real equal solutions.

Solution by factoring. The equation

$$(x - 3)(x + 2) = 0$$

is true if either the factor $x - 3 = 0$ or the factor $x + 2 = 0$ because 0 times any number equals zero. Hence, the solutions are $x_1 = 3$, $x_2 = -2$. Consequently, we can easily find the solutions if the quadratic is factorable. For example,

$$2x^2 + 7x - 15 = 0$$

can be factored to yield

$$(2x - 3)(x + 5) = 0.$$

Setting $2x - 3 = 0$ and $x + 5 = 0$, we have the solutions

$$x_1 = \frac{3}{2}, \ x_2 = -5.$$

Exercise. Solve $6x^2 + x - 2 = 0$ by factoring. Answer: $x_1 = 1/2$, $x_2 = -2/3$.

Finally, note that the factoring is elementary if the quadratic has no constant term. Thus,

$$2x^2 - 5x = 0$$

factors to

$$x(2x - 5) = 0,$$

so the solutions are $x_1 = 0$ and $x_2 = 5/2$.

A3.14 PROBLEM SET A3–3

Graph each of the following:

1. $y = 3x - 1$.
2. $y = x$.
3. $y = -5x + 6$.
4. $2y - 3x = 6$.
5. $y = \dfrac{x}{2} + 3$.
6. $y = 3x^2$.
7. $y = x^2 - 7$.
8. $y = -2x^2 + 10$.
9. $y = 2x^2 - x$.
10. $y = x^2 + 2x - 6$.

Solve by the quadratic formula:

11. $x^2 - 10x + 3 = 0$.
12. $2x^2 + 5x - 2 = 0$.
13. $4x^2 - 12x + 9 = 0$.
14. $3x^2 + 2x + 5 = 0$.
15. $x^2 - 2x = 0$.

Solve by factoring:

16. $x^2 - 3x + 2 = 0$.
17. $3x^2 - 6x = 0$.
18. $6x^2 + 7x - 20 = 0$.
19. $x^2 - 25 = 0$.
20. $2x^2 + 5x - 7 = 0$.

A3.15 REVIEW PROBLEMS

Solve for x, citing the operations performed:

1. $3x - 2 = 2x + 5$.

2. $\dfrac{2x}{3} - 1 = 5x$.

3. $\dfrac{3 + 2x}{4} - 2 = x$.

4. $\dfrac{2(a - x)}{b} + 3x = c$.

5. $2a = \dfrac{x}{3 - cx}$.

6. $\dfrac{3}{a(x - 1)} + \dfrac{1}{2} - b = 0$.

7. $5(x - b) = a + b[3 - 2(x + 1)]$.

8. $\dfrac{2}{3} - \dfrac{3x}{4} = 4x(2 - b)$.

9. $x - \dfrac{c}{d} = a - \dfrac{3[2 - b(x - 1)]}{4}$.

10. $\dfrac{1}{1 - x} + 1 = b$.

In the following compute x, given that $a = 6$, $b = 2$, and $c = 10$:

11. $x = \dfrac{ab - c}{b}$.

12. $x = \dfrac{a^2 b - b}{c}$.

13. $x = \dfrac{\dfrac{c^2}{b}}{a}$.

14. $x = \sqrt{c^2 - a^2}$.

15. $x = \left(3c - \dfrac{a}{2}\right)^{1/3}$.

16. $x = (3ab)^{-3/2}$.

17. $x = (c - a - b)[2b - a(c - 5b)]$.

18. $x = \dfrac{\dfrac{a^2}{b} - c}{b^2}$.

In the following, compute the exact value of y in the form of a fraction in the lowest terms, given that:

$$a = \frac{2}{7}; \ b = \frac{1}{3}; \ c = \frac{4}{9}; \ d = \frac{5}{12}; \ x = \frac{1}{8}; \ z = \frac{1}{5}.$$

19. $y = \dfrac{x}{a}$.

20. $y = \dfrac{ab}{2}$.

21. $y = d - \dfrac{1}{a}$.

22. $y = ax + bz$.

23. $y = \left(1 - \dfrac{b}{d}\right)^3$.

24. $y = (1 + 19x)^{1/3}$.

25. $y = \sqrt{c - b}$.

26. $y = \dfrac{\dfrac{b}{2} + 3x}{1 + a}$.

27. Compute y to the nearest cent:

$$y = at + b(t - c)$$
$$a = 4\tfrac{1}{4}; \ b = 22\%; \ c = \$18{,}000; \ t = \$28{,}000.$$

28. The equation relating the cost, C, of an item to its retail price, R, is $C + pR = R$, where p is the percent markup on the retail price.

a) What markup percent on retail is achieved if an item costs $10 and sells at a retail price of $15?

b) How much can a merchant pay for an item if he wishes to sell it for $20 and achieve a markup that is 30 percent of the retail price?

c) If the merchant seeks to achieve a markup that is 35 percent of the retail price, at what price should he sell an item that cost him $13?

29. Compute y accurate to three decimals if $x = 0.05$:

$$y = \frac{(1 + x)^{-3} + 1}{x}.$$

30. The amount of interest, I, earned on P dollars at simple interest of r percent per year for t years is $i = Prt$. Assuming a year to have 360 days,

a) How many days will it take for $2000 to earn $50 interest at 4 percent?

b) How many dollars must be invested at $5\frac{1}{3}$ percent if interest in the amount of $100 is to be earned in 90 days?

31. Temperature in degrees Fahrenheit, F, is related to temperature in degrees Celsius, C, by the formula

$$F = \frac{9}{5}C + 32.$$

Find the Celsius temperatures corresponding to the following Fahrenheit readings:

a) 0. b) 100. c) 51. d) −32.

32. When an asset is bought, its initial book value is $\$A$. The book value will decline over the N years of the asset's life to the salvage value, $\$S$. The ratio of salvage value to initial book value is $s = S/A$. In the straight-line method of computing book value from year to year, the appropriate formula is

$$B = A\left[1 - \frac{t}{N}(1 - s)\right],$$

where B is the book value after t years.

a) An asset with an initial book value of $1000 and a salvage value of $100 after 10 years of life will have what book value after 3 years?

b) Solve the formula for s.

c) If an asset with an initial book value of $1000 and a life of 10 years is quoted as having a book value of $650 after 5 years, find the ratio of the salvage value to the initial value, and find the salvage value.

d) What will the depreciation formula for book value be if the asset has no salvage value?

33. A customer buys an article on the installment plan. The price of the article (or the amount he still owes after making a down payment) is B dollars. The seller adds a carrying charge of C dollars and requires that the debt be paid off by n equal payments. Letting y be the number of payments that would be made in a year (that is, if payments are weekly,

y is 52; and if payments are monthly, y is 12), and letting r be the equivalent simple interest rate being paid by the customer, a formula states

$$r = \frac{2yC}{B(n+1)}.$$

a) An article priced at $200 is purchased. The customer pays $50 down. The seller adds a carrying charge of $28.50 and requires the debt to be repaid in 18 equal monthly installments. What is the amount of the installment and what is the interest rate being charged?

b) If the simple interest equivalent of the installment rate is to be 30 percent, what should be the carrying charge on an article priced at $300 if the debt is to be paid off in 12 monthly installments and no down payment is made?

34. If D (demand) is the number of units of an item that can be sold when the item is priced at p cents per unit, and if

$$D = \frac{20}{p-1} - 1$$

for values of p greater than 1 but less than 21, at what price will the demand be:

a) 19 units?   b) 9 units?   c) 3 units?   d) 1 unit?

35. Graph the following equations:

a) $y = 2x + 1.$ b) $5y + 4x - 20 = 0.$

c) $y = \dfrac{x}{3}.$ d) $x - y = 1.$

e) $y = 8 - x^2.$ f) $y = x^2 - 10x + 15.$

36. Solve by the quadratic formula:

a) $x^2 - 2x - 2 = 0.$ b) $4x^2 - 4x + 1 = 0.$

c) $x^2 + 2x + 5 = 0.$

37. Solve by factoring:

a) $6x^2 - 5x + 1 = 0.$ b) $16x^2 - 40x + 25 = 0.$

c) $3x^2 - 48x = 0.$ d) $x^2 - 100 = 0.$

Answers to problem sets

CHAPTER 1

Problem Set 1–1

1. See Figure A. $AB = 6$, $AC = 3$, $DB = 4$.

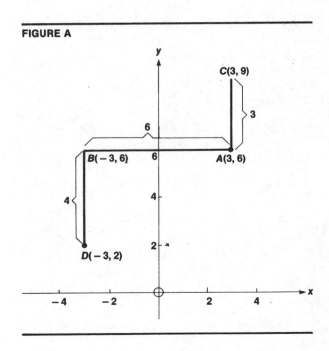

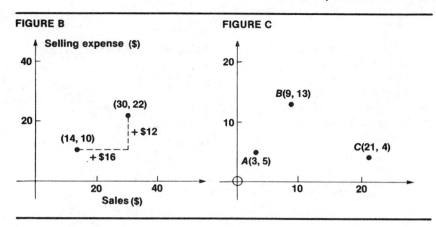

2. $AB = 2$, $AC = 24$, $AD = 17$, $BD = 15$.
3. AB is vertical, CD is horizontal.
4. $y_2 = y_1$, $x_1 = x_3$.
5. $y_2 = y_1$, $y_4 = y_3$.
6. *a)* 40 miles. *b)* 70 miles.
7. *a)* $16. *b)* $12. *c)* See Figure B.
8. Subscript notation preserves the letters x and y to mean abscissa and ordinate for *all* points.
9. 12 blocks.

Problem Set 1–2

1. *a)* (3, 8), (7, 5), 5.
   b) (−1, 6), (5, −2), 10.
   c) (−1, −14), (−10, −2), 15.
2. *a)* 10. *b)* 15. *c)* 5.10.
   d) 10. *e)* 3. *f)* 2.
3. $AB + BC + CA = 140 + 150 + 130 = 420$ miles.
4. *a)* See Figure C. *b)* 25 miles.
5. *a)* (50, 25). *b)* 55.9 feet. *c)* 50 feet.

Problem Set 1–3

1. *a)* The difference of the second *(y)* coordinates, $y_2 - y_1$.
   b) The difference of the first *(x)* coordinates, $x_2 - x_1$.
2. *a)* The steepest line is vertical.
   b) As we go from one point to another on the line, the run is zero so the slope ratio has a denominator of zero. A ratio with a denominator of zero is not a number.
3. 2/3.
4. Vertical. The plane evidently crashed.

854

5. 1.

6. *a)* 8.   *b)* $8.

7. The slope is 0.96 and $1 - 0.96 = 0.04$. The first is the marginal propensity to consume and the second the marginal propensity to save. The numbers 0.96 and 0.04 indicate that for an extra $1 of income, consumers spend $0.96 and save $0.04.

8. 1.   9. -1.   10. $-1/8$.

11. No slope number because segment is vertical.

12. 0.   13. 0.

14. No slope number because segment is vertical.

15. 1.   16. 21/32.   17. -1.   18. $-11/9$.

19. No slope number because segment is vertical.

20. 1.   21. $-1/13$.   22. See text.

23. $\dfrac{b-2}{a-3}, \dfrac{2-b}{3-a}$.

24. See Figure D.

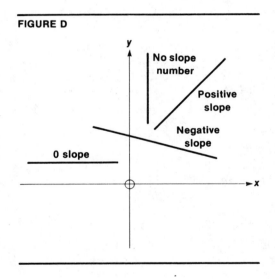

FIGURE D

25. If a is not zero, there is no number x such that $a/0 = x$; if a is 0, x is ambiguous because it could have any value.

26. F.   27. T.   28. T.
29. T.   30. T.   31. T.
32. T.   33. F.   34. T.

Problem Set 1–4

1. $y = 3x - 5$.   2. $y = -x + 1$.   3. $y = -\dfrac{2}{3}x + \dfrac{14}{3}$.

4. $y = \dfrac{1}{2}x - \dfrac{9}{2}$.

5. $y = -8$.

6. $y = -\dfrac{1}{6}x + \dfrac{22}{3}$.

7. $y = 13x + 57$.

8. $y = 0$.

9. $x = 5$.

10. $x = 0$.

11. $y = 4$.

12. $y = 5x - 23$.

13. $y = -\dfrac{1}{7}x + \dfrac{46}{7}$.

14. $y = \dfrac{6}{7}x + \dfrac{51}{7}$.

15. $y = x$.

16. $y = 9x + 14$.

17. $y = \dfrac{3}{2}x$.

18. $y = 4$.

19. $y = -\dfrac{3}{2}x$.

20. $x = -7$.

21. $y = \dfrac{1}{2}x + \dfrac{7}{2}$.

22. $y = 5$.

23. $y = 0$.

24. $y = -\dfrac{5}{7}x - \dfrac{29}{7}$.

25. $y = 0$.

26. $x = 0$.

27. $-\dfrac{24}{7}$.

28. -2.

29. $x = -6$.

30.   a) $y = 3$.

   b)   The tangent is the horizontal line $y = 500$ and shows the maximum (highest) profit is 500.

31.   a) $y = 5$.   b) $x = -10$.

32. $y = 0.25x + 50$.   33. $y = 3x + 50$.   34. $y = 3x + 10$.

35.   a) $y = 0.8m + 0.5$.

   b) $E = 0.1V + 50$. The slope is the rate of commission.

36.   F.

37.   T.

38.   T.

39.   F.

40.   T.

41.   T.

42.   See Figure E.

43.   See Figure F.

44.   See Figure G.

45.   See Figure H.

46.   See Figure I.

47.   3/2.

FIGURE E

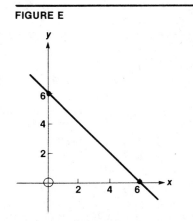

FIGURE F

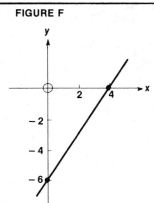

856

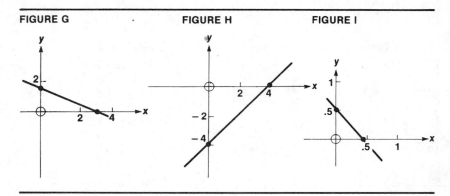

FIGURE G FIGURE H FIGURE I

48. −1.   49. 1/3.   50. 1.

51. *a)* $8x + 12y = 96$.   *b)* 1.5 pounds of A per pound of B.
    c) 2/3 pounds of B per pound of A.

52. $y = 2x − 6$.

53. If the equation is solved for y, the coefficient of x, which is the slope, turns out to be $−a/b$.

54. y/x is cost per unit; $y/x = 3$ or $y = 3x$ is the equation; slope is 3, and intercepts zero (line passes through the origin).

55. *a)* $3x − 2y + 8 = 0$.   *b)* $y + 3x + 12 = 0$.

56. $2200.

57. $y = 2x$.   58. $y = 4x/5$.

59. *a)* *D* is a right angle; hence, angle 1 is 90° −angle *BDC*. *DBC* is a right triangle; hence, angle 2 is 90° −angle *BDC*. Therefore, angle 1 = angle 2, and triangles *ABD* and *DBC* are similar because they have equal angles. It follows that corresponding sides are in proportion, so that $DB/AB = BC/DB = 1/(DB/BC)$, where DB/AB is the slope of l_2 and DB/BC is the slope of l_1.

    b) A horizontal and a vertical line do not have reciprocal slopes because 1/0 is not a number.

60. *a)* 1.25 miles to the right of the origin.
    b) Because the shortest distance from a point to a line is on a perpendicular to the line.

61. The coordinates of the origin (0, 0), satisfy any equation of the form $ax + by = 0$.

62. F.   63. T.   64. T.   65. T.   66. T.

67. T.   68. T.   69. T.   70. T.   71. T.

Problem Set 1–5

1. T.   2. T.   3. F.   4. T.   5. F.

6. T.   7. T.   8. T.   9. T.   10. T.

11. *a)* 150. *b)* 170. *c)* 3. *d)* 3.40. *e)* 3.
12. *a)* $10,000. *b)* $10,500. *c)* $10. *d)* $10.
13. *a)* $y = 5x + 2500.$ *b)* $10,000. *c)* $2500.
    d) $5. *e)* $6.25. *f)* $5.
14. *a), (b), (c)* See Figure J.
    d) $RS = 25 =$ fixed cost; $ST = 20 =$ variable cost for 10 units; $RT = 45 =$ total cost for 10 units.

FIGURE J

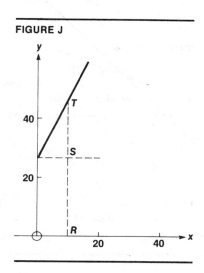

15. *a)* AB. *b)* AD. *c)* BD. *d)* BD/OA.
    e) AD/OA. *f)* AB/OA. *g)* EB. *h)* BD/OA.
16. *a)* Setup cost, or fixed cost.
    b) Variable cost if *OM* units are made.
    c) Variable cost per unit.
    d) Average cost per unit if *OM* units are made.

Problem Set 1–6

1. T. 2. F. 3. T. 4. T. 5. F. 6. T. 7. T.
8. F. 9. T. 10. F. 11. F. 12. T.
13. *a)* $R = 5q.$ $C = 2q + 60,000.$ *b)* $15,000. *c)* Loss of $30,000.
    d) 20,000 units. *e)* $100,000. *f)* See Figure K.
14. *a)* $R = 50q.$ $C = 20q + 120,000.$ *b)* $180,000.
    c) Loss of $90,000. *d)* 4000 units. *e)* $200,000.
    f) See Figure L.
15. 2500 units. 16. $5,500,000.
17. *a)* $200,000.
    b) The company should not shut down because if it does the loss will be $200,000 but if it produces and sells 1000 units, the loss will be $120,000, which is a smaller loss than $200,000.

FIGURE K

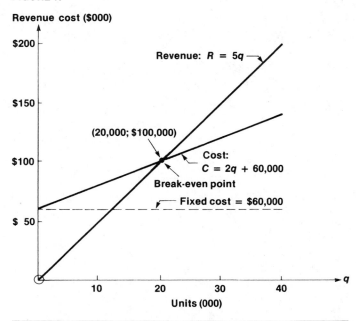

FIGURE L

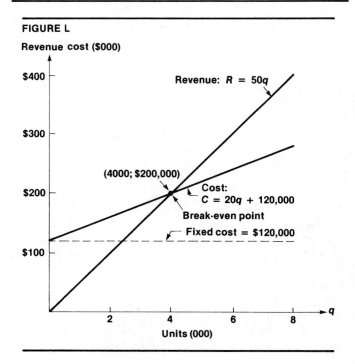

18. *a)* $500,000.
    b) The company should not shut down because if it does the loss will be $500,000 but if it produces and sells 100,000 units, the loss will be $450,000, which is a smaller loss than $500,000.

19. T.   20. T.   21. T.   22. T.   23. T.
24. T.   25. F.   26. F.   27. F.   28. T.
29. *a)* $60,000.   *b)* $y = 22,800 + 0.62x.$
    c) $5700.   *d)* See Figure M.
30. *a)* $80,000.   *b)* $y = 36,000 + 0.55x.$
    c) −$2250.   *d)* See Figure N.

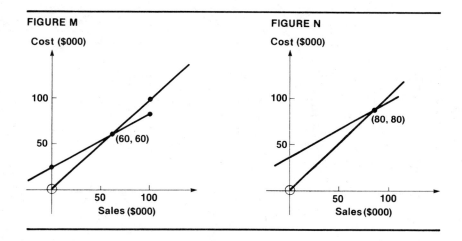

FIGURE M

Cost ($000)

(60, 60)

Sales ($000)

FIGURE N

Cost ($000)

(80, 80)

Sales ($000)

31. *a)* $0.47.   *b)* $29,786.   *c)* $63,626.
    d) $56,200.   *e)* $12,614.
32. *a)* $12,800.   *b)* $23,800.   *c)* $35,000.
    d) $3400 loss.
33. $1000.
34. *a), (b)* See Figures O and P.
35. The shift is 50 to the left horizontally, so demand is 50 units less at every price level. The shift is 5 downward vertically, so price is $5 per unit less at every level of demand.
36. The shift is 200 to the right horizontally, so demand is 200 units greater at every price level. The shift is 10 upward vertically, so price is $10 per unit higher at every level of demand.
37. *a)* Demand is a constant number of units at every level of price per unit.
    b) Price per unit is constant at every level of demand.
38. Demand is the constant function $q = 100$ tons per week, a vertical line.
39. The demand function is a horizontal line, $p =$ a constant.

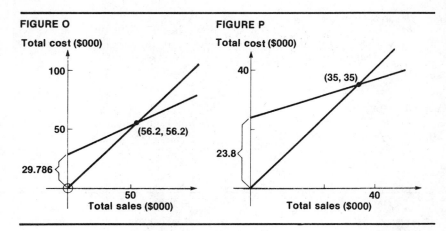

FIGURE O

Total cost ($000)

100

50   (56.2, 56.2)

29.786

50
Total sales ($000)

FIGURE P

Total cost ($000)

40   (35, 35)

23.8

40
Total sales ($000)

Problem Set 1–7

1.	F.	2.	T.	3.	T.	4.	T.	5.	F.
6.	T.	7.	T.	8.	T.	9.	F.	10.	T.
11.	T.	12.	F.	13.	T.	14.	T. ·	15.	T.
16.	T.	17.	T.	18.	T.	19.	T.	20.	T.

21. See Figures Q through S.

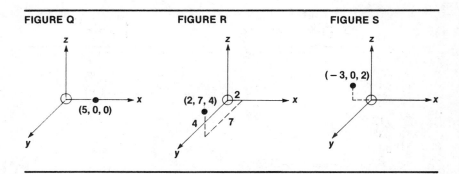

FIGURE Q

z

(5, 0, 0) → x

y

FIGURE R

z

(2, 7, 4) 2

4   7

→ x

y

FIGURE S

z

(−3, 0, 2)

→ x

y

22. $x + 0y + 3z = 6$.   23. See Figures T through Y.

24.   a) $w + 3x + 2y + 4z = 500$.
     b) $0.5w + 0.3x + 1.2y + 0.8z = 310$.

25. x arbitrary, $y = (6 - 3x)/2$; or y arbitrary, $x = (6 - 2y)/3$.

26. x and y arbitrary, $z = (2x + 3y - 18)/2$.
     x and z abritrary, $y = (18 + 2z - 2x)/3$.
     y and z arbitrary, $x = (18 + 2z - 3y)/2$.

27.   The number of pairs of values that satisfy the equation is without limit.

28. $3x + 4y = 12$.

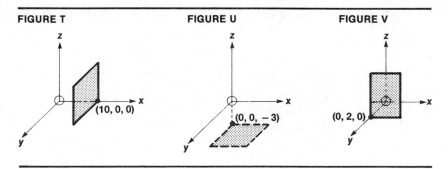

FIGURE T

(10, 0, 0)

FIGURE U

(0, 0, −3)

FIGURE V

(0, 2, 0)

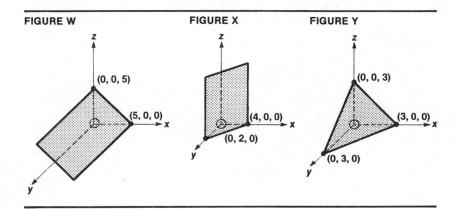

FIGURE W

(0, 0, 5)

(5, 0, 0)

FIGURE X

(4, 0, 0)

(0, 2, 0)

FIGURE Y

(0, 0, 3)

(3, 0, 0)

(0, 3, 0)

29. (4, 0) and (0, 3).
30. See Figure Z.
31. $x = 600$, $y = 400$ gallons.

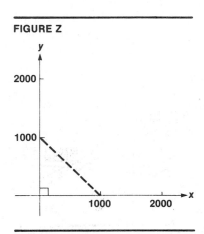

FIGURE Z

2000

1000

1000   2000

32. If per-gallon costs are not equal, and there is no restriction on the relative amounts of A and B in the tank.
33. $10x + 8y + 7z$.
34. *a)* $10x + 8y + 7z = 560$.
    b) The numbers must be positive integers, or 0.
35. The set of ordered pairs, (x, y), such that $y - 3x = 6$.
36. The set of ordered triples, (x, y, z), such that $2x + y - 3z = 15$.
37. The set of ordered quadruples, (w, x, y, z), such that $2w + x - y + 3z = 30$.
38. $\{(x, y, z): y = -2x + 3z + 15\}$.
39. $\{(w, x, y, z): y = 2w + x + 3z - 30\}$.
40. w, y, and z arbitrary; $x = 30 - 2w + y - 3z$.

CHAPTER 2

Problem Set 2–1

1. $(2, 3)$. 2. $(-1, 4)$. 3. $(1, 1)$. 4. $(3/2, 2)$. 5. $(1/2, 2/3)$.
6. $(1/2, 3/2)$.
7. 375 liters of Exall and 625 liters of Whyall.
8. Invest \$20,000 in Extron and \$30,000 in Whytron.
9. Make 15 Exbees and 10 Whybees.
10. Make 5 Exballs and 50 Whyballs.

Problem Set 2–2

1. *a)* See Figure A. *b)* $E(70, 15)$. *c)* See Figure A.

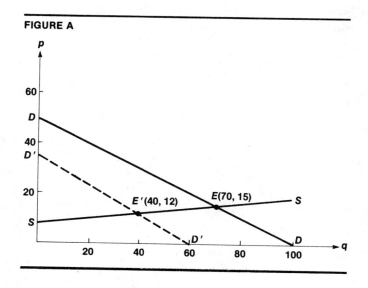

FIGURE A

d) E'(40, 12).

e) The decrease in demand was accompanied by a lower demand and a lower price per unit at equilibrium.

2. *a)* See Figure B.   *b)* E(100, 30).   *c)* See Figure B.

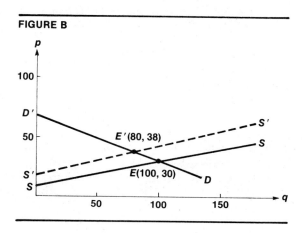

FIGURE B

d) E'(80, 38).

e) The decrease in supply was accompanied by a lower supply but a higher price per unit at equilibrium.

3. E'(125, 35). The increase in demand was accompanied by a higher demand and a higher price at equilibrium.

4. E'(80, 10). The increase in supply was accompanied by a higher supply but a lower price at equilibrium.

5. The demand function shifted to the right. At the new equilibrium, both demand and price are higher.

6. The supply function shifted to the left. At the new equilibrium, supply is lower but price is higher.

7. The supply function shifted to the right. At the new equilibrium, supply is higher but price is lower.

8. The demand function shifted to the left. At the new equilibrium, both demand and price are lower.

9. The supply function shifted to the right because a right shift of a supply function leads to a new equilibrium in which supply is *higher* but price is *lower.*

10. The demand function shifted to the left because with a left shift in demand *both* demand and price are lower at the new equilibrium.

Problem Set 2–3

1. *a)* A system in which the number of equations is *m* and the number of variables is *n.*

b) A system in which the number of equations is the same as the number of variables.

c) A system of three equations in two variables.

d) The total number of variables appearing in the equations of the system.

e) A linear combination of two equations is an equation formed by taking the sum of p times one equation plus q times the other, where p and q are any numbers.

2. a) x arbitrary, $y = (3 - 3x)/4$.

b) y and z arbitrary, $x = 15 - 2y + 4z$.

3. a) A false statement, such as $0 = 4$.

b) An identity, such as $0 = 0$.

4. See text.

5. a) See text.

b) Consistent means the set of three equations have one solution; that is, there is exactly one point that lies on all three planes. Inconsistent means there are no solutions; that is, there is no point that lies on all three planes. Dependent means the number of solutions is unlimited as would be the case if the three planes had a whole line in common.

6. See text.

7. a) $(3, -2)$. See Figure C.

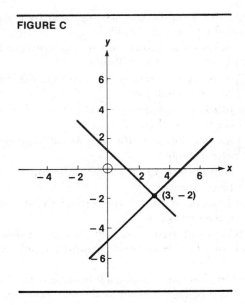

FIGURE C

b) $(3, -2)$. See Figure D.   c) $(32/17, 3/17)$. See Figure E.

8. a) Parallel lines, no solutions.   b) Lines have same graph.

FIGURE D

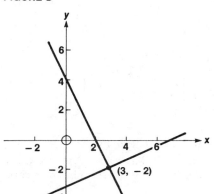

FIGURE E

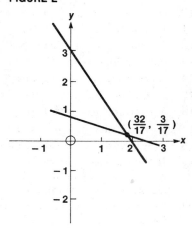

9. $(1/2, 1, 3/2)$.
10. x arbitrary, $y = x - 2$, $z = 5 - x$.
11. x arbitrary, $y = 3 - 2x$, $z = 5 - 3x$.
12. $(2/5, -3, 3/5)$.
13. No solutions.
14. No solutions.
15. x arbitrary, $y = 17 - 17x$, $z = 22 - 9x$.
16. $(1, 2, 3)$.
17. No solutions.
18. $(1/2, 1/3, 1/4, -2)$.
19. x and y arbitrary, $w = x + 2$, $z = y - x - 3$.
20. $(-10, -2, -5, 8)$.
21. $(1, 2, -2, -1)$.
22. y arbitrary, $w = 2y + 6$, $x = y + 3$, $z = 2y + 7$.

Problem Set 2–4

1. No solutions.
2. y arbitrary, $x = y + 2$, $z = 2y - 1$.
3. $(1, -1, 2/3)$.
4. No solutions.
5. No solutions.
6. $(2, 0, 4)$.
7. x arbitrary, $y = 2 - x$, $z = 2x$.
8. y, z arbitrary, $x = y + z + 1$, $w = 2y - z + 3$.

9. x and z arbitrary, $y = 2x + 3z - 3$.

10. y arbitrary, $x = 2 - y$, $z = 4 - 2y$.

11. No solutions.

12. x arbitary, $y = 1 - 2x$, $z = 3$.

13. No solutions.

14. y arbitrary, $x = y + 2$, $w = 7 + 2y$, $z = -3y - 5$.

15. x and z arbitrary, $w = 3x + z - 2$, $y = 5x + 2z - 6$.

16. x arbitrary, $w = x + 1$, $y = x$, $z = 3 - 2x$.

17. y, z arbitrary, $x = y + z$, $w = y - z$.

18. The inclusive intervals are 0 to 2 for x and y, 0 to 4 for z.

19. If the solution is stated with y arbitrary, we have $z = -3y - 5$. Consequently, for any nonnegative y, z is negative so the system has no solution if nonnegativity is imposed.

20. The inclusive intervals are: 1 to 5/2 for w; 0 to 3/2 for x and y; 0 to 3 for z.

Problem Set 2–5

1. $x = 10$, $y = 15$, $z = 20$.

2. z arbitrary in the range from 19 to 155/17, $x = 10z - 190$, $y = 155 - 7z$.

3. *a)* z arbitrary in the range from 0 to 3/2, $x = 4 + z$, $y = 3 - 2z$.
   b) Minimum is \$11.50, using 5.5 pounds of Exboy, 1.5 pounds of Zeeboy, and no Whyboy.

4. x arbitrary; $w = 40 - x$, $z = 70 - x$, $y = x - 10$.

5.
$w + x + y$		$= 90$	x, y arbitrary
$u + v$	$+ z = 60$		$w = 90 - x - y$
u $+ w$	$= 30$		$u = x + y - 60$
v $+ x$	$= 45$		$v = 45 - x$
$y + z = 75$.			$z = 75 - y$.

6. 10 grade A boxes, 15 grade B boxes.

7. The solution is z arbitrary, $x = (11z - 22)/2$, $y = (66 - 19z)/2$. It follows from these that z must be at least 2, but less than 66/19, so that $z = 2$, 3 are possibilities; however, if $z = 3$, x and y are not integers. The single solution is $z = 2$, $x = 0$, $y = 14$; that is, 0 boxes of grade A, 14 boxes of grade B, and 2 boxes of grade C.

8. The solution is y arbitrary, $x = 2 + y$, $z = 10 - 3y$; however, y must be between 0 and 3, inclusive.

9. $x = 2$, $y = 0$, $z = 10$.

10. The solution requires that $y = -2$, which is not permissible. The problem has no solution.

11. *a)* z units of C, arbitrary from 5 through 8; x units of A where $x = 80 - 10z$; y units of B where $y = z - 5$.
    b) Maximum is \$800, making 30 units of A, no B, and 5 units of C.

12. *a)* z units of C, arbitrary from 0 through 4; y units of B, where $y = z$; x units of A, where $x = 40 - 10z$.
    b) Maximum is \$800, making 40 units of A and no B or C.
13. *a)* $p_1 = \$415$, $q_1 = 65$ units; $p_2 = \$445$, $q_2 = 65$ units.
    b) $p_1 = \$474$, $q_1 = 96$ units; $p_2 = \$472$, $q_2 = 64$ units.
14. *a)* $p_1 = \$800$, $q_1 = 100$ units; $p_2 = \$400$, $q_2 = 100$ units.
    b) $p_1 = \$740$, $q_1 = 60$ units; $p_2 = \$420$, $q_2 = 160$ units.

CHAPTER 3

Problem Set 3–1

1. F.	2. T.	3. F.
4. F.	5. F.	6. F.
7. T.	8. F.	9. T.
10. T.	11. T.	12. T.
13. $x \le 2$.	14. $x \ge 3$.	15. $x < -3$.
16. $x > -2$.	17. $x \ge 0.5$.	18. $x \le -0.25$.

19. *a)* $y \le 6 - 1.5x$.   *b)* On and below the line.   *c)* $y \le 3$.
20. *a)* $y \ge 7 - 0.4x$.   *b)* On and above the line.   *c)* $y \ge 5$.
21. *a)* $y \le -9 + 1.4x$.   *b)* On and below the line.   *c)* $y \le 19$.
22. *a)* $y \ge -9 + 1.5x$.   *b)* On and above the line.   *c)* $y \ge 6$.
23. Yes.   24. No.   25. Yes.   26. No.

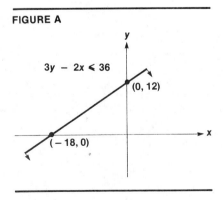

FIGURE A

$3y - 2x < 36$

(0, 12)

(−18, 0)

27. See Figure A.   28. See Figure B.   29. See Figure C.
30. See Figure D.   31. See Figure E.   32. See Figure F.
33. See Figure G.   34. See Figure H.   35. See Figure I.
36. See Figure J.

FIGURE B

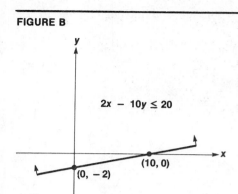

$2x - 10y \le 20$

(10, 0)

(0, -2)

FIGURE C

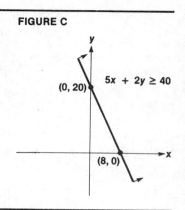

$5x + 2y \ge 40$

(0, 20)

(8, 0)

FIGURE D

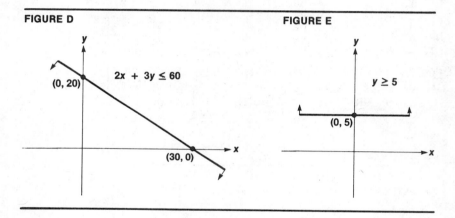

$2x + 3y \le 60$

(0, 20)

(30, 0)

FIGURE E

$y \ge 5$

(0, 5)

FIGURE F

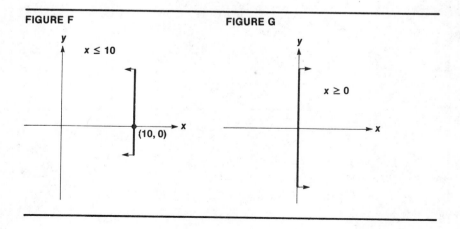

$x \le 10$

(10, 0)

FIGURE G

$x \ge 0$

FIGURE H

FIGURE I

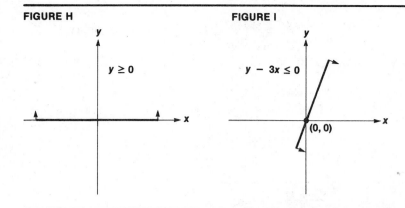

FIGURE J

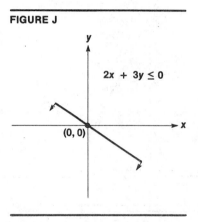

FIGURE K

FIGURE L

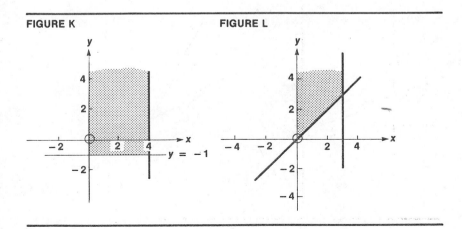

Problem Set 3–2

1. See Figure K.   2. See Figure L.   3. See Figure M.
4. See Figure N.   5. See Figure O.
6. If $2 \le x \le 30/7$, $0 \le y \le (x - 2)/2$; if $30/7 \le x \le 6$, $0 \le y \le (12 - 2x)/3$.
7. $0 \le x \le 52/19$, $(8 - x)/5 \le y \le 12 - 4x$.
8. If $52/19 \le x \le 3$, $12 - 4x \le y \le (8 - x)/5$; if $3 \le x \le 8$, $0 \le y \le (8 - x)/5$.
9. If $0 \le x \le 52/19$, $y \ge 12 - 4x$; if $52/19 \le x \le 8$, $y \ge (8 - x)/5$; if $x \ge 8$, $y \ge 0$.

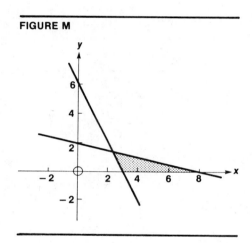

FIGURE M

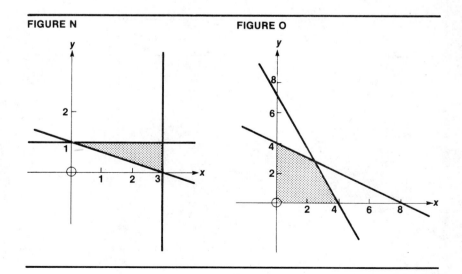

FIGURE N

FIGURE O

10. *a)* $h \le (76 - L)/2; 0 < L < 76.$
     b) $x \le (108 - L)/4; 0 < L < 108.$
11. $0 \le x \le 10, 8 - 0.5x \le y \le (45 - 3x)/5.$
12. $x = 0, y = 8.$
13. If $0 \le x \le 320, 0 \le y \le (2400 - 3x)/8;$ if $320 \le x \le 500, 0 \le y \le 500 - x.$
14. *a)* $0 \le x \le 2, 0 \le y \le 5 - x; 2 \le x \le 7/2, 0 \le y \le 7 - 2x.$

     b)

x	y
0	0, 1, 2, 3, 4, 5
1	0, 1, 2, 3, 4
2	0, 1, 2, 3
3	0, 1

15. Letting x be the number of Exballs, and y the number of Whyballs, $0 \le x \le 9/2, 0 \le y \le 2x; 9/2 \le x \le 18, 0 \le y \le (36 - 2x)/3.$
16. $0 \le x \le 13, x/4 \le y \le 4x; 13 \le x \le 88, x/4 \le y \le (286 - 2x)/5.$
17. $0 \le x \le 20, 0 \le y \le (40 - x)/2; 20 \le x \le 27, 0 \le y \le (190 - 6x)/7;$ $27 \le x \le 30, 0 \le y \le (120 - 4x)/3.$
18. $20 \le x \le 24, (190 - 6x)/7 \le y \le (40 - x)/2;$ $24 \le x \le 27, (190 - 6x)/7 \le y \le (120 - 4x)/3.$

CHAPTER 4

Problem Set 4–1

1. *a)* 7.   *b)* 16.5.   *c)* 18.
2. *a)* 12.   *b)* 19 and 3/7.
3. 20.5.
4. 28.1.
5. *a)* 2.3.   *b)* 7.   *c)* 15.
6. Maximum is 2.5 at $\left(\dfrac{3}{2}, \dfrac{1}{2}\right).$
7. Minimum is 12 at (0, 12).
8. Minimum is 12.8 at (1.6, 9.6).

Problem Set 4–2

1. Maximum profit of $19 by making 6 Extrans and 5 Whytrans.
2. Maximum profit of $33 by making and painting 12 Exbobs and 6 Whybobs.
3. *a)* No Exgrain and 9 ounces of Whygrain. Minimum cost is 18 cents.
     b) 3 ounces of Exgrain and 3 ounces of Whygrain. Minimum cost is 21 cents.
4. *a)* 450 Exballs and no Whyballs. Minimum cost is $450.
     b) 250 Exballs, 100 Whyballs. Minimum cost is $550.
5. 1000 units of A, 3000 units of B.
6. 9000 units of A, no units of B.

7. 800 each of types A and B.
8. $x = 30$ units of A, $y = 15$ units of B, $\theta_{max} = \$270$.
9. $x = 20$ units of A, $y = 20$ units of B, $\theta_{max} = \$200$.
10. $x = 30$ units of A, $y = 15$ units of B, $\theta_{max} = \$270$.
11. Because no part of $x + 2y = 60$ is part of the boundary of the solution space.
12. $\theta_{min} = \$27.3$ at 0 quarts of A and 9.1 quarts of B.
13. The new constraint is $y \geq 3x$. It is not binding because $y = 3x$ is not part of the boundary of the solution space.

Problem Set 4–3

1. $\theta_{max} = 580$ at $x = 520$, $y = 320$, $z = 0$.
2. $\theta_{min} = 1200$.
3. $\theta_{max} = 9$ and $27/82$.

Problem Set 4–4

1. No units of A or B, 1000 units of C.
2. $40/13$ units of A, $10/13$ units of B, $46/13$ units of C.
3. Maximum is $190 with no units of A, 40 units of B, and 75 units of C.

CHAPTER 5

Problem Set 5–1

1. T.	2. F.	3. F.	4. F.	5. T.
6. T.	7. F.	8. F.	9. F.	10. T.

11.
$\begin{array}{cc} 1/2 & 1/4 \\ -3/2 & 1/4. \end{array}$

12.
$\begin{array}{cc} -3/4 & 1/4 \\ -1/2 & 1/2. \end{array}$

13.
$\begin{array}{cc} 5 & -1 \\ 7 & -2. \end{array}$

14.
$\begin{array}{cc} 17/6 & 2/3 \\ 5/3 & 1/3. \end{array}$

15.
$\begin{array}{cc} 3/4 & -3 \\ -5/4 & -3. \end{array}$

16.

	y	
x	3	-2
w	5	-3^*

	w	
x	$-1/3$	$2/3$
y	$5/3$	$-1/3$

$x = -(1/3) + (2/3)w$
$y = (5/3) - (1/3)w.$

17.

	z	
v	10	4^*
y	5	-1

	v	
z	$-5/2$	$1/4$
y	$15/2$	$-1/4$

$z = -(5/2) + (1/4)v$
$y = (15/2) - (1/4)v.$

18.

	x	
y	8	-2
w	2	-1^*

	w	
y	4	2
x	2	-1

$y = 4 + 2w$
$x = 2 - w.$

19.

	v	
u	5	1^*
w	1	-1

	u	
v	-5	1
w	6	-1

$v = -5 + u$
$w = 6 - u.$

20.

	z	
x	3	2
y	4	-3^*

	y	
x	$17/3$	$-2/3$
z	$4/3$	$-1/3$

$x = (17/3) - (2/3)y$
$z = (4/3) - (1/3)y.$

Problem Set 5–2

1. *a)*

		w	*z*
y	−1/2	1/2	−1/2
x	7/2	3/2	−7/2.

b)

		x	*z*
w	−7/3	2/3	7/3
y	−5/3	1/3	2/3

c)

		y	*w*
z	−1	−2	1
x	7	7	−2.

d)

		y	*x*
w	7/2	7/2	−1/2
z	5/2	3/2	−1/2.

2. *a)*

		w	*u*	*y*
x	−3	−2	1	−2
v	−1	−5	2	−7.

b)

		v	*x*	*y*
u	13	−2	5	−4
w	5	−1	2	−3.

c)

		u	*x*	*y*
w	−3/2	1/2	−1/2	−1
v	13/2	−1/2	5/2	−2.

d)

		w	*x*	*u*
y	−3/2	−1	−1/2	1/2
v	19/2	2	7/2	−3/2.

e)

		w	*v*	*y*
u	1/2	5/2	1/2	7/2
x	−5/2	1/2	1/2	3/2.

f)

		w	*x*	*v*
u	19/3	4/3	7/3	−2/3
y	5/3	−1/3	2/3	−1/3.

3. *a)*

		v	*s*	*x*	*y*
r	16	2	−2	8	6
w	3	1	−1/2	2	0
t	−10	−6	3	−10	−4.

b)

$$r = 16 + 2v - 2s + 8x + 6y$$
$$w = 3 + v - (1/2)s + 2x$$
$$t = -10 - 6v + 3s - 10x - 4y.$$

4. *a)*

		w	*x*	*y*
u	4	3	2	1
v	2	−1	3*	−2.

b)

		w	*v*	*y*
u	8/3	11/3	2/3	7/3
x	−2/3	1/3	1/3	2/3

$$u = (8/3) + (11/3)w + (2/3)v + (7/3)y$$
$$x = -(2/3) + (1/3)w + (1/3)v + (2/3)y.$$

Problem Set 5–3

1.

		x	y
θ	0	2	1
r	20	−3	−2
s	12	−1	−1.

2.

		x	y	z
θ	0	1	2	1
r	8	−1	−1	−1
s	12	−2	−3	−1
t	10	−1	−2	−3.

3.

		w	x	y	z
θ	0	3	2.8	1	2
r	150	−2	−1	−2	−3
s	100	0	−2	−1	−2
t	200	−3	−1	−4	0.

Problem Set 5–4

NOTE: In the Problem Set, constants have been chosen so that hand calculations will not be excessively tedious. As a consequence, the values of x, y, and z that lead to the optimal θ may not be unique. The answers given have been obtained by the method developed in this chapter. Other methods will lead to the same θ, but the x, y, and z values may differ from those shown.

1. $\theta_{max} = 60$ at $x=0$, $y=20$, $z=0$.
2. $\theta_{max} = 35$ at $x=0$, $y=50/7$, $z=15/7$.
3. $\theta_{max} = 125/2$ at $x=3/2$, $y=29/2$, $z=1/2$.
4. $\theta_{max} = 35$ at $x=0$, $y=40/7$, $z=5/7$.
5. $\theta_{max} = 27$ at $x=0$, $y=3$, $z=1$.
6. $\theta_{max} = 54/5$ at $x=0$, $y=4$, $z=2/5$.
7. $\theta_{max} = 335/12$ at $x=0$, $y=5/6$, $z=95/12$.
8. $\theta_{max} = 22$ at $x=2$, $y=6$, $z=0$.
9. $\theta_{max} = 9$ at $x=0$, $y=3$, $z=3$.
10. $\theta_{max} = 79/7$ at $x=61/7$, $y=5/7$, $z=13/7$.
11. No units of A or B, 1000 units of C.
12. Maximum is $190 with no units of A, 40 units of B, and 75 units of C.

Problem Set 5–5

1.

		r	s
θ	0	20	10
x	5	−4	−6
y	3	−2	−1.

2.

		r	s
θ	0	10	20
x	4	−3	−1
y	2	−2	−2.

3.

		r	s	t
θ	0	2	4	6
x	1	−3	−2	−4
y	1	−3	−1	−2
z	1	−1	0	−2.

4.

		r	s	t
θ	0	10	15	20
x	2	−1	−2	−3
y	3	−1	−3	−1
z	1	−1	−1	−2.

5. $\theta_{min} = 10$ at $x = 3$, $y = 2$.
6. $\theta_{min} = 50$ at $x = 5$, $y = 0$, $z = 7$.
7. $\theta_{min} = 16$ at $x = 0$, $y = 8$.
8. $\theta_{min} = 42$ at $x = 6$, $y = 3$.
9. $\theta_{min} = 600$ at $x = 0$, $y = 600$, $z = 0$.
10. $\theta_{min} = 36$ at $x = 0$, $y = \dfrac{22}{19}$, $z = \dfrac{6}{19}$.
11. 3 quarts of Expop; 1 quart of Whypop. Minimum cost = $9.
12. No Excal, 3 pounds of Whycal, and 15 pounds of Zeecal. Minimum cost is 126 cents, or $1.26.

Problem Set 5–6

1. *a)* 100 Exglasses, no Whyglasses, and 80 Zeeglasses. Maximum profit is $125 thousand.
   b) Shadow price of grinding time is $30 per man-hour. Shadow price of polishing time is $50 per man-hour.
   c) 38 to 45 hours for grinding; 23 to 28 hours for polishing.
2. *a)* 4 superdeluxe, 9 deluxe, and no standard.
   b) Shadow price of a $1000 loan for land purchase is 1/5 thousand or $200. Shadow price of a $1000 loan for landscaping is 1/20 thousand or $50.
   c) 95 to 103 thousand dollars for land loan; 38 to 50 thousand dollars for landscaping loan.
3. *a)* Run the Exmill for 4 hours and the Whymill for 3 hours. Minimum cost $7 hundred.
   b) Shadow price of an A-Tran is 1/8 hundred or $12.50. Shadow price of a B-Tran is 1/4 hundred or $25.00.
   c) 12 to 36 A-Trans; 10 to 30 B-Trans.
4. *a)* X-Trucks for 60 days; Y-Trucks for 20 days. Minimum cost $9200.
   b) Shadow price for picking up and delivering a ton of gravel is 1/5 dollars, or 20 cents. Shadow price for picking up and delivering a ton of crushed rock is 2/5 dollars, or 40 cents.
   c) Gravel 17,000 to 20,000 thousand tons. Crushed rock 7000 to 12,000 thousand tons.
5. *a)* 10 Extrans, 12 Whytrans, no Zeetrans. Maximum profit is $58.
   b) Shadow price of a minute of forming time is $1/5 or 20 cents. Shadow price of a minute of painting time is $2/5 or 40 cents.
   c) Forming time, 140 to 155 minutes. Painting time, 60 to 75 minutes.
6. *a)* Run Exmix no hours, Whymix 2 hours, Zeemix 4 hours. Minimum cost is $1160.
   b) Shadow price of a barrel of A-oil is $20. Shadow price of a barrel of B-oil is $24.
   c) A-oil, 8 to 12 barrels. B-oil, 35 to 50 barrels.

CHAPTER 6

Problem Set 6–1

1. $(3 \quad 1 \quad 7)$.

2. $\begin{pmatrix} -2 \\ 2 \end{pmatrix}$.

3. $\begin{pmatrix} 10 \\ 13 \end{pmatrix}$.

4. $(8 \quad 0 \quad 4)$.

5. $\begin{pmatrix} 21 \\ 6 \end{pmatrix}$.

6. $(-10 \quad 18 \quad -6)$.

7. 41.

8. 5.

9. $\begin{pmatrix} 4 & -4 & 0 \\ -2 & 6 & 5 \end{pmatrix}$.

10. $\begin{pmatrix} 3 & 9 \\ 2 & -7 \end{pmatrix}$.

11. $\begin{pmatrix} -2 & -27 & 17 \\ 17 & -2 & -9 \end{pmatrix}$.

12. $\begin{pmatrix} 5 & -4 & 5 \\ 14 & -10 & 15 \end{pmatrix}$.

13. $(26 \quad 35)$.

14. $\begin{pmatrix} 3 & 18 \\ 3 & -30 \end{pmatrix}$.

15. $\begin{pmatrix} 13 & 16 \\ 8 & 7 \\ -5 & -9 \end{pmatrix}$.

16. $\begin{pmatrix} 7 & -1 & 5 & 6 \\ 11 & 3 & 8 & 13 \end{pmatrix}$.

17. $\begin{pmatrix} 1 & 4 \\ 2 & 5 \\ 3 & 6 \end{pmatrix}$.

18. $\begin{pmatrix} 1 & 2 & 3 \\ 3 & 2 & 1 \end{pmatrix}$.

19. $(0.06 \quad 0.07 \quad 0.08) \begin{pmatrix} 3000 \\ 2000 \\ 4000 \end{pmatrix} = \$640.$

20. *a)* $(10 \quad 20) \begin{pmatrix} 8 & 2 & 5 \\ 4 & 8 & 3 \end{pmatrix} = (160 \quad 180 \quad 110)$.

The first component of the right vector, 160, is 10 batches of Superior times 8 pounds of beef per batch, plus 20 batches of Regular times 4 pounds of beef per batch, for a total of 160 pounds of beef to make the required number of batches. Similarly, we see that 180 pounds of pork and 110 pounds of lamb will be required.

b) $\begin{pmatrix} 8 & 2 & 5 \\ 4 & 8 & 3 \end{pmatrix} \begin{pmatrix} 2.50 \\ 2.00 \\ 3.00 \end{pmatrix} = \begin{pmatrix} 39 \\ 35 \end{pmatrix}$.

The component in the answer, 39, is 8 pounds of beef times $2.50 per pound, plus 2 pounds of pork at $2 per pound, plus 5 pounds of lamb at $3 per pound, so the total cost of a batch of superior is $39. Similarly the total cost of a batch of regular is $35.

Problem Set 6–2

1. $\begin{pmatrix} 2 & 3 \\ 1 & 2 \end{pmatrix}\begin{pmatrix} x_1 \\ x_2 \end{pmatrix} = \begin{pmatrix} 5 \\ 3 \end{pmatrix}.$

2. $\begin{pmatrix} 1 & 2 \\ 2 & 3 \end{pmatrix}\begin{pmatrix} x_1 \\ x_2 \end{pmatrix} + \begin{pmatrix} y_1 \\ y_2 \end{pmatrix} = \begin{pmatrix} 10 \\ 12 \end{pmatrix}.$

3. $\begin{pmatrix} 3 & 0 & 1 & -1 \\ 2 & 1 & 0 & -5 \\ 0 & 1 & 3 & 1 \end{pmatrix}\begin{pmatrix} x_1 \\ x_2 \\ x_3 \\ x_4 \end{pmatrix} \le \begin{pmatrix} 5 \\ 10 \\ 8 \end{pmatrix}.$

4. $3x_1 + x_2 + 2x_3 = 5$
$x_1 + 4x_2 + x_3 = 4.$

5. $2x_1 + 3x_2 = 5$
$4x_1 + 6x_2 = 10$
$x_1 + 7x_2 = 6.$

6. $3x_1 + x_2 = 2$
$5x_1 + 4x_2 = 6.$

7. $p_{11}x_1 + p_{12}x_2 + p_{13}x_3 + p_{14}x_4 = q_1$
$p_{21}x_1 + p_{22}x_2 + p_{23}x_3 + p_{24}x_4 = q_2$
$p_{31}x_1 + p_{32}x_2 + p_{33}x_3 + p_{34}x_4 = q_3.$

8. $2x_1 + x_2 + 5x_3 + y_1 = 10$
$4x_1 + 6x_2 + 2x_3 + y_2 = 5.$

9. Maximize $3x_1 + 2x_2 + 4x_3$
subject to: $2x_1 + x_2 + 4x_3 \le 10$
$3x_1 + 5x_2 + 2x_3 \le 15$
$x_1 \ge 0$
$x_2 \ge 0$
$x_3 \ge 0.$

10. Maximize $2x_1 + 5x_2 + 4x_3 + 3x_4$
subject to: $x_1 + 3x_2 + 2x_3 + s_1 = 12 \quad x_1 \ge 0$
$x_2 + 3x_3 + 2x_4 + s_2 = 15 \quad x_2 \ge 0$
$5x_1 + 3x_2 + 2x_4 + s_3 = 10 \quad x_3 \ge 0$
$x_1 + 4x_2 + 6x_3 + s_4 = 20 \quad x_4 \ge 0$
$s_1 \ge 0$
$s_2 \ge 0$
$s_3 \ge 0$
$s_4 \ge 0.$

11. *a)* y must be q by 1 and g must be p by 1.
b) $Ay = g.$

Problem Set 6–3

1. *a)* 60% of customers who last purchased Eager will purchase Eager again and 40% will switch to Beaver.
b) (0.55 0.45) and (0.555 0.445). *c)* (5/9 Eager 4/9 Beaver).
2. *a)* (0.29 spender 0.71 saver). *b)* (1/8 spender 7/8 saver).

3. *a)* (0.01099 users   0.98901 non-users).

   b) This is an absorbing chain. The steady state is 100% users or the state vector (1   0).

4. The new state vector is the original *(a   b)* because the ones in the transition matrix mean that all in state #1 remain in #1 and all in #2 remain in #2.

5. (0.3   0.1   0.6); (0.1   0.6   0.3); (0.6   0.3   0.1).

Problem Set 6–4

1. $\begin{pmatrix} 1 & -3 \\ -2 & 7 \end{pmatrix}.$

2. $\begin{pmatrix} 1 & -4 \\ -2 & 9 \end{pmatrix}.$

3. $\begin{pmatrix} \dfrac{5}{4} & -\dfrac{1}{2} \\ -\dfrac{3}{4} & \dfrac{1}{2} \end{pmatrix}.$

4. $\begin{pmatrix} 2 & 1 \\ 1 & 1 \end{pmatrix}.$

5. Inverse does not exist.

6. Inverse does not exist.

7. $\begin{pmatrix} 0 & \dfrac{1}{2} \\ \dfrac{1}{3} & -\dfrac{1}{6} \end{pmatrix}.$

8. $\begin{pmatrix} -\dfrac{4}{7} & \dfrac{5}{7} \\ \dfrac{3}{7} & -\dfrac{2}{7} \end{pmatrix}.$

9. $\begin{pmatrix} -\dfrac{3}{2} & \dfrac{1}{2} \\ 1 & 0 \end{pmatrix}.$

10. $\begin{pmatrix} -\dfrac{5}{6} & \dfrac{3}{6} \\ \dfrac{2}{6} & 0 \end{pmatrix}.$

11. $\begin{pmatrix} 1 & 2 & 1 \\ 4 & 5 & -3 \\ 3 & 4 & -2 \end{pmatrix}.$

12. $\begin{pmatrix} \dfrac{3}{2} & -3 & -\dfrac{1}{2} \\ 2 & -3 & -1 \\ -2 & 4 & 1 \end{pmatrix}.$

13. Inverse does not exist.

14. $\begin{pmatrix} 2 & 2 & 3 \\ 0 & 1 & 1 \\ 1 & 1 & 1 \end{pmatrix}.$

15. $\begin{pmatrix} -\dfrac{4}{9} & \dfrac{5}{9} & \dfrac{2}{9} \\ -\dfrac{7}{9} & \dfrac{2}{9} & \dfrac{8}{9} \\ \dfrac{2}{3} & -\dfrac{1}{3} & -\dfrac{1}{3} \end{pmatrix}.$

16. Inverse does not exist.

Problem Set 6–5

1. *a)* $A = \begin{pmatrix} 8 & 5 \\ 3 & 2 \end{pmatrix}$; $x = \begin{pmatrix} x_1 \\ x_2 \end{pmatrix}$; $b = \begin{pmatrix} 2 \\ 1 \end{pmatrix}.$

   b) $A^{-1} = \begin{pmatrix} 2 & -5 \\ -3 & 8 \end{pmatrix}.$

   c) $\begin{pmatrix} x_1 \\ x_2 \end{pmatrix} = \begin{pmatrix} 2 & -5 \\ -3 & 8 \end{pmatrix}\begin{pmatrix} 2 \\ 1 \end{pmatrix}.$

d) (−1 2).

e) (1) (2 −3).   (2) (−5 8).   (3) (−3 5).

(4) (−14 23).   (5) (−11 17).

2. *a)* $A = \begin{pmatrix} 4 & 3 \\ 9 & 7 \end{pmatrix}$; $x = \begin{pmatrix} x_1 \\ x_2 \end{pmatrix}$; $b = \begin{pmatrix} 2 \\ 3 \end{pmatrix}$.

b) $A^{-1} = \begin{pmatrix} 7 & -3 \\ -9 & 4 \end{pmatrix}$.

c) $\begin{pmatrix} x_1 \\ x_2 \end{pmatrix} = \begin{pmatrix} 7 & -3 \\ -9 & 4 \end{pmatrix}\begin{pmatrix} 2 \\ 3 \end{pmatrix}$.

d) (5 −6).

e) (1) (7 −9).   (2) (−3 4).   (3) (4 −5).

(4) (11 −14).   (5) (−13 17).

3. *a)* $A = \begin{pmatrix} 6 & 8 \\ 2 & 3 \end{pmatrix}$; $x = \begin{pmatrix} x_1 \\ x_2 \end{pmatrix}$; $b = \begin{pmatrix} 3 \\ 1 \end{pmatrix}$.

b) $A^{-1} = \begin{pmatrix} 3/2 & -4 \\ -1 & 3 \end{pmatrix}$.

c) $\begin{pmatrix} x_1 \\ x_2 \end{pmatrix} = \begin{pmatrix} 3/2 & -4 \\ -1 & 3 \end{pmatrix}\begin{pmatrix} 3 \\ 1 \end{pmatrix}$.

d) (1/2 0).

e) (1) (−5/2 2).   (2) (−4 3).   (3) (3/2 −1).

(4) (−9 7).   (5) (−11/2 4).

4. $\begin{pmatrix} 1 & 7/5 \\ 1 & 8/5 \end{pmatrix}$.

5. *a)* $A = \begin{pmatrix} 3 & 0 & 5 \\ 2 & 2 & 5 \\ 0 & 1 & 1 \end{pmatrix}$; $x = \begin{pmatrix} x_1 \\ x_2 \\ x_3 \end{pmatrix}$; $b = \begin{pmatrix} 3 \\ 7 \\ 2 \end{pmatrix}$.

b) $A^{-1} = \begin{pmatrix} -3 & 5 & -10 \\ -2 & 3 & -5 \\ 2 & -3 & 6 \end{pmatrix}$.

c) $\begin{pmatrix} x_1 \\ x_2 \\ x_3 \end{pmatrix} = \begin{pmatrix} -3 & 5 & -10 \\ -2 & 3 & -5 \\ 2 & -3 & 6 \end{pmatrix}\begin{pmatrix} 3 \\ 7 \\ 2 \end{pmatrix}$.

d) (6 5 −3).

e) (1) (−3 −1 2).   (2) (−31 −15 19).   (3) (7 3 −4).

(4) (7 6 −4).   (5) (34 21 −20).

6. *a)* $A = \begin{pmatrix} 7 & 3 & 0 \\ 0 & 3 & 5 \\ 1 & 1 & 1 \end{pmatrix}$, $x = \begin{pmatrix} x_1 \\ x_2 \\ x_3 \end{pmatrix}$, $b = \begin{pmatrix} 1 \\ 2 \\ 3 \end{pmatrix}$.

b) $A^{-1} = \begin{pmatrix} -2 & -3 & 15 \\ 5 & 7 & -35 \\ -3 & -4 & 21 \end{pmatrix}$.

c) $\begin{pmatrix} x_1 \\ x_2 \\ x_3 \end{pmatrix} = \begin{pmatrix} -2 & -3 & 15 \\ 5 & 7 & -35 \\ -3 & -4 & 21 \end{pmatrix} \begin{pmatrix} 1 \\ 2 \\ 3 \end{pmatrix}.$

d) $(37 \quad -86 \quad 52).$

e) (1) $(16 \quad -37 \quad 22).$ (2) $(-13 \quad 31 \quad -18).$ (3) $(-34 \quad 79 \quad -47).$
    (4) $(-2 \quad 8 \quad -5).$ (5) $(3 \quad -7 \quad 4).$

7. $\begin{pmatrix} 3/2 & -3 & -1/2 \\ 2 & -3 & -1 \\ -2 & 4 & 1 \end{pmatrix}.$

8. b) (1) $(-3 \quad 0 \quad 6).$ (2) $(0 \quad 3 \quad 0).$

9. b) (1) $(1 \quad 0 \quad 0 \quad 0).$ (2) $(2 \quad -2 \quad 2 \quad -1).$

10. a) $x_1 + \qquad 2x_3 = H_1$
$x_2 + 3x_3 = H_2$
$x_1 + 2x_2 \qquad = H_3.$

b) $A = \begin{pmatrix} 1 & 0 & 2 \\ 0 & 1 & 3 \\ 1 & 2 & 0 \end{pmatrix}. \quad A^{-1} = \dfrac{1}{8}\begin{pmatrix} 6 & -4 & 2 \\ -3 & 2 & 3 \\ 1 & 2 & -1 \end{pmatrix}.$

c) $(130 \quad 35 \quad 15).$ d) $(200 \quad 100 \quad 100).$ e) $(208 \quad 24 \quad 56).$

Problem Set 6–6

1. a) $A = \begin{pmatrix} \dfrac{1}{5} & \dfrac{3}{10} \\ \dfrac{3}{5} & \dfrac{1}{10} \end{pmatrix}.$

b) One dollar of output of industry No. 1 requires ⅕ dollars of its own output and ⅗ dollars of industry No. 2 output.

c) $(I - A) = \begin{pmatrix} \dfrac{4}{5} & -\dfrac{3}{10} \\ -\dfrac{3}{5} & \dfrac{9}{10} \end{pmatrix}.$

d) $(I - A)^{-1} = \begin{pmatrix} \dfrac{5}{3} & \dfrac{15}{27} \\ \dfrac{10}{9} & \dfrac{40}{27} \end{pmatrix}.$

e) $X = \begin{pmatrix} \dfrac{5}{3} & \dfrac{15}{27} \\ \dfrac{10}{9} & \dfrac{40}{27} \end{pmatrix} \begin{pmatrix} d_1 \\ d_2 \end{pmatrix}.$

f) $x_1 = 675, \ x_2 = 900.$ g) $x_1 = 1350, \ x_2 = 1800.$

2. a) $\begin{pmatrix} x_1 \\ x_2 \end{pmatrix} = \begin{pmatrix} \dfrac{40}{27} & \dfrac{10}{9} \\ \dfrac{5}{9} & \dfrac{5}{3} \end{pmatrix} \begin{pmatrix} d_1 \\ d_2 \end{pmatrix}.$ b) $x_1 = 270, \ x_2 = 270.$

c) $x_1 = 540, \ x_2 = 540.$ d) $x_1 = 510, \ x_2 = 405.$

3. *a)* $x_1 = 2700, x_2 = 2720.$ *b)* $x_1 = 4500, x_2 = 4440.$

4. *a)* a_{ij} is the amount of industry i output required for \$1 output by industry j.

   b) a_{13} is the amount of industry No. 1 output required for \$1 output by industry No. 3.

   c) $A = \begin{pmatrix} a_{11} & a_{12} & a_{13} \\ a_{21} & a_{22} & a_{23} \\ a_{31} & a_{32} & a_{33} \end{pmatrix}.$

   d) $\begin{pmatrix} x_1 \\ x_2 \\ x_3 \end{pmatrix} = \left[\begin{pmatrix} 1 & 0 & 0 \\ 0 & 1 & 0 \\ 0 & 0 & 1 \end{pmatrix} - \begin{pmatrix} a_{11} & a_{12} & a_{13} \\ a_{21} & a_{22} & a_{23} \\ a_{31} & a_{32} & a_{33} \end{pmatrix} \right]^{-1} \begin{pmatrix} d_1 \\ d_2 \\ d_3 \end{pmatrix}$

   or

   $\begin{pmatrix} x_1 \\ x_2 \\ x_3 \end{pmatrix} = \begin{pmatrix} 1-a_{11} & -a_{12} & -a_{13} \\ -a_{21} & 1-a_{22} & -a_{23} \\ -a_{31} & -a_{32} & 1-a_{33} \end{pmatrix}^{-1} \begin{pmatrix} d_1 \\ d_2 \\ d_3 \end{pmatrix}.$

Problem Set 6–7

1. 22.
2. 15.
3. 230.

4. 45.
5. 36.
6. $\sum\limits_{j=1}^{q} j.$

7. $\sum\limits_{j=1}^{n} j^2.$
8. $\sum\limits_{j=1}^{4} j^3.$
9. $\sum\limits_{i=1}^{4} (i+2).$

10. $\sum\limits_{i=1}^{5} (3i).$

Problem Set 6–8

1. $y_1 + y_2 + y_3 + y_4 + y_5.$
2. $c_1 x_1 + c_2 x_2 + c_3 x_3.$
3. $b_1 y_1 + b_2 y_2 + \cdots + b_n y_n.$
4. $p_1 x_1 + p_2 x_2 + p_3 x_3 + p_4 x_4 + p_5 x_5 = 10.$
5. $a_1 x_1 + a_2 x_2 + \cdots + a_n x_n = c.$

6. $\sum\limits_{i=1}^{4} x_i.$

7. $\sum\limits_{i=1}^{6} a_i x_i = b.$

8. $\sum\limits_{i=1}^{9} c_i x_i.$

9. $\sum\limits_{i=1}^{q} a_i x_i.$

10. $n = 4, a_1 = 1, a_2 = 0, a_3 = 2, a_4 = 5, b = 7.$

Problem Set 6–9

1. $a_{11}x_1 + a_{12}x_2 + a_{13}x_3 + a_{14}x_4 = b_1$
    $a_{21}x_1 + a_{22}x_2 + a_{23}x_3 + a_{24}x_4 = b_2.$

2. Maximize $c_1x_1 + c_2x_2 + c_3x_3$
    subject to: $a_{11}x_1 + a_{12}x_2 + a_{13}x_3 \le b_1$
    $\qquad\qquad a_{21}x_1 + a_{22}x_2 + a_{23}x_3 \le b_2$
    $\qquad\qquad a_{31}x_1 + a_{32}x_2 + a_{33}x_3 \le b_3$
    $\qquad\qquad a_{41}x_1 + a_{42}x_2 + a_{43}x_3 \le b_4$
    $\qquad\qquad\qquad\qquad\qquad x_1 \ge 0$
    $\qquad\qquad\qquad\qquad\qquad x_2 \ge 0$
    $\qquad\qquad\qquad\qquad\qquad x_3 \ge 0.$

3. *a)* $a_{11}x_1 + a_{12}x_2 + \cdots + a_{1q}x_q = b_1$
    $\qquad a_{21}x_1 + a_{22}x_2 + \cdots + a_{2q}x_q = b_2$
    $\qquad \vdots \qquad\quad \vdots \qquad\qquad \vdots \qquad \vdots$
    $\qquad a_{p1}x_1 + a_{p2}x_2 + \cdots + a_{pq}x_q = b_p.$

    b) $\displaystyle\sum_{j=1}^{q} a_{ij}x_j = b_i; \quad i = 1, 2, \cdots, p.$

4. *a)* Maximize $c_1x_1 + c_2x_2 + \cdots + c_qx_q$
    subject to: $a_{11}x_1 + a_{12}x_2 + \cdots + a_{1q}x_q \le b_1$
    $\qquad\qquad a_{21}x_1 + a_{22}x_2 + \cdots + a_{2q}x_q \le b_2$
    $\qquad\qquad \quad \cdot \qquad\qquad \cdot \qquad\qquad\qquad \cdot$
    $\qquad\qquad a_{p1}x_1 + a_{p2}x_2 + \cdots + a_{pq}x_q \le b_p$
    $\qquad\qquad\qquad\qquad\qquad\qquad x_1 \ge 0$
    $\qquad\qquad\qquad\qquad\qquad\qquad x_2 \ge 0$
    $\qquad\qquad\qquad\qquad\qquad\qquad \cdot \qquad \cdot$
    $\qquad\qquad\qquad\qquad\qquad\qquad x_q \ge 0.$

    b) Maximize
    $$\sum_{j=1}^{q} c_j x_j$$
    subject to:
    $$\sum_{j=1}^{q} a_{ij}x_j \le b_i; \quad i = 1, 2, \cdots, p$$
    and $x_j \ge 0$ for all j.

5. $\displaystyle\sum_{j=1}^{8} a_{ij}x_j = b_i; \quad i = 1, 2, \cdots, 6.$

6. *a)* m equations, n variables; $m = 3$, $n = 4$.
    b) $a_{11} = 2$, $a_{12} = 3$, $a_{24} = 8$, $a_{23} = 9$, $a_{33} = 7$, $a_{34} = 6$.
    c) $a_{13} = 0$, $b_1 = 20$, $a_{31} = 5$, $a_{22} = 0$, $a_{32} = 0$.

7. *a)* $C = x_1d_1 + x_2d_2 + \cdots + x_{10}d_{10}.$

    b) $C = \displaystyle\sum_{i=1}^{10} x_i d_i.$

8. *a)* $C = x_1d_1 + x_2d_2 + \cdots + x_pd_p.$

    b) $C = \displaystyle\sum_{i=1}^{p} x_i d_i.$

Problem Set 6–10

1. *a)* 12. *b)* 62. *c)* 4.
   d) 14. *e)* 16. *f)* 62/3.
2. *a)* 24. *b)* 218. *c)* 8.
   d) 26. *e)* 64. *f)* 218/3.
3. *a)* 164. *b)* 560. *c)* 126.
   d) 400. *e)* 24.
4. *a)* 44. *b)* 108. *c)* 35.
   d) 81. *e)* 8.
5. *a)* See Figure A. *b)* $y = 1.9x - 4.5$. *c)* See Figure A.
6. *a)* See Figure B. *b)* $y = 4x/3$. *c)* See Figure B.

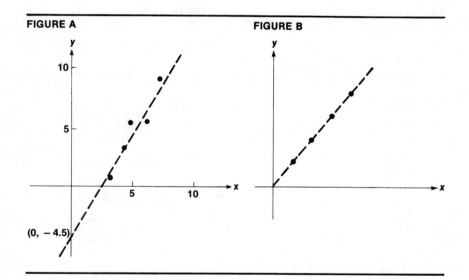

FIGURE A

(0, −4.5)

FIGURE B

7. $r = \dfrac{3(51) - 5(25)}{\sqrt{[3(13) - (5)^2][3(227) - (25)^2]}} = \dfrac{28}{\sqrt{784}} = 1$.

8. $r = 0.90$.
9. $\mu = 20$, $\sigma^2 = 4$, $\sigma = 2$.
10. $\mu = 10$, $\sigma^2 = 9$, $\sigma = 3$.
11. $\mu = 6$, $\sigma^2 = 100/9$, $\sigma = 10/3$.
12. *a)*

x	$x - \bar{x}$
3	−2
4	−1
8	3
15	0

$\bar{x} = \dfrac{15}{3} = 5$. $\sum (x - \bar{x}) = 0$.

b) $\sum_{i=1}^{n} (x_i - \bar{x}) = \sum x_i - \sum \bar{x} = \sum x_i - n\left(\dfrac{\sum x_i}{n}\right) = 0.$

13. a) $\sum_{i=1}^{n} (x_i - 5)^2 = \sum (x_i^2 - 10x_i + 25)$

$$= \sum x_i^2 - \sum 10x_i + \sum 25$$

$$= \sum x_i^2 - 10\sum x_i + 25n \qquad \text{because } \sum_{i=1}^{n} 25 = 25n$$

$$= \sum x_i^2 - 10(n\bar{x}) + 25n \qquad \text{because } \sum x_i = n\bar{x}.$$

 b) $(5-5)^2 + (4-5)^2 + (9-5)^2 = 25 + 16 + 81 - 10(3)(6) + 25(3)$
   $0 + 1 + 16 = 122 - 180 + 75$
   $17 = 17.$

14. a) $\bar{x} - b.$ \qquad\qquad b) $60 - 7 = 53.$

15. $\Sigma x_i^2 + 2n\bar{x} - 15n.$

16. At least $24/25$ or 96 percent of the grades were in the inclusive interval 57 to 87 percent.

17. At least 84 percent of the workers earn between \$4 and \$6, inclusive.

Problem Set 6–11

1. $\dfrac{1}{4}\begin{pmatrix} 2 & -4 \\ -3 & 8 \end{pmatrix}.$ \qquad 2. $-\dfrac{1}{11}\begin{pmatrix} -2 & -1 \\ 1 & 6 \end{pmatrix}.$ \qquad 3. $\dfrac{1}{2}\begin{pmatrix} 4 & -1 \\ -2 & 1 \end{pmatrix}.$

4. $\dfrac{1}{2}\begin{pmatrix} 4 & -5 \\ -2 & 3 \end{pmatrix}.$ \qquad\qquad 5. Matrix has no inverse.

6. Matrix has no inverse.

7. $\dfrac{1}{16}\begin{pmatrix} 5 & 9 & 8 \\ 7 & 3 & 8 \\ -1 & -5 & -8 \end{pmatrix}.$ \qquad 8. $\dfrac{1}{55}\begin{pmatrix} 6 & 14 & 3 \\ 4 & -9 & 2 \\ 9 & -34 & -23 \end{pmatrix}.$

9. $1\begin{pmatrix} 2 & 8 & -11 \\ -1 & -5 & 7 \\ 1 & 2 & -3 \end{pmatrix}.$ \qquad 10. $1\begin{pmatrix} 0 & -1 & 1 \\ -1 & 1 & 2 \\ 1 & 0 & -2 \end{pmatrix}.$

11. The matrix has no inverse. \qquad 12. $\dfrac{1}{4}\begin{pmatrix} 4 & 4 & 6 \\ 0 & 2 & 2 \\ 8 & 0 & 6 \end{pmatrix}.$

13. $-\dfrac{1}{58}\begin{pmatrix} 24 & -6 & -8 \\ -18 & -10 & 6 \\ -20 & 5 & -3 \end{pmatrix}.$ \qquad 14. The matrix has no inverse.

CHAPTER 7

Problem Set 7–1

Answers to Problems 1–10 were found on a hand calculator.

1. 1.58496. \qquad 2. 0.63093. \qquad 3. 0.69897.
4. 0.77815. \qquad 5. 0.11103. \qquad 6. 0.01230.

7.   18.85418.   8.   9.00647.   9.   0.84510.

10.   0.47712.

Answers to Problems 11–14 are based on Table 7–1.

11. $\log 125 = \log 5^3 = 3 \log 5 = 3(0.698970) = 2.096910$.

12. $\log 27 = \log 3^3 = 3 \log 3 = 3(0.477121) = 1.431363$.

13. $\log 81 = \log 9^2 = 2 \log 9 = 2(0.954243) = 1.908486$.

14. $\log 625 = \log 5^4 = 4 \log 5 = 4(0.698970) = 2.795880$.

Answers to Problems 15–20 were found on a hand calculator.

15.   1.53724.   16.   4.48289.   17.   −3.00000.

18.   −1.66096.   19.   14.27491.   20.   54.44869.

Problem Set 7–2

1. $\log_3 9 = 2$.   2. $\log_2 32 = 5$.   3. $3^x = N$.

4. $2^y = N$.   5. $\log_{1/2} 0.125 = 3$.   6. $\log_{0.2} 0.04 = 2$.

7. $16^{1/2} = 4$.   8. $27^{1/3} = 3$.   9. $\log_2 0.125 = -3$.

10. $\log_3 \left(\dfrac{1}{9}\right) = -2$.   11. $5^{-2} = 0.04$.   12. $2^{-4} = 0.0625$.

13. $10^2 = 100$.   14. $e^{2.9957} = 20$.   15. $e^{2.3026} = 10$.

16. $10^1 = 10$.   17. $10^{-0.3010} = 0.5$.   18. $e^{-0.6931} = 0.5$.

19.   1.   20.   1.   21.   2.

22.   3.   23.   1.   24.   1.

25.   1/2.   26.   1/3.

27. $\ln (e^{\ln x}) = \ln 2$ 28. $\log(10^{\log x}) = \log 3$

      $\ln x (\ln e) = \ln 2$ $\log x (\log 10) = \log 3$

         $\ln x = \ln 2$ $\log x = \log 3$

  antiln $(\ln x) =$ antiln $(\ln 2)$ antilog$(\log x) =$ antilog$(\log 3)$

           $x = 2$. $x = 3$.

29.   5.   30.   4.   31.   6.   32.   5.

33. $\ln \left(\dfrac{x^2 z^3}{y^2}\right)$.   34. $\log \left(\dfrac{x^5 y^{1/2}}{z^2}\right)$.

35.   4.4.   36.   −0.6.   37.   5.4.   38.   6.8.

39.   −1.8.   40.   −2.4.   41.   2.8.   42.   1.

43.   −3.5.   44.   −4.66667.   45. $\ln x^2 y^3$.

46. $\ln \left(\dfrac{x^3}{y^2}\right)$.   47. $\ln x$.   48. $2 \ln x$.

49. $\ln y^{1/x}$.   50. $\ln x^{1/y}$.   51.   1.5220.

52.   3.3201.   53.   0.3329.   54.   0.9900.

55.   0.3499.   56.   0.7183.   57.   5.2131.

58.   4.7795.   59.   11.23181.   60.   6.11626.

61.   0.14870.   62.   0.20112.   63.   0.09531.

64.   0.26246.

Problem Set 7–3 (Logarithms taken from Table I.)

1. 1.5391.	2. 0.5391.	3. 0.5391 − 1.
4. 0.5391 − 2.	5. 4.5391.	6. 0.9557 − 1.
7. 0.0043.	8. 2.4116.	9. −2.
10. 0.0969 − 2.	11. 3.	12. −1.
13. 0.786.	14. 132.	15. 2.3456.
16. 3.3692.	17. 0.6794 − 3.	18. 0.
19. 0.9943.	20. 2.7810.	21. 0.5051 − 2.
22. 0.8451 − 6.	23. 4.3892.	24. 0.7782.
25. −1.	26. 1.	27. 0.3075.
28. 2.7543.	29. 0.0043 − 4.	30. 0.5159.
31. −4.	32. 1.7559.	33. 0.6990.
34. 1.2041.	35. 0.2041.	36. 3.0934.
37. 2.0453.	38. 1.1931.	39. 1.0792.
40. 0.6021 − 2.	41. 0.1430 − 1.	42. 0.2304 − 1.
43. 0.2430.	44. 2.9186.	45. 0.7528 − 2.
46. 5.1903.	47. 0.3010 − 3.	48. 0.1367 − 1.
49. 0.7218.	50. 0.6794 − 9.	

Problem Set 7–4 (Logarithms taken from Table I.)

1. 28.	2. 280.	3. 0.08.	4. 0.8.
5. 7.	6. 2.9.	7. 25,800.	8. 5480.
9. 0.00407.	10. 0.000305.	11. 0.0174.	12. 0.174.
13. 1060.	14. 103,000.	15. 1.88.	16. 7.08.
17. 1.	18. 1.	19. 1.01.	20. 1.02.
21. 0.00102.	22. 0.0101.	23. 0.506.	24. 0.316.
25. 10,800.	26. 1040.		

Problem Set 7–5 (Logarithms taken from Table I.)

1. 3.5866.	2. 68.7.	3. 9140.	4. 0.7559 − 3.
5. 0.9731 − 1.	6. 5.37.	7. 8.71.	8. 0.8633 − 1.
9. 0.3010.	10. 1000.	11. 100.	12. 0.4771.
13. 0.	14. 10.	15. 1.95.	16. 0.4624 − 1.

Problem Set 7–6 (Logarithms taken from Table I.)

1. 1.3918.	2. 1.9536.	3. 0.5018 − 2.
4. 3.5148.	5. 0.6930 − 3.	6. 2.378.
7. 3.935.	8. 3.062.	9. 0.08072.
10. 0.05544.	11. 0.7839 − 1.	12. 4.055.
13. 0.1357.	14. 23.28.	15. 0.002328.
16. 0.0149.	17. 10.84.	18. 2.5413.
19. 4.093.	20. 0.6791 − 1.	21. 3.004.

22. 0.6868.	23. 72,780.	24. 0.5713 − 3.
25. 0.00838.	26. 0.7499.	27. 0.7068.
28. 1.0022.	29. 5517.	30. 0.002367.

Problem Set 7–7 (Logarithms taken from Table I.)

1. 934.	2. 2.356.	3. 47.36.
4. 1.454.	5. 0.00007002.	6. 145.3.
7. 1.673.	8. 0.002978.	9. 63.91.
10. 6.730.	11. 50.58.	12. 65.77.
13. 0.8898.	14. 0.00004106.	15. 2.154.
16. 49.46.	17. 5.773.	18. 1.817.
19. 5.241.	20. 0.9318.	21. 5.648.
22. 66.60.	23. 0.4810.	24. 0.004507.
25. 0.007792.	26. 7.905.	27. 0.6491.
28. 2.432.	29. 6.374.	30. 0.5409.
31. 4.644.	32. 1.5439.	33. −0.7924.
34. −0.6309.	35. −2.3223.	36. −1.2920.
37. 0.5373.	38. 0.5886.	39. 0.149.
40. 0.149.	41. −0.2752.	42. −0.0876.
43. 1.	44. 0.6309.	45. 0.246.
46. 0.072.	47. −0.1294.	48. −0.1398.
49. 0.324.	50. 3.17.	

CHAPTER 8

Problem Set 8–1

1. *a)* $35.	*b)* $535.	2. *a)* $80.	*b)* $1080.
3. *a)* $45.	*b)* $1045.	4. *a)* $60.	*b)* $2060.
5. *a)* $12.	*b)* $112.	6. *a)* $30.	*b)* $530.
7. *a)* $36.	*b)* $236.	8. *a)* $120.	*b)* $620.
9. *a)* $3600.	*b)* $8600.	10. *a)* $2400.	*b)* $6400.
11. 15 months.	12. 18%.	13. 8.82%.	14. $714.29.
15. 30 months.	16. 20 months.	17. 16%.	18. 5%.
19. $400.	20. $854.70.	21. $904.98.	22. $869.57.
23. $1,640.	24. $466.25.	25. $2580.65.	26. 12%.

Problem Set 8–2 (Answers computed on a hand calculator.)

1. $530.66.	2. $537.25.	3. $8689.81.	4. $1379.52.
5. $282.68.	6. $357.69.	7. $2880.61.	8. $4801.02.
9. 20.5 years.	10. 7.3 years.	11. 3.526%.	12. 11.25%.
13. $6851.15.	14. $2012.99.	15. 8.0 years.	16. 17.46%.

Problem Set 8–3 (Answers computed on a hand calculator.)

1. $214.55. 2. $1016.70. 3. $3069.57. 4. $2795.70.
5. $574.37. 6. $1747.31. 7. 78.4 units per man-hour.
8. $5000. 9. $852.16. 10. $815.87.

Problem Set 8–4 (Answers computed on a hand calculator.)

1. $21,538.44. 2. $60,401.98. 3. $81,990.98.
4. $396,566.67. 5. $6540.57. 6. $30,107.50.
7. $37,816.56. 8. $15,046.30. 9. $23,787.71.
10. $4307.69. 11. *a)* $6000. 11. *b)* $5287.67.
12. *a)* $53,373.88. 12. *b)* $125,000. 13. $312,232.31.
14. $86,126.35.

Problem Set 8–5 (All answers except 5 and 6 were computed on a hand calculator.)

1. $47.07. 2. $200.61.
3. *a)* $10,758.49.
   b) $6000 for interest; $4758.49 reduction of balance owed.
   c) $5714.49 for interest; $5044.00 reduction of balance owed.
4. *a)* $235.40. *b)* $120 interest; $115.40 reduction of balance owed.
   c) $117.69 interest; $117.71 reduction of balance owed.
5. *a)* $12,587.95.
   b) $11,250 interest; $1337.95 reduction of balance owed.
6. *a)* $275.35. *b)* $255.21 interest; $20.14 reduction of balance owed.
7. $4,828.51. 8. $1576.44. 9. $6902.95.
10. $687.30. 11. $200,443.11. 12. $50,886.21.

Problem Set 8–6 (All answers were computed on a hand calculator.)

1. $1146.39. 2. $232.50. 3. $11,040.20.
4. $800.80. 5. $2145.48. 6. $221,224.81.
7. 38,857. 8. $178.60. 9. $6074.34.
10. $115,573.86. 11. $3257.79. 12. $226.11.
13. $3888.44. 14. $12,942.29. 15. $10,119.86.
16. $39,296.11. 17. $1414.59. 18. $11,726.87.
19. $7357.48.

Problem Set 8–7

1. 6.3%. 2. 6.9%. 3. 7.8%. 4. 5.7%.
5. 6.29%. 6. 6.93%. 7. 7.75%. 8. 5.72%.
9. 8.24%. 10. 10.25%. 11. 12.68%. 12. 16.99%.
13. 10.47%. 14. 7.19%. 15. 19.56%. 16. 9.31%.
17. 9.38%. 18. 9.42%.

Problem Set 8–8 (All answers were computed on a hand calculator.)

1.	$1349.86.	2.	$1112.77.	3.	$7166.65.
4.	$5619.79.	5.	$341.93.	6.	$799.55.
7.	$913.93.	8.	$2582.12.	9.	0.0513.
10.	0.0618.	11.	0.0725.	12.	0.0833.
13.	0.1133.	14.	0.0862.	15.	0.0488.
16.	0.0953.	17.	$13,497.48.	18.	$15,348.58.
19.	$205.37.	20.	$19,480.87.	21.	$195.82.
22.	$4317.11.	23.	$23,085.36.	24.	$21,225.87.
25.	$31,267.68.	26.	$1185.51.	27.	$4782.21.
28.	$196.80.	29.	About $1,575,839,260.		
30.	$5444.39.	31.	0.0822.	32.	0.0790.
33.	0.0801.	34.	$86,697.75.	35.	$109.16.
36.	$217.32.	37.	$13,467.89.		

CHAPTER 9

Problem Set 9–1

1. *a)* 7. *b)* −8. *c)* $3a − 2$. *d)* $(3a − 2)^2 = 9a^2 − 12a + 4$.
   e) $3ab − 2$. *f)* $9y + 10$. *g)* $3x + 1$. *h)* 3.
2. *a)* 10. *b)* 0. *c)* 1.75. *d)* $(6a + 4)/a^2$.
   e) $x^2 + 4x + 1.75$. *f)* $a^2 + 5a + 4$. *g)* $a^2 + a − 2$.
   h) $2x + 2$.
3. *a)* $a^2y − y^2$. *b)* $x^2 − a^2$. 4. *a)* $2x + 15$. *b)* $6 + 5y$.
5. *a)* 0.25. *b)* 1/3. *c)* $(2/a) − (3/a^2) = (2a − 3)/a^2$.
   d) $[2/(x + 1)] − [3/(x + 1)^2] = (2x − 1)/(x + 1)^2$.
6. *a)* 5. *b)* 131/16. *c)* 13.
7. *a)* Δx means the change in x.
   b) $\Delta f(x)$ means the change in $f(x)$ when x changes by Δx.
8. 0.75. 9. −0.98. 10. $−3(\Delta x)$. 11. $m(\Delta x)$.
12. *a)* $2x(\Delta x) + (\Delta x)^2$. *b)* $4x(\Delta x) + 2(\Delta x)^2 − 3(\Delta x)$.
13. *a)* $4x(\Delta x) + 2(\Delta x)^2$. *b)* $2x(\Delta x) + (\Delta x)^2 + 2(\Delta x)$.
14. *a)* $0.02 + 2.99$. *b)* $3.19. *c)* $3.99.
15. *a)* $0.2g + 0.9$. *b)* $2.90. *c)* $10.90.

Problem Set 9–2

1. 1. 2. −13. 3. ab^2. 4. a^4. 5. 0. 6. 1.
7. Limit does not exist. 8. Limit does not exist. 9. $(a − 1)^{1/3}$.
10. 5. 11. $8a$. 12. $4b$. 13. 6. 14. 10. 15. 0.
16. Limit does not exist. 17. Limit does not exist. 18. 0.
19. $2x$. 20. $3x^2$. 21. $−1/4$. 22. $−1/25$.

Problem Set 9–3

1. $f'(x) = 3;\ f'(1) = 3.$
2. $f'(x) = -0.5;\ f'(1) = -0.5.$
3. $f'(x) = 2x - 2;\ f'(1) = 0.$
4. $f'(x) = 6x - 12;\ f'(3) = 6.$
5. $f'(x) = -1/x^2;\ f'(2) = -1/4.$
6. $f'(x) = -2/x^3;\ f'(-1) = 2.$

7. $f'(x) = \lim\limits_{\Delta x \to 0} \dfrac{(x + \Delta x)^5 - 2(x + \Delta x)^4 - x^5 + 2x^4}{\Delta x}.$

8. $f'(x) = \lim\limits_{\Delta x \to 0} \dfrac{3(x + \Delta x)^6 + 2(x + \Delta x)^3 - 3x^6 - 2x^3}{\Delta x}.$

9. $f'(x) = \lim\limits_{\Delta x \to 0} \dfrac{(x + \Delta x)^{1/3} - x^{1/3}}{\Delta x}.$

10. $f'(x) = \lim\limits_{\Delta x \to 0} \dfrac{(x + \Delta x)^{1/2} - x^{1/2}}{\Delta x}.$

Problem Set 9–4

1. 0.   2. 1.   3. 1/2.   4. 2.   5. 1/3.   6. $-3/2.$
7. $6x + 2.$ 8. $x^2 - x + 1.$ 9. $0.02x + 2.$ 10. $1.5x^2 - x.$
11. $m.$ 12. $2ax + b.$ 13. $(-2/x^2) + (2/x^3).$
14. $(-1/2x^2) - (3/x^4).$ 15. $3x^{1/2} - (2/x^{1/2}) + (2/x^{1/3}) + 2.$
16. $(2/3x^{2/3}) + (4/3)x^{1/3} + 6x^{0.2} + 1.$
17. $(-1/3x^2) - (1/x^{3/2}).$ 18. $(-1/x^{4/3}) + (3/x^{5/4}).$
19. $6x - 2.$ 20. $20y.$ 21. $a.$ 22. 0.
23. $2kh - (a/3h^{2/3}).$ 24. $(-a/m^2) - 6m.$
25. 0.   26. 0.   27. 2.   28. $-4.$ 29. 1.   30. 0.
31. 0.   32. $-1.$ 33. 5/3.   34. 1.
35. $x = 1/2.$ 36. $x = 100.$ 37. $x = 1/8.$
38. $x = 9/16.$ 39. $x = 3.$ 40. $x = 4.$
41. $f(x)$ is a parabola opening upward with vertex (minimum) at (0, 5).
42. $f(x)$ is a parabola opening downward with vertex (maximum) at (0, 10).
43. $f(x)$ is a parabola opening downward with vertex (maximum) at (5, 10).
44. $f(x)$ is a parabola opening upward with vertex (minimum) at (3, 4).
45. *a)* 5000 barrels.   *b)* $130,000.   *c)* $P(x)$ is a parabola opening downward, so its vertex is at the maximum.
46. *a)* 10 barrels.   *b)* $40 per barrel.   *c)* $c(y)$ is a parabola opening upward, so its vertex is at the minimum.
47. *a)* $4 per yard.   *b)* $6 per yard.
48. *a)* $55 per ton.   *b)* $40 per ton.   *c)* $55 per ton.

49. $\dfrac{d}{dx}[f(x) + g(x)] = \lim\limits_{\Delta x \to 0}\left[\dfrac{f(x + \Delta x) + g(x + \Delta x) - f(x) - g(x)}{\Delta x}\right]$

$$= \lim\limits_{\Delta x \to 0}\left[\dfrac{f(x + \Delta x) - f(x) + g(x + \Delta x) - g(x)}{\Delta x}\right]$$

$$= \lim_{\Delta x \to 0} \left[\frac{f(x + \Delta x) - f(x)}{\Delta x} + \frac{g(x + \Delta x) - g(x)}{\Delta x} \right]$$

$$= \lim_{\Delta x \to 0} \left[\frac{f(x + \Delta x) - f(x)}{\Delta x} \right] + \lim_{\Delta x \to 0} \left[\frac{g(x + \Delta x) - g(x)}{\Delta x} \right]$$

$$= f'(x) + g'(x).$$

Problem Set 9–5

1. *a)* 50 gallons.   *b)* $5 per gallon.   *c)* $A''(50) = 0.0016$ is positive, so we have a minimum.

2. *a)* $x = 25$ feet.   *b)* $A''(25) = -2$ is negative, so we have a maximum.
    c) 25 feet by 25 feet, a square.   *d)* 625 square feet.

3. *a)* $p = 0.50$.
    b) Because n is a number of people, it is positive. Therefore, $V''(0.5) = -2/n$ is negative, so we have a maximum.
    c) $V(0.5) = 0.025$. $V(0.1) = 0.009$.

4. *a)* 2000 gallons.
    b) $P''(2000) = -0.01$ is negative, so we have a maximum.
    c) $20,000.

5. $(2.5, 6.25)$ is a local maximum because $f''(2.5) = -2$ is negative.

6. $(3, 2)$ is a local minimum because $f''(3) = 4$ is positive.

7. $(0, 3)$ is a local maximum because $f''(0) = -2$ is negative.

8. $(0, 20)$ is a local minimum because $f''(0) = 4$ is positive.

9. $(5, 85)$ is a local maximum because $f''(5) = -6$ is negative.

10. $(0.5, 0.75)$ is a local minimum because $f''(0.5) = 2$ is positive.

11. $(0, 12)$ is a local maximum because $f''(0) = -24$ is negative. $(8, -244)$ is a local minimum because $f''(8) = 24$ is positive.

12. $(0, 2)$ is a local maximum because $f''(0) = -6$ is negative. $(2, -2)$ is a local minimum because $f''(2) = 6$ is positive.

13. $f(x)$ has no local extreme points.

14. $f(x)$ has no local extreme points.

15. $(5, -5)$ is a local maximum because $f''(5) = -18$ is negative. $(2, -32)$ is a local minimum because $f''(2) = 18$ is positive.

16. $(3, 0)$ is a local maximum because $f''(3) = -6$ is negative. $(1, -4)$ is a local minimum because $f''(1) = 6$ is positive.

17. $f(x)$ has no local extreme points.

18. $f(x)$ has no local extreme points.

19. $(2, 52)$ is a local minimum because $f''(2) = 48$ is positive.

20. $(3, 243)$ is a local maximum because $f''(3) = -108$ is negative.

21. $(1, 5)$ is a local maximum because $f''(1) = -12$ is negative.

22. $(3, 3)$ is a local minimum because $f''(3) = 36$ is positive.

23. $(4, -8)$ is a local minimum because $f''(4) = 0.25$ is positive.

24. $(27, 27)$ is a local maximum because $f''(27) = -2/81$ is negative.

25. $(4, -16)$ is a local minimum because $f''(4) = 0.75$ is positive.

26. (125, 25) is a local maximum because $f''(125) = -0.000355$ is negative.
27. (4, 32) is a local minimum because $f''(4) = 2$ is positive. $(-4, -32)$ is a local maximum because $f''(-4) = -2$ is negative.
28. (5, 30) is a local minimum because $f''(5) = 2.4$ is positive.

Problem Set 9–6

1. *a)* 50 by 66 feet.   *b)* $33,000.
2. *a)* 75 by 99 feet.   *b)* 7425 square feet.
3. *a)* $C(x) = \left(2x + 2\dfrac{A}{x}\right)e + ix.$ *b)* $x = \sqrt{\dfrac{2Ae}{i + 2e}}.$
4. *a)* $A(x) = x\left(\dfrac{D - 2ex - ix}{2e}\right).$ *b)* $x = \dfrac{D}{2(2e + i)}.$
5. *a)* 90 by 150 feet.   *b)* $90,000.
6. $210.   7. 500 square feet.
8. 18×12 inches; area = 216 square inches.
9. 1100 by 660 feet; increase would be 10,000 square feet.
10. *a)* $A = 2x^2 + 4xL.$ *b)* $A(x) = 432x - 14x^2.$
     c) $x = 108/7$; $A''(x) = -28$ is always negative, proving we have a maximum.
     d) 108/7 by 108/7 by 3(108/7), or about 15.4 by 15.4 by 46.3 inches.
11. $x = M/7$; $L = 3M/7.$
12. A cube 4 by 4 by 4 inches; area = 96 square inches.
13. 10 by 10 by 5 inches; area = 300 square inches.
14. 5 by 5 by 10 inches; minimum cost = $300.
15. $2.785.
16. One-inch squares.
17. 10/3 by 10/3 inches.   18. $r = 20$ feet; $h = 40$ feet.
19. $r = 24$ feet; $h = 250/9 = 27.8$ feet.
20. Maximum group charge $1512.50 for 55 persons or more.
21. *a)* $80r^2.$ *b)* $24 - 2r.$ *c)* $T(r) = 80r^2(24 - 2r).$
     d) $r = 8$ miles.   *e)* $40,960.
22. *a)* $N/Q.$ *b)* $cN/Q.$ *c)* $uQ.$
     d) $p(uQ)/2.$ *e)* $S(Q) = [cN/Q] + [p(uQ)/2].$
     f) $Q = \sqrt{2cN/pu}.$ *g)* 240 units.
23. *a)* $6Kx^2.$ *b)* $W/x^3.$ *c)* $(W/x^3)(6x^2d) = 6Wd/x.$
     d) $C(x) = 6Kx^2 + 6Wd/x.$
     e) $x = \sqrt[3]{\dfrac{Wd}{2K}}.$
     f) $x = 8$ feet.
24. *a)* $x^* = (k_1/k_2)^{1/2}.$ *b)* 35 miles per hour.
25. *a)* $x^* = [(p + k_1d)/k_2d]^{1/2}.$ *b)* 53.7 miles per hour.
26. 11,664 cubic inches.

Problem Set 9–7

1. $30(6x-5)^4$.
2. $10(2x+6)^4$.
3. $6(2x)^2$.
4. $2/(6x)^{2/3}$.
5. $2/(4x)^{1/2}=1/x^{1/2}$.
6. $12(9x)^{1/3}$.
7. $12(8x-3)^{1/2}$.
8. $20(12x-9)^{2/3}$.
9. $15(x-1)(3x^2-6x+2)^{3/2}$.
10. $4(x^2-2x+2)(x^3-3x^2+6x)^{1/3}$.
11. $1/(2x-3)^{1/2}$.
12. $4x/(3x^2+5)^{1/3}$.
13. $-8/(2x-3)^2$.
14. $-18/(3x-5)^2$.
15. $\left(\dfrac{-2}{x^2}\right)\left(\dfrac{1}{x}-2\right)$.
16. $\left(\dfrac{6}{x^3}\right)\left(5-\dfrac{1}{x^2}\right)^2$.
17. $-54/(3x-5)^3$.
18. $-72/(2x+10)^4$.
19. $5-30/(3x+2)^2$.
20. $0.1+1/(5-0.2x)^2$.
21. $3-1/(5+2x)^{3/2}$.
22. $-2-3/2(3x-7)^{3/2}$.
23. $28/3$.
24. $13/2$.

Problem Set 9–8

1. *a)* $y^*=400$ barrels.   *b)* \$164 per barrel.
2. *a)* $x^*=200$ barrels.   *b)* \$150 per barrel.
3. *a)* $x^*=52.5$ thousand gallons.   *b)* \$14.5 thousand.
4. *a)* $t^*=10$ hours.   *b)* 2.5 hundred pounds.
5. Q is $Q(0, 125)$.   6. Q is $Q(0, 25)$.
7. *a)* 4 miles.   *b)* \$24.5 hundred thousand, or \$2,450,000.
   c) \$1.60077 hundred thousand or \$160,077.
8. *a)* 8 miles.   *b)* \$47.64 hundred thousand, or \$4,764,000.
   c) \$2.21639 hundred thousand, or \$221,639.

9. *a)* $x^*=\dfrac{\left(\dfrac{ab}{k}\right)^{1/2}-c}{b}$.

   b) $f''(x)=\dfrac{2ab^2}{(bx+c)^3}$ is positive, so we have a local minimum.

10. *a)* $x^*=\dfrac{\left(\dfrac{a}{2k}\right)^2-b}{a}=\dfrac{a}{4k^2}-\dfrac{b}{a}$.

   b) $f''(x)=-\dfrac{a^2}{4(ax+b)^{3/2}}$ is negative, so we have a local maximum.

Problem Set 9–9

1. $12x+11$.   2. $19-42x$.
3. $9x^2-10x+6$.   4. $3(5-4x-5x^2)$.
5. $4x(x-1)^3+(x-1)^4=(x-1)^3(5x-1)$.
6. $3x^2(x+5)^2+2x(x+5)^3=5x(x+5)^2(x+2)$.
7. $\dfrac{3x^2(x+3)^{1/2}}{2}+2x(x+3)^{3/2}=x(x+3)^{1/2}\left(\dfrac{7x}{2}+6\right)$.
8. $\dfrac{4x^3}{(6x-1)^{1/3}}+3x^2(6x-1)^{2/3}=\dfrac{x^2(22x-3)}{(6x-1)^{1/3}}$.

9. $\dfrac{4x^2}{(3x^2+7)^{2/3}} + 2(3x^2+7)^{1/3} = \dfrac{2(5x^2+7)}{(3x^2+7)^{2/3}}$.

10. $\dfrac{9x^3}{(2x^3+5)^{1/2}} + 3(2x^3+5)^{1/2} = \dfrac{15(x^3+1)}{(2x^3+5)^{1/2}}$.

11. $-\dfrac{1}{(x-1)^2}$.

12. $\dfrac{3}{(x+1)^2}$.

13. $\dfrac{2x(x+3)}{(2x+3)^2}$.

14. $\dfrac{3x^2(2x+5)}{(3x+5)^2}$.

15. $\dfrac{3-2x^2}{(3+2x^2)^2}$.

16. $\dfrac{3(2x^2+1)}{(1-2x^2)^2}$.

17. $\dfrac{(3x+2)^{1/2} - \dfrac{3x(3x+2)^{-1/2}}{2}}{3x+2} = \dfrac{3x+4}{2(3x+2)^{3/2}}$.

18. $\dfrac{2(2x+3)^{1/2} - 2x(2x+3)^{-1/2}}{2x+3} = \dfrac{2(x+3)}{(2x+3)^{3/2}}$.

19. $\dfrac{2(x^2+5)^{1/3} - \dfrac{2x(2x+1)(x^2+5)^{-2/3}}{3}}{(x^2+5)^{2/3}} = \dfrac{2(x^2-x+15)}{3(x^2+5)^{4/3}}$.

20. $\dfrac{-5(x^3+2)^{1/3} - x^2(3-5x)(x^3+2)^{-2/3}}{(x^3+2)^{2/3}} = -\dfrac{3x^2+10}{(x^3+2)^{4/3}}$.

21. 40 miles per hour.

22. 30 miles per hour.

23. $(k_1/k_2)^{1/2}$ miles per hour.

24. $p^* = 0.10$ yields a maximum because $C'(0.1) = 0$ and $C''(0.1)$ is negative.

25. $p^* = 1/n$.

26. *a)* $x^* = 200$ pounds.   *b)* \$1000.

27. *a)* $x^* = 800$ pounds.   *b)* \$8000.

28. *a)* $(x-100)(y-60) = 150{,}000$.

   b) $y = \dfrac{150{,}000}{x-100} + 60$.

   c) $A(x) = \dfrac{150{,}000x}{x-100} + 60x$.   *d)* $x^* = 600$, $y^* = 360$ feet.

   e) 216,000 square feet.

29. Both sides should be 28.28 inches.

Problem Set 9–10

1. See Figure A.   2. See Figure B.   3. See Figure C.

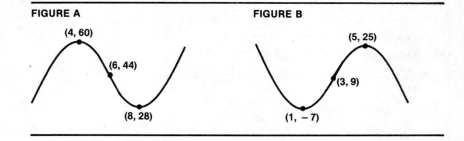

FIGURE A

(4, 60)

(6, 44)

(8, 28)

FIGURE B

(5, 25)

(3, 9)

(1, −7)

4. See Figure D.   5. See Figure E.   6. See Figure F.
7. See Figure G.   8. See Figure H.   9. See Figure I.
10. See Figure J.   11. See Figure K.   12. See Figure L.

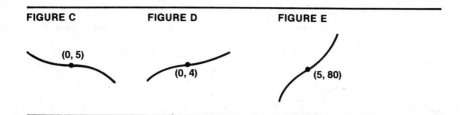

FIGURE C   (0, 5)

FIGURE D   (0, 4)

FIGURE E   (5, 80)

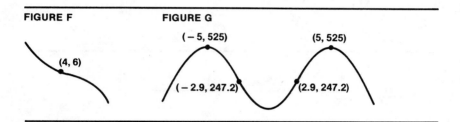

FIGURE F   (4, 6)

FIGURE G
(−5, 525)   (5, 525)
(−2.9, 247.2)   (2.9, 247.2)

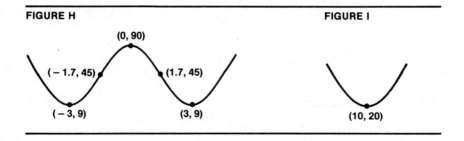

FIGURE H
(0, 90)
(−1.7, 45)   (1.7, 45)
(−3, 9)   (3, 9)

FIGURE I
(10, 20)

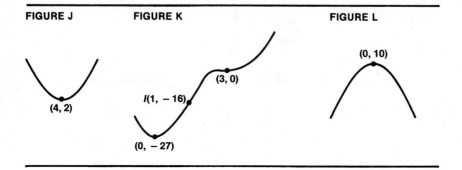

FIGURE J
(4, 2)

FIGURE K
(3, 0)
I(1, −16)
(0, −27)

FIGURE L
(0, 10)

13. See Figure M. 14. See Figure N. 15. See Figure O.
16. See Figure P. 17. See Figure Q. 18. See Figure R.

FIGURE M

FIGURE N

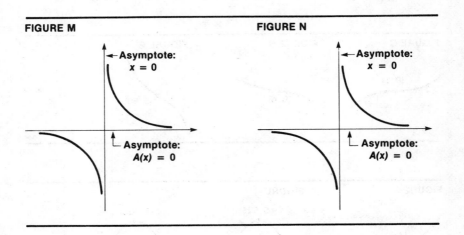

FIGURE O

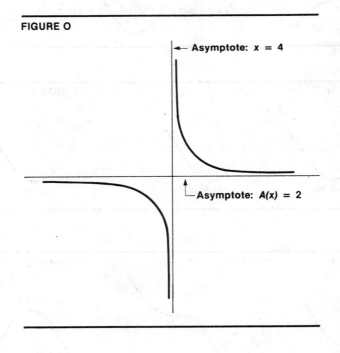

FIGURE P

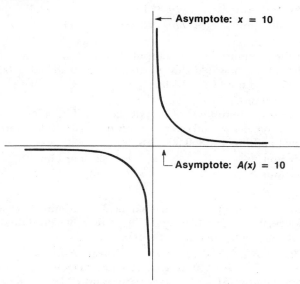

← Asymptote: x = 10

Asymptote: $A(x)$ = 10

FIGURE Q   **FIGURE R**

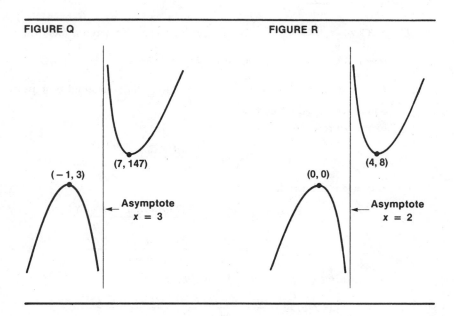

(7, 147)

(−1, 3)

← Asymptote
x = 3

(4, 8)

(0, 0)

← Asymptote
x = 2

Problem Set 9–11

1. *a)* $q^* = 100$ units.   *b)* $400.   *c)* $4.375 per unit.
2. *a)* $q^* = 400$ units.   *b)* $20,000.   *c)* $65 per unit.
3. *a)* $q^* = 70$ units.
    b) At the optimum level of output there will be a loss of $149.
    c) The producer would shut down and take a loss of $100 rather than operate with a loss of $149.
4. *a)* $q^* = 500$ units.
    b) At the optimum level of output there will be a loss of $10,000.
    c) The producer would operate with the $10,000 loss rather than shut down and incur a $60,000 loss.
5. $15 per unit.   6. $20 per unit.
7. $13 per unit.   8. $148.02 per unit.
9. *a)* $P(q) = R(q) - C(q)$.
    b) $P'(q) = R'(q) - C'(q)$. To have a local maximum, $P'(q)$ must be zero. Hence, $0 = R'(q) - C'(q)$ and $C'(q) = R'(q)$, which states that marginal cost must equal marginal revenue.
    c) $P''(q) = R''(q) - C''(q)$. If $R(q)$ is linear, $R''(q)$ is zero, and $P''(q) = -C''(q)$. To the right of a rising cubic inflection point, the curve is concave upward, so $C''(q)$ is positive and $P''(q) = -C''(q)$ is negative, proving that we have a local maximum.

10. *a)* $\text{ATC}(q) = \dfrac{C(q)}{q}$.

    b) $\text{ATC}'(q) = \dfrac{qC'(q) - C(q)}{q^2}$ by the product rule. To have a local mini-mum, $\text{ATC}'(q)$ must equal zero. Hence, $\dfrac{qC'(q) - C(q)}{q^2} = 0$; $qC'(q) - C(q) = 0$, and $C'(q) = \dfrac{C(q)}{q}$, which states that marginal cost must equal average total cost.

    c) From *(b)* rewrite

$$\text{ATC}'(q) = \frac{C'(q)}{q} - \frac{C(q)}{q^2}$$

and apply the product rule to get

$$\text{ATC}''(q) = \frac{qC''(q) - C'(q)}{q^2} - \frac{q^2 C'(q) - C(q)(2q)}{q^4}$$

$$= \frac{C''(q)}{q} - \frac{C'(q)}{q^2} - \frac{C'(q)}{q^2} + \frac{2C(q)}{q^3}.$$

Substituting $C'(q) = C(q)/q$ from *(b)* gives

$$\text{ATC}''(q) = \frac{C''(q)}{q} - \frac{C(q)}{q^3} - \frac{C(q)}{q^3} + \frac{2C(q)}{q^3} = \frac{C''(q)}{q}.$$

Since q and $C''(q)$ are positive, $\text{ATC}''(q) = C''(q)/q$ is positive and we have a local minimum.

CHAPTER 10

Problem Set 10–1

1. $f'(x) = 2e^x.$ $f''(x) = 2e^x.$
2. $f'(x) = 5^x \ln 5.$ $f''(x) = 5^x (\ln 5)^2.$
3. $f'(x) = 7^x \ln 7.$ $f''(x) = 7^x (\ln 7)^2.$
4. $f'(x) = 3e^x.$ $f''(x) = 3e^x.$
5. $f'(x) = -e^{-0.2x}.$ $f''(x) = 0.2e^{-0.2x}.$
6. $f'(x) = -e^{-0.4x}.$ $f''(x) = 0.4e^{-0.4x}.$
7. $f'(x) = -2e^{3-0.1x}.$ $f''(x) = 0.2e^{3-0.1x}.$
8. $f'(x) = 3e^{0.3x-6}.$ $f''(x) = 0.9e^{0.3x-6}.$
9. $f'(x) = 2xe^{x^2}.$ $f''(x) = 2e^{x^2}(2x^2 + 1).$
10. $f'(x) = -2xe^{-x^2}.$ $f''(x) = 2e^{-x^2}(2x^2 - 1).$
    In Problems 11–14, natural logarithms were taken from Table X.
11. $f'(x) = 2.07945(1/2)^{-3x} = 2.07945(2)^{3x}.$
    $f''(x) = 4.32411(1/2)^{-3x} = 4.32411(2)^{3x}.$
12. $f'(x) = 0.27489(0.4)^{-0.3x}.$ $f''(x) = 0.07556(0.4)^{-0.3x}.$
13. $f'(x) = 58.27(1.06)^x.$ $f''(x) = 3.395(1.06)^x.$
14. $f'(x) = -38.48(1.08)^{-x}.$ $f''(x) = 2.9615(1.08)^{-x}.$
15. See Figure A.   16. See Figure B.   17. See Figure C.
18. See Figure D.   19. See Figure E.   20. See Figure F.
21. See Figure G.   22. See Figure H.   23. 2.915.

FIGURE A   **FIGURE B**

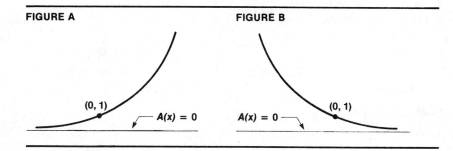

FIGURE C   **FIGURE D**

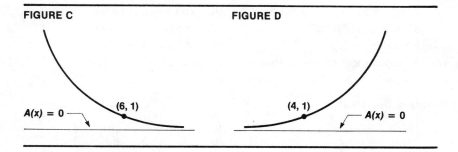

900

FIGURE E

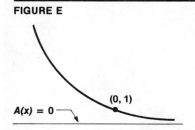

FIGURE F

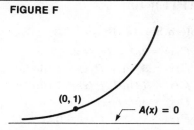

FIGURE G

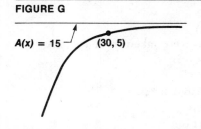

FIGURE H

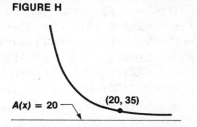

24. *a)* $t^* = 252.57286$ days.   *b)* $882,427.
25. *a)* $t^* = 160.94379$ days.   *b)* $1,024,056.
26. *a)* $t^* = 10.058987$ years.   *b)* $5,163,614.
27. *a)* $t^* = 11.453634$ years.   *b)* $7,381,624.
28. $t^* = 6.93$ years.   29. Maximum is 55.47.
30. Minimum is 1.   31. Minimum is 0.52.   32. Maximum is 12.18.
33. Maximum is 145.60.   34. Minimum is 4.85.

Problem Set 10–2

1. $1/x$.   2. $3/x$.   3. $2/(2x+3)$.   4. $-1/x$.
5. 1.   6. $2/3(2x+5)$.   7. $2(x+1)/x(x+2)$.
8. $(6x^2 - 6)/(2x^3 - 6x) = 3(x^2 - 1)/x(x^2 - 3)$.
9. $3/2(3x+2)$.   10. Minimum is 16.93.
11. Maximum is 180.26.   12. Minimum is -4.64.
13. Minimum is 2.30.   14. Minimum is 72.78.
15. Minimum is 9.45. (Note that x cannot be negative.)
16. 12 ounces.   17. 11.5 ounces.

Problem Set 10–3

1. $\dfrac{df(y)}{dy} \cdot \dfrac{dy}{dx}$.   2. $\dfrac{df(x)}{dx} \cdot \dfrac{dx}{dy}$.   3. $\dfrac{dp(q)}{dq} \cdot \dfrac{dq}{dh}$.

4. $\dfrac{dg(f)}{df} \cdot \dfrac{df}{dw}$.

5. $4y^3 \cdot \dfrac{dy}{dx}$.

6. $3x^2 \cdot \dfrac{dx}{dy}$.

7. $2e^{2w} \cdot \dfrac{dw}{dz}$.

8. $-e^{-0.5p} \cdot \dfrac{dp}{dq}$.

9. $\dfrac{2}{2y+3} \cdot \dfrac{dy}{dx}$.

10. $\dfrac{2z-3}{z^2-3z} \cdot \dfrac{dz}{dw}$.

11. $x + y \cdot \dfrac{dx}{dy}$.

12. $\left(y - x \cdot \dfrac{dy}{dx}\right)/y^2$.

13. $2x/3y^2$.

14. $9x^2/5y^4$.

15. $(x^5 + 1)/y(y-1)$.

16. $2/(20y^4 - 4y + 1)$.

17. $1/(1 - e^{-y}) = e^y/(e^y - 1)$.

18. $1/2(1 - ye^{y^2})$.

19. $-y \ln y$.

20. $1/x \ln x$.

21. $(2x - y^3)/3xy^2$.

22. $2(1 - xy^3)/3x^2y^2$.

23. $\dfrac{1}{1 - \dfrac{dg(y)}{dy}}$.

24. $\dfrac{2z}{2w - \dfrac{df(w)}{dw}}$.

25. *a)* 0.68.

  b) 68 cents of an additional dollar of income is spent.

  c) \$3.125.

  d) 0.85 or 85 percent.

26. *a)* 0.63.

  b) 63 cents of an additional dollar of income is spent.

  c) \$2.70.

  d) 0.90 or 90 percent.

27. *a)* 5.62.

  b) 3.51.

  28. *a)* 15.17.

  b) 5.26.

Problem Set 10–4

1. Cost increases at the (constant) rate of \$2.50 per additional book made.

2. At 500 miles, cost is increasing at the rate of \$0.15 per additional mile driven. At 1000 miles, cost is increasing at the rate of \$0.10 per additional mile driven.

3. At $t = 5$ years, the amount in the account is increasing at the rate of \$226.16 per additional year. At $t = 10$ years, the amount is increasing at the rate of \$332.31 per additional year.

4. At $t = 5$ years, the amount in the account is increasing at the rate of \$238.69 per additional year. At $t = 10$ years, the amount is increasing at the rate of \$356.09 per additional year.

5. *a)* $a(x) = \dfrac{16}{x} + 0.2$.

  b) $a'(x) = -\dfrac{16}{x^2}$.

  At 40 miles, the average cost per mile is decreasing at the rate of \$0.01 per additional mile driven.

6. *a)* $a(x) = 0.01x^2 - 3x + 300 + 10000/x$.

  b) $a'(x) = 0.02x - 3 - 10000/x^2$.

  c)   At 100 tons output, average cost per ton is decreasing at the rate of \$2 per additional ton produced. At 500 tons output, average cost per ton is increasing at the rate of \$6.96 per additional ton produced.

7. At 100 labor-hours worked, output is increasing at the rate of 0.359 tons per additional labor-hour worked.

8. At $t = 2$ years, sales are increasing at the rate of \$17.09 thousand per additional year.

9. At $t = 5$ years, the number of customers in the trading area is increasing at the rate of 39,923 per additional year.

10. At $t = 5$, the function is increasing at the rate of 10 percent per additional unit of x.

11. At $t = 5$ (or any number of) years, customers are increasing at the rate of 10 percent per additional year.

12. At $t = 2$ years, sales are increasing at the rate of 66.7 percent per additional year.

13. At $t = 10$ (or any number of) years, the amount in the account is increasing at the rate of 6.766 percent per additional year.

14. *a)* 100 units.   *b)* $\dfrac{N}{i}\left(\dfrac{2NF}{i}\right)^{-1/2} = \dfrac{N}{iL}.$ *c)* 2 units.

15. *a)* 200 units.   *b)* $\dfrac{F}{i}\left(\dfrac{2NF}{i}\right)^{-1/2} = \dfrac{F}{iL}.$

    c) Decrease by 10 units.

16. *a)* $x = 15.$ *b)* $-50/h^2.$ *c)* x would decrease by 0.25.

17. *a)* At $x = 5$, ln x changes at the rate of 0.2 per unit change in x.
    b) Increase by 0.0002.   *c)* 1.60964.

18. *a)* At $x = 4$, the square root of x changes at the rate of 0.25 per unit change in x.
    b) 0.0025.   *c)* 2.0025.

Problem Set 10–5

1. *a)* $e(200) = 1.2.$ *b)* At demand 200, a demand decrease of 1.2 percent accompanies a price increase of 1 percent.

2. *a)* $e(500) = 0.5.$ *b)* At demand 500, a demand decrease of 0.5 percent accompanies a price increase of 1 percent.

3. *a)* $q^* = 375.$ *b)* $e(q^*) = 1.$ 4. *a)* $q^* = 220.$ *b)* $e(q^*) = 1.$

5. *a)* $e(1000) = 1.25.$ *b)* $q^* = 1250.$

6. *a)* $e(500) = 0.80.$ *b)* $q^* = 400.$

7. *a)* $e(q) = 1$, always.   *b)* $R(q) = 1000.$
    c) Elasticity is one at all levels of demand, which means that proportionate changes in price and demand are the same but in different directions, so revenue remains constant. That is, $R(q) = qp(q) = 1000$. This case is called *unitary elasticity of demand*.

8. $e(q) = 2.$ 9. $e(q) = m.$

10. *a)* Yes. $p'(q) = -500/(0.5q + 1)^2$ is always negative so the curve slopes downward to the right, and $p''(q) = 500/(0.5q + 1)^3$ is positive so the curve is concave upward for $q \geq 0$.
    b) See *(a)*. Yes. $p'(q) = -2/q$ and $p''(q) = 2/q^2.$

Problem Set 10–6

1. *a)* 3. *b)* 0. *c)* −2. *d)* 0. *e)* 0.
2. *a)* 2. *b)* 0. *c)* 5. *d)* 0. *e)* 0.
3. *a)* $2x + 3$. *b)* 2. *c)* $2y - 2$. *d)* 2. *e)* 0.
4. *a)* $2x - 5$. *b)* 2. *c)* $-2y + 4$. *d)* −2. *e)* 0.
5. *a)* $6x - 2y$. *b)* 6. *c)* $-2x + 2y$. *d)* 2. *e)* −2.
6. *a)* $6x^2 + 3y$. *b)* $12x$. *c)* $3x + 4$. *d)* 0. *e)* 3.
7. *a)* $y^{1/2}/2x^{1/2}$. *b)* $-y^{1/2}/4x^{3/2}$. *c)* $x^{1/2}/2y^{1/2}$.
   d) $-x^{1/2}/4y^{3/2}$. *e)* $1/4x^{1/2}y^{1/2}$.
8. *a)* $2y^{1/3}/3x^{1/3}$. *b)* $-2y^{1/3}/9x^{4/3}$. *c)* $x^{2/3}/3y^{2/3}$.
   d) $-2x^{2/3}/9y^{5/3}$. *e)* $2/9x^{1/3}y^{2/3}$.
9. *a)* $2y^{1/2} - y/x^{2/3}$. *b)* $2y/3x^{5/3}$. *c)* $(x/y^{1/2}) - 3x^{1/3}$.
   d) $-x/2y^{3/2}$. *e)* $(1/y^{1/2}) - (1/x^{2/3})$.
10. *a)* $3x^{1/2}y^2/2$. *b)* $3y^2/4x^{1/2}$. *c)* $2x^{3/2}y$.
    d) $2x^{3/2}$. *e)* $3x^{1/2}y$.
11. *a)* $ye^x(x + 1)$. *b)* $ye^x(x + 2)$. *c)* xe^x.
    d) 0. *e)* $e^x(x + 1)$.
12. *a)* $y \ln y$. *b)* 0. *c)* $x(1 + \ln y)$.
    d) $x(1 + 1/y)$. *e)* $1 + \ln y$.
13. *a)* $2y/(x + y)^2$. *b)* $-4y/(x + y)^3$. *c)* $-2x/(x + y)^2$.
    d) $4x/(x + y)^3$. *e)* $2(x - y)/(x + y)^3$.
14. *a)* $-2y/(x - y)^2$. *b)* $4y/(x - y)^3$. *c)* $2x/(x - y)^2$.
    d) $4x/(x - y)^3$. *e)* $-2(x + y)/(x - y)^3$.
15. *a)* $3/(3x + 2y)$. *b)* $-9/(3x + 2y)^2$. *c)* $2/(3x + 2y)$.
    d) $-4/(3x + 2y)^2$. *e)* $-6/(3x + 2y)^2$.
16. *a)* $2e^{2x+3y}$. *b)* $4e^{2x+3y}$. *c)* $3e^{2x+3y}$.
    d) $9e^{2x+3y}$. *e)* $6e^{2x+3y}$.
17. $6z - 2x$. 18. $2w - 3z^2$. 19. 2. 20. 6. 21. −3. 22. 3.
23. *a)* 7. *b)* 3. *c)* 0. *d)* 2. *e)* 0. *f)* 0.
24. *a)* 7. *b)* 2. *c)* 0. *d)* 3. *e)* 0. *f)* 0.
25. *a)* 8. *b)* 5. *c)* 2. *d)* −4. *e)* −6. *f)* 1.
26. *a)* −14. *b)* 2. *c)* 4. *d)* −11. *e)* 2. *f)* −5.
27. At the production level of 3 gallons of Exall and 5 gallons of Whyall, cost is decreasing at the rate of $4 per additional gallon of Exall and cost is not changing (is stationary) with respect to a change in Whyall.
28. At the production level of 3 gallons of Exall and 7 gallons of Whyall, cost is decreasing at the rate of $8 per additional gallon of Exall and cost is increasing at the rate of $4 per additional gallon of Whyall.
29. At 400 gallons of Exall and 500 gallons of Whyall: *a)* Profit is increasing at the rate of $10 per additional gallon of Exall, and *b)* profit is increasing at the rate of $10 per additional gallon of Whyall. *c)* At 600 gallons of Exall and 500 gallons of Whyall, profit is decreasing at the rate of $10 per additional gallon of Exall. *d)* At 400 gallons of Exall and 600 gallons of Whyall, profit is decreasing at the rate of $10 per additional gallon of Whyall.

30. When labor expenditure and capital investment are, respectively, $64 and $27 million: *a)* Output is increasing at the rate of $1.50 per additional $1 of labor expenditure, and *b)* output is increasing at the rate of $7.11 per additional dollar of capital investment.

Problem Set 10-7

1. Minimum is 10.
2. Maximum is 23.
3. Maximum is 120.
4. Minimum is 50.
5. Minimum is 10.
6. Maximum is 52.
7. No extreme point.
8. No extreme point.
9. Maximum is 175.
10. No extreme point.
11. Maximum is 10.
12. Minimum is −5.
13. *a)* 500 gallons of Exall and 550 gallons of Whyall.
    b) Both P_{xx} and P_{yy} are negative and $D = (-0.1)(-0.2) - (0)^2$ is positive, so we have a maximum.
    c) $P_{max} = \$42,750$.
14. *a)* Use the Extron machine for 4 hours and the Whytron for 6 hours.
    b) Both C_{xx} and C_{yy} are positive and $D = 6(2) - (-2)^2$ is positive, so we have a minimum.
    c) $C_{min} = \$14$ per unit.
15. *a)* $L^* = 400$, $I^* = 250$. *b)* $C_{min} = \$3750$.
16. *a)* $L^* = 520$, $p^* = 0.8$. *b)* $C_{LL} = 0.0015$, $C_{pp} = 2600$ and $C_{pL} = 0$. Hence, $D > 0$ and both partials are positive so we have a minimum.
    c) $C_{min} = \$416$ thousand.
17. $L^* = \left[\dfrac{2cd(a+b)}{ab}\right]^{1/2}$. $I^* = \dfrac{bL^*}{a+b} = \left[\dfrac{2bcd}{a(a+b)}\right]^{1/2}$.

Problem Set 10-8

1. $y_f = 1.893 + 0.821x$.
2. $y_f = 9.571 + 0.357x$.
3. $y_f = 9.5 - 1.5x$.
4. $y_f = 11.25 - 2.5x$.
5. $y_f = 26.496 + 0.527x$.
6. $y_f = 32.0029 - 0.0516x$.

Problem Set 10-9

1. *a)* $h(x, y, \lambda) = 20x + 10y - x^2 - y^2 - \lambda(x + 2y - 10)$.
   b) (8, 1). *c)* Maximum is 105.
2. *a)* $h(x, y, \lambda) = x^2 + 2y^2 - 3x - 10y + 20.25 - \lambda(2x + 3y - 19)$.
   b) (3.5, 4). *c)* Minimum is 14.
3. *a)* $h(x, y, \lambda) = 3x^2 + 4y^2 - xy - 300 - \lambda(x + y - 16)$.
   b) (9, 7). *c)* Minimum is 76.
4. *a)* $h(x, y, \lambda) = 2xy - x^2 - 2y^2 + 200 - \lambda(2x - y - 30)$.
   b) (18, 6). *c)* Maximum is 20.
5. *a)* $h(x, y, \lambda) = 4xy - y^2 - 5x^2 + 16x + 10y - \lambda(2x + y - 60)$.
   b) (14, 32). *c)* Maximum is 332.
6. *a)* $h(x, y, \lambda) = 2x^2 + y^2 + 2x + 4y - \lambda(x + y - 8)$.
   b) (3, 5). *c)* Minimum is 69.

CHAPTER 11

Problem Set 11–1

1. $x + C.$

2. $z + C.$

3. $5y + C.$

4. $-7w + C.$

5. $x + \dfrac{x^2}{2} + C.$

6. $3y - \dfrac{y^2}{2} + C.$

7. $\dfrac{2x^3}{3} - \dfrac{3x^2}{2} + 4x + C.$

8. $\dfrac{3y^4}{4} + 2y^2 - y + C.$

9. $pq + C.$

10. $qp + C.$

11. $\dfrac{qp^2}{2} + C.$

12. $\dfrac{pq^2}{2} + C.$

13. $\dfrac{x^4}{4} + \dfrac{x^5}{5} - x + C.$

14. $\dfrac{y^3}{3} + \dfrac{y^6}{6} - y + C.$

15. $y^3 + y^5 + y + C.$

16. $x^4 - x^6 + x + C.$

17. $-\dfrac{1}{x} + C.$

18. $\dfrac{y}{343} + C.$

19. $-\dfrac{1}{2y^2} + C.$

20. $-\dfrac{1}{3x^3} + C.$

21. $5y + \dfrac{1}{y^2} + C.$

22. $7x + \dfrac{1}{x^3} + C.$

23. $x^2 + \dfrac{1}{x} + x + C.$

24. $\dfrac{y^2}{2} - \dfrac{1}{y^2} + y + C.$

25. $\dfrac{2p^{3/2}}{3} + C.$

26. $\dfrac{3q^{4/3}}{4} + C.$

27. $-\dfrac{9}{x^{1/3}} + C.$

28. $-\dfrac{10}{y^{1/2}} + C.$

29. $2x + \dfrac{1}{x} + \dfrac{1}{x^{2/3}} + C.$

30. $x^2 + \dfrac{1}{x^2} + \dfrac{3}{4x^{1/3}} + C.$

31. $4x^3 + 3x^4 + C.$

32. $x^3 + 3x^2 + C.$

33. $2(2x - 9)^4 + C.$

34. $\dfrac{5(3x + 5)^6}{3} + C.$

35. $-\dfrac{1}{3(3x - 9)} + C.$

36. $-\dfrac{1}{6(2x + 3)^3} + C.$

37. $-\dfrac{2(5 - 3x)^{1/2}}{3} + C.$

38. $\dfrac{1}{4(7 - 2w)^2} + C.$

39. $6(2x + 5)^{2/3} + C.$

40. $8(3x - 7)^{1/2} + C.$

Problem Set 11–2

1. 6.

2. 6.

3. 15.

4. 108.

5. 20.

6. 4.

7. 96/5.

8. 18.

9. 17/6.

10. 51/2.

11. 44.

12. 16.

13. 26/3.

14. 2.

15. 3.

16. 8.

17. $b^3 - a^3.$

18. $d^4 - c^4.$

19. $\dfrac{n^2}{2} + n - \dfrac{3}{2}.$

20. $n - n^2.$

Problem Set 11–3

1. 3.

2. 36.

3. 13/2.

4. 6.

5. 4.

6. 3/2.

7. 16.

8. 80.

9. 100.

10. 25/2.

11. 500.

12. 500/3.

13. *a)* 27/2. *b)* 23/2.

14. 8.

15. 36.

16. 500/3.

17. 125/6.

18. 9.

19. 64/3.

20. 3,375.

Problem Set 11–4

1. *a)* $13.8 thousand.   *b)* See Figure A.
   c) $17.4 thousand.   *d)* 20 years.
2. *a)* $180 thousand.   *b)* See Figure B.
   c) $140 thousand.   *d)* 25 years.

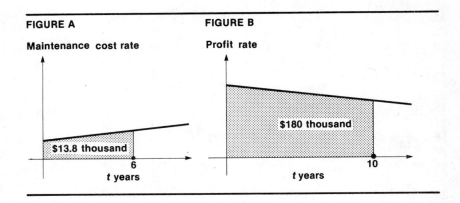

FIGURE A

Maintenance cost rate

$13.8 thousand

6

t years

FIGURE B

Profit rate

$180 thousand

10

t years

3. 100 million barrels.   4. 1944 barrels.
5. 13.5 years.   6. 18 years.   7. $100 + 80 = 180$ thousand.
8. $80 + 2 = \$82$ thousand.   9. $2600.   10. $20,000.
11. *a)* 50 months.   *b)* $200 thousand.
12. *a)* 36 months.   *b)* $17,500.
13. *a)* $3000 million.   *b)* See Figure C.

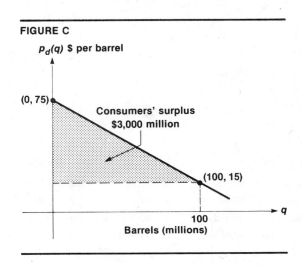

FIGURE C

$p_d(q)$ $ per barrel

(0, 75)

Consumers' surplus
$3,000 million

(100, 15)

100

Barrels (millions)

q

FIGURE D

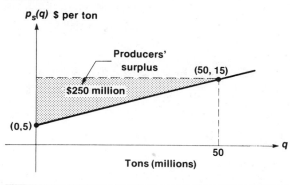

14. *a)* $250 million.   *b)* See Figure D.
15. $1856 thousand.   16. $3240 million.
17. *a)* $9000 thousand.   *b)* $4500 thousand.
18. *a)* $100 million.   *b)* $25 million.

Problem Set 11–5

1. $\ln x + C.$ 2. $\ln x + C.$ 3. $2 \ln x + C.$ 4. $3 \ln x + C.$

5. $-\dfrac{1}{x} + C.$ 6. $-\dfrac{1}{x} + C.$ 7. $\dfrac{\ln (5x + 4)}{5} + C.$

8. $-\ln (3 - x) + C.$ 9. $-5 \ln (3 - 0.2x) + C.$

10. $4 \ln (0.5x + 1) + C.$ 11. $-\dfrac{1}{2(2x - 1)} + C.$

12. $-\dfrac{2}{(5x + 3)} + C.$ 13. $2.30259.$

14. $1.$ 15. $0.44629.$ 16. $0.93269.$

17. 895.9 million barrels.   18. $140.3 million.

Problem Set 11–6

1. $e^x + C.$ 2. $(2^x/ \ln 2) + C.$
3. $-(3^{-x}/ \ln 3) + C.$ 4. $-e^{-x} + C.$
5. $2e^{0.5x} + C.$ 6. $[5(5^{0.2x})/ \ln 5] + C.$
7. $[-2.5(0.5)^{1-0.4x}/ \ln (0.5)] + C.$ 8. $-2e^{2-0.5x} + C.$
9. $-20e^{3-0.1x} + C.$ 10. $-20e^{5-0.2x} + C.$
11. $-[1/(0.8)^x \ln (0.8)] + C.$ 12. $(-2/ e^{0.5x}) + C.$
13. 17.18.   14. 14.11.   15. 3.61.   16. 9.02.
17. *a)* 343.7 million barrels.   *b)* 35.8 years.
18. *a)* $183.8 thousand.   *b)* 13.7 years.
19. *a)* $10,544.33.   *b)* 6.76 years.

20. *a)* 47.41 million pounds. *b)* 8.05 years.

21. 72.41 million pounds. 22. 93.23 million gallons.

23. $27.74 thousand. 24. $40 thousand.

Problem Set 11–7

1. $3x(\ln x - 1) + C.$ 2. $2x(\ln x - 1) + C.$

3. $x(\ln 2x - 1) + C.$ 4. $x(\ln 3x - 1) + C.$

5. $\dfrac{(2x+1)[\ln (2x+1)-1]}{2} + C.$

6. $\dfrac{(3x+5)[\ln (3x+5)-1]}{3} + C.$

7. $2(3x - 2)[\ln (3x - 2) - 1] + C.$

8. $2(2x - 1)[\ln (2x - 1) - 1] + C.$

9. $70.33 thousand. 10. 110.6 million bushels.

Problem Set 11–8

1. $\dfrac{x(2^x)}{\ln 2} - \dfrac{2^x}{(\ln 2)^2} + C.$ 2. $3x - (3/2) \ln (2x + 1) + C.$

3. $8\left[\dfrac{1}{(0.5x + 1)} + \ln (0.5x + 1)\right] + C.$

4. $e^x(x - 1) + C.$ 5. $-4e^{2-0.5x}(0.5x + 1) + C.$

6. $\dfrac{4}{3x + 4} + \ln (3x + 4) + C.$ 7. $2 \ln \left(\dfrac{x}{0.5x + 1}\right) + C.$

8. $12.5e^{0.4x-1}(0.4x - 1) + C.$ 9. $2\left[\dfrac{x}{5} + \dfrac{3}{25} \ln (5x - 3)\right] + C.$

10. $3 \ln \left(\dfrac{x}{3x + 2}\right) + C.$ 11. $5e^{0.1x^2-2} + C.$

12. $5[0.2x - \ln (1 + e^{0.2x})] + C.$

13. $x - 2 \ln (1 + 2e^{0.5x}) + C.$ 14. $-2e^{2-0.5x^2} + C.$

15. 8. 16. 3. 17. 2 18. 6.

19. Does not exist. 20. Does not exist. 21. 25. 22. 2.

Problem Set 11–9

1. 2.53. 2. 0.366. 3. 0.379. 4. 1.019.

5. 0.188. 6. 2.391. 7. 0.590. 8. 0.110.

9. 0.3411. 10. 0.3413. 11. 0.536. 12. 0.959.

13. $\dfrac{x^2 \ln x}{2} - \dfrac{x^2}{4} + C.$ 14. $\dfrac{x^4 \ln x}{4} - \dfrac{x^4}{16} + C.$

15. $-\left(\dfrac{\ln x + 1}{x}\right) + C.$ 16. $\ln x - \dfrac{\ln x}{x} - \dfrac{1}{x} + C.$

Problem Set 11–10

1. $y = x + C.$ 2. $y = x^3 + C.$ 3. $y = \ln x + C.$

4. $y = Ke^x.$ 5. $y = Kx.$ 6. $y = Ke^{-1/x}.$

7. $y = Ke^{x^{1/2}}$ 8. $y = K/x$.   9. $y = 5 \ln (0.2x + 3) + C$.

10. $y = 4 \ln (0.5x + 2) + C$.   11. $y = Ke^{0.5x} - 4$.

12. $y = Ke^{0.2x} - 15$.   13. $y = 0.5x^2 + x + 2$.

14. $y = x^2$.   15. $y = 0.4e^x$.   16. $y = 6/x$.

17. $y = 19e^{0.2x} - 15$.   18. $y = 5e^{0.5x} - 4$.

19. $y = 3x - 1$.   20. $y = (8/x) - 1$.

21.   a) $dS = (150 + 6t)\, dt$.   b) $S = 0$ when $t = 0$.
     c) $S(t) = 150t + 3t^2$.   d)   15,000.   e)   100 days.

22.   a) $dS = (100 + 2t)\, dt$.   b) $S = 0$ when $t = 0$.
     c) $S(t) = 100t + t^2$.   d)   7500.   e)   200 days.

23.   a) $dP = 0.1\, P dt$.   b) $P(t) = 100e^{0.1t}$.   c)   \$271.8 thousand.

24.   a) $dS = 0.05 S dt$.   b) $S(t) = 2000e^{0.05t}$.   c)   \$5436.56.

25.   a) $dS = 0.08 S dt + 500\, dt$.   b) $S(t) = 8250e^{0.08t} - 6250$.
     c)   \$12,110.71.   d)   13.7 years.

26.   a) $dS = 0.08 S dt - 500\, dt$.   b) $S(t) = 6250 - 1250e^{0.08t}$.
     c)   \$3468.07.   d)   20.1 years.

27.   a) $dC = 5e^{0.01t}\, dt$.   b) $C(t) = 500(e^{0.01t} - 1)$.
     c)   52.6 million barrels.   d)   33.6 years.

28.   a) $dC = 5e^{0.02t}\, dt$.   b) $C(t) = 250(e^{0.02t} - 1)$.
     c)   55.4 million barrels.   d)   29.4 years.

CHAPTER 12

Problem Set 12–1

1.   a)   0.30.   b)   0.55.   c)   0.   d)   0.125.   e)   0.
     f)   0.03.   g)   0.03.   h)   0.15.   i)   0.20.   j)   0.15.
     k)   0.50.   l)   0.50.   m)   0.775.   n)   0.47.   o)   0.85.
     p)   0.85.   q)   0.70.   r)   0.50.   s) $M \cup S$.   t) $B \cup C$.
     u) B or $A' \cap C'$.   v) $A' \cap M'$.   w) $B \cup M$.   x) $L \cup C$.

2.   a) $P(M | A) = 0.3 = P(M)$.
     b) M and A are independent in the probability sense.
     c)   Independent because $P(A | L) = 0.15 = P(A)$, or $P(L | A) = 0.20 = P(L)$.
     d)   Not independent because $P(B | L) = 0.75$ and $P(B) = 0.55$, so $P(B | L) \ne P(B)$.
     e) $P(A \cap B) = 0$. A and B are mutually exclusive.

3.   a)   0.42.   b)   0.60.

4.   a)   Yes. $P(WB) = 0$.   b)   No. $P(W | B) = 0$ does not equal $P(W)$,
       which is 0.20.   c)   No. $P(YW) \ne 0$.
     d)   Yes. $P(Y | W) = 78/120 = 0.65$, and $P(Y) = 390/600 = 0.65$.
     e)   No. $P(YR) \ne 0$.
     f)   No. $P(Y | R) = 270/300 = 0.9$ but $P(Y) = 390/600 = 0.65$, so $P(Y | R) \ne P(Y)$.

5.   a)   0.50.   b)   0.125.   c)   0.275.   d)   0.   e)   0.
     f)   0.   g)   0.25.   h)   0.625.   i)   0.10.   j)   0.25.

k) 0.40.   l) 0.   m) No. $P(YW) \neq 0$.

n) No because $P(W \mid Y) = 10/100 = 0.10$ does not equal $P(W) = 80/400 = 0.20$.   o) No. $P(WT) \neq 0$.

p) Yes, $P(W \mid T) = 50/250 = 0.2$ equals $P(W) = 80/400 = 0.2$.

q) Yes, $P(RG) = 0$.

r) No. $P(R \mid G) = 0$, which does not equal $P(R) = 0.50$.

6. $P(S) = 0.4$, and $P(S \mid H) = P(S \mid M) = P(S \mid L) = 0.4$.
$P(P) = 0.6$, and $P(H \mid P) = P(M \mid P) = P(L \mid P) = 0.6$.

7.

	H	M	L
Female	20	160	20
Male	10	80	10

Problem Set 12–2

1.

Area	L	M	S	
A	0.03	0.045	0.075	0.15
B	0.15	0.125	0.275	0.55
C	0.02	0.130	0.150	0.30
	0.20	0.30	0.50	1.00

2. a)

	F	F'	
C	0.24	0.36	0.60
C'	0.16	0.24	0.40
	0.40	0.60	1.00

b) 0.36   c) 0.76.   d) 0.

e) 0.40.   f) 0.60.   g) 0.

h) $P(CF) = 0.24$ and $P(C)P(F) = 0.60(0.40) = 0.24$. Hence, $P(CF) = P(C)P(F)$, so C and F are independent.

i) F and F' are complementary events.

3. a) 0.79.

b) As a consequence of these probability assignments, $P(A \cup B) = 1.09$, which is greater than 1, so the candidate should be advised to reassess her subjective probabilities.

4. a) $7/15 = 0.467$.   b) $1/15 = 0.0667$.   c) $7/15 = 0.467$.

d) $1 - P(0 \text{ defective}) = 1 - P(GGG) = 1 - (7/10)(6/9)(5/8) = 0.708$.

5. $P(CS) = 0.35$ because $P(CS) = P(C)P(S \mid C) = 0.5(0.7) = 0.35$.

6. $P(G \mid H) = 0.90$, where G means good job and H means honors. This follows from the rule $P(G \mid H) = P(GH)/P(H) = 0.09/0.10 = 0.90$.

7. a) 0.9025.   b) 0.0025.   c) 0.9975.

d) 0.095.   e) 0.95 because of independence.

8. *a)* *HHH, THH, HTH, HHT, TTH, THT, HTT, TTT.*
   b) 0, 1, 2, 3.
   c)

Event	*HHH*	*THH*	*HTH*	*HHT*	*TTH*	*THT*	*HTT*	*TTT.*
Probability	0.125	0.125	0.125	0.125	0.125	0.125	0.125	0.125.

   d)

Event	0	1	2	3
Probability	0.125	0.375	0.375	0.125.

   e) 0.875. *f)* 0.50. *g)* 0.375. *h)* 0.375.

9. *a)*

Event	W_1M	W_2M	W_1W_2	MW_1	MW_2	$W_2W_1.$
Probability	1/6	1/6	1/6	1/6	1/6	1/6.

   b) 1/3. *c)* 2/3. *d)* 1/3.

10. *a)* *WM.*
    b) $P(WM) = P(W)P(M \mid W) = (2/3)(1/2) = 1/3.$
    c) *WW.*
    d) $P(WW) = P(W)P(W \mid W) = (2/3)(1/2) = 1/3.$

Problem Set 12–3

1. 0.6. 2. 0.9975. 3. 0.3.
4. $P(X \mid Y) = 1/4;\ P(Y \mid X) = 1/6.$
5. $P(X) = 1/2;\ P(Y) = 2/5.$
6. 0.67.
7. *a)* 1/2. *b)* 4/7. *c)* 23/70. *d)* 47/70.
8. *a)* 0.35. *b)* 0.60. *c)* 0.55. *d)* 0.45.
9. *a)* 0.48. *b)* 0.16.
10. 0.61. 11. 0.4015. 12. 0.01099. 13. 0.07831.
14. *a)* $P(SC) = 0.5;\ P(S)P(C) = 0.48;\ P(SC) \neq P(S)P(C).$
    b) 0.12.
15. *a)* 1/4000. *b)* 3871/4000. *c)* 129/4000. *d)* 128/4000.
16. *a)* 1/12. *b)* 7/12. *c)* 1/4. *d)* 23/36.
17. 0.141.
18. 5/6. 19. 0.56. 20. 0.9. 21. 0.288. 22. 0.706.

Problem Set 12–4

1. 0.949. 2. 0.678. 3. 0.857. 4. *a)* 0.00336. *b)* 0.0198.
5. *a)* 0.998. *b)* 0.664.

Problem Set 12–5

1. *a)* 0.441. *b)* 0.657. *c)* 0.3087. *d)* 0.0513.
2. 0.992. 3. 0.913.
4. *a)* 0.886. *b)* 0.655.
5. *a)* 0.760. *b)* 0.0837.
6. 5.
7. *a)* 1/32 = 0.03125. *b)* 3/16 = 0.1875.
   c) 5/16, 0.3125. *d)* 1/2 = 0.50.
   e) 3/16 = 0.1875.

8. *a)* $1/1024 = 0.000977.$ *b)* $1/64 = 0.0156.$
   c) $45/512 = 0.0879.$ *d)* $53/512 = 0.104.$

9. *a)* $\sum\limits_{x=0}^{15} C_x^{100}(0.05)^x(0.95)^{100-x}.$ *b)* $\sum\limits_{x=11}^{200} C_x^{200}(0.05)^x(0.95)^{200-x}.$

10. *a)* $0.993.$ *b)* $0.421.$

11. *a)* $0.642.$ *b)* $0.027.$

12. *a)* $0.983.$ *b)* $0.006.$ *c)* $0.579.$
    d) $0.062.$ *e)* $0.377.$ *f)* $0.617.$

13. *a)* $0.115.$ *b)* $0.788.$ *c)* 1.000 to 3 decimals.
    d) $0.032.$ *e)* $0.210.$ *f)* $0.002.$

14. *a)* $0.349.$ *b)* $0.387.$ *c)* $0.651.$ *d)* $0.251.$

15. *a)* $0.074.$ *b)* $0.315.$ *c)* $0.012.$ *d)* $0.389.$

Problem Set 12–6

1. $0.$ 2. $\$0.8.$

3. Act 2.   4. $A_4.$

5. Make three. EMV $= \$8.35.$

6. *a)* $\$42.75.$
   b) $P(0 \text{ loss}) = 0.99825$ is not stated.

7. The author would choose A_1 in each case, even though A_2 presents no chance of loss. Individual readers may differ with the choice of A_1, especially in *c)*, where there is a 50 percent chance of a loss of $1000. However, $E(A_1)$ in *c)* is $1000, compared to $E(A_2) = \$250.$

CHAPTER 13

Problem Set 13–1

1. *a)* $p(x) = 0.1.$ *b)* $0.3.$
2. *a)* $p(x) = x/50.$ *b)* $0.09.$
3. *a)* $p(x) = (x^{1/2})/18.$ *b)* $0.259.$
4. *a)* $p(x) = (2 - x^{-2})/37.05.$ *b)* $0.462.$
5. *a)* $p(x) = (12x - 3x^2/32.$ *b)* $0.5.$

Problem Set 13–2

1. *a)* $\mu = 10.$ *b)* $\sigma^2 = 100/3.$ *c)* $\sigma = 5.77.$
2. *a)* $\mu = 50.$ *b)* $\sigma^2 = 2500/3.$ *c)* $\sigma = 28.87.$
3. *a)* $\mu = 4.$ *b)* $\sigma^2 = 2.$ *c)* $\sigma = 1.41.$
4. *a)* $\mu = 2.$ *b)* $\sigma^2 = 0.5.$ *c)* $\sigma = 0.707.$
5. *a)* $35/12$ hundred gallons.   *b)* 210 hundred gallons.
6. *a)* $17/15$ thousand gallons.   *b)* 102 thousand gallons.
7. *a)* See Figure **A.**

FIGURE A   **FIGURE B**

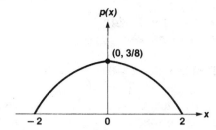

b) The function is symmetrical so its center of gravity is at the midpoint of the x-interval, which is zero.

c) $\int_{-1}^{1} x\left[\dfrac{3(1-x^2)}{4}\right]dx = \dfrac{3}{4}\left(\dfrac{x^2}{2} - \dfrac{x^4}{4}\right)\Big|_{-1}^{1} = 0.$

d) $(0.6)^{1/2} = 0.775.$

8. a) See Figure B.

  b) The function is symmetrical so its center of gravity is at the midpoint of the x-interval, which is zero.

  c) $\int_{-2}^{2} x\left[\dfrac{3(4-x^2)}{32}\right]dx = \dfrac{3}{32}\left(2x^2 - \dfrac{x^4}{4}\right)\Big|_{-2}^{2} = 0.$

  d) 0.894.

Problem Set 13–3

1. a) 0.7769.   b) 0.2325.   c) 0.0821.

2. a) 0.3935.   b) 0.1859.   c) 0.3679.

3. a) 0.5276. b) 0.6065.   4. a) 0.3935.   b) 0.4493.

5. a) 0.9933. b) 0.0821.   6. a) 0.1393.   b) 0.9512.

7. a) $[ke^{-4x}/4]\Big|_{1}^{\infty}.$ b) $k=4;\ p(x)=4e^{-4x}.$ c) $\mu=1/4.$

8. a) $[-k/2x^2]\Big|_{1}^{\infty}.$ b) $k=2.$ c) $\mu=2.$

Problem Set 13–4

1.		2.		3.	
a)	0.1915	a)	0.8400	a)	0.9772
b)	0.1915	b)	0.9544	b)	0.3023
c)	0.1587	c)	0.9974	c)	0.3085
d)	0.1498	d)	0.0456	d)	0.3413
e)	0.2266	e)	0.7881	e)	0.1498
f)	0.3721	f)	0.2295	f)	0.6826
g)	0.1554	g)	0.8413	g)	0.2266
h)	0.1464.	h)	0.1151.	h)	0.6247
				i)	0.0013.

Problem Set 13–5

1. $\bar{x}=7, s=2$.
2. $\bar{x}=15, s=2.94$.
3. $\bar{x}=11.2, s=2.17$.
4. $\bar{x}=3.5, s=2.07$.
5. $\bar{x}=0.58, s=0.0334$.

Problem Set 13–6

1. *a)* 52.2. *b)* 56.410. *c)* 41.775.
   d) 46.6 to 53.4. *e)* 58.225.
2. *a)* 0.1587. *b)* 0.9772. *c)* 23 parts.
   d) 4.56 percent. *e)* 1.5730 to 1.5770.
   f) 2.2484. *g)* \$53.44.
3. *a)* 0.3085. *b)* 69.15 percent. *c)* 134 cans.
   d) 10.56 percent. *e)* 16.465.
4. *a)* 1908 to 2092. *b)* 10.56 percent.
   c) 1868. *d)* 0.62 percent.
5. *a)* 0.13 percent. *b)* 5.48 percent. *c)* 183 to 217.
   d) 159. *e)* 99th percentile.

APPENDIX 1

Problem Set A1–1

1. *a)* "*B* is the set whose elements are 2, 4, and 6."
   b) "*C* is the set whose element is 0."
   c) "*Q* is the set whose members are the odd numbers 1 through 29."
   d) "*F* is the set whose elements are the lower case English vowels."
   e) "*L* is the set whose members are George and Charles."
   f) "*K* is the set of positive even integers starting with 2."
2. *a)* $F=$ {Lower case English vowels}.
   b) $S=$ {Positive integral multiples of 3}.
   c) $P=$ {First five positive integers}.
   d) $G=$ {First five upper case English letters}.
   e) $R=$ {Integers greater than 99}.
   f) $J=$ {Negative odd integers from -1 through -101}.
      $=$ {Negative odd integers greater than -102}.
3. *a)* {1, 3, 5, 7}.
   b) {January, February, March, April}.
   c) {14, 28, 42, . . .}.
   d) {5, 10, 15, . . . , 995}.
   e) ϕ.
   f) {$1^2, 2^2, 3^2, \ldots, 10^2$}.
4. ϕ has no elements, whereas {0} has the element 0.
5. *a)* $1/2 \notin$ {1, 2, 3, . . .}. *b)* $64 \in$ {4, 8, 12, . . .}.
   c) $a \in$ {a, e, i, o, u}. *d)* $4 \notin$ {1, 3, 5, . . .}.
   e) $\$ \notin$ {a, b, c, . . . , z}. *f)* $g \notin$ {a, b, c, d, e, f}.

6. a) {1, 3, 4, 7, 12}.   b) {3, 7}.
   c) φ.   d) {1, 4}.
   e) {1}.   f) φ.
7. a) {y : y + 2 = 10} = {8}.
   b) The set of y's such that y plus 5 equals 8 is the set with the single member, 3.
8. a) {m : m − 6 = 4} = {10}.
   b) The set of q's such that two times q is six is the set with the single member, 3.
9. φ, the empty set.   10. {11/4}.
11. a) No, because for each state there are two senators.
    b) Yes, because for each senator there is one and only one state.
12. a) Yes, because each apple grows on one and only one tree.
    b) No, because one tree may have 0, 1, or more than one apple.
13. Yes, because for each point on the horizontal line there is one and only one point on the slant line.
14. No, because for each point on the diameter there are two points on the circle.
15. a) The h function of z, or h of z.   b) 15.
    c) Multiply 4 by the value of z and add 7.
    d) Multiply 4 by (x + 2) and add 10.
16. a) The f function of x, or f of x.   b) 25.
    c) Multiply 3 by the value of x and add 10.
    d) Multiply 3 by (x + 1) and add 10.
17. {y : y = 2x + 7}.   18. {y : y = 5x}.
19. Yes, because for each value of x there is one and only one value for y.
20. No, because for the starting element, 0, there is no ending element; that is, 5/0 is undefined. See Appendix 2, Section 21.

Problem Set A1–2

1. a) {a, b, c, d, e, f, g, h}.   b) {a, b, c, d, e, i, j}.
   c) {d, e}.   d) φ.
   e) {d, e, f, g, h, i, j}.   f) φ.
2. a) {0, 1, 2, 3, ...}   b) φ.
   c) {5, 7, 9}.   d) φ.
   e) {1, 3, 5, ...}.
3. A ∪ B = {face cards}; A ∩ B = {queens}.
4. Assuming an ace is a face card, and using the notation 3D, jC to represent the 3 of diamonds and the jack of clubs, and so on:
   A ∪ B = {Face card, 2H, 3H, ... , 10H, 2D, 3D, ... , 10D}.
   A ∩ B = {jH, qH, kH, aH, jD, qD, kD, aD}.
   A ∩ C = φ.
   B ∩ C = {9H, 9D}.
5. a) φ.
   b) M ∩ N is the point at which the lines intersect.
6. a) Rain tomorrow, or warmer tomorrow, or rain and warmer tomorrow.
   b) Rain and warmer tomorrow.

APPENDIX 2

Problem Set A2–1

1. −14.	2. +13.	3. −12.	4. +3.
5. +5.34.	6. −24.	7. −23.	8. +3.
9. −4.	10. −4.	11. +3.	12. +1.
13. −2.	14. +13.	15. +11.	16. −45.
17. +15.	18. −3.	19. −7/2.	20. +3/2.
21. −32.	22. −28/5.	23. +12/7.	24. +2.
25. +1.	26. −5.		

27. $+3 + (−4) + (+5) + (−2) + (+6)$.   28. $+3 + (−2) + (+4) + (−5)$.

29. $−1.5 + (−3) + (−2.5)$.   30. $−3 + (+7) + (−4) + (+2)$.

31. $−10 + (+7) + (−4) + (+3)$.   32. $+12 + (−6) + (+5) + (−7)$.

33. −16.	34. 23.	35. −2.	36. 3.
37. 16.	38. 10.		

Problem Set A2–2

1. ab means a times b; cxy means c times x times y.
2. $100a + 10b + c$.
3. All.
4. If a and b are any odd integers, then $a + b$ is an even integer.
5. Whenever we add two clock numbers (in the time sense), the sum is another clock number.
6. By convention, it is agreed that three times the sum of a and c will be indicated by parentheses; thus, $3(a + c)$.
7. Any (whole) number.
8. In $a(b + c)$, we have the product of the number a by the number $(b + c)$. The distributive property says this product equals the sum of ab and ac.
9. Commutative, multiplication.
10. Commutative, addition.
11. Distributive.
12. Commutative, addition.
13. Convention; no sign means assume a plus sign.
14. Commutative, multiplication.
15. Convention; indicate addition of negative by minus sign.
16. Commutative, multiplication.
17. Associative, addition.
18. Commutative, multiplication.
19. Associative, multiplication.
20. Convention; b means $+b$; indicate addition of negative by minus sign.
21. Associative, addition.
22. Distributive property (inverse).

23. Commutative, addition.
24. Commutative, addition.
25. Associative, multiplication.
26. $-b$.
27. $7a - 5b$.
28. $5abc - 2d$.
29. $7 + x$.
30. $8abc$.
31. $8 + 2a - b$.
32. $-b + a$
  $= \quad a - b$ Commutative, addition.
33. $2b(3a)$
  $= 2b3(a)$ Associative, multiplication
  $= 2(3)ba$ Commutative, multiplication
  $= 6ba$ Associative, multiplication
  $= 6ab$ Commutative, multiplication.
34. $3 + xy + \ 5 + 3(2ab)$
  $= 3 + \ 5 + xy + 3(2ab)$ Commutative, addition
  $= \quad\quad 8 + xy + 3(2ab)$ Associative, addition
  $= \quad\quad 8 + xy + 6ab$ Associative, multiplication.
35. $3 + a + 2$
  $= 3 + 2 + a$ Commutative, addition
  $= \quad 5 + a$ Associative, addition.
36. $a + 3 + c + 2$
  $= a + c + 3 + 2$ Commutative, addition
  $= a + c + 5$ Associative, addition.
37. $acdb$
  $= adcb$ Commutative, multiplication
  $= adbc$ Commutative, multiplication.
38. $a + b$.
39. $a + (b + c)$.
40. $(a + b) + 2(cd)$.
41. $2(a + b) - 3(c + 2d)$.
42. $2d(a + b + c)$.

Problem Set A2–3

1.	$2abc - 4ab$.	2.	$ab + a - 2b - 2$.
3.	11.	4.	$ac - bc + 3c + 2a - 2b + 6$.
5.	$3a + 3b - 1$.	6.	$15a - 62$.
7.	$13a + 54$.	8.	$3ax - 6bx - 18x - 2a + 4b + 12$.
9.	$x + 2b - a$.	10.	$2x + 3y$.
11.	$a - 3b$.	12.	$a + 3x + 2$.
13.	$7b - 2ab$.	14.	$abc - 2abx + 10ab$.
15.	$ax - 2ab + 2b$.	16.	$3a - ac - 3bx + bcx$.

918

17. $ax + bx + x + ay + by + y$.
18. $ab + a - b - 1$.
19. -2.
20. $-2/5$.
21. $+4298.936$.
22. $-4/9$.
23. $+2333$.
24. See text.
25. $b(a - 2)$.
26. $a(3 + 5) = a(8) = 8a$.
27. $2a(2bc - b + 3)$.
28. $x(a - b + 1)$.
29. $a(3d - 5c + 1)$.
30. $2(2uv - xv + 1)$.
31. $ab(x + y - 1)$.
32. $2a(x - 3y + 2z)$.
33. $x(2 + a + b)$.
34. $-a(b + 3c + 1)$.
35. $(x + 1)(a + b)$.
36. $(x + 1)(1 + y)$.
37. $(x + y)(2 - a)$.
38. $(x + y)(a + b)$.
39. $(x - 2)(x + 1)$.
40. $(2x + 1)(x - 5)$.
41. $(4x - 3)(3x - 4)$.
42. $(5x - 1)(2x + 3)$.
43. $(x - 3)(x + 2)$.
44. $(x + 3)(x - 3)$.
45. $(x - y)(x + y)$.
46. $(2x + 3y)(2x - 3y)$.
47. F.
48. T.
49. T.
50. T.
51. F.
52. T.
53. F.
54. F.
55. T.
56. T.
57. F.
58. T.
59. F.
60. T.

Problem Set A2–4

1. F.
2. F.
3. T.
4. T.
5. T.
6. T.
7. F.
8. F.
9. T.
10. T.
11. T.
12. The product of a number and its reciprocal is one.
13. $3a/2c$.
14. Cancellation not possible as the expression stands.
15. 1.
16. $a + 3$.
17. $6c$.
18. Cancellation not possible as the expression stands.
19. $(2y + 6a + 1)/4y$.
20. $8x$.
21. See text.
22. $3ab/8$.
23. $\dfrac{2a + 2b}{15}$.
24. $-3a/4$.
25. $\dfrac{2ab + 4a}{21}$.
26. $\dfrac{-2}{3a + 3b}$.
27. $a + 6b + 2$.
28. $4a - 6b$.
29. $b(c - d)/2$.
30. $(b - 2)/(a + 3)$.
31. $[2 + (c - b)]/2a$.
32. $[2 + (c - b) + a]/2x(a + b)$.
33. $(x + y)/(y - x)$.
34. $1/3$.
35. $(5a - 6)/10$.
36. $(45b - 5ab + 12a)/30ab$.

37. $(a+24)/6a$.
38. $7/6$.
39. $(a-2ab+4b+bx-2)/b(a-2)$.
40. $(x-2ab+6)/2a$.
41. $(18abx-3ab+1)/6ab$.
42. $(11a+5b-3ab-5)/12(b-1)$.
43. $-(12x+17)/4(x+3)$.
44. $91/12$.
45. $27/176$.
46. $7/6$.
47. $(6b+2ab)/(5b-3a)$.
48. $(3abc-6b+18)/(2bc-6c)$.
49. $(4ac-6b+12bc)/(bc-12)$.
50. $(6a+4b-2c)/(12b-3a)$.

Problem Set A2–5

1. 16.
2. 7.
3. 1.041.
4. $4/3$.
5. $1/72$.
6. 1000.
7. Not a real number.
8. $31/12$.
9. $1/9$.
10. 1.
11. 8.
12. $1/2$.
13. 10.
14. $1/12$.
15. $5/4$.
16. 6.
17. 25.
18. 1.1025.
19. $1/8$.
20. $13/8$.
21. 5.
22. $9/4$.
23. $2/9$.
24. $5/4$.
25. $4/27$.
26. a^3.
27. $a^2b^2c^2$.
28. a^3b^7.
29. $a^3b^2c^3$.
30. x^4ab.
31. ab^2.
32. a^2/bc.
33. y^2b^2/x^2.
34. a^2/x.
35. x/y.
36. bc^3.
37. a^2b^3.
38. b^2/a^3c.
39. $-9bc$.
40. $-18/bc$.
41. ax.
42. a^3bc.
43. $1/ax^2$.
44. $2a/x^3$.
45. $1/ax^3$.
46. $3a^{7/3}b^{3/2}$.
47. $x^{3/2}/y^{1/6}$.
48. $a^{1/6}b^{7/6}$.
49. $2/3^{3/2}x^{1/2}$.
50. $2y^{11/6}/3x^2$.
51. $\dfrac{2x^2+1}{x^3+3x^2}$.
52. $\dfrac{2-x}{x-x^2}$.
53. $\dfrac{x+1}{3x^2}$.
54. $\dfrac{x^2}{a^{2x-2}}$.
55. Exponents in this expression cannot be combined.
56. a^2+ab.
57. $a^2-6ab+9b^2$.
58. $a^3-2a^2b+ab^2$.
59. x^2-y^2.
60. $1+a$.
61. $a-2a^{1/2}+1$.
62. $\dfrac{1+2a}{a^2}$.
63. 27.
64. F.
65. F.
66. T.
67. T.
68. F.
69. F.
70. T.
71. T.
72. F.
73. F.
74. T.
75. F.
76. F.
77. T.
78. F.
79. T.
80. F.
81. T.
82. T.
83. T.
84. F.
85. T.

APPENDIX 3

Problem Set A3–1

1. $2x - 3 = x + 4$ Add 3
   $2x = x + 7$ Subtract x
   $x = 7.$

2. $4x + 5 = 2x + 12$ Subtract 5
   $4x = 2x + 7$ Subtract $2x$
   $2x = 7$ Divide by 2
   $x = 7/2.$

3. $3x - 7 = 2x + 4$ Add 7
   $3x = 2x + 11$ Subtract $2x$
   $x = 11.$

4. $5 - 2x = 6$ Subtract 5
   $-2x = 1$ Divide by -2
   $x = -1/2.$

5. $7x - 5 = 3 - 4x$ Add 5
   $7x = 8 - 4x$ Add $4x$
   $11x = 8$ Divide by 11
   $x = 8/11.$

6. $3 - 2x = x - 4$ Subtract 3
   $-2x = x - 7$ Subtract x
   $-3x = -7$ Divide by -3
   $x = 7/3.$

7. $\dfrac{x}{3} + \dfrac{1}{2} = 3x$ Multiply by 6
   $2x + 3 = 18x$ Subtract 3
   $2x = 18x - 3$ Subtract $18x$
   $-16x = -3$ Divide by -16
   $x = 3/16.$

8. $\dfrac{x}{4} + \dfrac{x}{2} = 2 - \dfrac{5x}{8}$ Multiply by 8
   $2x + 4x = 16 - 5x$ Add $5x$
   $2x + 4x + 5x = 16$ Combine like terms
   $11x = 16$ Divide by 11
   $x = 16/11.$

9. $\dfrac{2x}{5} - \dfrac{3x}{2} = 4$ Multiply by 10
   $4x - 15x = 40$ Combine like terms
   $-11x = 40$ Divide by -11
   $x = -40/11.$

10. $1 - \dfrac{x}{3} + \dfrac{x}{2} = x - 4$ Multiply by 6
    $6 - 2x + 3x = 6x - 24$ Subtract 6
    $-2x + 3x = 6x - 30$ Subtract $6x$
    $-2x + 3x - 6x = -30$ Combine like terms
    $-5x = -30$ Divide by -5
    $x = 6.$

11. $\dfrac{1}{7} - \dfrac{x}{3} = x$ Multiply by 21

$3 - 7x = 21x$ Subtract 3

$-7x = 21x - 3$ Subtract $21x$

$-28x = -3$ Divide by -28

$x = 3/28.$

12. $1 - \dfrac{3x}{7} = 0$ Multiply by 7

$7 - 3x = 0$ Subtract 7

$-3x = -7$ Divide by -3

$x = 7/3.$

13. $\dfrac{x+3}{2} = x - \dfrac{1}{4}$ Multiply by 4

$2(x+3) = 4x - 1$ Apply distributive property

$2x + 6 = 4x - 1$ Subtract 6

$2x = 4x - 7$ Subtract $4x$

$-2x = -7$ Divide by -2

$x = 7/2.$

14. $\dfrac{5-2x}{3} + \dfrac{x}{2} = 1$ Multiply by 6

$2(5 - 2x) + 3x = 6$ Apply distributive property

$10 - 4x + 3x = 6$ Subtract 10

$-4x + 3x = -4$ Combine like terms

$-x = -4$ Divide by -1

$x = 4.$

15. $\dfrac{2x-1}{3} - \dfrac{1-x}{5} = 0$ Multiply by 15

$5(2x - 1) - 3(1 - x) = 0$ Apply distributive property

$10x - 5 - 3 + 3x = 0$ Add 8

$10x + 3x = 8$ Combine like terms

$13x = 8$ Divide by 13

$x = 8/13.$

16. $\dfrac{x+1}{2} - \dfrac{x-1}{3} = 5$ Multiply by 6

$3(x+1) - 2(x-1) = 30$ Apply distributive property.

$3x + 3 - 2x + 2 = 30$ Subtract 5

$3x - 2x = 25$ Combine like terms

$x = 25.$

17. $bx + 2 = c$ Subtract 2

$bx = c - 2$ Divide by b

$x = (c - 2)/b.$

18. $ax + 2 - x = 0$ Subtract 2

$ax - x = -2$ Factor

$x(a - 1) = -2$ Divide by $(a - 1)$

$x = -2/(a - 1).$

19. $\qquad ax + b = cx$ — Subtract cx

$\qquad ax - cx + b = 0$ — Subtract b

$\qquad ax - cx = -b$ — Factor

$\qquad x(a - c) = -b$ — Divide by $(a - c)$,

$\qquad x = b/(c - a).$ — change signs

20. $\quad ax + b = x - b$ — Subtract b

$\qquad ax = x - 2b$ — Subtract x

$\quad ax - x = -2b$ — Factor

$\quad x(a - 1) = -2b$ — Divide by $(a - 1)$

$\qquad x = -2b/(a - 1)$ — Change signs of numerator and denominator

$\qquad x = 2b/(1 - a).$

21. $\quad a(x - a) = 2x$ — Apply distributive property

$\quad ax - a^2 = 2x$ — Add a^2

$\qquad ax = 2x + a^2$ — Subtract $2x$

$\quad ax - 2x = a^2$ — Factor

$\quad x(a - 2) = a^2$ — Divide by $(a - 2)$

$\qquad x = a^2/(a - 2).$

22. $(x/a) - 1/2 = 3x$ — Multiply by $2a$.

$\qquad 2x - a = 4ax$ — Add a

$\qquad 2x = 4ax + a$ — Subtract $4ax$

$\quad 2x - 4ax = a$ — Factor

$\quad x(2 - 4a) = a$ — Divide by $(2 - 4a)$

$\qquad x = a/(2 - 4a).$

23. $\quad 2/(a - x) + 1/3 = 4$ — Multiply by $3(a - x)$

$\qquad 6 + a - x = 12(a - x)$ — Apply distributive property

$\qquad 6 + a - x = 12a - 12x$ — Add $12x$

$\qquad 6 + a + 11x = 12a$ — Subtract $(6 + a)$

$\qquad 11x = 11a - 6$ — Divide by 11

$\qquad x = (11a - 6)/11.$

24. $\qquad y = x/(b - cx)$ — Multiply by $(b - cx)$

$\quad y(b - cx) = x$ — Apply distributive property

$\quad yb - ycx = x$ — Add ycx

$\qquad yb = x + ycx$ — Factor

$\qquad yb = x(1 + yc)$ — Divide by $(1 + yc)$

$\quad yb/(1 + yc) = x$

or

$\qquad x = yb/(1 + yc).$

25. $(3/4) - (2x/3) = 2x(a - 1)$ — Multiply by 12

$\qquad 9 - 8x = 24x(a - 1)$ — Apply distributive property

$\qquad 9 - 8x = 24ax - 24x$ — Add $8x$

$\qquad 9 = 24ax - 16x$ — Factor

$\qquad 9 = x(24a - 16)$ — Divide by $(24a - 16)$

$\quad 9/(24a - 16) = x$

or

$\qquad x = 9/(24a - 16).$

26. $3(x-2) = 2 - a(x+2)$ Apply distributive property

        $3x - 6 = 2 - ax - 2a$ Add ax

   $3x + ax - 6 = 2 - 2a$ Add 6

     $3x + ax = 8 - 2a$ Factor

      $x(3 + a) = 8 - 2a$ Divide by $(3 + a)$

        $x = (8 - 2a)/(3 + a).$

27. $2/3(x-2) + 3/a - 1/2 = 0$ Multiply by $6a(x-2)$

   $4a + 18(x-2) - 3a(x-2) = 0$ Apply distributive property

   $4a + 18x - 36 - 3ax + 6a = 0$ Subtract $10a$

       $18x - 36 - 3ax = -10a$ Add 36

         $18x - 3ax = 36 - 10a$ Factor

        $x(18 - 3a) = 36 - 10a$ Divide by $(18 - 3a)$

           $x = (36 - 10a)/$

              $(18 - 3a).$

28. $ax - b/2 = c + 5[a - 2(b-x)]/6$ Multiply by 6

     $6ax - 3b = 6c + 5[a - 2(b-x)]$ Apply distributive property

     $6ax - 3b = 6c + 5[a - 2b + 2x]$ Apply distributive property

     $6ax - 3b = 6c + 5a - 10b + 10x$ Subtract $10x$

  $6ax - 10x - 3b = 6c + 5a - 10b$ Add $3b$

      $6ax - 10x = 6c + 5a - 7b$ Factor

     $x(6a - 10) = 6c + 5a - 7b$ Divide by $(6a - 10)$

         $x = (6c + 5a - 7b)/$

            $(6a - 10).$

29. $b/a - x = 2a(b - x)$ Multiply by a

      $b - ax = 2a^2(b - x)$ Apply distributive property

      $b - ax = 2a^2b - 2a^2x$ Add $2a^2x$

  $2a^2x + b - ax = 2a^2b$ Subtract b

    $2a^2x - ax = 2a^2b - b$ Factor

    $x(2a^2 - a) = 2a^2b - b$ Divide by $(2a^2 - a)$

         $x = (2a^2b - b)/(2a^2 - a).$

30. $x - a(b - x) = 2x - 3$ Apply distributive property

    $x - ab + ax = 2x - 3$ Subtract $2x$

  $-x - ab + ax = -3$ Add ab

     $-x + ax = -3 + ab$ Factor

   $x(-1 + a) = -3 + ab$ Divide by $(-1 + a)$

        $x = (-3 + ab)/(-1 + a)$ Use commutative property

        $x = (ab - 3)/(a - 1).$

31. $3(b - x) = 2 + b[x - (3 - x)]$ Apply distributive property

     $3b - 3x = 2 + b[x - 3 + x]$ Apply distributive property

     $3b - 3x = 2 + bx - 3b + bx$ Subtract $2bx$

  $3b - 3x - 2bx = 2 - 3b$ Subtract $3b$

   $-3x - 2bx = 2 - 6b$ Factor

   $x(-3 - 2b) = 2 - 6b$ Divide by $(-3 - 2b)$

        $x = (2 - 6b)/(-3 - 2b)$ Change signs of numerator and denominator

        $x = (6b - 2)/(2b + 3).$

32.
$$b(a + x) = a(b + x)$$
$$ba + bx = ab + ax$$
$$ba + bx - ax = ab$$
$$bx - ax = 0$$
$$x(b - a) = 0$$
$$x = 0.$$

Apply distributive property
Subtract ax
Subtract ba
Factor
Divide by $(b - a)$

Problem Set A3–2

1. 15.
2. 41.
3. 180.
4. 2.
5. 104.
6. 30.
7. 64.
8. 9.
9. 32.
10. 472.
11. 9/40.
12. 5/42.
13. 115/567.
14. −11/60.
15. 183/280.
16. −4435/192.
17. −27.
18. 2/3.
19. 16/21.
20. −122/453.
21. $33,083.33.
22. 0.50.
23. *a)* 37.5 percent. *b)* $6.00. *c)* $4.48.
24. *a)* 320 days. *b)* $16,200.
25. *a)* 212°F. *b)* 32°F. *c)* 12.92°F.
    d) 82.4°F. *e)* 14°F. *f)* 122°F.
26. *a)* $E = 0.2T + 0.125(T - 50,000)$.
    b) $50,000; $65,384.62; $80,769.23.
    c) $83,333.33.
27. *a)* $B = x[P - y(P - B)]$.
    b) $362,500.
28. *a)* 22.9 percent.
    b) $3.03.
29. *a)* $20,000. *b)* 0.2 *c)* 0.375.
30. *a)* $10.01. *b)* $5.01. *c)* 100. *d)* 250. *e)* 500.

Problem Set A3–3

Problems 1–10: See Figures A–J.

11. $x_1 = 9.69; x_2 = 0.31$.
12. $x_1 = 0.35; x_2 = -2.85$.
13. $x_1 = x_2 = 3/2$.
14. No real solutions.
15. $x_1 = 0; x_2 = 2$.
16. $x_1 = 2; x_2 = 1$.

FIGURE A

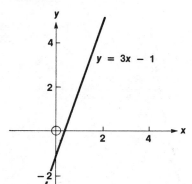

$y = 3x - 1$

FIGURE B

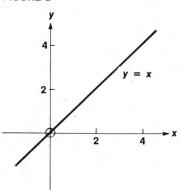

$y = x$

FIGURE C

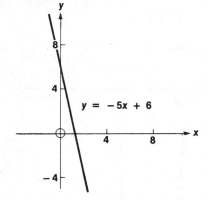

$y = -5x + 6$

FIGURE D

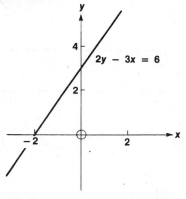

$2y - 3x = 6$

FIGURE E

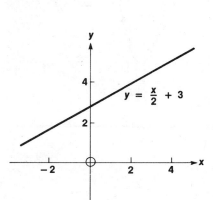

$y = \dfrac{x}{2} + 3$

FIGURE F

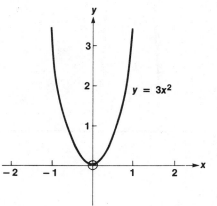

$y = 3x^2$

FIGURE G

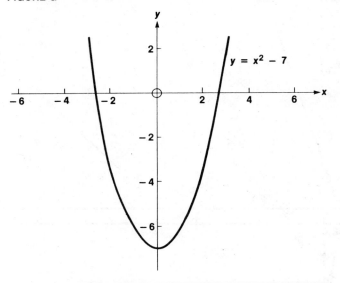

$y = x^2 - 7$

FIGURE H

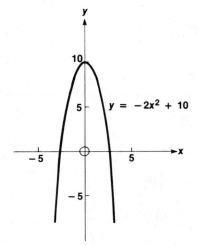

$y = -2x^2 + 10$

FIGURE I

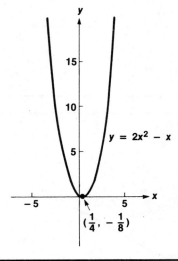

$y = 2x^2 - x$

$(\frac{1}{4}, -\frac{1}{8})$

FIGURE J

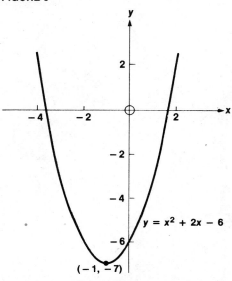

$y = x^2 + 2x - 6$

$(-1, -7)$

17. $x_1 = 0;\ x_2 = 2.$
18. $x_1 = -5/2;\ x_2 = 4/3.$
19. $x_1 = 5;\ x_2 = -5.$
20. $x_1 = -7/2;\ x_2 = 1.$

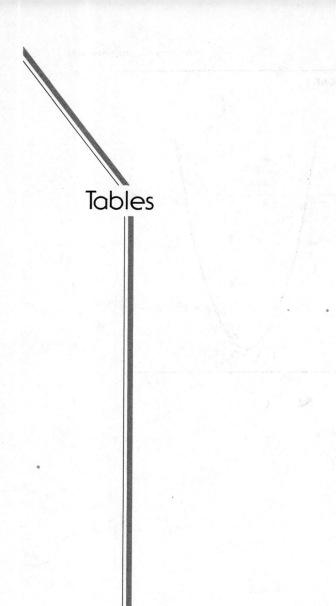

Tables

TABLE I
Common logarithms: 100–549

N	0	1	2	3	4	5	6	7	8	9
10	0000	0043	0086	0128	0170	0212	0253	0294	0334	0374
11	0414	0453	0492	0531	0569	0607	0645	0682	0719	0755
12	0792	0828	0864	0899	0934	0969	1004	1038	1072	1106
13	1139	1173	1206	1239	1271	1303	1335	1367	1399	1430
14	1461	1492	1523	1553	1584	1614	1644	1673	1703	1732
15	1761	1790	1818	1847	1875	1903	1931	1959	1987	2014
16	2041	2068	2095	2122	2148	2175	2201	2227	2253	2279
17	2304	2330	2355	2380	2405	2430	2455	2480	2504	2529
18	2553	2577	2601	2625	2648	2672	2695	2718	2742	2765
19	2788	2810	2833	2856	2878	2900	2923	2945	2967	2989
20	3010	3032	3054	3075	3096	3118	3139	3160	3181	3201
21	3222	3243	3263	3284	3304	3324	3345	3365	3385	3404
22	3424	3444	3464	3483	3502	3522	3541	3560	3579	3598
23	3617	3636	3655	3674	3692	3711	3729	3747	3766	3784
24	3802	3820	3838	3856	3874	3892	3909	3927	3945	3962
25	3979	3997	4014	4031	4048	4065	4082	4099	4116	4133
26	4150	4166	4183	4200	4216	4232	4249	4265	4281	4298
27	4314	4330	4346	4362	4378	4393	4409	4425	4440	4456
28	4472	4487	4502	4518	4533	4548	4564	4579	4594	4609
29	4624	4639	4654	4669	4683	4698	4713	4728	4742	4757
30	4771	4786	4800	4814	4829	4843	4857	4871	4886	4900
31	4914	4928	4942	4955	4969	4983	4997	5011	5024	5038
32	5051	5065	5079	5092	5105	5119	5132	5145	5159	5172
33	5185	5198	5211	5224	5237	5250	5263	5276	5289	5302
34	5315	5328	5340	5353	5366	5378	5391	5403	5416	5428
35	5441	5453	5465	5478	5490	5502	5514	5527	5539	5551
36	5563	5575	5587	5599	5611	5623	5635	5647	5658	5670
37	5682	5694	5705	5717	5729	5740	5752	5763	5775	5786
38	5798	5809	5821	5832	5843	5855	5866	5877	5888	5899
39	5911	5922	5933	5944	5955	5966	5977	5988	5999	6010
40	6021	6031	6042	6053	6064	6075	6085	6096	6107	6117
41	6128	6138	6149	6160	6170	6180	6191	6201	6212	6222
42	6232	6243	6253	6263	6274	6284	6294	6304	6314	6325
43	6335	6345	6355	6365	6375	6385	6395	6405	6415	6425
44	6435	6444	6454	6464	6474	6484	6493	6503	6513	6522
45	6532	6542	6551	6561	6571	6580	6590	6599	6609	6618
46	6628	6637	6646	6656	6665	6675	6684	6693	6702	6712
47	6721	6730	6739	6749	6758	6767	6776	6785	6794	6803
48	6812	6821	6830	6839	6848	6857	6866	6875	6884	6893
49	6902	6911	6920	6928	6937	6946	6955	6964	6972	6981
50	6990	6998	7007	7016	7024	7033	7042	7050	7059	7067
51	7076	7084	7093	7101	7110	7118	7126	7135	7143	7152
52	7160	7168	7177	7185	7193	7202	7210	7218	7226	7235
53	7243	7251	7259	7267	7275	7284	7292	7300	7308	7316
54	7324	7332	7340	7348	7356	7364	7372	7380	7388	7396

Taken by permission from Ernest Kurnow, Gerald J. Glasser, and Frederick R. Ottman, *Statistics for Business Decisions* (Homewood, Ill.: Richard D. Irwin, Inc., 1959), pp. 507–8.

TABLE I *(continued)*
Common logarithms: 550–999

N	0	1	2	3	4	5	6	7	8	9
55	7404	7412	7419	7427	7435	7443	7451	7459	7466	7474
56	7482	7490	7497	7505	7513	7520	7528	7536	7543	7551
57	7559	7566	7574	7582	7589	7597	7604	7612	7619	7627
58	7634	7642	7649	7657	7664	7672	7679	7686	7694	7701
59	7709	7716	7723	7731	7738	7745	7752	7760	7767	7774
60	7782	7789	7796	7803	7810	7818	7825	7832	7839	7846
61	7853	7860	7868	7875	7882	7889	7896	7903	7910	7917
62	7924	7931	7938	7945	7952	7959	7966	7973	7980	7987
63	7993	8000	8007	8014	8021	8028	8035	8041	8048	8055
64	8062	8069	8075	8082	8089	8096	8102	8109	8116	8122
65	8129	8136	8142	8149	8156	8162	8169	8176	8182	8189
66	8195	8202	8209	8215	8222	8228	8235	8241	8248	8254
67	8261	8267	8274	8280	8287	8293	8299	8306	8312	8319
68	8325	8331	8338	8344	8351	8357	8363	8370	8376	8382
69	8388	8395	8401	8407	8414	8420	8426	8432	8439	8445
70	8451	8457	8463	8470	8476	8482	8488	8494	8500	8506
71	8513	8519	8525	8531	8537	8543	8549	8555	8561	8567
72	8573	8579	8585	8591	8597	8603	8609	8615	8621	8627
73	8633	8639	8645	8651	8657	8663	8669	8675	8681	8686
74	8692	8698	8704	8710	8716	8722	8727	8733	8739	8745
75	8751	8756	8762	8768	8774	8779	8785	8791	8797	8802
76	8808	8814	8820	8825	8831	8837	8842	8848	8854	8859
77	8865	8871	8876	8882	8887	8893	8899	8904	8910	8915
78	8921	8927	8932	8938	8943	8949	8954	8960	8965	8971
79	8976	8982	8987	8993	8998	9004	9009	9015	9020	9025
80	9031	9036	9042	9047	9053	9058	9063	9069	9074	9079
81	9085	9090	9096	9101	9106	9112	9117	9122	9128	9133
82	9138	9143	9149	9154	9159	9165	9170	9175	9180	9186
83	9191	9196	9201	9206	9212	9217	9222	9227	9232	9238
84	9243	9248	9253	9258	9263	9269	9274	9279	9284	9289
85	9294	9299	9304	9309	9315	9320	9325	9330	9335	9340
86	9345	9350	9355	9360	9365	9370	9375	9380	9385	9390
87	9395	9400	9405	9410	9415	9420	9425	9430	9435	9440
88	9445	9450	9455	9460	9465	9469	9474	9479	9484	9489
89	9494	9499	9504	9509	9513	9518	9523	9528	9533	9538
90	9542	9547	9552	9557	9562	9566	9571	9576	9581	9586
91	9590	9595	9600	9605	9609	9614	9619	9624	9628	9633
92	9638	9643	9647	9652	9657	9661	9666	9671	9675	9680
93	9685	9689	9694	9699	9703	9708	9713	9717	9722	9727
94	9731	9736	9741	9745	9750	9754	9759	9763	9768	9773
95	9777	9782	9786	9791	9795	9800	9805	9809	9814	9818
96	9823	9827	9832	9836	9841	9845	9850	9854	9859	9863
97	9868	9872	9877	9881	9886	9890	9894	9899	9903	9908
98	9912	9917	9921	9926	9930	9934	9939	9943	9948	9952
99	9956	9961	9965	9969	9974	9978	9983	9987	9991	9996

TABLE II
Compound amount of $1 $(1 + i)^n$

n	1%	2%	3%	4%
1	1.01000	1.02000	1.03000	1.04000
2	1.02010	1.04040	1.06090	1.08160
3	1.03030	1.06121	1.09273	1.12486
4	1.04060	1.08243	1.12551	1.16986
5	1.05101	1.10408	1.15927	1.21665
6	1.06152	1.12616	1.19405	1.26532
7	1.07214	1.14869	1.22987	1.31593
8	1.08286	1.17166	1.26677	1.36857
9	1.09369	1.19509	1.30477	1.42331
10	1.10462	1.21899	1.34392	1.48024
11	1.11567	1.24337	1.38423	1.53945
12	1.12683	1.26824	1.42576	1.60103
13	1.13809	1.29361	1.46853	1.66507
14	1.14947	1.31948	1.51259	1.73168
15	1.16097	1.34587	1.55797	1.80094
16	1.17258	1.37279	1.60471	1.87298
17	1.18430	1.40024	1.65285	1.94790
18	1.19615	1.42825	1.70243	2.02582
19	1.20811	1.45681	1.75351	2.10685
20	1.22019	1.48595	1.80611	2.19112
21	1.23239	1.51567	1.86029	2.27877
22	1.24472	1.54598	1.91610	2.36992
23	1.25716	1.57690	1.97359	2.46472
24	1.26973	1.60844	2.03279	2.56330
25	1.28243	1.64061	2.09378	2.66584
26	1.29526	1.67342	2.15659	2.77247
27	1.30821	1.70689	2.22129	2.88337
28	1.32129	1.74102	2.28793	2.99870
29	1.33450	1.77584	2.35657	3.11865
30	1.34785	1.81136	2.42726	3.24340
31	1.36133	1.84759	2.50008	3.37313
32	1.37494	1.88454	2.57508	3.50806
33	1.38869	1.92223	2.65234	3.64838
34	1.40258	1.96068	2.73191	3.79432
35	1.41660	1.99989	2.81386	3.94609
36	1.43077	2.03989	2.89828	4.10393
37	1.44508	2.08069	2.98523	4.26809
38	1.45953	2.12230	3.07478	4.43881
39	1.47412	2.16474	3.16703	4.61637
40	1.48886	2.20804	3.26204	4.80102

TABLE II *(continued)*
Compound amount of $1 $(1 + i)^n$

n	5%	6%	7%	8%
1	1.05000	1.06000	1.07000	1.08000
2	1.10250	1.12360	1.14490	1.16640
3	1.15762	1.19102	1.22504	1.25971
4	1.21551	1.26248	1.31080	1.36049
5	1.27628	1.33823	1.40255	1.46933
6	1.34010	1.41852	1.50073	1.58687
7	1.40710	1.50363	1.60578	1.71382
8	1.47746	1.59385	1.71819	1.85093
9	1.55133	1.68975	1.83846	1.99900
10	1.62889	1.79085	1.96715	2.15892
11	1.71034	1.89830	2.10485	2.33164
12	1.79586	2.01220	2.25219	2.51817
13	1.88565	2.13293	2.40985	2.71962
14	1.97993	2.26090	2.57853	2.93719
15	2.07893	2.39656	2.75903	3.17217
16	2.18287	2.54035	2.95216	3.42594
17	2.29202	2.69277	3.15882	3.70002
18	2.40662	2.85434	3.37993	3.99602
19	2.52695	3.02560	3.61653	4.31570
20	2.65330	3.20714	3.86968	4.66096
21	2.78596	3.39956	4.14056	5.03383
22	2.92526	3.60354	4.43040	5.43654
23	3.07152	3.81975	4.74053	5.87146
24	3.22510	4.04893	5.07237	6.34118
25	3.38635	4.29187	5.42743	6.84848
26	3.55567	4.54938	5.80735	7.39635
27	3.73346	4.82235	6.21387	7.98806
28	3.92013	5.11169	6.64884	8.62711
29	4.11614	5.41839	7.11426	9.31727
30	4.32194	5.74349	7.61226	10.06266
31	4.53804	6.08810	8.14511	10.86767
32	4.76494	6.45339	8.71527	11.73708
33	5.00319	6.84059	9.32534	12.67605
34	5.25335	7.25103	9.97811	13.69013
35	5.51602	7.68609	10.67658	14.78534
36	5.79182	8.14725	11.42394	15.96817
37	6.08141	8.63608	12.22362	17.24563
38	6.38548	9.15425	13.07927	18.62528
39	6.70475	9.70351	13.99482	20.11530
40	7.03999	10.28572	14.97446	21.72452

TABLE III
Present value of $1 $(1 + i)^{-n}$

n	1%	2%	3%	4%
1	0.990099	0.980392	0.970874	0.961538
2	0.980296	0.961169	0.942596	0.924556
3	0.970590	0.942322	0.915142	0.888996
4	0.960980	0.923845	0.888487	0.854804
5	0.951466	0.905731	0.862609	0.821927
6	0.942045	0.887971	0.837484	0.790315
7	0.932718	0.870560	0.831092	0.759918
8	0.923483	0.853490	0.789409	0.730690
9	0.914340	0.836755	0.766417	0.702587
10	0.905287	0.820348	0.744094	0.675564
11	0.896324	0.804263	0.722421	0.649581
12	0.887449	0.788493	0.701380	0.624597
13	0.878663	0.773033	0.680951	0.600574
14	0.869963	0.757875	0.661118	0.577475
15	0.861349	0.743015	0.641862	0.555265
16	0.852821	0.728446	0.623167	0.533908
17	0.844377	0.714163	0.605016	0.513373
18	0.836017	0.700159	0.587395	0.493628
19	0.827740	0.686431	0.570286	0.474642
20	0.819544	0.672971	0.553676	0.456387
21	0.811430	0.659776	0.537549	0.438834
22	0.803396	0.646839	0.521893	0.421955
23	0.795442	0.634156	0.506692	0.405726
24	0.787566	0.621721	0.491934	0.390121
25	0.779768	0.609531	0.477606	0.375117
26	0.772048	0.597579	0.463695	0.360689
27	0.764404	0.585862	0.450189	0.346817
28	0.756836	0.574375	0.437077	0.333477
29	0.749342	0.563112	0.424346	0.320651
30	0.741923	0.552071	0.411987	0.308319
31	0.734577	0.541246	0.399987	0.296460
32	0.727304	0.530633	0.388337	0.285058
33	0.720103	0.520229	0.377026	0.274094
34	0.712973	0.510028	0.366045	0.263552
35	0.705914	0.500028	0.355383	0.253415
36	0.698925	0.490223	0.345032	0.243669
37	0.692005	0.480611	0.334983	0.234297
38	0.685153	0.471187	0.325226	0.225285
39	0.678370	0.461948	0.315754	0.216621
40	0.671653	0.452890	0.306557	0.208289

TABLE III *(continued)*
Present value of $1 $(1 + i)^{-n}$

n	5%	6%	7%	8%
1	0.952381	0.943396	0.934579	0.925926
2	0.907029	0.889996	0.873439	0.857339
3	0.863838	0.839619	0.816298	0.793832
4	0.822702	0.792094	0.762895	0.735030
5	0.783526	0.747258	0.712986	0.680583
6	0.746215	0.704961	0.666342	0.630170
7	0.710681	0.665057	0.622750	0.583490
8	0.676839	0.627412	0.582009	0.540269
9	0.644609	0.591898	0.543934	0.500249
10	0.613913	0.558395	0.508349	0.463193
11	0.584679	0.526788	0.475093	0.428883
12	0.556837	0.496969	0.444012	0.397114
13	0.530321	0.468839	0.414964	0.367698
14	0.505068	0.442301	0.387817	0.340461
15	0.481017	0.417265	0.362446	0.315242
16	0.458112	0.393646	0.338735	0.291890
17	0.436297	0.371364	0.316574	0.270269
18	0.415521	0.350344	0.295864	0.250249
19	0.395734	0.330513	0.276508	0.231712
20	0.376889	0.311805	0.258419	0.214548
21	0.358942	0.294155	0.241513	0.198656
22	0.341850	0.277505	0.225713	0.183941
23	0.325571	0.261797	0.210947	0.170315
24	0.310068	0.246979	0.197147	0.157699
25	0.295303	0.232999	0.184249	0.146018
26	0.281241	0.219810	0.172195	0.135202
27	0.267848	0.207368	0.160930	0.125187
28	0.255094	0.195630	0.150402	0.115914
29	0.242946	0.184557	0.140563	0.107328
30	0.231377	0.174110	0.131367	0.099377
31	0.220359	0.164255	0.122773	0.092016
32	0.209866	0.154957	0.114741	0.085200
33	0.199873	0.146186	0.107235	0.078889
34	0.190355	0.137912	0.100219	0.073045
35	0.181290	0.130105	0.093663	0.067635
36	0.172657	0.122741	0.087535	0.062625
37	0.164436	0.115793	0.081809	0.057986
38	0.156605	0.109239	0.076457	0.053690
39	0.149148	0.103056	0.071455	0.049713
40	0.142046	0.097222	0.066780	0.046031

TABLE IV

Amount of $1 per period: $s_{\overline{n}|i} = \dfrac{(1+i)^n - 1}{i}$

n	1%	2%	3%	4%
1	1.00000	1.00000	1.00000	1.00000
2	2.01000	2.02000	2.03000	2.04000
3	3.03010	3.06040	3.09090	3.12160
4	4.06040	4.12161	4.18363	4.24646
5	5.10101	5.20404	5.30914	5.41632
6	6.15202	6.30812	6.46841	6.63298
7	7.21354	7.43428	7.66246	7.89829
8	8.28567	8.58297	8.89234	9.21423
9	9.36853	9.75463	10.15911	10.58280
10	10.46221	10.94972	11.46388	12.00611
11	11.56683	12.16872	12.80780	13.48635
12	12.68250	13.41209	14.19203	15.02581
13	13.80933	14.68033	15.61779	16.62684
14	14.94742	15.97394	17.08632	18.29191
15	16.09690	17.29342	18.59891	20.02359
16	17.25786	18.63929	20.15688	21.82453
17	18.43044	20.01207	21.76159	23.69751
18	19.61475	21.41231	23.41444	25.64541
19	20.81090	22.84056	25.11687	27.67123
20	22.01900	24.29737	26.87037	29.77808
21	23.23919	25.78332	28.67649	31.96920
22	24.47159	27.29898	30.53678	34.24797
23	25.71630	28.84496	32.45288	36.61789
24	26.97346	30.42186	34.42647	39.08260
25	28.24320	32.03030	36.45926	41.64591
26	29.52563	33.67091	38.55304	44.31174
27	30.82089	35.34432	40.70963	47.08421
28	32.12910	37.05121	42.93092	49.96758
29	33.45039	38.79223	45.21885	52.96629
30	34.78489	40.56808	47.57542	56.08494
31	36.13274	42.37944	50.00268	59.32834
32	37.49407	44.22703	52.50276	62.70147
33	38.86901	46.11157	55.07784	66.20953
34	40.25770	48.03380	57.73018	69.85791
35	41.66028	49.99448	60.46208	73.65222
36	43.07688	51.99437	63.27594	77.59831
37	44.50765	54.03425	66.17422	81.70225
38	45.95272	56.11494	69.15945	85.97034
39	47.41225	58.23724	72.23423	90.40915
40	48.88637	60.40198	75.40126	95.02552

TABLE IV *(continued)*

Amount of \$1 per period: $s_{\overline{n}|i} = \dfrac{(1+i)^n - 1}{i}$

n	5%	6%	7%	8%
1	1.00000	1.00000	1.00000	1.00000
2	2.05000	2.06000	2.07000	2.08000
3	3.15250	3.18360	3.21490	3.24640
4	4.31012	4.37462	4.43994	4.50611
5	5.52563	5.63709	5.75074	5.86660
6	6.80191	6.97532	7.15329	7.33593
7	8.14201	8.39384	8.65402	8.92280
8	9.54911	9.89747	10.25980	10.63663
9	11.02656	11.49132	11.97799	12.48756
10	12.57789	13.18079	13.81645	14.48656
11	14.20679	14.97164	15.78360	16.64549
12	15.91713	16.86994	17.88845	18.97713
13	17.71298	18.88214	20.14064	21.49530
14	19.59863	21.01507	22.55049	24.21492
15	21.57856	23.27597	25.12902	27.15211
16	23.65749	25.67253	27.88805	30.32428
17	25.84037	28.21288	30.84022	33.75023
18	28.13238	30.90565	33.99903	37.45024
19	30.53900	33.75999	37.37896	41.44626
20	33.06595	36.78559	40.99549	45.76196
21	35.71925	39.99273	44.86518	50.42292
22	38.50521	43.39229	49.00574	55.45676
23	41.43048	46.99583	53.43614	60.89330
24	44.50200	50.81558	58.17667	66.76476
25	47.72710	54.86451	63.24904	73.10594
26	51.11345	59.15638	68.67647	79.95442
27	54.66913	63.70577	74.48382	87.35077
28	58.40258	68.52811	80.69769	95.33883
29	62.32271	73.63980	87.34653	103.96594
30	66.43885	79.05819	94.46079	113.28321
31	70.76079	84.80168	102.07304	123.34587
32	75.29883	90.88978	110.21815	134.21354
33	80.06377	97.34316	118.93343	145.95062
34	85.06696	104.18375	128.25876	158.62667
35	90.32031	111.43478	138.23688	172.31680
36	95.83632	119.12087	148.91346	187.10215
37	101.62814	127.26812	160.33740	203.07032
38	107.70955	135.90421	172.56102	220.31595
39	114.09502	145.05846	185.64029	238.94122
40	120.79977	154.76197	199.63511	259.05652

TABLE V

Present value of $1 per period: $a_{\overline{n}|i} = \dfrac{1-(1+i)^{-n}}{i}$

n	1%	2%	3%	4%
1	0.99010	0.98039	0.97087	0.96154
2	1.97040	1.94156	1.91347	1.88609
3	2.94099	2.88388	2.82861	2.77509
4	3.90197	3.80773	3.71710	3.62990
5·	4.85343	4.71346	4.57971	4.45182
6	5.79548	5.60143	5.41719	5.24214
7	6.72819	6.47199	6.23028	6.00205
8	7.65168	7.32548	7.01969	6.73274
9	8.56602	8.16224	7.78611	7.43533
10	9.47130	8.98259	8.53020	8.11090
11	10.36763	9.78685	9.25262	8.76048
12	11.25508	10.57534	9.95400	9.38507
13	12.13374	11.34837	10.63496	9.98565
14	13.00370	12.10625	11.29607	10.56312
15	13.86505	12.84926	11.93794	11.11839
16	14.71787	13.57771	12.56110	11.65230
17	15.56225	14.29187	13.16612	12.16567
18 . . : . .	16.39827	14.99203	13.75351	12.65930
19	17.22601	15.67846	14.32380	13.13394
20	18.04555	16.35143	14.87747	13.59033
21	18.85698	17.01121	15.41502	14.02916
22	19.66038	17.65805	15.93692	14.45112
23	20.45582	18.29220	16.44361	14.85684
24	21.24339	18.91393	16.93554	15.24696
25	22.02316	19.52346	17.41315	15.62208
26	22.79520	20.12104	17.87684	15.98277
27	23.55961	20.70690	18.32703	16.32959
28	24.31644	21.28127	18.76411	16.66306
29	25.06579	21.84438	19.18845	16.98371
30	25.80771	22.39646	19.60044	17.29203
31	26.54229	22.93770	20.00043	17.58849
32	27.26959	23.46833	20.38877	17.87355
33	27.98969	23.98856	20.76579	18.14765
34	28.70267	24.49859	21.13184	18.41120
35	29.40858	24.99862	21.48722	18.66461
36	30.10751	25.48884	21.83225	18.90828
37	30.79951	25.96945	22.16724	19.14258
38	31.48466	26.44064	22.49246	19.36786
39	32.16303	26.90259	22.80822	19.58448
40	32.83469	27.35548	23.11477	19.79277

TABLE V *(continued)*

Present value of $1 per period: $a_{\overline{n}|i} = \dfrac{1-(1+i)^{-n}}{i}$

n	5%	6%	7%	8%
1	0.95238	0.94340	0.93458	0.92593
2	1.85941	1.83339	1.80802	1.78326
3	2.72325	2.67301	2.62432	2.57710
4	3.54595	3.46511	3.38721	3.31213
5	4.32948	4.21236	4.10020	3.99271
6	5.07569	4.91732	4.76654	4.62288
7	5.78637	5.58238	5.38929	5.20637
8	6.46321	6.20979	5.97130	5.74664
9	7.10782	6.80169	6.51523	6.24689
10	7.72173	7.36009	7.02358	6.71008
11	8.30641	7.88687	7.49867	7.13896
12	8.86325	7.38384	7.94269	7.53608
13	9.39357	8.85268	8.35765	7.90378
14	9.89864	9.29498	8.74547	8.24424
15	10.37966	9.71225	9.10791	8.55948
16	10.83777	10.10590	9.44665	8.85137
17	11.27407	10.47726	9.76322	9.12164
18	11.68959	10.82760	10.05909	9.37189
19	12.08532	11.15812	10.33560	9.60360
20	12.46221	11.46992	10.59401	9.81815
21	12.82115	11.76408	10.83553	10.01680
22	13.16300	12.04158	11.06124	10.20074
23	13.48857	12.30338	11.27219	10.37106
24	13.79864	12.55036	11.46933	10.52876
25	14.09394	12.78336	11.65358	10.67478
26	14.37519	13.00317	11.82578	10.80998
27	14.64303	13.21053	11.98671	10.93516
28	14.89813	13.40616	12.13711	11.05108
29	15.14107	13.59072	12.27767	11.15841
30	15.37245	13.76483	12.40904	11.25778
31	15.59281	13.92909	12.53181	11.34980
32	15.80268	14.08404	12.64656	11.43500
33	16.00255	14.23023	12.75379	11.51389
34	16.19290	14.36814	12.85401	11.58693
35	16.37419	14.49825	12.94767	11.65457
36	16.54685	14.62099	13.03521	11.71719
37	16.71129	14.73678	13.11702	11.77518
38	16.86789	14.84602	13.19347	11.82887
39	17.01704	14.94907	13.26593	11.87858
40	17.15909	15.04630	13.33171	11.92461

TABLE VI

Per-period equivalent of $1 present value: $\dfrac{1}{a_{\overline{n}|i}} = \dfrac{i}{1-(1+i)^{-n}}$

n	1%	2%	3%	4%
1	1.010000	1.020000	1.030000	1.040000
2	0.507512	0.515050	0.522611	0.530196
3	0.340022	0.346755	0.353530	0.360349
4	0.256281	0.262624	0.269027	0.275490
5	0.206040	0.212158	0.218355	0.224627
6	0.172548	0.178526	0.184598	0.190762
7	0.148628	0.154512	0.160506	0.166610
8	0.130690	0.136510	0.142456	0.148528
9	0.116740	0.122515	0.128434	0.134493
10	0.105582	0.111327	0.117231	0.123291
11	0.096454	0.102178	0.108077	0.114149
12	0.088848	0.094560	0.100462	0.106552
13	0.082415	0.088118	0.094030	0.100144
14	0.076901	0.082602	0.088526	0.094669
15	0.072124	0.077825	0.083767	0.089941
16	0.067945	0.073650	0.079611	0.085820
17	0.064258	0.069970	0.075953	0.082199
18	0.060982	0.066702	0.072709	0.078993
19	0.058052	0.063782	0.069814	0.076139
20	0.055415	0.061157	0.067216	0.073582
21	0.053031	0.058785	0.064872	0.071280
22	0.050864	0.056631	0.062747	0.069199
23	0.048886	0.054668	0.060814	0.067309
24	0.047073	0.052871	0.059047	0.065587
25	0.045407	0.051220	0.057428	0.064012
26	0.043869	0.049699	0.055938	0.062567
27	0.042446	0.048293	0.054564	0.061239
28	0.041124	0.046990	0.053293	0.060013
29	0.039895	0.045778	0.052115	0.058880
30	0.038748	0.044650	0.051019	0.057830
31	0.037676	0.043596	0.049999	0.056855
32	0.036671	0.042611	0.049047	0.055949
33	0.035727	0.041687	0.048156	0.055104
34	0.034840	0.040819	0.047322	0.054315
35	0.034004	0.040002	0.046539	0.053577
36	0.033214	0.039233	0.045804	0.052887
37	0.032468	0.038507	0.045112	0.052240
38	0.031761	0.037821	0.044459	0.051632
39	0.031092	0.037171	0.043844	0.051061
40	0.030456	0.036556	0.043262	0.050523

TABLE VI *(continued)*

Per-period equivalent of $1 present value: $\dfrac{1}{a_{\overline{n}|i}} = \dfrac{i}{1 - (1+i)^{-n}}$

n	5%	6%	7%	8%
1	1.050000	1.060000	1.070000	1.080000
2	0.537805	0.545437	0.553092	0.560769
3	0.367209	0.374110	0.381052	0.388034
4	0.282012	0.288591	0.295228	0.301921
5	0.230975	0.237396	0.243891	0.250456
6	0.197017	0.203363	0.209796	0.216315
7	0.172820	0.179135	0.185553	0.192072
8	0.154722	0.161036	0.167468	0.174015
9	0.140690	0.147022	0.153486	0.161080
10	0.129505	0.135868	0.142378	0.149029
11	0.120389	0.126793	0.133357	0.140076
12	0.112825	0.119277	0.125902	0.132695
13	0.106456	0.112960	0.119651	0.126522
14	0.101024	0.107585	0.114345	0.121297
15	0.096342	0.102963	0.109795	0.116830
16	0.092270	0.098952	0.105858	0.112977
17	0.088699	0.095445	0.102425	0.109629
18	0.085546	0.092357	0.099412	0.106702
19	0.082745	0.089621	0.096753	0.104128
20	0.080243	0.087185	0.094393	0.101852
21	0.077996	0.085005	0.092289	0.099832
22	0.075971	0.083046	0.090406	0.098032
23	0.074137	0.081278	0.088714	0.096422
24	0.072471	0.079679	0.087189	0.094978
25	0.070952	0.078227	0.085811	0.093679
26	0.069564	0.076904	0.084561	0.092507
27	0.068292	0.075697	0.083426	0.091448
28	0.067123	0.074593	0.082392	0.090489
29	0.066046	0.073580	0.081449	0.089619
30	0.065051	0.072649	0.080586	0.088827
31	0.064132	0.071792	0.079797	0.088107
32	0.063280	0.071002	0.079073	0.087451
33	0.062490	0.070273	0.078408	0.086852
34	0.061755	0.069598	0.077797	0.086304
35	0.061072	0.068974	0.077234	0.085803
36	0.060434	0.068395	0.076715	0.085345
37	0.059840	0.067857	0.076237	0.084924
38	0.059284	0.067358	0.075795	0.084539
39	0.058765	0.066894	0.075387	0.084185
40	0.058278	0.066462	0.075009	0.083860

TABLE VII

Per-period equivalent of $1 future value: $\dfrac{1}{s_{\overline{n}|i}} = \dfrac{i}{(1+i)^n - 1} = \dfrac{1}{a_{\overline{n}|i}} - i$

n	1%	2%	3%	4%
1	1.0000000	1.0000000	1.0000000	1.0000000
2	0.4975124	0.4950495	0.4926108	0.4901961
3	0.3300221	0.3267547	0.3235304	0.3203485
4	0.2462811	0.2426238	0.2390270	0.2354900
5	0.1960398	0.1921584	0.1883546	0.1846271
6	0.1625484	0.1585258	0.1545975	0.1507619
7	0.1386283	0.1345120	0.1305064	0.1266096
8	0.1206903	0.1165098	0.1124564	0.1085278
9	0.1067404	0.1025154	0.0984339	0.0944930
10	0.0955821	0.0913265	0.0872305	0.0832909
11	0.0864541	0.0821779	0.0780774	0.0741490
12	0.0788488	0.0745596	0.0704621	0.0665522
13	0.0724148	0.0681184	0.0640295	0.0601437
14	0.0669012	0.0626020	0.0585263	0.0546690
15	0.0621238	0.0578255	0.0537666	0.0499411
16	0.0579446	0.0536501	0.0496108	0.0458200
17	0.0542581	0.0499698	0.0459525	0.0421985
18	0.0509820	0.0467021	0.0427087	0.0389933
19	0.0480518	0.0437818	0.0398139	0.0361386
20	0.0454153	0.0411567	0.0372157	0.0335818
21	0.0430308	0.0387848	0.0348718	0.0312801
22	0.0408637	0.0366314	0.0327474	0.0291988
23	0.0388858	0.0346681	0.0308139	0.0273091
24	0.0370735	0.0328711	0.0290474	0.0255868
25	0.0354068	0.0312204	0.0274279	0.0240120
26	0.0338689	0.0296992	0.0259383	0.0225674
27	0.0324455	0.0282931	0.0245642	0.0212385
28	0.0311244	0.0269897	0.0232932	0.0200130
29	0.0298950	0.0257784	0.0221147	0.0188799
30	0.0287481	0.0246499	0.0210193	0.0178301
31	0.0276757	0.0235963	0.0199989	0.0168554
32	0.0266709	0.0226106	0.0190466	0.0159486
33	0.0257274	0.0216865	0.0181561	0.0151036
34	0.0248400	0.0208187	0.0173220	0.0143148
35	0.0240037	0.0200022	0.0165393	0.0135773
36	0.0232143	0.0192329	0.0158038	0.0128869
37	0.0224680	0.0185068	0.0151116	0.0122396
38	0.0217615	0.0178206	0.0144593	0.0116319
39	0.0210916	0.0171711	0.0138439	0.0110608
40	0.0204556	0.0165557	0.0132624	0.0105235

TABLE VII *(continued)*

Per-period equivalent of $1 future value: $\dfrac{1}{s_{\overline{n}|i}} = \dfrac{i}{(1+i)^n - 1} = \dfrac{1}{a_{\overline{n}|i}} - i$

n	5%	6%	7%	8%
1 . . . . . 1.0000000	1.0000000	1.0000000	1.0000000
2 . . . . . 0.4878049	0.4854369	0.4830918	0.4807692
3 . . . . . 0.3172086	0.3141098	0.3110517	0.3080335
4 . . . . . 0.2320118	0.2285915	0.2252281	0.2219208
5 . . . . . 0.1809748	0.1773964	0.1738907	0.1704565
6 . . . . . 0.1470175	0.1433626	0.1397958	0.1363154
7 . . . . . 0.1228198	0.1191350	0.1155532	0.1120724
8 . . . . . 0.1047218	0.1010359	0.0974678	0.0940148
9 . . . . . 0.0906901	0.0870222	0.0834865	0.0800797
10 . . . . . 0.0795046	0.0758680	0.0723775	0.0690295
11 . . . . . 0.0703889	0.0667929	0.0633569	0.0600763
12 . . . . . 0.0628254	0.0592770	0.0559020	0.0526950
13 . . . . . 0.0564558	0.0529601	0.0496508	0.0465218
14 . . . . . 0.0510240	0.0475849	0.0443449	0.0412969
15 . . . . . 0.0463423	0.0429628	0.0397946	0.0368295
16 . . . . . 0.0422699	0.0389521	0.0358576	0.0329769
17 . . . . . 0.0386991	0.0354448	0.0324252	0.0296294
18 . . . . . 0.0355462	0.0323565	0.0294126	0.0267021
19 . . . . . 0.0327450	0.0296209	0.0267530	0.0241276
20 . . . . . 0.0302426	0.0271846	0.0243929	0.0218522
21 . . . . . 0.0279961	0.0250045	0.0222890	0.0198323
22 . . . . . 0.0259705	0.0230456	0.0204058	0.0180321
23 . . . . . 0.0241368	0.0212785	0.0187139	0.0164222
24 . . . . . 0.0224709	0.0196790	0.0171890	0.0149780
25 . . . . . 0.0209525	0.0182267	0.0158105	0.0136788
26 . . . . . 0.0195643	0.0169043	0.0145610	0.0125071
27 . . . . . 0.0182919	0.0156972	0.0134257	0.0114481
28 . . . . . 0.0171225	0.0145926	0.0123919	0.0104889
29 . . . . . 0.0160455	0.0135796	0.0114487	0.0096185
30 . . . . . 0.0150514	0.0126489	0.0105864	0.0088274
31 . . . . . 0.0141321	0.0117922	0.0097969	0.0081073
32 . . . . . 0.0132804	0.0110023	0.0090729	0.0074508
33 . . . . . 0.0124900	0.0102729	0.0084081	0.0068516
34 . . . . . 0.0117554	0.0095984	0.0077967	0.0063041
35 . . . . . 0.0110717	0.0089739	0.0072340	0.0058033
36 . . . . . 0.0104345	0.0083948	0.0067153	0.0053447
37 . . . . . 0.0098398	0.0078574	0.0062368	0.0049244
38 . . . . . 0.0092842	0.0073581	0.0057951	0.0045389
39 . . . . . 0.0087646	0.0068938	0.0053868	0.0041851
40 . . . . . 0.0082782	0.0064615	0.0050091	0.0038602

TABLE VIII
Areas under the normal curve

Normal Deviate, z	.00	.01	.02	.03	.04	.05	.06	.07	.08	.09
0.0	.0000	.0040	.0080	.0120	.0160	.0199	.0239	.0279	.0319	.0359
0.1	.0398	.0438	.0478	.0517	.0557	.0596	.0636	.0675	.0714	.0753
0.2	.0793	.0832	.0871	.0910	.0948	.0987	.1026	.1064	.1103	.1141
0.3	.1179	.1217	.1255	.1293	.1331	.1368	.1406	.1443	.1480	.1517
0.4	.1554	.1591	.1628	.1664	.1700	.1736	.1772	.1808	.1844	.1879
0.5	.1915	.1950	.1985	.2019	.2054	.2088	.2123	.2157	.2190	.2224
0.6	.2257	.2291	.2324	.2357	.2389	.2422	.2454	.2486	.2517	.2549
0.7	.2580	.2611	.2642	.2673	.2704	.2734	.2764	.2794	.2823	.2852
0.8	.2881	.2910	.2939	.2967	.2995	.3023	.3051	.3078	.3106	.3133
0.9	.3159	.3186	.3212	.3238	.3264	.3289	.3315	.3340	.3365	.3389
1.0	.3413	.3438	.3461	.3485	.3508	.3531	.3554	.3577	.3599	.3621
1.1	.3643	.3665	.3686	.3708	.3729	.3749	.3770	.3790	.3810	.3830
1.2	.3849	.3869	.3888	.3907	.3925	.3944	.3962	.3980	.3997	.4015
1.3	.4032	.4049	.4066	.4082	.4099	.4115	.4131	.4147	.4162	.4177
1.4	.4192	.4207	.4222	.4236	.4251	.4265	.4279	.4292	.4306	.4319
1.5	.4332	.4345	.4357	.4370	.4382	.4394	.4406	.4418	.4429	.4441
1.6	.4452	.4463	.4474	.4484	.4495	.4505	.4515	.4525	.4535	.4545
1.7	.4554	.4564	.4573	.4582	.4591	.4599	.4608	.4616	.4625	.4633
1.8	.4641	.4649	.4656	.4664	.4671	.4678	.4686	.4693	.4699	.4706
1.9	.4713	.4719	.4726	.4732	.4738	.4744	.4750	.4756	.4761	.4767
2.0	.4772	.4778	.4783	.4788	.4793	.4798	.4803	.4808	.4812	.4817
2.1	.4821	.4826	.4830	.4834	.4838	.4842	.4846	.4850	.4854	.4857
2.2	.4861	.4864	.4868	.4871	.4875	.4878	.4881	.4884	.4887	.4890
2.3	.4893	.4896	.4898	.4901	.4904	.4906	.4909	.4911	.4913	.4916
2.4	.4918	.4920	.4922	.4925	.4927	.4929	.4931	.4932	.4934	.4936
2.5	.4938	.4940	.4941	.4943	.4945	.4946	.4948	.4949	.4951	.4952
2.6	.4953	.4955	.4956	.4957	.4959	.4960	.4961	.4962	.4963	.4964
2.7	.4965	.4966	.4967	.4968	.4969	.4970	.4971	.4972	.4973	.4974
2.8	.4974	.4975	.4976	.4977	.4977	.4978	.4979	.4979	.4980	.4981
2.9	.4981	.4982	.4982	.4983	.4984	.4984	.4985	.4985	.4986	.4986
3.0	.4987	.4987	.4987	.4988	.4988	.4989	.4989	.4989	.4990	.4990

Adapted by permission from Ernest Kurnow, Gerald J. Glasser, and Frederick R. Ottman. *Statistics for Business Decisions* (Homewood, Ill.: Richard D. Irwin, 1959), p. 501. © 1959 by Richard D. Irwin, Inc.

TABLE IX
e^x and e^{-x}

x	e^x	e^{-x}	x	e^x	e^{-x}	x	e^x	e^{-x}	x	e^x	e^{-x}
0.01	1.0101	0.9900	0.31	1.3634	0.7334	0.61	1.8404	0.5434	0.91	2.4843	0.4025
0.02	1.0202	0.9802	0.32	1.3771	0.7261	0.62	1.8589	0.5379	0.92	2.5093	0.3985
0.03	1.0305	0.9704	0.33	1.3910	0.7189	0.63	1.8776	0.5326	0.93	2.5345	0.3946
0.04	1.0408	0.9608	0.34	1.4049	0.7118	0.64	1.8965	0.5273	0.94	2.5600	0.3906
0.05	1.0513	0.9512	0.35	1.4191	0.7047	0.65	1.9155	0.5220	0.95	2.5857	0.3867
0.06	1.0618	0.9418	0.36	1.4333	0.6977	0.66	1.9348	0.5169	0.96	2.6117	0.3829
0.07	1.0725	0.9324	0.37	1.4477	0.6907	0.67	1.9542	0.5117	0.97	2.6379	0.3791
0.08	1.0833	0.9231	0.38	1.4623	0.6839	0.68	1.9739	0.5066	0.98	2.6645	0.3753
0.09	1.0942	0.9139	0.39	1.4770	0.6771	0.69	1.9937	0.5016	0.99	2.6912	0.3716
0.10	1.1052	0.9048	0.40	1.4918	0.6703	0.70	2.0138	0.4966	1.0	2.7183	0.3679
0.11	1.1163	0.8958	0.41	1.5068	0.6637	0.71	2.0340	0.4916	1.1	3.0042	0.3329
0.12	1.1275	0.8869	0.42	1.5220	0.6570	0.72	2.0544	0.4868	1.2	3.3201	0.3012
0.13	1.1388	0.8781	0.43	1.5373	0.6505	0.73	2.0751	0.4819	1.3	3.6693	0.2725
0.14	1.1503	0.8694	0.44	1.5527	0.6440	0.74	2.0959	0.4771	1.4	4.0552	0.2466
0.15	1.1618	0.8607	0.45	1.5683	0.6376	0.75	2.1170	0.4724	1.5	4.4817	0.2231
0.16	1.1735	0.8521	0.46	1.5841	0.6313	0.76	2.1383	0.4677	1.6	4.9530	0.2019
0.17	1.1853	0.8437	0.47	1.6000	0.6250	0.77	2.1598	0.4630	1.7	5.4739	0.1827
0.18	1.1972	0.8353	0.48	1.6161	0.6188	0.78	2.1815	0.4584	1.8	6.0496	0.1653
0.19	1.2092	0.8270	0.49	1.6323	0.6126	0.79	2.2034	0.4538	1.9	6.6859	0.1496
0.20	1.2214	0.8187	0.50	1.6487	0.6065	0.80	2.2255	0.4493	2.0	7.3891	0.1353
0.21	1.2337	0.8106	0.51	1.6653	0.6005	0.81	2.2479	0.4449	2.1	8.1662	0.1225
0.22	1.2461	0.8025	0.52	1.6820	0.5945	0.82	2.2705	0.4404	2.2	9.0250	0.1108
0.23	1.2586	0.7945	0.53	1.6989	0.5886	0.83	2.2933	0.4360	2.3	9.9742	0.1003
0.24	1.2712	0.7866	0.54	1.7160	0.5827	0.84	2.3164	0.4317	2.4	11.0232	0.0907
0.25	1.2340	0.7788	0.55	1.7333	0.5769	0.85	2.3396	0.4274	2.5	12.1825	0.0821
0.26	1.2969	0.7711	0.56	1.7507	0.5712	0.86	2.3632	0.4232	2.6	13.4637	0.0743
0.27	1.3100	0.7634	0.57	1.7683	0.5655	0.87	2.3869	0.4190	2.7	14.8797	0.0672
0.28	1.3231	0.7558	0.58	1.7860	0.5599	0.88	2.4109	0.4148	2.8	16.4446	0.0608
0.29	1.3364	0.7483	0.59	1.8040	0.5543	0.89	2.4351	0.4107	2.9	18.1741	0.0550
0.30	1.3499	0.7408	0.60	1.8221	0.5488	0.90	2.4596	0.4066	3.0	20.0855	0.0498

TABLE X
Natural logarithms, 0.01 to 4.49. (insert minus sign before shaded numbers)

N	0	1	2	3	4	5	6	7	8	9
0.0		4.60517	3.91202	3.50656	3.21888	2.99573	2.81341	2.65926	2.52573	2.40795
0.1	2.30259	2.20727	2.12026	2.04022	1.96611	1.89712	1.83258	1.77196	1.71480	1.66073
0.2	1.60944	1.56065	1.51413	1.46968	1.42712	1.38629	1.34707	1.30933	1.27297	1.23787
0.3	1.20397	1.17118	1.13943	1.10866	1.07881	1.04982	1.02165	.99425	.96758	.94161
0.4	0.91629	.89160	.86750	.84397	.82098	.79851	.77653	.75502	.73397	.71335

Shaded numbers are negative. Insert minus sign.

N	0	1	2	3	4	5	6	7	8	9
0.5	0.69315	.67334	.65393	.63488	.61619	.59784	.57982	.56212	.54473	.52763
0.6	0.51083	.49430	.47804	.46204	.44629	.43078	.41552	.40048	.38566	.37106
0.7	0.35667	.34249	.32850	.31471	.30111	.28768	.27444	.26136	.24846	.23572
0.8	0.22314	.21072	.19845	.18633	.17435	.16252	.15082	.13926	.12783	.11653
0.9	0.10536	.09431	.08338	.07257	.06188	.05129	.04082	.03046	.02020	.01005

N	0	1	2	3	4	5	6	7	8	9
1.0	0.0 0000	0995	1980	2956	3922	4879	5827	6766	7696	8618
1.1	9531	*0436	*1333	*2222	*3103	*3976	*4842	*5700	*6551	*7395
1.2	0.1 8232	9062	9885	*0701	*1511	*2314	*3111	*3902	*4686	*5464
1.3	0.2 6236	7003	7763	8518	9267	*0010	*0748	*1481	*2208	*2930
1.4	0.3 3647	4359	5066	5767	6464	7156	7844	8526	9204	9878
1.5	0.4 0547	1211	1871	2527	3178	3825	4469	5108	5742	6373
1.6	7000	7623	8243	8858	9470	*0078	*0682	*1282	*1879	*2473
1.7	0.5 3063	3649	4232	4812	5389	5962	6531	7098	7661	8222
1.8	8779	9333	9884	*0432	*0977	*1519	*2058	*2594	*3127	*3658
1.9	0.6 4185	4710	5233	5752	6269	6783	7294	7803	8310	8813
2.0	9315	9813	*0310	*0804	*1295	*1784	*2271	*2755	*3237	*3716
2.1	0.7 4194	4669	5142	5612	6081	6547	7011	7473	7932	8390
2.2	8846	9299	9751	*0200	*0648	*1093	*1536	*1978	*2418	*2855
2.3	0.8 3291	3725	4157	4587	5015	5442	5866	6289	6710	7129
2.4	7547	7963	8377	8789	9200	9609	*0016	*0422	*0826	*1228
2.5	0.9 1629	2028	2426	2822	3216	3609	4001	4391	4779	5166
2.6	5551	5935	6317	6698	7078	7456	7833	8208	8582	8954
2.7	9325	9695	*0063	*0430	*0796	*1160	*1523	*1885	*2245	*2604
2.8	1.0 2962	3318	3674	4028	4380	4732	5082	5431	5779	6126
2.9	6471	6815	7158	7500	7841	8181	8519	8856	9192	9527
3.0	9861	*0194	*0526	*0856	*1186	*1514	*1841	*2168	*2493	*2817
3.1	1.1 3140	3462	3783	4103	4422	4740	5057	5373	5688	6002
3.2	6315	6627	6938	7248	7557	7865	8173	8479	8784	9089
3.3	9392	9695	9996	*0297	*0597	*0896	*1194	*1491	*1788	*2083
3.4	1.2 2378	2671	2964	3256	3547	3837	4127	4415	4703	4990
3.5	5276	5562	5846	6130	6413	6695	6976	7257	7536	7815
3.6	8093	8371	8647	8923	9198	9473	9746	*0019	*0291	*0563
3.7	1.3 0833	1103	1372	1641	1909	2176	2442	2708	2972	3237
3.8	3500	3763	4025	4286	4547	4807	5067	5325	5584	5841
3.9	6098	6354	6609	6864	7118	7372	7624	7877	8128	8379
4.0	8629	8879	9128	9377	9624	9872	*0118	*0364	*0610	*0854
4.1	1.4 1099	1342	1585	1828	2070	2311	2552	2792	3031	3270
4.2	3508	3746	3984	4220	4456	4692	4927	5161	5395	5629
4.3	5862	6094	6326	6557	6787	7018	7247	7476	7705	7933
4.4	8160	8387	8614	8840	9065	9290	9515	9739	9962	*0185

TABLE X *(continued)*
Natural logarithms, 4.50 to 8.99.

N	0	1	2	3	4	5	6	7	8	9
4.5	1.5 0408	0630	0851	1072	1293	1513	1732	1951	2170	2388
4.6	2606	2823	3039	3256	3471	3687	3902	4116	4330	4543
4.7	4756	4969	5181	5393	5604	5814	6025	6235	6444	6653
4.8	6862	7070	7277	7485	7691	7898	8104	8309	8515	8719
4.9	8924	9127	9331	9534	9737	9939	*0141	*0342	*0543	*0744
5.0	1.6 0944	1144	1343	1542	1741	1939	2137	2334	2531	2728
5.1	2924	3120	3315	3511	3705	3900	4094	4287	4481	4673
5.2	4866	5058	5250	5441	5632	5823	6013	6203	6393	6582
5.3	6771	6959	7147	7335	7523	7710	7896	8083	8269	8455
5.4	8640	8825	9010	9194	9378	9562	9745	9928	*0111	*0293
5.5	1.7 0475	0656	0838	1019	1199	1380	1560	1740	1919	2098
5.6	2277	2455	2633	2811	2988	3166	3342	3519	3695	3871
5.7	4047	4222	4397	4572	4746	4920	5094	5267	5440	5613
5.8	5786	5958	6130	6302	6473	6644	6815	6985	7156	7326
5.9	7495	7665	7834	8002	8171	8339	8507	8675	8842	9009
6.0	9176	9342	9509	9675	9840	*0006	*0171	*0336	*0500	*0665
6.1	1.8 0829	0993	1156	1319	1482	1645	1808	1970	2132	2294
6.2	2455	2616	2777	2938	3098	3258	3418	3578	3737	3896
6.3	4055	4214	4372	4530	4688	4845	5003	5160	5317	5473
6.4	5630	5786	5942	6097	6253	6408	6563	6718	6872	7026
6.5	7180	7334	7487	7641	7794	7947	8099	8251	8403	8555
6.6	8707	8858	9010	9160	9311	9462	9612	9762	9912	*0061
6.7	1.9 0211	0360	0509	0658	0806	0954	1102	1250	1398	1545
6.8	1692	1839	1986	2132	2279	2425	2571	2716	2862	3007
6.9	3152	3297	3442	3586	3730	3874	4018	4162	4305	4448
7.0	4591	4734	4876	5019	5161	5303	5445	5586	5727	5869
7.1	6009	6150	6291	6431	6571	6711	6851	6991	7130	7269
7.2	7408	7547	7685	7824	7962	8100	8238	8376	8513	8650
7.3	8787	8924	9061	9198	9334	9470	9606	9742	9877	*0013
7.4	2.0 0148	0283	0418	0553	0687	0821	0956	1089	1223	1357
7.5	1490	1624	1757	1890	2022	2155	2287	2419	2551	2683
7.6	2815	2946	3078	3209	3340	3471	3601	3732	3862	3992
7.7	4122	4252	4381	4511	4640	4769	4898	5027	5156	5284
7.8	5412	5540	5668	5796	5924	6051	6179	6306	6433	6560
7.9	6686	6813	6939	7065	7191	7317	7443	7568	7694	7819
8.0	7944	8069	8194	8318	8443	8567	8691	8815	8939	9063
8.1	9186	9310	9433	9556	9679	9802	9924	*0047	*0169	*0291
8.2	2.1 0413	0535	0657	0779	0900	1021	1142	1263	1384	1505
8.3	1626	1746	1866	1986	2106	2226	2346	2465	2585	2704
8.4	2823	2942	3061	3180	3298	3417	3535	3653	3771	3889
8.5	4007	4124	4242	4359	4476	4593	4710	4827	4943	5060
8.6	5176	5292	5409	5524	5640	5756	5871	5987	6102	6217
8.7	6332	6447	6562	6677	6791	6905	7020	7134	7248	7361
8.8	7475	7589	7702	7816	7929	8042	8155	8267	8380	8493
8.9	8605	8717	8830	8942	9054	9165	9277	9389	9500	9611

TABLE X *(continued)*
Natural logarithms, 9 to 9.99

N	0	1	2	3	4	5	6	7	8	9
9.0	9722	9834	9944	*0055	*0166	*0276	*0387	*0497	*0607	*0717
9.1	2.2 0827	0937	1047	1157	1266	1375	1485	1594	1703	1812
9.2	1920	2029	2138	2246	2354	2462	2570	2678	2786	2894
9.3	3001	3109	3216	3324	3431	3538	3645	3751	3858	3965
9.4	4071	4177	4284	4390	4496	4601	4707	4813	4918	5024
9.5	5129	5234	5339	5444	5549	5654	5759	5863	5968	6072
9.6	6176	6280	6384	6488	6592	6696	6799	6903	7006	7109
9.7	7213	7316	7419	7521	7624	7727	7829	7932	8034	8136
9.8	8238	8340	8442	8544	8646	8747	8849	8950	9051	9152
9.9	9253	9354	9455	9556	9657	9757	9858	9958	*0058	*0158

Natural logarithms, 10 to 99

N	0	1	2	3	4	5	6	7	8	9
1	2.30259	39790	48491	56495	63906	70805	77259	83321	89037	94444
2	99573	*04452	*09104	*13549	*17805	*21888	*25810	*29584	*33220	*36730
3	3.40120	43399	46574	49651	52636	55535	58352	61092	63759	66356
4	68888	71357	73767	76120	78419	80666	82864	85015	87120	89182
5	91202	93183	95124	97029	98898	*00733	*02535	*04305	*06044	*07754
6	4.09434	11087	12713	14313	15888	17439	18965	20469	21951	23411
7	24850	26268	27667	29046	30407	31749	33073	34381	35671	36945
8	38203	39445	40672	41884	43082	44265	45435	46591	47734	48864
9	49981	51086	52179	53260	54329	55388	56435	57471	58497	59512

Natural logarithms, 100 to 349

N	0	1	2	3	4	5	6	7	8	9
10	4.6 0517	1512	2497	3473	4439	5396	6344	7283	8213	9135
11	4.7 0048	0953	1850	2739	3620	4493	5359	6217	7068	7912
12	8749	9579	*0402	*1218	*2028	*2831	*3628	*4419	*5203	*5981
13	4.8 6753	7520	8280	9035	9784	*0527	*1265	*1998	*2725	*3447
14	4.9 4164	4876	5583	6284	6981	7673	8361	9043	9721	*0395
15	5.0 1064	1728	2388	3044	3695	4343	4986	5625	6260	6890
16	7517	8140	8760	9375	9987	*0595	*1199	*1799	*2396	*2990
17	5.1 3580	4166	4749	5329	5906	6479	7048	7615	8178	8739
18	9296	9850	*0401	*0949	*1494	*2036	*2575	*3111	*3644	*4175
19	5.2 4702	5227	5750	6269	6786	7300	7811	8320	8827	9330
20	9832	*0330	*0827	*1321	*1812	*2301	*2788	*3272	*3754	*4233
21	5.3 4711	5186	5659	6129	6598	7064	7528	7990	8450	8907
22	9363	9816	*0268	*0717	*1165	*1610	*2053	*2495	*2935	*3372
23	5.4 3808	4242	4674	5104	5532	5959	6383	6806	7227	7646
24	8064	8480	8894	9306	9717	*0126	*0533	*0939	*1343	*1745
25	5.5 2146	2545	2943	3339	3733	4126	4518	4908	5296	5683
26	6068	6452	6834	7215	7595	7973	8350	8725	9099	9471
27	9842	*0212	*0580	*0947	*1313	*1677	*2040	*2402	*2762	*3121
28	5.6 3479	3835	4191	4545	4897	5249	5599	5948	6296	6643
29	6988	7332	7675	8017	8358	8698	9036	9373	9709	*0044
30	5.7 0378	0711	1043	1373	1703	2031	2359	2685	3010	3334
31	3657	3979	4300	4620	4939	5257	5574	5890	6205	6519
32	6832	7144	7455	7765	8074	8383	8690	8996	9301	9606
33	9909	*0212	*0513	*0814	*1114	*1413	*1711	*2008	*2305	*2600
34	5.8 2895	3188	3481	3773	4064	4354	4644	4932	5220	5507

TABLE X *(continued)*
Natural logarithms, 350 to 799

N	0	1	2	3	4	5	6	7	8	9
35	5793	6079	6363	6647	6930	7212	7493	7774	8053	8332
36	8610	8888	9164	9440	9715	9990	*0263	*0536	*0808	*1080
37	5.9 1350	1620	1889	2158	2426	2693	2959	3225	3489	3754
38	4017	4280	4542	4803	5064	5324	5584	5842	6101	6358
39	6615	6871	7126	7381	7635	7889	8141	8394	8645	8896
40	9146	9396	9645	9894	*0141	*0389	*0635	*0881	*1127	*1372
41	6.0 1616	1859	2102	2345	2587	2828	3069	3309	3548	3787
42	4025	4263	4501	4737	4973	5209	5444	5678	5912	6146
43	6379	6611	6843	7074	7304	7535	7764	7993	8222	8450
44	8677	8904	9131	9357	9582	9807	*0032	*0256	*0479	*0702
45	6.1 0925	1147	1368	1589	1810	2030	2249	2468	2687	2905
46	3123	3340	3556	3773	3988	4204	4419	4633	4847	5060
47	5273	5486	5698	5910	6121	6331	6542	6752	6961	7170
48	7379	7587	7794	8002	8208	8415	8621	8826	9032	9236
49	9441	9644	9848	*0051	*0254	*0456	*0658	*0859	*1060	*1261
50	6.2 1461	1661	1860	2059	2258	2456	2654	2851	3048	3245
51	3441	3637	3832	4028	4222	4417	4611	4804	4998	5190
52	5383	5575	5767	5958	6149	6340	6530	6720	6910	7099
53	7288	7476	7664	7852	8040	8227	8413	8600	8786	8972
54	9157	9342	9527	9711	9895	*0079	*0262	*0445	*0628	*0810
55	6.3 0992	1173	1355	1536	1716	1897	2077	2257	2436	2615
56	2794	2972	3150	3328	3505	3683	3859	4036	4212	4388
57	4564	4739	4914	5089	5263	5437	5611	5784	5957	6130
58	6303	6475	6647	6819	6990	7161	7332	7502	7673	7843
59	8012	8182	8351	8519	8688	8856	9024	9192	9359	9526
60	6.3 9693	9859	*0026	*0192	*0357	*0523	*0688	*0853	*1017	*1182
61	6.4 1346	1510	1673	1836	1999	2162	2325	2487	2649	2811
62	2972	3133	3294	3455	3615	3775	3935	4095	4254	4413
63	4572	4731	4889	5047	5205	5362	5520	5677	5834	5990
64	6147	6303	6459	6614	6770	6925	7080	7235	7389	7543
65	7697	7851	8004	8158	8311	8464	8616	8768	8920	9072
66	9224	9375	9527	9677	9828	9979	*0129	*0279	*0429	*0578
67	6.5 0728	0877	1026	1175	1323	1471	1619	1767	1915	2062
68	2209	2356	2503	2649	2796	2942	3088	3233	3379	3524
69	3669	3814	3959	4103	4247	4391	4535	4679	4822	4965
70	5108	5251	5393	5536	5678	5820	5962	6103	6244	6386
71	6526	6667	6808	6948	7088	7228	7368	7508	7647	7786
72	7925	8064	8203	8341	8479	8617	8755	8893	9030	9167
73	9304	9441	9578	9715	9851	9987	*0123	*0259	*0394	*0530
74	6.6 0665	0800	0935	1070	1204	1338	1473	1607	1740	1874
75	2007	2141	2274	2407	2539	2672	2804	2936	3068	3200
76	3332	3463	3595	3726	3857	3988	4118	4249	4379	4509
77	4639	4769	4898	5028	5157	5286	5415	5544	5673	5801
78	5929	6058	6185	6313	6441	6568	6696	6823	6950	7077
79	7203	7330	7456	7582	7708	7834	7960	8085	8211	8336

TABLE X *(concluded)*
Natural logarithms, 800 to 1209

N	0	1	2	3	4	5	6	7	8	9
80	8461	8586	8711	8835	8960	9084	9208	9332	9456	9580
81	9703	9827	9950	*0073	*0196	*0319	*0441	*0564	*0686	*0808
82	6.7 0930	1052	1174	1296	1417	1538	1659	1780	1901	2022
83	2143	2263	2383	2503	2623	2743	2863	2982	3102	3221
84	3340	3459	3578	3697	3815	3934	4052	4170	4288	4406
85	4524	4641	4759	4876	4993	5110	5227	5344	5460	5577
86	5693	5809	5926	6041	6157	6273	6388	6504	6619	6734
87	6849	6964	7079	7194	7308	7422	7537	7651	7765	7878
88	7992	8106	8219	8333	8446	8559	8672	8784	8897	9010
89	9122	9234	9347	9459	9571	9682	9794	9906	*0017	*0128
90	6.8 0239	0351	0461	0572	0683	0793	0904	1014	1124	1235
91	1344	1454	1564	1674	1783	1892	2002	2111	2220	2329
92	2437	2546	2655	2763	2871	2979	3087	3195	3303	3411
93	3518	3626	3733	3841	3948	4055	4162	4268	4375	4482
94	4588	4694	4801	4907	5013	5118	5224	5330	5435	5541
95	5646	5751	5857	5961	6066	6171	6276	6380	6485	6589
96	6693	6797	6901	7005	7109	7213	7316	7420	7523	7626
97	7730	7833	7936	8038	8141	8244	8346	8449	8551	8653
98	8755	8857	8959	9061	9163	9264	9366	9467	9568	9669
99	9770	9871	9972	*0073	*0174	*0274	*0375	*0475	*0575	*0675
100	6.9 0776	0875	0975	1075	1175	1274	1374	1473	1572	1672
101	1771	1870	1968	2067	2166	2264	2363	2461	2560	2658
102	2756	2854	2952	3049	3147	3245	3342	3440	3537	3634
103	3731	3828	3925	4022	4119	4216	4312	4409	4505	4601
104	4698	4794	4890	4986	5081	5177	5273	5368	5464	5559
105	5655	5750	5845	5940	6035	6130	6224	6319	6414	6508
106	6602	6697	6791	6885	6979	7073	7167	7261	7354	7448
107	7541	7635	7728	7821	7915	8008	8101	8193	8286	8379
108	8472	8564	8657	8749	8841	8934	9026	9118	9210	9302
109	9393	9485	9577	9668	9760	9851	9942	*0033	*0125	*0216
110	7.0 0307	0397	0488	0579	0670	0760	0851	0941	1031	1121
111	1212	1302	1392	1481	1571	1661	1751	1840	1930	2019
112	2108	2198	2287	2376	2465	2554	2643	2731	2820	2909
113	2997	3086	3174	3262	3351	3439	3527	3615	3703	3791
114	3878	3966	4054	4141	4229	4316	4403	4491	4578	4665
115	4752	4839	4925	5012	5099	5186	5272	5359	5445	5531
116	5618	5704	5790	5876	5962	6048	6133	6219	6305	6390
117	6476	6561	6647	6732	6817	6902	6987	7072	7157	7242
118	7327	7412	7496	7581	7665	7750	7834	7918	8003	8087
119	8171	8255	8339	8423	8506	8590	8674	8757	8841	8924
120	9008	9091	9174	9257	9340	9423	9506	9589	9672	9755

TABLE XI

Monthly payment for $1 mortgage

$$\frac{\dfrac{j}{1200}}{1-\left(1+\dfrac{j}{1200}\right)^{-n}}$$

Annual Rate j%	Number of Months, n					
	120	180	240	300	360	420
7.00	.0116108479	.0089882827	.0077529894	.0070677920	.0066530250	.0063885636
7.25	.0117401041	.0091286288	.0079037598	.0072280686	.0068217628	.0065646724
7.50	.0118701769	.0092701236	.0080559319	.0073899118	.0069921451	.0067424260
7.75	.0120010631	.0094127575	.0082094856	.0075532876	.0071644225	.0069217594
8.00	.0121327594	.0095565208	.0083644007	.0077181622	.0073376457	.0071026088
8.25	.0122652625	.0097014036	.0085206565	.0078845013	.0075126660	.0072849114
8.50	.0123985689	.0098473956	.0086782324	.0080522708	.0076891348	.0074686057
8.75	.0125326750	.0099944865	.0088371071	.0082214364	.0078670041	.0076536314
9.00	.0126675774	.0101426658	.0089972596	.0083919636	.0080462262	.0078399297
9.25	.0128032722	.0102919229	.0091586683	.0085638184	.0082267543	.0080274432
9.50	.0129397558	.0104422468	.0093213119	.0087369666	.0084085421	.0082161160
9.75	.0130770242	.0105936266	.0094851685	.0089113742	.0085915441	.0084058939
10.00	.0132150737	.0107460512	.0096502165	.0090870075	.0087757157	.0085967243
10.25	.0133539002	.0108995092	.0098164339	.0092638328	.0089610130	.0087885561
10.50	.0134934997	.0110539892	.0099837989	.0094418171	.0091473929	.0089813402
10.75	.0136338680	.0112094798	.0101522895	.0096209272	.0093348136	.0091750290
11.00	.0137750011	.0113659693	.0103218839	.0098011308	.0095232340	.0093695765

TABLE XII–A
Rules for derivatives

1. $\dfrac{d}{dx}[f(x) + g(x)] = \dfrac{d}{dx} f(x) + \dfrac{d}{dx} g(x).$

2. $\dfrac{d}{dx}[f(x) - g(x)] = \dfrac{d}{dx} f(x) - \dfrac{d}{dx} g(x).$

3. $\dfrac{d}{dx} (a) = 0.$

4. $\dfrac{d}{dx} (x) = 1.$

5. $\dfrac{d}{dx}(ax) = a; \dfrac{d}{dx} af(x) = a\dfrac{d}{dx} f(x).$

6. $\dfrac{d}{dx} (x^n) = nx^{n-1}.$

7. $\dfrac{d}{dx}[f(x)]^n = n[f(x)]^{n-1}f'(x).$

8. $\dfrac{d}{dx} (e^x) = e^x.$

9. $\dfrac{d}{dx} (a^x) = a^x(\ln a).$

10. $\dfrac{d}{dx}[e^{f(x)}] = e^{f(x)}f'(x).$

11. $\dfrac{d}{dx}[a^{f(x)}] = a^{f(x)}[f'(x)](\ln a).$

12. $\dfrac{d}{dx} (\ln x) = \dfrac{1}{x}.$

13. $\dfrac{d}{dx} (\log x) = \dfrac{1}{x(\ln 10)} \doteq \dfrac{0.43429}{x}$

14. $\dfrac{d}{dx} [\ln f(x)] = \dfrac{1}{f(x)} [f'(x)] = \dfrac{f'(x)}{f(x)}.$

15. $\dfrac{d}{dx} [\log f(x)] = \dfrac{f'(x)}{f(x)(\ln 10)} \doteq \dfrac{(0.43429)f'(x)}{f(x)}.$

16. $\dfrac{d}{dx} f[g(x)] = \dfrac{d}{dg} [f(g)] \cdot \dfrac{d}{dx} g(x).$

17. $\dfrac{d}{dx} [f(x)g(x)] = f(x)g'(x) + g(x)f'(x).$

18. $\dfrac{d}{dx}\left[\dfrac{f(x)}{g(x)}\right] = \dfrac{g(x)f'(x) - f(x)g'(x)}{[g(x)]^2}.$

Note: *a* and *n* are constants;
$$\dfrac{d}{dx} f(x) = f'(x); \dfrac{d}{dx} g(x) = g'(x).$$

TABLE XII–B
Rules for integrals

1. $\int [f(x) + g(x)]dx = \int f(x)dx + \int g(x)dx.$

2. $\int [f(x) - g(x)]dx = \int f(x)dx - \int g(x)dx.$

3. $\int kf(x)dx = k\int f(x)dx.$

4. $\int dx = \int 1 dx = x + C.$

5. $\int kdx = kx + C.$

6. $\int x^n dx = \dfrac{x^{n+1}}{n+1} + C; \; n \neq -1.$

7. $\int x^{-1}dx = \int \dfrac{1}{x}dx = \int \dfrac{dx}{x} = \ln x + C.$

8. $\int (mx + b)^n dx = \dfrac{(mx + b)^{n+1}}{m(n+1)} + C; \; n \neq -1.$

9. $\int (mx + b)^{-1}dx = \int \dfrac{dx}{mx + b} = \dfrac{\ln (mx + b)}{m} + C.$

10. $\int \dfrac{xdx}{mx + b} = \dfrac{x}{m} - \dfrac{b}{m^2}\ln (mx + b) + C.$

11. $\int \dfrac{xdx}{(mx + b)^2} = \dfrac{b}{m^2(mx + b)} + \dfrac{\ln (mx + b)}{m^2} + C.$

12. $\int \dfrac{dx}{x(mx + b)} = \dfrac{1}{b}\ln \left(\dfrac{x}{mx + b}\right) + C.$

13. $\int e^x dx = e^x + C.$

14. $\int a^x dx = \dfrac{a^x}{\ln a} + C.$

15. $\int e^{mx+b}dx = \dfrac{e^{mx+b}}{m} + C.$

16. $\int a^{mx+b}dx = \dfrac{a^{mx+b}}{m(\ln a)} + C.$

17. $\int xe^{mx+b}dx = \dfrac{e^{mx+b}(mx - 1)}{m^2} + C.$

18. $\int xa^{mx+b}dx = \dfrac{xa^{mx+b}}{m(\ln a)} - \dfrac{a^{mx+b}}{(m \ln a)^2} + C.$

19. $\int xe^{ax^2+b}dx = \dfrac{e^{ax^2+b}}{2a} + C.$

20. $\int \dfrac{dx}{a + be^{mx}} = \dfrac{mx - \ln(a + be^{mx})}{am} + C.$

21. $\int \ln xdx = x(\ln x - 1) + C.$

22. $\int \log xdx = x\left(\log x - \dfrac{1}{\ln 10}\right) + C.$

23. $\int \ln (mx + b)dx = \dfrac{(mx + b)[\ln(mx + b) - 1]}{m} + C.$

24. $\int \log(mx + b)dx = \left(\dfrac{mx + b}{m}\right)\left[\log(mx + b) - \dfrac{1}{\ln 10}\right] + C.$

Note: a, b, k, m, n, and C are constants.

TABLE XIII–A

Cumulative binomial distribution for $n = 10$ (Tabulated values are $\sum\limits_{0}^{x} C_x^{10}\, p^x q^{10-x}$)

Values of x						Values of p							
	0.01	0.05	0.10	0.20	0.30	0.40	0.50	0.60	0.70	0.80	0.90	0.95	0.99
0 ...	.904	.599	.349	.107	.028	.006	.001	.000	.000	.000	.000	.000	.000
1 ...	.996	.914	.736	.376	.149	.046	.011	.002	.000	.000	.000	.000	.000
2 ...	1.000	.988	.930	.678	.383	.167	.055	.012	.002	.000	.000	.000	.000
3 ...	1.000	.999	.987	.879	.650	.382	.172	.055	.011	.001	.000	.000	.000
4 ...	1.000	1.000	.998	.967	.850	.633	.377	.166	.047	.006	.000	.000	.000
5 ...	1.000	1.000	1.000	.994	.953	.834	.623	.367	.150	.033	.002	.000	.000
6 ...	1.000	1.000	1.000	.999	.989	.945	.828	.618	.350	.121	.013	.001	.000
7 ...	1.000	1.000	1.000	1.000	.998	.988	.945	.833	.617	.322	.070	.012	.000
8 ...	1.000	1.000	1.000	1.000	1.000	.998	.989	.954	.851	.624	.264	.086	.004
9 ...	1.000	1.000	1.000	1.000	1.000	1.000	.999	.994	.972	.893	.651	.401	.096
10 ...	1.000	1.000	1.000	1.000	1.000	1.000	1.000	1.000	1.000	1.000	1.000	1.000	1.000

TABLE XIII–B

Cumulative binomial distribution for $n = 25$ (Tabulated values are $\sum\limits_{0}^{x} C_x^{25}\, p^x q^{25-x}$)

Values of x						Values of p							
	0.01	0.05	0.10	0.20	0.30	0.40	0.50	0.60	0.70	0.80	0.90	0.95	0.99
0	.778	.277	.072	.004	.000	.000	.000	.000	.000	.000	.000	.000	.000
1	.974	.642	.271	.027	.002	.000	.000	.000	.000	.000	.000	.000	.000
2	.998	.873	.537	.098	.009	.000	.000	.000	.000	.000	.000	.000	.000
3	1.000	.966	.764	.234	.033	.002	.000	.000	.000	.000	.000	.000	.000
4	1.000	.993	.902	.421	.090	.009	.000	.000	.000	.000	.000	.000	.000
5	1.000	.999	.967	.617	.193	.029	.002	.000	.000	.000	.000	.000	.000
6	1.000	1.000	.991	.780	.341	.074	.007	.000	.000	.000	.000	.000	.000
7	1.000	1.000	.998	.891	.512	.154	.022	.001	.000	.000	.000	.000	.000
8	1.000	1.000	1.000	.953	.677	.274	.054	.004	.000	.000	.000	.000	.000
9	1.000	1.000	1.000	.983	.811	.425	.115	.013	.000	.000	.000	.000	.000
10	1.000	1.000	1.000	.994	.902	.586	.212	.034	.002	.000	.000	.000	.000
11	1.000	1.000	1.000	.998	.956	.732	.345	.078	.006	.000	.000	.000	.000
12	1.000	1.000	1.000	1.000	.983	.846	.500	.154	.017	.000	.000	.000	.000
13	1.000	1.000	1.000	1.000	.994	.922	.655	.268	.044	.002	.000	.000	.000
14	1.000	1.000	1.000	1.000	.998	.966	.788	.414	.098	.006	.000	.000	.000
15	1.000	1.000	1.000	1.000	1.000	.987	.885	.575	.189	.017	.000	.000	.000
16	1.000	1.000	1.000	1.000	1.000	.996	.946	.726	.323	.047	.000	.000	.000
17	1.000	1.000	1.000	1.000	1.000	.999	.978	.846	.488	.109	.002	.000	.000
18	1.000	1.000	1.000	1.000	1.000	1.000	.993	.926	.659	.220	.009	.000	.000
19	1.000	1.000	1.000	1.000	1.000	1.000	.998	.971	.807	.383	.033	.001	.000
20	1.000	1.000	1.000	1.000	1.000	1.000	1.000	.991	.910	.579	.098	.007	.000
21	1.000	1.000	1.000	1.000	1.000	1.000	1.000	.998	.967	.766	.236	.034	.000
22	1.000	1.000	1.000	1.000	1.000	1.000	1.000	1.000	.991	.902	.463	.127	.002
23	1.000	1.000	1.000	1.000	1.000	1.000	1.000	1.000	.998	.973	.729	.358	.026
24	1.000	1.000	1.000	1.000	1.000	1.000	1.000	1.000	1.000	.996	.928	.723	.222

Index

*This book has been set in 10 point Gael, leaded
2 points. Chapter numbers are 48 point Gael
and chapter titles are 20 point Serif Gothic.
The size of the type page is 27 by 46½ picas.*